El Mirón Cave, Cantabrian Spain

El Mirón Cave, Cantabrian Spain

The Site and Its Holocene Archaeological Record

EDITED BY

LAWRENCE GUY STRAUS AND
MANUEL R. GONZÁLEZ MORALES

University of New Mexico Press | Albuquerque

Printed in the United States of America
17 16 15 14 13 12 1 2 3 4 5 6

LIBRARY OF CONGRESS CATALOGING-IN-PUBLICATION DATA

El Mirón Cave, Cantabrian Spain : the site and its Holocene archaeological record /
edited by Lawrence Guy Straus and Manuel R. González Morales.
p. cm.
Includes bibliographical references and index.
ISBN 978-0-8263-5148-7 (cloth : alk. paper) — ISBN 978-0-8263-5150-0 (electronic)
1. Mirón Cave (Spain) 2. Neolithic period—Spain—Cantabria. 3. Mesolithic period—Spain—Cantabria. 4. Excavations (Archaeology)—Spain—Cantabria. 5. Geology, Stratigraphic—Holocene. 6. Paleobotany—Holocene. 7. Cantabria (Spain)—Antiquities.
I. Straus, Lawrence Guy. II. González Morales, Manuel R. III. Title: Mirón Cave, Cantabrian Spain.
GN776.22.S7E5 2012
936.6′351—dc23

2011043902

DESIGN AND LAYOUT: MELISSA TANDYSH
Composed in 10.3/14 Minion Pro
Display type is Cronos Pro

Financing of color plates was provided by The Fund for Stone Age Research (UNM Foundation) whose principal donors are Jean and Ray Auel.

CONTENTS	AUTHOR	PAGE NO.
CHAPTER FIFTEEN: Taphonomic Study of the Large and Medium Mammals from the Post-Paleolithic Occupations in El Mirón Cave	Manuel Pérez Ripoll and María Dolores López Gila	262
CHAPTER SIXTEEN: Macromammalian Remains from the Holocene Levels of El Mirón Cave	Jesús Altuna and Koro Mariezkurrena	288
CHAPTER SEVENTEEN: Descriptions of the Post-Paleolithic Pit and Hearth Features of El Mirón Cave	Lawrence Guy Straus and Manuel R. González Morales	319
CHAPTER EIGHTEEN: The Post-Paleolithic Stone Industries of El Mirón Cave	Lawrence Guy Straus, Manuel R. González Morales, and John D. Rissetto	332
CHAPTER NINETEEN: The Ceramics of El Mirón Cave: Production, Morphology, and Discard	Cristina Vega Maeso	372
CHAPTER TWENTY: Conclusions: Preliminary Overview and Focus on the Post-Paleolithic Occupations of El Mirón Cave	Lawrence Guy Straus and Manuel R. González Morales	426
Contributors		443
Color Plates to Follow Page		196
Index		445

FIGURES, COLOR PLATES, AND TABLES

FIGURES, COLOR PLATES, AND TABLES

FIGURES, COLOR PLATES, AND TABLES

FIGURES, COLOR PLATES, AND TABLES

COLOR PLATES

FIGURES, COLOR PLATES, AND TABLES

TABLES

ACKNOWLEDGMENTS

THE EL MIRÓN CAVE PREHISTORIC PROJECT, LIKE ANY LARGE-SCALE, LONG-TERM archaeological excavation, is the fruit of the labor of many people—students, collaborating specialists, local residents, and government, foundation, and university administrators (in Spain and the United States). Words are not adequate to express our thanks to everyone who has helped our research in many different ways.

Since 1996 we have been the guests of the people of the town of Ramales de la Victoria, whose successive mayors (Fermín Gómez Seña and Jose Domingo San Emeterio) and councils have supported our research with lab space, storage, water, electricity, occasional manpower—and hospitality. Even more importantly, we have been digging in *their* cave on their communal land; El Mirón, like the facing San Vicente Peak, is a very prominent landmark, looking down as it does from the Monte Pando cliffside onto the town at the bottom of its valley. Ramales *is* caves: Cullalvera, Covalanas (both of which have bars named after them), El Mirón, and the legendary AER speleological society. Many "Ramaliegos" have been good friends—to us, to our student excavators, and to the project. Among them, two couples stand out: Cesar and Mary Carmen Gómez, who have a photography business and have done innumerable favors for us, from taking photographs and doing photocopying to allowing us use of their storeroom, and especially Joaquín ("Pencho") and Sara Eguizábal, guides extraordinary of the cave art site of Covalanas. Pencho, who has published on the Paleolithic paintings of Covalanas (and nearby La Haza), is in love with "his" cave and his town; he has transmitted his extraordinary enthusiasm and knowledge to countless visitors from throughout Spain, Europe, and America—including many of the world's most important paleoanthropologists and prehistorians. His unfailing support of the research in El Mirón is precious beyond words. For many of our students, their first (and unforgettable) experience with cave art has been Covalanas as shown and explained by Pencho.

The excavations in El Mirón have been authorized annually by the Consejería de Cultura (Cultural Affairs Department) of the Regional Government of Cantabria. The finds are curated in the Regional Museum of Prehistory and Archaeology (Santander), as required by law.

The research has been funded over the years by many private and public institutions. In Spain these have been the Fundación Marcelino Botín (Santander), the Gobierno de Cantabria (Consejerias de Cultura y de Agricultura y Pesca), and the Gobierno de España (Ministerio de Educación y Ciencia); in the United States they have been the L. S. B. Leakey Foundation (San Francisco), the National Geographic Society (Washington, D.C.), the National Science Foundation (Archaeology Program and Research Experiences for Undergraduates grants), the University of New Mexico (Research Allocations Committee and College of Arts and Sciences), and the Stone Age Research Fund (University of New Mexico Foundation; principal donors: Jean and Ray Auel). The generosity of all these granting agencies has been fundamental. Certain individuals therein deserve special credit, notably John Yellen at NSF, Esperanza Botella at Botín, the late Clark Howell and Richard Klein at Leakey, Henry Wright and Mindy Zeder at National Geographic, and Jose Antonio Cagigas and Justo Barreda in the Consejería de Cultura. Invaluable material support (equipment, laboratory and storage space) has been supplied by the Universidad de Cantabria (Departments of Historical Sciences, Physics, and Civil Engineering).

In addition to all the scientific collaborators whose chapters are in this volume, we wish to thank many others who have contributed their professional knowledge and expertise to the project in different ways. These include Alexander Cherkinsky (radiocarbon dating, formerly of Geochron Labs in Cambridge, Massachusetts, now at University of Georgia), Jaime Clapés (geophysics, Universitat Politécnica de Catalunya), Marie-Agnès Courty and Nicolai Fedoroff (micro-morphology/pedology, CNRS and INA, France), Brooks Ellwood (geophysics, Louisiana State University), Marcos García Diez (cave art, then of the Universidad del País Vasco), Carlos García Léaniz and Sonia Consuegra (fish DNA, Centro Ictiológico, Arredondo), F. Guitián (ceramic technology, Universidad de Santiago de Compostela), Salvador Rovira and Jesús Setien (archaeometallurgy, Museo Arqueológico Nacional in Madrid and Universidad de Cantabria, respectively), Enrique Serrano (glaciology, Universidad de Cantabria), Romualdo Seva and associates (archaeometry, Universidad de Alicante), Arantxa Aranburu (geological coring, Universidad del País Vasco), and Ruth Shahack-Gross and Steve Weiner (archaeochemistry, Weizmann Institute of Science, Israel). Many other specialist studies (notably DNA and stable isotopes) are currently under way.

Graduate and undergraduate students from the Universidad de Cantabria, the University of New Mexico, and universities throughout Spain, the rest of Europe, the Americas, and even Japan have diligently and intelligently excavated, piece-plotted, water-screened, floated, sorted, labeled, classified, and bagged hundreds of thousands of artifacts, manuports, and faunal remains. Many of them have participated in more than one campaign—some for several years, becoming field or lab supervisors who have been key to the success and quality of the research. There have been student specialists in living floor dissection; high-tech recording methods (digital photo-based photogrammetry, total station and computerized recording); bone and artifact cleaning and consolidation; faunal, osseous artifact, ceramic, and lithic sourcing analyses, and so on. Several have derived some or all of the data for their thesis or dissertation from the El Mirón excavations and have gone (or are going) on to become professional archaeologists in their own right (Rebecca Schwendler, Estebán Alvarez, John Rissetto, Ana Belen Marín Arroyo, Cristina Vega Maeso, Yuichi Nakazawa, Igor Gutierrez Zugasti, Alejandro García Moreno, Elisabeth Stone, David Cuenca). We are extremely thankful to and proud of all the many students who have participated in the research. Their good cheer, hard work, and enthusiasm have been inspirations to us and made it all the easier for us to direct them both in the field and back at our respective universities. The students—like the project itself—have become features of the summertime social life of Ramales, giving the town (and especially its many bars) the air of an international field school or study-abroad program, with "Pencho" as its social director and singing (etc.) master. Among student participants in the project there have been marriages, babies born, careers begun, lives changed.

Cave cartography was done by Eduardo Torres of Santander and modified by Manuel González Morales. Lithic artifact drafting was done by Natalia Suarez and Susana Ruiz. Redrafting of stratigraphic sections (originally drawn by Lawrence Straus) was done by Ron

Stauber of the University of New Mexico's Office of Contract Archaeology.

Finally, we wish to most warmly thank and acknowledge the many sacrifices of our respective spouses, Maria del Carmen Rapado Errazti and Yolanda Diaz Casado, and our respective daughters, Eva and Celia, who have all put up with a great deal as we spent countless hours obsessively digging, managing, e-mailing, talking, analyzing, and writing. We hope that they are not too jealous of our other "love," that remarkable cave, El Mirón.

We dedicate this volume to great paleoanthropologists/prehistorians who influenced our respective careers greatly and who died during the course of the project: Desmond Clark, Clark Howell, Francisco Jordá, Javier Fortea, Lew Binford, and now, tragically, our own William Farrand, who passed away in March 2011, having just given us his photograph negatives for this book. Bill was a key contributor to the El Mirón project for years. Joaquín González Echegaray and Les Freeman have been great inspirations for many years. ¡Mil gracias a todos!

The full roster of participants (mainly graduate and undergraduate students but also professionals and *aficionados*) is long and very international. The countries represented, in addition to Spain and the United States, include Argentina, Canada, Croatia, Cuba, France, Ireland, Japan, Mexico, Netherlands, Nicaragua, Portugal, and the United Kingdom (England and Scotland). Some worked for a few days or weeks, others for entire campaigns, generally two months in length. The core participants worked for several campaigns, some becoming excavation area or field laboratory supervisors (notably Maite Castanedo, Rebecca Schwendler, Ramiro Fernández Madrazo, John Rissetto, Hannah Dodd, Igor Gutierrez Zugasti, Elisabeth Stone, and Yuichi Nakazawa); others had special responsibilities for digital photogrammetry and EDM database organization (Angel Astorqui, Enrique Maeso, Jorge Caro, David Cuenca, Lisa Fontes, and Alejandro García Moreno) or conservation (Ziortza San Pedro). The crews are listed below by excavation campaign.

1996

Angel Astorqui
Gustavo Bengoechea
Maite (María Teresa) Castanedo
Francisco Javier Duarte
Pilar Mantecón
Marta Pascual Osaba
Susana Rodríguez
Ziortza San Pedro
Kim Sonderegger
David Vaughan
Christopher Warren

1997

Angel Astorqui
Gustavo Bengoechea
Maite Castanedo
Francisco Javier Duarte
Julia Farquhar
Ramiro Fernández Madrazo
Pascal Foucher
Nora Franco
Sébastian Lacombe
Pilar Mantecón
Marta Pascual Osaba
John Rissetto
Susana Rodríguez
Cristina San Juan
Ziortza San Pedro
Rebecca Schwendler
Kim Wilson
Ann Winegardner

1998

María Abellán
Domingo Alvarez Fernández
Angel Astorqui
Gustavo Bengoechea
David Cardo
Maite Castanedo
Richard Chatterton
Zoe Coughlan
Jose Antonio Cuervas
Francisco Javier Duarte
Elena Fernández García
Raquel Fernández García

1998 *(continued)*

Carlos Fernández González
Ramiro Fernández Madrazo
Marcos García Diez
Alicia Gómez Gómez
David Hernández
Katie Larsen
Sonia López Melón
Pilar Mantecón
Sean Maroney
Ramiro Pérez Milián
Ian Mylrea
Marta Pascual
John Rissetto
Silvia Ruiz Pedraza
Susana Ruiz Pedraza
Ziortza San Pedro
Vladimiro Sanz de la Fuente
Rebecca Schwendler
Natalia Suarez Rueda
Julián Talavera
Valentín Tejeiro
Amparo Valcarcel
Cristina Varela
Kim Wilson
Ann Winegardner

1999

María Abellán
Beatriz Acebo
Angel Astorqui
Sergio Barez del Cueto
Prof. Felipe Bate
Eduardo Carmona
Jorge Caro
Maite Castanedo
Nuño Castellano
Carolina Cortés
Alvaro Eguizabal
Raquel Fernández García
Pascal Foucher
Jorge González
Rafael González Riba
Antonio Jeronimo
Katie Larsen
Enrique Maeso
Pilar Mantecón
Ana Belen Marín
María Luisa Martínez López
Oscar Moro
Eric Nocerino
John Rissetto
Susana Rodríguez
Dennis Sandgathe
Tracey Sandgathe
Cristina San Juan
Ziortza San Pedro
Rebecca Schwendler
Julian Talavera
Jennifer Thompson
Cristina Vega Maeso
Ann Winegardner

2000

María Abellán
Beatriz Acebo
Sergio Aja
Domingo Alvarez Fernández
Angel Astorqui
Sergio Barez del Cueto
Clinton Blume
Eduardo Carmona
Jorge Caro
Maria del Mar Casas
Maite Castanedo
Nuño Castellano
Oscar de Castro
David Cuenca
Hannah Dodd
Alvaro Eguizabal
Raquel Fernández García
Carlos Fernández González
Ramiro Fernández Madrazo
Jesús Gancelo
Marcos García Diez
Elena Gil
William Hubbard
Rachel Kluendner
Enrique Maeso
Pilar Mantecón
Ana Belen Marín
Maria Soledad Martínez Avila
Oscar Mora
Eric Nocerino
Courtney Porreca
Susana Rodríguez
Silvia Ruiz
Vanesa Saiz
Ziortza San Pedro
Ana Maria Dos Santos Silva
Rebecca Schwendler
Jessica Thompson
Cristina Vega Maeso

2001

María Abellán
Beatriz Acebo
Estebán Alvarez Fernández
Angel Luis Arcal
Angel Astorqui
Sergio Barez del Cueto
María del Mar Casas
Nuño Castellano
David Cuenca
María Elvira Dato
Rene Diaz Girón
Ivan Diaz San Millán
Hannah Dodd
Sara Burns Duran
Ande Erce Domínguez
Ramiro Fernández Madrazo
Carmen García Escudero
Almudena García Rubio
Antonio García Tabernero
Elena Gil
Carlos Gil Ortiz
Ana Gómez Cabellud
Juliette Graziano
Igor Gutierrez Zugasti
David Hernández Rivero
Rachel Kluender
Kristin Koshgarian
Enrique Maeso
Pilar Mantecón
Juan Jose Marcos-Alberco
Ana Belen Marín
Ulises Mena
Iris Naïre Otte
Marián Presno
John Rissetto
Susana Ruiz Pedraza
Silvia Ruiz Tresgallo
María Sánchez Claramonte
Ziortza San Pedro
Teresa Steele
Deborah Stratmann
Natalia Suarez Rueda
Cristina Vega Maeso
Jose Manuel Vicente García
Antonio Villalpando
Debora Zurro

2002

María Abellán
Beatriz Acebo
Eduardo Carmona
Jorge Caro
Nuño Castellano
Stacey Chambliss
Rebecca Christy
David Cuenca
Hannah Dodd
Carlos Fernández González
Javier Fernández Irigoyen
Ramiro Fernández Madrazo
Marcos García Diez
Almudena García Rubio
Antonio García Tabernero
Carlos Gil
Elena Gil
Igor Gutierrez Zugasti
Cliodhna Ni Lionain
Marina Madarevic
Enrique Maeso
Pilar Mantecón
Yuichi Nakazawa
William Reitze
Aldara Rico
John Rissetto
María Sánchez Claramonte
Ziortza San Pedro
Jesús María Sola
Cristina Vega

2003

Jorge Caro
Nuño Castellano
Colby Phillips
John Rissetto
Alejandra Silva
Christina Sinkovec

2004

María Balde
Jorge Caro
Nuño Castellano
Matthew Dawson
Ivan Diaz San Millán
Irene García Lino
Angel González Gutierrez
Igor Gutierrez Zugasti
Laura Hoyo
Katherine Knauber-Ferriegel
Cliodhna Ni Lionain
Pilar Mantecón
Ana Belen Marín
Mihoko Nakazawa
Yuichi Nakazawa
John Rissetto
Oliva Rivero
Alejandra Silva
Jesús Sola
Elizabeth Stewart
Elisabeth Stone
Bea Viñas

2005

Feliciano Cadierno
Sara Cañarte
Jorge Caro
Aileen Cooper
David Cuenca
Matthew Dawson
Sara Diéz Getino
Irene García Lino
Sonia García Luque
Alejandro García Moreno
Daniel Garrido
Raquel González Gutierrez
Igor Gutierrez Zugasti
Jose Angel Honrado
Sarah King
Cliodhna Na Lionain
Miriam Llorens Liboy
Elena Mosquera
Yuichi Nakazawa
Zalema Rebollo
Olan Remontería
David Santamaría
Alejandra Silva
Elizabeth Stewart
Elisabeth Stone
Bea Viñas
Victoria Yannitto

2006

William Brown
Sofia Cadenas
Feliciano Cadierno
Sara Cañarte
Jorge Caro
Natalia Cortón
Sara Diéz Getino
Alejandro García Moreno
Nuria García Tuset
Maja Grguvic
Igor Gutierrez Zugasti
Jose Angel Honrado
Keiko Kitagawa
Jose López López
Ines López López-Doriga
Jose Luis Mazario
Elena Mosquera
Yuichi Nakazawa
Paula Ortega
Aixa Sanemeterio
Alejandra Silva
Elisabeth Stone

2007

Feliciano Cadierno
David Cuenca
Sara Diez
Michael Gibbons
Igor Gutierrez Zugasti
Erin Hegberg
Keiko Kitagawa
Zaira Marcos Gómez
Elena Mosquera
Aixa Sanemeterio
Elisabeth Stone
Joan Stone
Nikola Yukosavljevic

2008

Jorge Caro
David Cuenca
Michael Gibbons
Keiko Kitagawa
Elisabeth Stone

2010

Vitor Benito
Cristina Biele
Jorge Caro
David Cuenca
Laura DeFrank
Lisa Fontes
Carlos García Benito
Alejandro García Moreno
Igor Gutierrez Zugasti
Andrew Holmes
Zaira Marcos Gómez
Juan Martín Hafner
Alejandro Sierra
Joshua Vallejos

2011

Monica Alba Lozano
Victor Benito Hernández
David Cuenca Solano
Lisa Fontes
Alejandro García Moreno
Adrian Gurpegui Cotado
Igor Gutierrez Zugasti
Andrew Holmes
Margaret Homko
Laura Juanola Bosch
Shannon Landry
Mariana Nabais
Alejandro Sainz Sierra

We owe it all to all of you! ¡Gracias!

—Lawrence Guy Straus, Albuquerque
Manuel R. González Morales, Santander

Preamble

La Cueva del Mirón: A Grand Cave Site in the Cantabrian Cordillera of Northern Spain

Lawrence Guy Straus

According to legend, the name El Mirón has an ancient imperial origin.[1] In the fall of 1556 the king of Spain and Holy Roman emperor, Charles V, having abdicated in favor of his son Phillip II, was en route to his place of retirement, the monastery of Yuste in Extremadura (southwestern Spain). He had traveled by ship from Flanders to one of the habitual ports of both his imperial "commute" and the royal wool trade in northern Spain, Laredo, situated in what is today the autonomous region of Cantabria, between the cities of Santander and Bilbao. Ahead lay the steep climb up the Camino Real via the valleys of the Río Asón and its tributary, the Calera, and then over the cordillera by one of its lowest passes, Los Tornos, above the Vizcayan town of Lanestosa, up onto the high, dry northern Castilian Meseta of the ancient province of Burgos—the vast, hard lands of the Cid. But as his litter-bearers made their way up the Asón to the town of Ramales at the foot of the mountain chain, the emperor was still in the narrow, steep, humid world of Atlantic Spain, with its dense oak, hazel, elm, linden, and chestnut forests and verdant pastures. The Camino Real was probably little better than a rude track, especially as it ascended the grade from Ramales to Lanestosa, up the face of what the geologists call a hanging valley. At the confluence of the Calera with the Gándara, not far from where it joins the Asón in Ramales, its deep valley is surrounded by 1,000 m peaks: La Busta, El Moro, and most dramatic of all, San Vicente, at the eastern end of the jagged crest of the Sierra del Hornijo—visible on a clear day from the heights above Santander along the coast, some 35 km distant as the crow flies to the northwest.

The emperor had come some 20 km (about three and two-thirds Spanish leagues) from Laredo and his bearers were tired as they diagonally ascended the road up the scree at the base of El Haza cliff. Set down by his men so that they could rest, Charles drew open the curtains of the litter to let in the sun and was greeted with the spectacle of pyramidal Pico San Vicente framed by its rays. The old monarch is said to have pulled himself heavily to his feet, drawn out by the grandeur of the view. He then stood "mirón"—staring fixedly, awestruck—at the scene: the towering mountains before him, the deep canyon with its gushing torrent at his feet, and the sheer limestone wall of the cliff at his back. That cliff—the cirquelike western face of the Monte Pando shoulder of El Moro mountain—is filled with caves. The greatest of these is now known as El Mirón (plate I.1).

The legend speaks of the emperor's humility in the face of such a natural scene of wonderment and awe. He got back into his litter and his bearers continued their agonizing climb up to Lanestosa and Los Tornos—culminating in Yuste, where Charles died two years later. The man who had ruled much of Europe and America, the warrior who had defeated and captured the king of France and battled the Turks across the Mediterranean, the king in whose name Cortés and Pizarro laid claim for Spain to the Aztec and Inca Empires, respectively, had passed by and was greatly moved by the green and gray majesty of an overwhelming panorama that, unbeknownst to him, had been gazed upon for 40,000 years by anonymous women, men, and children of the ages of stone, copper, and bronze. And then he—like they—was gone.

Three and a half centuries were to pass. The great cave mouth in El Haza cliff would be used to house people, with their activities and their animals. Some of the inhabitants of the cave are reflected in the cave's alternative names: Cueva del Francés and Cueva de los Gitanos. Some (e.g., Alcalde del Río, Breuil, and Sierra 1911) were to think that the cave had even been used as an artillery emplacement during the 1839 Battle of Ramales in the First Carlist War (a Liberal victory, hence the qualifier added to the town's name). But this is probably incorrect, as the cannon was apparently sited in a cave on the opposite side of the gorge. (The error is important, however, because from the very beginning the cave's infilling was qualified as being "very mixed," supposedly as a result of military activities [Alcalde del Río, Breuil, and Sierra 1911]).

Then, again in fall—specifically mid-September 1903—two school principals explored the caves above Ramales. They were Father Lorenzo Sierra of the San Vincent de Paul high school in Limpias (not far from Laredo, at the head of the Asón estuary tidewater) and Hermilio Alcalde del Río of the vocational school in Torrelavega, a small, industrial city west of Santander. Both were avid amateur archaeologists. Two months later Alcalde would discover the monumental cave site of El Castillo and the important residential and art site of Hornos de la Peña, and in the same year he had begun to excavate in Altamira. Sierra discovered the cave art of El Salitre, the site (but not the art) of Cullalvera, and two entrances of the now-famous La Garma karstic system, all in 1903, and he was to discover Venta de la Perra in 1904 and El Valle in 1905. Naturally, given their respective bases, Alcalde tended to focus his explorations on the central and western parts of what was then the province of Santander and in eastern Asturias (El Pindal), while Sierra concentrated on eastern and central Santander. But the pair collaborated not only in their prospections around Ramales but also visited each other's discoveries together, often within just days of the finds. The list of their discoveries of sites and of cave art (including figures in caves that had already been explored by the first of this region's archaeologists, Sanz de Sautuola) during a five-year period is staggering (see Alcalde del Río 1906; Madariaga 1972; Sierra 1908; Madariaga 2000:71). One cave that they apparently explored together (although it may first have been visited by Sierra) was, however, for a long time to relegated to insignificance or simply forgotten: El Mirón.

Another half century was to pass before anyone was to make an attempt to investigate the archaeological potential of El Mirón Cave, which had been repeatedly but only very briefly suggested in the early literature (e.g., Sierra 1908:110; Alcalde del Río Breuil, and Sierra 1911:11; Cabré 1915:42; Obermaier 1924:157). And that attempt was anonymous and unpublished. A trench and a pit were dug across the width of the pitch-black inner cave, apparently by workmen engaged in the construction of a road up to Covalanas in the early 1950s. This apparent search for archaeological deposits in the interior gallery of El Mirón was likely directed by A. García Lorenzo, an amateur archaeologist and discoverer of several cave art loci (notably Las Chimeneas and Las Monedas near El Castillo), acting under the general direction of Father Jesús Carballo, founding director of the Provincial Museum of Prehistory in Santander. Probably having found little or nothing, this work was given up and soon forgotten. At some unknown time, a few artifacts probably collected from the El Mirón surface were donated to the Archaeological Museum of Cáceres—not far from Charles V's retreat in Yuste (Moure, González Sainz, and González Morales 1991).

And a hand ax or cleaver is said to have been found in El Mirón, but details are sketchy (e.g., Junta Directiva de la F.C.E. 1995).

Lawrence Straus (Straus 1975; 1983)—drawn by Cabré's (1915:47) mention of "Solutrean" artifacts—visited El Mirón in 1973, again in fall. He observed many cobbles and some flints and bones scattered on the surface of the vestibule, entered the inner cave with a feeble flashlight . . . and fell into the trench, which he assumed was a military affair, as mentioned by Alcalde del Río, Breuil, and Sierra (1911). The accident—far less magnificent than the legend of Carlos V—and the cave remained in his memory for nearly a quarter century. During this time, Manuel González Morales participated in a restudy of the paintings in nearby Covalanas and La Haza (Moure, González Sainz, and González Morales 1991) and later directed a research project centered at sites across the Asón estuary from Laredo, where the emperor had landed 450 years before (e.g., González Morales 2000; González Morales and Díaz Casado 2000; González Morales, Yudego, and Ituarte 2000). Now it is we who, year after year, have often paused to stare in awe at that grand view from that great cave, El Mirón, joining the hundreds of anonymous souls—and one emperor—to have done so for over 40 millennia. Now a monument and a sign stand below the cave along the old Camino Real: the first to Charles V and the second a panel briefly explaining what we are doing in "his" great cave. This book will begin to tell the latter story in greater detail.

Notes

1. The El Mirón legend was related to us by Joaquín ("Pencho") Eguizábal, lifelong resident of Ramales and Covalanas Cave guide. His source is an unpublished book manuscript by José María Haro titled "Consejas andróminas y falorias de Ramales de la Victoria," about which we have no further information.

References Cited

Alcalde del Río, H.

1906 *Las pinturas y grabados de las cavernas prehistóricas de la provincia de Santander.* Blanchard y Arce, Santander, Spain.

Alcalde del Río, H., H. Breuil, and L. Sierra

1911 *Les cavernes de la région Cantabrique.* A. Chêne, Monaco.

Cabré, J.

1915 *El arte rupestre en España.* Memoria 1. Comision de Investigaciones Paleontologicas y Prehistoricas, Madrid.

González Morales, M.

2000 La prehistoria de las Marismas: excavaciones en la Cueva de La Fragua (Santoña). Excavaciones de 1990, 1991, 1993, 1994, y 1996. In *Actuaciones arqueológicas en Cantabria, 1984–1999*, edited by R. Ontañón, pp. 177–180. Gobierno de Cantabria, Santander, Spain.

González Morales, M., and Y. Díaz Casado

2000 La prehistoria de las Marismas: excavaciones arqueológicas en los abrigos de la Peña del Perro. In *Actuaciones arqueológicas en Cantabria, 1984–1999*, edited by R. Ontañón, pp. 93–96. Gobierno de Cantabria, Santander, Spain.

González Morales, M., C. Yudego, and C. Ituarte

2000 La prehistoria de las Marismas: prospección arqueológica de la zona del bajo Asón y marismas de Santoña y toma de muestras en los yacimientos de las cuevas del Otero, La Chora, y El Valle. In *Actuaciones arqueológicas en Cantabria, 1984–1999*, edited by R. Ontañón, pp. 151–153. Gobierno de Cantabria, Santander, Spain.

Junta Directiva de la Federación Cántabra de Espeleología

1995 Informe de la Federación Cántabra de Espeleología sobre la zona de Ramales. *Asociación Cántabra para la Defensa del Patrimonio Subterraneo, memorias 1995*, pp. 132–146. Santander, Spain.

Madariaga, B.
1972 *Hermilio Alcalde del Río: Una escuela de Prehistoria en Santander*. Patronato de las Cuevas Prehistóricas, Santander, Spain.
2000 *Sanz de Sautuola y el descubrimiento de Altamira*. Patronato de las Cuevas Prehistoricas, Santander, Spain.

Moure, A., C. González Sainz, and M. González Morales
1991 *Las cuevas de Ramales de la Victoria (Cantabria). Arte rupestre en las cuevas de Covalanas y La Haza*. Universidad de Cantabria, Santander, Spain.

Obermaier, H.
1924 *Fossil Man in Spain*. Yale University Press, New Haven, CT.

Sierra, L.
1908 Notas para el mapa paletnográfico de la Provincia de Santander. In *Actas del Primer Congreso de Naturalistas Españoles*, pp. 103–117, Zaragoza, Spain.

Straus, L. G.
1975 *A Study of the Solutrean in vasco-cantabrian Spain*. University of Chicago.
1983 *El solutrense Vasco-Cantábrico: una nueva perspectiva*. Monografías 10. Centro de Investigación y Museo de Altamira, Madrid.

CHAPTER ONE

The Excavation of El Mirón Cave

Lawrence Guy Straus

It is a tradition in prehistoric archaeology that classic excavations were those done in *big* caves and rockshelters, with long stratigraphies. These became "type sites" for regions and time-slices or even for the entire record of continents. Sites such as Spy, Le Trou Magrite, La Quina, La Ferrassie, Laugerie, Isturitz, Mas d'Azil, Szeleta, Tabun, Haua Fteah, the caves at Grimaldi and at Klasies River mouth, and—closer to the subject of this book—El Castillo immediately come to mind. The traditional modus operandi of excavating large, deep sites was perfectly suited to the overall aims of the culture-historical approach to prehistory; these sites were likely to yield long sequences of distinctive industries exemplified by large, "representative" collections of diagnostic artifacts (and, sometimes, works of "art") that both allowed for reliable temporal/cultural attributions and provided the kinds of materials that museums (or private collectors) wanted to possess and display. These "mega-sites" were usually dug in extenso during the early decades of the discipline's existence (i.e., before World War II), with large crews of workers using picks, shovels, and sometimes dynamite and mine cars on rails. On the one hand, the volume of sediment removed—often in very short order—was staggering, making it hard for us today to imagine how the excavators (legendary figures such as Edouard Lartet, Denis Peyrony, Edouard Piette, Edouard Dupont, Dorothy Garrod, Réne and Suzanne de Saint-Périer, Hugo Obermaier, et al.) were able to recover as much as they did. On the other hand, the strategy of totally or substantially emptying large caves and rockshelters did provide ample coverage of broad areas, thereby "evening out" the effects of spatially segregated, activity-related debris and increasing the yields of rare finds, notably human burials and unusual works of portable art or adornment.

In recent decades, under the influence perhaps of processual archaeology and economic prehistory, and certainly because of ethical preservation considerations, financial constraints, and the increasing emphasis on great detail in interdisciplinary research, excavations of prehistoric sites have become smaller and much more painstaking—definitely much slower, while moving much less dirt. The profession has come to realize that prehistoric adaptive systems manifested themselves on the ground in the form of many different kinds of sites, from big, multipurpose sites (often long-term "residential" loci) to a wide variety of (often smaller) special-purpose, shorter-use sites, including "logistical" locations, to use Binford's (1980) terminology. These sites do not necessarily have long stratigraphic sequences and may only contain very partial samples of the full repertoire of a culture's material heritage. Yet they can sometimes give a much clearer picture of specific activities (e.g., specialized hunting camps, quarry-workshop sites, transit camps, shell-midden dumps) than the more complex palimpsests that generally characterize the "big" sites. A related development in recent prehistoric fieldwork has been the return to classic sites to conduct limited excavations designed to verify, refine, correct, and/or radiometrically date the old stratigraphic sequences, to collect samples for modern analyses (e.g., sedimentology, micromorphology, palynology, dental cementum seasonality studies, lithic microwear), and to gather artifact assemblages that are less heavily biased toward retouched tools (and at most large debris) to the detriment of microdebitage or even faunal remains. This strategy has the advantage of more-or-less guaranteed results; in re-excavating a known site (albeit with more modern, exhaustive methods and to ask "new" questions) the archaeologist knows in advance what periods and what range of materials (e.g., is there faunal preservation?) he or she is likely to find. The archaeologist can usually clean existing stratigraphic sections and dig back from them. In short, he or she is not "digging blind," which has the added practical advantage of allowing the archaeologist to tell funding agencies that success is practically guaranteed, at least when it is already known that intact sediments still remain in the targeted site.

Both of the directors of the Mirón project had had ample experience with the excavation of both previously researched and smaller, limited-function sites. They had collaborated (Straus as codirector and González Morales as Spanish inspector) with G. A. Clark in the excavation of a remnant of the classic Upper Paleolithic and Mesolithic site of La Riera Cave (Asturias), originally excavated by the Conde de la Vega del Sella in 1917. Straus had, over four three-month campaigns, excavated the remaining open-air portion of the large Magdalenian site of the Abri Dufaure (Les Landes, France), whose rockshelter had been entirely dug out by Abbé Henri Breuil and a colleague in 1900. He tested a number of sites of various periods in Portugal and then, in collaboration with Marcel Otte, dug small remnants of two classic Middle and Upper Paleolithic sites in Wallonia (Belgium)—le Trou Magrite and Hermitage-Huccorgne—as well as two small, specialized sites—l'Abri du Pape (Mesolithic) and la Grotte du Bois Laiterie (Magdalenian). González Morales had excavated another site originally discovered by Hermilio Alcalde del Río and first dug by Vega del Sella, namely an Asturian Mesolithic shell midden in Mazaculos Cave, and tested a number of other shell-midden sites. He also conducted a salvage excavation in the small Magdalenian rockshelter of Entrefoces (also in Asturias) and documented a surficial late Gravettian occupation immediately beneath rupestral art images in Fuente del Salín (western Cantabria). Finally, he had excavated the small Magdalenian-Azilian-Mesolithic sites of El Perro rockshelter and La Fragua Cave and the Mesolithic shell midden in La Trecha Cave and sampled for dating the earlier-excavated terminal Paleolithic and Mesolithic cave of La Chora, all in eastern Cantabria. All these excavations had helped answer interesting and sometimes important questions, but the samples and the coverages were generally limited and thus possibly biased.

In the present era, it is rare in western Europe to discover an essentially pristine cave site that is both large in area and deep in stratigraphy. The opportunity that El Mirón has presented to us is, although not unique (Franchthi Cave on the Argolid Peninsula of Greece is certainly a similar case), at least highly unusual, and it poses a great responsibility. From the outset, El Mirón has allowed (and obligated) us to try to "do it right," to excavate meticulously, to experiment and then establish thorough protocols for "total" recovery via flotation and fine screening and for computerized provenance recording. The great size of the cave vestibule, whose deposits proved to be largely undisturbed (despite superficial appearances and the opinions of some earlier archaeologists), would permit fairly broad exposure of "living surfaces" and therefore lend itself to analyses of structures and activity areas within this vast sheltered area.

From the outset in 1995, when Straus proposed to González Morales the idea of testing El Mirón, the goal of excavating *a large site in a montane setting* was clear. This would provide a contrast to the prevailing record for at least the Upper and Middle Paleolithic of Cantabria, which was based almost exclusively on sites that had been excavated for a century on or at the edge of the coastal plain (e.g., Altamira, El Pendo, Cueva Morín, El Juyo, El Castillo). It was also clear that El Mirón, with its large size, strategic location, panoramic view, and proximity to the major cave art sites of Covalanas, La Haza, and Cullalvera, was likely (among other things) to have been a significant Upper Paleolithic residential site. Few sites of that scale and depth had been excavated essentially de novo in this region since Obermaier's huge excavations at El Castillo in 1911–1914 or Vega del Sella's work in Cueto de la Mina (Asturias) in 1914–1915, since even Julio Martínez Santa-Olalla's 1953–1957 excavations in El Pendo had been preceded by years of diggings by Jesús Carballo and, at the beginning, by Marcelino Sanz de Sautuola.

But really *it was El Mirón that revealed its potential to us as we excavated it*, beginning with the startling discovery of intact Bronze Age layers immediately below a thin, surficial layer of dung and recent mixed fill within one of the cabin foundations near the front of the vestibule. As things developed, we discovered that this cave was to hold not only the major Paleolithic sequence that we hoped for and expected but also one of the most important post-Pleistocene sequences in the Cantabrian region. The sequence we uncovered in the initial test excavation in the Cabin area of the outer vestibule in 1996 showed itself to hold invaluable information on questions concerning the origins of food production and of complex societies in this peripheral, Atlantic fringe of the Iberian Peninsula (plate 1.1). The Holocene record would have to be carefully excavated, systematically recorded, and thoroughly analyzed. The first results of that effort are the main focus of this, the first volume in a projected series of monographs on the grand cave of El Mirón. In no other major excavation of a Paleolithic site in the Cantabrian region had either much evidence been found of or much attention paid to post-Paleolithic cultural evidence. Often, when promising Paleolithic cave sites were to be dug, surface layers with "modern" or "recent" artifacts and fauna represented impediments that had to be removed expeditiously. (For example, Breuil and Obermaier were quite vague on the subject of the contents and age ["Neolithic" or "Eneolithic"] of the topmost stratum in El Castillo, in which, nonetheless, they had found a great mixture of ceramic types and a number of rather spectacular metal artifacts, now known to be mid–Bronze Age [Cabrera 1984].) In recent years, some excavations of Neolithic and Chalcolithic sites have been conducted and there has been renewed interest in the study and dating of megalithic monuments along the Atlantic facade of northern Spain (for example, see papers in Rodríguez Casal 1997). However, it is an important and rare opportunity to study how El Mirón Cave, as a particular place that had been a major hunter-gatherer site for the exploitation of a specific local environment during thousands of years, had been transformed into a locus for pastoral and perhaps agricultural food production, with all the material and social ramifications.

In El Mirón we are constrained neither by a small cave area nor by the limits imposed by earlier excavations. With an area of about 250 m^2, the vestibule of El Mirón is almost like an open-air site, but with a roof and walls—a far cry from some of the smaller caves that we have excavated in recent years, such as Le Bois Laiterie or La Fragua. Excavations in such a large site—if sustained for substantial amounts of time, year after year—can potentially sample a variety of activity areas for any given level. In a large cave the archaeological signatures of different activities may be more clearly segregated than in more physically constrained cavities, where palimpsests of residues from many distinct activities may be far more common and thus difficult to "read." Since the use of space in any particular period is likely to have been organized in relation to the structural features of the cave (e.g., terrace at the mouth, north versus south zones with differing exposure to winds and drafts both from outside and inside the cave, exterior zone with sunlight all day versus rear zone with bright natural light only at the end of the afternoon, limited roof-fall blocks that could serve as "site furniture"), the archaeological sampling of different zones within the vestibule should, in principle, lead to a fuller picture of the gamut of activities conducted within it during any given period. Unlike the modus operandi of traditional excavations (including even ones done in recent decades), the goal of exposing broad areas of a large site such as El Mirón should not be just to "lump" together all the finds from individual layers to give a general picture of contents in (or "typical" of) a period. Rather, it should be to try to examine possible synchronic similarities and differences of a spatial character in order to approach the problem of the organization of space. Then, through diachronic, inter-level comparisons, one can study how the structured use of a particular place may or may not have changed in relation to changes in human activities and in the physical configuration of the cave, all within the context of a potentially dynamic environment. *Both* "space" and "time" are of significance in the detailed study of archaeological places, neither to the detriment of the other.

Thus, the opportunity provided by El Mirón Cave is to conduct research along both of the strategic axes by which archaeological excavation can be conducted: vertical and horizontal (plate 1.2). Many excavations (including our own) have emphasized—generally by dint of circumstances (e.g., the size and/or nature of the site, the intact area remaining to be dug, available time, funding, and other resources)—one or the other of these axes. Many excavations of open-air sites obviously emphasize the study of the "horizontal," that is, broad activity area exposures and behavioral analyses, with little or no stratigraphic depth and hence scant time depth (Pincevent, Gönnersdorf and the other Paris Basin and Neuwied Basin Magdalenian sites are classic examples). Most cave/rockshelter excavations have naturally tended to stress the "vertical" (i.e., stratigraphic depth, chronology and the study of change through time in environments, diagnostic artifacts or assemblages, and faunal indicators). The behavioral interpretations derived from "vertical" excavations suffer from stringent spatial sampling limitations (i.e., small, only partially representative areas, usually in only one part of a site, often dug back from old excavation sections). The exquisite "snapshots" of near-moments of time from some of the most spectacular "horizontal" excavations nonetheless usually do not deal with issues of how the role of particular places may change through time as (global, regional, and local) environments, resources, and human survival strategies change. But the often-segregated emphases on timelessness and time do not have to be separated in the context of archaeological excavations, if both horizontal and vertical axes of research can be combined as integral components of individual projects.

There is a significant problem that affects the excavation and interpretation of "rich" sites such as El Mirón: the often continuous accumulation of vast amounts of cultural debris with few or no sterile layers, and hence the practical difficulty of determining strict contemporaneity of finds across space. It is a common characteristic of many of the caves of the Cantabrian region that were repeatedly favored for human use (i.e., ones that combined strategic location with good shelter and solar orientation and access to food and fuel) to be "stuffed" with dark, organic-rich deposits continuously littered with artifacts and faunal remains. This is particularly typical of major residential sites of the Magdalenian period, when the problem is often further complicated by prehistoric trampling, fires, and frequent construction activity (pits, walls, and possibly surface leveling) that had the repeated effect of "blurring" the distinctions among occupation layers. Such sites as Altamira, El Castillo, El Juyo, La Riera, Cova Rosa, Tito Bustillo, and so on have major Magdalenian palimpsests that are usually very dark in color and extraordinarily rich—and hard to meaningfully subdivide stratigraphically, at least over significant distances within the sites. Similar problems of possible construction- and residential intensity–related "churning" of culturally very rich deposits can affect the records of late Neolithic, Chalcolithic, and Bronze Age times in the Cantabrian caves—a fact compounded by the frequent use of suitable cavities as livestock stables and human burial places (even ossuaries) as well as residential locations. Stabling, frequent and massive burning (plate 1.3) (including on-site metallurgy in the Bronze Age), and intensive digging (and re-digging) of (often large) pits (plate 1.4) during the post-Mesolithic prehistory of El Mirón Cave clearly affected the nature of the depositional record in the upper zones of the site's stratigraphies. Yet the "fresh" aspect of the sharply bounded pits; the often pristine, relatively unmixed nature of ash and charcoal layers and pockets; and the clear color and texture contrasts among sediments of different origins or composition generally made the job of observing and maintaining stratigraphic and depositional distinctions in the upper levels of El Mirón relatively straightforward, compared with the more "time-averaged" Late Glacial strata.

Faced with the complexity of such a "dense" deposit—particularly in the mid-Holocene and Tardiglacial strata—we (like other excavators of rich cave sites) have had to combine the advantages of both vertical and horizontal excavation methods. In the outer vestibule area (the Cabin), where there were no preexisting holes that could provide "window" to the stratigraphy underlying the dirt floor of the submodern structures, excavation would always be conducted by first carefully digging small areas (1–4 m^2), which would later be expanded to a full 10 m^2 as the work proceeded downward. This allowed us to observe (and, if needed, correct for errors in our interpretation of) the stratigraphy before excavating larger areas of individual levels, which were uncovered by following the slope trends revealed in both nearby east-west and north-south sections. Below the incredibly complex topmost (post-Mesolithic) layers—with their mounds of ash and many pits within pits—the strata in the Cabin area lie very close to horizontal, making relatively straightforward the task of following surfaces, although there are problems of sedimentary and cultural intergradation among the Magdalenian levels exposed to date.

In the inner vestibule (the Corral area), we were able to use the existence of a large, refilled pothole at the northeastern (left rear) corner of the chamber to provide us with a "preview" of the Tardiglacial and Early Postglacial stratigraphy. By emptying (and screening) the contents of this "crater" we exposed its western and southwestern edges. Excavation of a maximally 1 m^2 test pit (Square V8) back from the southwestern edge of the pothole did provide a valuable window on and platform for radiocarbon dating of the Magdalenian levels to be exposed over another about 10 m^2 during successive excavation campaigns in the Corral area. That excavation, as it proceeded downward, created stratigraphic sections ("after the fact") along its southern, western, and northern edges. But the actual dissection of the Magdalenian levels (plate 1.2) was in reality guided not only by the (west- and north-trending) slopes of overlying levels displayed in those stratigraphic sections but also by following the inclinations of long bones, blades, cobbles, and sedimentologically discrete lenses (such as patches of ash and charcoal, red ochre, or flour-fine gray-green silt) as well as the presence of apparent features such as hearths, pits, and a possible rock wall. Despite all our best efforts to follow and dissect individual living floors in this area, the dark (grayish or blackish brown), organically and culturally rich nature of the sediments, the evidence for considerable prehistoric digging and other alterations (e.g., trampling, scuffing) of surfaces, the lack of continuous sterile (or even "poor") layers, their frequent indistinctiveness even in the sections, and the distance from those sections in some areas, all make it very difficult to be absolutely sure of the discrete integrity of individual levels. Movements of objects from the size of a bone splinter to that of a charcoal lump among these levels are clearly suggested by inconsistencies among the radiocarbon determinations. This problem could—in principle—be addressed by refitting, but this is made very problematic by the extremely small size of the flint artifacts (including the usually diminutive and "exhausted" beach pebble cores) and the triturated (marrow-cracked?) nature of the bone assemblages. The reality is that the Magdalenian deposits are time-averaged palimpsests that can be fairly reliably unraveled only in the immediate vicinity of either stratigraphic sections or definite prehistoric features such as hearths.

In order to be able to correlate the Upper/Epi-Paleolithic levels of the Cabin and Corral areas, a dog-leg connecting trench (1 m wide) was dug between the two. This excavation reached the top of the Lower Magdalenian complex of levels in three squares and (as of 2004) penetrated it and the Solutrean horizon in one (P6) (plate 1.5). It was also extremely fruitful because early Neolithic deposits were found in its three westernmost meters, increasing the sample sizes of artifacts and faunal and botanical remains for this period.

Despite the palimpsest problems introduced above, El Mirón Cave provides a spectacular opportunity to combine both axes of excavation—vertical (plate 1.6) and horizontal (plate 1.7)—in the study of change through time and spatial organization within a major place in the montane landscapes of late Quaternary Cantabria. While the cave and the hard limestone physical relief surrounding it remained essentially stable during the Upper Pleistocene and Holocene, surface geomorphology, climate, and vegetation changed substantially during the tens of millennia represented by the sedimentary infilling of the cave. It is our job to explore the interplay between humans and those aspects of stability and change through the course of the last glacial-interglacial cycle.

References Cited

Binford, L.

1980 Willow smoke and dogs' tails: hunter-gatherer settlement systems and archaeological site formation. *American Antiquity* 45:4–20.

Cabrera, V.

1984 *El yacimiento de la cueva del Castillo*. Bibliotheca Praehistorica Hispana 22. Instituto Español de Prehistoria, Madrid.

Rodríguez Casal, A.

1997 *O Neolitico Atlantico e as orixes do Megalitismo*. Universidade de Santiago de Compostela, Santiago, Spain.

CHAPTER TWO

The Río Asón Drainage in Eastern Cantabria

Lawrence Guy Straus

The Vasco-Cantabrian Region

Vasco-Cantabrian Spain is a distinctive, high-relief geographical region characterized by a coast closely abutted by a mountain chain, the Cantabrian Cordillera. It lies astride the forty-third parallel of latitude, forming the north-central margin of the Iberian Peninsula along the broad Atlantic embayment known alternatively as the Mar Cantábrico (by the Spaniards), the Golfe de Gascogne (by the French) or the Bay of Biscay (by the British). From east to west, it is made up of the modern provinces of Guipúzcoa, Vizcaya, Cantabria, and Asturias, which total no more than about 14,000 km^2. A physiographically well-defined region, it is about 350 km long but narrow: never more than about 50 km (and sometimes a mere 25 km or so) wide between the present shore and the cordilleran crestline (figure 2.1). The cordillera forms in part the continental divide between the Atlantic and the Mediterranean basins (the Río Ebro drains into the Mediterranean to the south of the mountains in their eastern—Basque and east Cantabrian—sector). The hyphenated name—Vasco-Cantabria—is used here to stress some important topographic differences between the eastern and western parts of the region. The coastal Spanish Basque Country (provinces of Guipúzcoa and Vizcaya) lies at the right-angle corner of the Cantabrian Sea, where the Pyrenees and Cantabrian Cordillera meet in a "chaos" of mountains that often plunge down to the sea with deep, narrow gorges but few broad valleys or stretches of coastal plain. The Basque Mountains—although including some dramatically steep, localized massifs such as the Sierras de Amboto, Gorbeia, Urbasa, Aitzkorri, or Aralar—are generally relatively low (maximum elevation 1,400–1,500 m, but usually much less) and have the most accessible passes between the Atlantic coast and the interior of Spain (namely Navarra, Alava, and eastern Burgos in Old Castile). These characteristics continue westward into the easternmost part of the province of Cantabria, which is now an autonomous administrative region but was historically part of Old Castile.

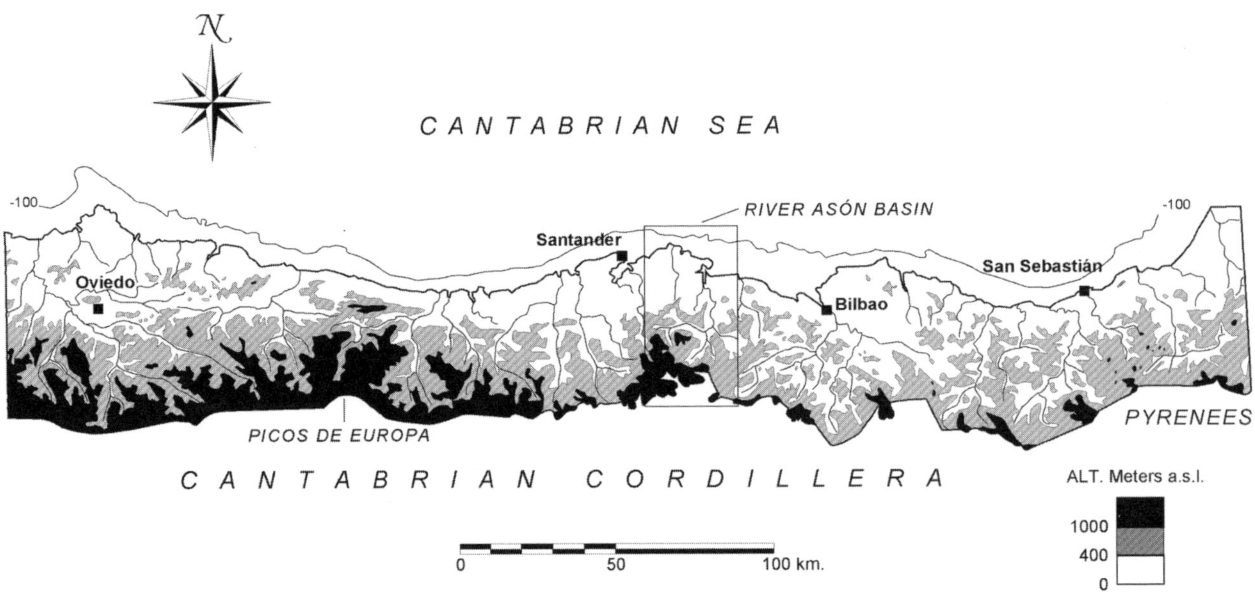

Fig. 2.1. Map of Vasco-Cantabria, showing the Cantabrian Sea (Bay of Biscay), mountain ranges, major rivers, and cities. Plan by L. G. Straus, redrafted by Ronald Stauber.

Westward, the cordilleran summits rise and, in the area where the coastal provinces of Cantabria and Asturias meet the interior province of León, they are abutted to the north by an awesome series of massifs: the Picos de Europa, whose summits—at or slightly over 2,600 m above present sea level—are the highest in northwest–north central Spain and as high as the highest peaks of the western Pyrenees, despite their proximity to the shore. The topography of central and western Cantabria and Asturias is, however, more regularly structured than that of the Basque Country. There is usually a coastal lowland zone (albeit often hilly), which is relatively broad in central Cantabria (around the present-day bay and city of Santander) and central Asturias (the triangle north of Oviedo between Cabo Peñas and the cities of Avilés and Gijón). This strip is generally flanked by coastal hill ranges rising from several hundred to slightly over 1,000 m above present sea level. In many sectors, south of the coastal ranges are intermontane valleys that are also oriented east-west (i.e., parallel to the shore). Still farther south lie the ever-higher, usually parallel chains of the cordillera itself and finally the 800–1,000 m high Meseta of Old Castile. The Vasco-Cantabrian region—dominated by Cretaceous and Carboniferous limestone bedrock—transitions into the region of ancient shield rock and lower relief that characterizes the northwestern corner of the Iberian Peninsula: Galicia. This area generally lacks caves and has a somewhat different prehistoric record.

Coastal and mountainous Vasco-Cantabrian Spain is also a land of many short, steep rivers in deeply entrenched valleys. From east to west, the major rivers include the Bidasoa, Urumea, Oria, Urola, Deba, Ibaizabal, Asón, Miera, Pas-Pisueña, Besaya-Saja, Deva-Cares, Sella-Piloña, and Nalón-Narcea. The Basque rivers are separated from one another all (or almost all) the way down to the Holocene shore, while the lower courses of rivers in much of Cantabria and Asturias are more readily connected along the broader coastal lowland zone. Under full glacial conditions of sea level regression, the coastal plain throughout the whole Vasco-Cantabrian region was only about 4–12 km wider than it is today under interglacial conditions—far less than off the Gulf of Gascony shore of Aquitaine or Charentes or even off the Gulf of Lions shores of Catalonia or Languedoc. So, intervalley communications within the Vasco-Cantabrian region in both the Last Glacial and in the Holocene were easiest in the lower courses and, in the former period, along the now-inundated continental shelf. Today the only continuous avenues of west-to-east communication (narrow-gauge railroad, national road, and the new divided highway) also closely follow the shore, except where mountain chains plunge down to it, as is most generally the case from eastern Cantabria eastward to the French border. Otherwise, the rest of the main roads and the wide-gauge railroads linking Madrid to the provincial capitals of northern Spain all run from the Meseta to the shore via Cordilleran passes and the main river valleys that cut northward from them. Above the coastal zone (where it is present at all), roads between valleys (generally only fairly recently paved) are steep, narrow, and switchbacked. Despite short distances on the map, travel over intervalley passes is far from easy, even today (and especially in a snowy winter).

Traditional life in northern Spain has always in historic times been organized along the lines of valleys. While of course never completely isolated—and indeed often connected by upland pastures and grazing practices (including various forms of transhumance and mutualism, trade, and intermarriage)—the valleys of Vasco-Cantabria have constituted fundamental social, economic, and even administrative units (as reflected by the modern boundaries of the rough equivalents of townships and counties). Until the (relatively recent) advent of mass access to automobiles, many people (unless drawn to industrial cities for employment—something that really became common only in the twentieth century) spent their lives within the valley from which their families had stemmed, visiting or marrying into adjacent valleys but rarely straying far afield except for commercial or administrative visits to the provincial capital or more distant places, for reasons of obligatory military service, for example. The traditional pastoral economy in the valleys of northern Spain was often largely self-sufficient, although there were long-standing trade connections with Castile and Andalucia for such basic products as olive oil or wheat, not at all or little produced in cool, humid Atlantic Spain (e.g., Caro Baroja 1943; Freeman 1979).

The present climate of Vasco-Cantabria is a temperate, oceanic one but with great differences in temperature between the coastal zone and the cordillera, although striking microclimates sometimes exist within major valleys of the montane interior (such as in the well-known case of the relatively broad valley of Liébana, surrounded by the Picos de Europa). Heavily influenced by the ocean, particularly because of the presence of a branch of the Gulf Stream in the Cantabrian Sea, the regional climate is highly pluvial and temperature ranges along the coast are equable (all climatic information is from Terán and Sole Sabaris 1968). Average annual rainfall totals in the coastal cities of Santander and Bilbao are 1,172 mm and 1,142 mm, respectively. Precipitation (including considerable snow, which, however, is very rare in the coastal lowlands) is even higher in the mountains—up to 1,500–2,000 mm per year. In contrast, just barely south of the cordillera in the Ebro (i.e., Mediterranean) drainage of the interior Basque province of Alava, average annual rainfall is already significantly lower (828 mm in the capital city, Vitoria-Gasteiz, but in the 600 mm range farther south in the province). In northern Castilla-León (Duero drainage) the aridity is even more marked, with average annual rainfall of 478, 380, 408, and 665 mm in the cities of Burgos, Aranda de Duero, Palencia, and León, respectively. Average annual temperature in Santander and Bilbao (both on the coast) is 13.6°C and 14.3°C, respectively, and the annual range is very limited. Average annual temperatures in the mountains are much colder (e.g., only 9.1°C at Reinosa in south-central Cantabria, whose elevation is 850 m above sea level) and annual variation in temperature is far greater in the mountains than along the coast. The deep, narrow valley floors of the montane zone (especially the south-north-oriented ones) are characterized by shortened periods of open sunshine (due to shade) and nightly cold air drainage. The annual variation in temperature on the northern Meseta of Old Castile is brutal, with long winters and mean January temperatures of only 3.3°C and 3.9°C in León and Burgos, respectively. Even Vitoria, with an average annual temperature of 11°C, has an average January temperature of slightly less than 5°C.

Touted in tourist propaganda as "Green Spain," Vasco-Cantabria is sometimes described as being like Ireland with mountains and like Switzerland with a seacoast. Prior to modern times, the temperate, pluvial climatic regime in the lower elevations favored the growth of dense, luxuriant, oak-dominated deciduous forests, whose other components included such trees as hazel, chestnut, elm, linden, maple, willow, poplar, and alder, often with a rich understory of shrubs. The upper elevations were dominated by beech and pine, with high mountain slopes covered with alpine meadow plants.

Today, rocky cliffs are often bare or characterized by *matorral*. Ericaceous heathlands are also found along exposed stretches of the coast, which includes raised platforms (*rasas*) in many sectors. The modern vegetation is, of course, significantly the product of human activity (charcoal burning, forest clearance for pasture and construction), particularly as this is a region whose economy has long depended heavily on ovicaprine and/or bovine pastoralism. The result is that the coastal zone and lower valleys are usually characterized by lush (and heavily manured) grasslands. Now that the dairy industry has been severely curtailed as a result of European Union milk quotas, it is possible that interglacial woodlands will once again spread. This is especially likely in the steep uplands, which are less easy or profitable to exploit from the point of view of rationalized large-scale cow herding but were better suited for the small-scale (and often transhumant) family-level production units so overwhelmingly characteristic of the valleys of northern Spain until recently. Native arboreal vegetation has largely been replaced by siviculture plantations of eucalyptus and Monterrey pine, especially in the hills and lower mountain slopes. This wood is used mainly for paper production. The north-south contrast between the vegetation of the Vasco-Cantabrian region and that of the upper Ebro Basin in the eastern sector and the northern Castilian Meseta in the west is stark and was also in the Pleistocene. The main factor in determining this difference is the year-round humidity so characteristic of the narrow Atlantic facade as opposed to the aridity of the Mediterranean world on the other site of the cordillera. This sharp ecological break is clearly perceivable as soon as one crosses any one of the cordilleran passes, such as Los Tornos. With the difference in climate and vegetation come immediately visible distinctions in the human land-use and settlement patterns even today.

Despite its overall characteristic of having a humid, equable, oceanic climate, and despite being a relatively small natural region, Vasco-Cantabria harbors within itself significant contrasts in elevation and relief, local climate, and hence native vegetation. In addition, although highly contrasted in many respects, its coastal and montane zones are located in very close proximity to one another and thus are ecologically (and economically) complementary. It is a region of short, narrow valleys sloping steeply down from a high cordillera across several fore-ranges to a narrow coastal lowland or—especially in the Basque Country—directly to the shore. The valleys connect the complementary habitats of the mountains and the coast and provide for the possibility of relatively high degrees of self-sufficiency, while the difficulty of intervalley communication could lead to significant social isolation. (The existence of the Basques—with their intervalley differences in dialect and customs—as well as such other ancient traditional groups as the Pasiegos in Cantabria and the Vaqueiros of Asturias, is clear evidence of this fact.) East-west communication is governed by the regional topographic structure and is thus mainly channeled along the coastal zone, and even there it is highly restricted in areas where the mountains plunge down to the sea, such as in the border areas between Cantabria and Vizcaya and between Vizcaya and Guipúzcoa. Although the modern administrative units into which this natural region is divided are four provinces (distributed among three autonomous regions)—which, given the physiography of northern Spain, are not surprisingly among the very smallest in the whole country—it is the valleys that constitute the historic entities in which, until the present, people have traditionally lived their lives and made their living. This way of life probably had very deep roots imposed by the bedrock environment of the region—back to the Neolithic and even back to the Paleolithic.

The Río Asón Drainage

The Asón is the easternmost major river of Cantabria. Its course is only 40 km long, but this is relatively long for the region (figure 2.2). Directly or through its tributaries (notably the Clarón and Clarín in the coastal zone, the Silencio and Carranza in the mid-valley, and the Calera, Gándara, and Bustablado in the upper valley) drains about 800 km^2 of eastern Cantabria plus a small area of westernmost Vizcaya. The Asón itself is born in a dramatic waterfall that gushes from the top of a U-shaped, moraine-filled glacial valley below 1,160 m high Sia Pass east of 1,707 m high Castro Valnera peak in the cordillera, whose other summits in this sector range around 1,300–1,500 m above sea level. The Calera flows down from Los Tornos Pass near where the provinces of

Cantabria, Vizcaya, and Burgos meet; at only 920 m, it is one of the lowest cordilleran passes. The source of the Gándara is in the high mountain valley of Soba between the first and second cordilleran ranges and between Sia and Tornos passes. The Bustablado descends from the west, down from 575 m high Alisas Pass, which divides the Asón drainage from that of the Río Miera and its estuarine mouth, the Bay of Santander. The upper Asón is separated from the upper Miera by a steep, south-north interfluve ridge whose summits range from 1,500 to 500 m.

The lower Asón drainage is separated from that of the Bay of Santander/lower Miera by 200–300 m hills that cut across the coastal strip inland of the Cabo de Ajo. The lower course of the Asón is separated from Vizcaya by a zone of particularly rugged relief, notably the 731 m high massif of Campo Ventoso, whose northern slope plunges straight down to the shore. It is this massif that forced the route of the coastal narrow-gauge railroad from Santander to go up the Asón and then up the Carranza into the interior of Vizcaya before going northward again to Bilbao. The old national road hugs the coastal cliffs via switchbacks in this sector around the Cantabria-Vizcaya border, and the new highway goes through a long tunnel.

The tidewater part of the lower Asón (known as the Ría de Treto) reaches as far inland as the town of Limpias—historically a port—about 9 km inland of the present river mouth between the cities of Laredo and Santoña. The estuarine zone of the Asón, although not very large in absolute terms because of the narrowness of the coastal strip and the high relief of this area, is one of the most extensive in Cantabria (figure 2.3). At Laredo the river presently turns eastward from its northerly course and flows into the ocean between the huge El Puntal sand spit and the southern base of the 376 m high limestone massif of Monte Buciero, which dominates Santoña. In earlier Holocene (and presumably Pleistocene) times, however, the river almost certainly flowed straight north, entering the sea to the west, between Buciero and 239 m high Cueto Cabrero Brusco. That paleo-channel is choked with dune-covered peat dated to medieval times (González Morales, Yudego, and Ituarte 2000). Monte Buciero contains two known Late Paleolithic–Mesolithic sites (El Perro and La Fragua) and a minor cave art locus (San Carlos). Other Mesolithic

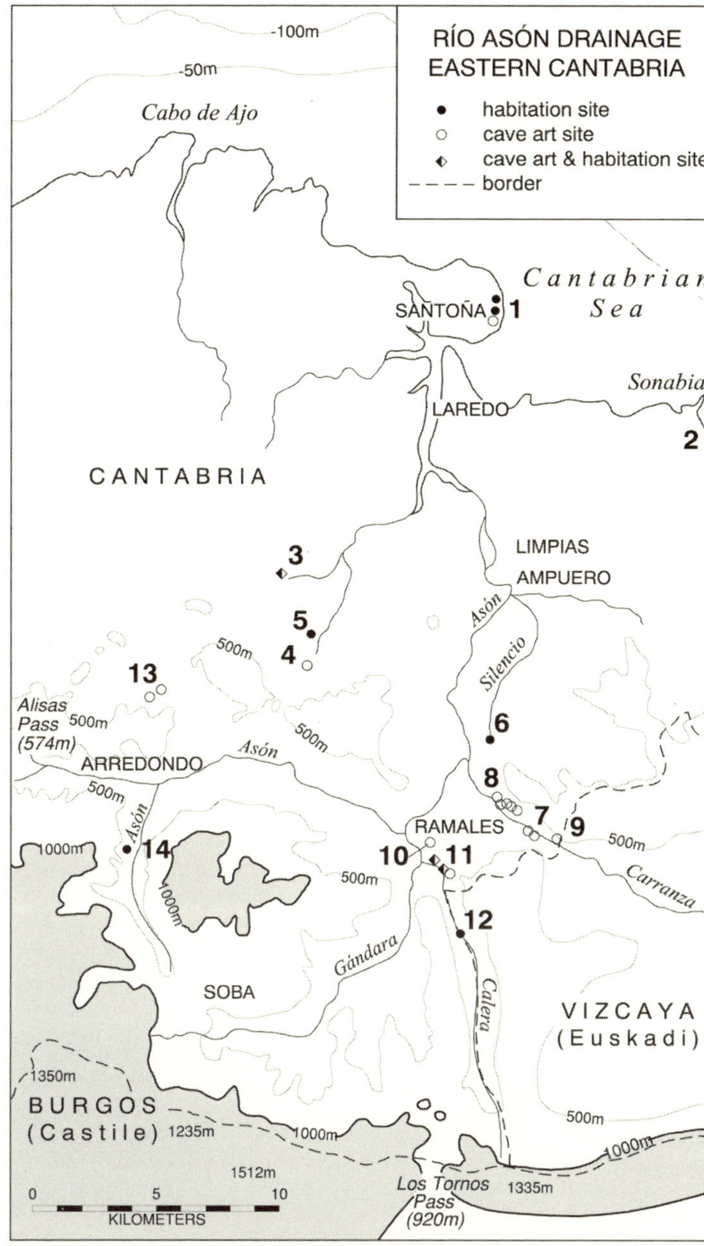

Figure 2.2. Map of Río Asón valley, with major prehistoric archaeological sites. 1. La Fragua, El Perro, San Carlos; 2. La Trencha; 3. El Otero; 4. Cobrante; 5. La Chora; 6. El Valle; 7. Sotarriza, Cova Negra; 8. Morro, Pondra, Areo A, B, and C; 9. Venta Laperra, Polvorín; 10. Cullalvera; 11. El Miron, Covalanas, La Haza, La Luz, El Horno; 12. Tarrerón; 13. Las Emboscados, Patatal, Confresnedo, Cubio Redondo; 14. Cubera. Plan by L. G. Straus, redrafted by Ronald Stauber.

Fig. 2.3.
Photo of the lower Asón valley (present-day tidewater sector). Photo by L. G. Straus.

Fig. 2.4.
Photo of menhir of Las Hayas on ridge crest east of the lower Asón valley. Photo by L. G. Straus.

sites once existed in and around Laredo (e.g., La Baja Cave) (González Morales 1990).

The estuarine zone is today characterized by extensive marshes drained by a maze of channels. Indeed, the area west of the Ría de Treto is a 5 km wide, low-lying plain with gently rolling hills. In the inland (Voto-Aras) part of this zone are three Upper and Epi-Paleolithic cave sites with occupation debris and art: El Otero (Gonzalez Echegaray, Garcia Guinea, and Begines 1966), La Chora (Gonzalez Echegaray et al. 1963), and, on the northern flank of the first coastal range, Cobrante (Rasines 2008).

In contrast, the eastern edge of the lower Asón is flanked by a steep range of hills, with summit elevations typically ranging between 200 and nearly 400 m. The ridge-crests both here and farther upstream are dotted with megalithic monuments (tumuli, menhirs, dolmens) in dominant spots that often have panoramic views of the valley (figure 2.4). There are other short inlets and valleys between this range and the Nervion estuary at Bilbao, but generally the landscape between the lower Asón and west-central Vizcaya is very rugged and difficult to cross, in either the west-east or north-south directions.

The Asón valley narrows quickly in the upstream direction, notably above the town of Ampuero. The river cuts through a series of foothill ranges with elevations up to about 200–550 m, as close as 8 km from the present shore at Laredo. Its middle course is closely paralleled by its tributary the Río Silencio, which emerges from the vast cave of El Valle at the foot of a 500–600 m range in the town of Rasines. This cave is one of the most classic, richest Upper Magdalenian-Azilian sites in the Vasco-Cantabrian region (Garcia-Gelabert and Talavera 2004). Immediately upstream of El Valle, the Asón is joined by its main eastern tributary, the Río Carranza (plate 2.1). This river drains a large area of westernmost Vizcaya and rises in the Montes de Ordunte (maximum elevation 1,104 m) at the common border of Vizcaya, Burgos, and the Cantabria enclave of Trucios. Some 16 km farther to the east, beyond the headwaters of the Carranza (a total of 30 km from its confluence with the Asón) and near the natural interior avenue of communication between Cantabria and the Bilbao area, is the major Paleolithic and post-Paleolithic cave site of Arenaza, with rupestral paintings very similar to those of the upper Asón caves of Covalanas and La Haza. The Carranza gorge, between the Cantabria-Vizcaya border and the Asón confluence, is lined with an extraordinary concentration of Paleolithic occupation and cave art sites, notably Venta de la Perra, Polvorín, Arco, Pondra, and Sotarriza—none but the second of which, unfortunately, has been excavated with modern methods (González Sainz and San Miguel 2001). Polvorín has recently yielded two Upper Magdalenian harpoons, a travertine slab engraved with a horse image, and other artifacts (Ruiz Idarraga 2007).

The Asón passes through a deep, narrow, 3 km long gorge (El Portillo) immediately above its confluence with the Carranza. This dramatic chokepoint effectively separates the middle from the upper Asón, as at this point the river cuts through the northernmost and lowest of the cordilleran ranges, with summits in the 600–900 m range. At the head of this gorge, at the town of Ramales, the valley opens up into the relatively broad (ca. 1 km), long (12 km), east-west-trending Ruesga Valley. This is the intermontane, dramatically tectonic segment of the Asón's course, flanked to the south by the precipitously steep, jagged crest of the Sierra del Hornijo, whose summits range from 1,000 to over 1,400 m above sea level (plate 2.2). The mountain valley of Matienzo, high above the north side of the Ruesga Valley, contains numerous caves, at least two of which (Patatal and Los Emboscados) have Upper Paleolithic art and one (Cubío Redondo) a Mesolithic shell midden site (Ruiz Cobo and Smith 2001). At the town of Arredondo (quixotically self-styled as "the Capital of the World"), the tributary Bustablado, flowing down from the west below Alisas Pass, joins the Asón, which itself turns a right angle south. The highest part of the Asón's course is a very narrow, mountain gorge (in part glacially U-shaped in profile), 7 km long and dominated by peaks and ridges over 1,000 m in elevation. There is a little-known site of possible Magdalenian age (Cubera) at the base of the west face of this gorge (Chaline 1965).

At Ramales two mountain streams join the Asón: the Gándara and the Calera (figure 2.5). On the edge of the town is the mouth of the huge cave of Cullalvera, with paintings and engravings of Magdalenian style about 1 km from the gaping mouth. The Gándara drains the high, isolated Soba valley, which lies between the Sierra del Hornijo and the main chain of the cordillera, whose peaks here range between 1,200 and 1,300 m. The Gándara spills out into the Calera above Ramales through an extremely narrow gorge between 723 m high La Mortera and 957 m high Pico San Vicente at the eastern end of the Sierra del Hornijo. The Calera flows down from Los Tornos Pass, forming the border between Vizcaya and Cantabria down to the eastern flank of La Mortera. At that point—the top of its "hanging valley" segment—it disappears into a karstic system to resurge near its confluence with the Gándara, at the foot of the dramatic, cirquelike El Haza cliff, below the caves of El Mirón and Covalanas—the latter a UNESCO World Heritage rupestral art site (Garcia Diez and Eguizábal 2003). Only after prolonged, intense rains does the Calera also flow in its thalweg over the edge of this cliff between La Mortera and Monte Pando. At this montane "barrier" there are not only the two above-mentioned cave sites plus La Haza Cave (with a small Solutrean site and paintings nearly identical to those of Covalanas) on the western side (Garcia Diez and Eguizábal 2007) but also El Horno (Upper Magdalenian and Azilian—a small cave but rich in mobile art and ornaments, like El Valle) at the base of the absolutely vertical La Pared cliff (Fano 2008) (overlooked by several gigantic caves such as Cuevamur, which may have been

Fig. 2.5. Photo of the upper-middle Asón valley (Ruesga Valley) from the north, showing the Sierra del Hornijo (*right*), La Busta Peak (*center*), and Monte Pando (*left*). Photo by L. G. Straus.

connected to El Mirón) and nearby Cueva de la Luz (with two Solutrean points and rupestral engravings), as well as Tarrerón and Las Pajucas caves (Mesolithic, Chalcolithic) on the southern valley side above the town of Lanestosa (Apellaniz 1971). Solutrean artifacts have also been found during house construction at the upper end of the town of Ramales, at the base of the cliff near La Haza Cave (Bohigas et al. n.d.).

The drop at the northwest edge of the upper, "hanging" segment of the Calera valley down to its confluence with the Gándara is about 100 m. Overall, the El Haza cliff is about 200 m high, the top half of which is sheer limestone rock, the bottom half being a scree talus. The dramatic relief of the area around Ramales (the deep Asón valley, the steep mountains and vertical cliffs that loom above it, the "hanging valley" of the Calera between Lanestosa and Monte Pando) is the product of major tectonic movements that included an uplifted wedge and various fault blocks (Prieto and Vidal 1995) (plate 2.3).

The old Camino Real ascends in ramp-like fashion along the top of the talus from Ramales to the edge of the Calera's hanging valley at the narrow notch between Pando and Mortera mountains. In contrast, the modern paved road climbs from the Calera-Gándara confluence up the steep, lower, northern slope of Mortera in a series of sharp switchbacks. The El Haza cliff presents a very dramatic facade to the inner cordillera (plate 2.4). Besides being riddled with caves, it is certainly no accident that the strategic, ecotonal Ramales-Carranza-Lanestosa area has such an impressive concentration of Paleolithic and post-Paleolithic sites. The juxtaposition of the major east-west Ruesga Valley with the high mountain habitats of the cordillera, together with the avenues of communication that lead down from both Vizcaya and Burgos across the mountains down to the Asón, make this area both a natural complement to the coastal zone and a significant "meeting" place between the Cantabrian environment and those of the Basque Country and Castile. El Mirón and the other sites are clearly located at a crossroads of different Iberian "worlds."

References Cited

Apellaniz, J. M.
1971 El Mesolítico de la cueva de Tarrerón y du datación por el C14. *Munibe* 23:91–104.

Bohigas, R., F. López, R. Martinez, J. A. Moure, E. Munoz, and R. de Prado
n.d. Informe espeleológico sobre el proyecto de trazado de la Variante de Ramales de la Victoria. Unpublished contract archaeology report to the Gobierno de Cantabria, Santander, Spain.

Caro Baroja, J.
1943 *Los pueblos del norte de la Peninsula Iberica*. Consejo Superior de Investigaciones Prehistóricas, Madrid.

Chaline, J.
1965 Observaciones preliminaries sobre los terrenos cuaternarios en los alrededores de Arredondo. *Cuadernos de Espeleología* 1:21–26.

Fano, M.
2008 La Prehistoria del Valle del Asón. Excavaciones en la cueva de El Horno (Ramales de la Victoria), 2000–2001. In *Actuaciones arqueológicas en Cantabria: 2000–2003*, edited by R. Ontañón, pp. 73–78. Gobierno de Cantabria, Santander, Spain.

Freeman, S. Tax
1979 *The Pasiegos: Spaniards in No Man's Land*. University of Chicago Press, Chicago.

García Diez, M., and J. Eguizábal
2003 *La cueva de Covalanas*. Gobierno de Cantabria, Santander, Spain.
2007 Los dibujos rojos de estilo paleolítico de la Cueva de La Haza (Ramales de la Victoria, Cantabria): estudio monografico. *Munibe* 58:177–222.

García-Gelabert, M. P., and J. Talavera
2004 *La Cueva del Valle, Rasines, Cantabria, España*. British Archaeological Reports S-1252. Oxford, UK.

González Echegaray, J., M. A. Garcia Guinea, and A. Begines
1966 *Cueva del Otero*. Excavaciones Arqueológicas en España 53, Madrid.

Gonzalez Echegaray, J., M. A. Garcia Guinea, A. Begines, and B. Madariaga
1963 *Cueva de la Chora*. Excavaciones Arqueológicas en España 26, Madrid.

González Morales, M.
1990 La prehistoria de las Marismas: excavaciones en el Abrigo de la Peña del Perro (Santoña, Cantabria). *Cuadernos de Trasmiera* 2:13–28.

González Morales, M., C. Yudego, and C. Ituarte
2000 La prehistoria de las Marismas: prospección arqueológica de la zona del bajo Asón y Marismas de Santoña y toma de muestras en los yacimientos de las cuevas del Otero, La Chora y El Valle. In *Actuaciones arqueológicas en Cantabria, 1984–1999*, edited by R. Ontañón, pp. 151–153. Gobierno de Cantabria, Santander, Spain.

González Sainz, C., and C. San Miguel
2001 *Las cuevas del desfiladero*. Universidad de Cantabria and Gobierno de Cantabria, Santander, Spain.

Prieto, J., and R. Vidal
1995 Nuevo modelo tectónico interpretivo para la Cuña de Ramales (Cantabria) y su relación con la zona noroccidental de la Cuenca Vasco-Cantábrica. *Munibe (Ciencias Naturales)* 47:29–36.

Rasines, P.
2008 Excavaciones arqueológicas en la cueva de Cobrante (San Miguel de Aras, Voto). In *Actuaciones arqueológicas en Cantabria, 2000–2003*, edited by R. Ontañón, pp. 61–66. Gobierno de Cantabria, Santander, Spain.

Ruiz Cobo, J., and P. Smith
2001 El yacimiento del Cubío Redondo (Matienzo, Ruesga): una estación mesolítica en Cantabria. *Munibe* 53:31–55.

Ruiz Idarraga, R.
2007 Cueva de El Polvorín (Valle de Carranza). *Arkeoikuska* 06:136–137.

Terán, M. de, and L. Sole Sabaris (eds.)
1968 *Geografía regional de España*. Ediciones Ariel, Esplugues de Llobregat, Spain.

CHAPTER THREE

El Mirón Cave

Lawrence Guy Straus

El Mirón Cave is located at the eastern edge of Cantabria province, in the township of Ramales de la Victoria, within a kilometer of the border of the province of Vizcaya (Basque Autonomous Region). It is also close (12 km south) to the border of the province of Burgos (Old Castile). In terms of longitude and latitude, the cave is at 43°14'48" North × 3°27'05" West of the Greenwich Meridian. In UTM this translates to x = 463353.83, y = 4788392.82 (range 30, zone 30T). The huge mouth of El Mirón Cave is imposing because of its size and location (plate 3.1). At 260 m above sea level (a.s.l.), it is eminently visible from the valley floor at the Gándara-Calera confluence (100 m a.s.l.), looming from high on the rampart-like eastern end of the Ruesga Valley at Ramales (70 m a.s.l.). It is, of course, not the only vast, gaping cavern entrance in this area of highly karstic Aptian (Lower Cretaceous) limestone.

Immediately east of El Mirón on the opposite, equally vertical face of Monte Pando (La Pared del Eco) are the immense mouths of Cuevamur (plate 3.2) and Cueva de la Luz. Far less visible, opening out onto a low-lying dolina on the edge of Ramales, is the gigantic mouth of Cullalvera (figure 3.1), but unlike the caves high up on Monte Pando, it is just the end of a very long cavern—to date measuring 16 mapped km. Despite its paintings and engravings located nearly a kilometer from the mouth (and a faint rupestral painting recently discovered near the mouth), not much remains of a once-extant Paleolithic living site in Cullalvera (attested during recent walkway construction near the mouth by the finding of an engraved bone whose style, like that of the rock art, is suggestive of late Magdalenian age), since a violent current flows from its mouth after periods of prolonged, intensive rain, actually causing floods in areas of Ramales in front of this imposing cave (González Sainz, Muñoz, and Morlote 1997). Although not nearly as long as Cullalvera, in terms of easily accessible area El Mirón—at about 130 m deep—is still an impressive cavern. In contrast, Cuevamur has essentially no level habitable area, although, through an interior chimney, it does lead back to an extensive series of galleries that have been explored by experienced speleologists. Much less extensive, Cueva de la Luz opens out

Fig. 3.1. Photo of Cullalvera Cave mouth. Photo by L. G. Straus.

slightly above and to the south of the base of the sheer, 200 m high La Pared del Eco cliff and contains an unexcavated archaeological site whose surface finds suggest an Upper Paleolithic component (Junta Directiva de la F.C.E. 1995)—including two Solutrean points and rupestral engravings (Montes, Muñoz, and Morlote et al. 2003; Gómez et al., 2006). (We recently obtained a ^{14}C date on a bone from the exposed Cueva de la Luz deposit that tends to confirm a Solutrean age.)

Immediate Geographic Location and Location

El Mirón is at the southern end of the vertical sector of the cirquelike, 150 m high El Haza cliff (figure 3.2). Beginning immediately to the south of the cave the slope of Monte Pando (475 m a.s.l.) becomes gentler, leading down to the notch between it and La Mortera mountain (719 m a.s.l.) at the upper end of the "hanging" upper Calera Valley. Thus, although the scree-covered talus immediately below El Mirón is extremely steep, the cave is readily accessible from the Calera notch. This is in fact the route chosen by civil engineer Alfredo García Lorenzo when he directed construction of the road up to the cave art site of Covalanas, directly above El Mirón, in the early 1950s. Ever since then the access to El Mirón has been via an approximately 125 m path that runs northward, straight across the Pando slope from the first switchback of the unpaved Covalanas road, about 50 m above the elevation of the Calera notch. Since it is unlikely that the scree talus below El Mirón was any less steep in Pleistocene times (and may well have been even steeper), it is reasonable to conclude that the southern slope was also the easiest way into the cave during the Last Glacial. From El Mirón's location adjacent to the notch, prehistoric humans (foragers and pastoralists) could—with varying degrees of ease—exploit the surrounding slopes, the upper Calera Valley, the Soba Valley via the Gándara, and the Ruesga Valley of the Asón. And they could see animals at a great distance toward the west from the mouth of the cave, especially during times when arboreal vegetation was scarce or nearly absent in the Ruesga Valley and on its bordering slopes (plate 3.3). This cave is an "eagle's nest," but it is a large one, and it is not hard to reach on foot. Its surroundings—the sheer cliffs of Monte Pando, itself a shoulder of El Moro (818 m a.s.l.); the towering slopes of pyramidal Pico San Vicente (909 m a.s.l.) (plate 3.4); and the serrate crest of the Sierra del Hornijo (with peaks up to 1,200–1,400 m a.s.l.)—are grandiose. The cavern is spacious, generally dry, and well-oriented vis-à-vis the sun and the north winds, and its vestibule is apparently unencumbered by blocks. All of these qualities—with varying degrees of relative importance—would have made El Mirón exceptionally attractive to humans throughout millennia as a site both for long-term residence and for other functions. Until recently El Mirón was used as a stable and—within the living memory of townspeople—as a human residence (putatively that of gypsies). The Haza and Pared del Eco cliffs, and indeed the El Mirón Cave arch itself, are currently very popular rock-climbing venues, with numerous established piton routes.

Fig. 3.2. Photo of El Mirón Cave mouth from the west. Covalanos is above it, slightly to the left, behind trees. Photo by L. G. Straus.

Physical Description

The El Mirón mouth is 16 m wide just inside the dripline and (as measured by laser scanning) about 19 m high (plate 3.5). There is presently a boulder-strewn terrace at the foot of the vertical cliff-face in front of El Mirón; it is no more than about a dozen meters wide and not much longer than the cave mouth itself. It is marked at its northern edge by a large mound of sediments that are believed to be the contents of the mouth of Covalanas that were shoveled out by García Lorenzo's workmen to make that cave more accessible for tourists (Garcia Diez and Eguizábal 2003). The size and configuration of the terrace in front El Mirón during the Last Glacial are unknown, but one can assume that it was no larger than it is now. At present there is a shallow rockshelter at the base of the sheer cliff directly north of the cave mouth. It has not yet been archaeologically tested.

El Mirón Cave is formed at the intersection of a vertical and a horizontal diaclase, both spectacularly straight and obviously following structural planes of weakness within the otherwise extremely hard limestone. Neither on the surface nor in the areas we have excavated have there been many large roof- or wall-fall blocks. A striking exception is the large block with engravings that outcrops at the center-rear of the vestibule. Inspection of the vestibule ceiling reveals it to be smooth, with no obvious recent collapse scars. It also bears the stubs of several thick but *dry* (i.e., inactive) stalactites, as well as a few large active ones (at the very front and midpoint of the vestibule). These stalactites are clearly of great antiquity, again indicating the long-term stability of the ceiling—not surprising given the hardness of the limestone. Beginning shortly inside the mouth per se, the height of the vestibule ceiling is very consistently about 13 m above the nearly level present ground surface of this chamber (as measured by E. Torres, a speleological topographer).

The vestibule measures 30 m deep from the dripline to its rear wall (excluding a narrow, 2 m deep alcove in the southeast corner) (figure 3.3). The outer part of the vestibule is slightly funnel-shaped, quickly narrowing from 16 to 11 and then to 9 m wide, at which point (midway back from the dripline) it becomes nearly parallel-sided, with a width of 7 m. At the right rear (northeast corner) of the vestibule the cave narrows by half, to a width of 3–4 m. This narrow passage is 20 m long and has a colluvial ramp sloping steeply (ca. 30°) upward toward the inner cave (plate 3.6). This slope represents the erosional face of an ancient alluvial deposit that probably had originally filled all or most of the vestibule as well as the inner cave (plate 3.7). The slope continues (slightly less steeply) back another 10 m, but the cave in this stretch expands again to a width of 7–8 m, which it then maintains all the way back at least to the end of its accessible interior.

Sunlight penetrates essentially to the top of this slope, where the surface of the alluvial in-filling levels out, at about 10 m above the ground surface in the vestibule. There are remnants of breccia adhering to the cave walls above the slope (plate 3.8). A bone encased in this capping travertine has been radiocarbon dated to about

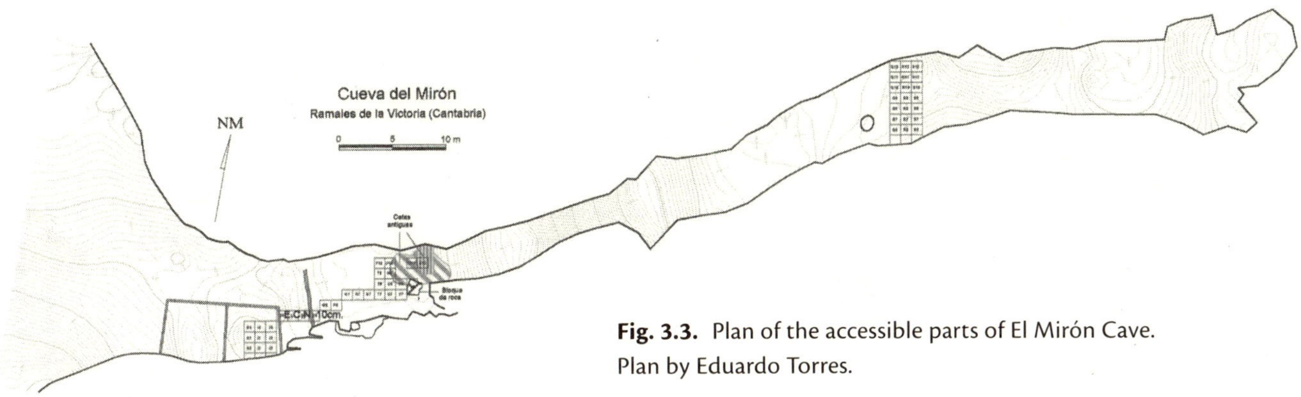

Fig. 3.3. Plan of the accessible parts of El Mirón Cave. Plan by Eduardo Torres.

10,500 BP. Flints are also visible in the breccia. This breccia attests to a westward prolongation of the horizontal top of the alluvial infilling as recently as the very end of the Pleistocene and to the presence of Azilian-era visits to the inner cave. Under this component and in direct contact on the slope with the now-exposed ancient colluvial-alluvial fill there had once been a wedge of dark brown, organic- and artifact-rich Magdalenian deposits, a basal remnant of which was preserved in a small concavity of the southern cave wall about one-third of the way up the slope (Niche A) (plate 3.9). This remnant has been ^{14}C-dated to 16,600 BP. The erosional face of the slope has obviously retreated significantly during the Holocene (attested by the presence of fossilized bones and Upper Paleolithic–style osseous and lithic artifacts in post-Mesolithic levels in the Cabin area), probably due to a combination of running water (which still trickles down the slope today after prolonged periods of rain [notably in winter] that must enter the inner cave through some small, as-yet-undiscovered chimney or crack) and movements up and down the slope by humans and animals. Earlier and much more violent, powerful flows of water out of the cave are indicated by the presence of deep, wide grooves gouged on a downward angle (roughly parallel to the ramp slope) in both cave walls of the narrow ramp passage. These flows are also attested by the very large water-worn cobbles that are major components of the colluvial deposits of the slope (as documented by a trench dug midslope by the project sedimentologists).

The inner cave is dark and can be explored only with artificial lighting. It is accessible for about 60 m beyond the top of the slope. At this point, since the cave ceiling remains nearly level throughout its length, the alluvial deposit (capped by travertine) reaches to (or almost to) the ceiling. The cave seems to bifurcate, and a long, impenetrable passage that trends northeastwardly can be perceived by lying on the ground surface. The alluvial infilling has clearly undergone water erosion within the inner cave as well as on the slope (and, presumably, catastrophically in the vestibule). The center of the inner cave has a channel running down the middle, with "terrace" remnants, capped by travertine, lining sectors of the cave walls. The channel was obviously once considerably deeper than it is now in the center of the inner cave, since the 1950s trench that cuts across it (straightened and deepened by us) has revealed a stratigraphic sequence of at least medieval, Bronze Age, and Magdalenian levels, all dated by radiocarbon and totaling nearly 1 m in depth (figure 3.4). This sector of the inner cave (about 16 m from the top of the slope and about 33 m from the end of the accessible area) is extremely wet, with very active ceiling drips forming mondmilch (unconsolidated calcium carbonate precipitate) so quickly as to have already covered the 1950s backdirt mound parallel to the west side of the trench. After a period or periods of intensive flow, the channel that was excavated in the ancient alluvium—so clearly marked by the fossilized terrace remnants adhering to the cave walls along its sides—refilled with alternating layers of clay and mondmilch, interrupted by at least one spectacular episode of burning in the Middle Ages, around AD 1100. All this attests to an active inner cave that is certain to have had geomorphological repercussions on the more habitable vestibule area, namely in the form of the continuing transport of alluvial clays, silts, sands, and cobbles from inside toward the exterior.

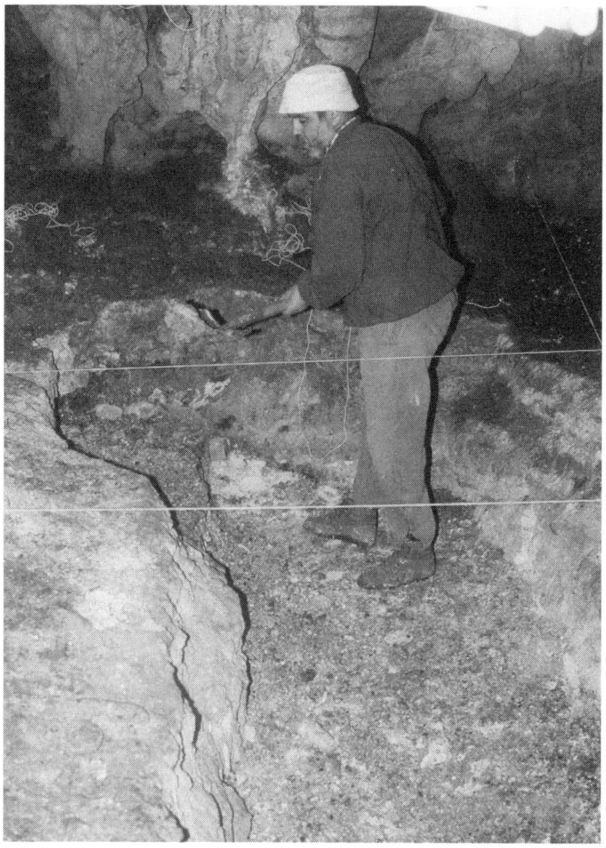

Fig. 3.4. Photo of the Inner Cave Trench. Photo by L. G. Straus.

Recent Uses of the Cave and Conditions at the Beginning of the Excavations

El Mirón Cave was being used to stable livestock (probably mostly goats) in the late twentieth century. Local residents still living during the time of our excavations were among those who kept animals in the cave. Besides the obvious wooden and stone corral walls and feeding troughs, the existence of stone foundations of two contiguous square structures, together with the presence of domestic cultural debris on the surface and in mixed fill in the vestibule attest to residential uses of the cave in the recent past. Such debris includes pottery (even glazed ware), glass, remnants of shoes, tins, and so on. It also includes an automatic pistol clip (with bullets) found apparently hidden and covered with a thin level of ashy silt in a small alcove, two-thirds of the way back along the south wall of the vestibule (figure 3.5). This last artifact could conceivably be testimony to use of the cave during the Spanish Civil War (1936–1939) or even shortly afterward, when guerillas (known under Franco's regime as "bandits") continued to be active in the Cantabrian Mountains and elsewhere. Sadly, the recent residential use of El Mirón is also evidenced by our discovery of human remains in mixed fill just below the surface behind a block in the corral area at the south rear of the vestibule. These remains could be of gypsies whose use of the cave was related to us by local inhabitants.

The stone structures of the outer vestibule are well-built, dry-wall foundations made with large, naturally square or rectangular blocks of limestone (probably from the talus scree at the base of the cliff just outside the cave mouth) (plate 3.10). The walls are straight and—at least for the inner of the two "rooms" or "cabins" that we emptied of debris—today rise only about a meter above the packed dirt floor. The lowest course seems to have been at least in part (especially the south wall segment of the inner room) set into the ground by way of a foundation. In short, preparation by digging was required to make these rooms. The outer of the rooms measures slightly more than 5 m square, while the inner one (where we excavated) is slightly less than 5 m square. Each structure is formed by a thick, common north-south wall, two other right-angle walls, and the

Fig. 3.5. Photo of pistol clip hidden in an alcove of the cave vestibule. Photo by L. G. Straus.

slightly bowed southern wall of the cave. There is no door or passage between the two rooms. The rest of the structures had collapsed some time ago, with many of their wall blocks filling the room interiors. Many others may have been used subsequently to construct the wall that bisected the vestibule, undoubtedly to contain livestock within its rear area. It is unknown whether the "cabins" originally had been roofed or not. If they were, the beams and other wooden elements of the structures had already disappeared (perhaps scavenged for reuse elsewhere) by the time we began our work. The outer of the two rooms is placed several meters back from the dripline, in a highly sheltered location, but with daylong sunshine.

We chose to excavate within the walls of the inner of these two rooms on the hunch that the rooms might have been spared from major interior hole-digging perhaps during the past century or so. They were in addition full of wall-collapse blocks, which would have made recent pothunting in those places neither enticing nor easy. Finally, observation of the vestibule ceiling revealed it to be dry, smooth, unscarred by block-falls, and marked by the presence of the stubs of large, old stalactites, all of which testified to long-term stability and the likelihood that this area would not contain large blocks that could hinder our excavation. Significantly, the rooms were built just to the south of two areas of permanent dripping from the center of the cave ceiling. These drips can become very intense during periods of prolonged rainfall, with water obviously percolating through the sediments immediately below their points of impact. They come from large, active stalactites on the ceiling, probably along a crack in the limestone bedrock in the axis of the vertical diaclase. Obviously, not only from the standpoint of commodity of excavation but also from that of likely preservation problems underground, we wanted to avoid digging in these constant drip zones.

The other stone structure we encountered in the El Mirón vestibule—which we eventually dismantled—was a 5.5 m long wall, about 1 m high, running roughly south-north across the point at which the vestibule narrows to a 6–7 m width, about two-thirds of the way back from the dripline. This wall was constructed of mainly large blocks (possibly scavenged from the cabins), but not as carefully coursed as the room walls. It had no foundation; the lowest course lay on top of (or just barely below) the 1996 ground surface, suggesting that the wall had been built not too long ago (although it was present in 1973, when Straus first visited the cave). This wall stopped about 1 m short of the north cave wall, permitting access to the vestibule rear. No gate was preserved. This wall probably served to keep livestock in the vestibule rear area.

The ground surface of the vestibule as we found it in 1996 was approximately level all the way back to its rear, with a very sharp inflection upward at the base of the colluvial-alluvial slope at the north rear. The absence of Chalcolithic and Bronze Age deposits in the mid-vestibule and of Neolithic, Chalcolithic, and Bronze Age ones in the vestibule rear suggests that these Holocene-age sediments may have been removed in sub-modern times to provide a level surface for livestock stabling and perhaps as a source of fertilizer for fields in the valley floor below the cave. Certainly, recent human alterations to the uppermost prehistoric archaeological sequence were far more serious in the vestibule rear than in the front, where a Bronze Age horizon lay immediately below the ovicaprine excrement–covered floor of at least the innermost of the two historic house rooms or cabins.

References Cited

Garcia Diez, M., and J. Eguizábal
2003 *La cueva de Covalanas*. Gobierno de Cantabria, Santander, Spain.

Gómez, A., A. Chauvin, F. Savanti, E. Gutiérrez, and M. L. Serna
2006 El registro arqueológico de la Cueva de la Luz: nuevas aportaciones. *Sautuola* 12:75–88.

González Sainz, C., E. Muñoz, and J. M. Morlote
1997 De nuevo en La Cullalvera (Ramales, Cantabria). Una revision de su conjunto rupestre paleolítico. *Veleia* 14:73–100.

Junta Directiva de la Federación Cántabra de Espeleología
1995 Informe de la Federación Cántabra de Espeleología sobre la zona de Ramales. *Asociación Cántabra para la Defensa del Patrimonio Subterráneo, memorias 1995*, pp. 132–146. Santander, Spain.

Montes, R., E. Muñoz, and J. Morlote
2003 Découverte de gravures pariétales à la Cueva de la Luz (Ramales de la Victoria, Cantabrie, Espagne). *International Newsletter on Rock Art* 37:1–7.

CHAPTER FOUR

Stratigraphy of El Mirón Cave (1996–2007)

Lawrence Guy Straus
and Manuel R. González Morales

Given its very large size, we obviously wanted to sample different areas of El Mirón Cave—especially the sunlit vestibule, but also the dark inner cavern. However, we are fully cognizant of the complexity and internal variation that generally characterize the stratigraphy of caves, particularly when their infillings are deep and have been subject to a wide range of natural and anthropic processes. Correlations between stratigraphic units in separate excavation areas—even if they are relatively close in space—can often be tenuous and problematic (Farrand 1975, 1985, 1988, 2000; Laville, Rigaud, and Sackett 1980; L. Straus 1997; L. G. Straus 1979, 1990, 2000; Woodward and Goldberg 2001). Stratigraphy should be first and foremost descriptive of what has been empirically observed in each area of excavation—independent of hypotheses about inter-area relationships or potential explanations of formation processes. "Genetic" and climatic interpretation of the deposits is the job of the sedimentologist and micromorphologist (in this case, William Farrand [Farrand 2001a, b, see also this volume; Straus et al. 2001] and Marie-Agnès Courty [Courty and Vallverdu 2001], respectively, together with their students [e.g., Hubbard 2000] and colleagues), as well as the paleomagnetism specialist (Ellwood et al. 2001). The responsibility of the archaeologist is to attempt to excavate by "natural" levels and to describe them as they reveal themselves, making a concerted and continuous effort not to mix materials from what appear to be distinct stratigraphic units. The key is to follow an inductive methodology in dealing with stratigraphy, allowing the lenses and levels to show themselves and trying deliberately to be as objective as possible in their discernment and description. And—something that is often very difficult, especially when sterile layers are absent—the archaeologist must try, as he or she digs down, to assure that assemblages of finds represent objects that were really deposited *relatively* contemporaneously. The problem of palimpsests is serious enough without the archaeologist compounding it by cutting across and therefore "lumping" the contents of distinct layers. But there is no guarantee that we have not done so, especially in the dark, organic, archaeologically rich Early Magdalenian layers of the cave. Stratigraphic definition is a

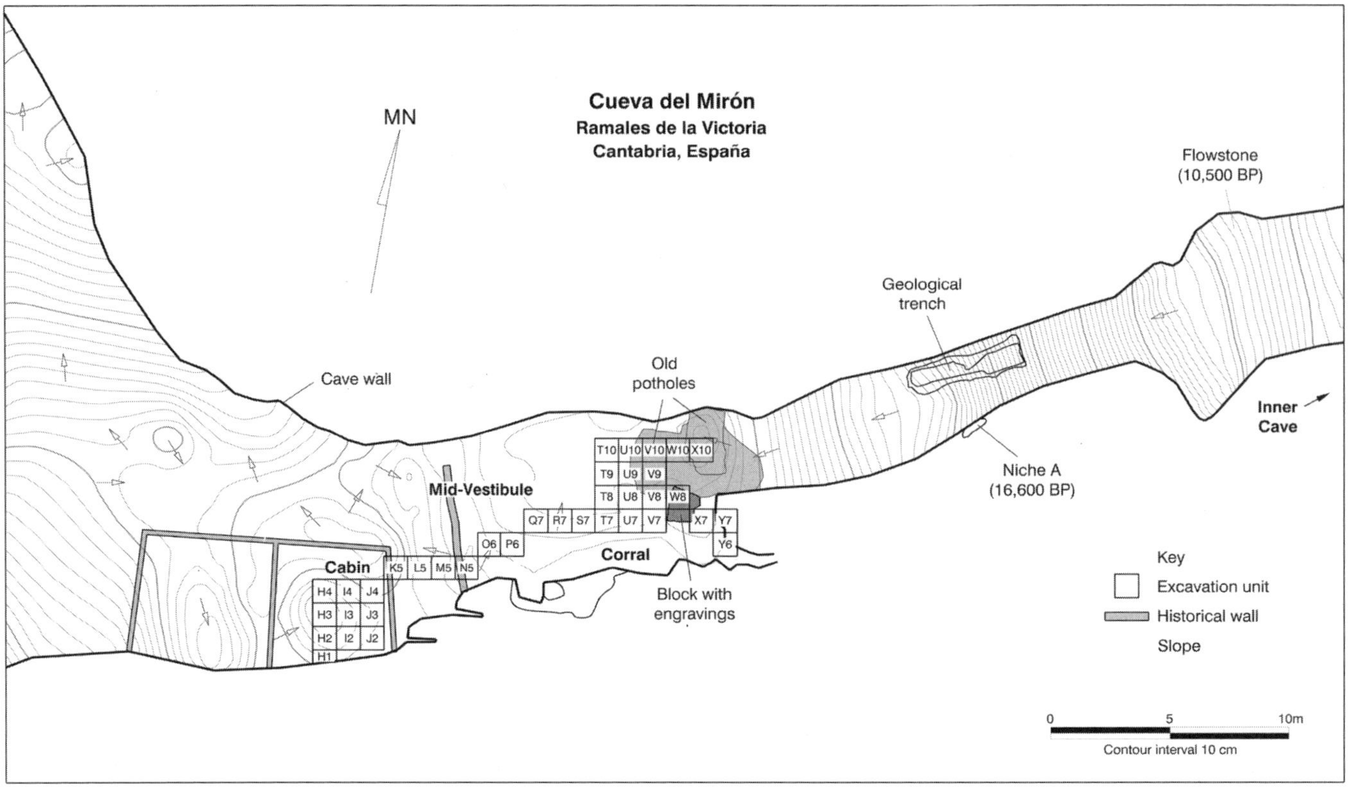

Fig. 4.1. Plan of El Mirón Cave vestibule. Plan by Eduardo Torres, redrafted by Ronald Stauber.

daily work in progress during the course of any excavation—especially in a cave with thick, complex deposits. The conduct of excavation in very thin "spits" ("dissection" layers) within palimpsest levels can help mitigate the problem of artificial "archaeological mixing," but only to the extent that we manage to accurately follow the prehistoric "lay of the land" within each meter and quarter-meter square, taking our clues from the inclination of long or flat bones, stone slabs, and artifacts.

In the case of El Mirón Cave, two excavation areas—each measuring 9–10 m^2, depending on depth—were opened: one (the Cabin) in the inner (eastern)-most of two historic-era cabins near the front of the south side of the vestibule and the other (the Corral) in a recent stable area at the vestibule rear (figure 4.1). (Exact surface areas excavated varied slightly: larger with depth in the Corral because of the outward-sloping sides of the old pothole, but slightly smaller with depth in the Cabin because of the generally inward-sloping cave wall.) We described a stratigraphy separately for each area, and for at least the uppermost parts of the deposits the two are totally different. In order to understand the stratigraphic relationships between the Cabin and Corral areas, a narrow, dogleg trench was dug between them down to the top of the early-middle Magdalenian levels. This Mid-Vestibule Trench provides strong evidence for stratigraphic continuity below the levels pertaining to the time of the middle Holocene, but with significant lateral facies variation in the 15 m between the western end of the Cabin area and the eastern edge of the Corral excavation. The stratigraphies of the various excavation areas (from the cave mouth toward its rear) are described as we encountered them, in reverse stratigraphic order (i.e., from top to bottom). For geological and cultural purposes, they are described sedimentologically, paleobotanically, faunistically, and archaeologically, respectively, in the other chapters of this volume. The present chapter is a chronicle of the levels as we dug and defined them. The stratigraphic sections are labeled with reference to the walls (e.g., north, west) of the excavation area that they represent and specifically the meter squares between (or within) which they lie (e.g., J–K/1–4: section between Squares J1–4 and K1–4; W–X/8: section midway through Squares W8 and X8). All the stratigraphic sections have been drawn and photographed (both on film and digitally). Most of the vestibule sections were drawn by Straus but the inner cave ones were drawn by González Morales, in both cases assisted by various students. All were re-drafted by Ronald Stauber (cartographer at the University of New Mexico Office of Contract Archeology) under Straus's supervision.

Cabin Stratigraphy

As of 2002 we had excavated the Cabin area (averaging about 10 m²) to a depth of about 2.75 m below the 1996 ground surface. At the time we began work, both cabin rooms were filled with stones (collapsed wall blocks), some modern garbage, loose dung, and sediments. This surficial fill was removed from the whole inner (eastern) cabin in early June 1996, exposing a relatively compact, grayish surface rich in ash and charcoal. This mottled, about 3–10 cm thick layer probably constituted

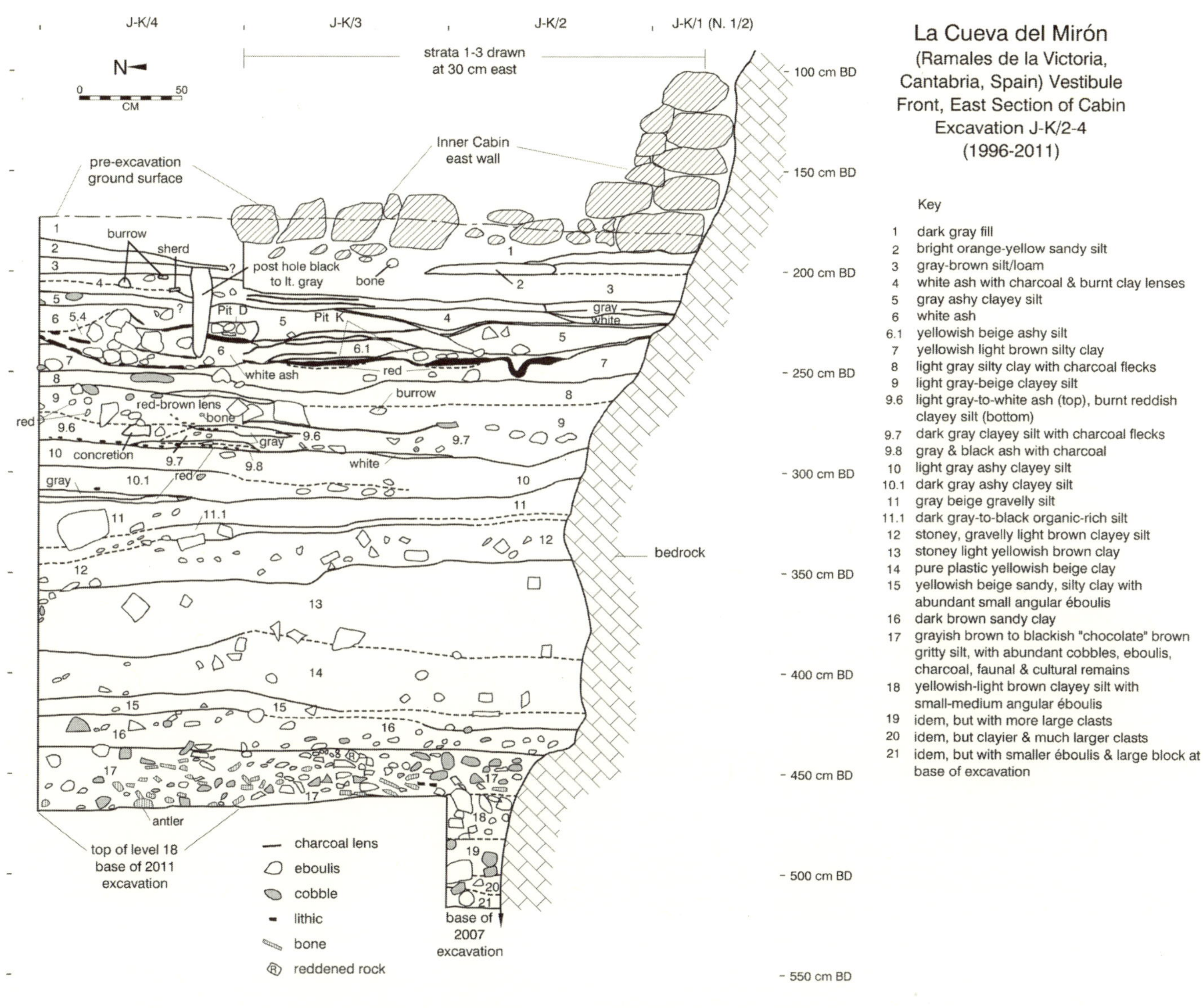

Fig. 4.2. East section of the Cabin excavation area (J–K/1–4). Section by L. G. Straus, redrafted by Ronald Stauber.

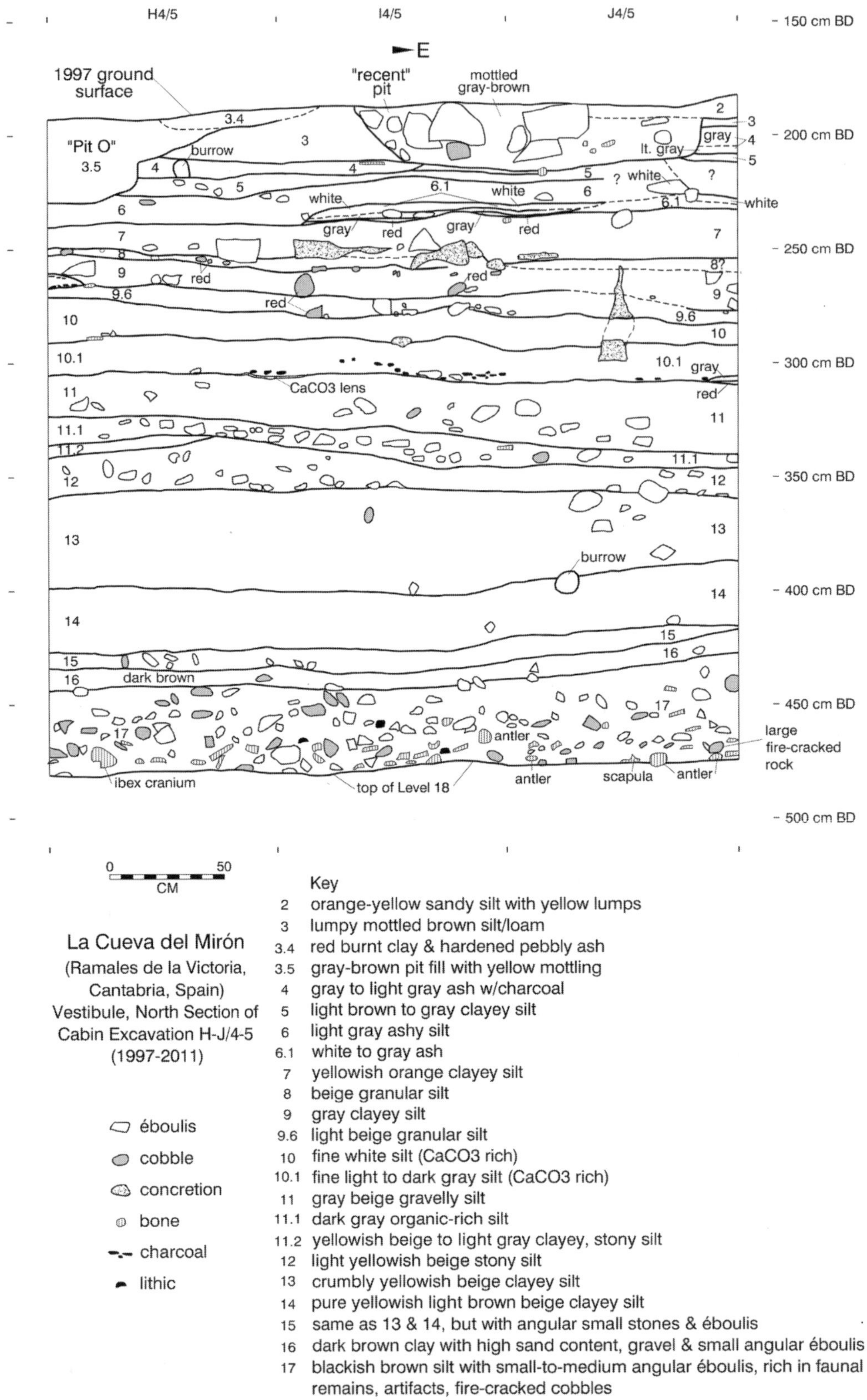

Fig. 4.3. North section of the Cabin excavation area (H–J/4–5). Section by L. G. Straus, redrafted by Ronald Stauber.

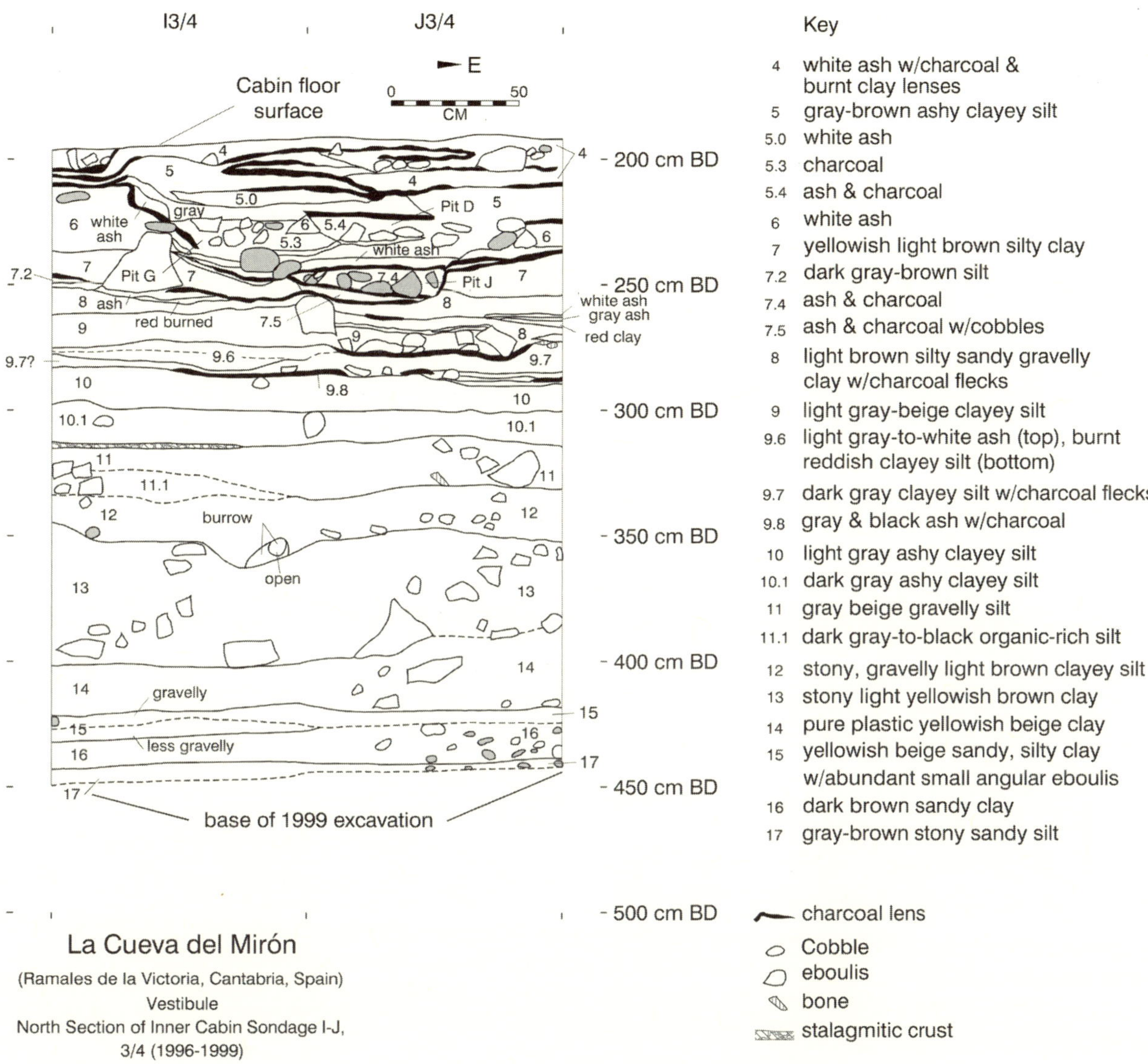

Fig. 4.4. North section of the Cabin excavation area sondage (I–J/3–4). Section by L. G. Straus, redrafted by Ronald Stauber.

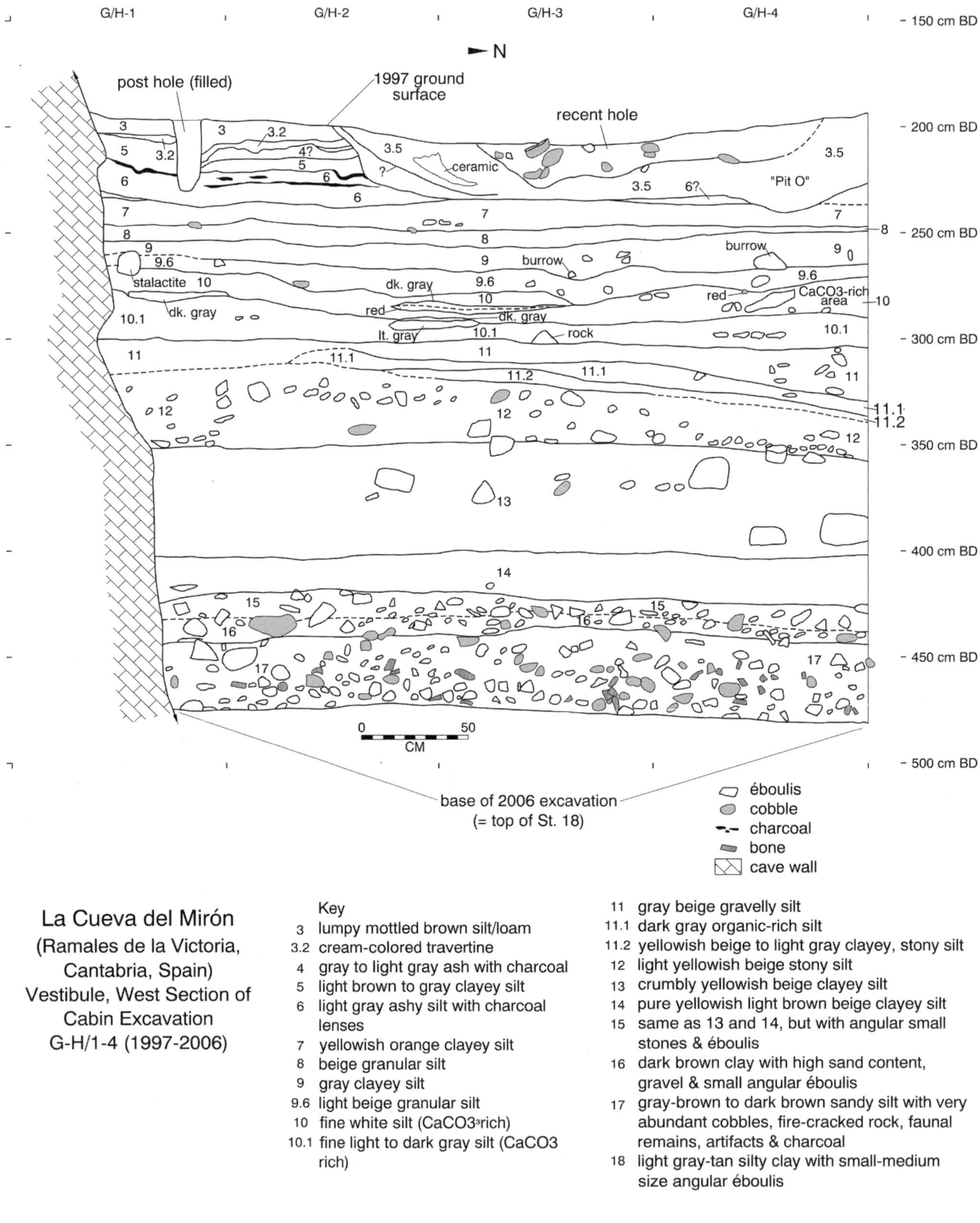

Fig. 4.5. West section of the Cabin excavation area (H–G/1–4). Section by L. G. Straus, redrafted by Ronald Stauber.

the de facto floor of the cabin and contained light (in weight and color), recent-appearing bones of domesticated animals, although at least the lower part of this layer (1) might not be significantly mixed with modern material. Once this Stratum 1 was dug through in a small, ad hoc test trench in the southeast corner of the Cabin, we encountered a compact, beige (locally yellowish), gritty, clayey silt layer (2) with domestic animal remains, together with sherds and flints, but no modern artifacts. It quickly became apparent that the contents of this layer were at least fairly homogeneous and probably pertained to the Bronze Age, based on the characteristics of the ceramics. (However, it was sometimes impossible, when first encountered, to separate in practice from Stratum 1—hence the many finds labeled "1–2.") Stratum 2 seemed to be discontinuous, consisting of beige or yellow patches embedded in a larger, more continuous stratum (3), which is a dark gray or grayish-brown, clayey silt rich in charcoal, bones, and organic matter as well as non-modern (thick, grit-tempered) sherds and flints. Although perforated by postholes and a couple of broader pits, all of which may have pertained to the recent cabin, these Bronze Age levels are essentially intact. In their fill there are no extraneous modern finds. The staggering conclusion that we reached is that modern/recent people and their animals had been living essentially atop an ancient prehistoric surface that had nevertheless remained largely undisturbed for three millennia.

Since the Cabin was the first area where we began systematic excavation, we decided to label the levels in the Cabin area from 1 to 99. Decimal numbers were assigned to localized lenses, pit fills, and levels perceived ex posteriori (or after the fact, as in the case of Level 10.1, whose darker gray color relative to Level 10 was seen in sections only after an L-shaped, 3 m² test pit had cut through it). Normally, the decimal-designated lens(es) lie(s) below the level with the same numerical designation. Each different excavation area was assigned a different range of 100 numbers for stratigraphic designations (i.e., 1–99, 100–199, 200–299 . . . Roman I–C). Criteria for distinguishing new levels and lenses are of a common sense, ad hoc nature. They include color, texture, sedimentary content (including clast amount and size), and archaeological (cultural) content, individually or (more often) in combination.

By far, the visibly most complex part of the stratigraphy that we have uncovered to date in El Mirón is the upper meter or so of the Cabin sequence, corresponding to Levels 1 to 9.8: Bronze Age, Chalcolithic, and late Neolithic. Within this 1 m thickness, which is riddled with small and large pits (sometimes pits dug within pits), we defined 33 distinct levels and lenses (many being limited-area pit fills). It was practically very difficult to follow many of these units continuously in excavation, owing to the mass of pits. The more significant strata in the mid-Holocene time period are numbered Levels 3, 4, 5, 6, 7, 8, and 9, but most of these are interrupted to varying extent (notably by a large pit, trench, or channel [3.5] of possibly late Bronze or Iron Age that cuts through early Bronze and Chalcolithic Levels 3–6 and down to the top of Level 7 in the northwest corner of the Cabin excavation area and that in turn was cut into by a "recent" pit). Some horizons (especially Level 4) are clearly very complex palimpsests of burning episodes (pockets, piles, lenses of ash with patches and flecks of charcoal). Typically, in the ash lenses or zones there are very few surviving cultural or faunal remains, perhaps in part due to the effects of fire, although it should be noted that while bones are often charred, they are not calcined, indicating low-temperature fires (Shahack-Gross and Weiner, n.d.). Such deposits may well represent the firing of dung and straw to cleanse the cave areas being used for stabling animals (i.e., sheet trash bonfires).

Many of the pits (some being meter-wide circular holes, notably in Levels 4, 5, and 7) are filled with fire-cracked rocks and rounded cobbles. Some of the latter were probably used as grinding stones. There are also numerous possible postholes, often associated with the large pits in Level 4. Other kinds of combustion activity is indicated by Lens 3.4, which is bright red with white ash patches and small white pebbles, possibly related to metallurgy (?), since surrounding Level 3 yielded pieces of apparent slag and a copper pin. In short, the top meter of the Cabin stratigraphy is a *highly* anthropogenic deposit; humans and their domesticated animals were the main agents of accumulation and alteration of these sub-recent sediments. However, apparent episodes of localized ponding water and calcium carbonate precipitation are attested by such lenses as 3.1 and 3.2—the first with, but the second without, any cultural remains. Excluding Level 1, the top meter of the Cabin stratigraphic sequence dates between slightly more than 5250 BP (Level 9.6, late Neolithic) and slightly less than 3700 BP (Level 3, early Bronze Age). Despite the complications of intensive burning and digging by post–Atlantic phase human occupants of the cave, these strata and those that underlie them in the Cabin area are bedded essentially horizontally.

Underlying the complex and extraordinarily anthropogenic uppermost levels, there is a 10–20 cm thick, continuous layer (10) of fine, whitish silt that is rich with calcium carbonate but less abundant in cultural and ungulate faunal remains. This unit has been dated by two nearly identical AMS determinations of around 5700 BP on charcoal samples. Level 10 also lacks the wealth of pits, dug-out hearths, bonfire ash mounds, fire-cracked rocks, and postholes that characterize most of the overlying levels. Before its formation there seems to have been a significant hiatus (perhaps nondepositional in nature, due to the lack of any perceptible erosional unconformity), since immediately underlying Level 10.1 (darker gray, intensively $CaCO_3$-enriched silt) is a layer dated by three determinations on charcoal to between 8380 and 9550 BP. This 12–18 cm thick layer is also continuous and lacks anthropogenic features. Indeed, it is extremely poor in any cultural or macro-faunal remains. In turn, there was another depositional hiatus sometime before the formation of Level 10.1, either beneath undated Level 11 or directly below 10.1. The latter hypothesis is perhaps more likely, based on the cultural similarities (i.e., Azilian-type artifacts) between Levels 11 and 11.1, the latter level being dated to 11,720 BP. Each of the hiati seems to represent more than 2,000 "missing" years. The causes for these gaps in deposition are unclear.

Level 11 is also a grayish-beige, gravelly silt, and 11.1 is a dark gray silt rich in organic matter and cultural remains. Clearly the latter is a relatively significant occupation lens centered on a hearth with some fire-cracked rocks and whose charcoal has been dispersed. Level 11.2, with small numbers of similar Azilian-like artifacts, is another thin, localized lens of clayey, stony silt that is lighter in color than 11.1. These two lenses are underlain by a more substantial but archaeologically rather poor stratum (12) consisting of light brown-beige, clayier silt with some angular *éboulis* (limestone rock fragments). Most of the artifacts and faunal remains were found in the upper part of this 10–35 cm thick level. Among the cultural finds is a fragment of a unilaterally barbed antler harpoon closely associated with a large lump of charcoal that was radiocarbon dated to 12,970 BP. Level 12 is thus attributable both archaeologically and chronometrically to the Upper Magdalenian. Continuing downward from Level 12, as we excavated the Cabin area we found that the sediments tended to get clayier with depth. Levels 13 and 14 intergrade in terms of their yellowish-beige, clayey silt matrix, but there is a zone of more abundant, medium-size éboulis at the base of 13 that defines the interface between the two strata. Otherwise, stones are rare in both. Level 13 is some 25–50 cm thick and Level 14, 20–25 cm. Archaeological finds are extremely scarce in Level 13 (14,930 BP) and in the upper part of 14. There is a small concentration of artifacts together with some animal teeth and corroded bones near the base of Level 14 (14,600±190 BP). The process of excavation, flotation, and screening of these two clayey levels was extremely time-consuming and arduous. Not enough organic matter was found to reasonably permit radiocarbon dating. Clearly, during the period of time corresponding to the Middle Magdalenian, the southwestern (front) part of the Mirón vestibule was little used by people, following a time of intensive occupation of this area in the Early Magdalenian that had concluded with the formation of Level 15.

The top of Level 15 also intergrades with these overlying clayey layers but contains many more gravels and small éboulis. The base of the level is less clayey, darker brown, loose, grainy silt. Thin (5–10 cm), Level 15 is culturally rich (lithic artifacts—mainly debitage—but very little bone) and is dated by two conventional ^{14}C determinations to around 15,100 BP. Level 16 (15–20 cm thick) is a dark brown, sandy, clayey silt with gravel, water-worn cobbles, and small éboulis. Lithic artifacts are abundant, but the heavily fragmented faunal remains are in poor condition. It is dated by AMS to 15,180 BP. Level 17 (30–40 cm thick) is a less sandy, gritty silt generally of somewhat lighter brown color and even less clayey than overlying Level 16, with which it intergrades. The big difference between these two, otherwise rather similar, strata is the enormous wealth of occupation debris in Level 17, notably well-preserved faunal remains, including many large bones. This level is also characterized by areas of densely packed, medium-size rocks (both cobbles and éboulis), some or many of which may have been "manuports" (including hammerstones and anvils, or at least humanly arranged pavement and hearth elements [fire-cracked rocks], even if the sources of the cobbles were probably close or very close at hand). There are several hearths rich in charcoal-stained sediments in just the 9 m² excavated, making for a Level 17 that is locally quite black. Three conventional and two AMS ^{14}C determinations that are in generally coherent stratigraphic order place Level 17 between 15,370 and 15,700 BP.

At the base of Level 17 there is a sharp stratigraphic break. The underlying levels (18–21) are lighter (yellowish-brown), clayey silt–silty clay, with angular limestone éboulis that tend to be increasingly large with depth. These levels (excavated in 2007 in Square J2, a 1 m² sondage adjacent to the south cave wall) are far poorer in cultural remains (artifacts, manuports, and fauna) than Level 17, although none is completely sterile. Level 21, at the (arbitrary) base of this test pit, was dated on a lump of charcoal (identified as *Juniperus* sp. by M. Ruiz under the supervision of L. Zapata) to 16,050±40 BP. There are, as yet, no clear temporally diagnostic artifacts from Levels 18–21.

Mid-Vestibule Stratigraphy

WESTERN SECTION (K–N/5)

The doglegged Mid-Vestibule Trench (1 m wide) consists of Squares L–N/5, O–P/6 and Q–S/7. Only the top 10–20 cm of Square K5 was dug, for fear of collapsing the tall, standing eastern (J–K/1–4) stratigraphic section of the Cabin area. Likewise, S7 was only very partially excavated to avoid damage to the southwestern corner of the Corral area stratigraphic sections. The

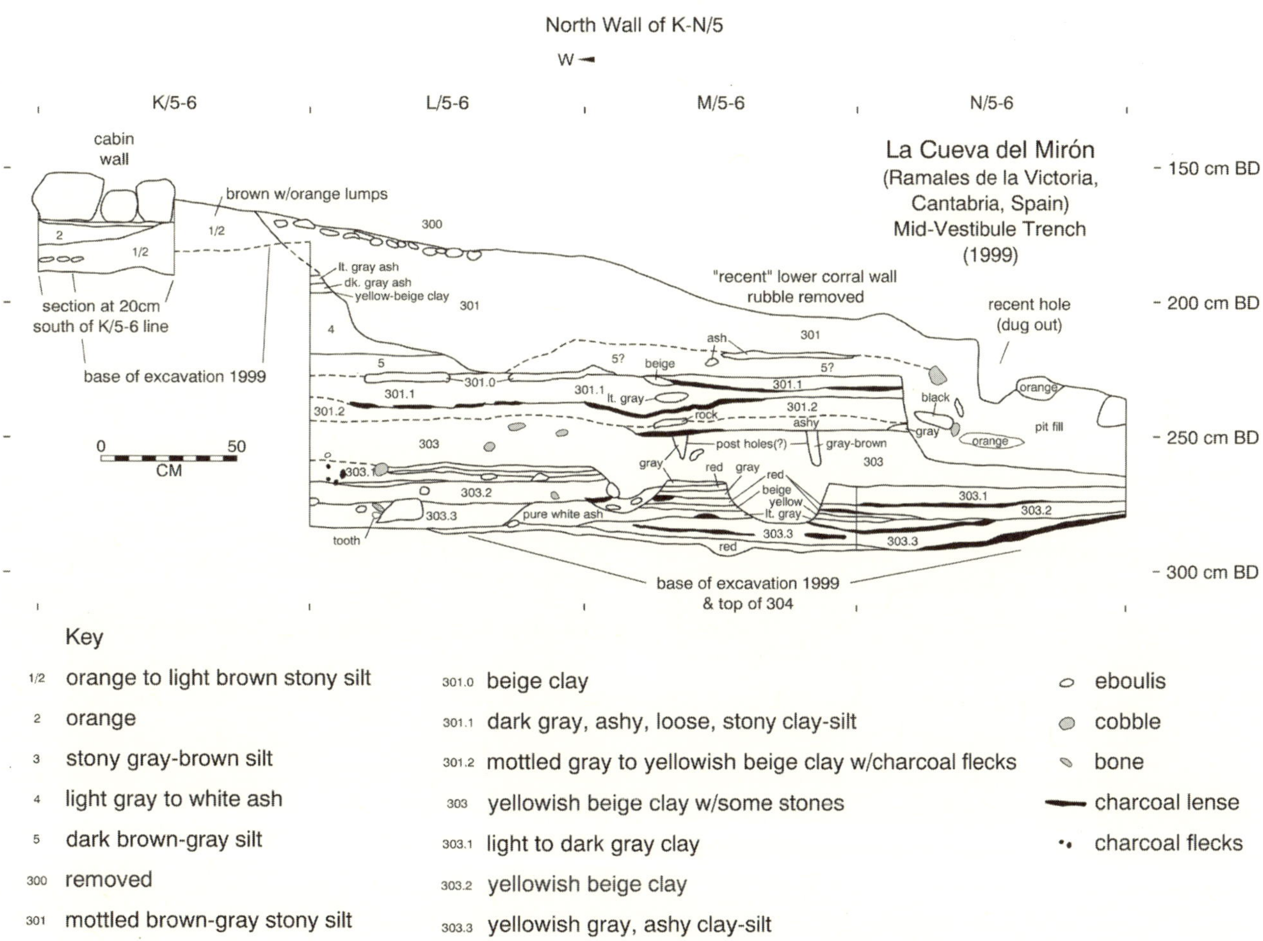

Fig. 4.6. North section of the Mid-Vestibule Trench, western end (K–N/5–6). Plan by L. G. Straus, redrafted by Ronald Stauber.

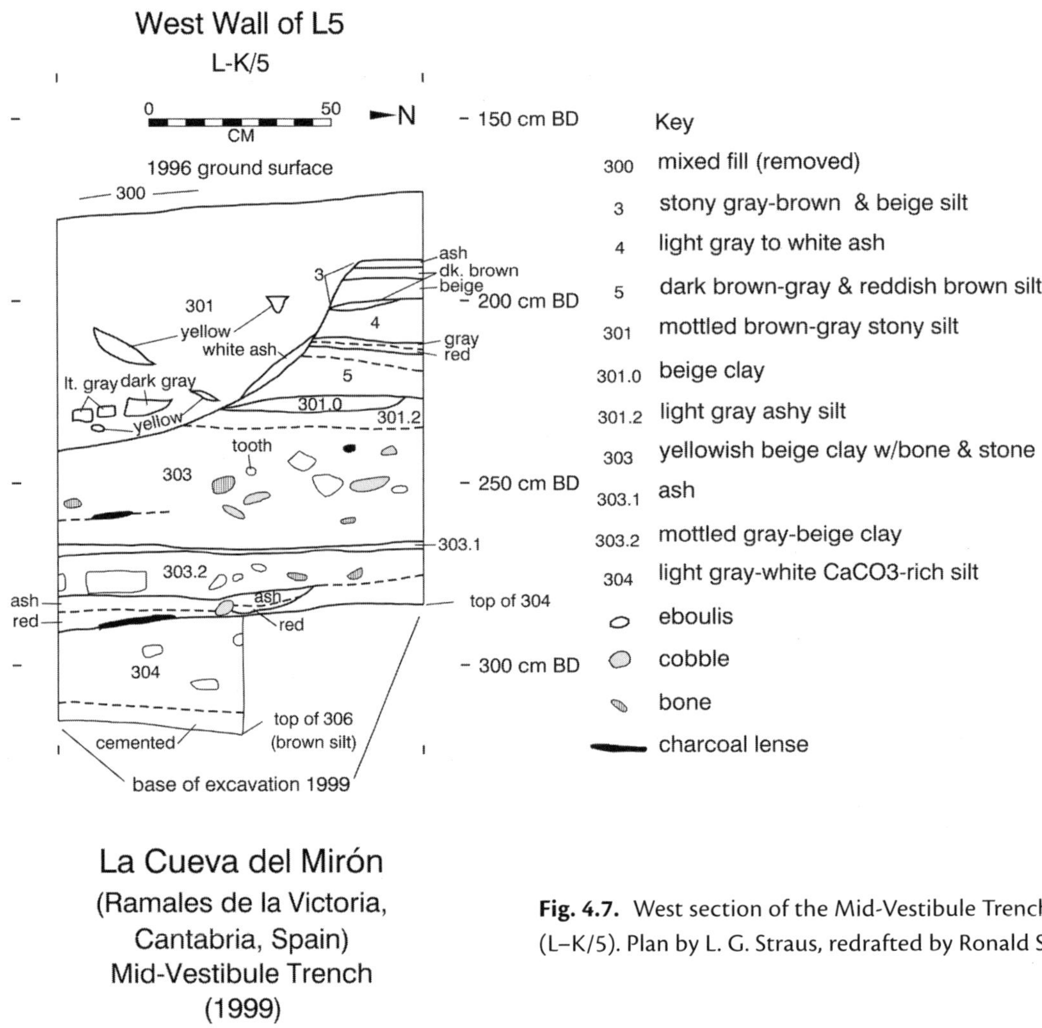

Fig. 4.7. West section of the Mid-Vestibule Trench (L–K/5). Plan by L. G. Straus, redrafted by Ronald Stauber.

Mid-Vestibule Trench seems to show the most complex recent depositional and disturbance history in the site, at least insofar as we know it.

At the far western end of the trench (K5 and the western edge of L5), there is a continuation of the adjacent Cabin area stratigraphy in the top 75 cm (at most), with remnants of Strata 1, 2, 3, 4, and 5 fairly clearly represented, albeit only locally. However, beginning near the K–L/5 square boundary it is apparent that the topmost (Bronze Age) levels had been dug out in more recent times and replaced by a fill with mixed (ancient and modern) archaeological contents (Level 301). This digging cut into ash- and charcoal-rich Level 4 at the western end of L5 but seems to have stopped in some places at a deposit possibly equivalent to Level 5 but for which no continuous, physical connection with the Cabin area has been made. Farther east (in M–N/5), hypothetical Level 5 has also been cut into by pit digging of uncertain age. The fill (partly Level 301 but also other, older—and more likely intact/unmixed—deposits called Lenses 301.0 and 301.1—which definitely seem to be sandwiched *between* Strata 5 and 303, unlike 301 per se, which is not found to actually underlie Stratum 5) contains several charcoal patches, alternating with beige or gray clay, some stones, bones, and possible slag. Along the northern edge of L–M/5 this is underlain by another lens (301.2), which is also gray, clayey silt rich in ash and speckled with charcoal. Archaeological finds are rare in these lenses—all attesting to considerable burning as well as digging, presumably during the Chalcolithic. These lenses (301.0–301.2) are probably roughly equivalent to Cabin Levels 6–7. The pit(s) filled

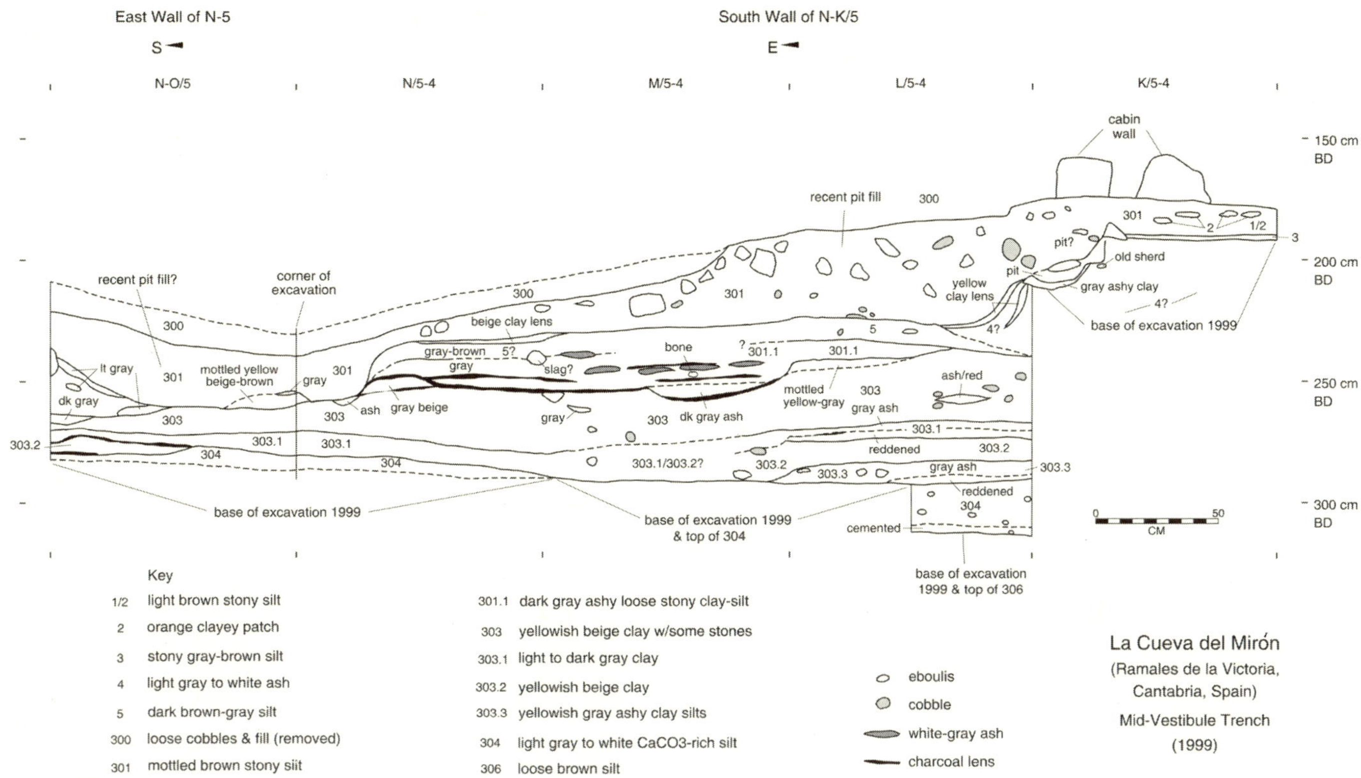

Fig. 4.8. South and east sections of the Mid-Vestibule Trench (N–K/5–4, N–O/5). Plan by L. G. Straus, redrafted by Ronald Stauber.

with mixed Level 301 material seems to have cut partially through the 301.0–301.2 lenses as well as through Levels 2–5. This complex situation is made all the more difficult to decipher because all the levels are ashy, clayey silts, with many thin, discontinuous patches and lenses, often visible only ex post facto in stratigraphic section. Below the upper section disturbances of 301, the strata in L–N/5 are bedded virtually horizontally.

Continuing the stratigraphic description in the L–N/5 sector, there is a thick (ca. 10–60 cm) stratum of yellowish-beige clay with isolated stones (Level 303). It contains relatively abundant faunal remains, undecorated ceramic sherds, and a limited number of lithics. Level 303 is cut into by a few possible postholes, apparently dug down from one of the overlying 301 subdivisions. In turn, 303 is underlain by a series of thin lenses (themselves often internally subdivided by charcoal, pure white ash, and burnt red clay lenses and patches) that alternate between ashy, yellowish-gray to yellowish-beige, clayey silt (303.1–303.3). Strata 303 and 303.1 continue laterally into Squares O–P/6 (albeit with interruptions by posterior pits) before pinching out probably near the P–Q/6-square border.

Levels 303, 303.1, and 303.3 are radiocarbon dated respectively to 5500, 5520, and 5790 BP. These dates are equivalent to those from Cabin Stratum 10, and the ceramics are identical (plain, well made, thin walled)—early Cantabrian Neolithic. (With ^{14}C standard deviations included, the stratigraphic equivalence of 303.1 and 303 could be extended to Strata 9.6–9.8 in the Cabin.) Strata 303.1–303.3 lie above pasty, light gray-white, clayey silt, calcium carbonate–rich Stratum 304, which is virtually devoid of artifacts and faunal remains. Its base is concreted. It is about 35–40 cm thick and, in a small (0.25 m^2) sondage dug in L5 subsquare C (the southwest corner), it was shown to lie atop a loose brown silt layer (306). Dated to 11,650 BP in Square P6, Level 306 is probably equivalent to Level 11 in the Cabin, which lies immediately above 11.1 (itself dated to 11,720 BP).

CENTRAL SECTION (O–P/6)

Squares O6 and P6 were originally dug as an isolated 2 × 1 m sondage. At its western end (hence 1 m north of and diagonally adjacent to the K–N/5 trench), excavation revealed another very complex situation caused by apparently repeated pit digging. Underneath surficial mixed deposits (loose rubble, Level 300, and more compact, stony brown silt, Level 301—both with mixtures of "old" and modern artifacts and bones), essentially all of Square O6 was occupied by a large pit that showed in excavation and in section to have been dug and filled on at least two occasions. As shown in both the north and west sections of O6, we essentially had "quartered" what was originally a huge "crater"-like

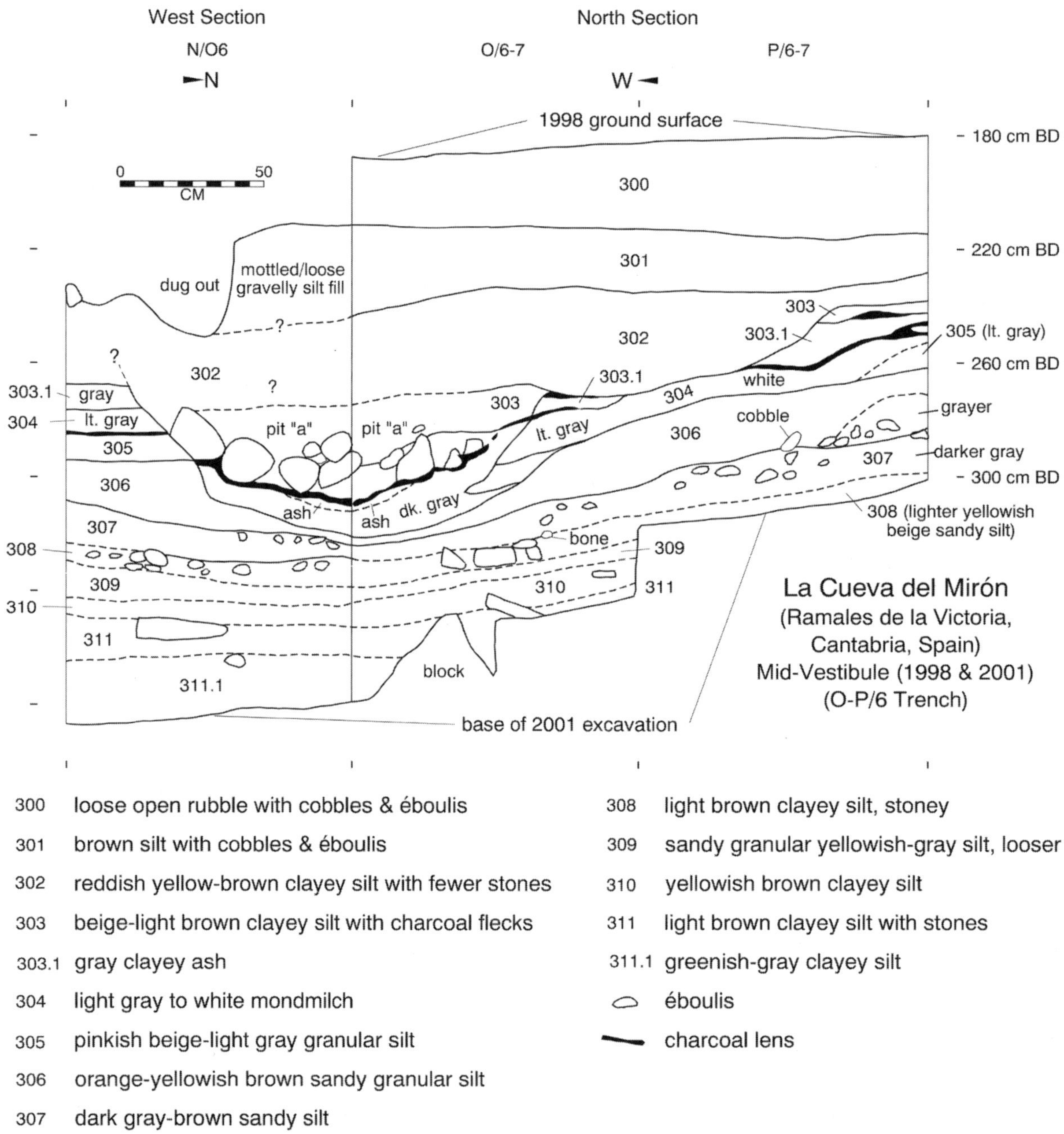

Fig. 4.9. West and north sections of the Mid-Vestibule Trench, central part (N–O/6, O–P/6–7). Plan by L. G. Straus, redrafted by Ronald Stauber.

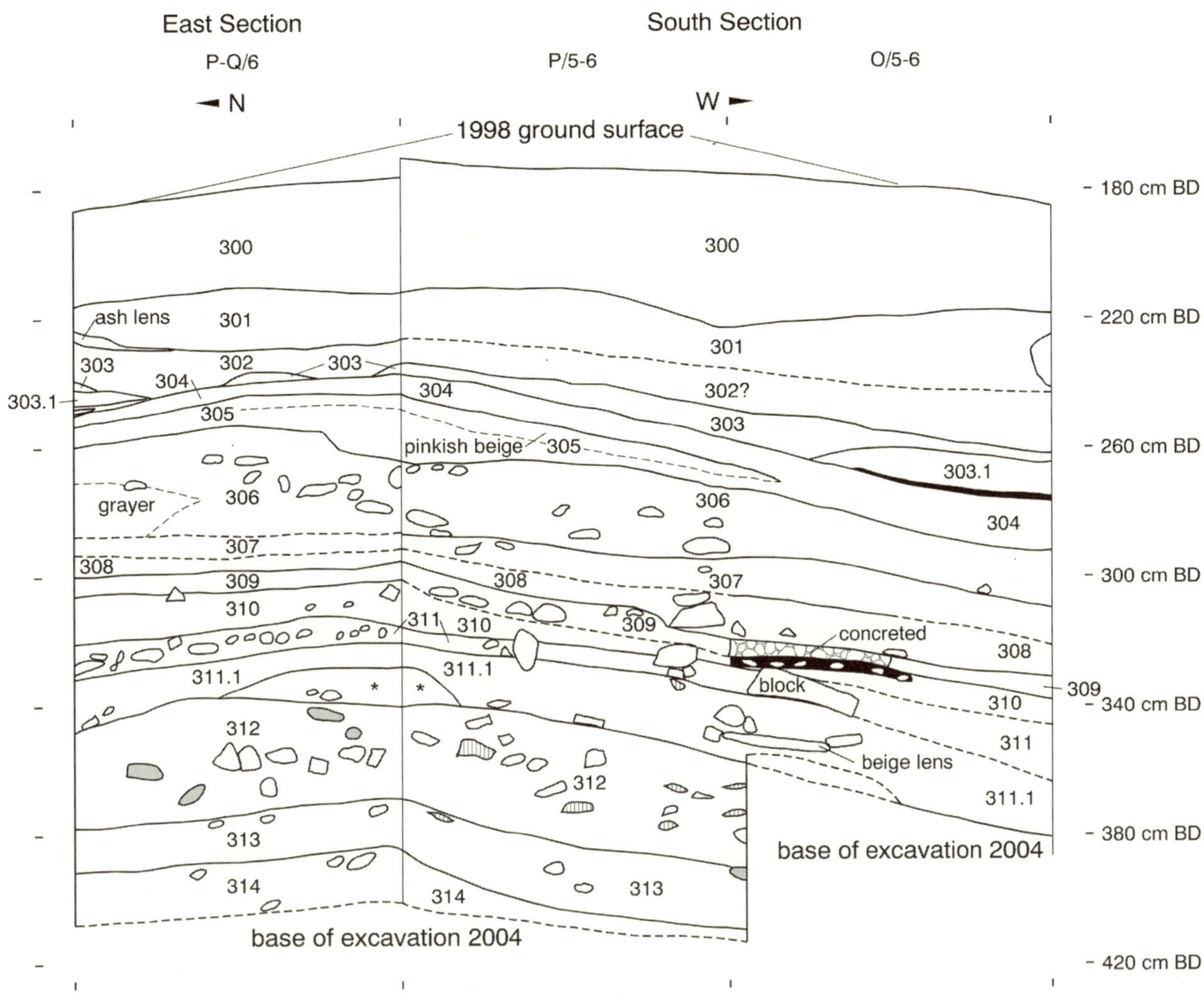

Key

300 loose open rubble with cobbles & éboulis
301 brown silt with cobbles & éboulis
302 reddish yellow-brown clayey silt with fewer stones
303 yellowish beige clayey silt
303.1 gray clayey ash
304 white mondmilch
305 pinkish beige-light gray granular silt
306 orange-yellowish brown sandy, granular silt
307 dark gray loose, gritty, sandy silt with angular éboulis & cobbles
308 light gray clayey granular silt
309 light yellow-beige sandy silt with small white stones
310 dark gray-light brown, pasty, clayey silt with angular éboulis & cobbles
311 light brown compact clayey silt, with some éboulis
311.1 greenish gray fine clayey silt
312 chocolate brown to black organic silt with many bones & some éboulis
313 alternating yellowish orange-brown and dark brown silt with some bones and weathered white éboulis
314 loose gravelly silt, light yellowish gray with angular éboulis
* unnumbered wedge of light yellowish beige sandy silt

La Cueva del Mirón
(Ramales de la Victoria,
Cantabria, Spain)
Mid-Vestibule (2001 & 2004)
(O-P/6 Trench)

éboulis
cobble
bone
charcoal lens

Fig. 4.10. East and south sections of the Mid-Vestibule Trench, central part (P–Q/6, O–P/5–6). Plan by L. G. Straus, redrafted by Ronald Stauber.

feature. The uppermost and very broad fill included a mass of fire-cracked rocks and several lumps of what appeared to be "slag" in a jet-black charcoal, white ash, or gray, clayey silt matrix. This both includes and was overlain by reddish or yellow-brown clayey silt that is poor in stones (Level 302). This may be the level from which at least the upper part of the pit may have been (re-)dug, as it apparently cuts through Level 303. The narrower, lower part of the pit (labeled "Feature 1998-a") seems to have been dug earlier, however, possibly from the top of Level 303 and cutting through Levels 304, 305, and the top of 306. Charcoal from the base of the pit has been dated to 4910 BP, which would place the first phase of this large-structure construction in the very late Neolithic. If, however, the slag is confirmed as metallurgically derived, then the reopening, widening, and last reuse of the pit as a kiln would have occurred in the early Bronze Age, more than a millennium later and perhaps equivalent to Level 3 in the Cabin.

Level 304 is a particularly striking pasty-white, mondmilch-like layer in O–P/6 ("mondmilch" is a porous, soft, white travertine). Extremely rich in $CaCO_3$, it is undoubtedly equivalent to Level 10.1 in the Cabin, underlying as it does Levels 303.1–303.3 (that date to ca. 5800–5500 BP) and overlying 305.

Level 305 (not present in K–N/5) is a pinkish-beige–light gray, granular silt dated to 10,270 BP. It is thus probable that (as in the Cabin) there was a depositional hiatus between 304 and the 303 series of levels. However (unlike in the Cabin) there may have been less of a gap (or none) below 304, since 305 seems to fall right at the traditional end of the Last Glacial, temporally equivalent to the late Azilian. As noted above, Level 306 (35–75 cm thick, orange-yellowish brown, sandy granular silt in O–P/6—apparently more granular than in the mini-sondage in L5) dates to 11,650 BP, making it a probable correlate of Cabin Level 11. In contrast to the Cabin and western sector of the Mid-Vestibule Trench, all the levels below 302 in O–P/6 have a definite slope downward (westward) in the direction of the cave mouth. The slope steepens the closer one gets to the rear of the vestibule and the face of the colluvial-alluvial ramp, as one would expect.

Stratum 307 is a dark gray, loose, gritty, sandy silt with angular éboulis and water-worn cobbles. It is about 20–40 cm thick and grades downward into 308, which is a gray–light brown, clayey silt that is also stony and measures some 15–20 cm thick. Stratum 308 is distinguished by being relatively rich in artifacts and faunal remains (including numerous salmonid vertebrae). It had a small feature that consisted of a concentration of limestone slabs and cobbles in P6 (hearth?). Charcoal from this level was dated to 12,350 BP, making it approximately equivalent to Cabin Level 12, which yielded a typical Upper Magdalenian harpoon.

Stratum 309, which is partially concreted, is a light brown, clayey, sandy silt with small "pea" gravels. It is only 10–25 cm thick. Stratum 310, of about the same thickness as 309, is a dark gray, pasty, clayey silt with angular éboulis. Stratum 311 (ca. 20–25 cm thick) is a light brown, more compact, clayey silt with fewer rocks. Stratum 311.1 is a greenish-gray, clayey silt, also about 20–25 cm thick. None of these levels immediately below 308 yielded much cultural or faunal material.

Below 311.1, however, there is an abrupt contact with a very dark brown, charcoal-rich, silty loam layer containing bones, called 312, which at the time of writing has been excavated in only the 1 m^2 sondage of Square P6. Small- to medium-size, white, corroded éboulis are abundant at the base of this thick, complex stratum. Level 312, by its stratigraphic position, color, granulometry, and richness in bones and artifacts is undoubtedly equivalent to at least Corral Levels 108–115 and probably in turn equivalent to Cabin Levels 16–17. At least the first of these hypothetical correlations is supported by a sondage dug within the eastern section of the Mid-Vestibule Trench (R–S/7). Level 312 is clearly an amalgam of at least three distinct units (darker, lighter, darker), recognized in the stratigraphic sections after this deep, narrow test pit had been dug. The middle spit of this complex unit has been dated to 15,850±170 BP. Below Level 312 is a sharply distinct, compact, yellowish-orange, granular, clayey silt that is far poorer in faunal remains and artifacts: Level 313—dated to 17,400±270 BP. It is probably equivalent to Level 120 at the base of the pothole in the Corral area. It lies atop an even lighter, apparently culturally even poorer level (314) that has not yet been excavated.

EASTERN SECTION (Q–S/7)

This is the least-developed section of the Mid-Vestibule because of a large amount of recent disturbance in the area between the 7 row and the south wall of the vestibule (the area where we found relatively recent human remains just below the 1996 surface as we removed dung and other modern debris from the corral). It is also the case that the upper stratigraphy is partially interrupted by a large limestone block that crops out partially in Square Q7. However, the sequence is very clear and identical to that of the upper part of the Corral stratigraphy in a 0.25 m^2 mini-test pit dug in Square R7d.

The upper deposits are missing or totally disturbed along the 7 row down to the top of a white, calcium carbonate–rich, sandy silt, 5 cm thick. This levels grades

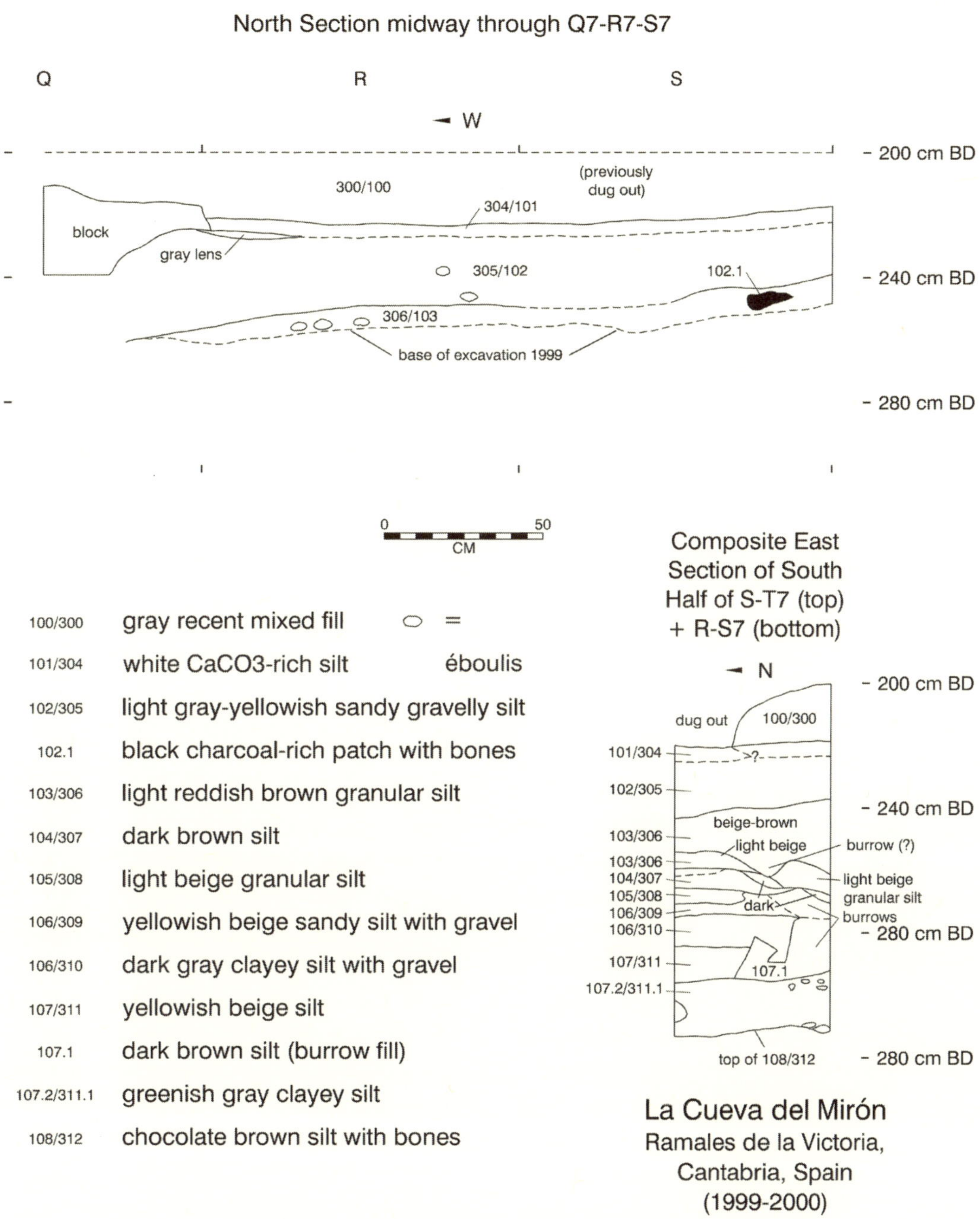

Fig. 4.11. North and east sections of the Mid-Vestibule Trench, eastern end (Q–S/7, R–S/7, and S–T/7). Plan by L. G. Straus, redrafted by Ronald Stauber.

into Level 304 toward the west and Level 101 toward the east and serves as an excellent "marker horizon." It is, in effect, continuous from the Corral area at the rear of the vestibule—where it (101) is a $CaCO_3$-rich, whitish-gray/brown, stony silt—through the Mid-Vestibule Trench. There this level (locally called 304) becomes increasingly less granular and more like a calcium carbonate mondmilch. This horizon continues into the Cabin area toward the front of the vestibule—where it (locally called 10.1) is a fine silt, rich in $CaCO_3$ and actually locally cemented. This horizon slopes gently down westward, toward the mouth of the cave, from the V row to the N row, where it levels out, becoming quite flat in the J–H rows.

Below 304/101 in Q–S/7 is a light yellowish-gray, sandy, gravelly silt: Level 305, which is equivalent to 102 in the Corral. This is a thick horizon: 18–30 cm. It is underlain by a 5–10 cm thick layer of light reddish-brown, sandy, gravelly silt with a few stones (306); it is equivalent to 103 in the Corral area. Within it, in S7, there is a charcoal-rich patch with bone that is at the same depth as Lens 102.1 in the adjacent Corral area, where this lens has been radiocarbon dated to 11,950 BP. It should be recalled that Level 306 in P6 is dated to 11,650 BP. But for the interruption of the large block in Q7, this unit is continuous with 306 in R–S/7 and with 103 in T–V/7. Furthermore, we can suppose (by stratigraphic position, depth, and extrapolated slope) that 103/306 is equivalent to 11 and/or 11.1 in the Cabin—the latter dated to 11,720 BP. These are probably classifiable as "Azilian," while the 102.1 fire lens is somewhat earlier and assignable in time to the terminal Magdalenian.

Below this point the eastern end of the Mid-Vestibule stratigraphy was exposed only in the small R7 sondage. Level 307 here is a dark gray-brown silt (5–7 cm); 308 a light beige, granular silt (3–5 cm); 309 a yellowish-beige, sandy silt with pea gravels (5–8 cm); 310 a darker gray silt (10–15 cm); and 311 a fine, yellowish-beige silt (10 cm). The latter is disturbed by rodent burrows that are equivalent to burrows that riddle the Corral area stratigraphy at exactly this depth and that we designated as 107.1. Below 311 and 107.1 in R7 is a greenish-gray, clayey silt (311.1) that is about 15 cm thick. It in turn lies atop a dark ("chocolate") brown silt of "flour" texture, called 312 in the P6 sondage and here in R7 clearly equivalent to Stratum 108 (and underlying levels) in the Corral area stratigraphic section, 1 m to the east.

The tentative, approximate equivalences of the Mid-Vestibule strata with those of the Corral and Cabin are thus as follows:

314=121=18–21 313–312=110–119=16–17
311.1=107–108=14–15 308–310=105–106=12–13
306–307=102.1–104=11.1–11.2 305=102=11
304=101–102=10.1

The Cabin-area correlations are not based on physical connections but rather on depth, slope, sedimentological characteristics, and radiocarbon dates. They are less solid than the correlations to the Corral area. The Neolithic levels of the western and central sectors of the Mid-Vestibule Trench are missing in the eastern sector and in the Corral area. The correlation of those early Neolithic levels (303–303.3) to Cabin Level 10 is quite plausible. And there is a rough equivalence between the remnants of more recent layers and pit fills (albeit some partially mixed) that overlie the 303 series and the late Neolithic and Chalcolithic levels of the Cabin (9–40).

Vestibule Rear Stratigraphy

CORRAL EXCAVATION AREA: UPPER PART

The ground surface of the corral (bounded by a crude wooden fence and containing wooden feeding troughs) in the southeast corner of the vestibule rear was understandably covered with some 15–25 cm of dung, straw, and modern refuse. This was removed down to the top of compact, apparently intact sediments in the area to the east, south, and west of a large roof-fall block. The top of this block protruded above the ground surface and was soon discovered (by González Morales) to bear linear engravings that continued downward below that surface.

In contrast, there were no compact, homogeneous sediments in the northeast corner of the vestibule rear located at the foot of the steep colluvial-alluvial slope that leads up and back toward the inner cave. Here it was immediately apparent to us that there was a large, filled-in pothole. Ultimately, by digging out (and dry-screening) its mixed fill over several years, we determined its full extent (ca. 15 m^2) and volume (ca. 25 m^3). It first served the purpose of providing sections (albeit somewhat irregular) back from which to excavate

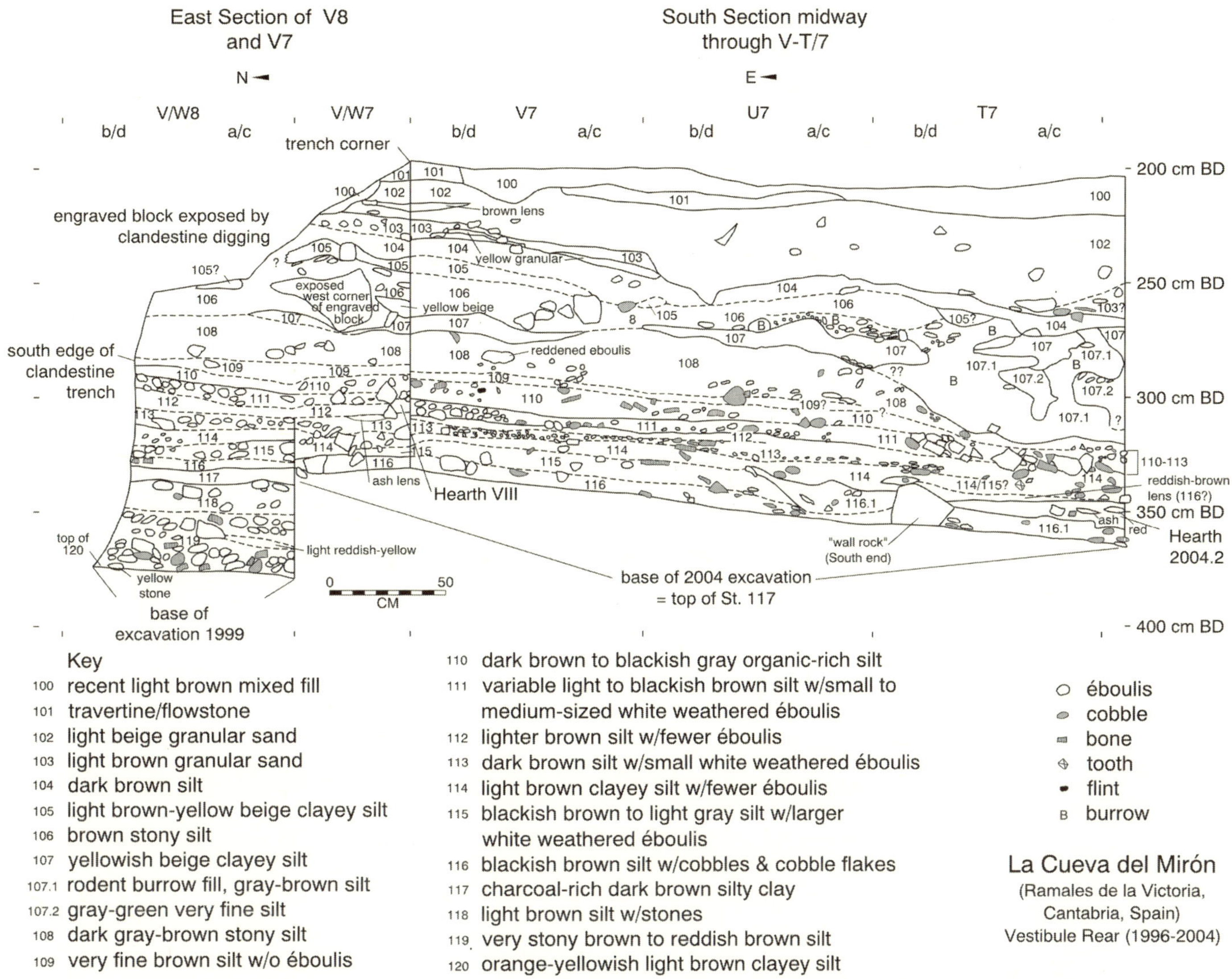

Fig. 4.12. Southeast corner and south sections of the Corral excavation area (V–W/8–7, V–T/7). Plan by L. G. Straus, redrafted by Ronald Stauber.

and to determine our initial stratigraphic units in the Corral area. Later, once completely emptied down to its base (which coincided with a sharp stratigraphic break—the top of orange-yellowish-brown, clayey silt Level 120, which the pothunters had probably wrongly interpreted as sterile basal "cave clay"), the pothole allowed us to dig a 2 m² sondage (W–X/10) down into pre-Magdalenian levels, all the way into the Middle Paleolithic. There had also been "recent" (but luckily much shallower) pitting in the northwest corner of the Corral area excavation (in Square T10), which cut down as far as Stratum 106.1

Thus, the main Corral area stratigraphy was basically defined in Squares V–U/7–8 (especially V8, which was dug down to the top of Level 120 on the south edge of the pothole immediately adjacent to the engraved block) for its upper sequence and in W–X/10 for the lower strata. Since these two key stratigraphic sections can be physically linked by the continuity of highly distinctive Stratum 120 across the less than 2 m

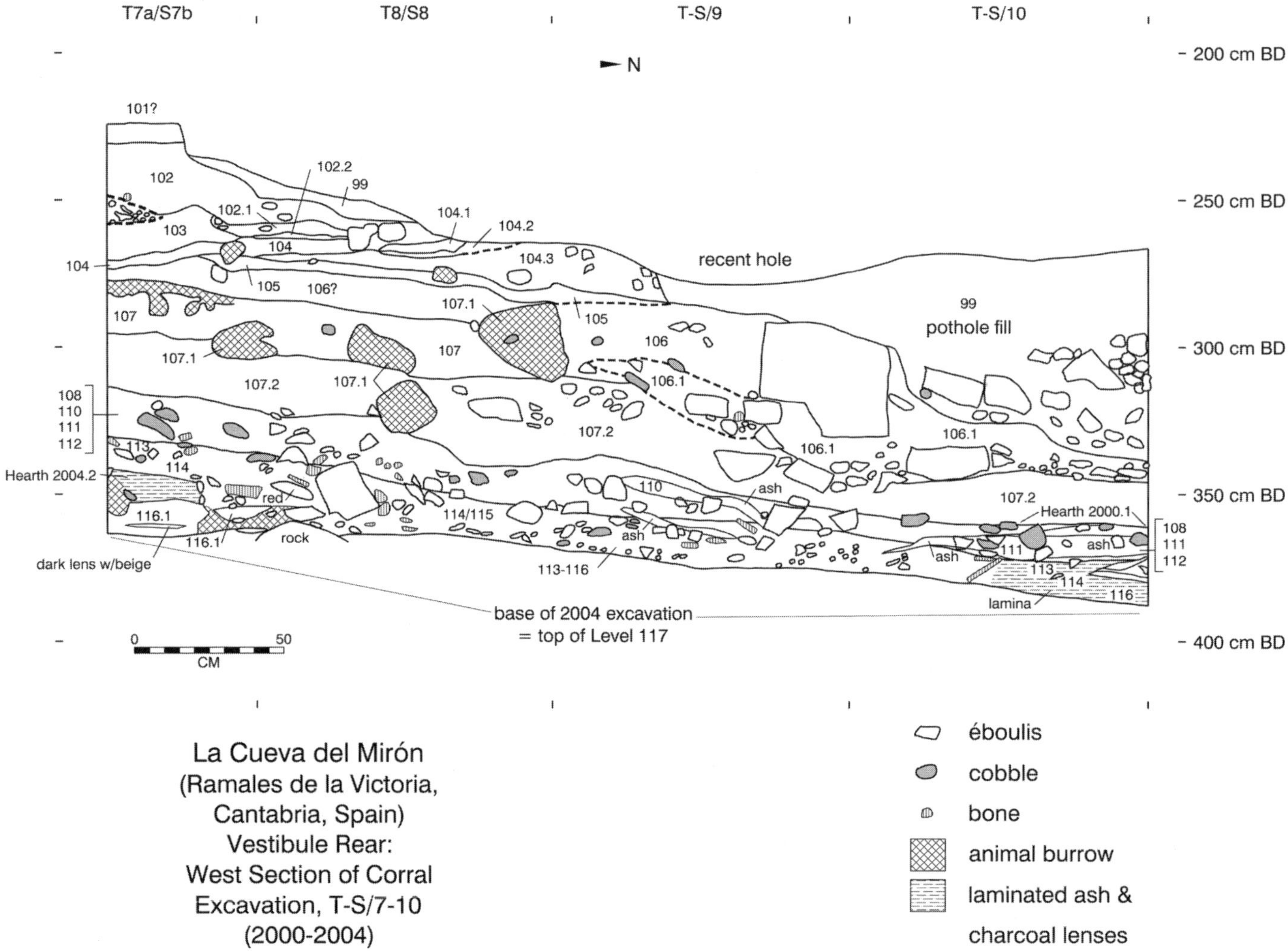

Fig. 4.13. West section of the Corral excavation area (T–S/7–10). Plan by L. G. Straus, redrafted by Ronald Stauber.

of horizontal space that separates them at the base of the pothole, the numeric designation series (originally two separate ones: the 100s [i.e., 100–120] and the 400s [400–407]) were eventually merged into one series that presently runs from 100 at the top to 130 at the base of the "pothole sondage."

The uppermost part of the intact Corral sequence is characterized by a series of sandy silts. In general, grain size of the upper strata is coarser in the Corral than in the Cabin, which makes sense if the source of much of the vestibule in-filling is the colluvial-alluvial slope at the rear of the vestibule, from which sediments are being transported (by running water in great part). The sediments of Strata 101, 102, and 103 are granular, sandy silts without visible clay content. They vary in color. Stratum 101 is gray–light brown and is rich in precipitated calcium carbonate; 102 is yellowish-orange and grades into 103, which is grayish-brown but with a reddish-yellow lens. Strata 102 and 103 are locally separated in the southwest part of the Corral excavation area (T/7–8) by a thin lens of ash and charcoal with a base of reddened (burnt) sediments that clearly represents a fire episode (a simple hearth laid on a surface corresponding to the beginning of Stratum 102 deposition [the lens of yellowish-beige, granular, sandy-gravelly silt below 102.1 is designated 102.2]). Strata 101–103 are poor in finds, but lithic artifacts (mainly small flakes and bladelets but few retouched tools) were found in

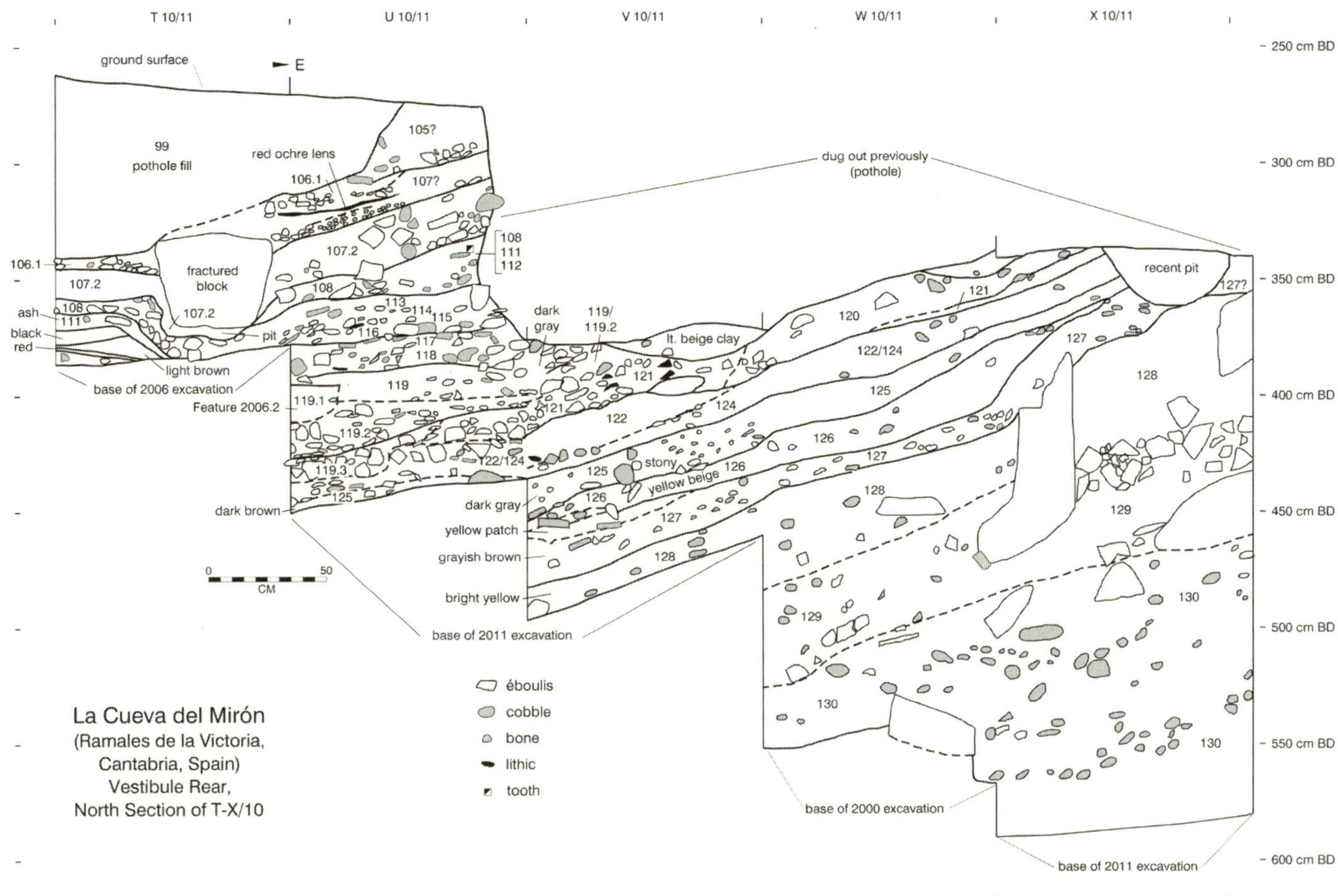

Fig. 4.14. North section of the Corral excavation area (T–X/10–11). Plan by L. G. Straus, redrafted by Ronald Stauber.

small numbers in all of them, together with fragmentary faunal remains (mainly bone splinters). If there is any "concentration" of artifacts among these uppermost levels in the Corral, it is the debitage (but no formal tools) associated with the 102.1 hearth. This is reminiscent of the situation in the Cabin, where it is the Stratum 11 complex (particularly the fire Lens 11.1) that yielded the only real concentration of artifacts in the terminal Late Glacial and initial Post-Glacial part of the stratigraphic sequence. These "hearth" lenses (as well as intervening Stratum 306, which has a small number of lithic artifacts) are penecontemporaneous in radiocarbon terms and correspond to the time of transition between the classic terminal Magdalenian and Azilian cultural phases.

The uppermost (i.e., most recent) truly culturally "rich" layer in the Corral excavation area is Stratum 104: a dark, gray to blackish-brown, clayey silt with fewer gravels than in the overlying levels (thus perhaps signifying less flowing water). In Square T8 (eastern end of the area), Stratum 104 is underlain by series of lenses: the upper two (104.1, 104.2) are clearly fire-feature layers (ash and reddened burnt silt, respectively); the lowest (104.3) is a light gray silt. Stratum 105 is a patchy, clayey silt unit in terms of color (greenish-yellow, beige-brown, and dark grayish-brown). This patchiness could suggest some disturbance process such as cryoturbation or burrowing, but the sediments are compact and seem "intact" compared to the obvious burrow fills of Stratum 107.1.

In contrast, underlying Stratum 106 is a homogeneous, dark brown to gray-black, fine, unconsolidated silt with some flints and blackened (charred?) bones (as in 105). This layer contains several medium-size

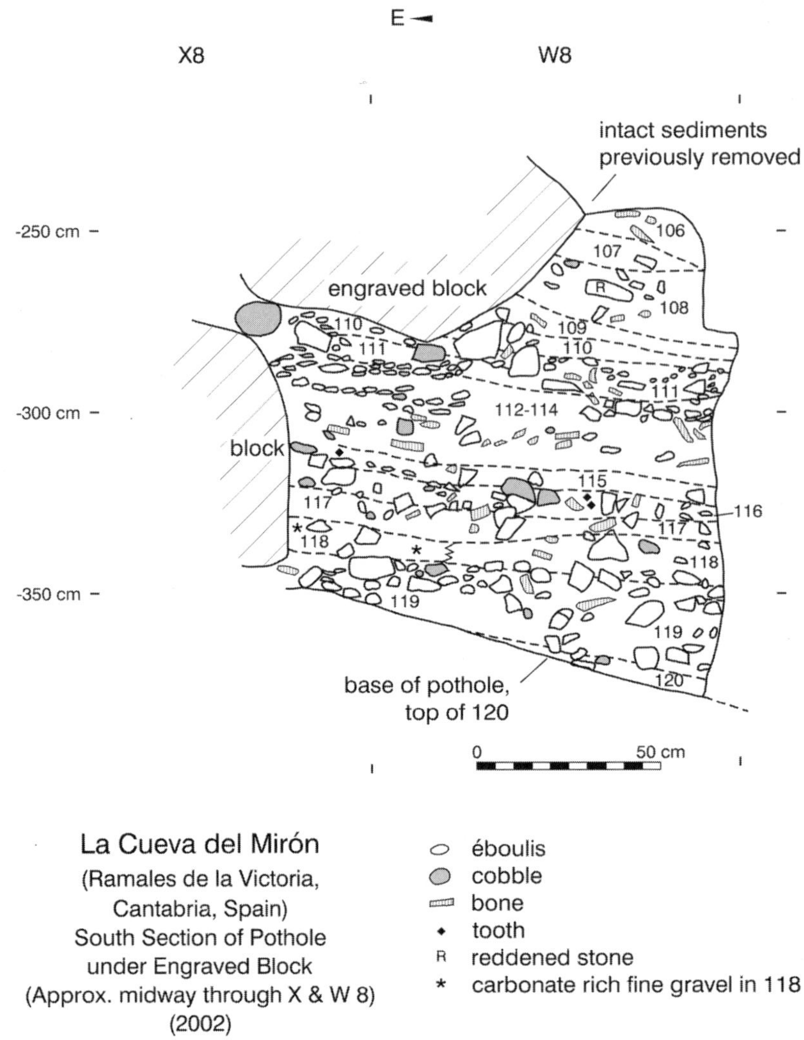

Fig. 4.15. South section of the Corral area pothole under the engraved block (W–X/8). Plan by L. G. Straus, redrafted by Ronald Stauber.

limestone blocks, whose fall may have locally distorted underlying Stratum 107, which is a yellowish-beige, compact silt with numerous finds. In some areas (such as Square V8 in the east) there is no clear Stratum 107 and Strata 106 and 108 merge, both being fine silt of light to dark brown color. There is a large pit in T8 dug from 106, cut into 107, and filled with light brown, fine silt with small pebbles and éboulis (but without clay), labeled 106.1. Stratum 108 is extremely rich in flint artifacts as well as blackened bones and some fire-cracked cobbles, attesting to considerable burning, even though no hearth centers were found in the area excavated. There are also some white limestone éboulis. In the western part of the Corral excavation area, 107 and 106 are separated by a maze of light brown silt pockets with few stones. These interconnected patches of loose fill (labeled 107.1) clearly represent rodent burrows (probably made by voles, according to G. Cuenca Bescós [pers. comm., 2001]). Their fill is rich in artifacts and bones, probably mainly derived from Stratum 107, into which the burrows seem to have been excavated. This disturbed unit is not ubiquitous in the Corral excavation area; in other spots 107 is underlain by a lens of very fine silt of distinctive gray-green color. This clear marker lens (perhaps attesting to localized ponding) is cut into by the bottom of some burrows. At the western end of the

Corral excavation area, Strata 107.1 or 107.2 lie atop a distinctive, continuous layer of chocolate-brown fine silt, Stratum 108. This level contains small, white, weathered éboulis and many blackened bones, flint artifacts, and some cobbles (including some that are fire-cracked or fire-blackened). Where 107.1 and 107.2 are missing—at the eastern end of the area (notably V/7–8)—Strata 106 and 108 intergrade in terms of sedimentological characteristics and archaeological contents.

Below this horizon (end of the mid-Magdalenian), distinction among the stratigraphic units originally defined in Square V8 becomes increasingly difficult (and indeed, in places perhaps illusory or even arbitrary to some extent). In V8 at least there seems to be an alternation between more and less stony (i.e., éboulis-enriched) levels, the matrix of all of which is more or less fine, brown to grayish-brown silt with abundant artifactual and faunal remains and "manuports" (i.e., cobbles, whose origin in fact was likely the adjacent colluvial ramp at the rear of the vestibule). This alternation is sometimes less clear and less continuous as one digs westward and northward from Square V8. Nonetheless, the excavation of these early to mid-Magdalenian deposits in the Corral (in reality a kind of "dissection" by spits 1–2 cm in thickness at most) was able to follow the local "lay of the land" (i.e., a double slope down toward the north and especially down toward the cave mouth in the west). This was done by uncovering and following the slope of long bone shafts, elongated cobbles, large éboulis, localized lenses of ochre pigment or colored silt, blades, and large flakes. The ability to at least locally (i.e., more or less within 1 m^2 areas) follow "real" stratigraphic units that may approximate "occupation surfaces" is also assisted by the existence of features such as hearths, pits, and even a possible "wall" within these deposits. Thus, although the lack of more-or-less culturally sterile horizons below Stratum 109 (though there do seem to be culturally poorer and richer zones) means that this dark brown, organically- and culturally-rich deposit (Strata 110–119) is undoubtedly a palimpsest of residues from many, repeated reoccupations of the vestibule rear, it can be—at least partially—teased apart and assigned relatively convincing level designations. Obviously, however, the confidence level in claiming strict stratigraphic continuity within the so-called strata declines as distance increases laterally, the greatest problems being correlation between V/7–8 in the southeast and T–U/10 in the northwest. At this time we have the impression that the levels are thicker in the southeast (as observed in sondage V8) and thinner at least in the northeast (as observed in the cleaned western edge of the old pothole).

Blackened bones and éboulis are (nearly or completely) absent in Stratum 109, which is otherwise similar to 108. Stratum 109 may represent an episode of lesser (or nil) human occupation of at least the rear vestibule area. Stratum 110 (14.8–16.1 kya) is also sedimentologically similar to the overlying levels (dark brown to blackish-gray, fine silt) but again contains many blackened bones, small éboulis, and flint artifacts. It was atop Stratum 110 that the later-to-be-engraved block seems to have fallen (Straus, González Morales, and Carretero 2011). Stratum 111 is distinguished by having a dense concentration of small- to medium-size, white, highly weathered éboulis, while 112 is defined by having far fewer. However, this is a distinction based on qualitative judgment and 111 can intergrade with 113 where there is no clear intervening "stoneless" zone. Color is not a very good discriminating characteristic of any of these levels, since it ranges more laterally than vertically from dark brown to light gray-brown in all of them. Certain zones (notably the southwest and northwest corners of the Corral area, i.e., T/7–8 and T10, respectively) are always chocolate brown, but other patches of darker or lighter silt are found in different areas of individual levels.

Stratum 114 is again generally characterized by having fewer éboulis in most areas, while 115 seems to have more and larger stones. Stratum 116 forms a rather well-defined "living surface": the 2 m long, south-north trending rock wall along the border between the T and U rows seems to have been "built" atop this level, and there are a few pits/hearths associated with it as well. Otherwise, the éboulis are less dense in this level. Stratum 116 is also distinguished by its archaeological contents, which seem particularly to include many non-flint cobbles, cobble cores, and large flakes struck from them. All the above-described levels have been excavated in the whole Corral area. At the time of this writing, however, underlying Strata 117–119 have been exposed only in the 1 m^2 area of the V8 sondage.

Stratum 117 revealed another clear, hard surface sloping down toward the west and south. It is characterized by a large amount of fairly large, white éboulis, possibly representing a significant rockfall episode (perhaps not surprising, given an age corresponding to the end of the Last Glacial Maximum). Stratum 118 in V8 slopes even more steeply down toward the west, which makes sense, since it is close to the underlying and steeply sloping base of the colluvial ramp. The matrix of 118 is clayier and it is even harder than 117, at least in the small area exposed. Éboulis of various sizes are enveloped in this clayey silt matrix. Underlying Stratum 119 in V8 is a massive layer of medium-large, rounded éboulis in an orange-brown, very clayey silt matrix like that of 118, with which it intergrades. Both levels are rich in bones. Stratum 119 is also very steeply sloping. This unit is cut into by a pit that we partially excavated in the northwestern corner of Square U10. The loose, dark grayish-black, silty fill of this pit was called 119.1. Rather arbitrarily, the lower part of Stratum 119 abutting this pit fill was called 119.2. The contents of this unit are rich in artifacts, manuports (cobbles, fire-cracked rocks), and faunal remains. Like the overlying levels, 119 and 119.2 contain very thin and/or localized lenses of fine silt of various colors, including red ochre. Stratum 119/119.2 lies atop the radically distinct (and problematic) Stratum 120, which is also exposed only across the whole base of the pothole and which we excavated in the W-X/10 sondage, just slightly more than a meter northeast of V8.

CORRAL EXCAVATION AREA: LOWER PART ("FOOTSLOPE SONDAGE")

On its surface, as exposed at the base of the large pothole at the foot of the colluvial ramp, Stratum 120 looks like a typical basal "cave clay." In reality, it is yellowish-beige, clayey silt with rock fragments and pebbles that contrasts sharply with the generally very dark or grayish-brown sediments of the early to mid-Magdalenian levels (108–119/119.2, as well as 16–17 and 312). These visibly organic, culturally rich layers that overlie 120 are also often rich in white limestone éboulis; they are typical of the kinds of dense Magdalenian habitation layers found in many Cantabrian sites such as El Juyo, El Rascaño, Altamira, El Castillo, La Riera, and Tito Bustillo. In contrast to such layers, Stratum 120 presents an appearance of organic and cultural sterility, which is probably why the clandestine excavators always stopped as soon as they reached its surface, despite having gone to the effort of breaking blocks (including the northern end of the huge engraved block) in the obviously rich Magdalenian-age deposit. Once dug into, however, Stratum 120 (originally called "400" until the physical correlation with the sedimentary unit directly underlying Stratum 119 in Square V8 was made) revealed itself to include angular éboulis and water-worn cobbles as well as flint artifacts and numerous faunal remains beneath the hard, surficial veneer of compact, clayey silt matrix. Since Strata 116–119/119.2 date to around 17 ka and Strata 125–126 to around 19 ka, deposition of the yellowish clayey silts (presumably directly derived from the adjacent colluvial ramp) seems to have terminated with the formation of Stratum 120 around 18 kya. Indeed, immediately underlying Stratum 121 is dated to 18,390±300 BP.

Had the "pothunters" cut through the surface veneer of Stratum 120, they would have encountered a series of alternating, yellowish and brown (presumably organic-rich) lenses (120 per se, 121 [ex-400.1], 122 [ex-401], 123 [ex-401.1], 124 [ex-401.2], 125 [ex-402], 126 [ex-403], and 127 [ex-404]). The brown levels generally have more lithic artifacts and faunal remains (including both ungulate bones and teeth and fish vertebrae). The yellowish-beige levels are all compact, clayey silt. All these levels, which present a spectacular banded appearance in section, slope very steeply down toward the west, that is, following the trend of the colluvial slope down toward the cave mouth. These layers (120–127), dating between 18,390 (Stratum 121) and 19,230 BP (Stratum 127), are assigned to the Solutrean technocomplex, with characteristic invasively retouched foliate and shouldered points in some of them. Stratum 127.1 is a localized lens of fine, yellowish-gray silt that seems to have been laminated (by water gently running down the slope?).

Stratum 128 (ex-405) is a bright yellow, smooth, clayey silt. The lower part of this unit is stonier, and its base is defined by a number of large limestone blocks. It is not culturally sterile, but lithic artifacts (mainly flakes, some large blades), faunal remains, and charcoal fragments are scarce and scattered throughout a considerable thickness, suggesting only occasional human visits at least to this part of the cave during a period that an AMS radiocarbon assay dates to about 27.6 ka, that is, early Upper Paleolithic. In contrast, underlying

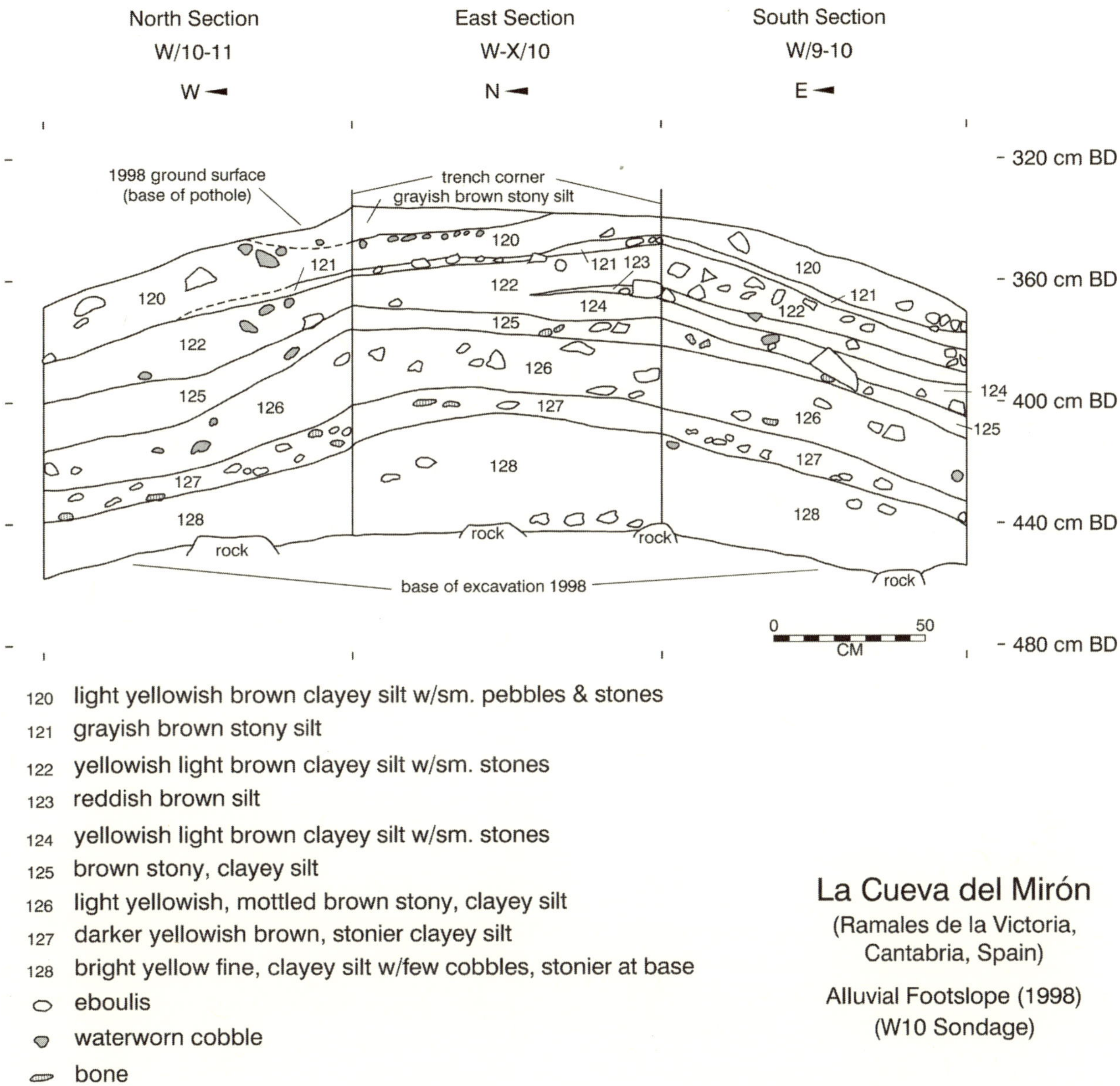

Fig. 4.16. North, east, and south sections of W10 sondage at base of pothole (W/10–11, W–X/10, W/9–10). Plan by L. G. Straus, redrafted by Ronald Stauber.

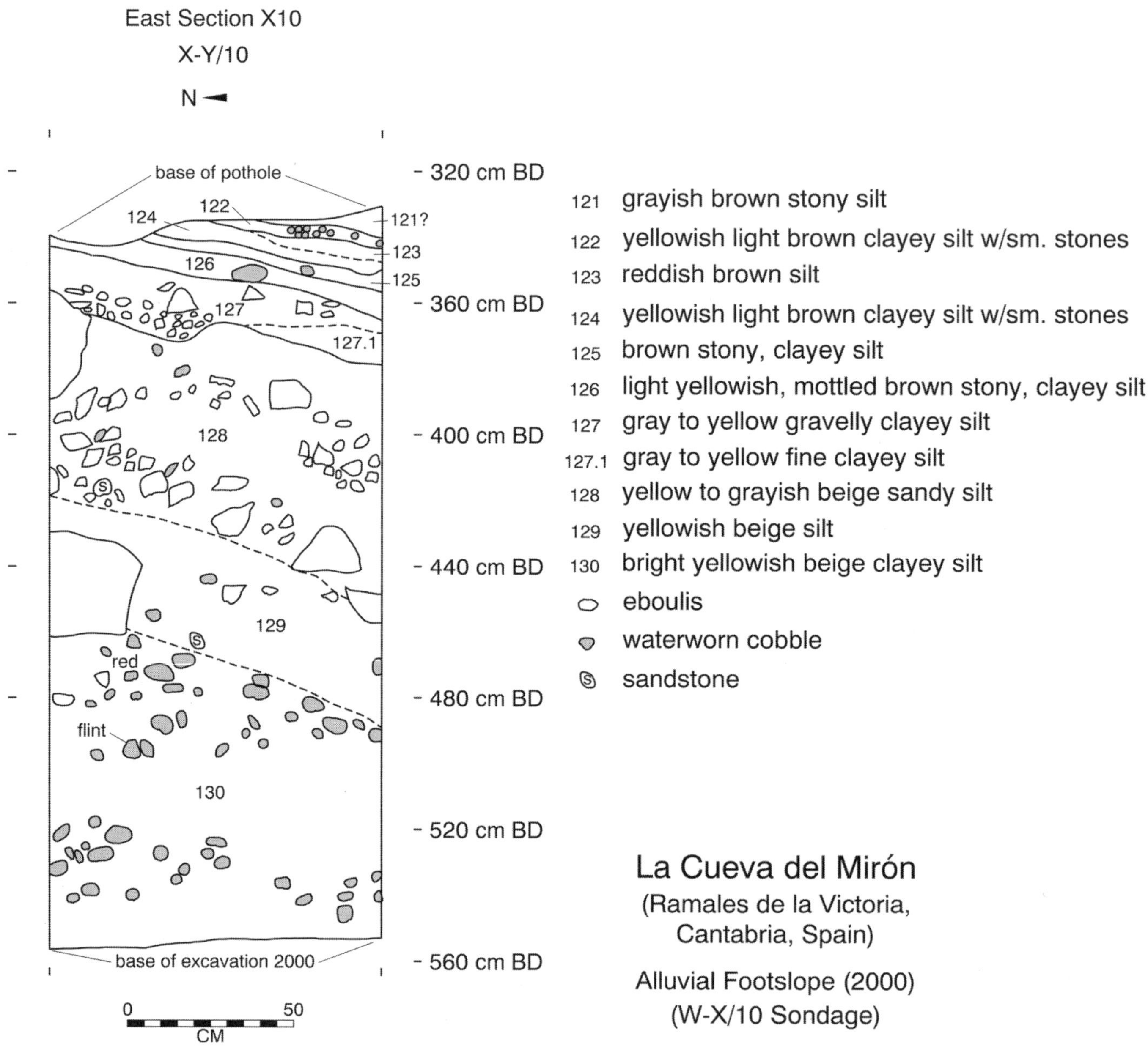

Fig. 4.17. East section of X10 sondage at base of pothole (X–Y/10). Plan by L. G. Straus, redrafted by Ronald Stauber.

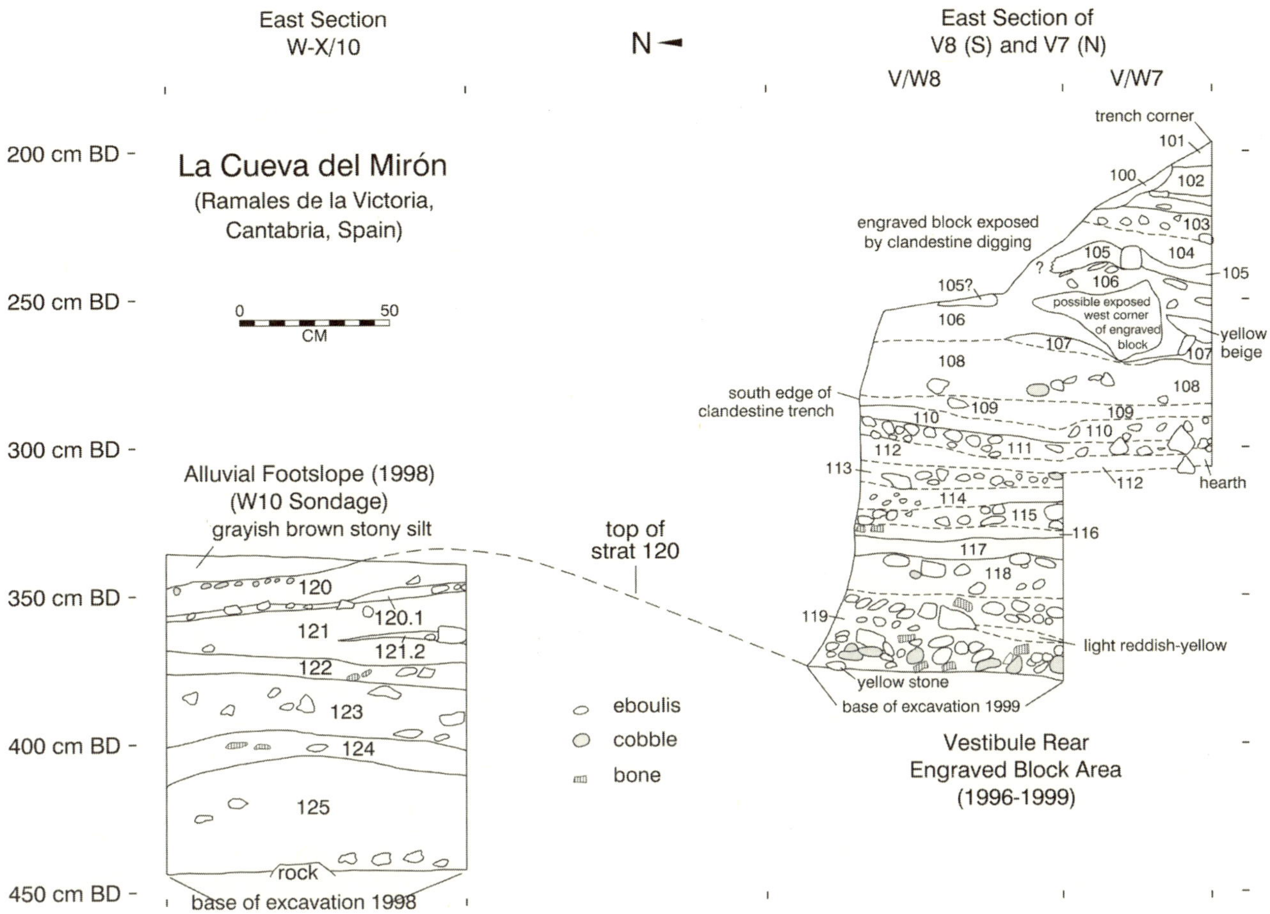

Key

100 recent light brown mixed fill
101 travertine/flowstone
102 light beige granular
103 light brown granular
104 dark brown silt
105 light brown-yellow beige clayey silt
106 brown stony silt
107 beige clayey silt
107.1 rodent burrow fill, gray-brown silt
107.2 gray-green very fine silt
108 dark gray-brown stony silt
109 very fine brown silt without eboulis
110 dark brown to blackish gray organic-rich silt
111 variable light to blackish brown silt w/small to medium-sized white weathered eboulis
112 lighter brown silt w/fewer eboulis
113 dark brown silt w/small white weathered eboulis
114 light brown clayey silt w/fewer eboulis
115 blackish brown to light gray silt w/larger white weathered eboulis
116 blackish brown silt w/cobbles & cobble flakes
117 charcoal-rich dark brown silty clay
118 light brown silt w/stones
119 very stony brown to reddish brown silt
120 yellow-orange to light brown clayey silt w/sm. stones
120.1 grayish brown stony silt
121 yellowish light brown clayey silt w/sm. stones
121.1 reddish brown silt
121.2 same as 121
122 brown stony, clayey silt
123 light yellowish, mottled brown stony, clayey silt
124 darker yellowish brown, stonier clayey silt
125 bright yellow fine, clayey silt w/few cobbles, stonier at base

Fig. 4.18. Correlation of the V7 and W10 sections.
Plan by L. G. Straus, redrafted by Ronald Stauber.

Stratum 129 is almost completely sterile. It intergrades sedimentologically with 128 (being an even brighter, pasty, "pure" clayey silt); the boundary between the two units we defined as the blocks at the base of 128. Indeed, there are also many large blocks within 129. The deepest level excavated to date in the W–X/10 sondage is 130, which is a reddish-yellow, sandy silt (i.e., the clay content is far less than that of the overlying levels). There are virtually no angular éboulis (which are present in the overlying levels), but there are many pebbles and large, water-worn cobbles in 130. This seems to have been a water-laid deposit, possibly the top of the steep colluvial ramp at this depth. There are a very few artifacts (including a pair of flake denticulates and a notch) and charcoal fragments, a few of which were dated to around 41.3 ka: late Middle Paleolithic.

CORRAL EXCAVATION AREA: THE DEPOSIT BEHIND THE ENGRAVED BLOCK

Perhaps one of the most extraordinary discoveries in the vestibule rear has been the existence of an apparently intact Upper Paleolithic deposit in the approximately 1 m wide area between the huge engraved rock and the cave wall. Recall that before we began systematic excavation in the Corral area we removed about 15–25 cm of dung and recently mixed fill. Below this loose debris in the area behind the engraved block, where we excavated a few subsquares distributed between Squares X–Y/7–8, we encountered a light gray-brown silt with stones, lithic artifacts, and old-appearing bones. This layer (500) may be somewhat mixed but did not yield modern artifacts.

Below 500 is a gray or gray-brown, fine, powdery silt rich in ash, burnt flints, and bones. This level (501) is clearly an intact fire residue unit, again with no trace of modern intrusives, but now with no obvious traces of disturbance. Beneath this "hearth" is a fine, white, semi-cemented, calcium carbonate–enriched silt with small, angular éboulis and some cobbles. This level (502) is clayey at its base. The 501 hearth was apparently made atop Level 502. Level 503 is a brown silt with small, white éboulis, with abundant bones. It is also carbonate-rich and may be a localized lens. Level 503.1 is an ash- and charcoal-rich layer dated on charcoal to 15,120±40 BP.

Level 504 is a reddish, grayish-brown, loose, powdery silt in direct contact with the sloping bedrock of the rear wall of the cave. It has abundant flecks of red ochre and galena or mica. Like 502 and 503, this level is clearly banked up against this sloping wall at the back of the vestibule. In 2010 and 2011 Level 504 yielded the ochre-stained mandible and numerous bones of a secondary human burial preliminarily dated on associated animal bones to 15,740±40 BP—Lower Magdalenian (Straus, González Morales, and Carretero 2011). The burial layer (504) is underlain in the northern half of X7 by a bedrock shelf and in the southern half (and in Y7) by hearth layer 505 and culturally rich occupation layer 506, which is probably equivalent to Level 110.

With increasing depth, each of these levels becomes narrower, since each is a wedge underlain by the sloping bedrock. All these levels behind the engraved block also slope very steeply down toward the west because of that underlying bedrock. They contain striking numbers of lithic and osseous artifacts of Magdalenian aspects (e.g., burins, antler *sagaies*). As the intact series of levels 501–504 is not overtopped here by any layer (with only the uncertain possibility of Stratum 500?) clearly equivalent in characteristics to Strata 101–102, which exist farther toward the west in the Corral excavation (and which are culturally nearly sterile, but of an age that is—by correlation with the Cabin stratigraphy—equivalent to the early Holocene Mesolithic), it is most likely that 501–503 are in fact of late Magdalenian age. Extrapolating their slope (which is, of course, interrupted by the presence of the engraved block, which makes physical correlation impossible), one could hypothesize that they could be at least approximately equivalent to Corral Strata 103–106. This was confirmed by digging Square W7 in 2011. Indeed, projection of the slope of 103 in V7 upward toward the rear of the vestibule would have it running up to the present ground surface midway through Y7, which is about where 501 contacts the cave wall. More recent terminal Pleistocene and early Holocene levels may have been removed from the back of the vestibule by the corral activities of humans (perhaps including deliberate leveling) and/or animals. This possibility is further suggested by the presence of small remnants of travertine crust cemented to the walls of a small alcove at the extreme southeast rear of the vestibule a couple of meters farther back from the X–Y/7 excavation area and about 1.5 m above the 1996 ground surface. This travertine would have originally capped the stratigraphic deposit in this alcove.

Inner Cave Stratigraphy

The 5 m long by 2 m wide trench across the inner cave (now designated as 8–12/Q–R) was dug in the 1950s; we cleaned and straightened its sections in 1996. We also dug a small sondage below the base of that trench in Square 11R. (South of and aligned with the old trench, there is an approximately 2 × 1 m pit—now called 5–6/R and probably of the same vintage—which we did not touch.) The trench and pit were dug into sediments that seem to fill a channel cut longitudinally through the ancient alluvial infilling of the inner cave. The uppermost layer (I) visible in the trench sections is apparently consolidated backdirt from the 1950s dig; it is mixed clay and travertine. Stratum II is the pre-1950 cave surface, consisting of a thin lens of homogeneous clay. Stratum III is a thin layer of porous, soft, white travertine, analogous to what is being precipitated currently in the area immediately west of the trench, where there is a very active drip from the cave ceiling. Stratum IV is a jet black to gray-brown clay-silt layer with pulverized charcoal that has been radiocarbon dated to AD 1040–1200 (calibrated)—that is, late Middle Ages. No artifacts and only a few faunal remains were recovered, although obviously a large fire had been made at this time.

Stratum V is another mondmilch layer, but it is discontinuous, that is, patches of more or less pure but porous calcium carbonate. Stratum VI is a yellowish-orange to light brown clay layer, which is culturally sterile. Stratum VII is a thick (ca. 40 cm.) mondmilch layer that is gravelly at its base. Charcoal from Stratum VII is radiocarbon dated to 3200 BP, that is, Bronze Age, perhaps roughly equivalent in age to Stratum 2 in the Cabin area. Stratum VIII in the 11R sondage is another yellowish-orange to light brown clay. Right at the contact between VII and VIII there is a concentration of lithic artifacts, faunal remains, and charcoal. This evidence of human use of this part of the dark inner cave continues within VIII itself and has been radiocarbon dated to 14.6 ka, that is, mid-Magdalenian, roughly equivalent in age to Strata 108 in the Corral and 14 in the Cabin and perhaps 110–111.1 in the Mid-Vestibule. This clay is underlain by an orange-brown, sandy, gravelly layer with virtually no clay. This layer (IX) is archaeologically sterile, at least in the small area in which it was dug. It could be a testament to an episode of flowing water within the inner cave channel.

Finally, along the cave walls in the area immediately above the top of the present colluvial ramp there are

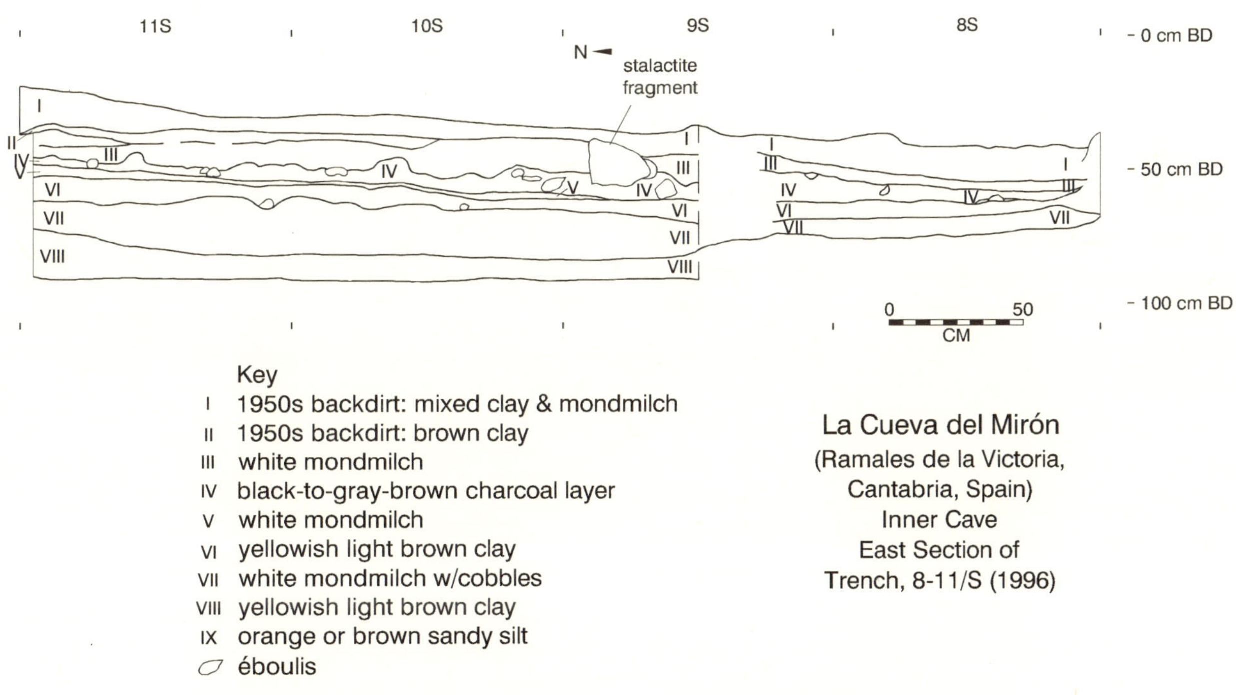

Fig. 4.19. East and north sections of the Inner Cave Trench (9–12/S, S12/13). Plan by M. González Morales, redrafted by Ronald Stauber.

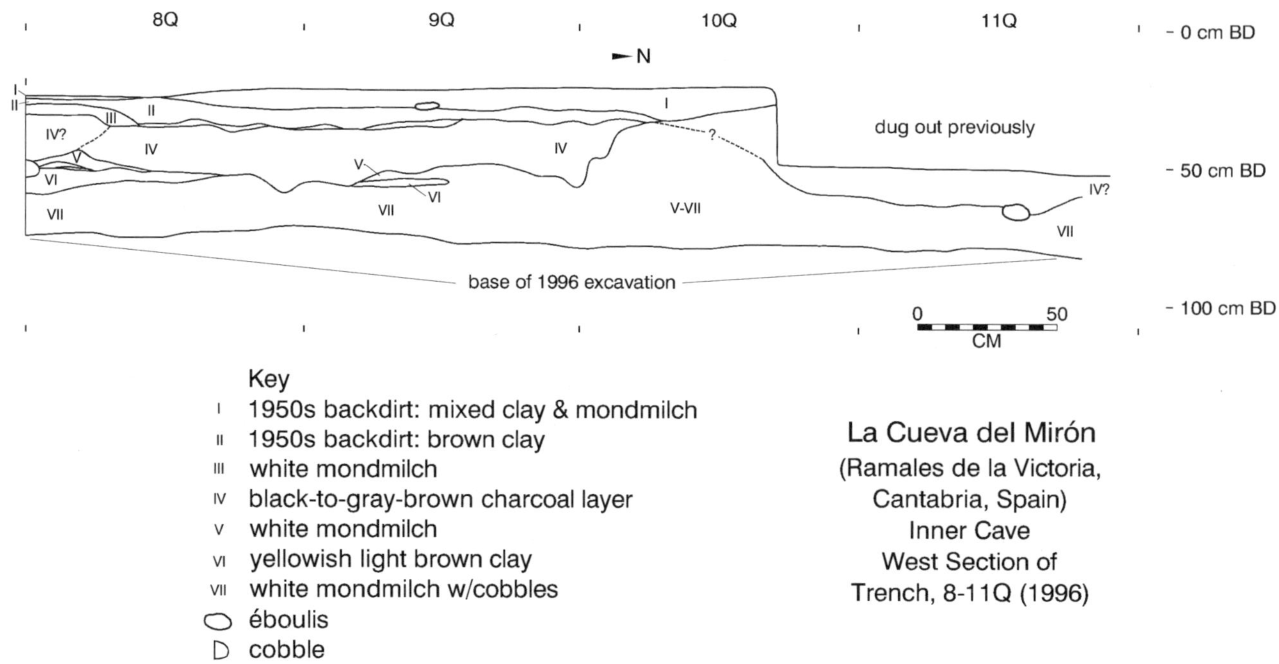

Fig. 4.20. West section of the Inner Cave Trench (9–12/Q). Plan by M. González Morales, redrafted by Ronald Stauber.

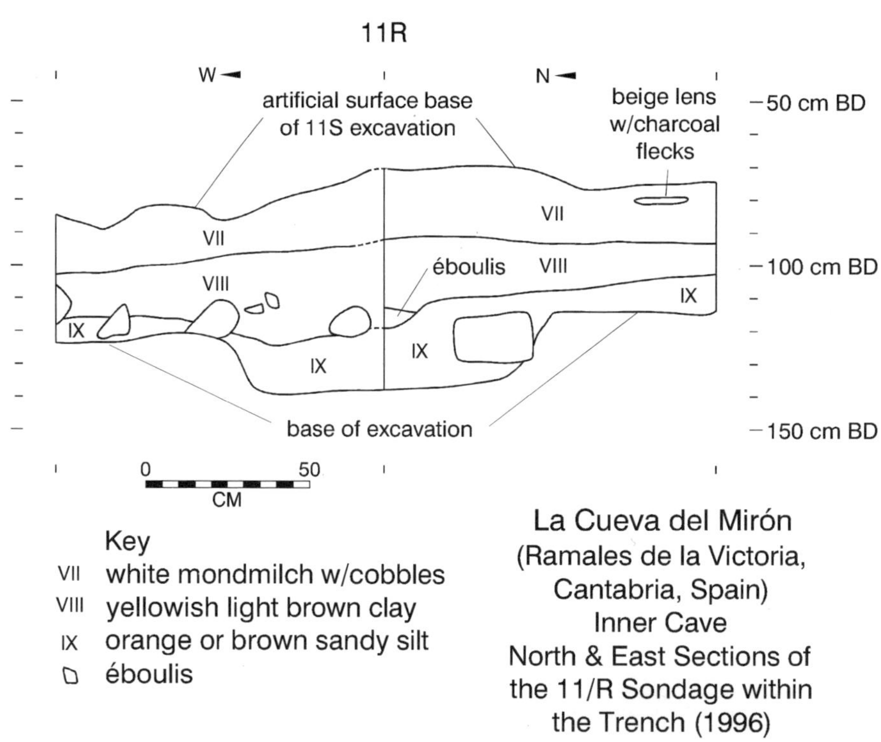

Fig. 4.21. North and east sections of the inner cave 11R sondage (11–12R, 11/R–S). Plan by M. González Morales, redrafted by Ronald Stauber.

remnants of a horizontal travertine crust that are adhering to the bedrock. These attest to where the surface of the inner cave alluvial fill had been before the ramp was eroded eastward, as a result of water flow and the movements of animals and humans during the Holocene. Breccia adhering to the base of the travertine crust displays a few flints and bone/tooth fragments. One of the latter was dated (twice) to around 10.5 ka, equivalent to an Azilian age, like Stratum 305 in the Mid-Vestibule Trench. These breccia remnants show that there has been significant erosion from the colluvial ramp and that it has been retreating eastward toward the inside of the cave throughout the course of the Holocene (as well as far earlier, before the cave vestibule had seen any human occupation or Upper Pleistocene in-filling) and that there had been human uses of the narrow corridor area connecting the vestibule with the inner cave at least at the time of the Pleistocene-Holocene transition. Exploration of the inner cave thus occurred at least during the mid-Magdalenian, Azilian, Bronze, and Middle Ages. Indeed, about a quarter of the way up the ramp there is one of several "niches" (Niche A) in the south bedrock wall of the cave that was filled with bone- and artifact-rich, chocolate-brown, silty loam very similar to the early to mid-Magdalenian deposits in the Corral, Mid-Vestibule and Cabin excavation areas. It is radiocarbon dated to 16,600 BP. A wedge of Magdalenian occupation residues had apparently been banked up against at least the lower- to mid-slope surface of the colluvial-alluvial ramp before it was eroded away, presumably in the early to mid-Holocene, perhaps after the capping Azilian level and travertine had been broken under conditions of increased temperature and humidity.

Appendix: Stratigraphic Designations and Informal Descriptions

The terms *stratum* and *level* are interchangeably used equivalently and are abbreviated as "St." Decimals generally refer to "lenses," which are thin, localized, patchlike stratigraphic units that are often channel, pit, or hearth fills or fire remnants. They usually lie under the main stratum from which their numeric designation is derived. Major exceptions to the "localized patch" definition for decimalized units are Strata 10.1 and 107.1. The former is in fact a major, separate level that—because of the subtle (i.e., shades of gray) nature of its sedimentological distinction from overlying Stratum 10—was not recognized in the initial, 3 m² test pit dug in 1996. The latter consists of the fill of rodent burrows dug into Stratum 107 and that probably mix materials from (at least) Strata 107 and 108. Stratum 3.5 is a large, late-prehistoric pit or channel fill that postdates Stratum 3, whose true (intrusive) stratigraphic position was established after it had been named.

The units listed here are those that were recognized by the excavation codirectors (Straus and González Morales). Further distinctions—both vertically and horizontally within levels—were recognized by the project sedimentologist (Farrand); these additional stratigraphic subdivisions are incorporated into table 5.1 of his chapter in this volume.

STRATUM OR LENS DESCRIPTIONS AND CHARACTERISTIC CONTENTS

Outer Vestibule (Cabin area: rows H–K/1–4; from top downward)

1 Surficial ovicaprine dung and rubble from collapsed cabin stone walls.

2 Beige, stony, sandy clay; locally light brown or, in K1–2, bright orange-yellow. Floor of cabin, with cobbles and other small stones in places. Rich in bones (cow, cervid/ovicaprid, pig).

3.5 Yellow-orangish, gritty (decomposed sandstone?), mottled basin fill (possibly equivalent to St. 2 in H3–4). Lumpy and somewhat discontinuous; in places clayey. Cuts through and is thus posterior to Levels 3–6 and part of 7. Also known as "pit O." In turn, a smaller pit was dug into it "recently."

3.0 Small, flat, 1–2 cm thick, white-gray ash lens. Separates St. 1 and 3 in H–I/2 section. It is 14 cm. higher (at -206 cm B.D.) than St. 3.1 (which is a horizontal travertine). There is a very thin (ca. 1 cm) black charcoal lens atop the white.

3 Dark gray → gray brown, clayey silt, with charcoal flecks; rich in bones and organic matter. Contains ceramics of "old" appearance, flints, etc.

3.1 White, gritty travertine lens in I2c+d, I1a+b, and parts of J2c + J1a. This patch is absolutely flat (at -220 cm B.D.) and is probably a result of a puddle forming at bottom of an already partly St. 5–refilled pit that had cut through St. 4 in center of I2.

3.2 Cream-color calcium carbonate lens sloping slightly down from north to south in the H–I/2–3 section; 1–3 cm. thick. A gours: ponded travertine, brecciated with sherds and bones. Exists in H1, but not in I–H/1 section.

3.3 Dark gray → brown layer under 3.2 in I–H/2–3 section. 4–10 cm thick. (Defined ex post facto in section on July 3, 1997. Not individualized in 1996.)

3.4 Red (burnt) and white (indurated ash?) packed with small pebbles, below surface fill sediments (i.e., St. 1 or 3?) in H4b only. A combustion feature. Possibly part of the late fill of "pit O"/St. 3.5 and, if so, posterior to St. 3.

4 Hearth horizon. Alternating lenses of light gray-white ash (locally cemented) and jet black charcoal with chunks of burnt wood, plus rubified + fire-cracked rocks. Two fire foci found in 1996: "A" (earlier/lower) in J3c+d and "B" (higher/later) in J3a+b, with limestone rocks, cobbles, and sherds. Surrounded by possible postholes (or burrows), with a definite posthole in J2. Rich in ceramics around hearth foci and some burnt (but not calcined) bones.

5 Dark brown → gray → black (with charcoal). Greasy, clayey silt. Below St. 4, but locally (in J2a + J3c) below both 4 and a lower lens of 4 ash.

5.0 White ash lens in I2, I3, J3 west (+ probably also in J2 west).

5.1 Wedge of dark brown silt between ash lenses of St. 4 in J2a and J3c.

5.2 Black charcoal patches. Loose, dark brown-gray loam with charcoal, fire-cracked rock, and cobbles in-filling pit H in I2a–d and north edges of I1a+b.

5.3 Dark gray, loose organic fill of pit G in I3a+b, cut into St. 6.

7.4 Upper charcoal + ash-rich, dark organic fill of pit J base in I3a.

7.5 First (basal) generation of charcoal, ash, and cobbles within pit J in J4.

5.4 Ash and charcoal fill of pit D (called simply St. 5 in 1996) in N3/4 of pit in J4d. This matrix surrounds rocks, some quite large. Pit D and therefore lens 5.4 predate pit J and also therefore predate lenses 7.4 and 7.5 (despite order inversion of strata numbers). Pit J was dug into pit D.

5.5 Grayz → white ash with charcoal and burnt wood; fill of small pit R dug into pit G in J4c.

6 White ash layer clearly earlier (lower) than St. 4 and separated from it by 5 (+5.1). Separation is thicker toward the south and east than toward the north and west. It is a gritty, compact ash with charcoal flecks. It is cut through in J2c + southern part of J2a by a large, rock-filled pit (pit C). (Base of pit C in J2 + fill of pit D in J3 are called St. 5., dug through St. 6 ash.)

6.1 Clay lens (confused with top of St. 7 in I3b+d and I2b) plus an ash + charcoal lens under it and atop the real summit of St. 7. Maximum thickness of this complex in I3b = 12 cm; minimum in I2b = 2 cm (where it is only a simple ashy clay lens with charcoal flecks). Contains bones. Cut by base of pit D (St. 5+6) in south edge of I3b; in turn it overlies pit J at the western edge of I3b. Includes last (upper) level of pit J.

7 Light brown → beige, homogeneous silty "cave clay." Humid, plastic, and found throughout J1–3. Not pure clay. Contains dispersed charcoal lumps and flecks, plus bones, cobbles, and fire-cracked rock.

7.1 Localized gray, hardened ash lens within St. 7 in I2a+b and I3c+d.

7.2 Pit I fill: dark gray-brown, organic-rich. In I3a+c.

7.3 Gray, hardened ash lens/patch (1–2 cm thick) under (and separated by an unnumbered clay lens from) pit I in I3a+b and a bit of I3c+d. It underlies the two limestone blocks that formed the eastern wall of pit I.

8 Light gray-beige, pure clay. Very plastic and humid, crumbly and greasy; contains charcoal flecks. A fairly clear textural break with St. 7.

8.1 Ash and charcoal lens with rocks along the northern edge of I+J3, under continuous St. 8 clay, which separates 8.1 from the base of pits D and J. This ash lens is in the north half only of I+J3 but does not continue southward. It might represent a ground surface exposed at the top of St. 9. Contains some sherds.

9 Darker gray clay with charcoal flecks.

9.1 Brown → gray → black sequence of Structure L roasting/fire pit fill in I2a+b. Ash and charcoal at base with cobble + fire-cracked rock fill. Contains sherds, flints.

9.2 White → light gray-yellowish ash lens in middle of and atop circular stone feature M in J3. It is both under and above St. 9 fill.

9.3 Charcoal-rich, brown clayey fill among the rock fill of pit M in J3a+b.

9.4 Ash lens at base of pit M in J3a.

9.6 Light gray → white ash atop reddish burnt clayey silt, fire lens (ca. 2 cm thick); localized in I3.

9.7 Dark gray, clayey silt with charcoal flecks (like St. 9) in I3. 6–9 cm thick.

9.8 Another thin (1–2 cm) ash and charcoal lens at base of 9.6 in I3.

10 Light gray-brown silty clay. Poor in flints, but with some bones.

10.1 Dark gray ashy, clayey silt. 8–9 cm thick. In I3 this subdivision was made, where St. 10 per se is 10–11 cm thick.

11 Orangish-light brown → beige gravelly, silty clay with medium → large (10–20 cm) éboulis, generally angular. Loose gravel. Contains lenses defined in 1997, but not individualized in 1996, are alternately dark, charcoal-rich or light-color, water-washed gravels, especially in Square I2a. St. 11 was subdivided in Square I3 into:

11.1 Darker gray level or zone at base of St. 11 in I3 (at least south part): 3–4 cm thick.

11.2 Light yellowish-beige, stony, clayey silt level at base of 11.1 in I3 and in I–H/2 section; 2–4 cm thick. Not very clayey like underlying St. 12. Resembles St. 11, so where 11.1 is absent, the two (11+11.2) merge. (Note: clay content increases with depth from St. 11.1 through St. 13.)

12 Purer clay; orangish light brown/beige. Abundant gravels, some rounded, and larger éboulis. Contains bones and artifacts, especially at top.

13 Yellowish-orange to light brown, more or less pure clay, with few gravels and occasional, isolated, large, weathered éboulis. Gradational color and textural changes vis-à-vis St. 12 in I-J/2. The basis for distinction is the purer clay matrix of 13. It gets clayier and less stony with depth. Archaeologically sterile save for rare bones and flints (and these may be due to downward percolation along the cave wall in I and J2, given the existence of a void there). Larger éboulis at base, but still without gravels. A few flints right at base of blocks.

14 Pure clay: plastic, dense, compact, yellowish-beige/khaki, very similar to St. 13. Several cm below the éboulis (and below the flints at the base of these St. 13 blocks), after a zone of sterile clay, there is a concentration of flints in J3, with some corroded bone fragments and teeth. St. 14 is individualized more on an archaeological than on a sedimentological basis.

15 Same clay as in 13 and 14, but with small stones and éboulis in J2+3 at least. Stones are angular and continuous through J3 and J2 (except for the zone immediately adjacent to the cave wall in J2c+d, which is loose or void). Lower part of the matrix is darker brown, grainier, and siltier (not pure, plastic, dense, compact clay as at top of 15). Culturally rich.

16 Dark brown, crumbly clay with a high sand content, plus gravel and small (</=5 cm) angular éboulis (in J3). Not dense, compact, pure, plastic clay. Contains artifacts, triturated bones, and cobbles. Intergrades with St. 15 and there is lateral variation between J3 and J2 with much darker sediments in J3 (= St. 16) and much lighter toward the cave wall (esp. in J2) (= St. 15). In short, the distinction is ± arbitrary in practice.

17 Lighter gray-brown, but locally blackish or chocolate-brown, sandy, gritty silt with many stones (angular éboulis, but fewer cobbles); looser and with a lower clay content than overlying St. 16, with which it intergrades. Rich in flints and especially large bones (better preserved than in

St. 16). Several hearths, with areas dense in fire-cracked rocks and charcoal-stained sediments.

18 Lighter, yellowish-beige, clayey silt with angular limestone éboulis of small-moderate size (up to ca. 10 × 10 cm). Relatively poor in artifacts, faunal remains, and organic/cultural matter in general, especially in (stark) contrast with overlying 17, though not sterile. More yellow near the south cave wall; more brown toward the north. 20 cm thick.

19 Similar to 18, but somewhat clayier and with larger limestone clasts (? 10 cm). Contains cultural and faunal remains in rather low density. 10 cm thick.

20 Similar matrix to the overlying two layers, but even clayier (and moister) with even larger éboulis (? 20 cm). Concreted in the area adjacent to the cave wall. Poor, but not sterile in cultural remains. 10 cm thick.

21 Similar fine, clayey silt sedimentary matrix to St. 18–20, with smaller éboulis (5–10 cm), but at least one very large block at the arbitrary base of excavation in 2007. Somewhat richer in cultural remains (including an exceptionally large chunk of juniper charcoal). The distinctions among St. 18–21 are rather arbitrary, especially since these deposits have only been excavated in a 1 × 1 m^2 sondage (J2) adjacent to the south cave wall, where the sediments are naturally very humid and spalled clasts relatively abundant.

Mid-Vestibule (R7, O–P/6 sondages + westward continuation in N–K5; from top downward)

300 Loose, open-work surface fill: gravel, pebbles, cobbles, and éboulis with varying amounts (usually slight) of dark brown, organic-rich silt. Contains wood, mixed modern and late prehistoric (sherds), and old ("Magdalenian-aspect") artifacts.

301 More compact, but still mixed fill: dark brown silt with cobbles (more silt than clasts, in contrast to St. 300). Fill of recent holes (with some glass in places).

301.0 Light yellowish-beige, clayey lens between St. 5 (gray-brown) and St. 301.1 (dark gray, ashy) in L5 and north section of M5 (where it was not recognized during excavation). 2–3 cm.

301.1 Gray ash layer with black charcoal lens at base. Ca. 6–10 cm. Recognized ex post facto in M and N5 in 1999. Not distinguished or labeled as such during excavation of surficial level (301). Underlies St. 5 and might = St. 6. It is cut by a recent hole filled with St. 301 sediments in Squares N5–6.

301.2 Mixture of yellowish-beige clay at top with darker gray ash and charcoal lens at base. Also cut by a "recent" hole in N5–6. Ca. 5–10 cm. Recognized ex post facto in N and M5 in 1999. Might = St. 7 (+7.1/7.3).

302 Lighter reddish-brown silt with gritty sand, gravel, and some more isolated cobbles. In P6 it is rather compact but looser in O6. It lies atop a compact and very distinctive layer of mondmilch (304) in P6 and silt layer (303) in O6. Evidence of fires (ash, charcoal) at its base. Distinction between 301 and 302 is ± arbitrary in many places. 302 includes fill of a pit that cut thru 303 in O6 and W part of P6. Not present in N–L/5.

303 Yellowish-beige, clayey silt (2–6 cm thick in P6). Intact, but cut by a pit filled with 302. In turn, upper 303 fills at least the base of pit "a." St. 303–303.3 are equivalent to St. 10 (+ perhaps St. 9.6–9.8).

303.1 Fill of a hearth pit or a bonfire layer under 303. Gray and black (ash + charcoal). Cut by pit "98a" in O6; also cut by a later pit filled with Level 302 sediment in P6.

303.2 Yellowish-beige, clayey layer below ash and charcoal fire layer of 303.1 in N–M5. Ca. 5–6 cm.

303.3 Paired ash-charcoal (and sometimes ash-charcoal-ash-charcoal) lenses between 303.2 and 304 in N and M5. Ca. 2–10 cm thick. Thickens from east to west. Defined in 1999.

304 White, greasy, wet, pure mondmilch. Sterile. (= St. 101).

305 Light beige, granular, sandy silt (= St. 102). Pinkish/orangish lenses and light gray/whitish lenses, all granular, but whitish ones being $CaCO_3$-enriched. Sterile. Few stones except travertine slabs (especially in active drip area in P6c+d). This is a lens within St. 304 which feathers out in O6. 304+305 probably = St. 10.1.

306 Dark brown-grayish, granular, sandy silt with flint artifacts, bones, cobbles, and éboulis (including angular fragments) (= St. 102.1–103 = St. 11–11.2?).

307 Loose, gritty, sandy silt matrix (as in 306), but with more stones (angular éboulis and cobbles) and, especially in 06, dark gray in color (as opposed to the yellowish-orange-beige of some zones of 306). Culturally richer than 306 (= 104).

308 Pasty, clayey silt, generally without sand or grit. Light beige/yellowish, but darker brown in P6. There is some sand and angular éboulis + cobbles and it is wet in the downslope area of 06a+c, due to the constant ceiling drip in P6d (= 105–106 = 12).

309 Yellowish-beige, sandy silt with small granular ("pea") gravels.

310 Darker gray, pasty, clayey silt (wet in P6) with angular gravels and larger éboulis. Fine matrix is locally (in 06c+d) washed out due to the drip, with a localized concretion in P6. More stones and gravels than clayey silts in washed-out area. Patches of sandy silt with a more yellowish-beige color within the gray, clayey silt may be rodent burrow fills. Culturally poor. Intergrades with 309 and may also be equivalent to 13(?).

311 Light brown, clayey silt, compact, fewer rocks (= 107+107.1 = 14?).

311.1 Slightly darker, reddish-brown, compact silt with some rocks, especially at base. Very slight clay content; mostly silt (= 108–109 = 15?). Lies directly atop dark, chocolate-brown silt (312).

312 Excavated only in Square P6, but top also reached in small sondages in Squares L5 and R7. Dark, chocolate-brown to black, charcoal-rich, silty loam (but with a lighter-color midsection), with abundant bones and artifacts; many small, weathered, white éboulis at base (= 110–116 = 16–17). Could be subdivided into at least three distinct levels.

313 A horizon consisting of an alternation of yellowish, orange-brown lenses and a gray, charcoal-rich lens. Fewer cultural remains than in St. 312 (= 117–119.2).

314 Light grayish-brown silt with abundant, limestone gravels and small éboulis, more angular than in 313. Fewer bones; little or no charcoal. Possibly equivalent to Footslope St. 120 and Cabin St. 18–21.

Colluvial-Alluvial Footslope (X–W10 sondage, partially continued into Squares U and V10 in 2010 and 2011; from top downward)

400 = 120 Mottled yellow-beige, clayey silt with many stones and cobbles, underlying mixed pothole fill (12 cm). There is a surficial, unnumbered wedge of gray silt with many small pebbles in W10d that could be mixed fill. But 400 per se lacks ceramics or other "modern" artifacts, though at base of pothole.

400.1 = 121 Thin (3–7 cm), brown, stony silt lens with organic content and cultural material at base of 400 in W10c and d only. 1–3 cm.

401 = 122 Yellowish-beige, homogeneous, compact, clayey silt with stones, gravels, and cobbles. Contains a few flints and bones.

401.1 = 123 Thin, brown, stony silt lens within 401 in W10c+d only. Contains flints and bones. 3–7 cm.

401.2 = 124 Yellowish-orange beige, stony silt below 401.1 in W10c+d. Merges with 401 in W10a+b. ±Sterile. 5 cm.

402 = 125 Fairly dark brown silt with flints and bones, steeply sloping down to W. 10–12 cm.

403 = 126 Somewhat more yellowish, lighter (beige), clayey, sandy silt with cobbles, éboulis, many bones, and flint + antler artifacts. Steeply sloping. 20–25 cm.

404 = 127 Darker brown-gray, stony, loose "organic-cultural" lens (10 cm) with large flake and other artifacts + abundant bones and teeth (including ibex). Steeply sloping down to west. Distinction with 403 is not clear or sharp.

127.1 Yellow-gray, fine, apparently laminated silt. A little channel fill within 127, found only in SE (upslope) corner of X10. 1–10 cm.

405 = 128 *Bright* yellow, clayey silt, smooth, compact, sloping steeply down to W. Sharp contact with 404. Base reached in 1998 is at layer of large limestone blocks. Lower part of

405 is stonier and grayer beige than top. (Core boring shows that similar yellow silt continues at least 20 cm below the rocks.) Culturally poor, though there are macro- and micro-mammal and fish bones, as well as small numbers of flakes. 25–45 cm.

406 = 129 Gradational below 128, but underlies its stony lower part. An apparently purer, pasty, clayey, bright-yellow silt. It lies among huge limestone blocks.

130 Gradational below 129, but redder and brighter yellow silt and less clayey. Contains many water-worn cobbles and pebbles. Generally underlies large blocks and seems to continue at least another 1.5 m below base of the year 2000 excavation in X10, albeit with some blocks (based on augering and core-boring). Contains flakes, bones, and charcoal, but dispersed.

Vestibule Rear (Corral area: rows T–V/7–8, plus S–Q/7; from top downward)

99 Loose, light reddish-brown silt (recent fill/"backdirt") with cobbles and stones, atop which there is a sort of wall below 1996 surface dung in U–T/8. This lies atop St. 102 and is recent rubble below corral dung.

100 Top 15–25 cm of ovicaprine dung, recent mixed fill, cobbles, bones, etc. in U–V–W/6–8.

101 Base of ±modern surface fill. Gray-brown silt with stones and local calcium carbonate-enriched patches caused by roof drips.

102 Yellowish-orange, sandy silt without clay content; compact but loose to excavate. Contains a few dispersed flints and bones.

102.1 Thin (2–5 cm) ash and charcoal burnt (red) fire lens in a basin in T8c+d and T7a+b.

102.2 Like St. 102 in T8/7 section. Light yellowish-beige, granular, sandy, but slightly more gravelly (with larger clasts) than 102 per se. It exists under 102.1 but merges with it on sides of the fire pit fill. It is ca. 1–5 cm thick. Exists also as a very thin lens in T–S/7 section.

103 Gray-brown sandy silt (as in 102). Contains a reddish-yellow localized lens of the same granular sandy silt that divides 103 laterally and horizontally. Contains some flints and bones.

104 Dark gray/blackish-brown (darker to the west, lighter brown to the east), compact and clayier silt with fewer gravels. Contains small → medium angular éboulis.

104.1 Gray ash overlying reddened (burnt) sediments in T8a+c at west edge of pit III. Ca. 1–3 cm thick.

104.2 Reddened burnt sediments (probably altered St. 107 surface) under ash patch of 104.1 in T8a + ca. 1–2 cm thick.

104.3 Light gray silt layer between 104.2 and 105 in S-T/8 section. Defined ex post facto in 1997.

105 Greenish-yellow, clayey silt, but heterogeneous, with large patches of beige-brown → dark grayish-brown silt, with lateral patches, perhaps due to some mixing processes such as cryoturbation (?). But these are intact sediments.

106 Homogeneous dark brown (with gray-black zones), loose silt, with burnt bones and flints, as in St. 105. Large éboulis blocks lie on and/or in this dark, culturally rich level in V7+8. Very rich in blackened bones, especially in V8, with some continuity in V7b.

106.1 Pit fill in T9–10. Fine, light brown silt with small pebbles and some small limestone éboulis. No clay content. Fill of pit E, dug from 106 (or 105?). It cuts through 107 and is disturbed at south end by extensive (large and small) animal burrows.

107 Yellowish-beige, compact silt (±like the yellowish patches within 105). May lie under 106 in V7, but is convoluted by limestone block falls (which also may explain the heterogeneity of St. 105). Best represented in U7.

107.1 Light brown silt with a few large stones, ca. 20 cm thick. Below yellow-beige sterile 107 in T7b. Contains artifacts and bones (unlike 107). Burrow fill.

107.2 Gray-green, very fine silt; sterile. In T7–9, under St. 107. Ponding sediments(?). Riddled by burrows filled with 107.1. Lies atop St. 108. Top tinted with red ochre in T10.

108 Light → darker brown silt. Heterogeneous. Contains small weathered (white) éboulis. Less rich in bones or flints than St. 106, but intergrades with it and is basically not separated from 106 by any

continuous St. 107 in V8 (and barely so in V7). St. 108 gets darker toward its base, but seems to vary in color both through space and depth. It is full of blackened bones, many flints, white éboulis, some cobbles (some of which are burnt). It is stony. Overlain by large blocks fallen from cave ceiling.

109 Fine, flour-like brown silt without stones or burnt bones. A localized lens.

110 Another dark brown/blackish-gray, organic-rich silt full of blackened bones, stones, flints. Under "sterile" St. 109. Intergrades with St. 108 where 109 is very thin or absent.

111 Alternating patches of darker and lighter brown and blackish-brown silt, rich in blackened bones and flints (as in 108 and 110), but with small → medium size, white, highly weathered éboulis. A distinct stony lens, but with silt matrix identical to that of the above levels. About 3–5 cm thick. Seems to fill a depression or north-south channel in V8a–c. Hearth in V7b.

112 Lighter brown silt with many fewer éboulis than in St. 111, but still rich in artifacts and fauna. (There is better separation between St. 111 and 113 in V8d than in subsquare c.)

113 Another lens of small, white, corroded limestone gravels/stones in V8. Matrix is dark brown silt with many bones, but fewer flints apparently. Not as dark as St. 111, but patchy in color again. It is ± continuous with St. 111 in V8c+a, but separated from it by stoneless St. 112 in V8d.

114 Light brown, clayey silt with fewer éboulis (some small white ones, especially in V8d) under the stones of St. 113.

115 Many larger (medium-size), corroded, white limestone éboulis in V8, as in St. 111 and 113. Matrix varies from blackish-brown to lighter gray, clayey silt. Still ±rich in bones and flints.

116 Blacker silt (burning evidenced by blackened bones, but few burnt lithics). Presence of many water-worn cobbles, cobble cores, and flakes with cobble cortex. Fewer éboulis than in St. 115.

117 Larger white éboulis in a compact, gray-brown silt matrix, lighter at the top and darker at the base. A clear, hard surface, sloping down toward south and west. A sharp break in texture and a major rockfall episode. Matrix is rich in burning evidence (a few calcined bones and fire-cracked rocks + charcoal) and consists of loose, blackish sediments below surface.

118 Hard, very compact, light brown (or orangish-brown), clayey silt with gravels and small limestone "flecks" dispersed throughout the hard, clayey-silt matrix, but no larger éboulis. Slopes much more steeply down toward the west than the overlying levels.

119 Massive layer of medium-large, rounded, weathered éboulis in an orange-brown, clayey silt essentially like 118, with which it intergrades. Contains abundant bones like 118. Slopes down steeply to W and N, like 118.

119.1 Dark gray, loose, silty loam fill of a pit (2006.1) partially exposed in U10. Small white éboulis at base.

119.2 Continuation of 119 at its base, surrounding the 119.1 pit fill. Dark brown (but highly variegated), clayey silt with abundant éboulis and cobbles, with many fine silt lenses/patches of various colors (beige, greenish-gray, reddish), as well as ashy areas. Culturally rich. Lies directly atop 121/122.

119.3 Dark brown, loose, silty loam with relatively few éboulis in U10 A + north part of C. Fill of pit 2011.1. Contains blackened bones. Surrounded by more compact, stonier but otherwise similar sediments called St. 122 in 2011. Possible continuation of pit 2006.2 (119.1).

Vestibule Rear between Engraved Block and Cave Wall (floor of the Corral; from top downward)

500 Loose, light gray-brown silt with stones, artifacts, and bones. Probably mixed at surface (as of 2001) in Y–X/7–8, W7.

501 Fine, powdery, lumpy silt with ash and small angular éboulis; locally (in X7b) with charcoal, burnt bones, and flints. Gray to light gray or gray-brown.

502 Fine, white silt, semi-cemented ($CaCO_3$-rich), with small, angular éboulis in X7b. Some cobbles; clayey at base.

502.1 Compact yellowish clayey silt; edge of hearth pit in X7c.

503 Brown silt with small white limestone éboulis and many bones with carbonates. May be just a

localized lens in X7. Culturally rich with tools of "Magdalenian" aspect.

503.1 Mottled yellowish silt with ash, charcoal, and reddened patches/lenses in X7c+d.

504 Dark, reddish-gray-brown, loose, powdery silt in direct contact with inward-sloping bedrock of rear wall of cave vestibule in V7. Successive wedges of 502 and 503 may be banked up against the cave wall rear against 504. Rich in ochre and galena or mica. Contains a human secondary burial, partially in a pit in X7.

505 Light gray, very fine, ashy silt with charcoal flecks and patches. A hearth layer in X7 south and Y7 southwest. (Dug as part of 504 in X7 in 2010; defined in 2011.)

506 Yellowish-beige, granular to fine silt with gray lenses and abundant small clasts in south half of X7; crushed and cemented atop bedrock ledge in center of square; rich in artifacts and faunal remains; disturbed by rodent burrows (defined in 2011 below human burial). May be equivalent to Level 110 and was hit by the block that was later engraved and painted.

Inner Cave ("Trinchera": rows 5–12/Q–S; from top downward)

I Backdirt from 1959 excavation flanking the east and west sides of the trench. Mixed clay and mondmilch.

II Pre-1959, brown, clayey cave floor surface. 3–5 cm.

III Thin, white mondmilch layer. 5–7 cm.

IV Black → gray-brown, charcoal-rich layer ("the black layer"). 6–9 cm. Discontinuous in 8–12/Q+R (west) but continuous in 8–12/S+R (east).

V Mondmilch layer; discontinuous and completely absent toward south within Trinchera. 0–5 cm.

VI Yellowish-orange/light brown clay layer. 5–10 cm.

VII Thick mondmilch layer </=40 cm thick; gravelly at base, with cobbles, flints, bones, and teeth at base (in 10–11/R).

VIII Yellowish-orange/light brown clay. Objects are found right at the contact between base of VII and top of VIII. Bones, teeth, and flints continue within VIII.

IX Orange-brown, sandy, gritty sediment with little or no clay. Archaeologically sterile (found only in 0.25 m^2 mini-test pit: 11Rb).

References Cited

Courty, M. A., and J. Vallverdu

2001 The microstratigraphic record of abrupt climatic changes in cave sediments of the western Mediterranean. *Geoarchaeology* 16:467–500.

Ellwood, B., F. Harrold, S. Benoist, L. Straus, M. González Morales, K. Petruso, N. Bicho, J. Zilhão, and N. Soler

2001 Paleoclimate and intersite correlations from late Pleistocene/Holocene cave sites: results from southern Europe. *Geoarchaeology* 16:433–463.

Farrand, W. R.

1975 Sediment analysis of a prehistoric rockshelter: the Abri Pataud. *Quaternary Research* 5(1):1–26.

1985 Rockshelter and cave sediments. In *Archaeological Sediments in Context*, edited by J. Stein and W. Farrand, pp. 21–39. Center for the Study of Early Man, University of Maine at Orono.

1988 Integration of Late Quaternary climatic records from France and Greece: cave sediments, pollen, and marine events. In *Upper Pleistocene Prehistory of Western Eurasia*, edited by H. Dibble and A. Montet-White, pp. 305–320. University Museum Monograph, vol. 54. The University Museum, University of Pennsylvania, Philadelphia.

2000 *Depositional History of Franchthi Cave: Sediments, Stratigraphy and Chronology*. Excavations at Franchthi Cave, Greece, fascicle 12. Indiana University Press, Bloomington.

2001a Archaeological sediments in rockshelters and caves. In *Sediments in Archaeological Context*, edited by J. K. Stein and W. R. Farrand, pp. 29–66. University of Utah Press, Salt Lake City.

2001b Sediments and stratigraphy in rockshelters and caves: a personal perspective on principles and pragmatics. *Geoarchaeology* 16:537–557.

Hubbard, W.

2000 Geoarchaeological study of El Mirón Cave, Ramales de la Victoria, Cantabria, Spain. M.A. thesis, University of Michigan.

Laville, H., J.-P. Rigaud, and J. R. Sackett

1980 *Rockshelters of the Perigord*. Academic Press, New York.

Shahack-Gross, R., and S. Weiner
n.d. Geochemical Investigation of black bones from El Mirón Cave. Unpublished report submitted to L. G. Straus and M. R. González Morales.

Straus, L.
1997 Convenient cavities: some human uses of caves and rockshelters. In *The Human Use of Caves*, edited by C. Bonsall and C. Tolan-Smith, pp. 1–8. British Archaeological Reports S-667. Oxford, UK.

Straus, L. G.
1979 Caves: a paleoanthropological resource. *World Archaeology* 10:331–339.
1990 Underground archaeology: perspectives on caves and rockshelters. In *Archaeological Method and Theory*, vol. 2, edited by M. Schiffer, pp. 255–304. University of Arizona Press, Tucson.
2000 Caves and rockshelters, excavation. In *Archaeological Method and Theory: An Encyclopedia*, edited by L. Ellis, pp. 89–94. Garland, New York.

Straus, L. G., M. R. González Morales, and J. M. Carretero
2011 Lower Magdalenian secondary human burial in El Mirón Cave, Cantabria, Spain. *Antiquity* 85:1151–1164.

Straus, L. G., M. González Morales, W. Farrand, and W. Hubbard
2001 Sedimentological and stratigraphic observations in El Mirón, a late Quaternary cave site in the Cantabrian Cordillera, northern Spain. *Geoarchaeology* 16:603–630.

Woodward, J., and P. Goldberg
2001 Rockshelter sediment records and environmental change in the Mediterranean region. *Geoarchaeology* 16:223–631.

CHAPTER FIVE

Sedimentology of El Mirón Cave

William R. Farrand

At the invitation of L. G. Straus and M. González Morales, I began sedimentological study of El Mirón in 1997. The objectives of this study were the description of the sediments enclosing the archaeological finds, understanding their stratigraphic relations, deciphering their origins and paleoenvironmental settings, and ultimately evaluating the role of sediments in reconstructing human cultural activities in and around the site. The philosophical approach and methodology in this study follow those of W. R. Farrand (2001), adapted as necessary to the idiosyncratic aspects of this particular site.

Fieldwork

GENERAL

The studies reported here are based on field observations in 1997, 1998, 2000, and 2002 by the author, with the assistance of William Hubbard in the field in 2000. In each of these seasons we examined the sections exposed by the excavation team and selectively sampled the strata following the identifications of the excavators. Also, Hubbard and I excavated a geologic trench partway up the ramp ("colluvial slope," as described by Straus [chap. 3, this volume]) that leads from the vestibule up into the dark inner cave. No archaeological remains were found in the ramp sediments, which were poorly sorted, heterogeneous fluvial deposits (plate 5.1).

In addition, we explored the surrounding area for evidence of sources of sediment external to the cave, possible river terraces, and related geomorphic features. We were guided in the latter effort by Enrique Serrano, then of Universidad de Cantabria, who was studying the glacial history of the upper Río Asón and the eastern Cantabrian Cordillera. Of special interest was Cuevamur, a cave near El Mirón with a sedimentary fill similar to the older sediments in El Mirón.

SAMPLING

Seventy-seven stratigraphic samples, listed in table 5.1, were collected from the Cabin, Mid-Vestibule, Corral, and Footslope excavation areas of the sunlit vestibule, as well as from the Trinchera in the dark, inner cave and from the ramp. Miscellaneous samples included pieces of the cave bedrock and terra rossa soil from outside the cave. The samples were shipped to the University of Michigan for analysis. Samples were selected by visual inspection of the exposed strata in the excavated areas with the goal of taking at least one sample from each archaeological stratum or two or more samples if the stratum was rather thick or seemed to include some internal variation. In a few cases a given stratum was resampled in successive seasons or in different excavation squares to evaluate lateral variations. Each sample was labeled with the number of the stratum and the square from which it was collected and in some cases with the addition of "upper," "mid," or "lower" to indicate multiple samples from a given stratum.

Samples commonly weighed between 100 g and 1000 g, with larger samples from beds of coarser sediment. In most cases the Munsell color of the sampled sediment was recorded in the field under moist conditions. The remainder were color-coded in the laboratory under dry conditions, as noted in table 5.1. The moisture content was measured upon arrival in the laboratory, only a few weeks after the samples were collected. They had been wrapped in double plastic bags that preserved the original moisture state reasonably well.

The remainders of these samples are archived in the Museum of Anthropology at the University of Michigan.

Methodology

GRANULOMETRY

The samples were dried and weighed again during initial processing in the laboratory. Then various fractions were separated for further analysis. The coarse fraction, from 128 mm to 10 mm, was measured by hand. The sand and granule fraction, 10 to 0.063 mm, was sorted by wet sieving, and the finest fraction was measured with a Spectrex™ laser counter, with verification of some samples by a Coulter™ counter.

In addition, Hubbard (2000, figs. 12–15) determined the granulometry of the fine fraction after decalcification with 10 percent HCl and mild heating.

ROUNDNESS

The degree of rounding of the larger clasts—that is, the extent of rounding of originally angular corners of the rock fragments—was observed visually and indexed according to the system of W. R. Farrand (1975, 2001). Clasts of a selected size range, in this case between 8 mm and 32 mm, were classified as follows:

- **Class 0:** only angular corners and edges; no rounding
- **Class 1:** more angular corners and edges than rounded ones—subangular
- **Class 2:** more rounded corners than angular ones—subrounded
- **Class 3:** no angular corners or edges—rounded

Then the weight percentage of each class was calculated and a Roundness Index (R.I.) was determined with the equation:

$$\text{R.I.} = \frac{[(Wgt\ Class\ 0)0 + (Wgt\ Class\ 1)1 + (Wgt\ Class\ 2)2 + (Wgt\ Class\ 3)3] \times 100}{3}$$

Thus, a sample with an R.I. = 0 contains only angular rock fragments and a sample with an R.I. = 100 contains only completely rounded fragments. For clarity, it should be pointed out that a completely rounded rock fragment is not necessarily spherical in its overall form. It may be brick-shaped, or cubic, or platy, but all its edges and corners are rounded.

CALCIUM CARBONATE AND ORGANIC MATTER

These measures were determined on the fine fraction (<2 mm) by means of loss-on-ignition analysis in a muffle furnace. Dried samples were combusted first at 550°C to burn off organic matter and then again at 1,000°C to burn off CO_2, which was then calculated as $CaCO_3$. Duplicate samples and laboratory standards were measured to assure constant results.

PH

The pH of the fine fraction (<2 mm) was measured in a 1:1 slurry of sediment and distilled water by means of an electronic pH meter. The slurry was allowed to stabilize for one hour before the reading was made.

ROCK FRAGMENT LITHOLOGY

It was quickly obvious that rocks of various types occurred in the sediments, so we decided to tabulate the fragments in two categories—(a) limestone fragments derived from the cave bedrock and (b) all other lithologies, that is, lumped as "non-limestone." The latter comprised micaceous sandstone, siltstone, chert, and quartzite, none of which occur in the bedrock near El Mirón and which tend to be well-rounded cobbles and pebbles. The non-limestone rocks occurred abundantly in the sediments of the narrow corridor ramp at the rear of the vestibule and in the lower levels of the Corral but rarely in the Mid-Vestibule Trench or the Cabin area.

CULTURAL REMAINS

Most of the sediment samples contained residues of human activity, such as bone fragments, lithic debitage, charcoal, small sherds, and shell fragments (table 5.2). In almost all cases these are small fragments. They were weighed and their aggregate percentage by weight of the total sample was calculated, labeled "Cultural Debris," "D fraction," or "D%" (terminology of W. R. Farrand 1975). In most cases the percentage is small (<2 percent) but rises to 15–20 percent in culturally rich strata. These values give only a rough indication of the intensity of human activity associated with a given stratum. More accurate estimates of cultural richness are given by tabulation of artifacts and other human debris by the archaeologists. As seen in table 5.2, all sediment samples contained some cultural material, except some of the deeper layers in the Footslope excavation, notably Stratum 129.

CLAY MINERALOGY

Hubbard (2000) analyzed the clay minerals in the fine fraction by X-ray diffractometry (XRD), following the methods of Moore and Reynolds (1997), for the purpose of distinguishing differences in provenance of sedimentary components. As explained by Hubbard (2000:15), "Although the methodology of preparing the samples included measures to remove $CaCO_3$, all was not removed owing to the massive amount of $CaCO_3$ in [some] samples. Reducing the samples any further would have resulted in complete dissolution of the clay-sized particles, yet enough $CaCO_3$ was removed to allow for the measurements of the peaks of other minerals." Hubbard did not sample the lower levels in the Cabin or Footslope trench for clay minerals.

Stratigraphic Considerations

GENERAL

The stratigraphy followed in this chapter (figure 5.1) is that established by the archaeologists (Straus and González Morales, chap. 4, this volume), and the samples are labeled according to the archaeologists' designations. Thus, samples 2 through 16/17 come from the Cabin area; samples 102 through 113/114 come from the Corral area. (Levels 115 through 119 were not sampled.) Samples 400 (=120) through 405 (=128) and 129 through 130 come from the Footslope sondage and samples IV through IX were collected from the Trinchera in the inner cave.

INTER-TRENCH CORRELATION

Clearly some of the strata continue from one excavation area to another. This was obvious in a general way from visual inspection alone, but in a number of cases it could be confirmed by radiocarbon dating and was supported by sedimentological similarities, as in the case of the carbonate-rich, pale gray–colored Levels 10, 10.1, 304–305, and 101–102 (all of early Holocene age), which is presumably a single mondmilch layer (plate 5.2a, b). On the basis of these correlations, the entire sequence in the Mid-Vestibule Trench matches the lower half of the Cabin strata, Levels 10 through 17, and matches Levels 101 through 109 in the Corral area. Correlation based on the

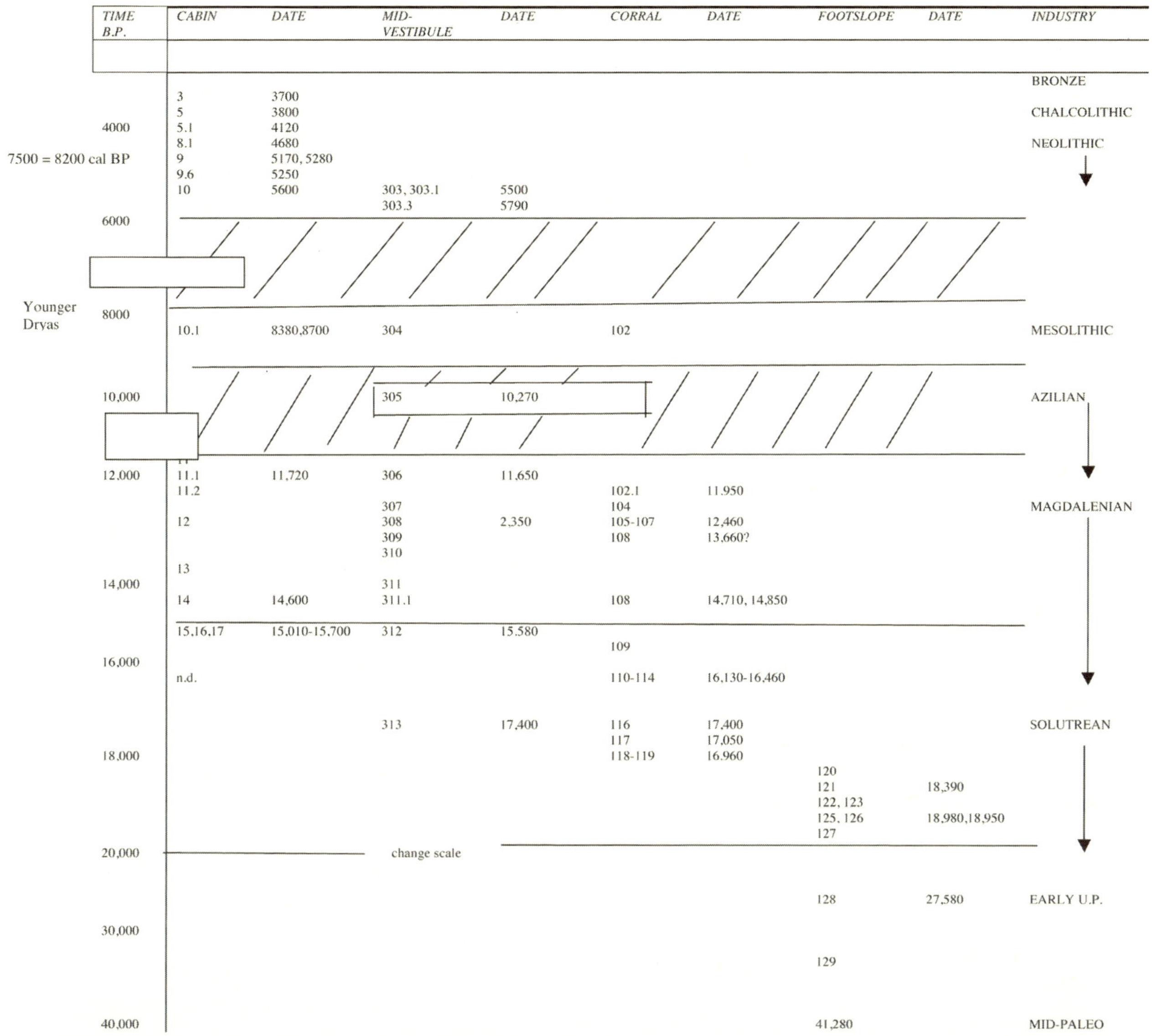

Fig. 5.1. Stratigraphic correlation chart. Areas with diagonal lines indicate major unconformities. Two especially prominent, widespread cold spells are indicated by the boxes on the left margin. Drawing by William Farrand.

radiocarbon chronology is given in figure 5.1. In accord with this correlation, I have arranged the sequence of sedimentological samples in a composite stratigraphy in table 5.2 in order to bring together and compare data from a single geological stratum or stratum cluster. Thus, for example, Levels 13, 106, 310, and 14 are clustered on the horizontal axis of those figures and in the left column of table 5.2, and so on. The results of these correlations are presented in the stratigraphic summary below.

Cabin levels above Level 10 (which is early Neolithic) apparently have no equivalents in the mid- or rear vestibule, except the very surface level of modern disturbed sediment (Levels 1, 300–301, 100). Levels 303–303.3 in the Mid-Vestibule Trench are continuous with Cabin Level 10. Levels below Level 109 in the Corral and Footslope areas have not yet been reached in the Cabin area, although excavation is continuing and borings as well as a 1 m² sondage indicate more sediments are present there (Levels 18–21: 16,000–16,600 BP). In the Mid-Vestibule Trench excavation has now reached Level 313, the ¹⁴C dates of which match those of 116 or 117 in the Corral area.

Stratigraphic Summary

The highlights of each stratum or stratum complex in table 5.2 are given here in terms of granulometry, roundness, $CaCO_3$ content, presence of exotic rocks, existence of angular rock fragments presumably produced by frost action, and other occasionally pertinent characters, combining field and laboratory observations. As mentioned above, the sediment color was determined under controlled conditions either in the field or in the laboratory, according to the Munsell color scheme. The colors given in this chapter do not necessarily correspond to those of the excavators, as seen in section drawings elsewhere in this volume, which were assigned empirically according to the impressions of the artist drawing the sections in the field.

A stratum complex is a group of strata that is considered to be a single geological layer. A stratum complex may combine laterally and/or vertically adjacent strata as defined during excavation. The sedimentological character of a given stratum complex may vary somewhat laterally (e.g., in color, grain size, $CaCO_3$ content, or cultural content), but it is considered to have been deposited simultaneously throughout the cave vestibule. See table 5.1 for depths and Munsell colors and table 5.2 for numerical values. The uppermost strata that are considered to be modern and/or mixed sediments are not included in the discussion below.

Stratum 2: Dark brown, fine sand with granules and some angular limestone rocks and occasional exotic pebbles; low in $CaCO_3$ and organic matter; modern bones associated with old cabin floor.

Stratum 3: Dark brown, silty clay with abundant limestone rocks and some yellow sandstone pebbles; moderate amount $CaCO_3$ (29 percent) and relatively high organic matter (8.29 percent); charcoal present. Levels 3.5, 3.0, 3.1, 3.2, 3.3, and 3.43, which were not sampled for sediments, were largely hearth and ash complexes.

Stratum 4: Pinkish-gray, silty ash (hearth), very rich in $CaCO_3$ (73 percent) from ash, but low organic matter; abundant angular, fire-cracked limestone rocks.

Stratum 5: Mottled grayish-brown to brown, silty clay with a few granules but no large rocks; abundant $CaCO_3$ (60 percent) and moderate amount of organic matter (6 percent). Sublevels 5.1 through 5.5 not sampled.

Stratum 6: White ash, not sampled.

Stratum 6.1: Brown, silty clay with a few angular limestone fragments and charcoal; lower part of hearth Stratum 6.

Stratum 7: Brown to dark brown, silty clay with abundant limestone fragments (fire-cracked) and some charcoal; moderately rich in $CaCO_3$ (50 percent) and organic matter (7.5 percent); some cobbles noted by excavators (not in sediment sample).

Stratum 8: Grayish-brown, silty clay, fissured at top, with some granules but no large rock fragments; very rich in $CaCO_3$ (76 percent); moderate organic matter; some charcoal present; clear upper contact with Stratum 7.

Stratum 9: Brown, very moist, silty clay with some large angular limestone fragments at top, just below contact with Stratum 8; high $CaCO_3$ (71.5 percent), moderate organic matter; some charcoal.

Stratum Complex 10 and 303–303.3: Pale brown to light brown, silty clay ("grayish-white" mondmilch), very compact at top and very moist and vesicular at base (in Cabin); very rich in $CaCO_3$ (80 percent in Cabin and 60 percent in Mid-Vestibule); abundant angular limestone fragments in upper part but only granules with depth and none in 303 (Mid-Vestibule); abundant microfauna, especially in lower part of Stratum 10.

Stratigraphic unconformity

Stratum Complex 10.1, 304, and 102: Mixed dark reddish-gray and brown to dark brown, silty clay in Cabin but pale yellowish-brown silt in Mid-Vestibule and Corral; very rich in $CaCO_3$ (57–86 percent) but low organic matter; some rounded granules in 10.1, which is a thin, ashy hearth residue; few rock fragments except at inner end of Mid-Vestibule Trench (Square P–O/6), which has abundant angular limestone fragments; Stratum 102 in Corral breaks into granular fragments (see histogram) but also contains some angular limestone rocks; Stratum 102 clearly fills a depression in the Corral area with a clear lower contact.

Stratigraphic unconformity

Stratum 305: Pale yellowish-brown silt, similar to 304 but distinctly darker at top under clear contact with 304, suggesting a weathering interval (figure 5.1a); some granules and angular rock fragments; 305 is cut out by later pits on north side of Mid-Vestibule Trench.

Stratigraphic unconformity

Stratum Complex 11–11.2, 306, and 102.1: Stratum 11 is light brown to reddish-brown, silty clay, gritty in upper part; abundant angular limestone fragments in middle and lower part; 306 is yellowish-brown and sandier, also with angular limestones (plate 5.2a); Stratum 11 in much richer in $CaCO_3$ (78 percent) than 306 (24 percent); clear to sharp contact with 305; 102.1 is a hearth remnant. All these levels appear to be truncated at the top.

Stratum Complex 104 and 307: Dark brown, very fine sandy loam with rare angular limestone fragments; low to moderate $CaCO_3$ (12–31 percent) and moderate organic matter; very clear contact with 306 above (plate 5.2b).

Stratum Complex 12, 308, 309, and 105(?): Brown to dark brown, silty clay with abundant angular limestone fragments, some up to 15–16 cm (plate 5.2b); moderate amount of $CaCO_3$ at top but decreases with depth; exotic rocks rare; Stratum 105 appears mixed.

Stratum Complex 13, 14, 106–108, 310, 311, and 311.1: Brown to dark brown, silty clay in Cabin and Mid-Vestibule (plate 5.2b) but fine sand in Corral, with small angular rock fragments abundant but some 8–9 cm (in Corral) and up to 25–30 cm (in Cabin); exotic pebbles rare; $CaCO_3$ moderate to low—26 percent in Mid-Vestibule but less in Cabin and Corral.

Stratigraphic unconformity

Stratum Complex 15–17 and 312: Brown to dark brown, silty clay with numerous angular limestone fragments and bones (plate 5.3); distinct color contrast with 311.1 and sharp upper contact with 14; low $CaCO_3$ and organic matter; no exotic rocks.

(No sampling below Stratum 17 or 312 in Cabin and Mid-Vestibule.)

Stratum Complex 109 and 110: Soft, dark brown loam with bones and occasional white (weathered) limestone fragments and exotic pebbles; low $CaCO_3$, but relatively high organic matter (6–7 percent).

Stratum Complex 111–114: Dark reddish-brown loam with variable amount of angular limestone fragments and rounded exotic cobbles; clasts are chalky white; low to moderate $CaCO_3$ but relatively rich in organic matter; bones common.

(Strata 115–119 not sampled but similar to Strata 111–114, with white rocks common and abundant organic matter; culturally rich.)

Stratum 120 (400 in Sq. W10): Dark yellowish-brown, loamy matrix with abundant exotic pebbles and chalky limestone fragments and some bone; low $CaCO_3$ and organic matter.

Stratum 121 (400.1 in Sq. W10): Thin, dark brown layer at base of 400 with fewer rocks; limestone fragments are chalky; low $CaCO_3$ and organic matter.

Stratum Complex 122, 123, and 124 (401, 401.1, and 401.2 in Sq. W10): Medium yellowish-brown, sandy loam with pebbles and limestone fragments, similar to 400; 124 is a bit more yellow (plate 5.4).

Stratum 125 (402 in Sq. W10): Distinct olive-brown, sandy loam with about equal numbers of limestone fragments and exotic pebbles; 402 especially rich in bones and charcoal; low $CaCO_3$ but relatively rich in organic matter.

Stratum 126 (403 in Sq. W10): Medium yellowish-brown, sandy loam with abundant limestone fragments, exotic pebbles, flint, and bones, especially in 403; low $CaCO_3$ and organic matter; clear color contact with 125 above but gradational with 127 below (plate 5.4).

Stratum 127 and 127.1 (404 and upper 405 in Sq. W10): Yellowish-brown to brownish-yellow, sandy and gravelly loam including a thin, fine-grained lens (127.1) in middle; limestone fragments more abundant than exotic pebbles; modest amount of $CaCO_3$ (17–22 percent); 404 and 405 relatively rich in organic matter and cultural debris; lower contact wavy and clear, probably erosional.

Stratigraphic unconformity?: Approximately 9,000-year gap between 126 and 128.

Stratum 128 (405 lower in Sq. W10): Brownish-yellow, sandy loam, not as gravelly as 127; limestone fragments abundant and some exotic pebbles; large limestone blocks begin to appear—up to 30 cm diameter; modest amount of $CaCO_3$ (17–20 percent); as above, 405 richer in cultural debris than 128 (plate 5.4).

Stratum 129 (406 in Sq. W10): Strong brown (moist), sandy loam with abundant exotic pebbles and some angular limestone fragments; low $CaCO_3$ and organic matter; cultural material rare; limestone blocks up to 1 m diameter at this level; lower boundary arbitrary.

Stratum 130: Strong brown (moist), sandy loam, same as 129; limestone fragments and exotic pebbles common; large limestone roof-fall blocks present.

Results

GRANULOMETRY

The particle-size distribution for each sample is presented in the form of a histogram (figures 5.2–5.6) that displays the weight percentage of each phi size class from -7 phi (128 mm) to smaller than 10 phi (1 μm). The phi scale is the standard in sedimentology, being a logarithmic transform of the grain size in millimeters, where Phi = $-\log_2 10$ (grain size in mm). Thus, 1 mm = 0 phi; larger grain sizes have negative values and smaller ones are positive. The upper limit of -7 phi is an arbitrary but practical limit for the collecting and shipping of samples. A greater upper limit would require a much larger total sample for statistical purposes. Some histograms were previously published by Hubbard (2000; Straus et al. 2001) but, owing to recalculations, some of them differ slightly from those presented herein.

It is immediately obvious that the size distribution is polymodal with two or three modes (peaks) for each sample. (This is the usual pattern found in cave and rockshelter sediments [W. R. Farrand 2001], although the peak values of the different modes vary in different sites.) In the case of El Mirón samples, there are generally three modes: (a) coarse (rock fragments and pebbles), (b) granules and sand, and (c) fine silt and clay.

Fig. 5.2. Granulometric histograms for the Cabin area. Graph by William Farrand.

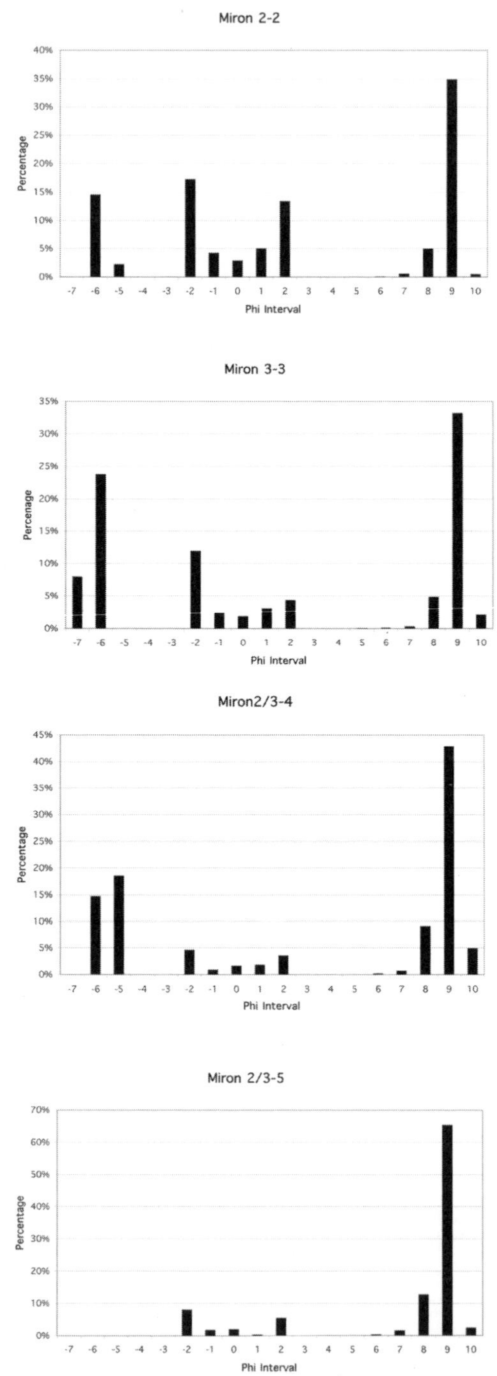

Fig. 5.2. *(continued)*

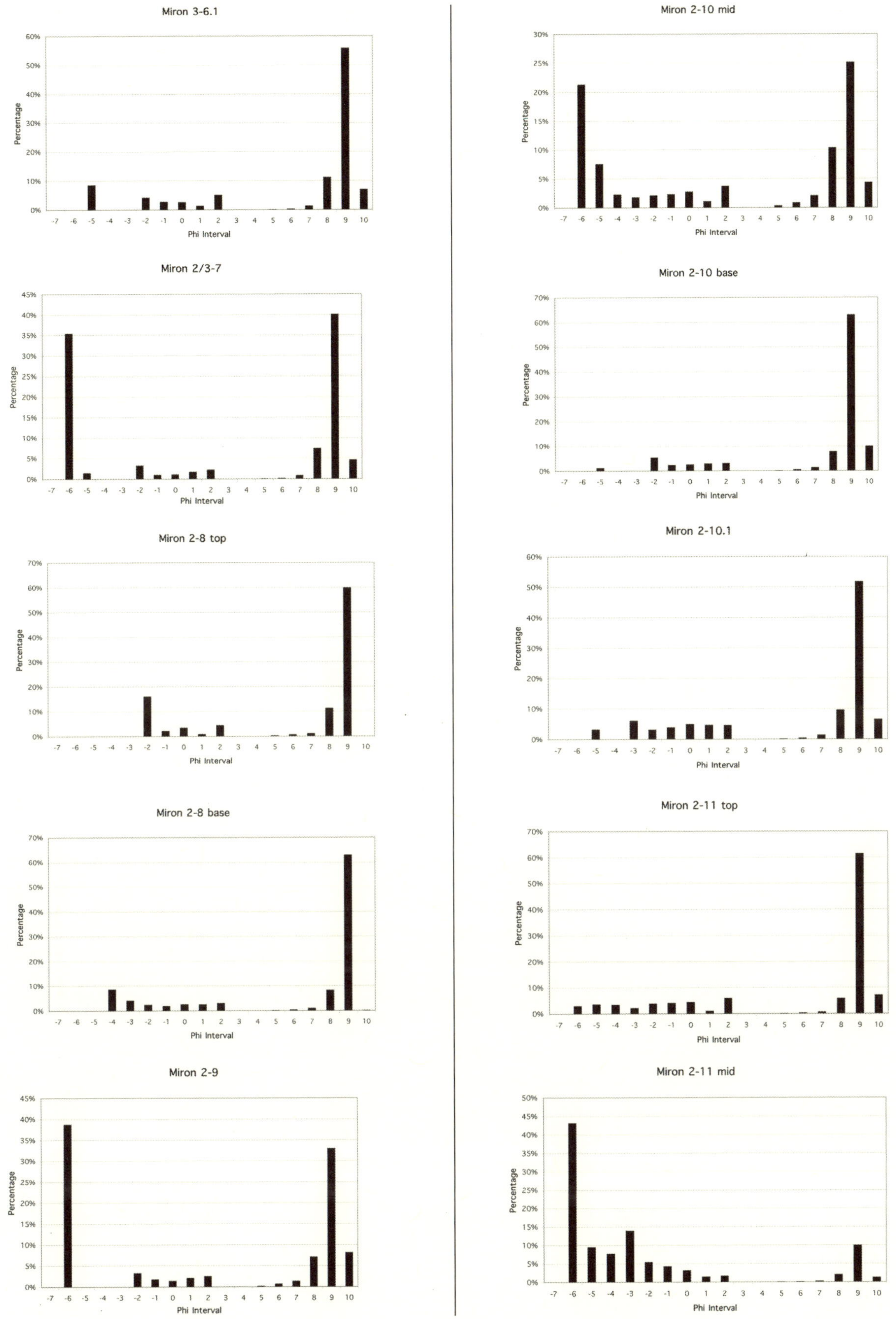

Fig. 5.2. *(continued)*

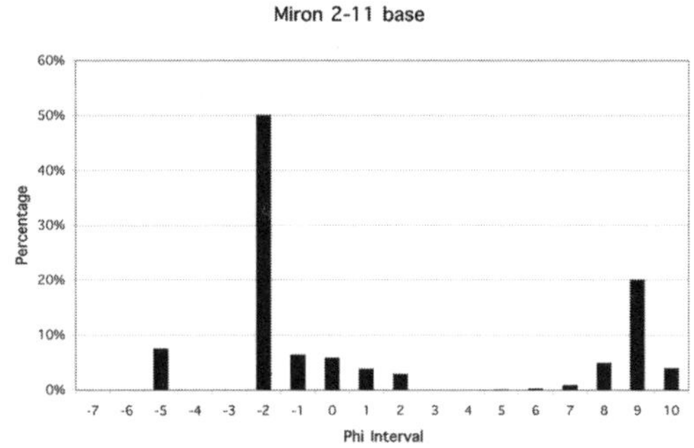

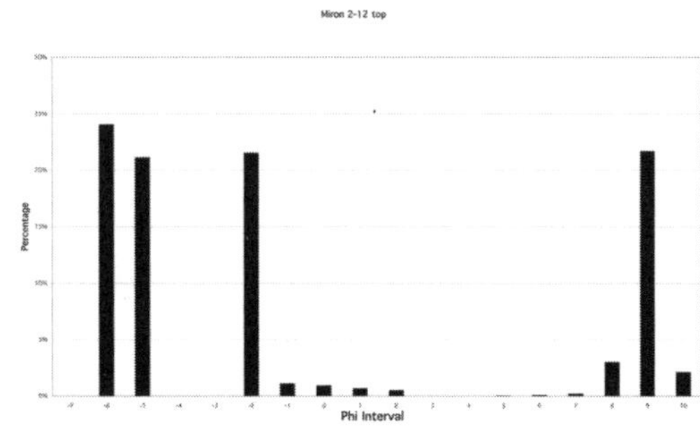

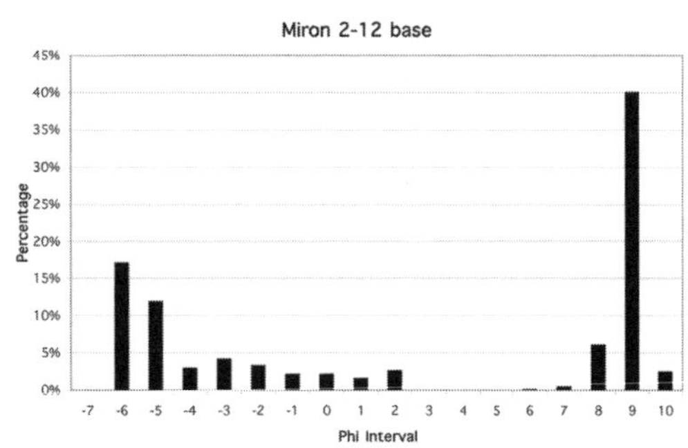

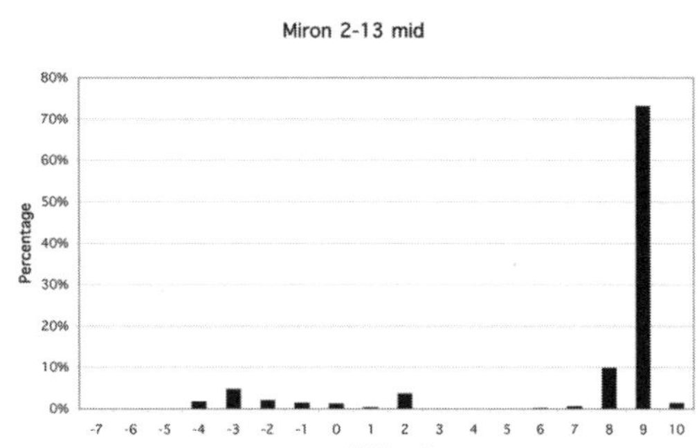

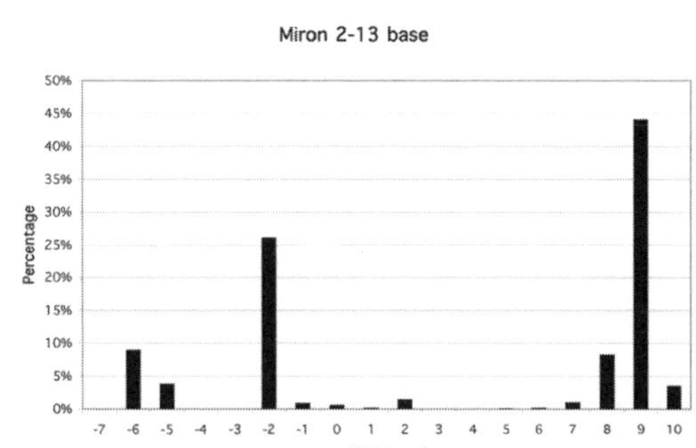

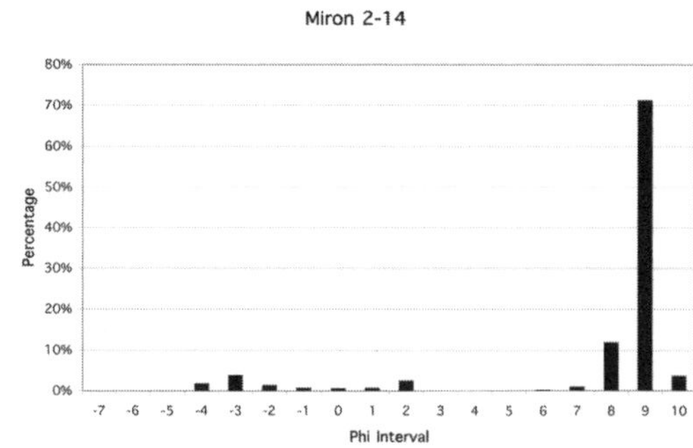

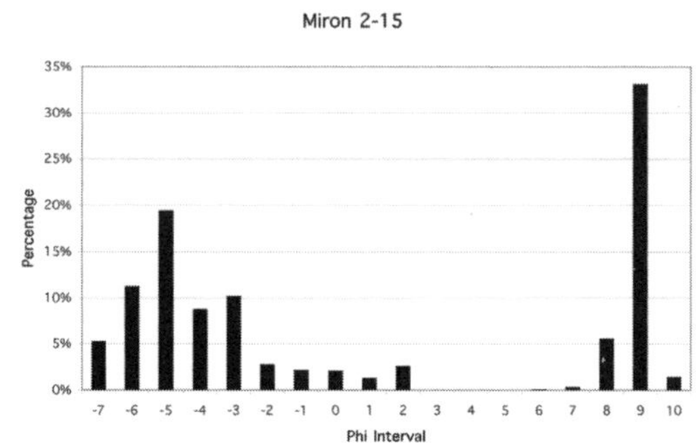

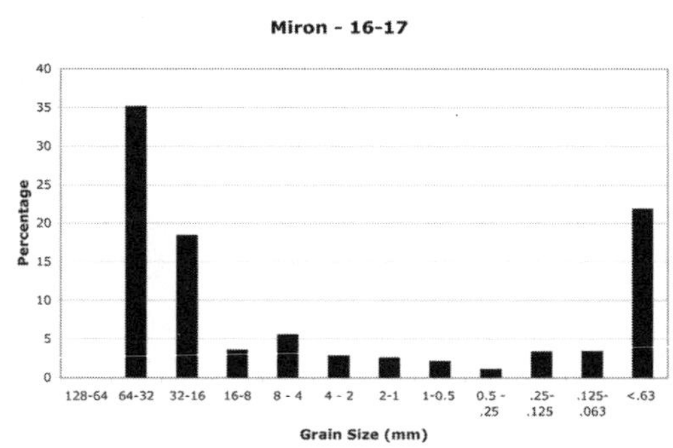

Fig. 5.3. Granulometric histograms for the Mid-Vestibule Trench. Graph by William Farrand.

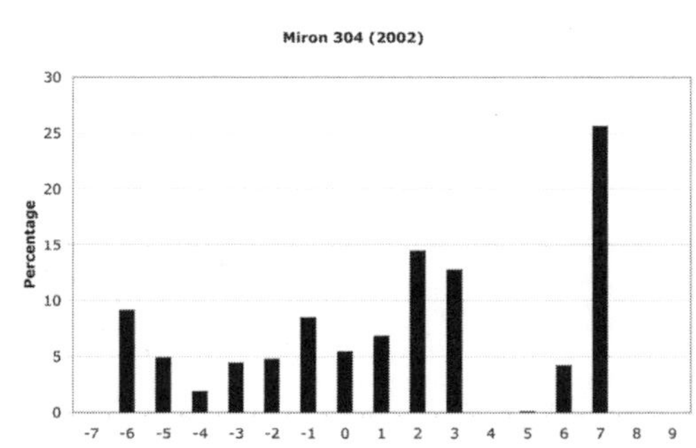

Fig. 5.3. *(continued)*

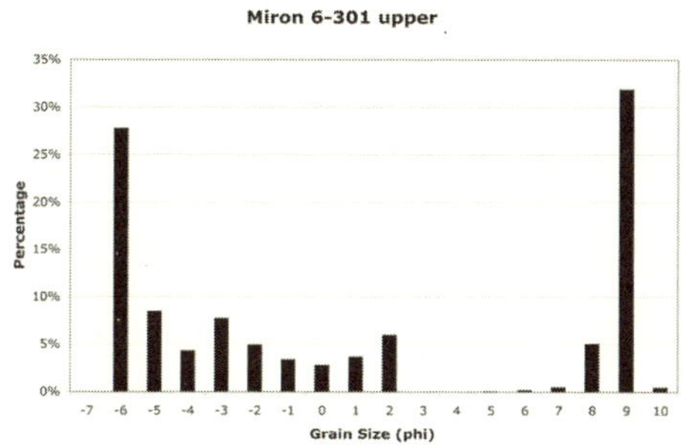

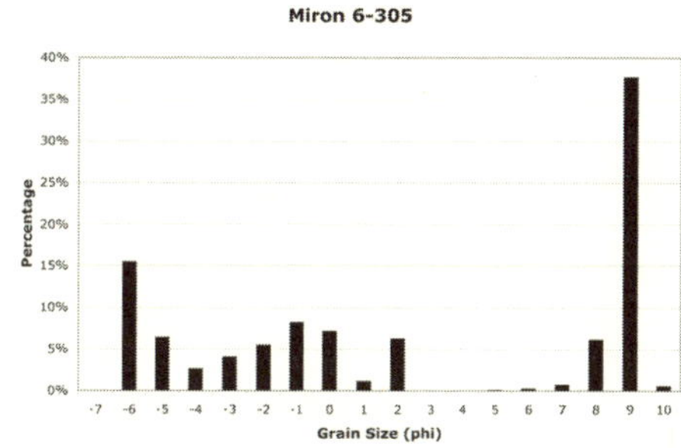

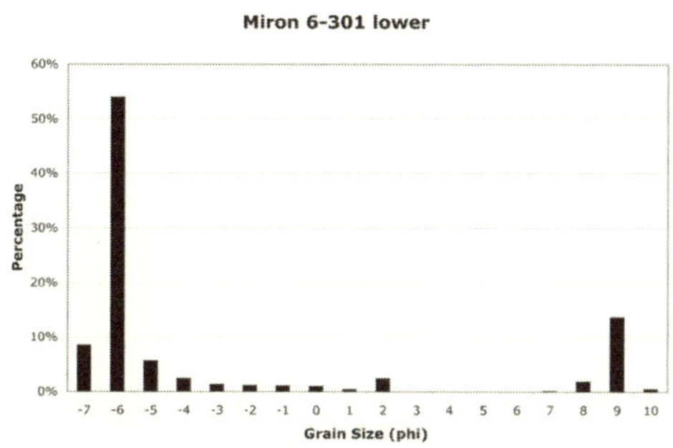

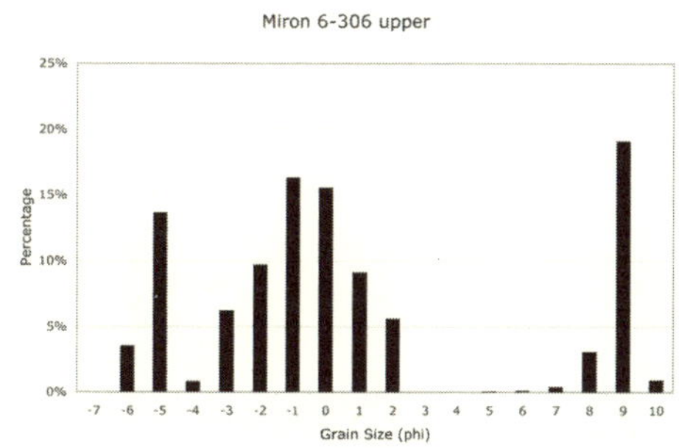

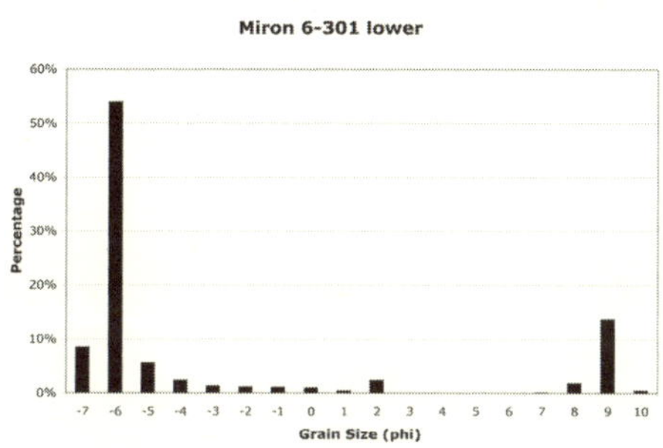

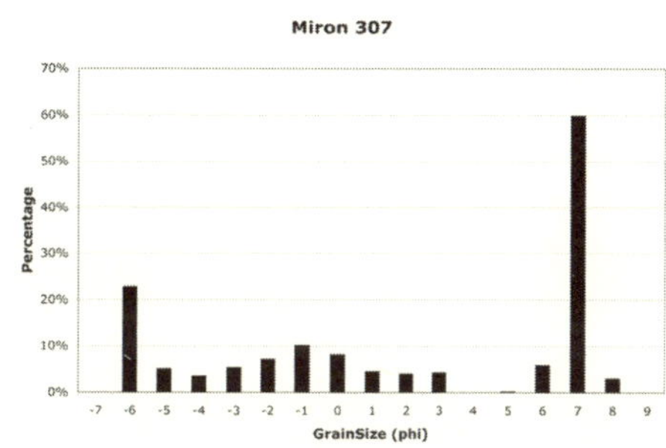

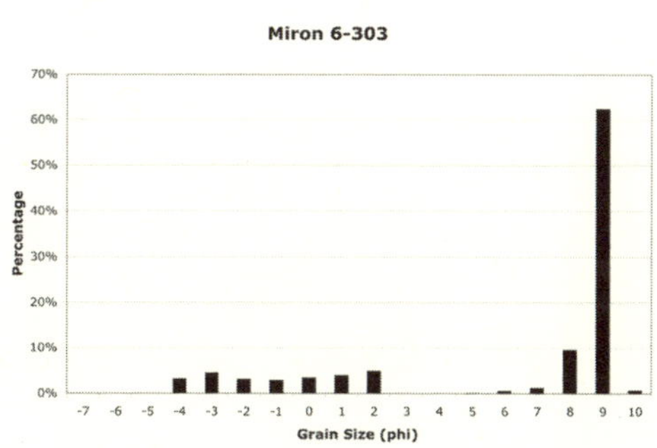

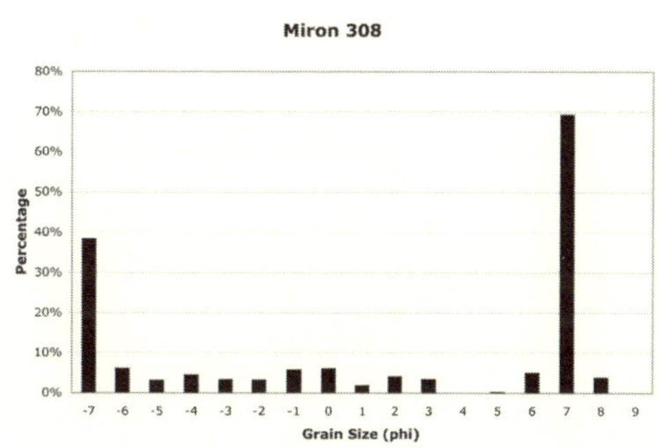

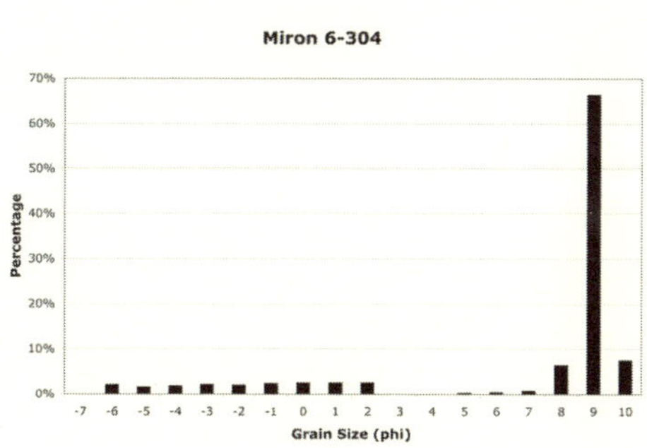

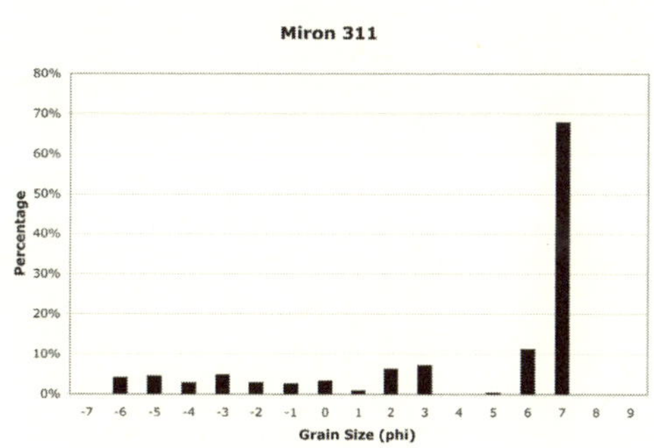

Fig. 5.3. *(continued)*

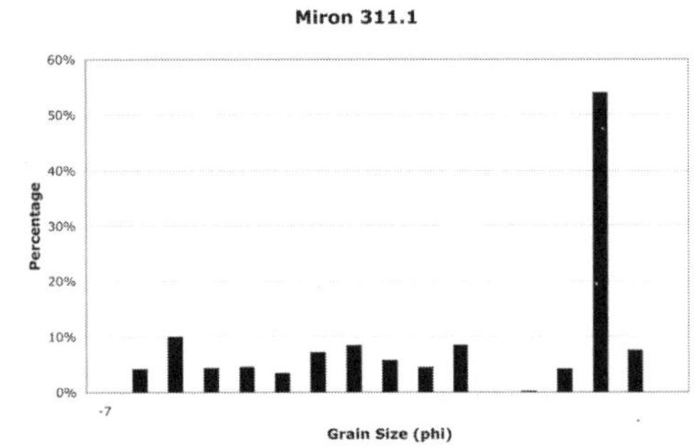

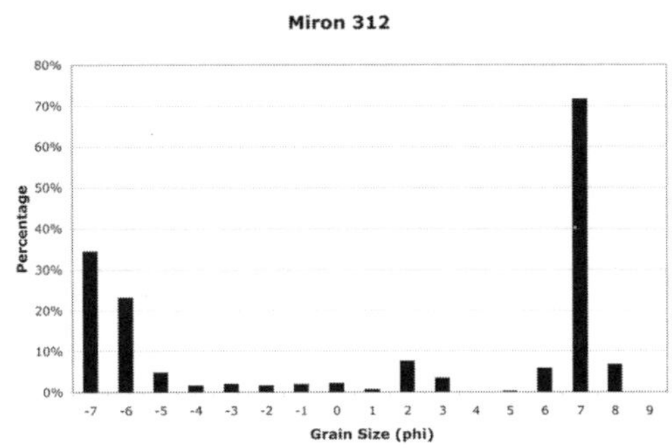

Fig. 5.4. Granulometric histograms for the Corral area. Graph by William Farrand.

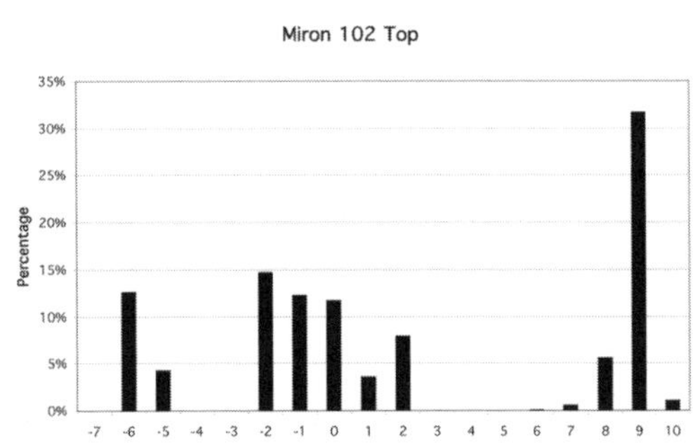

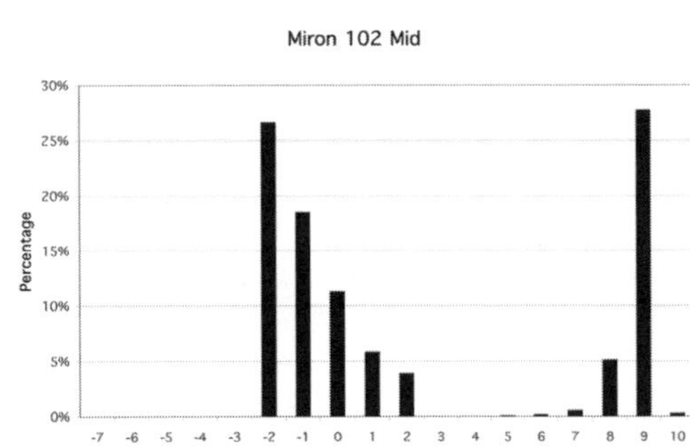

Fig. 5.4. *(continued)*

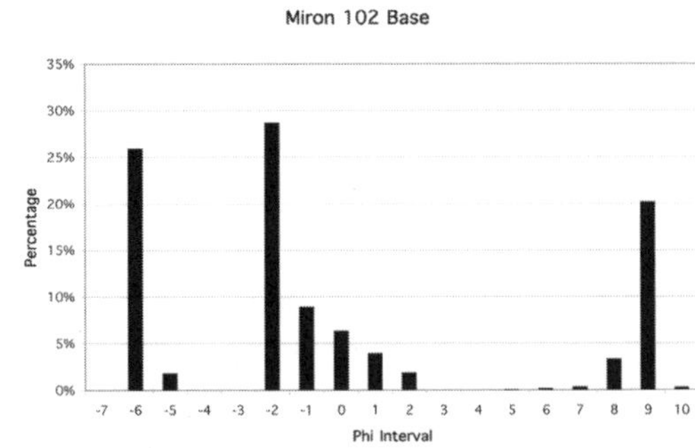

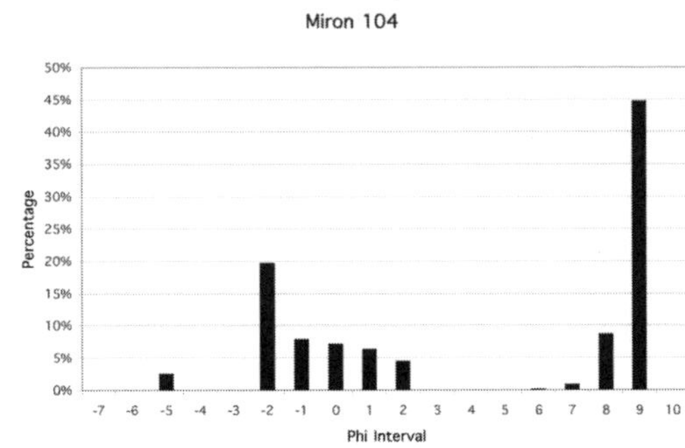

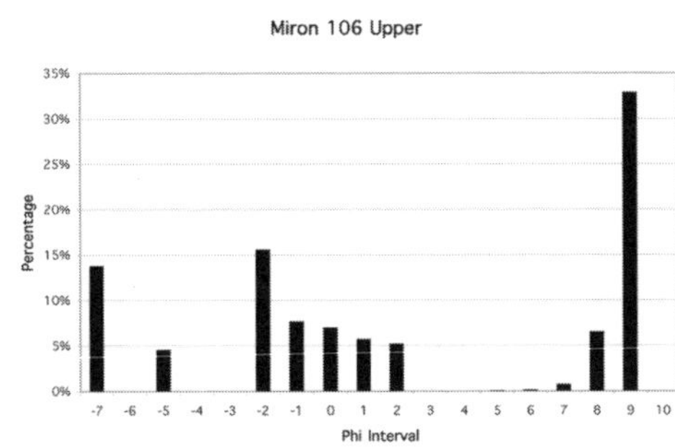

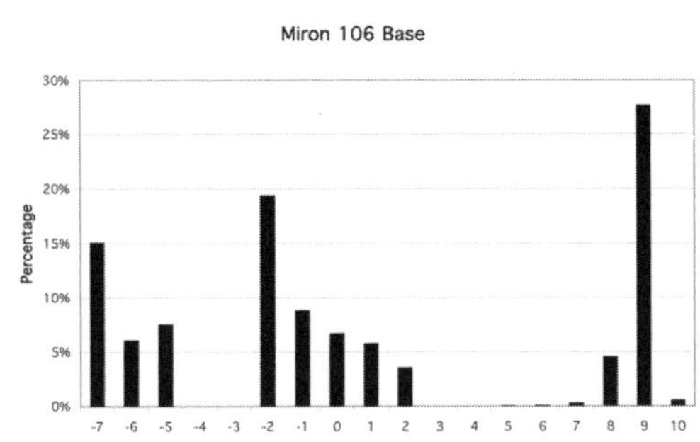

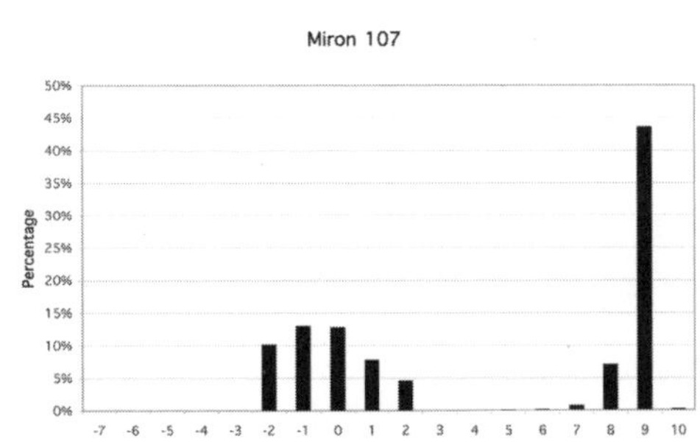

Fig. 5.4. *(continued)*

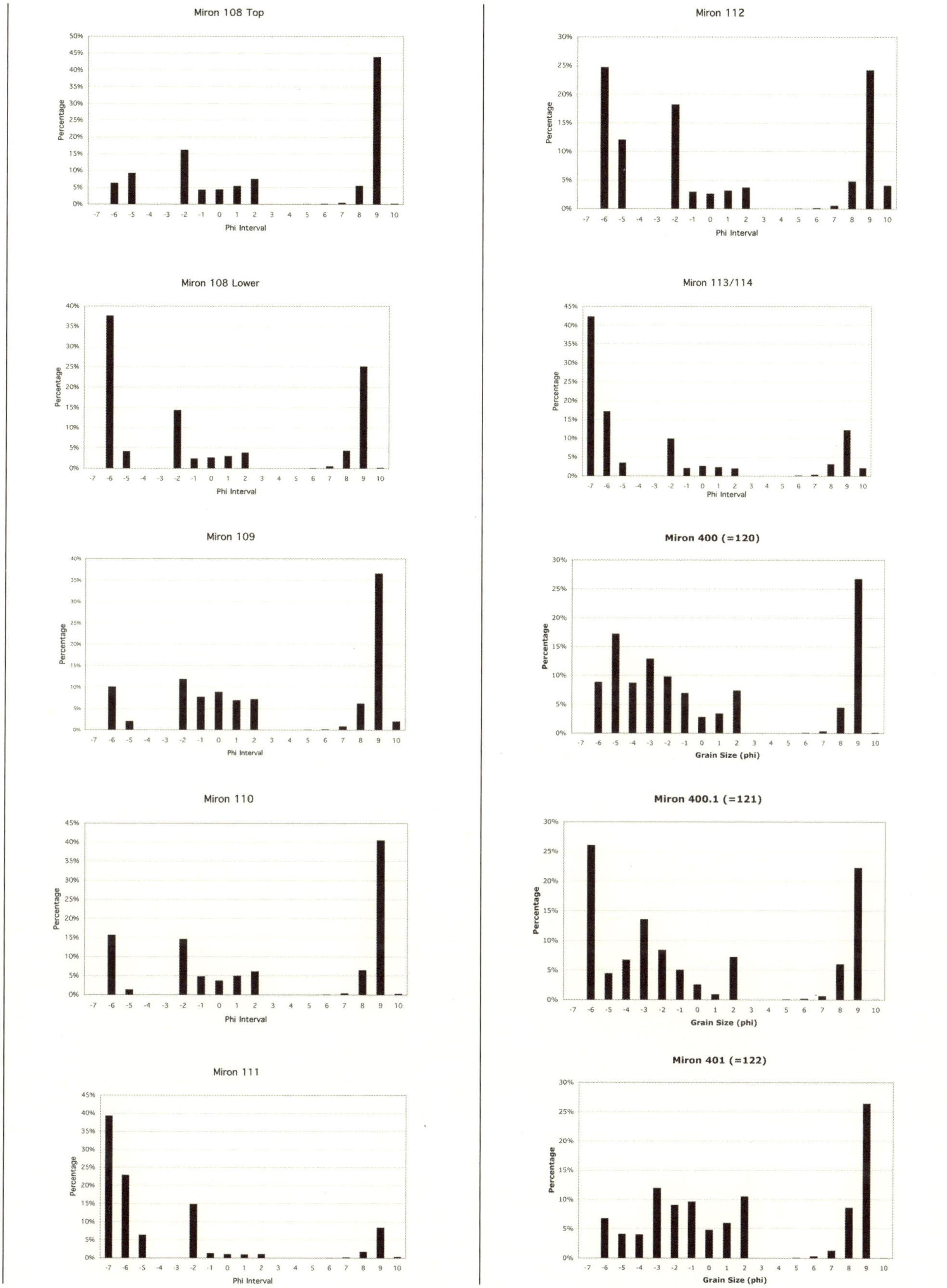

Fig. 5.4. *(continued)*

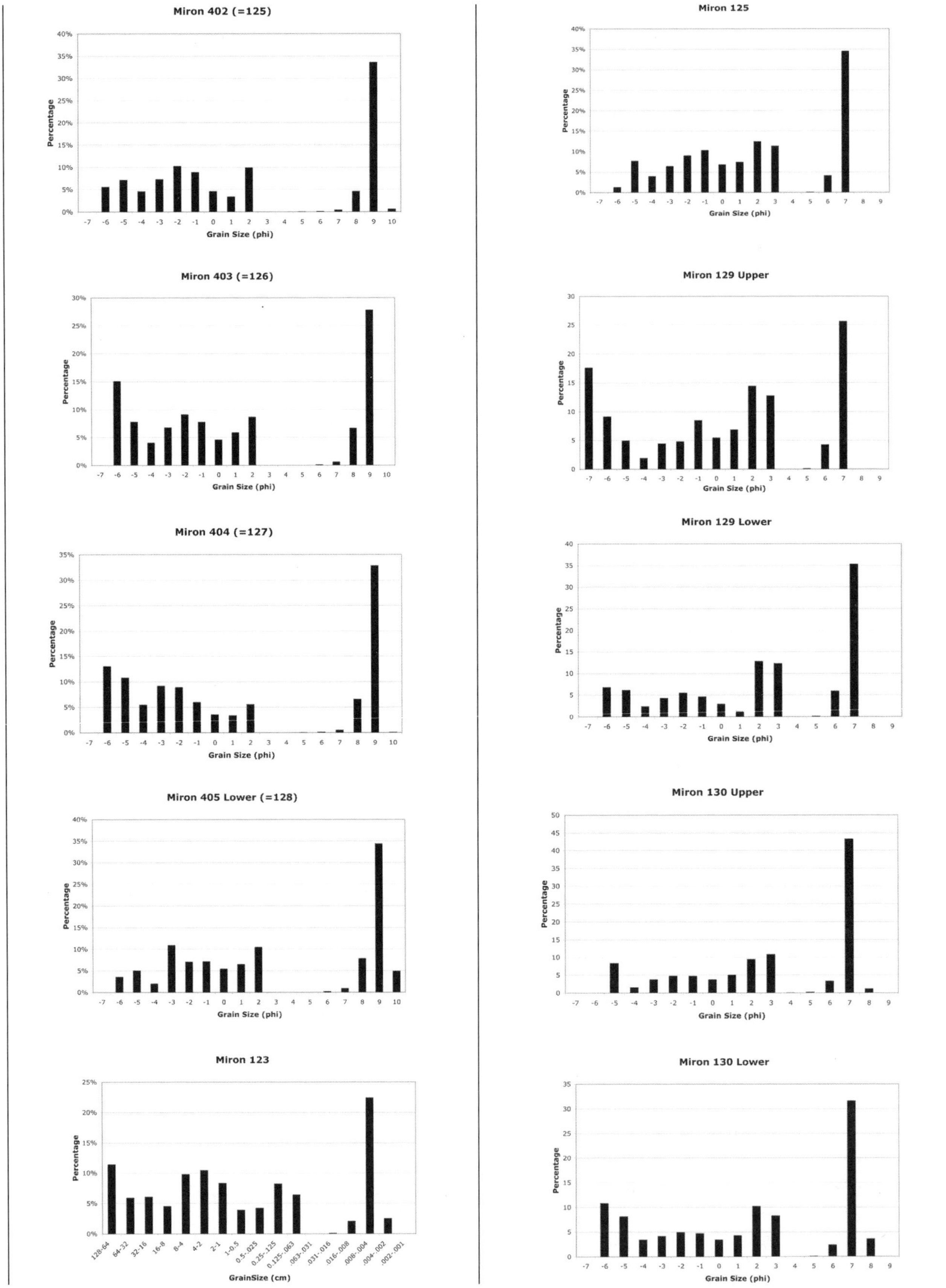

Fig. 5.5. Granulometric histograms for Cuevamur and the El Mirón Cave colluvial-alluvial ramp deposit. Graph by William Farrand.

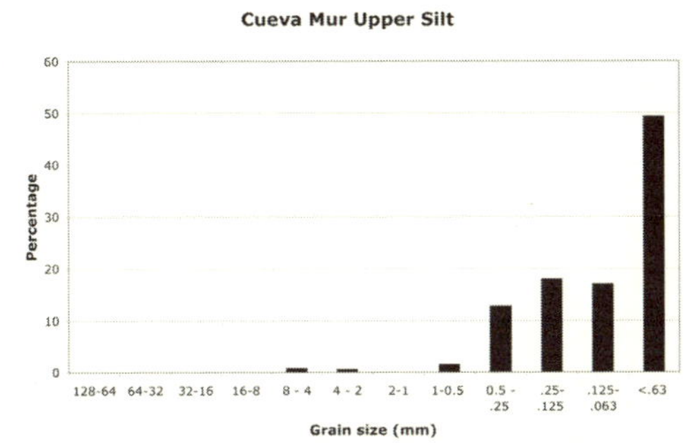

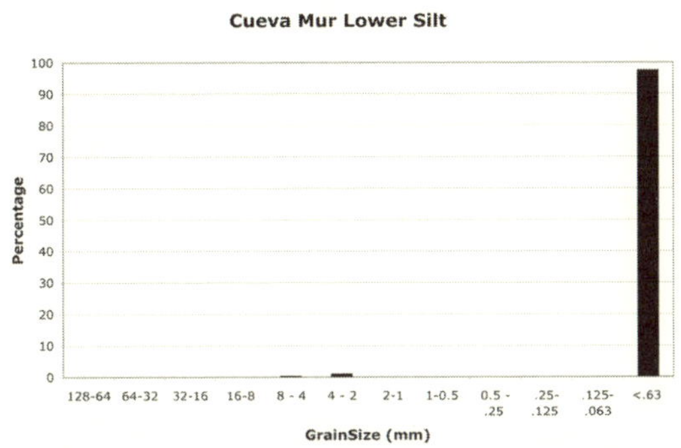

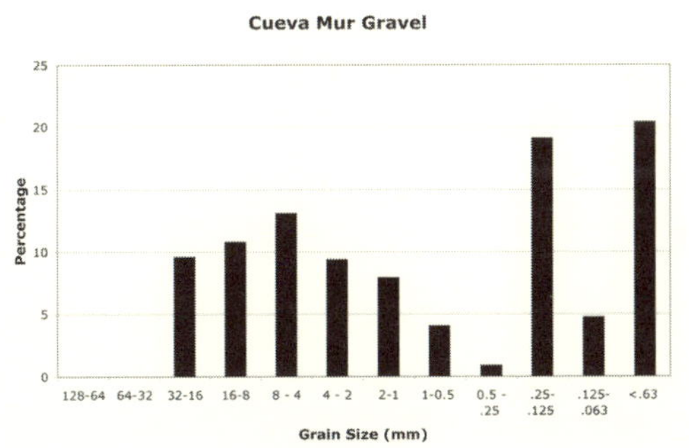

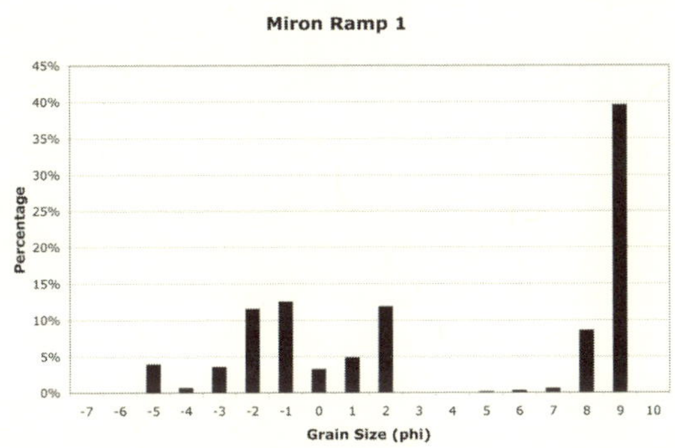

Fig. 5.5. (*continued*)

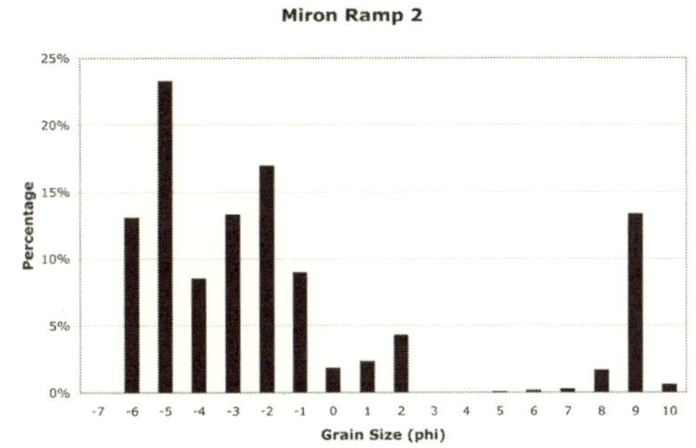

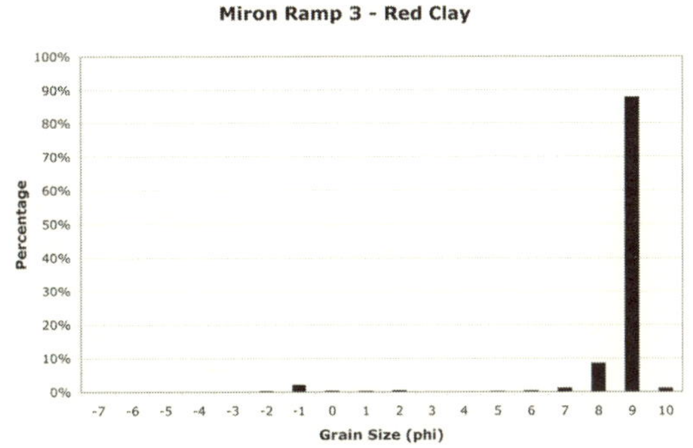

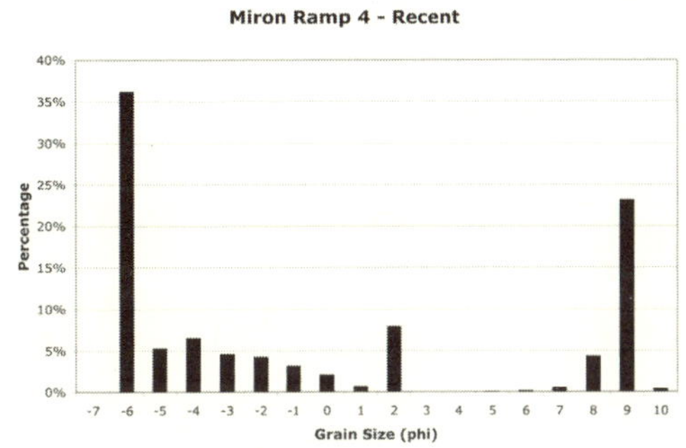

Fig. 5.6. Granulometric histograms for the Inner Cave Trench and interior wall deposit. Graph by William Farrand.

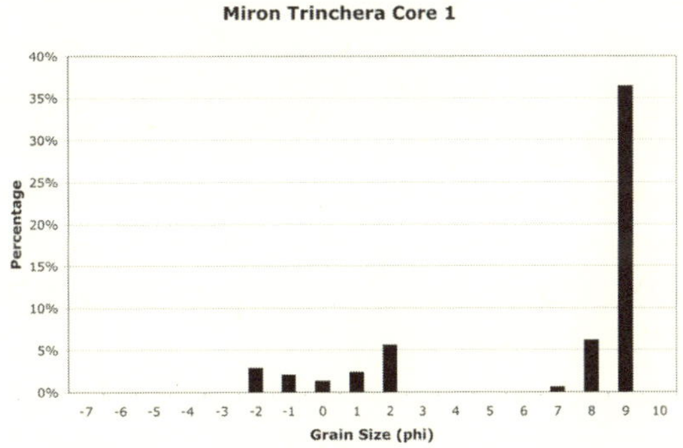

Fig. 5.6. *(continued)*

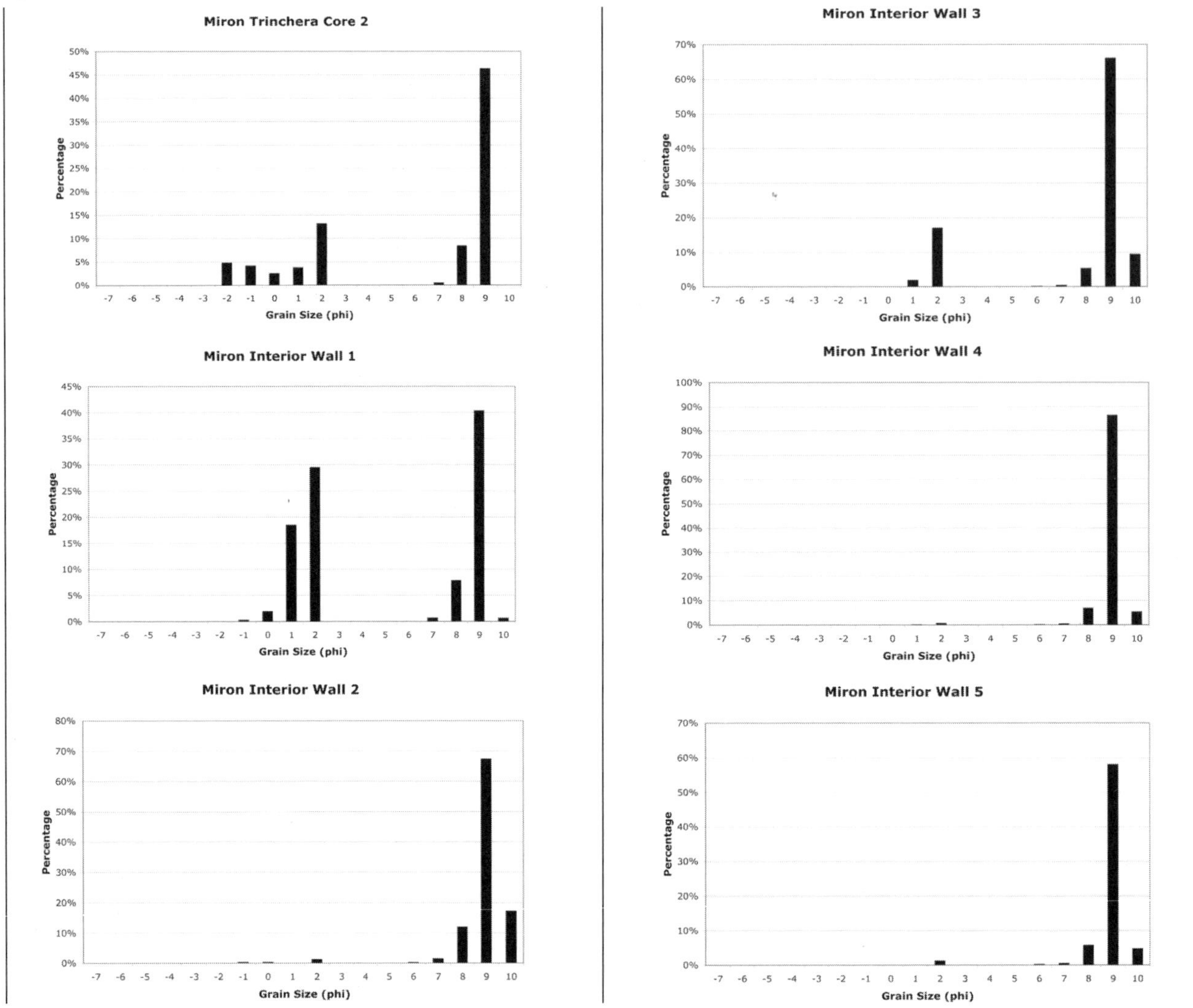

Almost all samples have a very prominent fine mode between 7 and 10 phi with a strong peak at 7–8 phi or 9–10 phi—that is, fine silt or coarse clay. This peak commonly reaches 25–30 percent and in some cases constitutes 60–70 percent of the total sample weight. Only two samples are exceptions to this general situation—samples 111 and 113/114, with ≤ 12 percent (figure 5.4). In these cases the samples have a very strong coarse peak (-7 phi) that suppresses the relative proportions of other peaks. In several samples the fine mode stands alone, with minimal presence of any coarser grains, for example, samples 303 and 304 in the Mid-Vestibule Trench (plate 5.2a) and samples 5, 8 base, 11 top, and 14 in the Cabin excavation (figure 5.2).

Without exception, an absence or near absence of grains exists in the interval between 4 and 6 phi (0.063–0.016 mm). In some cases the gap is wider—from 3 to 7 phi (0.125–0.008 mm). This indicates an absence of very fine sand and coarse, silt-sized grains. The implication is that the very fine fraction originated independently of the coarser sediment; in other words, it is not simply the fine tail of a normal distribution.

Overall, two general patterns occur among the histograms of the El Mirón samples—bimodal and trimodal distributions. Considering the trimodal pattern first, these samples have a mode centered on -2 to +3 phi (granules to medium sand) and another centered on -4 to -5 phi (pebbles), in addition to the commonly occurring mode in the fine silt range. Examples are samples 2-2, 3-3, 11 base, 102, and 106, among others (figures 5.6, 5.8), thus mostly in the upper levels of the Corral and the Cabin excavation areas.

Fig. 5.7. Decalcification histograms (after Hubbard 2000).

The bimodal pattern consists of a broad and commonly irregular mode, without a prominent peak, spanning the -7 to +3 phi range (cobbles, pebbles, and coarse to medium sand) along with the usual very fine (silt) mode. Examples are samples 15 and 16/17 in the Cabin (figure 5.2) and almost all samples from 120 (=400) and lower in the Footslope excavation (figure 5.4), thus from stratigraphically lower levels in the cave. This pattern mirrors the granulometry of the colluvial ramp sediments—Ramp 1 and Ramp 2 (figures 5.1, 5.5)—which are the most probable major source of the lower sediments in the vestibule.

A consistent difference in the finest fractions occurs between the bimodal and trimodal patterns. For samples with a bimodal distribution, the very fine peak is usually between 7 and 8 phi, whereas in the trimodal pattern the finest peak is distinctly finer, at 8–9 phi. However, this difference may be just a statistical artifact. If the true peak lies very close to 8 phi and the distribution is slightly skewed to finer or coarser sizes, the histogram bar may shift slightly right or left, and the difference is not significant.

Hubbard (2000) compared the granulometry of the <2 mm fraction of the original (raw) samples with that of decalcified samples. The results showed several trends (figure 5.7). First, commonly there is a reduction or even near absence of the coarse sand fraction (-1 to +1 phi, or 2 to 0.5 mm) and a concomitant increase in the finest silt and clay peaks (8 to 10 phi, or 4 to 1 μm). This suggests that the coarse sand "grains" are in fact aggregates of silt and clay cemented by $CaCO_3$ and that the grains released by acid solution are added to the fine peaks. This situation prevails in the Corral and Footslope samples. In other cases decalcification strongly reduces or removes the finest clay (9 to 10 phi, or 2 to 1 μm) fraction and also reduces the size of the fine silt peak (8 to 9 phi), suggesting that the silt and clay fraction is largely $CaCO_3$. This trend is found in the carbonate-rich samples of the Cabin excavation. These relations point up a distinct difference between the sediments in the front and rear of the vestibule.

LITHOLOGY

The numbers of limestone and non-limestone rock fragments in each sample are given in table 5.2 and graphed in figure 5.8. It is striking that almost no non-limestone fragments occur in the Cabin area or in the Mid-Vestibule Trench, and they are rare above Stratum 113 of the Corral area. However, in the lower levels of the Corral and especially in the Footslope trench non-limestone rocks not only dominate but are numerous.

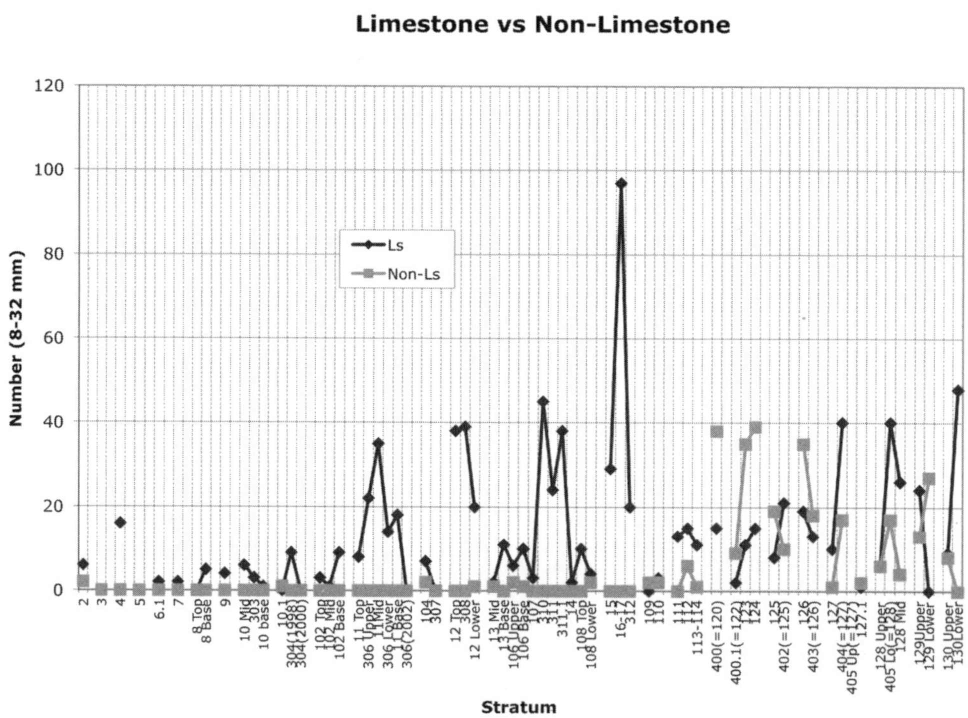

Fig. 5.8. Distribution of limestone vs. non-limestone (foreign) rock fragments (8–32 mm size class) vs. composite stratigraphy. Note absence or near absence of non-limestone fragments in the Cabin and Mid-Vestibule areas (Strata 2 through 312) but increasing number in the Corral area and the W–X10 sondage. Graph by William Farrand.

Moreover, the two samples from the colluvial slope, Ramp 1 and Ramp 2, contain only non-limestone rocks. The decreasing abundance of these non-limestone rocks from the inner cave into the outer vestibule is an indication of lateral transport toward the cave mouth of sediment derived from the ramp area.

Limestone fragments, usually angular, occur in almost every sample, but they are particularly abundant from Stratum 11 and lower in the Cabin area, below Stratum 104 in the Corral area, and throughout the Footslope strata. In contrast, they are absent in the ramp deposit. The scarcity of limestone fragments in the upper levels of the Cabin, Mid-Vestibule, and Corral areas coincides with the Holocene time period, with a climate presumably similar to that of the present. This suggests that greater abundance of limestone fragments in the sediments occurred during the colder climates of the Late Pleistocene epoch under a frost-dominated climate.

ROUNDNESS

As described above, the roundness of the rock fragments between 8 and 32 mm was evaluated. The results are tabulated in table 5.2 and graphed in figure 5.9, where they are compared with the abundance of non-limestone rocks. It is obvious that roundness and non-limestone rocks are positively correlated in the Corral and Mid-Vestibule areas. This relation results from the fact that the non-limestone rocks are, without exception, rounded pebbles and cobbles apparently of fluvial origin, whereas limestone rock fragments are broken and angular pieces detached from the cave walls and ceiling. In the Cabin area the relation is not as clear, but here one is dealing with very small numbers of rock fragments and thus questionable statistics. One outlier, in Stratum 10.1, is readily explained by the presence of one subrounded, non-limestone cobble, perhaps a manuport, but there are no other limestone clasts in that sample.

CALCIUM CARBONATE

The $CaCO_3$ content of all samples is listed in table 5.2 and plotted in figure 5.9. The curve in figure 5.9 shows a clear difference between the upper levels in the Cabin, Stratum 11 and above, and the rest of the samples. Values of 60 percent or more prevail in these upper levels. There is a steady decline in values below Stratum 12 to 10–20 percent $CaCO_3$. In the Mid-Vestibule and Corral areas, $CaCO_3$ is high only

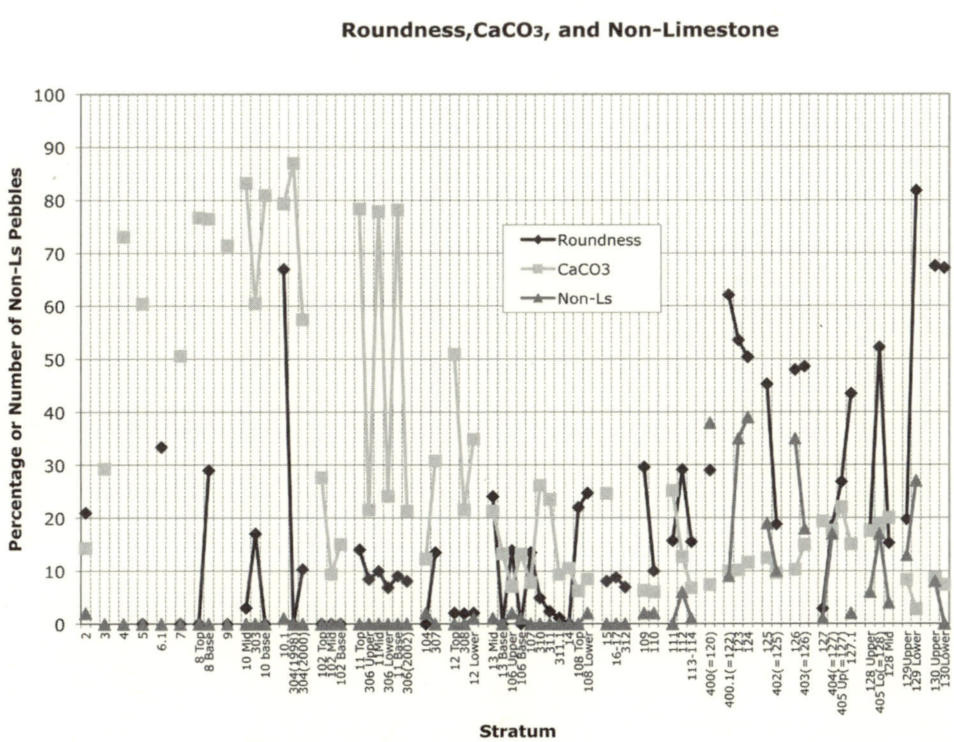

Fig. 5.9. Comparison of Roundness Index, percentage of $CaCO_3$, and number of non-limestone rocks vs. composite stratigraphy. Graph by William Farrand.

in the topmost levels (101–102 top, 303, 304, and 305), which are pale-colored sediments with very little clastic material. Below these carbonate-rich deposits the $CaCO_3$ values remain low, in the 10–20 percent range. The carbonate-rich layers superficially resemble mondmilch deposits that are currently forming in the dark inner cave, discussed further below.

pH, ORGANIC MATTER, AND CULTURAL DEBRIS

These variables are listed in table 5.2 and plotted in figure 5.10. The pH does not vary greatly throughout the sequence, with values commonly between 8 and 9. Extreme pH values are 7.69 and 9.34, and a distinct plateau occurs in Stratum Complex 16/17–312–108 and Strata 109 through 113–114. The pH curve does not follow the $CaCO_3$ curve, as one might expect if the pH were controlled primarily by the alkalinity of the carbonate. On the other hand, the highest pH values (above 9) occur in levels richest in organic matter, namely the middle strata in the Corral area, especially 108 through 113/114. (This same relation probably holds for Strata 115 through 119, which were not sampled but have the same dark-colored, organic matrix.) Thus, it appears that pH is controlled more directly by the amount of organic matter than by the amount of fine-grained $CaCO_3$ in the sediment. The increase in organic matter in these strata probably reflects increased human activity. Note the peaks of cultural debris in Stratum Complex 16/17–302–108 and again in 109, 110, and 112, coincident with high pH and organic matter. I have found similar relations between pH and human activity at the Abri Pataud in the Dordogne (W. R. Farrand 1975), although in that case the increased human activity coincided with lowered (i.e., more acid) pH. Another source of organic matter is guano from roosting birds. Courty (n.d.) finds evidence of abundant bird droppings in Corral strata, especially in 101 through 107, and to a lesser degree in 108 through 116.

CLAY MINERALS

As mentioned above, the clay mineralogy was determined by Hubbard (2000) who identified the main classes of clay-sized minerals present in terms of their relative abundances on a semi-quantitative basis. The three major excavation areas show distinctively different suites of clay-sized minerals (figure 5.11). In Square W10 at the rear of the vestibule one sees rather uniform and

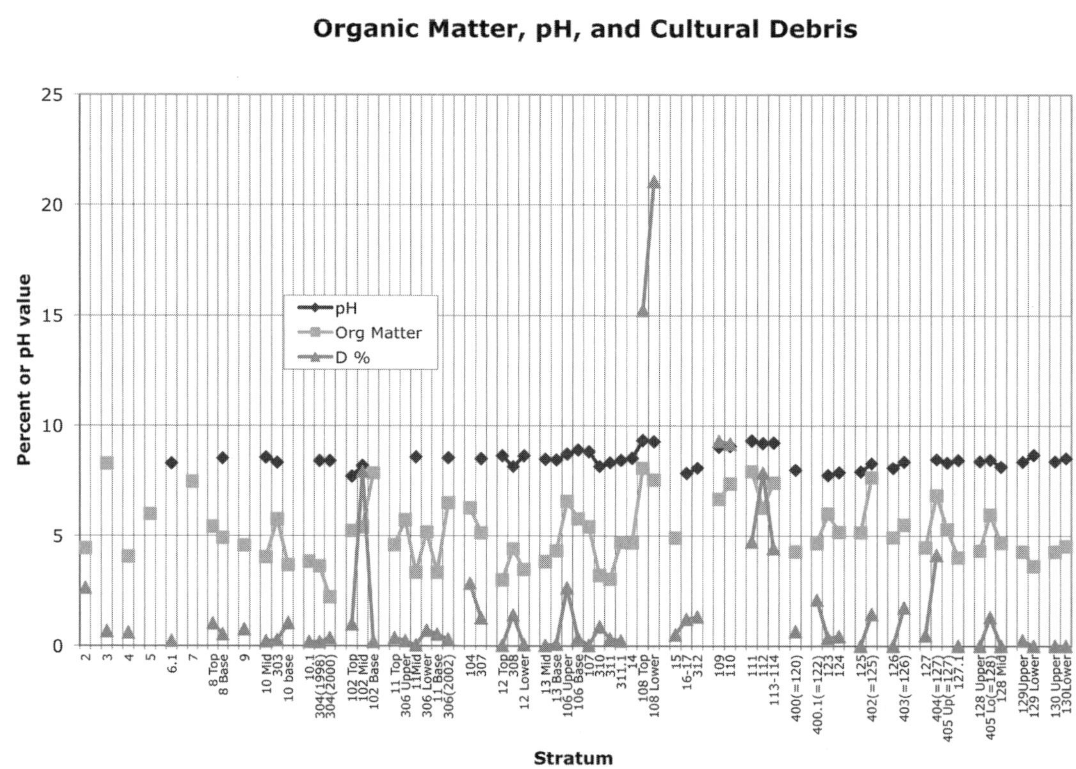

Fig. 5.10. Percentage of organic matter and cultural debris and pH vs. composite stratigraphy. Graph by William Farrand.

nearly equal percentages of quartz, illite, and kaolinite but no calcite. At the other extreme, in the Cabin area, the samples above Level 12 are strongly dominated by calcite, although some quartz and illite appear in Level 11. This overwhelming amount of calcite remained even after Hubbard had removed $CaCO_3$ in pretreatment of the XRD samples, which testifies to the great abundance of $CaCO_3$ in the Cabin sediments, as discussed above. The lowest level in the Cabin tested by Hubbard, Level 12, shows a pattern more closely similar to that in Square W10, with increased amounts of illite and kaolinite. In between these end points the Corral area clay mineral suites are intermediate mixtures, mostly showing a decreasing trend of quartz accompanied by increasing kaolinite with depth and a constant percentage of illite. An exception is seen in the uppermost level (102), where calcite strongly dominates the clay minerals, as it does in the upper Cabin strata. This matches the high abundance of $CaCO_3$ found in Level 102 (figure 5.9).

Thus, the clay mineral analysis reinforces the patterns found in the $CaCO_3$ results and in the lithology and roundness of rock fragments. A consistent gradient of mineralogy and lithology is found from the Footslope area outward through the vestibule to the Cabin area, showing an increase in bulk $CaCO_3$ and calcite among the clay-sized minerals, along with an increase in angular limestone rock fragments and a decrease in rounded non-limestone pebbles.

Discussion

SEDIMENT SOURCES—EXTERNAL, INTERNAL, ANTHROPIC

Sediment that accumulates in a cave may be derived from several sources successively or simultaneously, commonly leading to a mélange of materials (W. R. Farrand 2001). In the case of the inhabited areas of El Mirón the possible sources are

- Rock fragments detached from the cave ceiling and walls
- Material derived from the same areas by solution and precipitated on the cave floor
- Silt and fine sand blown into the cave from the outside
- Sediment from the interior of the cave that has been eroded and redeposited in the vestibule area,

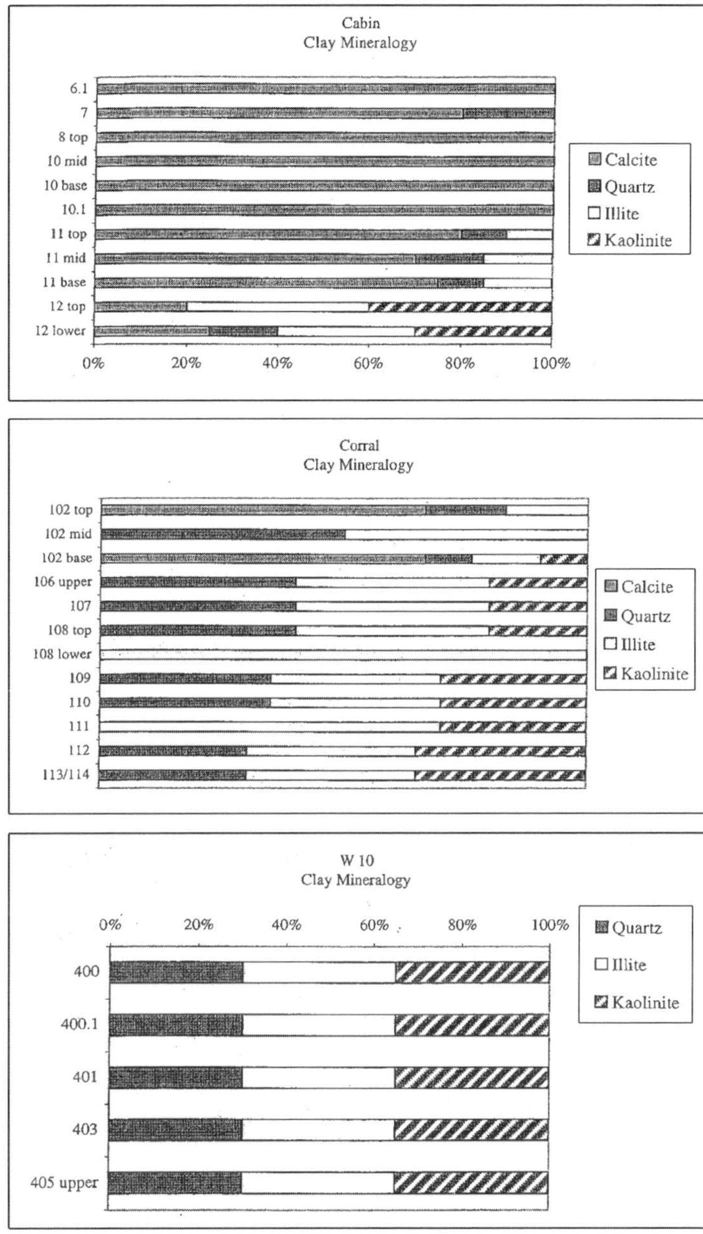

Fig. 5.11. Approximate abundances of minerals in the clay-size fraction of selected samples from the Cabin and Corral excavation areas and the W10 sondage (after Hubbard 2000). Graphs by William Hubbard.

which could be reworked older cave sediments or dissolution clays derived from the bedrock
- Sediment, artifacts, faunal remains and vegetal debris, excrement, and other debris brought into the cave by animals, birds, and humans, including sediment (mud) on their feet, fur, or clothing

Another conceivable source would be river sediment that floods into the cave mouth, but in the case of El Mirón this is highly unlikely during the period of habitation (the last 50,000 years or so) because the cave sits some 100 m above the Río Calera, where there is no indication of river terraces nearly as high as El Mirón. In any case, there are no sediments in the Cabin, Mid-Vestibule, or Corral areas that resemble fluvial deposits similar to those of the modern Calera. Moreover, the vestibule deposits slope outward toward the cave mouth, not inward, as would be expected if river sediments had washed in.

On the other hand, the ramp deposits in the passage leading from the vestibule into the dark inner cave do appear to be of fluvial origin. They consist of well-rounded pebbles, cobbles, and head-sized boulders of rock types exotic to the cave bedrock in a sandy, silty matrix (plate 5.1) and contain no angular limestone fragments that might have been derived from inside the cave. No stratification was seen in our limited excavation (about 150 cm deep) in these deposits, and the lack of good sorting gives the impression that the ramp deposit, as we see it, is probably colluvium reworked from an original fluvial deposit that had washed into the cave well before any human occupation documented in our excavation. The steep gradient of the ramp appears to be an erosional slope down which some of the old alluvium was washed into the vestibule, where it was mixed with other sediments accumulating in the Corral and Cabin areas. This hypothesis will be discussed further below.

ROCK FRAGMENTS

Rocks detached from the cave walls and ceiling are angular to subangular limestone clasts (table 5.2) that resemble fragments detached by frost action. If so, this implies that the ambient winter temperatures in the cave dropped at least to -5°C (W. R. Farrand 1975). In addition, strata with greater numbers of limestone fragments record times of greater numbers of freeze-thaw cycles per year (W. R. Farrand 1975, 2001). Thus in the Mirón sequence abundant limestone fragments in Strata 11 through 17 in the Cabin and in 102 through 108 in the Corral (and correlative levels in the Mid-Vestibule Trench) must reflect cold temperatures during the period from 15,000 to about 11,000 BP, which was the final major cold episode of the Last Glacial (figure 5.1). This is in agreement with the results of Courty and Vallverdu (2001) for Strata 10.1 through 12 based on their micromorphological observations. Reduced numbers of limestone fragments in Strata 109 through 125 in the Corral and Footslope areas may indicate somewhat less rigorous conditions, although those areas toward the rear of the vestibule are more sheltered from the outside weather. But again, the number of limestone fragments increases abruptly in the lower levels of the Footslope sondage, Strata 127 through 130, presumably indicating more rigorous climatic conditions in the period from more than 40,000 to roughly 20,000 BP.

FINE-GRAINED CARBONATE SEDIMENT

Certain levels are very rich in fine-grained carbonate, even as much as 80–90 percent $CaCO_3$ by weight—for example, Strata 4 through 11 in the Cabin and Stratum 304 in the Mid-Vestibule Trench (figure 5.9). Stratum 102 in the Corral area is also very pale colored but contains only 15–28 percent $CaCO_3$ by loss-on-ignition analysis, although calcite dominates the clay mineral fraction (ca. 75 percent). The source of this carbonate appears to be $CaCO_3$ dissolved by water filtering through the bedrock above the cave and precipitated around seeps on the cave ceiling and side walls, ultimately disintegrating and falling to the cave floor, only to be dissolved again and incorporated into the sediments. The role of vegetation in this process appears to be critical, as hypothesized by M. A. Courty (pers. comm.) for those areas within reach of daylight in the vestibule.

This process is continuing today, as evidenced by masses of fine, soft $CaCO_3$ hanging on the ceiling and cascading down the cave walls (figure 5.12), and commonly associated with clinging vegetation. These areas drip today at times of heavy rainfall, but this activity must have been more intense in the past, especially at the time of Stratum Complex 10–304–102 around 5500 BP in the mid-Holocene. At that time very little other sedimentation was occurring, such as rockfall or

Fig. 5.12. Photo of recent travertine accumulation draped on south bedrock wall of vestibule. Photo by William Farrand.

wind-borne input, given the purity of these carbonate layers. Moreover, those wet conditions may have discouraged human occupation at that time.

The dripping process can be observed also in the dark cave interior, where it has resulted in thick layers of mondmilch, such as was encountered in the Trinchera excavation. Some of the mondmilch was dated to very late Holocene time (e.g., Stratum VII = 3230±40 BP and Stratum IV = AD 1040–1204) and is even occurring today, such as on top of backdirt piles dating to the 1950s. Vegetation is not involved in this case since no daylight reaches this area.

WIND-TRANSPORTED FINE SEDIMENT

Examination of the granulometric histograms for samples from the Cabin area reveals a particularly strong, well-sorted peak in fine silt and coarse clay (especially between 7 and 9 phi, or 0.008 and 0.002 mm). This mode is clearly separated from the mode in fine to coarse sand by the absence or near-absence of any grains in the coarse silt range (3 to 7 phi, or 0.125 to 0.008 mm), suggesting a separate and distinct source of the finest sediment. As mentioned above, it is not merely the fine tail of a broader (normal) granulometric distribution.

There is, however, a difference in the composition of those silt peaks between the upper and lower parts of the Cabin strata. In Stratum 11 and above, the sediments are greatly enriched in $CaCO_3$, as much as 75–85 percent by weight (figure 5.9), whereas in Stratum 12 $CaCO_3$ drops to 50 percent, then to 34 percent, and below Stratum 12 $CaCO_3$ is never higher than 25 percent, although the silt-sized mode remains high—25 to 70 percent, depending on the amount of coarse rock fragments in a given stratum. (More coarse material reduces the *relative* size of the silt peak.)

A similar fine, silty peak occurs in histograms from the rear of the vestibule, but the clay mineral assemblages suggest a different source in that area. Even after acid pretreatment, the clay minerals in the Cabin area above Stratum 12 are overwhelmingly calcite, whereas in the Footslope area (Square W10) there is no calcite, only equal parts of quartz, kaolinite, and illite. It is the same in the Cabin area below Stratum 11. Similarly, in the Corral area strata below the uppermost Stratum 102, no calcite occurs among the clay-sized minerals (figure 5.11). (See discussion of clay mineralogy above.) The fine peak in the sediments at the rear of the vestibule is derived largely from reworking of the old colluvial ramp deposits inside the cave, as can be seen by the similarity of the histograms from the ramp (figure 5.5) and from Footslope Strata 120–130 (figure 5.4), as well as the abundance of rounded, non-limestone pebbles and cobbles. In contrast, non-limestone rock fragments are rarely found in the Cabin area, suggesting that the silty fraction in the lower Cabin sediments is derived from a different source, likely an eolian input of loess from outside the cave. This interpretation is supported by the observations of Courty and Vallverdu (2001:493), who propose that the fine, loamy sediment matrix of the Cabin strata was blown into the cave mouth by winds blowing up the Río Calera valley. On the other hand,

Courty (n.d.) finds that the eolian input was considerably weaker in the Corral area, and I would suggest that the fine, silty peak in Stratum 108 and lower is mostly derived from reworked colluvium, as indicated by the presence of rounded non-limestone rock fragments and the clay minerals.

Specifically, I suggest that the loessic sediment was derived from the lower Río Gándara and the upper Río Asón (Ruesga) valley in the vicinity of Ramales de la Victoria, where broad floodplain terraces exist. There is no floodplain or terrace in the Río Calera valley on the slopes below El Mirón, where the valley is deeply incised into bedrock. Thus, the winds would have been from the northwest, more or less directly into the cave, likely rendering the cave mouth an unpleasant area at times and accounting for the scarcity of cultural debris in Strata 11 through 15 (figure 5.10).

CUEVAMUR

About 250–300 m southeast of El Mirón, and about 60 m higher, other caves open along the face of a vertical limestone wall, La Pared del Eco, which is about 200 m high. The most prominent of these caves is Cuevamur, or Cueva del Muro (plates 5.5a, b, 5.6a, b). It has a square opening, about 15 m on a side, exposing a considerable fill of rocky and loamy sediment (that slopes steeply upward toward the cave's rear wall, where it is capped in places by an ancient-appearing travertine [see below]). No archaeological remains were seen, presumably because the present surface is quite steep and heavily eroded. In contrast to El Mirón, the outer chamber or vestibule of Cuevamur is connected only by a very small opening (*una pequeña gatera*) in the upper rear corner of the cave, which leads to a much larger karstic network that has been extensively explored by AER speleologists (AER 1968). This entrance was initially quite small but was enlarged with dynamite, according to AER, and now is bordered by an iron frame.

These two caves are part of an extensive karstic network in the Monte Pando–El Moro massif. To understand the development of this network or complex, one needs to visualize the progressive growth of the caves as a function of uplift of the mountains and/or the down-cutting of streams. Karstic cavities develop by groundwater solution, generally following bedding planes, joints, and faults, at the level of the groundwater table (i.e., the top of the zone of groundwater saturation). The groundwater table, in turn, is controlled by the local base level of nearby streams, in this case the Río Calera. In the present case, caves such as Cuevamur and Covalanas, at 340 and 330 m/sl, respectively, would have developed when the local water table and the Río Calera stood at approximately their level. As the mountain massif was uplifted or the river cut more deeply into its canyon (or both occurring together), the water table would be lowered and caves would begin to form at lower levels, such as the level of El Mirón (280 m/sl), and the Cuevamur/Covalanas caves would have been high and dry. "El resto de las cuevas del muro del Eco así como la C. del Francés [El Mirón] son posteriores a Cuevamur, formadas por el mismo río subterráneo el emigrar este a niveles inferiores" (AER 1968). As the river cut more deeply, the cave passages of El Mirón would have become dry, and lower-level caves would have developed, such as La Haza and Cueva del Río and others. It is after the water table falls below the cave level that stalactites and stalagmites begin to form. Also, after the formation of the caves at a given level and while the river is still close to the level of the cave, it is possible for the river to flood into the cave mouth and deposit sediment. Therefore, one needs to visualize the bed of the Río Calera nearly as high as the mouth of El Mirón at the time that the coarse alluvial sediments washed into the vestibule and beyond into the cave interior. This is in accord with the hypothesis of Butzer (1981) that such fluvial sediments were washed into the mouths of a number of other caves in the Monte Castillo area, some 45 km west of El Mirón, probably in Early Pleistocene or even Mio-Pliocene times. By comparison, fluvial sediments penetrated as much as 200 m or so into some Castillo caves, for example, Las Monedas. Subsequently the river has cut down to its present level.

An interesting study has been published of river down-cutting dated by cave formations in the Sierra Nevada of California (Stock, Anderson, and Finkel 2004). The caves are dated by speleothems that formed in them after the local water table dropped below the cave level, thus providing a minimum age for the karstic activity in each cave. Some 400 m of bedrock erosion occurred there in about two million years. Thus, this is on the same scale in terms of relief and age that we have in the Calera Valley, if Butzer (1981) is correct that the fluvial influx in the complex caves of El Castillo hill in central Cantabria is early Pleistocene or even Mio-Pliocene. Geologists studying the formation of the Cantabrian Cordillera have concluded that, while the bedrock structures developed in Early Tertiary times, uplift and subsequent erosion of the mountain chain occurred in the Late Tertiary: "Avec la dernière phase tectonique miocène, le pays, s'il avait sa structure, n'avait pas encore son relief et sa configuration actuels. . . . Après une période d'aplanissement (Mio-pliocène) dont témoignent des restes de surfaces de nivellement, une puissante reprise d'érosion, sous climat vraisemblablement semi-aride, est attributable à un soulèvement assez recent (Pliocène final? Quaternaire ancien?)" (Feuillée and Rat 1971:41).

The importance of Cuevamur for our study of El Mirón lies in the resemblance of its sediments to those in the ramp in El Mirón. The sandy loam sediments enclose well-rounded pebbles, cobbles, and boulders of the same non-limestone lithologies that occur in El Mirón (plates 5.5b, 5.6a). There are no limestone fragments except some meter-sized limestone blocks that lie loosely on top of the Cuevamur fill and appear to be relatively recent roof-fall. In addition, the Cuevamur fill is clearly stratified, with pebbly-cobbly layers alternating with fine-grained sandy and silty beds. See the histograms in figure 5.5.

The fill in Cuevamur has the appearance of considerable antiquity, and a tentative U/Th age estimate for the capping travertine (plate 5.6b) is minimally 200,000 years (J. L. Bischoff, pers. comm., June 11, 2004). Moreover, the absence of angular limestone fragments in the matrix suggests a non-glacial (even preglacial?) environment at the time of its emplacement.

Similarity in composition and in height above the valley floor strongly suggest that the filling of Cuevamur and the old ramp deposit in El Mirón were deposited by the same mechanism, namely fluvial floods of the ancient Río Calera, which was then much higher than the present, that washed these sediments into existing cave mouths. It should also be noted that we found traces of similar cobble deposits on the upper slopes of Monte Pando, 150–200 m (estimated) above El Mirón and Cuevamur, which are additional evidence of a long history of down-cutting by the Río Calera, long predating the human occupation revealed in El Mirón.

An alternative suggestion, that fluvial sediment entering Cuevamur upstream could have transited through an internal karstic passage and exited downstream through El Mirón, is considered highly unlikely on several counts. First, there is no way that the volume of fluvial/alluvial sediment that we encounter in El Mirón could have passed through the very small entrance in the rear of Cuevamur. Second, inside that entrance is a narrow passage that descends initially some 13 m into the interior of the Cuevamur system. Still farther down is a large, complex series of chambers (Gran Sima, etc.) and passages that presumably would have trapped any fluvial sediment that entered from the Cuevamur mouth. Moreover, the AER (1968) map of the Cuevamur system does not show any connections leading in the direction of El Mirón. Finally, the velocity of flow necessary to transport the large cobbles that we find in the ramp deposits would have had to have been tremendous, appropriate for a mountain torrent but not for an internal karstic river. For example, a cobble 10 cm in diameter requires a flow velocity on the order of 3 m/sec, or roughly 10 km/hr, and we found cobbles at least twice that size (plate 5.1).

In summary, the filling of Cuevamur provides an important clue to understanding the origin of ramp deposits in El Mirón. The fluvial deposits in Cuevamur could have been emplaced only from outside the cave, by an ancient Río Calera washing exotic sediments into the cave mouth. The same must, therefore, be true for the old ramp deposits in El Mirón. In El Mirón the fluvial sediments penetrated well into the dark inner-cave chamber, as evidenced by cemented cobble deposits hanging on the walls above the Trinchera excavation area. The inner cave must have been filled nearly to the ceiling, as was the case in some of the Monte Castillo caves (Butzer 1981). After the cobbles were deposited and cemented but before subsequent erosion (see next paragraph), gently flowing water from the cave interior deposited laminated fine sand, silt, and clay, rather typical karstic residues, on top of the cobble deposit. A remnant of these fine-grained sediments preserved just below the ceiling on the north wall of the inner cave near the Trinchera is about 65 cm thick, comprising, from top to bottom:

• Upper very fine-grained sand, cross-bedded?
• Pale brown, laminated silt, ca. 1 cm thick
• Lower very fine-grained sand
• Reddish-brown clay with manganese flecks
• Silt with manganese nodules

Subsequently, these fluvial deposits were partially eroded by waters discharging through the karstic system. This erosion cut a channel in the inner cave floor, at least 2 m deep in the area of the Trinchera excavation. The top of the Trinchera excavation is about 1 m below the residual cobble level, and Magdalenian artifacts occur about 50–60 cm deep in the Trinchera Stratum VIII (14,620 BP). An auger hole revealed another 1.25 m of micaceous sandy clay loam sediment with small pebbles below the lowest point of the Trinchera excavation, but it is not clear whether this sediment is part of the channel fill or part of the original fluvial deposit.

In any case, erosion of the old fluvial sediments from the inner cave presumably continued down the ramp area, reworking the old fluvial sediments into a colluvial (mudflow?) deposit that one sees today—that is, a coarse, randomly sorted, crudely stratified sediment of head-sized boulders in a sandy loam matrix. Some erosion of the ramp deposits continued during the times of habitation in the vestibule, as seen in the sandy loam matrix and rounded, non-limestone pebbles in the Footslope trench and the Corral area. Also, a remnant of sediment with Magdalenian artifacts, dated 16,600 BP, is preserved in a niche on the south wall of the ramp as a relic of slope deposits that have been otherwise entirely removed since Magdalenian times, after the destruction of the 10,500 BP capping travertine.

TOPOGRAPHY OF THE SEDIMENTARY FILL

Visual inspection of the several excavation areas indicates a general slope of the strata from the rear of the vestibule toward the cave mouth, but at a lesser gradient than the surface of the ramp. If one plots the depths below datum (B.D.) of certain well-defined strata in each excavation area and their correlates in adjacent areas, one finds a persistent outward gradient of about 7° (figure 5.13). The gradient is a bit steeper in the Mid-Vestibule Trench (ca. 11°) and then flattens out in the Cabin area. This configuration reinforces the idea that the bulk of the vestibule sediment was derived by erosion of the ancient colluvium in the ramp. There is no indication of a gradient reversal, that is, of strata sloping from the mouth into the cave, confirming the hypothesis that there was no significant input of sediment external to the cave, except for windblown loess.

A question remains concerning the disposition of the old colluvial deposits in the vestibule. Were they completely removed by erosion, or do they occur at depth below the archaeological layers? Geophysical soundings by Clapés (1997) indicate a change of sediments from finer to coarser at about 5 m below the present surface, reaching bedrock at about 9 m throughout the vestibule. Bedrock was also encountered at about 9 m depth below the erosional surface of the ramp. Only the upper 2.75 m has been excavated so far in the Cabin

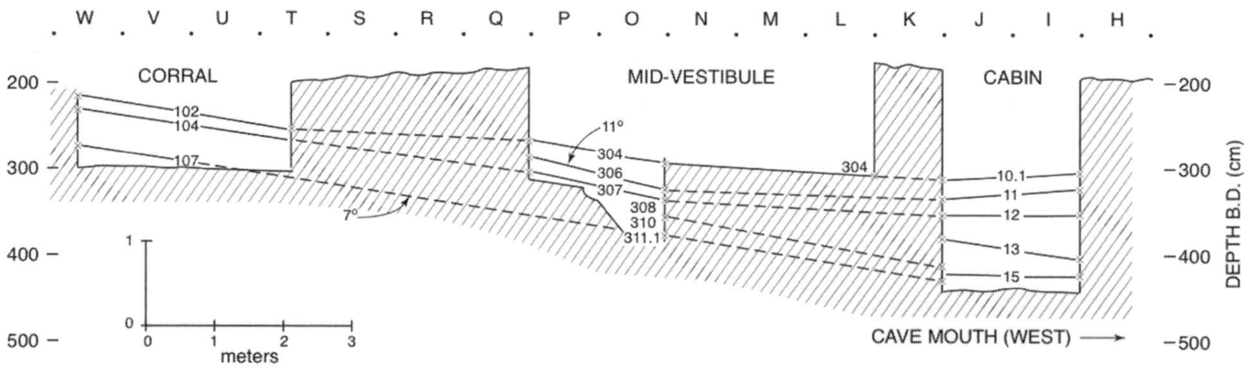

Fig. 5.13. Schematic sagittal profile in El Mirón Cave illustrating the three excavation areas and key strata as of 2000. Vertical exaggeration = 1.25x. Plan by William Farrand.

area, but another 2.35 m was cored below the base of Stratum 17, revealing "yellowish brown sandy clayey silt" with some rock fragments and "charcoal, flints and bone fragments" (Straus and González Morales 2003:135 and pers. comm.). I have not seen those core samples, but the description is similar to the lower layers of the Footslope trench. A sondage (Square J2) against the south cave wall revealed about 60 cm of clayey silt deposits with small/medium limestone clasts and variable amounts of cultural and faunal remains (Levels 18–21). The geophysical pattern at depth resembles the signal at or near the surface in the ramp area, suggesting that the old colluvium does continue to the cave mouth, where it has been buried by younger (archaeological) deposits.

SEDIMENTATION RATE

Sufficient radiocarbon dates are available for the reconstruction of sedimentation rate curves for different areas of the cave (figure 5.14). The best control exists for the Cabin area, where two segments show rather uniform rates—about 3.18 cm/100 yrs from Stratum 17 to Stratum 11.1 and about 4.55 cm/100 yrs from Stratum 10 to the top. Between these two segments the apparent rate is much less, and in fact it may have been nil except for a brief episode at the time of Stratum 10.1, meaning that unconformities, or hiatuses, occur between Strata 11 and 10.1 and again between 10.1 and 10, each perhaps as much as 2,000 years in duration. Three radiocarbon dates are attributed to Stratum 10.1, which is only about 8–9 cm thick and is discontinuous. Two dates, 8380±175 BP (at 296–308 cm B.D.) and 8700±40 BP (at 310–320 cm B.D.) lie within two standard deviations of each other and conceivably can be averaged to 8540±180 BP. The third sample, 9550±50 BP (at 309–320 cm B.D.), was collected at about the same level as the 8380 BP sample and may be spurious, and is rejected here in the absence of any evidence of a hiatus at this level. Thus the two hiatuses are inferred to occur between about 11,000 and 8500 BP and 8500 and 6000 BP. These intervals are centered, respectively, on two widely recognized, important climatic episodes, the Younger Dryas cold reversal and a younger, brief cold snap centered on 7500 BP, which lasted only one century or less (Alley et al., 1993). The latter event is commonly referred to as the "8200 cal BP event."

Radiocarbon dates for the Mid-Vestibule Trench determine rate curves that are roughly parallel to those of the Cabin area, although there are fewer control points. Sedimentation between Levels 313 and 308 is somewhat slower than that between Levels 11.1 and 17, implying faster sedimentation in the Cabin area, which is logical because the Cabin area is closer to the area of input of sediments (mostly loess) from outside the cave.

In contrast, some of the radiocarbon dates from the Corral area are confused, indicating either some spurious dates or turbation of the sediments by human or animal activities. In figure 5.14 the data points are connected in stratigraphic order to emphasize that the disorder is stratigraphic, not statistical. The dashed line through these points suggests the average sedimentation rate, about 2.0 cm/100 yrs, thus slightly slower than in the Cabin area, which is reasonable but certainly not to be taken as definitive.

The Footslope curve, which joins the Corral curve at Stratum 126, is controlled by only four dates, from Stratum 125 to 130, yielding an average rate of about 0.75 cm/100 yrs. This seems reasonable in view of the difference in sediment type between the Corral and Footslope areas but must be considered only a first approximation. However, the possibility of one or more hiatuses in the interval between 40,000 and 20,000 BP, a generally mild period prior to the Last Glacial Maximum (LGM), cannot be excluded. Specifically, a hiatus of as much as 9,000 years is possible, as shown in figure 5.14, as suggested by the difference in radiocarbon ages for Strata 126 and 128 and the sedimentary evidence of an erosional surface at the base of Stratum 127 (plate 5.4).

These sedimentation rates are similar to those of other sites, such as the Abri Pataud (Dordogne, France) and Franchthi Cave (Argolid, Greece). At the Abri Pataud, which is located in a more intense freeze-thaw climate, the rates range from 6.9 to 9.3 cm/100 yrs (W. R. Farrand 1975:41). Franchthi Cave, in a frost-free, Mediterranean climate, has a wide range of rates, generally from 2 cm/100 yrs to 23.3 cm/100 yrs, even as much as 250 cm/100 yrs, the latter during a period of intense human habitation (W. Farrand 2000:fig. 6.2). In general, the sedimentation rate appears positively correlated with the intensity of habitation in Franchthi Cave but less so in the Abri Pataud, where occupation horizons are clearly separated by barren layers of

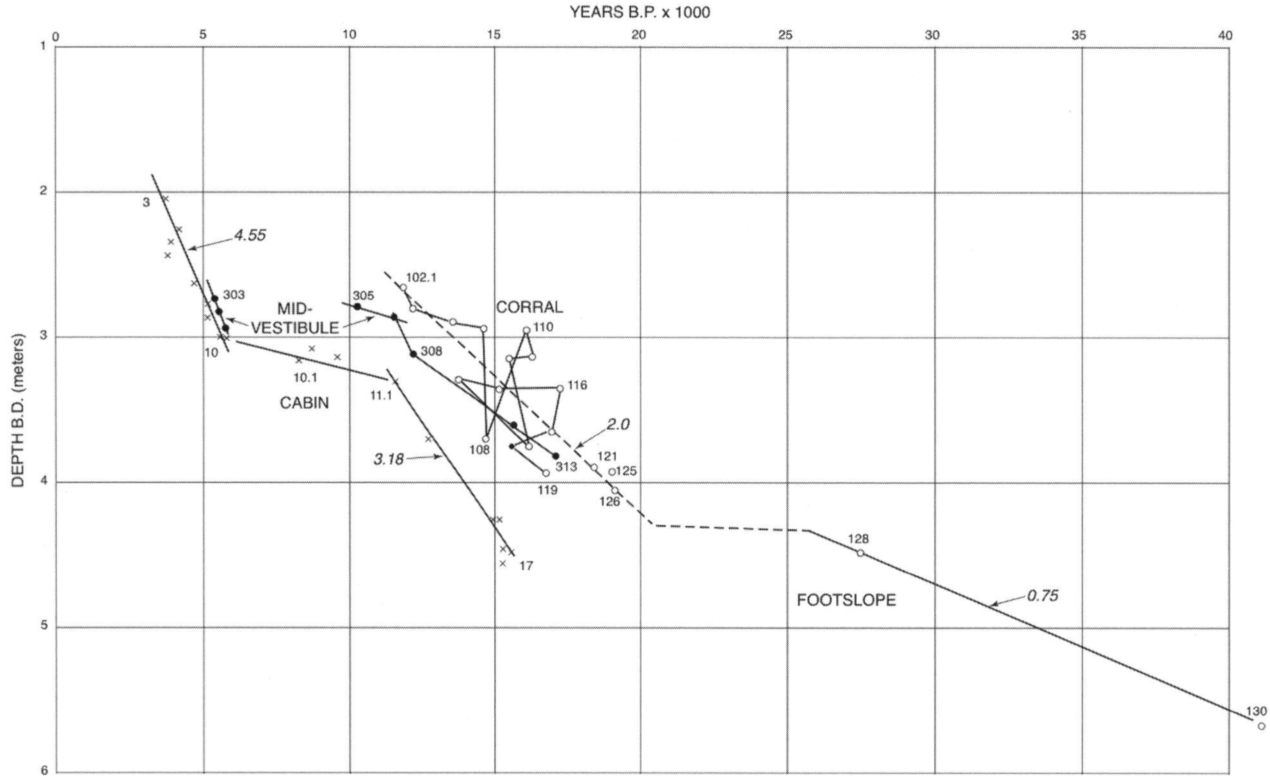

Fig. 5.14. Time-depth curves for the three excavation areas, showing the approximate rates of sediment accumulation (cm/100 yrs), figures in italics. Control points are C14 dates. Plan by William Farrand.

rock fragments (*éboulis*). In El Mirón these tendencies seem to be mixed. In the Cabin area sedimentation was faster during the Holocene (4.55 cm/100 yrs) than during the time of sparser Magdalenian occupation (3.18 cm/100 yrs), and in the Footslope area with very sparse habitation the rate was very slow (0.75 cm/100 yrs). Unfortunately the stratigraphically incoherent radiocarbon dates in the Corral excavation do not allow a firm comparison with the other areas, although a general trend is suggested by the dashed line in figure 5.14 that has approximately the same slope as for the Cabin area.

REGIONAL COMPARISONS

Other studies of cave sedimentology in Cantabria have been conducted by Hoyos Gómez (1995) and his colleagues and by Butzer (1981). In both cases those researchers studied a number of caves and rockshelters, too numerous to review in detail here. Thus they were able to synthesize and integrate their paleoclimatic conclusions by comparing a given site with the others. In the present study we are dealing with a single cave with all its idiosyncrasies, and we must sort out the local controls on sedimentation from the effects of regional influences. It turns out that in El Mirón the vestibule sedimentation is strongly influenced by the presence of the ancient alluvial-colluvial filling in the cave interior that shed sediments into the Corral, Mid-Vestibule, and to a lesser extent the Cabin areas. This component of the sedimentation was conditioned largely by hydrologic conditions in the cave interior that can be related to the external environment only indirectly, if at all. Another complication is human activity that masks environmental impacts. This is particularly evident in the Corral area, which is rich in cultural debris and where turbation is indicated by the incoherent radiocarbon dates.

A comparison of the climatic syntheses of Hoyos Gómez and Butzer for the period from 20,000 to 10,000 BP is given in figure 5.15. Their interpretations agree well in general, although Hoyos Gómez's sequence is more detailed. I have interpolated the interpretations for El Mirón in figure 5.15 to the extent possible, including indications of unconformities at about 20,000, 15,000, and 11,000–10,000 BP. Following the hiatus at 20,000 BP, the clearly stratified Strata 125 and 126 suggest a greater influence of running water in the cave and thus a moister climate. The next clear climatic indication occurs in Strata 17, 16, and 15, where an abundance of angular limestone suggests intense freeze-thaw but less eolian activity than above the 15,000 BP hiatus. From Stratum 14 upward through Stratum 11 the Cabin-area sediments are dominantly loessic, indicating an increase in wind action relative to frost action. Finally, the Younger Dryas cold episode is marked by an absence of sediments in El Mirón, except for some sterile mondmilch in Stratum 305 in the Mid-Vestibule.

El Mirón matches the interpretations of Hoyos Gómez and Butzer reasonably well for the humid episode around 20,000–19,000 BP and for the Younger Dryas cold 11,000–10,000 BP. It is suggestive that the "humid, fresh" phase of Hoyos Gómez and the "cold, wet, unstable" environment of Butzer (figure 5.15) correlates with the erosional unconformity around 15,000 BP in El Mirón. Otherwise, the El Mirón sequence does not match Hoyos Gómez's phases in detail.

Summary

The sedimentary filling of El Mirón consists of four major components that accumulated more or less simultaneously but in varying proportions, which is typical of most cave and rockshelter deposits. Specifically, in

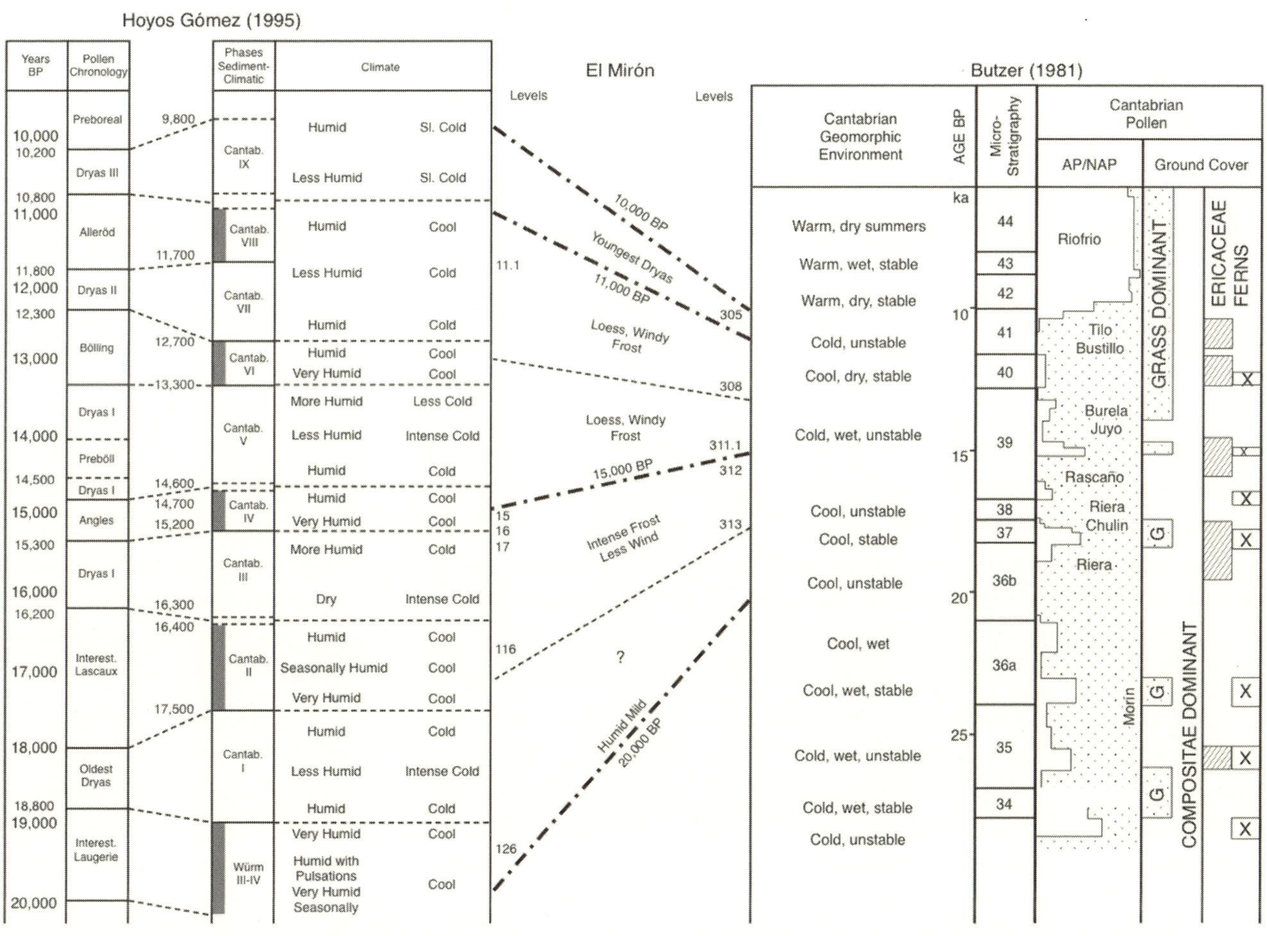

Fig. 5.15. Comparison of El Mirón Cave stratigraphy and suggested climatic conditions with sequences in other Cantabrian caves proposed by Butzer (1981) and Hoyos Gómez (1995). Heavy dot-dash lines in the Mirón column indicate major unconformities in the cave sequence. Plan by William Farrand.

El Mirón these components are (a) rockfall and carbonate drip from the cave ceiling and walls, (b) eolian loess blown into the cave mouth, (c) washing of old fluvial/colluvial sediments from the cave interior outward into the vestibule area, and (d) anthropogenic input resulting intentionally or unintentionally from human activities. Each of these inputs operated differently in a spatial sense. Eolian loess is more important near the cave mouth. Reworked fluvial/colluvial components are more abundant in the rear of the vestibule near the ramp from which they were derived and are not easy to distinguish in the area of the cave mouth (Cabin area). Human influence is more pronounced in the mid- to rear vestibule, apparently because the inhabitants shunned the more wide-open cave mouth area, at least after the time of Stratum 14. Earlier, however, the cave mouth was heavily utilized, for example, in Levels 16 and 17. And finally, the rock debris falling from the ceiling and walls, presumably largely due to freeze-thaw action, is more abundant toward the cave mouth than in the rear of the vestibule, which is consistent with the diminution of the influence of outside weather as one penetrates farther into the cave. However, Straus reports that Levels 118–119 in the Corral area (sediments not sampled) do contain abundant angular rock fragments.

It is found that the cave floor formerly sloped distinctly down toward the cave mouth (figure 5.13). Subsequent to the time of Level 15 sediments have accumulated more rapidly in the Cabin area, which has tended to level the present cave floor, although some leveling of the floor may be due to historic activities associated with the construction of the former cabins and corral. Given the former slope of the floor, it is clear that there was no significant input of sediment into the cave mouth, except windblown material.

The rate of sediment accumulation can be readily determined for the Cabin area based on the consistent set of radiocarbon dates (figure 5.14). The rate was about 3.18 cm/100 yrs from Stratum 17 to Stratum 11 and somewhat faster, about 4.55 cm/100 yrs, from Stratum 10 to the surface. However, between Strata 11 and 10 the available dates reveal two prominent hiatuses that correspond to two well-known cold interruptions during the Late Glacial climatic amelioration. At these times the cave appears to have been largely abandoned, at least the cave mouth area. Unfortunately, no sediments of this age are preserved in the Corral area in the rear of the vestibule. These are the Younger Dryas cold episode, about 11,000–10,000 BP, and the century-long cold event at about 7500 BP (8200 cal BP). It was during these times that the carbonate-rich, mondmilch-like sediments accumulated in the vestibule (e.g., Strata 10.1, 304, 305, and 102), presumably indicating a wetter climate, and probably colder as well, judging from regional climate information and the scarcity of human occupation in the cave.

Acknowledgments

This research was made possible by the physical, financial, and moral support of Lawrence Straus and Manual González Morales, who were very generous with their time and advice in the field. They arranged financial support through the U.S. National Science Foundation, National Geographic Society, L. S. B. Leakey Foundation, and Fundación M. Botín. Laboratory space and supplies were available in the Museum of Anthropology at the University of Michigan. Invaluable laboratory assistance was rendered by William J. Hubbard, whose master's thesis in geological sciences focused on sediments sampled through 1999 (Hubbard 2000), and by Samantha Sands, of the Michigan Undergraduate Research Opportunity Program, in 2000 through 2004. Katherine Clahassey drafted several of the figures.

References Cited

AER (Agupación Espeleológica Ramaliega)
1968 Cuevas de Cuevamur y del Mirón o Francés. *Cuadernos de Espeleología* 3:119–125.

Alley, R., D. Meese, C. Shuman, A. Gow et al.
1993 Abrupt increase in Greenland snow accumulation at the end of the Younger Dryas event. *Nature* 362: 527–529.

Butzer, K. W.
1981 Cave sediments, Upper Pleistocene stratigraphy, and Mousterian facies in Cantabrian Spain. *Journal of Archaeological Science* 8:133–183.

Clapés, J.
1997 Estudio mediante técnicas geofisicas del subsuelo de la Cueva del Mirón. Unpublished report.

Courty, M. A.
n.d. Microstratigraphic study of El Mirón Cave deposits, 1997 sampling: preliminary report. Unpublished report submitted to L. G. Straus and M. R. González Morales.

Courty, M. A., and J. Vallverdu
2001 The microstratigraphic record of abrupt climatic changes in cave sediments of the western Mediterranean. *Geoarchaeology* 16:467–500.

Farrand, W.
2000 *Depositional History of Franchthi Cave: Sediments, Stratigraphy and Chronology*. Excavations at Franchthi Cave, Greece, fascicle 12. Indiana University Press, Bloomington.

Farrand, W. R.
1975 Analysis of the Abri Pataud sediments. In *Excavation of the Abri Pataud, Les Eyzies, (Dordogne)*, edited by H. L. Movius Jr., pp. 27–68. Vol. ASPRB 30. Peabody Museum, Harvard University, Cambridge, MA.
2001 Archaeological sediments in rockshelters and caves. In *Sediments in Archaeological Context*, edited by J. K. Stein and W. R. Farrand, pp. 29–66. University of Utah Press, Salt Lake City.

Feuillée, P., and P. Rat
1971 Structures et paléogéographie des pyrénéo-cantabriques. In *Histoire structurale du Golfe de Gascogne*, pp. 1–48. Publications de l'Institut Français du Pétrole, vol. 2. Editions Technip, Paris.

Hoyos Gómez, M.
1995 Paleoclimatología del Tardiglacial enla Cornisa Cantábrica Basada en los Resultados Sedimentológicos de Yacimientos Arqueológicos Karsticos. In *El final del Paleolítico Cantábrico*, edited by A. Moure and C. Gonzalez Sainz, pp. 15–75. Universidad de Cantabria, Santander, Spain.

Hubbard, W.
2000 Geoarchaeological study of El Mirón Cave, Ramales de la Victoria, Cantabria, Spain. M.A. thesis, University of Michigan.

Moore, D. E., and R. C. Reynolds
1997 *X-Ray Diffraction and the Identification and Analysis of Clay Minerals*. Oxford University Press, Oxford, UK.

Stock, G. M., R. S. Anderson, and R. C. Finkel
2004 Pace of landscape evolution in the Sierra Nevada, California, revealed by cosmogenic dating of cave sediments. *Geology* 32:193–196.

Straus, L., M. González Morales, W. Farrand, and W. Hubbard
2001 Sedimentological and stratigraphic observations in El Mirón, a Late Quaternary cave site in the Cantabrian Cordillera, northern Spain. *Geoarchaeology* 16:603–630.

Straus, L. G., and M. González Morales
2003 Early-Mid Magdalenian excavations in El Mirón Cave (Ramales, Cantabria, Spain): Report on the VII campaign (2002). *Eurasian Prehistory* 1:117–137.

Table 5.1. Descriptions of sampled strata

Outer vestibule (Cabin area: rows H–K/1–4, from top downward)

Level 2, Sq. K2 (cut back 30 cm E of drawn section): 193 to 197 bd; dark yellowish-brown (10YR 3–4/6 m), very fine sand with small granules and a few rocks.

Level 3, Sq. K3: 197 to 202 bd; dark brown (7.5YR 3/2 m) silty clay with common to abundant rock fragments, including yellow sandstone pebbles; also charcoal.

Level 4, Sq. J–K/2–3: 217 to 220 bd; pinkish-gray (7.5YR 7/2 m) silt (ash), with many angular rock fragments; some charcoal.

Level 5, Sq. J–K/2–3: 232 to 238 bd; mottled grayish-brown and brown (7.5YR 3–4/2 and 4/4 m) silty clay.

Level 6.1, Sq. J–K/2–3: 232 to 238 bd; lower "yellow-beige" part of 6.1; brown (7.5 YR 5/4 m) silty clay with some charcoal and rock fragments, perhaps burnt.

Level 7, Sq. J–K/2–3: 246 to 251 bd; brown to dark brown (7.5 YR 4/3 m) silty clay, with some charcoal and angular rock fragments.

Level 8 top, Sq. H–I/2: 259 to 263 bd; (grayish) brown (7.5YR 4-5/2 m) silty clay, fissured, with some small rock fragments.

Level 8 base, Sq. H–I/2: 264 to 268 bd; same as above, but not fissured; some charcoal present.

Level 9, Sq. H–I/2: 271 to 276 bd; (light) brown (7/5YR 5/2 vm) silty clay with some angular limestone fragments at top just below contact with level 8.

Level 10 top, Sq. H–I/2: 283 to 285 bd; light brown (7.5YR 6/2 m) very compact silty clay, with some ash(?) and granules(?).

Level 10 middle, Sq. H–I/2: 285 to 187 bd; brown (7.5YR 5/2 m) silty clay, softer than level 10 top; contains microfauna.

Level 10 base, Sq. H–I/2: 289 to 292 bd; pale brown (10YR 6/3 vm) wet clay, vesicular, with abundant microfauna, and some small rock chips.

Level 10.1, Sq. H–I/2: 294 to 296 bd; thin layer of mixed colors—dark reddish-gray (5YR 4/2 m) and (lighter) brown to dark brown (7.5YR 4/2 m) silty clay with small rock chips and charcoal and bone fragments.

Level 11 top, Sq. H–I/2: 297 to 300 bd; light brown (7/5YR 5/2 m) gritty clay, with some angular rock fragments and some charcoal.

Level 11 middle, Sq. H–I/2: 308 to 312 bd; reddish-brown (5YR 4/3 m) silty clay(?), with abundant small rock chips (granules) and rock fragments up to 3 cm.

Level 11 base, Sq. H–I/2: 315 to 318 bd; reddish-brown (5YR 5/2 m) clayey silt with abundant rock chips (but less than in Level 11 middle).

Level 11.2, Sq. H–I/2: 322 to 325 bd; strong brown (7/5YR 4–5/2 m), rocky clayey silt, a distinctly brighter layer.

Level 12 top, Sq. H–J/2: 330 to 334 bd; brown to dark brown (7.5YR 4/4 m) silty clay with abundant rock fragments.

Level 12 lower, Sq. H–I/2: 344 to 348 bd; brown to dark brown (10YR 4/3 m) clayey silt with fewer rock fragments than upper part of 12; the level includes some larger limestone blocks up to 15–16 cm.

Level 13 middle, Sq. H–J/2: 365 to 370 bd; brown to dark brown (7.5YR 4/2 m) smooth, silty clay, drier than levels above; almost no rock fragments but an occasional large block (6–7 cm).

Level 13 base, Sq. H–J/2: 390 to 395 bd; brown to dark brown (7.5YR 4/3 m) silty clay, similar to 13 middle but more rock fragments; several large limestone blocks in this level, 25–30 cm across.

Level 14 middle, Sq. H–J/2: 406 to 411 bd; dark brown to brown (7/5YR 4/3 m) silty clay with few rock fragments and rare large rocks; no apparent difference between 13 and 14 except more large rocks in 13.

Level 15, Sq. H–I/2: 418 to 423 bd; brown to dark brown (7.5YR 4/4 m) silty clay matrix with many small, angular rock fragments; includes small bone fragments and sand-size shell fragments (snails?); matrix like 13 and 14, but sharp upper contact with 14.

Level 16 base and 17 top: (dark) brown (10YR 5/3 d) sandy, loamy matrix with angular rock fragments, <10 cm but mostly <5 cm and one 20 cm very well-rounded cobble.

Mid-Vestibule Trench (Sq. R/7, O–P/6 sondages plus westward continuation in N–K/5, from top downward)

Level 301 upper, Sq. P6: bottom of sample 220 cm bd; possibly old (Bronze Age?) disturbance; dark brown, slightly sandy loam, ca. 6 cm thick.

Level 301 lower, Sq. P6: bottom 237 cm bd; similar to 301 upper but with many more rock fragments.

Level 303, Sq. P6: 241 cm bd; light (yellowish-brown) gray (10YR 7/2 d) clay loam, but no cobbles; sharp contact with 301.

Level 303, Sq. M/5–6: 260–270 cm bd; homogeneous yellowish-gray silt with some bone, charcoal chunks, angular limestone clasts, and rare very rounded cobbles.

Level 304, Sq. L5: 295 cm bd; light tan, $CaC0_3$-rich mondmilch with occasional very angular limestone clasts.

Level 304, Sq. P–O/6: 248 cm bd; pale yellowish-brown (white 10YR 8/1 d) silt; smooth, greasy feel; a few rock fragments, chert, and flecks of charcoal; clear to sharp contact with 303. Note: 304 plunges west decidedly, toward cave mouth.

Level 305: 257 cm bd; sampled upper half of 305; similar to 304, but not as smooth and a little browner, especially at the top; occasional flint, some bone; clear contact with 304.

Table 5.1. *(continued)*

Mid-Vestibule Trench (*continued*)

Level 306 upper: 270 cm bd; medium yellowish-brown (10YR 6.5/4 d), sandy loam with abundant flint and bones and small decayed limestone granules; sharp to clear contact with 305.

Level 306 lower: 280 cm bd; similar to 306 upper but more pebbles and rock fragments.

Level 306: moved to Sq. N–O/6: 300 to 308 cm bd; light yellowish-brown (10YR 6.5/4 d) granular sediment, slightly darker in top 2 cm; lower contact with 307 very clear.

Level 307: 381–325 cm bd; dark brown (pale brown 10YR 6/3.5 d) rocky and granular sediment.

Level 308: 331–338 cm bd; dark brown (pale brown 10YR 6.3/5 d) rocky sediment with granular matrix; like 307 but more rocks; contains bones.

Level 310: 345–353 cm bd; grayer (pale brown 10YR 5.5/3 d) and clayier, some small rocks, but no pebbles; includes a black rocky lens on south wall.

Level 311: 358–363 cm bd; light yellowish-brown (2.5Y 6/3 d) silty loam with some small rocks.

Level 311.1: 375–380 cm bd; similar to 311 but more sandy, light yellowish-brown (10YR 6/4 d).

Level 312: ca. 385 cm bd (= floor of mini-sondage); dark brown (10YR 3/3 d) sediment with rocks and bones; distinct color contrast with 311.1; correlates with Level 15 or 16.

Vestibule rear (Corral area: rows T–V/7–8, plus S–Q/7; from top downward)

Level 102 top: 215–218 cm bd; crumbly (light) brown (7.5YR 5/4 m) silt, with particles or granules of pure silt, which are lighter brown (7.5YR 6/4 sl m).

Level 102 mid: 233–238 bd; same as above; occasional bone fragments and rock clasts.

Level 102 base: 243–248 bd; somewhat coarser silt granules; also rock fragments; immediately above sharp contact with level 104.

Level 104: 249–253 bd; dark brown (7.5YR 3/2 m) very fine, sandy loam, with rare rock fragments.

Level 106 top: 258–263 bd; dark brown (7.5YR 3/4 m) very fine, sandy loam, with small rock chips and fragments.

Level 106 base: 266–270 bd. Brown to dark brown (7.5YR 4/4 m) stony loam with small rock chips.

Level 107: 269–275 bd; dark brown to strong brown (7.5YR 4/4–6 m) fine, sandy loam.

Level 108 top: 287–289 bd, on shelf at base of section wall; dark reddish-brown (5YR 2.5/2 m) silty, clay loam, with bones, charcoal, and white limestone fragments.

Level 108 lower, Sq. V/7–8: 291–294 bd; dark (reddish) brown (5YR 3/3 or 7.5YR 3/4 m) loam (?), with white-coated, angular limestone fragments up to 8–9 cm (not sampled).

Level 109, Sq. V/7–8: 296–298 bd; soft, dark brown (5YR 3/3 or 7.5YR 3/4 m) loam with bones and occasional white rock fragments.

Level 110, Sq. V/7–8: 300–305 bd; similar to level 109; dark (reddish) brown (5YR–7.5YR 3/2 m).

Level 111, Sq. V/7–8: 315–321 bd; rocky, dark reddish-brown (5YR 2.5/2 m) loam with bones.

Level 112, Sq. V/7–8: 323–328 bd; same soft, dark reddish-brown (5YR 2.5/2 m) loam, with fewer rocks and fewer(?) bones; sample immediately on top of large, flat, rounded cobble (of level 113).

Level 113 and/or 114, Sq. V/7–8: 329–333 bd; same dark reddish-brown (5YR 2.5/2–3 m) sediment, but includes rounded (broken) cobbles.

Colluvial terrace footslope (X–W/10 sondage, from top downward)

Level 120 (= 400): top of sample 331 cm bd and bottom of sample 333 cm bd; dark yellowish-brown matrix with abundant pebbles, rock fragments, and some bone; limestone fragments chalky (weathered).

Level 121 (= 400.1): bottom of sample 343 cm bd; dark brown layer at base of level 400, ca. 2 cm thick; limestone fragments chalky (weathered).

Level 122 (= 401): bottom 351 cm bd; medium yellowish-brown (10YR 4.5/6 d) sandy loam with rock fragments and pebbles, similar to 400.

Level 123: 345–349 cm bd; medium yellowish-brown (10YR 4.5/6 d) loam with small pebbles and rock fragments.

Level 124: 353–359 cm bd; like 123 but a bit more yellow (yellowish-brown 10YR 5/5 d)

Level 402: bottom 357 cm bd; olive brown (2.5Y 4/4 d) sandy loam with rock fragments and bone; some pebbles; and some charcoal (not sampled).

Level 125: 360–363 bd; very distinct layer, not much different from 124 except color, olive brown (2.5Y 4/4 d); many flecks of charcoal, some bone fragments.

Level 126 (= 403): 367 cm bd; medium yellowish-brown (10YR 5/8 d) sandy loam with rock fragments, pebbles, bone, and flint fragments.

Table 5.1. *(continued)*

Colluvial terrace footslope (*continued*)

Level 126: 366–371 cm bd; medium yellowish-brown (10YR 5/8 d); clear contact with 125 above (color contrast) but merges with 127 below; one worked flint.

Level 404: top 374 cm bd and bottom 382 cm bd; medium to dark yellowish-brown (10YR 5.5/4 d) sandy loam with pebbles, rocks fragments, abundant bone fragments, and some charcoal.

Level 127: 380–383 cm bd; yellowish-brown (10YR 5.5/4 d) gravelly facies of 127, below 127.1 on south wall; lower contact is sharp cut-and-fill rupture with 128.

Level 127.1: 375–378 cm bd; yellowish-brown (10YR 51/7 d) lens of finer sediment in middle of level 127, divides 127 into 3 facies.

Level 128 (= 405) upper: bottom 392 cm bd; brownish-yellow (10YR 6/6, d) fine, sandy loam; not as sandy as upper levels; some pebbles and bone, but not as rich as above; sharp contact with 404.

Level 128 (= 405) lower: bottom 403 cm bd; like 405 upper; brownish-yellow (10yr 6/6 d) fine, sandy loam with pebbles and bone fragments.

Level 128: 377–382 cm bd; brownish-yellow (10YR 6/6 d) sandy loam; not as gravelly as 127, but large limestone blocks begin to appear; a 30 cm block in NE corner and 8 x 12 cm block just below this sample.

Level 128: 407–419 cm bd; brownish-yellow (10YR 6/6 d) sandy loam with a little charcoal and some bone fragments.

Level 129 upper: 455–462 cm bd; strong brown (7.5YR 5/6 m) sandy loam.

Level 129 lower: 474–479 cm bd; same sediment and same color (strong brown 7.5YR 5/6 m; yellowish-brown 10YR 5/8 d) as 128.

Level 130: 502–508 cm bd; strong brown (7.5YR 5/6 m; yellowish-brown 10YR 5/8 d) sandy loam, as in 129; arbitrary boundary between 129 and 130.

Level 130: 544–552 cm bd; same sandy loam, strong brown (7.5YR 4/6 m), yellow-brown (10YR 7/4 d).

Inner cave (Trinchera: rows 5–12/Q–S, from top downward)

Level IV: brown (10YR 5/3 m) silty clay, with abundant charcoal fragments, some pebbles and fragments of mondmilch; on the whole, rather mixed; = 900 BP.

Level V: very pale brown (10YR 8/3 m) mondmilch mottled or mixed with dark reddish-brown sediment.

Level VI upper: yellowish-brown (10YR 5/5 m) silty clay, with numerous small, white granules (=ripped up mondmilch?), small charcoal chips, and maybe Mn nodules.

Level VII top: cream-colored (10YR 9?/3 m) mondmilch; slightly crunchy but essentially quite "clean."

Level VII lower (moved to R/S line): mondmilch (10YR 8–9/3 and 10YR 7/3 m) and some thin, browner streaks and/or nodules, some as yellow as 10YR 8/6; marbled in places; some very small Mn? spots.

Level VIII top: yellowish-brown (10YR 5/5 m) silty clay with some very fine sand; abundant small granules and small, rounded cobbles of exotic types; much wetter than above; abundant, very pale gray, fibrous growth in interstices—mineral or vegetal?

Level IX: strong brown (7.5YR 5/7 m) very fine, sandy clay with abundant small granules, pebbles, and some small cobbles; small, white granules adhering to clumps in sample (= surface contamination?); sandier and redder than level VIII.

Inner cave (interior wall above travertine)

1. Upper very fine sand
2. Pale brown laminated silt
3. Lower very fine sand
4. Reddish-brown clay
5. Silt with Mn nodules

Cuevamur

1. Sandy silt
2. Bedded gravel with rounded pebbles
3. Sandy silt

Note: bd = below datum. For Munsell colors, d = dry, m = moist, sm = slightly moist, vm = very moist.

Table 5.2. Data by excavation area

Sample	Depth B.D.	Roundness Index	pH	$CaCo_3$	Organic Matter	LS	Non-Ls	D%	Bone/Flint/Shell/Other[a]
Outer vestibule (Cabin area)									
2	197	21.0	—	14.3	4.44	6	2	2.64	32/1/0/0
3	202	—	—	29.3	8.29	0	0	0.67	52/2/1/0
4	220	0.8	—	73.03	4.06	16	0	0.62	9/0/0/0
5	238	0.0	—	60.39	5.98	0	0	—	—
6.1	238	33.3	8.31	—	—	2	0	0.23	0/0/0/3
7	251	0.0	—	50.63	7.46	2	0	—	—
8 top	263	0.0	—	76.63	5.42	0	0	1.02	18/0/7/0
8 base	268	29.0	8.53	76.44	4.91	5	0	0.54	15/0/3/0
9	276	0.0	—	71.43	4.57	4	0	0.77	8/1/5/1
10 mid	287	3.0	8.56	83.22	4.05	6	0	0.21	19/0/2/0
10 base	292	0.0	—	80.83	3.68	1	0	1.05	76/0/5/0
10.1	296	67.0	—	79.25	3.83	0	1	0.19	21/0/5/0
11 top	300	14.0	—	78.29	4.59	8	0	0.36	23/0/10/2
11 mid	312	10.0	8.58	77.85	3.34	35	0	0.04	5/1/2/0
11 base	318	9.0	—	78.15	3.35	18	0	0.53	24/2/6/0
12 top	334	2.0	8.64	50.88	3.00	38	0	0.02	1/0/2/0
12 lower	348	2.0	8.65	34.89	3.47	20	1	0.06	4/6/2/0
13 mid	370	24.0	8.49	21.28	3.83	2	1	0.03	3/0/0/0
13 base	395	0.0	8.46	13.29	4.33	11	0	0.08	1/0/2/0
14	411	0.0	8.52	10.49	4.69	2	0	—	0/0/0/0
15	423	8.0	—	24.60	4.91	29	0	0.52	1/0/0/0
16–17	440	8.8	7.83	—	—	97	0	1.22	+/0/0/1
Mid-Vestibule Trench									
303	241	17.0	8.34	60.44	5.76	3	0	0.28	21/10/0/0
304(2000)	248	0.0	8.42	86.92	3.63	9	0	0.18	54/1/0/0
304(2002)	293	10.3	8.42	57.45	2.23	all	0	0.36	+/0/+/1
306 upper	270	8.4	—	21.50	5.71	22	0	0.23	34/2/9/0
306 lower	280	6.8	—	24.10	5.17	14	0	0.70	68/4/0/0
306(2002)	308	8.0	8.55	21.29	6.50	n.d.	0	0.31	+/0/+/2
307	325	9.2	8.51	30.77	5.13	all	0	1.28	+/0/+/1
308	338	1.9	8.15	21.51	4.42	39	0	1.40	+/0/0/0
310	353	4.9	8.16	26.11	3.21	45	0	0.88	+/0/+/1
311	363	2.3	8.32	23.46	3.05	24	0	0.33	+/0/+/2
311.1	380	1.1	8.43	9.27	4.70	38	0	0.23	+/0/0/0
312	385	6.9	8.07	—	—	20	0	1.33	+/0/0/0
Vestibule rear (Corral area)									
102 top	218	0.0	7.69	27.59	5.22	3	0	0.98	8/1/1/0
102 mid	238	0.0	8.18	9.34	5.39	1	0	7.93	33/0/0/0
102 base	248	0.0	—	14.98	7.84	9	0	0.19	24/0/0/0
104	253	0.0	—	12.32	6.27	7	2	2.85	83/18/0/0
106 upper	263	13.9	8.72	7.01	6.57	6	2	2.64	39/5/0/0

Table 5.2. *(continued)*

Sample	Depth B.D.	Roundness Index	pH	$CaCo_3$	Organic Matter	LS	Non-Ls	D%	Bone/Flint/Shell/Other[a]
Vestibule rear (Corral area) *(continued)*									
106 base	270	0.0	8.9	13.19	5.78	10	1	0.34	19/9/0/0
107	275	13.5	8.83	7.82	5.42	3	0	0.03	3/0/0/0
108 top	289	22.0	9.34	6.24	8.08	10	0	15.23	376/55/0/0
108 lower	294	24.7	9.28	8.48	7.53	4	2	21.09	223/24/0/0
109	298	29.6	9.04	6.28	6.67	0	2	9.30	157/16/0/0
110	305	10.0	9.07	5.99	7.35	3	2	9.19	179/22/0/0
111	321	15.7	9.32	25.15	7.92	13	0	4.74	182/12/0/0
112	328	29.1	9.21	12.71	6.26	15	6	7.84	156/24/0/0
113–114	333	15.5	9.23	6.71	7.39	11	1	4.44	60/13/0/0
Colluvial terrace footslope									
400(=120)	331	29.0	7.99	7.32	4.28	15	38	0.67	68/4/0/0
400.1(=121)	343	62.1	—	9.95	4.67	2	9	2.10	40/2/0/0
401(=122)	351	30.8	8.19	—	—	8	20	0.99	162/9/0/0
123	349	61.6	7.74	10.09	6.00	11	35	0.37	5/1/0/0
124	359	50.4	7.86	11.63	5.18	15	39	0.45	6/1/0/0
125	363	45.3	7.91	12.43	5.15	8	19	0.00	0/0/0/0
402(=125)	357	18.8	8.28	—	—	21	10	1.44	195/5/4/0
126	371	48.0	8.08	10.28	4.92	19	35	0.00	0/0/0/0
403(=126)	367	48.6	8.35	—	—	13	18	1.74	115/7/5/0
127	383	2.8	—	19.33	4.47	10	1	0.47	4/1/0/0
404(=127)	382	18.6	8.48	—		40	17	4.14	180/4/12/0
405 upper(=128)	392	26.8	8.32	—	—	—	—	—	—
127.1	378	43.5	8.43	15.06	4.02	1	2	0.00	0/0/0/0
128 upper	382	18.0	8.37	17.55	4.33	6	6	0.00	0/0/0/0
405 lower(=128)	403	52.2	8.45	—	—	40	17	1.31	257/4/12/0
128 mid	419	15.3	8.14	20.02	4.69	26	4	0.00	0/0/0/0
129 upper	462	19.7	8.35	8.33	4.28	24	13	0.26	0/1/0/0
129 lower	479	81.8	8.67	2.75	3.62	0	27	0.00	0/0/0/0
130 upper	508	67.6	8.38	8.68	4.28	9	8	0.00	0/0/0/0
130 lower	552	67.2	8.52	7.32	4.54	48	0	0.00	0/0/0/0
Ramp									
1	—	—	—	—	—	0	8	0.00	—
2	—	—	—	—	—	0	63	0.00	—
2000	—	83.7	8.56	4.73	4.02	3	47	0.00	—
Cuevamur									
Upper silt	—	0	8.29	1.97	2.62	0	0	0.00	—
Lower silt	—	0	8.37	2.48	3.47	0	0	0.00	—
Gravel	—	33.3	8.63	2.10	5.14	—	—	0.00	—

[a]+ = present

CHAPTER SIX

El Mirón Cave Radiocarbon Dates

Lawrence Guy Straus
and Manuel R. González Morales

In order to develop a chronology for the El Mirón site that is at least partially independent of temporally/culturally "diagnostic" artifacts and in order to be able to objectively correlate and compare levels both among the different excavation areas within the cave and between El Mirón and other regional sites, a major effort and financial commitment was made to date the site by the radiocarbon method. To avoid possible problems of inter-laboratory comparability, almost all the dates were processed by Dr. Alexander Cherkinsky, almost all at Geochron Laboratories (GX) in Cambridge, Massachusetts. After Cherkinsky moved to the Center for Applied Istotope Studies at the University of Georgia (Athens, Georgia, USA), he ran the final seven dates (UG) there, using the same rigorous pretreatment protocols (including filtering) as at Geochron. Three specific methods were used: conventional, conventional with extended radioactivity counting, and accelerator mass spectrometry (AMS). The targets for accelerator mass spectrometry were prepared by Cherkinsky and the AMS measurements were made at a variety of accelerator facilities (notably the Lawrence Livermore Laboratory at Berkeley, California; the University of Arizona; and the University of Georgia). Seven AMS dates on bone were commissioned at the Oxford Accelerator (OxA) for red deer samples from which DNA was extracted by Professor Adrian Lister and Meirav Meiri.

Between 1996 and 2011 a total of 76 radiocarbon dates were obtained from El Mirón, the largest number from any single early prehistoric site in the Cantabrian region (or probably from Spain or western Europe, for that matter). Thirty-five dates were run on wood charcoal, 1 on a grain of wheat, 38 on bone collagen, and 2 on two different fractions (bioapatite and collagen) from a single ungulate tooth. All determinations were determined to have acceptable $\delta^{13}C_{PDB}$ values (ranging between ca. -20 and -28‰). All dates are expressed in radiocarbon years before AD 1950 (i.e., BP), using the ^{14}C half-life of 5568 years. All but the oldest two dates (27,590 and 41,280 BP) have been calibrated according to the CALIB programs of Stuiver et al. (1998, as periodically updated), giving a range of $\pm 1\delta$ (one standard deviation).

The bone collagen dates were almost all done on multiple bones from single spits in one or more subsquares of single-meter squares. The bones selected were usually large, not readily identifiable, long-bone splinters from medium- to large-size ungulates (almost certainly red deer and/or ibex in most cases). Charcoal was used when there were relatively abundant, spatially concentrated large chunks, although occasional isolated chunks were also used as dating samples. In two cases, Square X7, Level 503.1, and Square J2, Level 21, individual chunks of charcoal could be identified (*Juniperus* sp., determined by M. Ruiz and L. Zapata) and in another case (Square M5, Level 303.3) an individual emmer wheat grain was identified during study by L. Peña-Chocarro (*Triticum dicoccum*). Care was taken to not select bones or charcoal from spits where the excavation notes mentioned any obvious or suspected human or natural disturbance (e.g., rodent burrows); however, some charcoal pieces were deliberated sampled from obvious hearths, and these could have been subjected to prehistoric disturbance by digging into or through underlying levels, leading to dates that appear either too young or too old relative to the levels in which the hearths were encountered during excavation. And in one case (GX-24465), an unusually large and well-preserved chunk of charcoal (not seen in situ by either one of the directors or by the Corral excavation area crew chief) was collected by an excavator on the surface of a Magdalenian level but almost certainly had fallen from the top of an adjacent stratigraphic section, hence explaining its medieval date.

All the post-Paleolithic/post-Pleistocene dates were run on charcoal (or on the wheat grain). Most of the Paleolithic dates were run on bone collagen, as charcoal lumps are rare in those levels, even though charcoal "dust" or staining is fairly common both in apparent hearths and spread throughout occupation levels in general.

Results

The full list of El Mirón radiocarbon dates is given in table 6.1. There are 26 dates from the Cabin (outer vestibule) excavation area, 10 from the Mid-Vestibule Trench, 23 from the Corral (vestibule rear), and 6 from the ramp and inner cave areas.

From bottom to top of the cultural occupation sequence, the dates are distributed as follows:

Mousterian (1 date from the Corral, sondage W–X/10): 41,280±1120 BP;

Early Upper Paleolithic (1 date from the Corral, sondage W–X/10): 27,580±210 BP;

Solutrean (4 dates from the Corral, V/9–10, sondage W–X/10): 19.0–18.4 kya, possibly plus 1 date of 17,400±270 BP from Mid-Vestibule Level 313 that has no associated Solutrean points (18,310–21,105 cal BC);

Early (i.e., Initial, Lower, and Middle) **Magdalenian** (13 dates from the Cabin, 1 date from the Mid-Vestibule, 18 dates from the Corral, 2 from the human burial area behind the engraved block, and 2 dates from the ramp and inner cave): 17–13.7 kya (18,750–14,210 cal BC);

Upper Magdalenian (4 dates from the Corral, Mid-Vestibule, and Cabin): 13–12.3 kya (13,990–12,165 cal BC);

Final Magdalenian/Early Azilian (4 dates from Corral, Mid-Vestibule, and Cabin): 12–12.7 kya (12,135–11,525 cal BC);

Azilian (2 dates from the ramp, 1 from the Mid-Vestibule, and 1 from the Cabin): 10.7–10.3 kya (10,095–9820 cal BC);

Mesolithic (3 dates from the Cabin): 9.6–8.4 kya (9120–7180 cal BC);

Neolithic (5 dates from the Mid-Vestibule and 6 from the Cabin): 5.8–4.7 kya (4770–3370 cal BC);

Chalcolithic (3 dates from the Cabin): 4.1–3.7 kya (2860–1930 cal BC);

Early Bronze Age (1 date from the inner cave and 1 from the Cabin): 3.7–3.2 kya (2140–1440 cal BC);

Middle Ages (1 date each from the inner cave and the Corral): 900–540 BP (cal AD 1040–1440).

There is a certain amount of circular reasoning involved in the cultural phase attributions of many levels (especially in the Magdalenian), as unambiguous "diagnostic" artifacts (e.g., Solutrean points, Lower Magdalenian striation engraved scapulae, Upper Magdalenian harpoons, Neolithic double-bevel retouched geometic microliths, Chalcolithic arrowheads) are not always present (at least in the areas of many occupation levels that we excavated). (Even El Mirón's ceramics are undecorated and temporally rather banal; the only diagnostic decorated [chevron incised] sherd was found in mixed surface deposits and might represent a Late Bronze Age occupation not represented in the excavations [Vega Maeso, chap. 19, this volume].) The transitions between what archaeologists call "Solutrean" and "Initial Magdalenian," "Upper Magdalenian" and "Azilian," and "Chalcolithic" and "Early Bronze Age" are particularly ambiguous and arbitrary. In any excavation (especially one that does not even come close to covering 100 percent of the once-occupied site area), there are always going to be questions of sampling effect, as well as prehistoric behavioral reasons (spatially discrete activity area functions, inter-site functional differences, artifact curation effects, etc.) for the presence or absence of certain temporally indicative "fossil director" artifacts. This is clearly the case with El Mirón.

Anomalies

In general, especially if single or double standard deviations are taken into consideration, the radiocarbon dates are in coherent stratigraphic order within their respective excavation areas. The post-Pleistocene dates are particularly coherent. Most inversions in the order of the central tendencies of dates from stratigraphically contiguous or "nearby" levels are easily covered by the standard deviations of the individual assays (e.g., Chalcolithic Levels 7 and 5.1 in the Cabin; spits within Lower Magdalenian Level 17 of the same area). However, there are some egregious anomalies among the Late and Early Magdalenian dates from the Corral.

Level 108 in the Corral has four dates, and one can either choose a long (ca. 14.8 kya) or a short (ca. 13.7 kya) chronology separated by a radiocarbon millennium; to make matters even less understandable, one of the "old" dates and one of the "recent" dates come from the same square, V8, but (in reverse stratigraphic order) from contiguous spits. The new AMS date (UB-9286) for Level 105 on bone from a single spit (7) in a single W7 subsquare (C) is probably more reliable than the conventional date for underlying Level 106 (GX-32382), which had been done on bones from several different squares and spits, especially as the stratigraphic relationship between Levels 106 and 105 is not always clear, the latter being a discontinuous entity.

There are two dates from Level 111 that differ by nearly a radiocarbon millennium, but both have large standard deviations. However, the "old" date centered on 16,370 BP makes sense vis-à-vis the 16,130 BP date for overlying Level 110 (though now contradicted by two OxA dates of ca. 14,800 BP) and the single date of 16,460 BP from underlying Level 114. The "worst" date in the Corral sequence—13,900±840 BP—is from underlying Level 115; even at +2δ this date would still be out of stratigraphic order at 15,580 BP. This date should be discarded. Unfortunately, however, it is underlain by a pair of contradictory dates from adjacent squares (V7 and V8) in Level 116; again, one can choose a long (17,400 BP) or a short (15,220 BP) chronology for this level, but the latter would contradict the choice of dates in the 16–16.5 kya range for overlying Levels 110–114. The same choice applies to dates from Level 117 ("old" at 17,050 BP) and Level 118 ("young" at 15,460 BP, albeit with a large standard deviation of ±190 that is not enough to bridge the gap with the "old" date, even at +2δ).

Underlying Levels 119 and 119.2 are dated by inverted assays that nonetheless seem to converge at about 16.5 kya, but these are "young" vis-à-vis overlying levels 116 and 117 that may be dated to around 17 kya. There is no apparent pattern as to where "old" and "young" dates in the Corral came from, despite the existence of a double slope (down toward both the west and north) and despite the practical difficulty experienced in "following" thin, individual levels with certainty across this 10 m^2 excavation area. There are both "old" (GX-27114, GX-28209) and "young" (GX-31933, GX-32656) dates in the double-downslope part of the Corral area (T–U/10, U9) that is farthest from the initial sections used to define the Corral stratigraphy in square V8. And there are dating contradictions within the same or contiguous levels in "upslope" squares (notably V8). All this suggests that in this area of intensive, relatively continuous human use, including repeated hearth-pit digging, where the levels basically constitute a massive palimpsest of Early Magdalenian occupation deposits with no sterile lenses, dated materials (bone splinters and small lumps of charcoal) may have moved both up and down within the stratigraphy. Although we excavated these deposits by trying to follow the (relatively gentle) slopes indicated by the northern, western, and southern stratigraphic sections of the Corral, as well as the local "lay-of-the-land" as indicated by the inclinations of long bones, limestone *éboulis*, cobbles, and localized patches of ochre and distinctive sediments such as greenish-gray silt, it is also possible (though not proven) that we may have unconsciously passed from one "level" to another, given the color and content similarities of all the Early Magdalenian deposits. This would lead to incorrect equivalences being made between particular levels upslope with others downslope.

There is one case of date inversion in the Cabin excavation area that is also difficult to explain: the charcoal AMS date from Level 21 (UG-03364), which is >2δ younger than the bone collagen date from overlying Level 19 but statistically identical to the bone collagen date from Level 18, which is even higher up in the square J2 sondage. The sediments of all these levels are dense, compact, clayey silt/silty clay, and the excavator (Cabin area crew chief Dr. Igor Gutierrez Zugasti) reported observing no voids along the south cave wall, something that, however, had been encountered higher up in the stratigraphy in squares J2, I2, and H1 that directly abut the cave wall. There was no apparent evidence of disturbance, and the large chunk of charcoal was uncovered in situ by Gutierrez in Level 21. The date of 16,050 BP seems young relative to the immediately overlying levels of this "cave clay" deposit, as well as with regard to the classic Lower Magdalenian Level 17, the oldest date for which is 15,700 BP. There does not seem to be any pattern of charcoal dates being *younger* than bone collagen dates in the El Mirón series, although there are certainly cases of *both* charcoal and bone dates (e.g., GX-24469: Level 111; GX-23394: Level 115; GX-31933: Level 118) seeming to be too young vis-à-vis other coherent dates on both materials from the same or contiguous levels. Such problems plague all archaeological sites (especially perhaps in caves and rockshelters) with large series of radiocarbon dates. Thus the emphasis on radiocarbon-based dating of long, complex culture-stratigraphic sequences is a double-edged sword: it is preferable to date many, most, or even all levels and to try to obtain multiple dates per level when sample materials and funding are adequate, but with the multiplication of assays comes the multiplication of "problem dates." The resolution of which dates are "good" and hence the ones that are to be accepted is difficult and fraught with the problem of a priori assumptions and circular reasoning. But where the archaeology of the terminal Quaternary would be without radiocarbon dating is hard to imagine in this day and age. So El Mirón displays a good overall pattern of radiocarbon ages, but it is one that is dotted with specific problems and therefore difficult individual archaeological choices.

Conclusions

Nonetheless, there seem to be some major subdivisions of cultural time after the Solutrean (with foliate and shouldered points) in the Corral area: Initial Magdalenian (Levels 119.3–116) ca. 17 kya; Lower Magdalenian (Levels 115–110) ca. 16.5–16 kya; and Middle Magdalenian (based on Level 108, but possibly including Levels 109 and 107) ca. 14.5–13 kya. However, whether there are really significant, standardized differences in terms of lithic and osseous industries among these three traditional Magdalenian subdivisions in El Mirón remains to be seen. Are there essential temporal distinctions beyond the many, rather stochastic, inter- (or even intra-) level variations that are probably the products of functional variations in time and space related to how people used the cave (or different areas within the cave) during different visits (and kinds of visits) to the site as well as the products of sampling accidents? This we do not know (but see Straus and González Morales 2008; Straus, González Morales, and Stewart 2008). It will be the subject of future analyses and discussions.

The general, inter-area, stratigraphic correlations, based on sedimentary and archaeological content similarities, physical (near-) contiguity, and radiocarbon dates are as follows:

Early Magdalenian: Inner cave Level VIII, ramp Niche A, Corral Levels 119.2–108, Mid-Vestibule Levels 310–312, Cabin Levels 13–21;
Upper Magdalenian: Corral Levels 105–107, Mid-Vestibule Levels 307–308, Cabin Level 12;
Final Magdalenian/Early Azilian: Corral Level 102.1, Mid-Vestibule Level 306, Cabin Level 11.1;
Azilian: Ramp Breccia, Mid-Vestibule Level 305, Cabin Level 11 (?);
Mesolithic: Corral Levels 101–102, Mid-Vestibule Level 304, Cabin Level 10.1;
Neolithic: Mid-Vestibule Levels 303–303.3 + pit 98a base, Cabin Levels 8–10;
Chalcolithic: Cabin Levels 7–4;
Bronze Age: Inner cave Level VI; Cabin Levels 2–3.

While not perfect, the El Mirón radiocarbon series is a very significant source of chronological information for the Late Quaternary (Late Middle Paleolithic, Upper Paleolithic, Mesolithic, and post-Paleolithic) record of Cantabrian Spain. It will serve as a key element in the ongoing construction of the chronostratigraphic framework for the region for years to come, much as the far shorter (Late Upper Paleolithic and Mesolithic) radiocarbon series from La Riera Cave in eastern Asturias (Straus and Clark 1986) has done since the late 1970s. It is joined by the spectacular series of radiocarbon dates from the 11 cave sites of the La Garma karstic system near Santander in coastal Cantabria (Arias and Ontañón 2008).

References Cited

Arias, P., and R. Ontañón
2008 Zona arqueológica de La Garma (Omoño, Ribamontán al Monte): Campañas 2000–2003. In *Actuaciones arqueológicas en Cantabria, 2000–2003*, edited by R. Ontañón, pp. 43–60. Gobierno de Cantabria, Santander, Spain.

Straus, L. G., and G. A. Clark
1986 *La Riera Cave: Stone Age Hunter-Gatherer Adaptations in Northern Spain.* Anthropological Research Papers 36. Arizona State University, Tempe.

Straus, L. G., and M. González Morales
2008 Addendum and correction to "Early Magdalenian variability: new evidence from El Mirón Cave, Cantabria, Spain." *Journal of Field Archaeology* 33:367–369.

Straus, L. G., M. González Morales, and E. B. Stewart
2008 Early Magdalenian variability: new evidence from El Mirón Cave, Cantabria, Spain. *Journal of Field Archaeology* 33:197–218.

Stuiver, M., P. Reimer, E. Bard, J. Beck, G. Burr, K. Hughen, B. Kromer, G. McCormac, J. van der Plicht, and M. Spark
1998 INTCAL 98 radiocarbon calibration, 24000–0 cal BP. *Radiocarbon* 40:1041–1083.

Table 6.1. El Mirón Cave radiocarbon dates (1996–2011)

Square	Zone[a]	Level[b]	Spit[c]	Period[d]	Date BP	1 SD	Material[e]	Lab no.	Method[f]	Calibrated date[g]
H2	OV	3	4	BA	3700	40	Ch	GX-25851	AMS	2140–2030 BC
J2	OV	5	3	Chal	3820	240	Ch	GX-22127	Conv.	2580–1930 BC
I3	OV	5.1	4	Chal	4120	50	Ch	GX-22130	AMS	2860–2590 BC
H2	OV	7	14	Chal	3740	120	Ch	GX-24460	Cxcnt	2310–1960 BC
I3	OV	8.1	13	Neo?	4680	60	Ch	GX-22131	AMS	3610–3370 BC
J2	OV	9	8	Neo	5170	170	Ch	GX-22128	Conv.	4220–3790 BC
H4	OV	9	22	Neo	5280	40	Ch	GX-24461	AMS	4220–4000 BC
I4	OV	9.6	22	Neo	5250	150	Ch	GX-24462	Cxcnt	4320–3950 BC
I3	OV	10	19	Neo	5570	50	Ch	GX-23414	AMS	4450–4360 BC
I3	OV	10	19	Neo	5690	50	Ch	GX-23413	AMS	4580–4460 BC
J4	OV	10.1	34	Mes	8380	175	Ch	GX-24463	Cxcnt	7590–7180 BC
I4	OV	10.1	28	Mes	8700	40	Ch	GX-25852	AMS	7745–7609 BC
I4	OV	10.1	29	Mes	9550	50	Ch	GX-24464	AMS	9120–8790 BC
J3	OV	11		AZ	11,785	55	Bc	OxA-22087	AMS	11,790–11,620 BC
J4	OV	11.1		AZ/UM	11,205	55	Bc	OxA-22088	AMS	11,260–11,120 BC
I3	OV	11.1	25	AZ/UM	11,720	140	Bc	GX-23391	Conv.	12,040–11,520 BC
J2	OV	12	12	UM	12,970	70	Ch	GX-22132	AMS	13,990–13,300 BC
J4	OV	13		MM	14,930	70	Bc	OxA-22089	AMS	16,540–16,320 BC
H–J	OV	14	sev.	MM	14,600	190	Bc	GX-32383	Conv.	15,870–15,210 BC
I3	OV	15	43	LM	15,010	260	Bc	GX-23392	Conv.	16,400–15,610 BC
I3	OV	15	43	LM	15,220	300	Bc	GX-23393	Conv.	16,680–15,810 BC
I3	OV	16	44	LM	15,180	100	Bc	GX-23415	AMS	16,490–15,910 BC
J3	OV	17	32	LM	15,470	240	Bc	GX-24466	Conv.	16,920–16,150 BC
J3	OV	17	39	LM	15,450	160	Bc	GX-27115	Conv.	16,850–16,180 BC
J2	OV	17	20	LM	15,700	190	Ch	GX-25853	Cxcnt.	17,150–16,440 BC
H3	OV	17	74	LM	15,610	90	Bc	OxA-22093	AMS	16,900–16,730 BC
H3	OV	17	79	LM	15,370	80	Ch	GX-32654	AMS	16,720–16,140 BC
J2	OV	18	35	LM	16,080	40	Bc	UG-3366r	AMS	17,410–17,230 BC
J2	OV	19	36	LM	16,600	40	Bc	UG-3365r	AMS	17,740–17,640 BC
J2	OV	21	38	LM	16,050	40	Ch	UG-3364r	AMS	17,370–17,180 BC
O6	MV	302?	Pit 98-a	Neo?	4910	80	Ch	GX-28211	Conv.	3770–3640 BC
L5	MV	303	13	Neo	5500	90	Ch	GX-25854	Conv.	4450–4250 BC
L5	MV	303.1	14	Neo	5520	70	Ch	GX-25855	Conv.	4450–4260 BC
M5	MV	303.3	10	Neo	5550	40	Grain	GX-30910	AMS	4450–4340 BC

Table 6.1. *(continued)*

Square	Zone[a]	Level[b]	Spit[c]	Period[d]	Date BP	1 SD	Material[e]	Lab no.	Method[f]	Calibrated date[g]
L5	MV	303.3	16	Neo	5790	90	Ch	GX-25856	Cxcnt.	4770–4540 BC
P6	MV	305	9	AZ	10,270	50	Bc	GX-24467	AMS	10,360–9820 BC
P6	MV	306	11	AZ/UM	11,650	50	Bc	GX-24468	AMS	11,860–11,530 BC
P6	MV	308	16	UM	12,350	180	Ch	GX-28210	Cxcnt.	13,400–12,170 BC
P6	MV	312	25	LM	15,850	170	Bc	GX-31932	Conv.	17,330–16,630 BC
P6	MV	313	29	Sol?	17,400	270	Bc	GX-31194	Conv.	19,200–18,310 BC
U7	VR	surface	#	Med	540	100	Ch	GX-24465	Cxcnt.	AD 1300–1440
T8	VR	102.1	3	AZ/UM	11,950	70	Bc	GX-23417	AMS	12,140–11,890 BC
W7	VR	105	7	UM	13,490	70	Bc	UG-9286	AMS	14,850–14,630 BC
U–V	VR	106	sev.	UM	12,460	180	CBc	GX-32382	Cxcnt.	13,480–12,210 BC
V8	VR	108	4	MM	13,660	70	Bc	GX-22703	AMS	14,690–14,210 BC
U7	VR	108	10	MM	13,710	70	Bc	GX-32381	AMS	14,750–14,270 BC
V8	VR	108	5	MM	14,710	160	Bc	GX-23397	Conv.	15,970–15,350 BC
T10	VR	108”	8	MM	14,850	60	Ch	GX-27114	AMS	16,090–15,560 BC
U8	VR	110	18	LM	14,760	70	Bc’	OxA-22091	AMS	16,100–15,830 BC
U8	VR	110	18	LM	14,795	75	Bc’	OxA-22092	AMS	16,120–15,890 BC
V8	VR	110	8	LM	16,130	250	Bc	GX-23396	Conv.	17,700–16,880 BC
V8	VR	111	14	LM	16,370	190	Bc	GX-23395	Conv.	17,940–17,200 BC
U7	VR	111	20	LM	15,530	230	Bc	GX-24469	Conv.	16,980–16,220 BC
T8	VR	112	26	LM	15,430	75	Bc	OxA-22090	AMS	16,780–16,630 BC
T10	VR	114	17	LM	16,460	50	Bc	GX-28209	AMS	17,990–17,360 BC
V8	VR	115	19	LM	13,800	840	Bc	GX-23394	Conv.	15,730–13,520 BC
V8	VR	116	20	LM	15,220	100	Bc	GX-23416	AMS	16,530–15,960 BC
V7	VR	116»	26	LM	17,400	80	Ch	GX-29439	AMS	19,090–18,420 BC
V8	VR	117	24	LM	17,050	60	Ch	GX-25857	AMS	18,760–18,030 BC
U10	VR	118	38	LM	15,460	190	Ch	GX-31933	Cxcnt.	16,880–16,170 BC
V8	VR	119	28	LM	16,960	80	Ch	GX-25858	AMS	18,570–17,920 BC
U9	VR	119.2	52	LM	16,320	160	Bc	GX-32656	Conv.	17,870–18,160 BC
V9	VR	121	3	Sol	18,390	300	Bc	GX-32655	Cxcnt.	19,420–20,380 BC
W10	VR	125	4	Sol	18,980	360	Bc	GX-24470	Conv.	21,110–20,040 BC
W10	VR	126	5	Sol	18,950	350	Bc	GX-24471	Conv.	21,070–20,010 BC
V10	VR	127	15	Sol	19,230	50	Ch	UG-7216	AMS	21,170–20,680 BC
X10	VR	128	20	EUP	27,580	210	Ch	GX-27113	AMS	29,940–29,460 BC
X10	VR	130	36	MP	41,280	1120	Ch	GX-27112	AMS	43,770–42,150 BC

Table 6.1. *(continued)*

Square	Zone[a]	Level[b]	Spit[c]	Period[d]	Date BP	1 SD	Material[e]	Lab no.	Method[f]	Calibrated date[g]
X7	BB	503.1	5.1	LM	15,120	40	Ch	UG-7799	AMS	16,340–16,160 BC
X7	BB	504	7	LM	15,740	40	Bc	UG-7217	AMS	16,990–16,820 BC
Mid-slope	Niche A			LM	16,600	90	Bc	GX-30398	AMS	18,160–17,510 BC
Slope top IC	Flow-stone			AZ	10,740	40	Ta'	GX-27521a	AMS	10,980–10,700 BC
Slope top IC	Flow-stone			AZ	10,390	50	Tc'	GX-27521c	AMS	10,670–10,020 BC
8–9/Q	IC	IV		Med	900	80	Ch	GX-22129	Conv.	AD 1040–1200
11Q	IC	VII		BA	3230	40	Ch	GX-28013	AMS	1520–1440 BC
11R	IC	VIII	10	MM	14,620	80	Ch	GX-22347	AMS	15,820–15,290 BC

Note: Total number of dates = 76

[a]IC = inner cave; OV = outer vestibule (Cabin); VR = vestibule rear (Corral); MV = Mid-Vestibule (trench); BB = behind block

[b]": charcoal from hearth pit dug in or from the indicated stratum

[c]sev. = several lumped spits due to low bone density; # = lg. charcoal lump fallen into the excavation from Corral surface after removal of top ca. 20 cm of recent dirt and dung

[d]Med = medieval; BA = Bronze Age; Chal = Chalcolithic; Neo = Neolithic; Mes = Mesolithic; AZ = Azilian; UM = Upper Magdalenian; MM = Middle Magdalenian; LM = Lower Magdalenian; Sol = Solutrean; EUP = Early Upper Paleolithic; MP = Middle Paleolithic

[e]Ch = charcoal; Ta = tooth bioapatite; Tc = tooth collagen; Bc = bone collagen; Cbc = carbonized bone collagen; ': same tooth or bone

[f]AMS = accelerator; Conv. = conventional; Cxcnt. = extended count

[g]Stuiver et al. 1998: CALIB 4.1.2 and 6.0.0 (range @ 1 sigma, rounded to nearest 10 yrs)

CHAPTER SEVEN

Preliminary Archaeomagnetic and Rock-Magnetic Results from the Holocene Fire Lenses in El Mirón Cave

Ángel Carrancho and Juan José Villalaín

TRANSLATED BY LAWRENCE GUY STRAUS

Analysis of the magnetic record of burnt archaeological materials can provide interesting data, especially with regard to chronology. The Earth's magnetic field undergoes changes in direction and intensity that vary both spatially and temporally around the surface of the Earth and that can end up being registered via various mechanisms (mainly thermal) in diverse archaeological materials. One of the phenomena responsible for these changes, of direct interest to archaeology, is the one known as *secular variation* of the earth's magnetic field (SV), the principle on which archaeomagnetic dating is based.

Archaeomagnetism is an interdisciplinary branch of geophysics that studies directional variations (declination [D] and inclination [I]) and changes in intensity (F) in the magnetic field in given areas by means of the analysis of the magnetic record in archaeological materials (Aitken 1990). SV measures regional changes in direction and intensity of the magnetic field on the order of 10^2 to 10^3 years. Thanks to direct recordings of SV in western Europe over the past four centuries (Alexandrescu, Courtillot, and Le Mouël 1997), we know that the magnetic field has varied over the past 150 years at a rate of ~0.20°/yr for middle latitudes (McElhinny and McFadden 2000). In addition to being an erratic phenomenon, SV has a regional character, considered inhomogeneous for regions of greater than 500 km in radius.

Logically, in order to determine past changes in the Earth's magnetic field, it is necessary to use geological and/or archaeological materials susceptible to having registered these variations in magnetization over time. The archaeomagnetic method is based on the fact that a material heated to temperatures over the Curie point of ferromagnetic materials present in its composition (e.g., 580°C for magnetite), acquires a thermoremanence (TRM) at the moment of cooling. This TRM represents an instant in the history of the Earth's magnetic field and can be very stable through time if the material remains in situ and undergoes no further thermal and/or chemical events that would reset the direction and intensity of the original magnetism. However, secondary magnetizations (e.g., viscous remanent magnetization [VRM]) acquired after heating can be superimposed on the original TRM recorded during the last firing of the material. Occasionally these secondary magnetizations can partially or totally destroy the direction of the primary or characteristic remanent magnetization (ChRM). However, the demagnetization procedures used in the laboratory often permit elimination of these secondary magnetic components.

One cannot obtain an archaeomagnetic date if there is no previously established, region-specific master curve of SV with which to compare the direction and/or intensity of the combustion structure (hearth, bonfire, etc.) being analyzed. Such a curve is created with the largest possible number of archaeomagnetic data (D, I, F) that can be obtained by analyzing remanent magnetism in burnt archaeological materials that were previously dated by other methods (notably radiocarbon). The larger the number of archaeomagnetic points there are in a master SV curve, the more accurate will be the archaeomagnetic dates that can be produced for a region.

In recent years, European archaeomagnetic specialists have made considerable efforts to develop master SV curves in several regions of the continent. At present, there exist master SV curves in Austria (Schnepp and Lanos 2006), Bulgaria (Kovacheva, Jordanova, and Karloukovshi 1998), France (Thellier 1981; Bucur 1994; Gallet, Genevey, and Le Goff 2002), Germany (Schnepp and Lanos 2005), Hungary (Márton 2003; Márton and Ferencz 2006), Iberia (Gómez-Paccard et al. 2006), Italy (Tema, Hedley, and Lanos 2006), and the United Kingdom (Batt 1997; Zananiri et al. 2007). With the exception of Bulgaria, whose curve covers the past eight millennia, the rest of the records do not go beyond the past three millennia. For Spain in particular the extant curve spans the period between 850 BC and AD 1960 (Gómez-Paccard et al. 2006). The curve is open to the addition of new data, but specialists are already beginning to obtain the first archaeomagnetic dates for Iberia (e.g., Catanzariti et al. 2008; Gómez-Paccard and Beamud 2008; Ruiz-Martínez, Pavón-Carrasco, and Catanzariti 2008). It is thus important to obtain more and especially older data on SV in Iberia in order to be able to extend the SV curve back in time and thus to date older archaeological sites by archaeomagnetism as soon as possible.

With the goal of extending the application of archaeomagnetic methods to older archaeological materials and adding new SV data to the Iberian curve, we present here the results from a study of burnt Holocene levels in El Mirón Cave (Ramales de la Victoria, Cantabria). The presence of many burnt levels that are well dated by radiocarbon and apparently well preserved indicates that this is a good context in which to obtain new archaeomagnetic data. In addition we conducted detailed rock-magnetic experiments to determine the properties of natural remanent magnetization (NRM) as well as the nature and stability of mineralogical phases that carry the archaeomagnetic signal. The study of the magnetic properties (composition, concentration, and granulometry of the magnetic minerals) of these burnt facies can provide information relevant to the use and preservation of the fire structures and lenses in the Holocene strata of the site.

The Archaeological Site and Chronological Framework

El Mirón Cave (43°14'48" N; 3°27'5" W) is located in the upper valley of the Río Asón on the northern edge of the Cantabrian Cordillera (Spain) at about 260 m above present sea level. The archaeological excavations directed by L. G. Straus and M. R. González Morales since 1996 have revealed a long cultural sequence with levels spanning the late Middle Paleolithic through the early Bronze Age (plus traces of medieval and modern human activity). The site has 65 radiocarbon dates, ranging from 41,000 BP (uncal.) to AD 1400 (Straus and González Morales 2003, 2007, and unpublished data), making it one of the most complete and best-dated cave site sequences for the late Upper Pleistocene and Holocene in western Europe.

The excavations have been centered on two areas, each about 9–10 m², in the vestibule of the cave, connected by a 9 × 1 m trench (figure 7.1). Our interest in this work lies in the major presence of burning episodes in the Holocene stratigraphy of the cave, specifically in the Neolithic, Chalcolithic, and early Bronze Age levels. The Neolithic has been identified in the outer vestibule (OV) Cabin area (Levels 10–8) and in the western half of the Mid-Vestibule (Levels 303.3–303) Trench and consists principally of clearly burned layers defined by

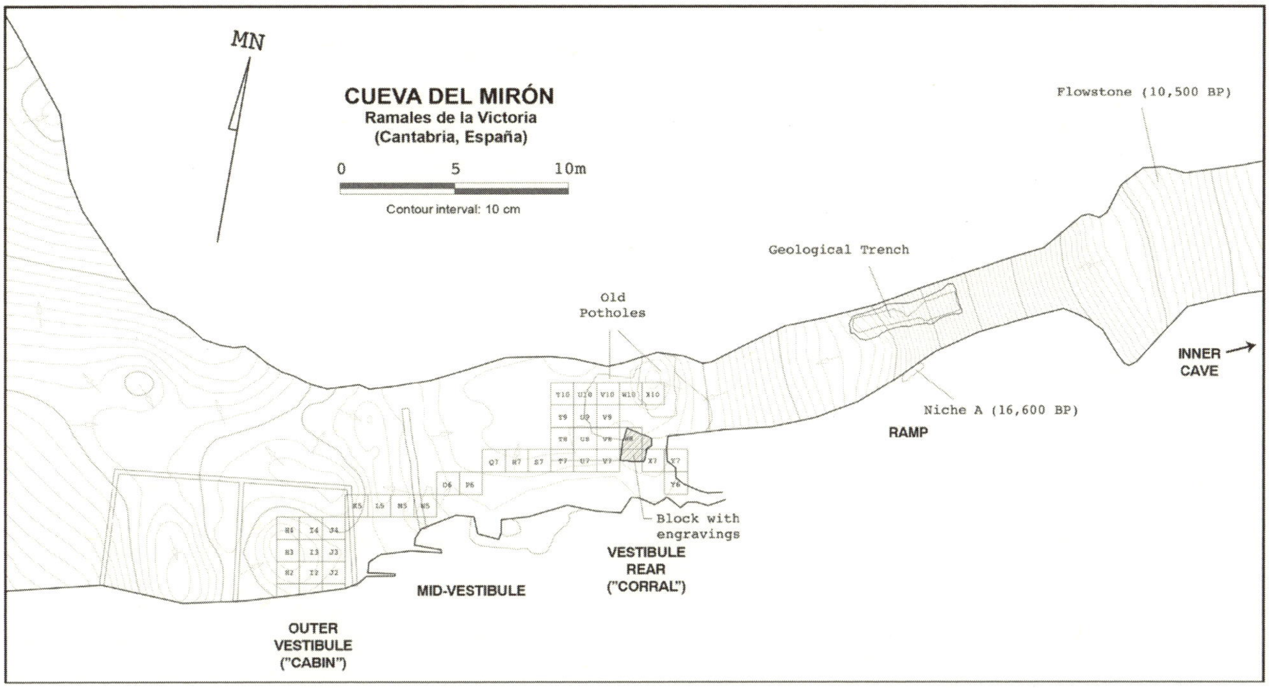

Fig. 7.1. Plan of El Mirón Cave vestibule, showing excavation areas. Topography by E. Torres, modified by Ronald Stauber.

the presence of ceramic and lithic artifacts and faunal remains. In Level 303.3 an individual carbonized grain of wheat was dated to 5500 BP (ca. 4400 cal BC), constituting among the oldest evidence of agriculture in the Cantabrian region (Peña-Chocarro et al. 2005). The Mid-Vestibule Trench levels physically correlate with OV Levels 10–8, dated between about 4600 and 3500 cal BC (Straus and González Morales 2003). The upper part of the Neolithic sequence is not clearly dated, and the only date (ca. 3700 cal BC [GX-28211]; see table 7.1) is only approximate and indeed problematic, since it was obtained from charcoal at the base of a large pit (Feature 98a), possibly corresponding to the base of Level 302, which itself is disturbed. This pit may have continued to be used and modified during Chalcolithic or Bronze Age times. For these reasons, this date is not included in this study.

The Chalcolithic levels (7–4) in the OV consist of a massive succession of ash and charcoal lenses. They are dated to the Chalcolithic by radiocarbon assays and by the typology of artifacts (including characteristic arrowheads), with an age of around 2500 cal BC (Straus and González Morales 2003). In specific parts of these levels, as exposed in the three stratigraphic sections of the Cabin area, it is possible to observe the presence of pits or alteration surfaces, thereby possibly explaining the inversion of dates GX-22130 and 24460 (see table 7.1; Straus and González Morales 2003). Finally, OV Levels 2 and 3 correspond to the early Bronze Age, with an age estimated at around 2100 cal BC (Straus and González Morales 2003). They contain ceramic sherds, lithic artifacts, remains of domesticated (and wild) animals, possible lumps of slag, a copper pin, and ample evidence of in situ burning, some of which might have been related to in situ metallurgical activities.

In the fire features studied here one can often observe lenses of white or gray ash, 2–10 cm thick, lying atop rubified facies of dark color containing charcoal and other organic matter. This kind of layered sedimentation is characteristic of caves that served as stables, where sheet burning of livestock (primarily ovicaprine) dung and straw was repeatedly done to clean out parasites (Brochier, Villa, and Giacomarra 1992). In the archaeological literature such fire lenses are known as *fumiers* (literally "dung" in French) or "animal burnt layers" (in English), and their presence is well known in Mediterranean Europe beginning in the Neolithic (e.g., Brochier 1983; Brochier, Villa, and Giacomarra

1992; Courty, Macphail, and Wattez 1991; Boschian and Montagnari-Kokelj 2000; Alday, Fernández, and Yusta 2003; Karkanas 2006; Vergès et al. 2008).

With the objective of being able to suggest an age for each fumier (or ash-rubefaction couplet), the available radiocarbon dates for the respective study levels are listed in table 7.1. Knowledge of the age of material is critically necessary in order to be able to add information to the master SV curve. The *terminus post* and *ante quem* dates provided by radiocarbon assays establish maximum and minimum limits for the last heating episodes of the burning lenses and hearths. As a chronological criterion we took into consideration the upper and lower calibrated dates in BC at ±1δ for the levels bracketing each fumier (table 7.3). According to this criterion, several sampled fumiers between Mid-Vestibule Levels 303.3 and 303 share the same upper and lower ages. Fortunately, the lateral continuity of some of these facies permits establishment of stratigraphic correlations or differentiations among them. Thus the six dates corresponding to the initial Neolithic levels (303.3–303 and 10) suggest rapid sedimentation (Straus and González Morales 2007), with formation in certainly less than 500 years. This fact, together with the overall high rate of sedimentation calculated for all the Holocene levels (3 cm/100 yrs. [Straus et al. 2001]), favored the preservation of discrete fire lenses.

Materials and Methods

SAMPLING

We took 191 oriented archaeomagnetic samples from a total of 15 burning features (fumiers, hearths), for an average of 10–15 samples per feature (table 7.2). These features correspond to Bronze Age, Chalcolithic, and Neolithic levels exposed in the sections of the Cabin and Mid-Vestibule Trench (figure 7.1; table 7.2). The samples were taken in both the ash and underlying rubified lenses, which were generally well preserved. Sampling was done by means of a nonmagnetic cylindrical piston specially designed for soft lithologies and incorporating a built-in orientation system. The oriented samples were saved in cylindrical plastic capsules (Ø 16.5 mm, L 17 mm, vol. ~3.6 cm^3) and stored under cold conditions (~3°C) to avoid chemical alterations. Other samples representative of each facies were injected by the same means into plaster capsules for thermal demagnetization experiments. The NRM of the plaster recipient is virtual nonexistent compared to the magnetization of the archaeological samples (normally ranging between 1 percent and less than 4.7 percent in the worst cases). The plaster samples were impregnated with a solution of sodium silicate (water glass) to consolidate them.

METHODS

NRM and bulk magnetic susceptibility were measured with a 2G SQUID magnetometer (noise level ~5 × 10^{-12} Am^2) and a Kappabridge KLY-4 (AGICO, noise level ~3 × 10^{-8} SI), respectively, at the Paleomagnetism Laboratory of the University of Burgos (Spain). The alternating field (AF) demagnetization of the NRM was carried out in 15–17 steps until maximum fields of 100–120 mT. The thermal demagnetization of NRM was carried out in 16 steps until temperatures of 600–615°C with a thermal demagnetizer TD48-SC (ASC).

Several rock-magnetic experiments were performed on pilot samples from each facies present in the fire levels at the Fort Hoofddijk Paleomagnetic Laboratory of the University of Utrecht (Netherlands). The temperature dependence of magnetic susceptibility was determined for representative samples up to 700°C in air with a KLY3-CS3 system at heating and cooling rates of 6–10°C min^{-1}. An alternating gradient magnetometer (MicroMag Model 2900, noise level ~2 × 10^{-9} Am^2) was used to measure hysteresis loops, isothermal remanent magnetization (IRM) acquisition, and back-field demagnetization curves (∀ 1.6 T) as well as FORC (First Order Reversal Curves) diagrams on bulk sample (~50 mg).

Results and Discussion

NRM STABILITY

The representation of diagrams with archaeomagnetic directions has been done by means of orthogonal demagnetization NRM plots (figure 7.2) and stereographic projections (figures 7.6 and 7.7). The resultant diagrams are generally of good quality and thus are interpretable. The characteristic directions (ChRM) and their statistical parameters were calculated using linear regression and Fisher (1953) statistics.

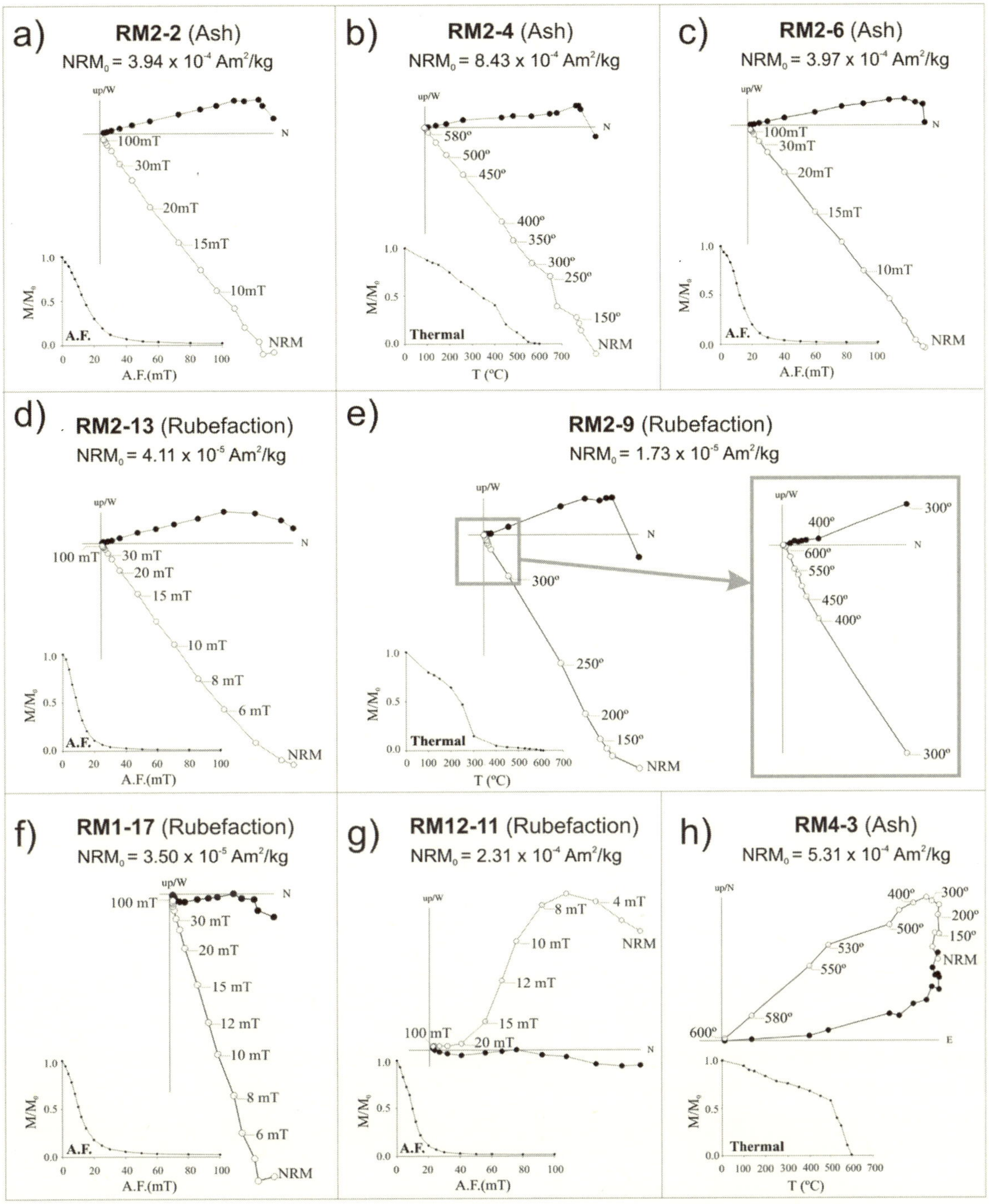

Fig. 7.2. Orthogonal NRM demagnetization plots for representative ashes and rubefactions from the following structures: (a–e) *fumier* RM2 (outer vestibule [Cabin] Level 9.8); (f) *fumier* RM1 (between outer vestibule Levels 6.1–7); (g) *fumier* RM12 (Mid-Vestibule Levels 301.2–301.0); (h) *fumier* RM4 (between outer vestibule Levels 3–5). AF demagnetization diagrams include two ashes (a,c) and three rubefactions (d,f,g). Thermal demagnetization plots for two ashes (b,h) and a rubefaction (e) are also shown. Open circles show projections of vector endpoints onto the vertical plane. The intensity (NRMo), sample code, facies, and demagnetization spectra are indicated for each sample. The decay curves represent the normalized intensity (M/Mo) after their respective demagnetization step. In the right part of Figure 7.2e, the final steps of the diagram are blown up to denote the presence of a high-temperature component between 400 and 600°C. Figures 7.2g and h represent two specimens with anomalous NRM behavior. See text for explanation. Graphs by A. Carrancho.

Two types of magnetic behavior have been distinguished among the samples as functions of the structure of the NRM and of the different sedimentary facies analyzed. All the samples display a secondary component of viscous character that was easily removed in the first steps of magnetic cleaning (<10–15 mT per A.F; <150–250°C by thermal demagnetization). This secondary component is especially notable in the rubefactions, and part of it is probably acquired during manipulation of the samples in the laboratory. In any event, it is lacking in interest for our purposes. The ashes, save for some exceptions that we will discuss later, are characterized as presenting a well-defined magnetic component that decreases univectorially toward the origin point and has a relatively high intensity that is practically demagnetized at 80–100 mT (figure 7.2a, c). The ashes thermally demagnetized define their ChRM between 250°C and 580–600°C (figure 7.2b). The maximum unblocking temperatures (T_{ub}) in the NRM thermal demagnetization curves are produced around 580–600°C, suggesting the presence of magnetite or slightly oxidized magnetite (maghemite).

Rubefactions demagnetized by alternating fields (AF) display either a single magnetic component (figure 7.2d) or several partially overlapping components (figure 7.2f). As in the case of ashes, the greater part of the NRM is carried by minerals of low coercivity (magnetite/maghemite). The thermal demagnetization revealed the occasional presence of partial thermoremanences (pTRM) in the rubefactions. If the heating temperature does not exceed the Curie point of the ferromagnetic material present (e.g., 580°C for magnetite; 680°C for hematite), the sample acquires a pTRM. The interesting thing from the archaeological viewpoint is that the unblocking temperature (T_{ub}) of the pTRMs allows us to estimate the maximum heating temperature that the rubified sedimentary facies underwent during burning. This is visible in the abrupt fall in intensity accompanied by the change in direction of the vector magnetization, which in figure 7.2e is produced at 400°C and which in other rubefactions we have observed between 350–450°C. The magnetic component of high temperature (~400–600°C [figure 7.2e]) can be interpreted as the original substrate magnetization.

We have identified anomalous behaviors in the demagnetization of the NRM of some ashes and rubefactions. These have to do with samples whose direction is badly defined, those with an NRM structure that is not univectorial, or those that have an anomalous direction (e.g., figures 7.2g and 7.2h). These samples often present the lowest Koenigsberger ratio (Q_n) values and have mainly (but not exclusively) been identified in fumiers of the outer vestibule area. Although these samples have not been included in the calculation of the mean direction, in the section of this chapter on archaeomagnetic directions we will discuss the possible causes of this behavior.

ROCK-MAGNETIC PROPERTIES

NRM and Magnetic Susceptibility

In figure 7.3 we plot the initial intensities of the NRM (A/m) against magnetic susceptibility (SI) on a logarithmic scale. The values of the NRM fall between 0.016 and 1.2 A/m (between 2.2×10^{-5} and 1.6×10^{-3} Am^2kg^{-1} expressed by unit of mass), while the susceptibility oscillates between 2.3×10^{-4} and 1.7×10^{-2} SI (between 1.7×10^{-5} and 2.2×10^{-7} m^3kg^{-1}). Within its range of variability the highest values are found among ashes, in contrast to the more moderate values of the rubefactions. The Koenigsberger ratio (Q_n [cf. Stacey 1967]), which quantifies the ratio between remanent and induced magnetization (Q_n = NRM / χH, where H.40 Am^{-1} is the present, local geomagnetic field strength), displays values that oscillate between 1 and 10 (figure 7.3). Despite the relatively low values in comparison with those obtained in studies of ovens or adobe bricks between 1 and 100 (e.g., Gómez-Paccard et al. 2006), the remanent magnetization is still higher than the induced magnetization (Q_n ratio >1; Q_n mean of ashes =3.15; Q_n mean for rubefactions =2.003). The resultant values are within the range for hearths and burned floors (Linford and Canti 2001; Schnepp et al. 2003) and are very similar to those obtained by Herries, Kovacheva, and Kostadinova (2008) in a medieval oven in Bulgaria and practically identical to Neolithic fumiers from Mirador Cave in the Sierra de Atapuerca (Burgos), some 125 km south of El Mirón on the uplands of Old Castile (Carrancho et al. 2009). Except for samples in which we could observe a relationship between low values for Q_n with

Fig. 7.3. NRM intensity (A/m) plotted versus bulk magnetic susceptibility for three representative SI on a logarithmic scale for ashes (black circles) and rubefactions (gray circles). Koenigsberger ratio isolines are also shown. Graphs by A. Carrancho.

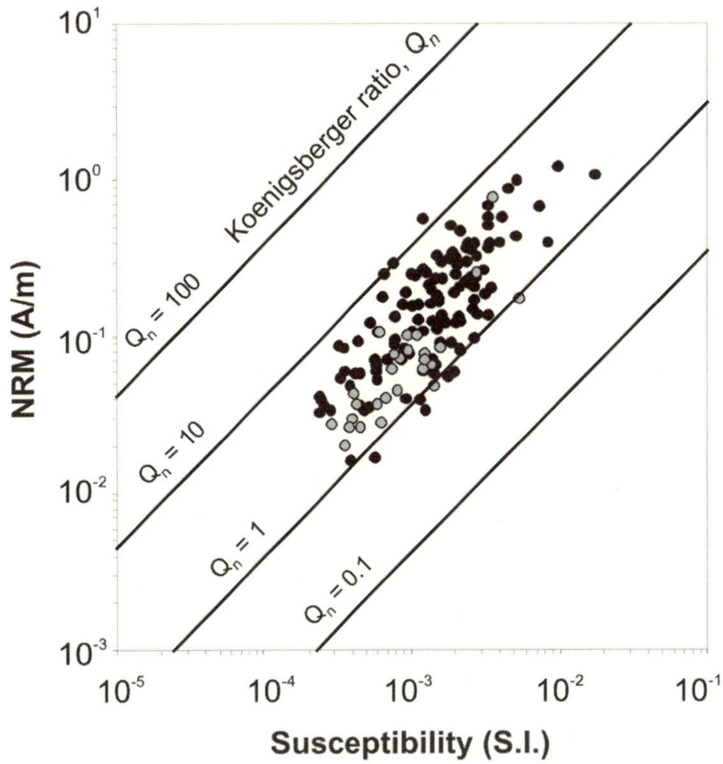

evidence of mechanical alteration of the sediments, the values presented here can in most cases be considered to be carriers of a thermoremanent magnetization.

Thermomagnetic Curves (K-T)

In pilot samples of ashes and rubefactions we measured the temperature dependence of the magnetic susceptibility (figure 7.4). This experiment permitted us to determine the ferromagnetic mineralogy that is present and gave us information on the conditions that produced the combustion in these fires.

The thermomagnetic curves we studied display heterogeneous behavior. Figures 7.4a–c show examples of the thermomagnetic curves of three ash lenses and figures 7.4d–f show examples for three rubefactions.

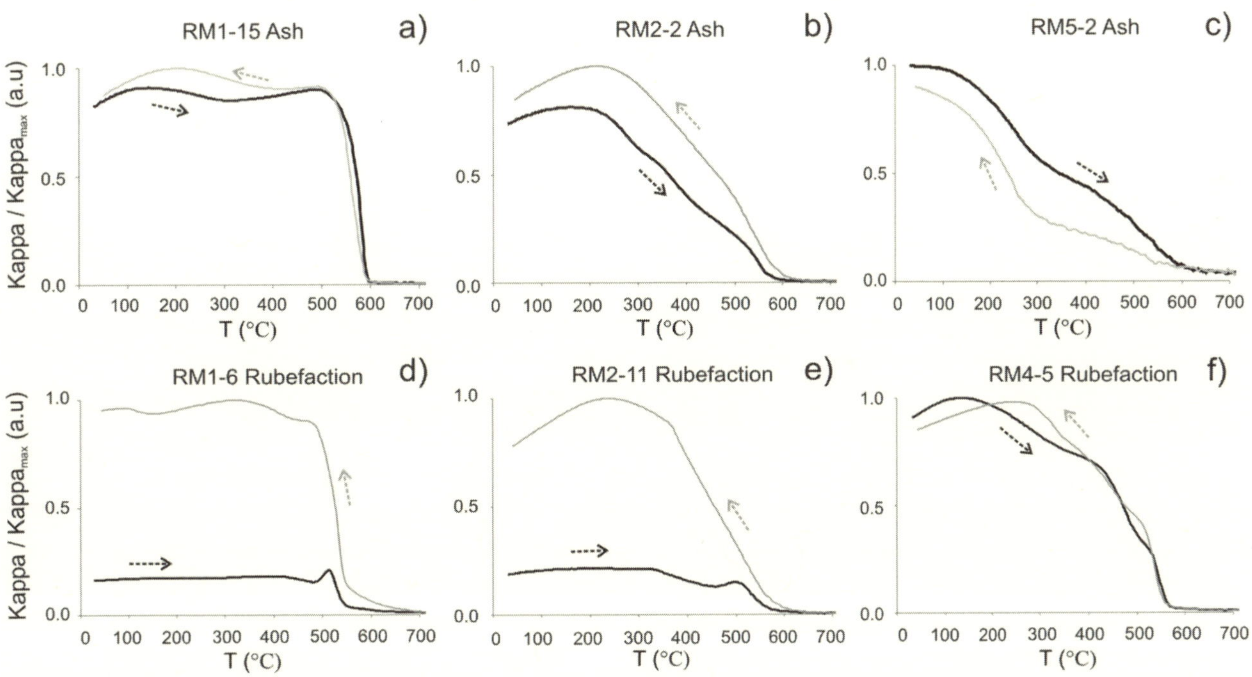

Fig. 7.4. High-temperature variations of low-field magnetic susceptibility for three representative ashes (a–c) and three rubefactions (d–f). Heating curves are plotted with a thick black line and cooling with a thin gray line, with their respective arrows. Sample code is also indicated. Curves are normalized by their maximum susceptibility. Graphs by A. Carrancho.

A general characteristic is that the greatest fall in susceptibility during heating took place around 580°C, which suggests that the main carrier of magnetization is magnetite or a Ti-poor titanomagnetite. In ash sample RM5–2 (figure 7.4c) this fall occurred at temperatures slightly above 580–600°C, probably indicating the presence of a slightly oxidized magnetite (maghemite). Such Curie points maintain coherence with maximum unblocking temperatures observed in the thermal demagnetization of the NRM. Among the heating curves of the ashes and rubefactions one can see falls at intermediate temperatures (~250–350°C), suggesting the presence of unstable maghemite or, less likely, ferromagnetic sulphurs and/or magnetite with high Ti, Al, or Mg content.

The degree of reversibility among the heating and cooling curves, indicative of the thermal stability of a sample, is greater among the ashes than among the rubefactions. The most irreversible samples are rubefactions RM1–6 and RM2–11 (figures 7.4d–e), where one can see a substantial increase in susceptibility during cooling between 580°C and 400°C. This lack of reversibility, caused by chemical changes responsible for the formation of magnetite, is certainly the result of the samples (figure 7.4d–e) not having undergone high heating temperatures, at least not the temperatures we used in this experiment (700°C). Rubefaction RM4–5 (figure 7.4f) has a much more reversible character, probably because this sample also contains ash. The ash samples, for their part, display a more stable and relatively reversible character, especially when compared with the rubefaction samples. Ash sample RM1–15 (figure 7.4a) is highly reversible, and the cooling curve of ash sample RM5–2 (figure 7.4c) has susceptibility values lower than in its heating curve, no doubt because it was heated to high temperatures.

Hysteresis Measurements

The form of the hysteresis cycles that were obtained—practically closed at around ∀ 150 mT (figure 7.5a)—indicates the dominant presence of a mineral with low coercivity (magnetite/maghemite). The intensity of magnetization is notably greater in the most-burned facies (ashes) compared to the rubified substrate. In the example shown in figure 7.5a, which corresponds to Neolithic fumier Lens 9.8, ash sample RM2–2 has a degree of magnetization four to five times greater than its associated rubefaction sample (RM2–11). This phenomenon is characteristic of all the samples studied.

After correcting for the effect of the paramagnetic signal, we plotted the parameters of the Day diagram (M_{rs}/M_s vs. B_{cr}/B_c [Day, Fuller, and Schmidt 1977]) in

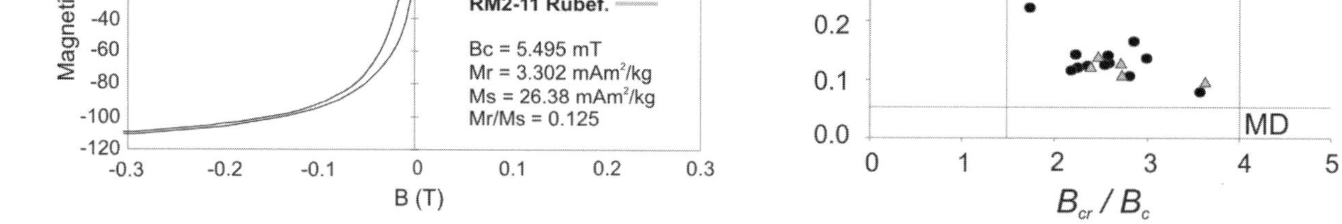

Fig. 7.5. (a) Hysteresis loops for two characteristic samples from the outer vestibule Neolithic Level 9.8. The central parts of the loops are shown up to ±300 mT. The loops are displayed after paramagnetic correction and on a mass-specific basis. Sample code and hysteresis parameters are listed as well for both examples. (b) M_{rs}/M_s vs. B_{cr}/B_c for representative ashes (black circles) and rubefactions (gray triangles) from different Holocene burnt levels. SD (single domain), PSD (pseudo-single domain) and MD (multi-domain) regions are indicated according to Day, Fuller, and Schmidt (1977). Graphs by A. Carrancho.

order to determine the magnetic domain state of the samples (figure 7.5b). Both the ashes and the rubefactions are grouped together in the area of the pseudo-single domain (PSD), although clear differences cannot be perceived between the facies. This result was confirmed by FORC diagrams (not shown here).

Considering the similarity of the magnetic properties of these burnt facies, it is highly probable that both the conditions of combustion and the types of fuel used were similar among the fumiers we studied. The presence of organic matter favored the necessary conditions for a reducing atmosphere, thereby causing the formation of magnetite. This magnetite can present itself as partially oxidized (maghemite), insofar as there is an increased intake of oxygen during cooling. Similar studies done on stabling levels in El Mirador Cave at Atapuerca (Burgos) (Carrancho et al. 2009) reflect very similar composition and magnetic domains states.

According to archaeological and ethnographic observations (Shahack-Gross et al. 2003; Karkanas 2006), the fuel used in this kind of fire is a mixture of grasses (straw), wood (sticks), and animal dung. It would be interesting to determine whether different fuels have sufficiently distinctive magnetic properties to be able to differentiate among them in archaeological settings. What is clear is that the ash lenses show higher concentrations of magnetic minerals than the rubified lenses below them. These magnetic results have potential value for the study of the formation of fire-rich archaeological deposits and for evaluating the degree of preservation vs. alteration that they may have undergone.

Archaeomagnetic Directions

Of the 15 features or lenses sampled, we report 8 well-defined directional data with values that fall within the expected range for the SV. The mean directions we obtained, their associated statistics, and the respective age intervals are all detailed in table 7.3. In figure 7.6 the archaeomagnetic directions are represented in equal-area projections together with their mean directions and error circles at the 95 percent confidence level (α_{95}). In figure 7.7, we show these directions together with their confidence circles and their corresponding age intervals. The mean direction is calculated with a minimum of 7 samples; the dispersion parameter (k) is reasonably acceptable, and in 7 of 8 sites the α_{95} is between 4.3 and 6.80 (table 7.3). In 7 of the 15 studied features or lenses it was not possible to obtain an archaeomagnetic result because of anomalous magnetic behavior among the majority of their samples, as will be explained below. We also analyze the archaeomagnetic data and evaluate their implications, both archaeomagnetic and archaeological.

Our research strategy at the site involved sampling every feature that had been potentially burned or shown signs of having undergone heating, even though the intensity and degree of preservation thereof were variable. However, there is a clear correlation between

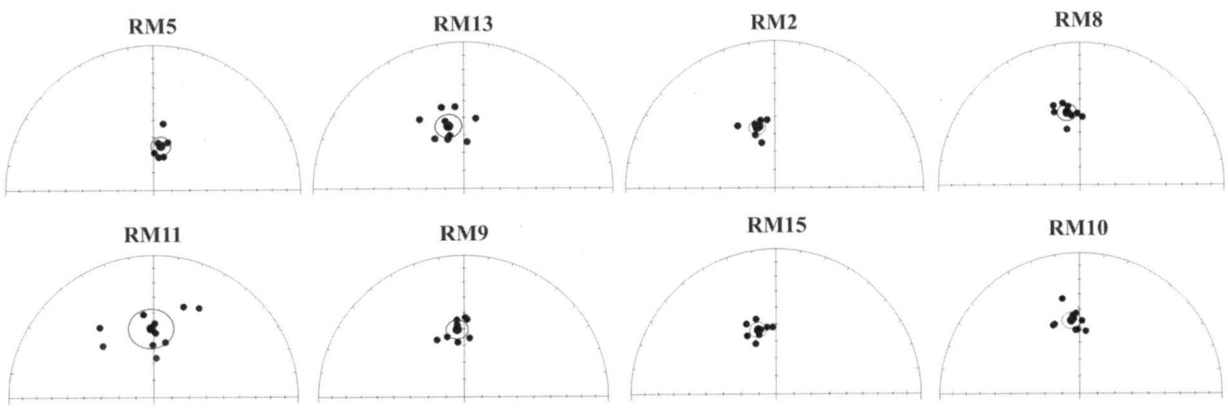

Fig. 7.6. Directions of the ChRM determined per sample, mean direction, and error circle (α95) calculated for the structures compiled in table 7.3 and considered representative of the secular variation. Graphs by A. Carrancho.

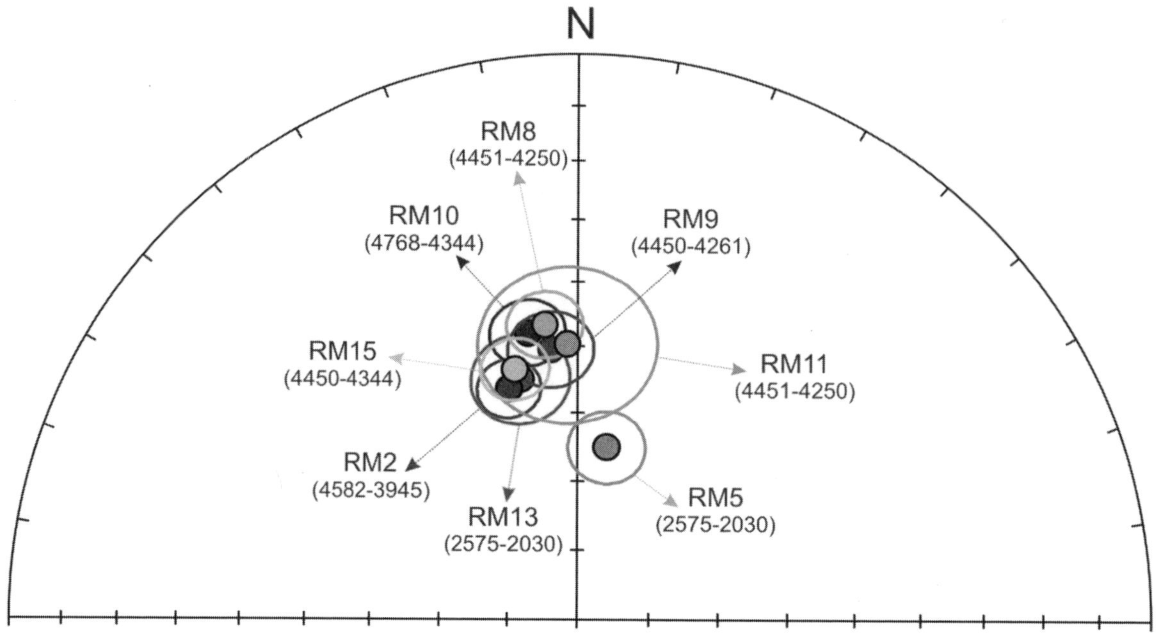

Fig. 7.7. Equal-area projection showing the mean directions, the confidence circles (α95), and the temporal interval considered (cal yr BC) for every structure reported in table 7.3. Graphs by A. Carrancho.

the quality of the archaeomagnetic data obtained and good feature preservation. Among the features in which anomalous magnetic behavior was identified (these being omitted from table 7.3 and figures 7.6 and 7.7, namely ashes with demagnetization diagrams displaying non-univectorial NRM, anomalous archaeomagnetic directions, or high directional dispersion), we have observed evidence of mechanical alteration of the sediments, such that the material was no longer in situ. There are several indications of mechanical disturbance: the presence of ashes mixed with non-burned sediments, the occasional absence of or discontinuities in rubified facies, a lesser thickness of the ashes, and especially low values for the Q_n ratio, indicating that these do not carry a stable TRM. All these factors are particularly influential in being able to obtain archaeomagnetic data from these materials.

Some fumiers with anomalous directions (e.g., RM3, between Cabin Levels 9.6 and 9.7) do not have their corresponding underlying rubefaction facies. In other cases the rubefaction facies was discontinuous or badly delimited (e.g., RM4, also in the Cabin area) or the ashes (or at least some of them) also contained unburned sedimentary matrix material (e.g., RM6, 7, 12, and 14 in the Mid-Vestibule Trench). All these observations are indicative of alteration that occurred after the TRM had been acquired, probably by mechanical disturbance of the sediments. This interpretation is justified by the fact that the samples that are not pure ash display the lowest values of the Q_n ratio, due to their having lost their TRM through disorganization of their magnetic moments. The anomalous directions in apparently stable diagrams (e.g., figures 7.2g–h) can be explained by mechanical movements of particles, which, although produced on a small scale, can perfectly well generate spurious directions. This seems to be the case with sampling site RM1 (between Cabin Levels 6.1 and 7), where the majority of the samples display anomalous directions despite the presence of well-preserved rubefaction facies.

In this sense it is logical to imagine that some alteration process had affected Cabin area features RM1, 3, and 4, where we were unable to obtain any archaeomagnetic results. This is particularly visible in RM4, where in fact we could clearly see some evidence of reworking of the ashes. In Mid-Vestibule sampling sites RM6, 7, 12, and 14 (see table 7.2) the alteration was at least partial, with significant portions of the ashes mixed together with unburned sediments and giving Q_n ratio values that are particularly low and where the numbers of individual samples accepted for the calculation of the mean direction (*N*) does not exceed 50 percent of the number of samples taken from the studied feature (*N*=). Considering that these are non-lithified materials and that the magnetization resides in micrometric particles with a certain degree of mobility, there probably had been some deformation effect and/or mechanical alteration after acquisition of the TRM that was able to totally or partially affect the remanence (NRM) of some of these features.

In table 7.3 we show the results obtained from those features that display coherent directional behavior and that we consider to be representative of SV. Specifically, sampling sites RM8, 9, 10, and 11, located among Mid-Vestibule Neolithic levels 303.3–303, all display very similar directional results in keeping with their temporal range (table 7.3; figures 7.6 and 7.7). These fumiers are stratigraphically almost continuous, contain thick (~10 cm) ash facies, and present very clear rubefaction facies. Although the statistics for RM11 could be better, its mean direction is well defined and its confidence circle (α_{95}) encompasses the rest of the mean directions from features of similar age (figure 7.7). Feature RM15 (Mid-Vestibule Level 303.3) displays a very similar direction, although its ash facies is no more than 4–5 cm thick, and feature RM2 (Cabin Level 9.8) has a direction that is very compatible with the contemporaneous loci studied in the Mid-Vestibule Trench (table 7.3). We have no reason to suspect that the magnetic directions of the stratigraphically lower fumiers would reflect remagnetizations caused by heating among overlying ones. Linford and Canti (2001) show experimentally, with a system of thermocouples to control for temperature, that the downward penetration of heat is not significant beyond just a few centimeters of depth. Even after repeated burning episodes after which the ashes were removed to eliminate a potential insulation effect, Carrancho and Villalaín (2008) show in similar experiments that heating does not exceed 150–200°C below 4 cm of depth underneath a fire surface.

For fumier RM10 (under Mid-Vestibule Level 303.3), we took the minimum age (t_{min} = 4768 cal BC) to be the lower limit of radiocarbon date GX-25856 (see table 7.1). This fumier lies directly atop Mesolithic Level 304, whose age is about 3 kyr older than the initial Neolithic due to the existence of a sedimentary hiatus (Straus and González Morales 2003). The proposed $t_{\min}$ seems fairly reasonable given the standard deviations of the other available dates (see table 7.1).

The directions we obtained for features RM5 and RM13 (between Levels 3–5) deserve some special comment. What one would expect upon sampling distinct features in the same level in different stratigraphic sections, or ones that are at least similar in age, would be that they have very similar magnetic directions. However, fumiers RM5 and RM13 display directions that are clearly different (table 7.3; figures 7.6 and 7.7). The wide temporal range (~500 yr [table 7.3]) that is provided by the chronological intervals conceivable for these features does not allow us to determine whether or not we are dealing with two temporally separate burning episodes despite the fact that they share the same mean age (t_{mean} = 2302 cal yr BC). It would be desirable to have another absolute date that would permit us to determine if the two features are synchronous or, if not, to specify the amount of time that might have separated the two burning episodes in question. Unable at present to clarify this chronological uncertainty, for now we have no solid basis for eliminating either one of the two archaeomagnetic directions. Nevertheless, it is true that sampling locus RM13 displays a degree of directional dispersion (figure 7.6), something that is perhaps related to a somewhat irregular geometry in its outlines, but the Fisherian distribution of its directions indicates that its mean direction is well defined.

The implications of these analyses are of great utility from both archaeomagnetic and archaeological perspectives. If we set aside Chalcolithic sampling loci RM5 and RM13, the other six archaeomagnetic directions we obtained correspond to Neolithic features that had formed during a fairly short period of time (table 7.3). Although this is something already indicated by the radiocarbon dates from these levels (table 7.1), the overlap reflected in the mean directions from the Neolithic features in figure 7.7 also suggests formation during a short period of SV. It is important to stress that these data are preliminary and that our intention is to take more samples from specific targeted levels to refine the directions and to reduce the confidence circles (α_{95}).

A more detailed discussion in which the El Mirón data would be compared with contemporaneous records of SV would seem premature at this point given the small number of available data and their limited chronological distribution. We must keep in mind that these are the oldest directional archaeomagnetic results yet obtained from burnt archaeological materials on the Iberian Peninsula. In the future and as we obtain new archaeomagnetic data of sufficiently good quality, from geographically nearby contexts and similar time periods, we will be able to integrate these results into the master curve for Iberia in order to be able to model SV and thus to actually date other sites by archaeomagnetism.

From the archaeological viewpoint, the vectorial analysis of these kinds of burning features permits us to determine their in situ (or disturbed) character and thus their level of preservation. This work has allowed us to determine the types of criteria that these kinds of fire lenses must meet in order to be able to get good archaeomagnetic results, namely: (1) high values for the Q_n ratio, (2) presence of all the obligatory facies in a combustion feature, (3) absence of evidence of mechanical alteration of the sediments, (4) greatest possible thickness of the ash lenses, and (5) a majority of demagnetization diagrams with univectorial NRM among the ash facies. For archaeomagnetic studies, any process susceptible to causing directional distortions in a studied feature, no matter how minimal, is critical to the validity of the results. Nevertheless, if post-depositional processes are not severe and do not provoke major displacements among the archaeological artifacts within the stratigraphy, there is no reason why the cultural interpretation of the archaeological record should be adversely affected. Even though the statistical parameters of the features presented in table 7.3 could be improved upon, the coherency of the mean directions and their scarce dispersion suggest a relatively acceptable level of preservation.

The analysis of NRM stability and the rock-magnetic experiments that were done on the burnt Holocene levels from El Mirón Cave have permitted us to magnetically characterize these facies. In general, they are dominated by magnetite produced by combustion in an essentially reducing atmosphere favored by the presence of organic matter. The identification of p-TRMs in the rubified facies underlying the ash lenses has permitted us to determine final combustion temperature (ca. 350–450°C). The greater instability observed in the thermomagnetic curves of the rubefactions confirms the partial heating that these facies underwent relative to the ash lenses.

This work has also allowed us to determine the conditions that must co-occur in order to obtain valid archaeomagnetic results. These are: high values of the Q_n ratio indicating a stable TRM, presence of all the facies, the greatest possible thickness of ash lenses with univectorial NRM demagnetization diagrams, and clear absence of any mechanical or chemical alteration processes that could have distorted the original archaeomagnetic direction. It has been possible to identify a stable magnetic component both in ash lenses and in rubefactions, thereby recording the Earth's magnetic field among these facies. Except for a few cases that did not meet the above-stated conditions, we have obtained eight satisfactory archaeomagnetic directions that are representative of SV. Although the number of data is low and their statistics could be improved, at the time of writing these are the oldest directional archaeomagnetic data obtained on burnt archaeological materials from the Iberian Peninsula. Finally, we would like to stress the utility of magnetic analyses for determining the in situ (or disturbed) character of burning features such as fumiers (ash + reddened sediment lenses from the sheet-burning of animal dung and straw) and for evaluating the processes of formation and preservation thereof in archaeological sites.

Acknowledgments

Ángel Carrancho acknowledges the financial support of the Fundación del Patrimonio Histórico de Castilla y León. This research has also been funded by the CGL2009-10840 project of the Spanish Ministry of Education and Science and the BU004A09 project of the Junta de Castilla y León. Thanks are also expressed to the University of Utrecht paleomagnetic laboratory, where some of these analyses were carried out. This chapter was translated into English by Lawrence Straus, who invited us to contribute it.

References Cited

Aitken, M. J.
1990 *Science-Based Dating in Archaeology*. Longman Archaeology Series. Longman, London.

Alday, A., J. Fernández, and I. Yusta
2003 Suelos de habitación-suelos de corrales: los casos de Atxoste y Los Husos. *Veleia* 20:183–225.

Alexandrescu, M., V. Courtillot, and J. L. Le Mouël
1997 High-resolution secular variation of the geomagnetic field in western Europe over the last four centuries: comparison and integration of historical data from Paris and London. *Journal of Geophysical Research* 102:20245–20258.

Batt, C. M.
1997 The British archaeomagnetic calibration curve: an objective treatment. *Archaeometry* 39:153–168.

Boschian, G., and E. Montagnari-Kokelj
2000 Prehistoric shepherds and caves in Trieste Karst (northeastern Italy). *Geoarchaeology* 15:332–371.

Brochier, J. E.
1983 Combustion et parcage des herbivores domestiques: le point de vue du sédimentologue. *Bulletin de la Societé Préhistorique Française* 80(5):143–145.

Brochier, J. E., P. Villa, and M. Giacomarra
1992 Shepherds and sediments: geo-ethnoarchaeology of pastoral sites. *Journal of Anthropological Archaeology* 11:47–102.

Bucur, I.
1994 The direction of the terrestrial magnetic field in France during the last 21 centuries. *Physics of the Earth and Planetary Interiors* 87:95–109.

Carrancho, Á., and J. J. Villalaín
2008 Estudio experimental de las transformaciones en la mineralogía magnética en fuegos prehistóricos y sus implicaciones en arqueomagnetismo. *Geotemas* 10:1151–1155.

Carrancho, Á., J. J. Villalaín, D. E. Angelucci, M. J. Dekkers, J. Vallverdú, and J. M. Vergès
2009 Rock-magnetic analyses as a tool to investigate archaeological fired sediments. A case study of Mirador Cave (Sierra de Atapuerca, Spain). *Geophysical Journal International*. 179:79–96.

Catanzariti, G., G. McIntosh, A. M. Monge Soares, E. Díaz-Martínez, P. Kresten, and M. L. Osete
2008 Archaeomagnetic dating of a vitrified wall at the Late Bronze Age settlement of Misericordia (Serpa, Portugal). *Journal of Archaeological Science* 35:1399–1407.

Courty, M. A., R. I. Macphail, and J. Wattez
1991 Soil micromorphological indicators of pastoralism; with special reference to Arene Candide, Finale Ligure, Italy. In *Archeologia della pastorizia nell'Europa Meridionale*, edited by R. Maggi, R. Nisbet, and G. Barker, pp. 1–4. Rivista Studi Liguri, vol. 57. Bordighera, Italy.

Day, R., M. Fuller, and V. A. Schmidt
1977 Hysteresis properties of titanomagnetites: grain size and composition dependence. *Physics of the Earth and Planetary Interiors* 13:260–267.

Fisher, R. A.
1953 Dispersion on a sphere. *Proceedings of the Royal Society, London* 217:295–305.

Gallet, Y., A. Genevey, and M. Le Goff
2002 Three millennia of directional variations of the Earth´s magnetic field in western Europe as revealed by archaeological artefacts. *Physics of the Earth and Planetary Interiors* 131:81–89.

Gómez-Paccard, M., and E. Beamud
2008 Recent achievements in archaeomagnetic dating in the Iberian Peninsula: application to Roman and mediaeval Spanish structures. *Journal of Archaeological Science* 35:1389–1398.

Gómez-Paccard, M., G. Catanzariti, V. C. Ruiz-Martínez, G. McIntosh, J. I. Núñez, M. L. Osete, A. Chauvin, P. Lanos, D. H. Tarling, D. Bernal-Casasola, J. Thiriot, and A. W. Group
2006 A catalogue of Spanish archaeomagnetic data. *Geophysical Journal International* 166:1125–1143.

Herries, A. I. R., M. Kovacheva, and M. Kostadinova
2008 Mineral magnetism and archaeomagnetic dating of a mediaeval oven from Zlatna Livada, Bulgaria. *Physics and Chemistry of the Earth A* 33(6–7):496–510.

Karkanas, P.
2006 Late Neolithic household activities in marginal areas: the micromorphological evidence from the Kouveleiki Caves, Peloponnese, Greece. *Journal of Archaeological Science* 33(11):1628–1641.

Kovacheva, M., N. Jordanova, and Karloukovshi
1998 Geomagnetic field variations as determined from Bulgaria archaomagnetic data. Part II: the last 8000 years. *Surveys in Geophysics* 19:431–460.

Linford, N. T., and M. G. Canti
2001 Geophysical evidence for fires in antiquity: preliminary results from an experimental study. *Archaeological Prospection* 8:211–225.

Márton, P.
2003 Recent achievements in archaeomagnetism in Hungary. *Geophysical Journal International* 153:675–690.

Márton, P., and E. Ferencz
2006 Hierarchical versus stratification statistical analysis of archaeomagnetic directions: the secular variation curve for Hungary. *Geophysical Journal International* 164:484–489.

McElhinny, M. W., and McFadden, P. L.
2000 *Paleomagnetism: Continents and Oceans*. Academic Press, London.

Peña-Chocarro, L., L. Zapata, M. J. Iriarte, M. González Morales, and L. G. Straus
2005 The oldest agriculture in northern Atlantic Spain: new evidence from El Mirón Cave (Ramales de la Victoria, Cantabria). *Journal of Archaeological Science* 32:579–587.

Ruiz-Martínez, V. C., F. J. Pavón-Carrasco, and G. Catanzariti
2008 First archaeomagnetic data from northern Iberia. *Physics of the Earth and Planetary Interiors* 33:566–577.

Schnepp, E., and P. Lanos
2005 Archaeomagnetic secular variation in Germany during the past 2500 years. *Geophysical Journal International* 163:479–490.
2006 A preliminary secular variation reference curve for archaeomagnetic dating in Austria. *Geophysical Journal International* 166:91–96.

Schnepp, E., R. Pucher, C. Goedicke, A. Manzano, U. Müller, and P. Lanos
2003 Paleomagnetic directions and thermoluminescence dating from a bread oven-floor sequence in Lübeck (Germany): a record of 450 years of geomagnetic secular variation. *Journal of Geophysical Research* 108:2078–2092. doi:10.1029/2002JB001975.

Shahack-Gross, R., F. Marshall, and S. Weiner
2003 Geo-ethnoarchaeology of pastoral sites: the identification of livestock enclosures in abandoned Maasai settlements. *Journal of Archaeological Science* 30:439–459.

Stacey, F. D.
1967 The Koenigsberger ratio and the nature of thermoremanence in igneous rocks. *Earth and Planetary Science Letters* 2:67–68.

Straus, L. G., and M. González Morales
2003 El Mirón Cave and the ^{14}C Chronology of Cantabrian Spain. *Radiocarbon* 45(1):41–58.
2007 Further radiocarbon dates for the Upper Paleolithic of El Mirón Cave (Ramales de la Victoria, Cantabria, Spain). *Radiocarbon* 49:1205–1214.

Straus, L., M. González Morales, W. Farrand, and W. Hubbard
2001 Sedimentological and stratigraphic observations in El Mirón, a late Quaternary cave site in the Cantabrian Cordillera, northern Spain. *Geoarchaeology* 16:603–630.

Stuiver, M., P. Reimer, E. Bard, J. Beck, G. Burr, K. Hughen, B. Kromer, G. McCormac, J. van der Plicht, and M. Spurk
1998 INTCAL98 radiocarbon age calibration, 24,000–0 cal BP. *Radiocarbon* 40:1041–1083.

Tema, E., I. Hedley, and P. Lanos
2006 Archaeomagnetism in Italy: a compilation of data including new results and a preliminary Italian secular variation curve. *Geophysical Journal International* 167:1160–1171.

Thellier, E.
1981 Sur la direction du champ magnétique terrestre en France durant les deux derniers millenaires. *Physics of the Earth and Planetary Interiors* 24:89–132.

Vergès, J. M., E. Allué, D. E. Angelucci, F. Burjachs, Á. Carrancho, A. Cebrià, I. Expósito, M. Fontanals, S. Moral, A. Rodríguez, and M. Vaquero
2008 Los niveles neolíticos de la cueva de El Mirador (Sierra de Atapuerca, Burgos): nuevos datos sobre la implantación y el desarrollo de la economía agropecuaria en la submeseta norte. In *Actas del IV Congreso del Neolítico en la Península Ibérica, Alicante, Spain*, vol. 1, pp. 418–427. Alicante, Spain.

Zananiri, I., C. M. Batt, P. Lanos, D. H. Tarling, and P. Linford
2007 Archaeomagnetic secular variation in the UK during the past 4000 years and its application to archaeomagnetic dating. *Physics of the Earth and Planetary Interiors* 160:97–107.

Table 7.1. ^{14}C dates from the sampled mid-Holocene levels at El Mirón Cave (from Straus and González Morales 2003, 2007)

Square	Zone[a]	Level	Period[b]	Date BP	SD	Lab nr.	Material	Method[c]	Calibrated date (BC)[d]
H-2	OV	3	BA	3700	40	GX-25851	Charcoal	AMS	2140–2030 BC
J-2	OV	5	Chal	3820	240	GX-22127	Charcoal	Conv	2575–1931 BC
I-3	OV	5.1	Chal	4120	50	GX-22130	Charcoal	AMS	2858–2586 BC
H-2	OV	7	Chal	3740	120	GX-24460	Charcoal	Cxcnt	2305–1963 BC
I-3	OV	8.1	Neo	4680	60	GX-22131	Charcoal	AMS	3612–3371 BC
J-2	OV	9	Neo	5170	170	GX-22128	Charcoal	Conv	4221–3789 BC
H-4	OV	9	Neo	5280	40	GX-24461	Charcoal	AMS	4217–4001 BC
I-4	OV	9.6	Neo	5250	150	GX-24462	Charcoal	Cxcnt	4318–3945 BC
I-3	OV	10	Neo	5570	50	GX-23414	Charcoal	AMS	4449–4359 BC
I-3	OV	10	Neo	5690	50	GX-23413	Charcoal	AMS	4582–4458 BC
O-6	MV	Pit 98a	Neo	4910	80	GX-28211	Charcoal	Conv	3773–3641 BC
L-5	MV	303	Neo	5500	90	GX-25854	Charcoal	Conv	4451–4250 BC
L-5	MV	303.1	Neo	5520	70	GX-25855	Charcoal	Conv	4451–4261 BC
M-5	MV	303.3	Neo	5550	40	GX-30910	Grain	AMS	4450–4344 BC
L-5	MV	303.3	Neo	5790	90	GX-25856	Charcoal	Cxcnt	4768–4540 BC

[a]OV = outer vestibule (Cabin); MV = Mid-Vestibule (trench)
[b]BA = Bronze Age; Chal = Chalcolithic; Neo = Neolithic
[c]AMS = accelerator; Conv = conventional; Cxcnt = extended count
[d]Stuiver et al. (1998); CALIB 4.1.2 (range at 1δ)

Table 7.2. Sampling and measurement details of the archaeomagnetic samples collected from El Mirón Cave Holocene burnt levels

Fumier code	Level	Zone[a]	N'[b]	Ashes	Rubefactions[c]	AF[d]	TH[e]
RM4	between 3 and 5	OV	7	6	1	5	2
RM5	between 3 and 5	OV	11	11	—	9	2
RM13	4	MV	11	11	—	9	2
RM1	between 6.1and 7	OV	19	11	8	12	7
RM12	301.2–301.0	MV	11	8	3	9	2
RM3	between 9.6 and 9.7	OV	10	10	n.p.	8	2
RM2	9.8	OV	13	7	6	11	2
RM6	303.1–303	MV	17	9	8	12	5
RM7	303.1–303	MV	18	9	9	14	4

Table 7.2. *(continued)*

Fumier code	Level	Zone[a]	N′[b]	Ashes	Rubefactions[c]	AF[d]	TH[e]
RM8	between 303 and 303.2	MV	12	12	—	12	—
RM11	303.1–303	MV	15	11	4	12	3
RM14	303.1	MV	12	12	—	10	2
RM9	303.3–303.2	MV	10	10	—	8	2
RM15	303.3	MV	14	14	—	11	3
RM10	beneath 303.3	MV	11	11	—	9	2
TOTAL			**191**	**152**	**39**	**151**	**40**

[a]OV = outer vestibule (Cabin); MV = Mid-Vestibule (trench)
[b]Number of independently oriented samples taken and analyzed from the site
[c]n.p = not present; — = no sample collected
[d]Alternating field demagnetization of the natural remanent magnetization (NRM)
[e]Thermal demagnetization of the natural remanent magnetization (NRM)

Table 7.3. Mean archaeomagnetic directions from El Mirón Cave Holocene burnt levels

Fumier code	Level	Zonea	t_{minb}	t_{maxc}	t_{meand}	N´/Ne	Declination (°)	Inclination (°)	a95 (°)f	kg
RM5	between 3 and 5	OV	2575	2030	2302	11/7	9.3	64.8	5.3	130.1
RM13	4	MV	2575	2030	2302	11/10	346.7	54.0	6.8	51.9
RM2	9.8	OV	4582	3945	4263	13/8	343.5	54.9	4.3	90.0
RM8	between 303 and 303.2	MV	4451	4250	4350	12/8	350.1	47.4	5.1	163.4
RM11	303.1–303	MV	4451	4250	4350	15/10	357.8	49.6	12.0	17.2
RM9	303.3–303.2	MV	4450	4261	4355	10/8	354.3	50.3	5.8	92.4
RM15	303.3	MV	4450	4344	4397	14/9	345.9	52.4	4.7	123.5
RM10	beneath 303.3	MV	4768	4344	4556	11/10	353.6	46.5	5.1	90.0

[a]OV = outer vestibule (Cabin); MV = Mid-Vestibule (trench)
[b]Minimum age of the structure in cal. years BC
[c]Maximum age of the structure in cal. years BC
[d]Mean age of the structure in cal. years BC
[e]Number of independently oriented samples taken and analysed from the site (*N'*) / samples taken into account in the calculation of the mean site direction (*N*)
[f]Precision parameter of characteristic remanent magnetization (Fisher 1953). See text for explanation.
[g]95 percent confidence limit of characteristic remanent magnetization (Fisher 1953). See text for explanation.

CHAPTER EIGHT

The Vegetational Contexts of El Mirón Cave throughout the Early to Middle Holocene

Contributions from the Palynological Study

María José Iriarte Chiapusso

TRANSLATED BY LAWRENCE GUY STRAUS

The cave of El Mirón (Ramales de la Victoria, Cantabria) contains an interesting archaeological deposit with important stratigraphic sections that attest to a human presence from at least the late Upper Pleistocene up to the present. This cave, with a vestibule that measures between about 16–18 m wide by 30 m deep by 13 m high and with a mouth that may be as high as about 20 m, opens out onto the limestone cliff that forms the western face of Monte Pando. It dominates a steep-sided valley where the rivers Gándara and Calera join and then in turn flow into the Río Asón at Ramales. At present, the cave is about 150 m above the valley floor at Ramales and about 260 m above sea level. The geographical location of the cave is manifestly strategic, dominating major avenues of communication between the Cantabrian coast and mountains and between the intermontane sector of the Asón Valley and the Meseta of Old Castile and the interior of Vizcaya as well as the high, enclosed mountain valley of Soba above Ramales (figure 8.1).

The Asón basin, in which El Mirón is located, currently harbors an interesting vegetational landscape because of its high, abrupt relief, diversity of morphological forms, and lithological and pedological substrates. The coastal zone (the present-day Atlantic shore is no more than 25 km from the cave) has been intensely affected by human activity, and the autochthonous arboreal taxa have been decimated by the progressive spread of plantations of eucalyptus, which are now prominent in this area of Atlantic heaths. As one goes up, following the altitudinal gradient, the arboreal vegetation becomes progressively dominated by Cantabrian oaklands in areas of limestone bedrock. It is those areas of the interior that are not specifically rocky uplands—namely the river valley floors and gentle lower slopes—that are preferentially used by humans for agriculture and especially livestock raising, which of course affect the kinds of deciduous woods that can survive there. Finally, in the highest elevations calcicolous beech woods replace the oaks.

Throughout this article we will refer to information obtained by means of palynological analysis of the vegetational environments that existed in the surroundings of El Mirón during its human occupations in the early to middle Holocene. Thus, the results

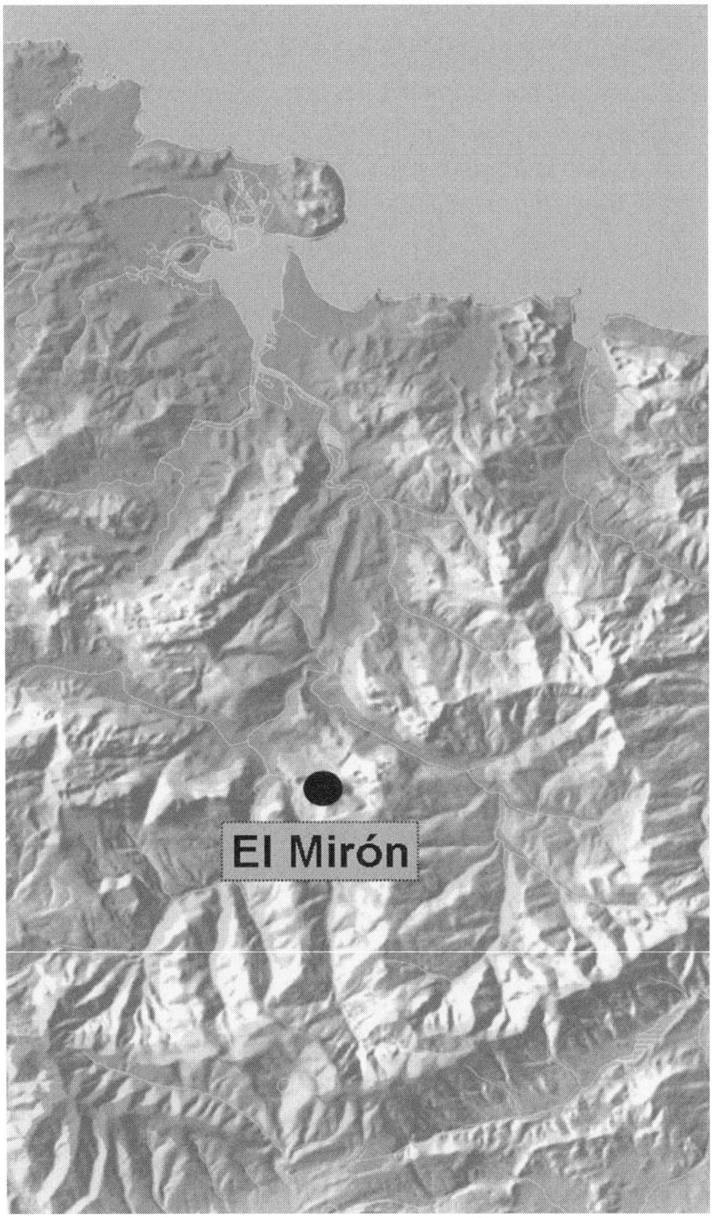

Fig. 8.1. Location of El Mirón Cave. Digital topography from the Servicio Geográfico del Ejercito, used with permission. Modified by María José Iriarte.

given below represent only the relatively short, most recent period of time relative to the long archaeological sequence in this site.

Palynological Methodology

The methodology of pollen analysis consists of several phases. The first step, fieldwork, involves the taking of samples by means that vary according to conditions and characteristics that are functions of the type of sedimentary deposit that one is going to study. Next, in the laboratory, the sediment samples are subjected to physical and chemical treatment by means of which we extract the pollens and spores that are preserved therein. The next step involves the identification of the pollen grains and spores by means of an optical microscope and with reference to comparative collections of pollens and published drawings. After statistical analysis of the resultant pollen counts, they are represented graphically in the form of pollen diagrams or histograms, depending on the type of deposit and/or the kinds of results that had been obtained (e.g., sample size and diversity). The final step consists of interpretation and comparison of the pollen spectrum with other relevant spectra from other contemporaneous sites and cores in the same or nearby regions, as well as with continental or even global records. What follows is a description of the different phases in the palynological analysis of the archaeological site in El Mirón Cave.

EL MIRÓN SAMPLING

Ever since L. G. Straus and M. González Morales began the excavation of this cave in 1996, we were involved collaboratively in developing the strategy for palynologically sampling the sediments exposed. The strategy has been adapted to the differing characteristics of the deposits, and the samples are thus representative of the sequence and its lateral variations. Given the dimensions of the cave, its long stratigraphic sequence, its sedimentary characteristics, and the course of the archaeological campaigns, the palynological sampling has been done in several different columns, in all of which the sampling interval has been 5 cm except when sedimentary characteristics (e.g., presence of large stones or animal burrows) have prevented it. Among the columns sampled to date, information on the Holocene part of the sequence has been obtained from two (figures 8.2, 8.3). First of all we studied the twenty most recent samples from the pollen column collected in the excavation area known as the Cabin, toward the exterior and near the south wall of the cave vestibule (specifically, from the section between grid Squares H3 and G3). Information from here was supplemented

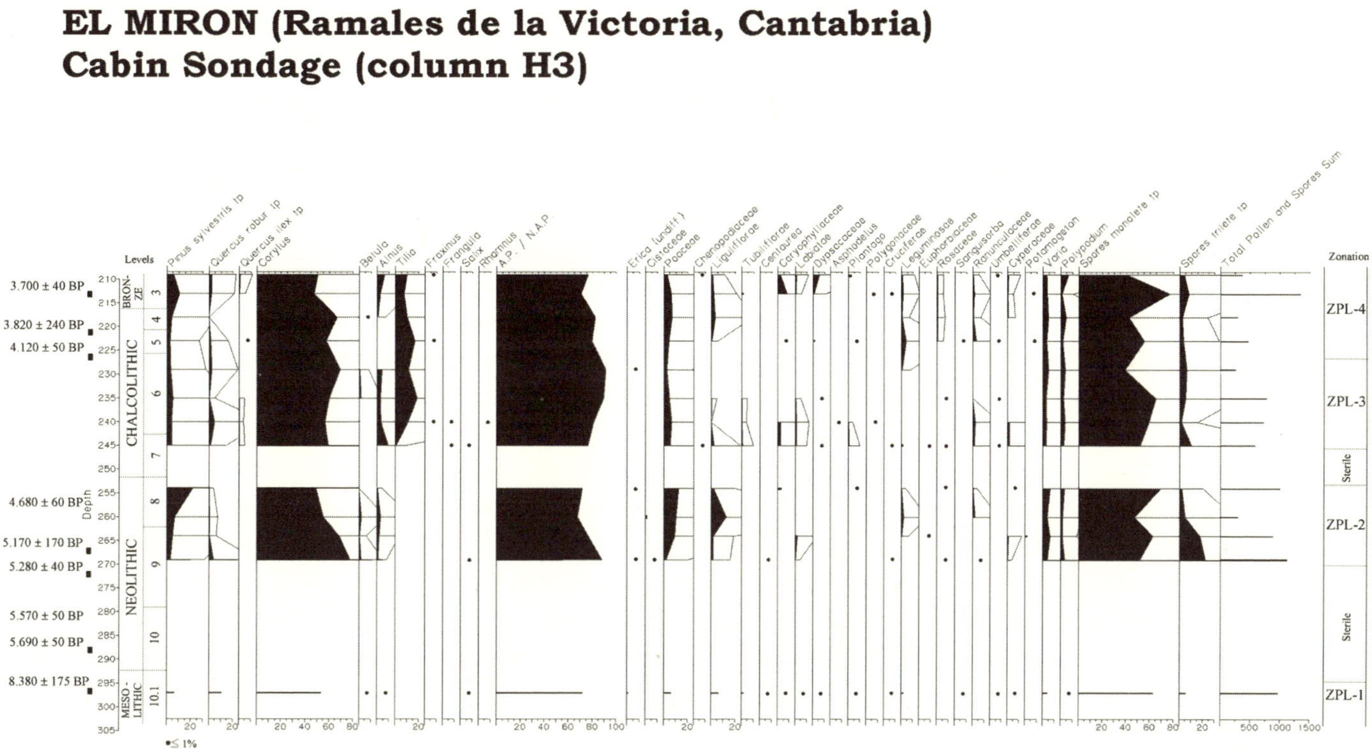

Fig. 8.2. Pollen diagram from the Cabin (outer vestibule) excavation area. Graph by María José Iriarte.

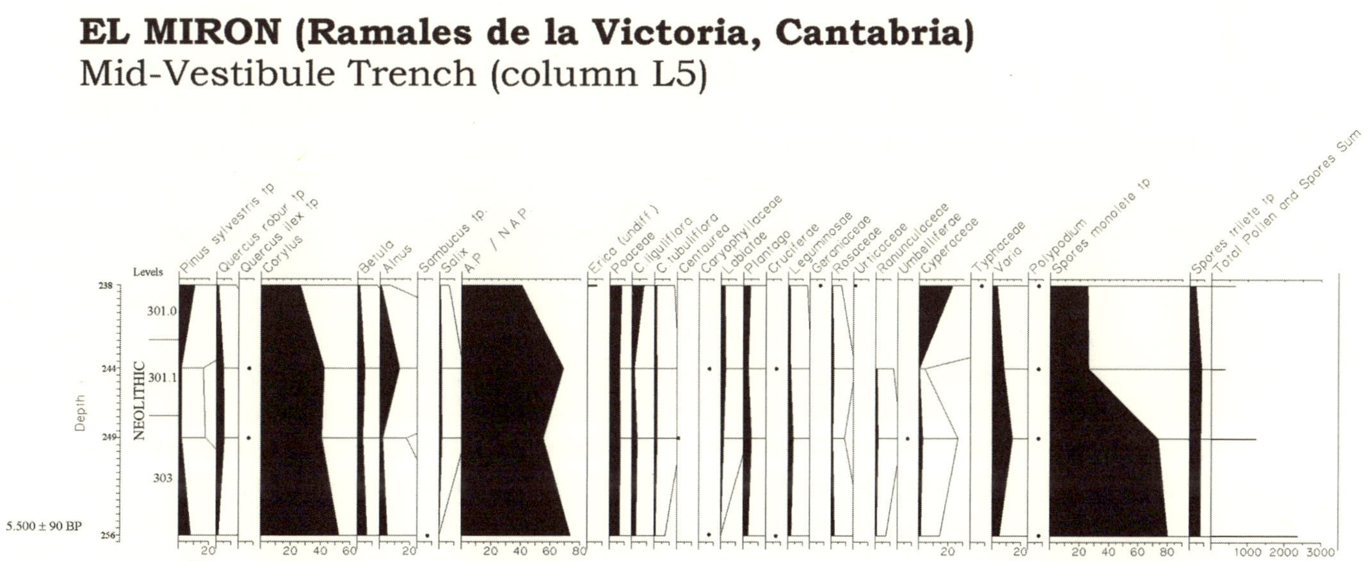

Fig. 8.3. Pollen diagram from the Mid-Vestibule Trench. Diagram by María José Iriarte.

by a sample column taken at the eastern end of the mid-vestibule trench that connects the Cabin with the Corral excavation area at the rear of the vestibule (specifically, from the section between Squares L5 and K5). The trench samples were actually taken in two columns (starting at 290 cm below the site datum plane) because of the possibility of disturbance in the sediments; the older part of the sequence is called column B and the younger, column A. The stratigraphic assignments of the samples from all three columns are given in tables 8.1 and 8.2, although for the purposes of this publication (exclusively on the Holocene levels), we include only samples 15–18 from L15 column B in our analyses. The enumeration of the samples presented here is inverted as a result of requirements of the computer program used to prepare the pollen diagrams. In reality, sampling is always from bottom to top to avoid pollen contamination, so sample 1 is always the oldest.

PHYSICAL AND CHEMICAL TREATMENT OF THE SEDIMENTS

The physical treatment of the sampled sediments (i.e., washing with distilled water and/or screening) prepares them for being attacked chemically. This treatment begins with the use of HCl to eliminate carbonates. After neutralization of the HCl, a 20 percent solution of NaOH (hot reaction, using double-boiling) is added to act on silicates. Then the residue is neutralized again. In order to separate the pollen content from this residue we use the concentration method of dense (Thoulet) liquor described by Goeury and Beaulieu (1979). To do this the sediment must be slightly acid, so a drop of HCl is added. After shaking, the Thoulet liquor is filtered through carbonate filters, which collect the existing pollen content from the sample. After a new HCl attack to eliminate carbonates from the filtering, a drop of 10 percent KOH solution is added to eliminate non-pollen, non-spore organic compounds. Then the residue is neutralized once more.

THE POLLEN COUNT

Data obtained in the pollen count are considered to be statistically representative on the basis of the relationship between the number of grains and the number of taxa present in each sample. By this criterion, sediment samples are considered to be palynologically sterile if they have fewer than 100 pollens and spores and if there are fewer than 10 taxa represented. Also not considered valid are samples with more than 100 grains (or even close to 200) *of which the majority* belong to a single taxon that is accompanied with few other taxa. We consider those samples that have a large number of pollens (200 or more) and spores and that include more than 15 plant taxa to be representative.

POLLEN IDENTIFICATION

Pollen identification and counting were done using two microscopes: a Nikon (Alphaphot-2 YS2 model) and an Optiphot with 40x and 100x objectives and oil immersion. Of critical importance were the comparative collections of pollen grains, along with the palynological photographic archives at the Palynological Laboratory of the Sociedad de Ciencias Aranzadi in San Sebastián and in the Prehistory Subdepartment at the Universidad del País Vasco in Vitoria. In addition, we consulted various pollen atlases (Erdtman 1971; Moore, Webb, and Collinson 1991; Punt 1976; Punt and Blackmore 1991; Punt, Blackmore, and Clarke 1988; Punt and Clarke 1980, 1981, 1984; Valdes, Diez, and Fernández 1987), as well as other works that refer to more specific topics in pollen identification.

The nomenclature used in this study for various taxa is the one presented in *Flora Europaea* (Tutin 1980) and *Flora Ibérica* (Castroviejo 1986, 1990). The terminologies used by various authors in their pollen analyses do vary somewhat with respect to particular taxa. This fact can cause confusion among readers from other specialties who obviously may not be completely familiar with the systematics used by palynologists.

The notation "tp." after the Latin name for a species indicates that the pollen grains in question are most likely of the "type" in question, even though absolutely positive identification cannot be assured. All radiocarbon dates are given as uncalibrated ages before present.

REPRESENTATION OF THE RESULTS

The pollen diagram is the graphic representation of the results (expressed in percentages) obtained from a pollen column. Arrayed along the abscissa are the various taxa and their index of presence and along the ordinal axis one displays at real scale the points along the stratigraphic column where the samples were taken (together with the appropriate references to both radiometric and cultural chronologies for the site area in question). The various taxa included in the study are shown in vertical columns in the graph, with their percentage values being calculated relative to the sum total of the AP (arboreal pollen), NAP (non-arboreal pollen), and "other" categories (here including those pollen grains that are either indeterminate or indeterminable). Those values that do not surpass 1 percent are shown as dots on the graph, in order to provide a better overall vision of the spectrum. For the same reason, in the case of El Mirón, Level 20, which comes after a pollen hiatus, has been represented in the form of a histogram. Following the same criterion, except for the column showing the global AP/NAP relationship together with the total sum of pollens and spores, we have added an exaggerated curve (represented by hatching), which gives a better picture of the progression of those taxa that have low percentage values.

INTERPRETATION OF THE POLLEN ANALYSIS

The basic postulates for the interpretation of a pollen analysis are as follows:

1. The credibility of the results depends principally on the precision of and care taken in the methods used for sampling, laboratory treatment, and so on.
2. The fossil pollen rain reflects ancient vegetation, in the same way that the modern pollen rain corresponds (albeit indirectly) to modern vegetation.
3. The evolution of the various vegetal species during the Quaternary did not involve substantial changes either in the morphological characteristics of the plants or in their ecology, thereby permitting comparisons with modern species.
4. The geographic situation of a site would give it particular characteristics as a function of its altitude, latitude, and orientation and of its potential uses by humans and animals.

The existence of these postulates must not allow us to forget that palynological interpretation is conditioned by a series of limitations and possibilities inherent in the method. First of all, one has to take into consideration such criteria for representativeness of pollens and spores as the differential production and transport mechanisms of pollen among taxa (for example, 10 percent pine pollen does not mean the same thing as 10 percent elm pollen, the former being a well-known "overproducer" and the latter a proven "underproducer"), among other factors (including the fact that only certain taxa are wind-pollinated, while others are pollinated by insects or birds, for example). Thus one cannot establish the evolution of a vegetational landscape based solely on the dynamics of a single taxon, it being necessary to relate that phenomenon with the rest of the taxa represented in the diagram. Any results must be complemented by comparison with any other existing palynological studies from the same region and time periods, whether they are from archaeological or natural sites (e.g., bogs, lakes, marshes), as well as with results from other disciplines (e.g., sedimentology, paleontology, charcoal analyses, and carpology).

Another important factor is related to the particular characteristics of each archaeological site studied (e.g., location and orientation of the site, type of site and deposit, soil chemistry, stratigraphic sequence, disturbance agencies). It is not the same thing to interpret a Pleistocene sequence in a cave as a Neolithic open-air site. In the first instance, a decline in the arboreal portion of the spectrum general corresponds to climatic deterioration, while in the second, the origin of such an arboreal decline could be either climatic downturn or human activity linked to the development of a food production economy, in which case once again it is the *ensemble* of the pollen spectrum that will help us determine the root cause of the process of vegetational change. An incomplete understanding of the process of construction of a megalithic monument or the disturbance of sediments as a consequence of the digging of a grave, can, for example, lead to an incorrect interpretation of palynological results.

It is also fundamental to have a broad knowledge of the geographical context of the site being studied. The geological evolution of the surroundings and of any karstic setting, the sedimentology, the distribution of the modern vegetation, the historical and present-day human exploitation of the territory around the site, the dominant climatology, and so on are all indispensable for better understanding the paleobotanical development.

Palynological Analysis of El Mirón Cave

What follows is the analysis of the pollens extracted from three columns of samples in the Holocene deposits of the El Mirón vestibule. Analysis of the Upper Pleistocene samples will be the subject of future publications.

REPRESENTATION OF POLLENS AND SPORES

In both areas sampled for Holocene-age pollen (Cabin H3 and Mid-Vestibule Trench L5) there are samples with deficient pollen preservation. Although the chemical characteristics of the substances that make up the external layer or exine of pollen grains do confer a great resistance capacity for a long time, several factors can alter that capacity and thus cause problems such as those experienced in this palynological analysis. Among these factors, the ones that have the greatest potential for destroying the pollen/spore content of a deposit in a cave are the chemical composition of the sediment, oxidation of the pollen grains associated with processes such as combustion—attested by the presence of hearths—and local sediment leaching processes.

In the Cabin excavation area (table 8.3), the layers that are either sterile or poor in pollen (and thus statistically not valid) are concentrated in early Neolithic Levels 10 and 9 (in addition to sample number 10 from Chalcolithic Level 7 and sample number 1 from Bronze Age Level 3). In the Mid-Vestibule Trench the pollen sample preservation problems are even more serious, possibly because of the intensity of burning. In this area, between the two columns analyzed (tables 8.4 and 8.5) only four samples from column A (Neolithic Levels 301.0, 301.1, and the upper half of 303) turned out to be palynologically fertile (samples 5, 6, 7, and 8). The rest of the samples either are sterile (samples 3–1 from column A and samples 15 and 18 from column B) or the numbers of pollen grains preserved are not numerous enough to be considered statistically valid.

PALYNOLOGICAL ANALYSIS OF THE CABIN AREA

The reader is reminded that the pollen sequence for this analysis (figure 8.2) begins with Level 10.1, since we are reporting on the Holocene deposits only here. It ends with Level 3, hence covering the period from the early Mesolithic through the early Bronze Age, albeit with several hiati. The discussion of the results has been organized in terms of pollen zones (e.g., ZPL-1 = Mirón pollen zone 1) established as a function of the development of the different taxa that make up the pollen record, following chronological order from oldest to most recent:

ZPL-1: Level 10.1, Mesolithic
ZPL-2: Levels 9 and 8, Neolithic
ZPL-3: Levels 7 and 6, Chalcolithic
ZPL-4: Levels 5–3, Chalcolithic and Bronze Age

Around 8380 ^{14}C BP (ZPL-1) the arboreal fraction dominated the vegetational landscape at the time of this Mesolithic-age level. The principal component of these woods was *Corylus* (54 percent), followed by *Quercus robur* tp. (10 percent, which is the highest value for English oak in the whole sequence) and *Pinus sylvestris* tp. (6 percent), while the rest of the arboreal taxa have values of less than or equal to 1 percent: *Betula, Alnus, Salix.* For their part, among the herbaceous and arbustive taxa the main components of the spectrum are *Compositae liguliflora* (7 percent), *Poaceae* (5 percent), *Centaurea, Leguminosae*, and *Ericaceae.* The representation of fern spores is numerically important; the sum of spores of *Monolete* tp., *Trilete* tp., and *Polypodium* comes to 72 percent of the global pollen + spore total for this sample.

Following this sample upward, the sequence is interrupted by both palynological and sedimentary hiati, the former of which includes Level 10 and the lower half of 9, until about 5200 ^{14}C BP. The first fertile sample (14) is from upper Level 9, culturally attributed to the Neolithic; it represents the beginning of ZPL-2 and is characterized by the first arboreal pollen maximum (AP = 88 percent). Comparing this sample with the earlier one, one observes that there is now an even greater disproportion between values for *Corylus* (78 percent) and those for *Quercus robur* tp., *Pinus sylvestris* tp., and *Betula* (4,3, and 2 percent, respectively), while the herbaceous-arbustive fraction sees a decline in the representations of *Poaceae, Compositae liguliflora*, and *Ericaceae* (not exceeding 2 percent each), although the fern spores do increase slightly to 77 percent of the global pollen + spore total.

Throughout the second pollen zone (ZPL-2) and up to another palynological hiatus that interrupts the sequence at the beginning of the first Chalcolithic level (7) there is a progressive decrease in the arboreal fraction, especially in terms of *Corylus* (a reduction of 29 percent) and *Quercus robur* tp. (in sample 11 it reaches its lowest value in the whole Holocene sequence). In spite of this decline, at no time in ZPL-2 does the arboreal pollen fraction (AP) fall below 67 percent, and in fact, in the last (uppermost) sample it recuperates somewhat. This is due to the curve for *Pinus sylvestris* tp., which runs contrary to those of the two main deciduous taxa mentioned above, possibly because of the capacity for "opportunism" that characterizes the Scots pine as a colonizer of open spaces. In this way, in Level 11 Scots pine reaches its highest percentage value in the whole sequence (21.8 percent). The representations of *Betula* and *Alnus* are fairly stable at between 1and 2 percent.

In this period of woodland shrinkage, as is logical, the herbaceous-arbustive fraction increases its representation, particularly in terms of the higher percentages of *Poaceae* and *Compositae liguliflora*. The pollen curve of the *Poaceae* has a gradual, progressive upward trend (from 2 to 12.5 percent), while that of the *Compositae liguliflora* jumps up sharply in sample 12 (to 12.5 percent), outnumbering the *Poaceae*, but then falling back to its earlier values (samples 14 and 13) and then falling further to 2 percent in sample 11. The rest of the components of this spectrum have discontinuous curves in the samples from the most recent Neolithic levels in El Mirón: *Ericaceae*, *Cistaceae*, *Centaurea*, *Caryophyllaceae*, *Labiatae*, *Rosaceae*, *Leguminosae*, *Euphorbiaceae*, *Ranunculaceae*, *Cyperaceae*, etc. The dynamic of the spores is first to increase in percentage, then to fall back in sample 12 (from 81.6 percent to 53.7 percent of the pollen + spore sum), and finally to increase again at the end of this period (74 percent).

After the palynological hiatus, in respect to the arboreal fraction, beginning with the first Chalcolithic sample corresponding to the upper part of Level 7 one can detect certain differences in the vegetation after about 4680 ^{14}C BP, signaling the start of ZPL-3. From this time on the woodlands around El Mirón are enriched with two new taxa: *Tilia* and *Quercus ilex* tp. Linden develops truly great significance in the vegetation since, beginning with sample 8, it becomes the second-best-represented tree after hazel. (The pollen curve of holm oak, on the other hand, is discontinuous and of lesser importance.)

In the first (lowest) three Chalcolithic samples (9, 8, and 7) one observes *Tilia*'s clear tendency for increase (from 1 to 18.6 percent), which coincides with the stable presence of *Pinus* (4–5 percent) and of *Quercus ilex* tp. (<1.5 percent), while *Quercus robur* tp. reaches 4.5 percent in sample 8. In the oldest of these samples *Corylus* and *Alnus* have higher percentage values than at the end of the Neolithic (59 and 9 percent, respectively), although in the next sample they decrease, especially in the case of alder, which goes down by more than 50 percent. The further evolution of these taxa from this time on varies in that in the case of hazel a recuperation seems to have taken place, while alder stayed at around 3 percent of the pollen total for the rest of the sequence. Another noteworthy aspect of these Chalcolithic samples is that in the first (lowest) two there is a greater diversity of taxa than in any other part of the entire Holocene sequence, since, along with the taxa mentioned above, there are also pollens of *Fraxinus*, *Frangula*, *Salix*, and *Rhamnus*, all of whose values are less than 1 percent. The main component of the herbaceous-shrub fraction continues to be the *Poaceae*, although with a dynamic that is different than among the Neolithic levels, namely a slight decreasing tendency that bottoms out in sample 6 at less than 7 percent. In addition, at this time and given discontinuous series that are displayed by the other herbaceous taxa, there are nonetheless continuous representations of *Compositae*, *Labiatae*, *Ranunculaceae*, *Caryophyllaceae*, *Plantago*, and *Cyperaceae*. For their part, the spores as a group rise from 64.7 to 72.4 percent of the global pollen + spore sum, although the individual curves of the various components (*Polypodium*, *Spores Monolete* tp., and *Spores trilete* tp.) are highly variable.

From 4120 ^{14}C BP on (ZPL-4) there is another decline (this time in "sawtooth" fashion) in the arboreal fraction that mainly affects *Corylus* and *Tilia*. However, *Pinus*, in its role as a colonizer of open spaces, does begin a contrary increasing tendency that culminates in Level 3, early Bronze Age, around 3700 ^{14}C BP. In this same level *Quercus robur* tp. recuperates, ending up at values around 2 percent, and *Q. ilex* tp. again achieves values slightly above 1 percent. At the top of Chalcolithic Level 4 the alder curve starts up again (having been interrupted in sample 5), with a progressive advance that culminates in the second maximum in the sequence for *Alnus* (6 percent) in the most recent sample from the Bronze Age. In this period the only other trees accompanying the above-mentioned taxa are *Fraxinus* and *Betula*.

In the herbaceous-shrub fractions there is some recuperation in the diversity of taxa as well as in the percentage representations of these plants. The *Poaceae* continue to dominate (with an increasing trend, except in the last sample of the sequence), accompanied by *Compositae liguliflora*, *Leguminosae*, *Rosaceae*, *Ranunculaceae*, and *Cyperaceae*. In the samples from Level 3 (Bronze Age) the *Caryophyllaceae* reappear with values of 8 and 2 percent, together with *Dipsacaceae* (1–5.6 percent), *Labiatae* (around 1 percent), and *Plantago* (2.8 percent). For their part, the dynamic of the fern spores is in a curve of sawtooth form, with ups and downs in terms of percentages of the global pollen + spore sum (60.5, 49.5, 85.8, and 51.9 percent).

PALYNOLOGICAL ANALYSIS OF THE MID-VESTIBULE TRENCH

From the whole stratigraphic sequence at the western end of the Mid-Vestibule Trench we have palynological information only from the Neolithic levels dating to after 5500±90 BP, specifically from the upper part of Level 303 and from Lenses 301.1 and 301.0 (pollen column A, samples 8–5).

This limitation in the pollen record makes it difficult to define any pollen zonation. In this sedimentary interval (figure 8.3) arboreal cover in the landscape witnessed a considerable decline (from 73.7 percent of the pollen total in sample 8 down to 40.6 percent in sample 5), although there is a slight recuperation in intervening sample 6. This same dynamic is reflected in the curves for *Corylus* (from 52.5 to 26 percent) and *Betula* (from 6.5 to 0.3 percent), while the curves for *Pinus*, *Quercus robur* tp., *Alnus*, and *Salix* all show different trends. In the first case there is a decline in the representation of pine in the middle samples of this short sequence, followed by an increase in the last (highest) sample, up to 10.5 percent. There is a totally opposite dynamic for *Quercus robur* tp. and *Salix*, both taxa whose maximum percentages are found in the central part of the sequence, then falling back to their initial values in the last sample. The curve for *Alnus* displays a major increase in sample 6 (from 1.8 to 13 percent). There are only traces of *Quercus ilex* tp. and *Sambucus*.

In the herbaceous-shrub fractions the dominant taxonomic group is *Poaceae*, with a fairly stable curve that oscillates between 6.5 and 8 percent. The grasses are accompanied by *Compositae liguliflora* (in fact, in sample 5 this group has the same percentage representation as the *Poaceae*: 8 percent), *C. tubuliflora* (with values below 1.6 percent), *Labiatae* (with an increasing trend), *Plantago* (3–5 percent), *Leguminosae*, *Rosaceae*, and *Cyperaceae*, the latter a taxon that sees a major increase in its percentage values in the last (highest) sample (up to 23 percent). The rest of the herbaceous taxa are present in trace form only, with the exception of *Ranunculaceae*, which interrupts its evenly continuous curve (with a bit more than "traces") only in sample 5, where it is absent.

The representation of the spores, like that of the trees, has a clear declining trend. In the two oldest samples (8 and 7) their values are 88.3 and 81.7 percent, respectively, declining significantly in sample 6 (35.7 percent) and yet again, but more moderately, in sample 5 (to 30.5 percent).

The sterility of the rest of the pollen sequence prevents us from providing any information on the Mesolithic or early Neolithic levels on the one hand or, on the other hand, for the Chalcolithic. Thus correlations between the H3 and L5 columns can be made only for the later Neolithic period.

CORRELATION OF THE TWO POLLEN SEQUENCES: DISCUSSION

The possible correlation between the pollen spectra of the Cabin and Mid-Vestibule Trench (based on samples that are separated by only about 5 m) concerns the period after about 5500 ^{14}C BP and before about 4900 ^{14}C BP. The Cabin area ZPL-2 would seem to relate to the short sequence from the Mid-Vestibule Trench, since both diagrams coincide in terms of the main tendencies for the taxa represented. The first characteristic element is the declining trend for arboreal vegetation, mainly *Corylus*, as well as, to a lesser extent, *Quercus*. The two diagrams also coincide in terms of stability in the percentages of *Betula* and *Alnus*, in the progression witnessed by *Pinus* toward the end of the Neolithic, in the increase and overall dominance of the *Poaceae* within the herbaceous fraction, in the pulse of *Compositae liguliflora* during the second half of this period, and in the reappearance of *Ericaceae* in the most recent samples of both spectra.

However, there are some differences between the two records in terms of the specific percentage values, both relative to the AP/NAP ratios (those of the Mid-Vestibule Trench being lower than those of the Cabin—not more than 74 percent), and as regards the main taxa. In the Mid-Vestibule Trench it is noteworthy that there is a continuous curve of some of the herbaceous taxa (*Plantago*, *Ranunculaceae*, *Leguminosae*, *Compositae tubuliflora*, *Cyperaceae*), as well as of willow (*Salix*), which is less abundantly and less continuously represented in the Cabin. Finally, another element that differentiates the two records is the absence from the Cabin of pollens of *Quercus ilex* tp., a taxon that is, in contrast, present (albeit discontinuously) in the central samples from the Mid-Vestibule Trench. The Cabin also lacks *Sambucus* (elder) and *Compositae tubuliflora*, both of which are minimally present in the Mid-Vestibule Trench (the former as only a trace in the earliest sample).

Lacking a complete geochronological record from each of these two sequences, it is difficult to specify how the two spectra relate. At first sight, based on the similarities in composition and trends, one could say that the same chronological sequence is found in both columns, although in the case of the Mid-Vestibule Trench it starts with somewhat older dates. However, within the same global dynamic the differences mentioned above are striking. Various hypotheses could be suggested to explain these differences, but none of them is conclusive:

1. Asynchrony in the horizontal placement of the mid- to late Neolithic occupations could have occurred, such that the pollen columns from the two areas (despite their nearness) would represent phases that were slightly different in time.
2. Topographic phenomena related to the different locations of the pollen columns within the vestibule (H3 being closer to the cave wall and mouth than L5) could have caused distortions in the pollen rain within the vestibule (perhaps due to variations in the penetration and course of wind currents).
3. Distortion in the pollen record generated by anthropic factors, such as, for example, the possible use of certain areas within the cave vestibule as a stable during this time, could lead to a situation whereby *Plantago* (used as fodder?) presents a continuous representation in the Mid-Vestibule Trench area but only occasional traces in the Cabin. The peak representation of *Cyperaceae* in sample 5 of the trench could be due to humans bringing sedges into the cave (also as fodder and/or bedding?).

THE VEGETATIONAL LANDSCAPE OF EL MIRÓN CAVE IN ITS REGIONAL CONTEXT

Given the location of El Mirón on the Atlantic slope of the Vasco-Cantabrian Mountains (a series of chains that link the Pyrenees to the east with the higher ranges of the Cantabrian Cordillera and Picos de Europa to the west), in this section we will include for comparative purposes a series of palynological studies of natural and archaeological sites that are also located in the Atlantic zone of the northern face of these mountains.

IMMEDIATE SURROUNDINGS OF THE SITE

The pollen records that are geographically closest to El Mirón Cave (plate 8.1) are at the western end of the Montes de Ordunte, specifically at Los Tornos Pass (Soba, Cantabria). In the bog of the same name, at an elevation of 920 m above sea level, Peñalba (1989) and Muñoz Sobrino (2001) have conducted several pollen analyses from cores. Information on the Holocene from this montane setting at the continental divide between the Atlantic and the Ebro River drainage (Mediterranean) is complemented by studies of the bog at El Escudo Pass farther west in Cantabria (1,011 m above sea level) (Muñoz Sobrino 2001) and at the Saldropo bog farther east in Vizcaya (García-Antón, Ruiz, and Ugarte 1989; Peñalba 1989; Ugarte, Garcia, and Ruiz 1986). These deposits have provided Holocene pollen records for the specific region that is of particular interest in relation to El Mirón. Given the geographic area and the time period that are of concern to this study, the most closely relevant pollen analysis is that of Muñoz (2001) on core samples from the bogs at the passes of Los Tornos and El Escudo. The base of the sequence from the latter shows an open landscape characterized by dominant *Ericaceae*, which would be related to climatic phases prior to the regional expansion of woodlands at the beginning of the Holocene (El Escudo: ZPL-1).

The initial Holocene is represented both in the El Escudo bog (ZPL-2 and ZPL-3) and in the Los Tornos one (ZPL-1 and ZPL-2) and first displays arboreal colonization by the pioneer taxa *Betula* and *Corylus*, accompanied to a lesser extent by *Quercus robur* tp. Progressively the woods took hold and became established as *Corylus* became numerically more significant, along with *Quercus robur* tp., while *Betula* decreased. It is in this period—seventh millennium BC—that the Holocene pollen sequence from El Mirón picks up (H3 column, ZPL-1). The hiati before and after the sole, isolated, valid sample from Level 10.1, dated to around 8300 ^{14}C BP, do not permit us to correlate this sample precisely within the regional sequence, but we can situate it within zone 2b (CAN-2b) of the regional vegetational evolution (Muñoz Sobrino 2001). Neither for the end of the initial phase of the Holocene (in which has been detected a retreat in the woodland expansion that permitted the development of *Pinus*, *Ericaceae*, and *Poaceae*), nor for the first pollen zone of the middle Holocene (Los Tornos I: ZPL-3) is there any pollen record in El Mirón Cave. Similarly, there seem to be sedimentary gaps in the study of the Los Tornos 1 core (Peñalba 1989).

After the expansion of *Corylus* throughout the initial Holocene, during the middle Holocene there should have been a progression of various arboreal formations corresponding to the Post-Glacial climatic optimum. However, in the regional sequence throughout the fourth millennium BC, contrary to the expected increase in the representation of woods, the arboreal pollen values actually declined (Los Tornos I: ZPL-4). This phenomenon is also reflected in the pollen sequence from El Mirón. After the sedimentary and palynological hiati, the diagrams from the cave start again in Neolithic levels postdating 5500 ^{14}C BP. Both the dynamic of the Cabin column (H3, Levels 9 and 8) and the Mid-Vestibule Trench column (L5, Levels 303, 301.1, and 301.0), there is a progressive decline in deciduous arboreal pollen (mainly concerning *Corylus* and *Quercus robur* tp.), while there were important increases in *Pinus* and especially *Poaceae*. Once again pollen hiati cut the evidence for the evolution of the vegetational landscape in the El Mirón deposit. In the nearby bog core Los Tornos I there is a notable decline in trees at the end of this period and the first appearance of *Cerealia*, which continued—albeit with fluctuations in percentage—until the end of that high mountain sequence. And yet *Cerealia* pollen are absent from the lower elevation site of El Mirón.

From this time on, in what remained of the middle Holocene, the vegetational landscape of Los Tornos Pass was characterized by a rather equal codominance of *Poaceae*, *Quercus robur* tp., and *Corylus* (according to the analyses of the Los Tornos I and Los Tornos 1 cores by Muñoz Sobrino and C. Peñalba, respectively). One characteristic of the vegetation recorded at El Mirón—nearly 700 m lower in elevation—is the clear predominance of *Corylus*, a fact that does not coincide with the similarity of percentage values of *Corylus* and *Quercus* at Los Tornos. Relative to the Neolithic levels, there is some evidence for regrowth of woodlands, based in part on the expansion of hazel. Another aspect that differentiates the Los Tornos bog from El Mirón Cave is the dynamic of *Tilia*. In both deposits this taxon appears at about the same time—in the middle of the fifth millennium BP—but its development is different, since the presence of lime is quite sporadic at Los Tornos but at El Mirón it manifests a significant development over time, becoming the second-best-represented arboreal taxon. For their parts, *Poaceae*, *Quercus*, and *Alnus* show slight decreasing trends—more notable in the case of alder—that contrast with the stable representation of *Pinus* throughout this period. After about 4120 ^{14}C BP, in the last (uppermost) Chalcolithic levels, the deciduous arboreal fraction starts another regression that, after 3700 ^{14}C BP, is taken advantage of by other taxa, such as *Pinus*, to colonize open spaces in the environs of the cave.

REGIONAL CONTEXT

In the wider Vasco-Cantabrian regional geographical context, the available palynological information (see the chapter appendix for the list of pollen analyses with primary references) comes from both archaeological sites and non-anthropic sources (mainly bog and lake cores) (figure 8.4). However, several aspects, such as the scarcity of pollen sequences that lack sedimentary and/or palynological hiati, put limits on that information. The pollen analyses from the Cantabrian deposits at Tina Mayor, Liencres, Pico Ano, and El Pendo Cave,

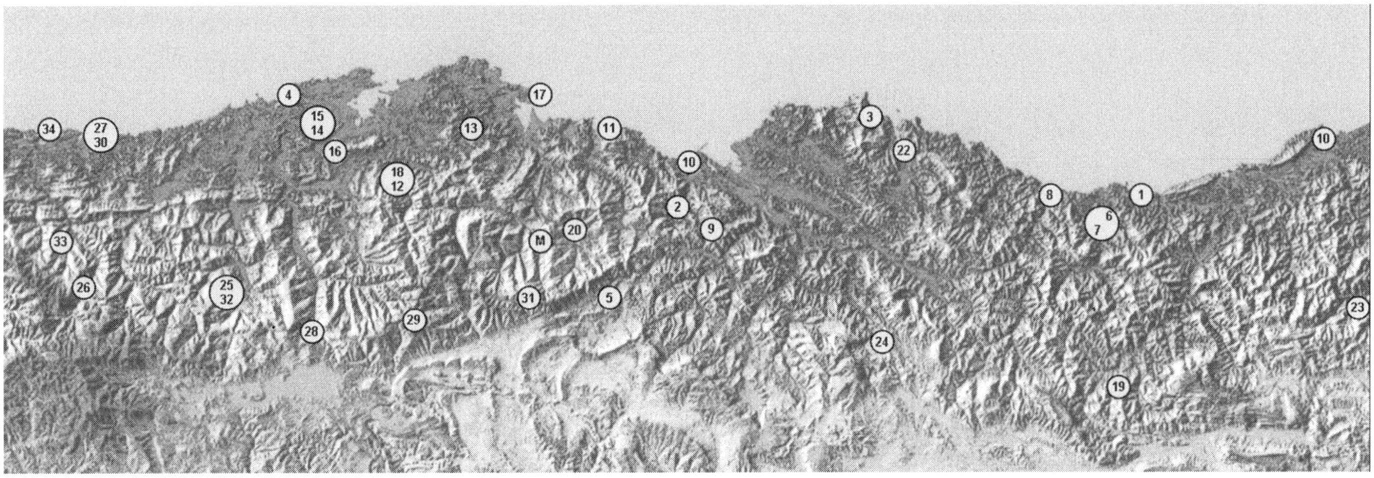

Fig. 8.4. Archaeological sites and geological loci with palynological records in Cantabria and the coastal Basque provinces (Vizcaya and Guipúzcoa) (see chapter 8 appendix for numbered list of sites). Digital topography from the Servicio Geográfico del Ejercito, used with permission. Modified by María José Iriarte.

together with the Vizcayan locus of La Boheriza 2, are not included in the comparisons made here because of their very low counts of preserved pollens and spores, for geological reasons that invalidated the pollen sampling, or because only certain taxa were identified from among the whole pollen + spore assemblage in the samples.

The vegetational references for the beginning of the Holocene come only from loci in the coastal zone or in sublittoral valleys (i.e., lower courses). The stratigraphic records from the Cantabrian sites of Cueva Morín, El Otero, and El Rascaño end in levels associated with the Azilian or with the transition between the Upper Magdalenian and Azilian. But although there are very few pollen samples, those that exist do provide evidence for the Pleistocene-Holocene transition in the form of a definite increase in trees, especially *Corylus*. This same process, although with a more significant increase in *Quercus* in this case, is evidenced among the Magdalenian and Azilian levels of the coastal site of El Perro rockshelter at the present mouth of the Río Asón (12,140–10,160 ^{14}C BP). Both in cave deposits (El Salitre [montane Cantabria], Ekain [coastal Guipúzcoa]—9460 and 9540 ^{14}C BP, Urtiaga [coastal Guipúzcoa], and Arenaza [interior Vizcaya]—9600 ^{14}C BP) and in the open-air site of Pareko Landa (coastal Vizcaya), beginning with levels dated to the eighth millennium BC one observes a progressive woodland expansion (Iriarte 2002). In these sequences it is fundamentally *Corylus* that dominates this arboreal advance, although there are cases such as Urtiaga (8700 ^{14}C BP: upper part of level C) in which the expansion of *Quercus* came before that of *Corylus*. A clear predominance of forest is observed in the marsh of the Urdaibai (Vizcaya) from the beginning of the Holocene. In the initial phases *Pinus sylvestris* tp., *Quercus robur* tp., *Corylus*, and *Quercus ilex* tp. pollens constitute the basis of the arboreal fraction. There is also a continuous presence of *Betula*, *Alnus*, and *Fagus*, as well as a discontinuous record of other tree mesophytes or mesotermophiles (*Castanea*, *Tilia*, *Juglans*, *Carpinus*, *Ulmus*, *Ilex*, *Fraxinus*, *Salix*, and *Populus*) and sclerophyllous trees (*Olea*, *Phillyrea*). In the case of the Peña del Perro rockshelter this is evident beginning even with the earlier phases, although by 10,000 BP *Quercus* started to share its dominant role with *Corylus*, and then with *Betula*, *Corylus*, and *Populus* in a much more mixed deciduous woodland at the time of the early Mesolithic occupation (9260 ^{14}C BP).

The vision we have of the vegetational landscape for the period between about 8000 and 6000 ^{14}C BP in the eastern area of the Vasco-Cantabrian coastal zone and lower valleys (based on spectra from the Bidasoa estuary on the French border, Pareko Landa, Urdaibai, and Arenaza I) continues to reflect the process started earlier and—taking into account differences caused by specific geographical locations and types of deposits—this coincides with what we also know from highland sites (El Escudo, Los Tornos [both on the Cantabrian-Burgos border at the Cordilleran crest], and Belate in Navarra).

The sixth millennium BC provides more information and a broader geographic dispersion of sampled loci, all of which gives us a better understanding of this interesting period, during which the new economic activities developed by humans (i.e., food production) implied the introduction of and/or increase in a major new factor in the evolution of the landscape: "anthropization," or human impact.

In this area of the coast in the first half of the sixth millennium the arboreal representation varies among the deposits sampled from AP values above 50 percent (Herriko Barra, Covacho Arenillas—5580 ^{14}C BP) to ones below 15 percent (Pico Ramos, not far from El Mirón on the western coast of Vizcaya—5860 ^{14}C BP). The main arboreal components of the vegetation continue to be *Corylus* and *Quercus robur* tp., although on occasion *Alnus* (at Herriko Barra on the shore of Guipúzcoa, with 30–60 percent) and *Tilia* (at Covacho Arenillas on the eastern shore of Cantabria, with 18 percent) can attain high values, higher especially than those of *Quercus*. For their part, the principal representatives of the herbaceous plants are the *Poaceae* and *Compositae*, especially *C. liguliflora*. At the open-air site of Herriko Barra (6010 and 5960 ^{14}C BP) the presence of cereal pollens is noteworthy and attests to agricultural activities in the immediate surroundings of the site (Iriarte et al. 2006). At the other end of the region, near San Vicente de la Barquera in western Cantabria, in a bog at La Jerra Beach (ca. 5800 BP) there is evidence of dominantly wooded cover, albeit on the decline, and with the *Corylus* and *Quercus* curves going in opposite directions from one another, ending up with the latter outnumbering both the former and *Pinus*. Later (ca. 5300 ^{14}C BP), in a different core at the same beach-side bog (La Jerra 2), there is a clear dominance of the herbaceous fraction (*Poaceae* and *Ericaceae*), followed in the uppermost sample by evidence of a return to more wooded conditions. In the opinion of the authors of the study of the La Jerra cores, the absence of cereal pollens in these records does *not* mean that the deforestation had not been caused by humans (Mary and Medus 1993; Mary, Medus, and Delibrias 1975). In the Urdaibai marsh there is also a decline in the values of arboreal pollen and indications of human activity during the middle Holocene (Iriarte et al. 2006).

In the montane interior, and especially during the second half of the sixth millennium, at the Saldropo bog (in the foothills of the Sierra de Gorbea between Vizcaya and Alava) beginning around 5630 ^{14}C BP the landscape is dominated by *Poaceae*, *Corylus*, and *Quercus*, while on the edge of the Montes de Ordunte—specifically at the megalithic site of Cotobasero-Basorogane (La Cabaña 4 dolmen), relatively close to El Mirón in western Vizcaya—the nearly equal representations of *Corylus* and *Quercus* are accompanied by *Fagus* and *Alnus*, among other arboreal components, along with *Ericaceae* and *Poaceae* among the herbaceous plants.

This evolution of the mixed-oak forest continued throughout the fifth millennium BP, albeit with variations in the degree of representation of its principal component taxa. The pollen record in the littoral zone is restricted to the coastal deposit at Bederna (San Vicente de la Barquera, Cantabria) and in the cave of Pico Ramos (Muzkiz, Vizcaya). The function of the latter site had changed from a Neolithic habitation to a Chalcolithic burial cave (4790, 4100 ^{14}C BP). As in its earlier occupation, the arboreal pollen in the Chalcolithic period total is less than 20 percent and there are no great changes among the main components of the spectra, although there is evidence of agricultural activities in the vicinity of the cave. At the Bederna estuary (4770 ^{14}C BP) the slight presence of *Alnus* together with the behavior of the *Corylus* and *Pinus* curves led Mary, Medus, and Delibrias (1975) to think that the landscape was undergoing the effects of significant human activities.

In the lower valleys of Guipúzcoa the stratigraphic sequences of the caves of Amalda and Urtiaga pick up again in this (fifth) millennium with Chalcolithic levels, after several sedimentary hiati. In the first site *Corylus* and *Alnus* outnumber *Pinus*, while at Urtiaga, after a first phase of codominance by *Quercus* and *Corylus*, the main arboreal components of the mixed-oak forest that develops are *Quercus* and *Tilia*. Unfortunately, the absence of records both earlier and later than the mid-Holocene spectra from these sites does not allow us to define the existence and magnitude of possible processes of landscape humanization. In the Cave of Arenaza I (western Vizcaya, not far from El Mirón), there is a palynological hiatus that conditions the Holocene record. The arboreal vegetation attested for the upper part of the Chalcolithic level (4270 ^{14}C BP) displays a slight retreat relative to the last AP values known from before the hiatus (nearly 30 percent), with the main difference concerning the fact that *Corylus* is now virtually the only taxon present. This same phenomenon is also seen at the megalithic monument of Praalata, situated near the head of the basin of the Oria River in Guipúzcoa (4310 ^{14}C BP).

From the pollen record at the Saldropo bog during this period we can deduce that throughout the Post-Glacial climatic optimum there was a taxonomic enrichment of the mixed-oak forest relative to earlier periods of the Holocene. Together with *Quercus*, *Ulmus*, *Tilia*, *Fraxinus*, and *Hedera* become well established. On the edge of the bog there seem to have been abundant hazels (*Corylus*), while *Taxus* (yew) was limited to a few isolated individuals scattered at high elevations throughout the understory of the mixed-oak forest and alders (*Alnus*) would have lined the streams that fed into the bogs. Beginning at the base of the sequence studied by B. Ruiz Zapata (dating to 4510 ^{14}C BP) there is a continuous curve of *Fagus* with values that fluctuate between 5 and 10 percent. In the western Vizcayan valley of Karrantza, near El Mirón, at the open-air site of Ilso Betaio, the evolution of *Fagus* (10–12 percent) is also noteworthy, together with that of *Alnus*, in a rather open vegetational landscape in which *Corylus* was progressively declining while the mixed-oak woods maintained fairly stable values. For its part, the Alsa bog displays an arboreal fraction dominated by *Pinus*, *Betula*, and *Corylus* at this time.

The landscape evidence we have for the area of and near Guipúzcoa is from the Basque Mountains. In the pre-Pyrenees the vegetation at Belate (northwest Navarra) includes a major arboreal component, in which most notable are *Corylus*, *Quercus*, and *Pinus*. In this time period—whose upper limit is dated to 2960 ^{14}C BP—there are sporadic cereal pollens in some of the records. The information available from the dolmen of Ataun-Burunda in the mountains of Guipúzcoa shows variations as a function of the different deposits analyzed, while at the dolmen of Napalatza the arboreal representation is significant (based mainly on *Corylus*). At the Guipuzcoan dolmens of Aitxu, Unanabi, and Zorroztarri the landscape was more open and *Corylus* had percentages that were closer to those of the other arboreal taxa: *Fagus* and *Quercus* in the first case, *Quercus* and *Pinus* in the second, and *Fagus*, *Pinus*, and *Alnus* in the third. At Zorroztarri (3280 and 3030 ^{14}C BP) the presence of cereal pollen is continuous. Continuing in the mountainous area, near the continental divide between the Atlantic and Mediterranean watersheds in the Saldropo bog the processes of anthropization (woodland clearance) are quite apparent in the vegetation around the deposit, with the first actual signs of agriculture per se appearing after 3590 ^{14}C BP. Finally, in upland Cantabria in the top level of El Salitre Cave (one valley to the west of the Asón) there is now evidence of an open landscape among whose few arboreal taxa are *Corylus*, *Pinus*, and *Betula*.

SUMMARY

The evolution of the vegetational landscape as a response to the ameliorating climatic conditions after 10,000 ^{14}C BP (beginning of the Holocene) is attested by the progressive increase in trees throughout the region (as in the rest of southwestern Europe). The first reflection of Holocene-type vegetation in the area around El Mirón Cave, dating to around 8380 ^{14}C BP, is characterized by a clear dominance of the arboreal fraction (73 percent) in the early Mesolithic pollen spectrum. This fits well with what we know from the available records both in the Vasco-Cantabrian Mountains (at the passes of El Escudo, Los Tornos, and Belate), and in the coastal zone and lower valleys (Bidasoa estuary, Pareko Landa, Urdaibai, and Arenaza I). This development of the woodlands is mainly based on the expansion of three arboreal taxa—*Corylus*, *Betula*, and *Quercus robur* tp.—although the specific representation of each of these varies by site. The greatest disproportions between hazel and English oak are found at Arenaza I and El Mirón, with a clear dominance of the former at these two sites, which are only about 28 air km apart.

The existence of two hiati (sedimentary and palynological) in the El Mirón sequence forces us to go up to the fourth millennium to continue our creation of a picture of the vegetational landscape around this cave. The period between about 5300 and 4600 ^{14}C BP sees the continuation of the main characteristics of the start of the Holocene, albeit with a few subtle differences. Although at first the arboreal pollen index had been even higher than in the Mesolithic, throughout the later Neolithic occupation there was a decline in woodland cover. In addition, the disproportion between *Corylus* and *Quercus* increased. Nonetheless, the *Pinus* percentages grew at the expense of *Corylus* and both *Betula* and *Alnus* augmented their respective representations.

It is interesting to note that the decline in arboreal cover occurred during a period defined by the adoption of a food production economy. From this time on the new necessities created by animal husbandry and agriculture, among other economic activities, interfered significantly with the evolution of the vegetation. In the coastal zone during the first half of this millennium, along with records showing a continued clear dominance of the arboreal fraction (as at Covacho Arenillas) are other records with few tree pollens (namely at Picio Ramos). Nevertheless, in both of these sites the sedimentary and/or palynological data are too scarce to really give us an idea as to the grade of impact that anthropization was having on the landscape. In the case of La Jerra Beach, farther west, around 5800 ^{14}C BP there is evidence of an important arboreal fraction in which one can nevertheless detect a slight decrease at the end of the sequence. In addition, just as at El Mirón, hazel shows a declining trend from the very beginning. In a second pollen sequence from the same coast bog, La Jerra 2, at a time closer to the mid-Neolithic samples we have from El Mirón (5300 ^{14}C BP), the reduction in arboreal pollen is considerable, although it increases again in the last sample back to the AP values that had characterized the beginning of the millennium. Thus, except for the case of Covacho Arenillas for circumstances mentioned earlier, the majority of the littoral deposits confirm a thinning of the woodlands related to human activities, most notably agricultural practices in the surroundings of the open-air site of Herriko Barra. The decline in trees and expansion of *Poaceae* at El Mirón are not discordant with the record for this time in the Vasco-Cantabrian Mountains and central Cantabrian Cordillera (e.g., Cueto de la Avellanosa). In addition, specifically at the bog of Los Tornos (along with the La Cabaña dolmen in Karrantza, the pollen record closest to El Mirón), the deterioration of the woodland cover during this period is confirmed by the presence of cereal pollens. The existence of agricultural activities has also been detected at the foot of the Ordunte Mountains (Burgos) in the area where a megalithic necropolis is adjacent to various living sites that have been attributed to the final Neolithic and Chalcolithic (Iriarte, unpublished data).

With the definitive recommencement of the El Mirón Holocene pollen sequence in the Chalcolithic, there is a change in the composition of the arboreal fraction relative to the earlier Neolithic and Mesolithic levels. First is the notable appearance and expansion of *Tilia* and second is the increase in the representation of *Alnus*, which becomes one of the most important woodland species, outnumbering *Pinus*, *Quercus*, and *Betula*, although it is always palynologically less numerous than *Corylus*. In the Vasco-Cantabrian Mountains (El Escudo, Los Tornos, Saldropo, Belate) this lime expansion phenomenon is not observed, although the species is either occasionally attested in trace quantities or is found continuously in very low numbers as in Karrantza, both in the La Cabaña 4 and in the open-air site of Ilso Betaio. In the coastal zone and lower valleys the two spectra that have the largest representations of *Tilia* are Covacho Arenillas and Urtiaga. In the first case, the low values of lime are quite similar to those of El Mirón, but the age is supposedly about a millennium older. With respect to Urtiaga, there is a coincidence with El Mirón in that *Tilia* appears for the first time in the Chalcolithic and develops rapidly, but in Urtiaga the values always remain below 7 percent. Throughout the Holocene we have evidence of an interesting evolution in the representation of *Alnus*: in coastal sites—at least along the Basque shore—the alder expansion seems to have been related to episodes of minor sea level regression (Bidasoa estuary, Herriko Barra). In the records that we have from the Cantabrian shore (La Jerra, Covacho Arenillas, Bederna) the representation of *Alnus* does not exceed 1 percent, but this could be due to the fact that the levels sampled in these sequences correspond to a period of marine *transgression*. In the rest of the analyses the presence of alder is constant but with percentage values that do not exceed 10 percent in most cases.

Another fact that should be noted at El Mirón is the appearance of *Quercus ilex* tp.—first as traces, but then later in the Chalcolithic levels in larger quantities—coinciding with the presence of this sclerophile element in the eastern sector of the Cantabrian Cordillera in specific kinds of settings and with a limited distribution (Muñoz Sobrino 2001). In the littoral-sublittoral zone there are also records containing this species in the Bederna deposit and in Urtiaga Cave.

Around 4120 ^{14}C BP the El Mirón general arboreal pollen curve presents a sawtooth form with a clear descending trend that even interrupts the individual curves for *Quercus ilex* tp. and *Alnus* (which recommence during the Bronze Age). The arboreal decline ends with the last sample in the sequence, after 3700 ^{14}C BP. The only taxon that departs from this trend is *Pinus*, a fact that is related to its opportunistic nature as a colonizer of open spaces, just as also happened during the Neolithic occupations. Around 3700 ^{14}C BP, during the early Bronze Age occupation of the cave, pine reached its second-highest value in the sequence, coinciding with minimum Holocene values for *Corylus* and *Tilia*; then it fell back again at the end of this period.

Final Reflections

THE VEGETATIONAL LANDSCAPE

In synthesizing the results obtained in the palynological study of the Holocene levels of El Mirón Cave, the first aspect to be noted is the high representation of arboreal taxa throughout the whole sequence: at no time is the AP less than 65 percent. The dominant woodland type in this record is the mixed deciduous one, with an evolution from the initial Holocene onward being consistent with that reflected in the upper valleys of the Gándara and Calera rivers. However, one must not forget that this pollen spectrum reflects a landscape in which human intervention is patent, for which reason the trends in the curves of its component taxa were conditioned by the process of anthropization.

Within the vegetational spectrum defined in the El Mirón sequence, the two principal types of forest that are evidenced are mixed deciduous woods and riparian or gallery woods. However, taking into consideration the ecological characteristics of the component taxa of the arboreal fraction, it is not easy to determine with precision to which of these forest types each taxon belongs. The main component of the gallery woods is the alder, a species that inhabits the waterlogged soils of the banks of rivers and other water bodies. In this geographic context alder is continuously present from the Neolithic levels onward (except in Level 5—Chalcolithic), in contrast to the sporadic representation of the willow, which normally would line the primary river course or even could play a role in the stages of degradation of the alder woods. Among the herbaceous taxa associable with this type of woods, the best represented are the *Cyperaceae*, although these can also be found on slopes and around springs at footslopes, as is the case today (Herrera 1995). With the recommencement of the pollen sequence in the Chalcolithic after a hiatus, *Frangula* appears, coinciding with the period of greatest representation of alder, which could indicate that this taxon also formed part of the gallery woodland. However, one should not forget that it can also form part of the shrubby fringes of mixed woods. One also detects the occasional presence of *Typhaceae* and *Potagometon*, both of which prefer riverine backwaters and ponded areas.

The main bioclimatic surface reflected in this landscape is the hillslope, characterized fundamentally by mixed deciduous woods. However, within the composition of this type of woods one can observe a clear difference between the part of the sequence that corresponds to the oldest levels (Mesolithic and Neolithic) and the upper parts, which correspond to the Chalcolithic and the Bronze Age, the more recent levels being those in which the mixed woods are enriched with taxa such as *Tilia* and *Fraxinus*.

Within the framework of the increasingly solid benignity of the climatic conditions of the Holocene, the distribution of various vegetational landscape types has also been conditioned by other factors such as the morphology of the relief and the nature and instability of the substrate. The location of El Mirón Cave in a limestone mountain (plate 8.1) with very steep slopes at the base of which is entrenched the Río Calera has been a constant conditioning factor in the types of vegetation that developed in the immediate surroundings of the site. At the present time there is a clear predominance of oak, especially dense along the footslope nearest to the river and becoming more dispersed higher up, where it is capable of subsisting in tiny patches of soil among rocky outcrops. Together with this species we find others such as *Crataegus monogyma* (hawthorn), *Cornus sanguinea* (cherry), and even *Fraxinus* (ash) located in areas where the soil is a little deeper and with some degree of preference for those zones that are relatively close to the river. In the low shrub and herbaceous strata (mainly in the area closest to the cave), most prominent are the *Rosaceae*, *Gramineae*, and ferns such as *Pteridium aquilinum*.

One constant throughout the whole vegetational sequence at El Mirón is the dominance of the arboreal fraction by *Corylus*. The dispersion of hazel is a characteristic of the beginning of the woodland expansion during the Holocene. However, especially during the climatic optimum, hazel loses ground in competition with other taxa, such as oak. In the regional pollen sequence established by Muñoz Sobrino (2001), once the initial phase of the Holocene is finished the values for *Corylus* and *Quercus robur* tp. are closer to equal and oak even exceeds hazel in ZPL-5 at Los Tornos Pass. However, in the environmental context of the cave, as noted earlier, hazel simply dominates in all the samples from the early and middle Holocene. This could have to do with a conjunction of several local factors, not just climatic ones.

We have already succinctly discussed the main characteristics of the relief surrounding the cave. They translate into a lack of areas with soils adequate for the development of mixed-oak woods or the other kinds of trees and shrubs that usually accompany them. In this environment, unfavorable for the development of mature, leafy forest, the extension of oaks is limited to those small areas harboring cool, deep soils (and, in addition, the deeply entrenched valley of the Calera and Gándara rivers below the cave does not permit the existence of terraces of any significance). Thus hazel would naturally play an important role in a kind of deciduous woods conditioned by the scarcity of soil and by a karstified and often bare rock substrate, occupying steep slopes and limestone outcrops, as well as areas with gentler slopes that interrupt the elevated ridges of the limestone mountain massifs and even the cliff bases, where it is able to survive despite rockfalls. For all these reasons, it seems that the predominant factor in the major representation of hazel at this site is the steep, rocky setting of El Mirón Cave. In the same way one can explain the important representation of lime in the Chalcolithic and the Bronze Age, since this taxon is fully adapted to zones at the foot of cliffs and, given atmospheric humidity sufficient to even occupy rocky walls, it could share this difficult habitat with hazel. In this type of woods, conditioned by rocky habitat, it would also have been possible to encompass other deciduous arboreal species such as *Fraxinus*. However, given the adaptability of ash, it is not possible to be sure that its discontinuous presence in the pollen record is associated only with the rocky nature of the habitat. For its part, the most favorable places for the development of *Quercus ilex* tp. are in sunny spots among the rocks, with carbonate-rich, well-drained soils, where it would cohabit with other sclerophile species together with deciduous elements. During the Neolithic occupations the presence of *Quercus ilex* was scarce, but it increased beginning in the Chalcolithic, a time during which there are also traces of *Rhamnus*, a sclerophile taxon also associated with this kind of rocky substrate woods. One must not forget that the orography of the terrain conditions the development of both the tree and shrub fractions and the herbaceous one, with a limitation in the extension of herbaceous taxa and ferns to the level areas, cracks, and fissures of the rocks, thereby generating a discontinuous herbaceous fraction.

If we consider that the representation of hazel is tied to a kind of deciduous woodland that is strongly conditioned by rocky habitat, we cannot forget that, given its ecological characteristics, hazel can also form a part of the shrubby fringes of mixed-oak (*Quercus robur* tp.), deciduous woods that develop on deeper soils and that could exist somewhat farther from El Mirón, namely in the form of riparian woods.

THE HUMANIZATION OF THE LANDSCAPE

Another important characteristic of the El Mirón pollen sequence is its evidence for the anthropization of the landscape. The first period of Holocene woodland retreat at El Mirón, after 5280 ^{14}C BP, took place during the Neolithic (although we lack pollen from the oldest Neolithic levels at the site, which go back to 5800 ^{14}C BP) and coincides with processes of similar characteristics that have been detected throughout the

Cantabrian region (Peñalba 1989; Ramil-Rego 1992) and in the pollen spectra located closest to El Mirón, such as Los Tornos I and Ordunte. In Los Tornos I one would have normally expected the major development of mature mixed-deciduous woods during the Post-Glacial climatic optimum. However, what one observes in the diagram is a gradual *reduction* in the arboreal fraction during the period of greatest arboreal diversity, with an increase in *Gramineae* and evidence for the existence of agricultural activities in the area near the bog. In this last aspect there would seem to be corroboration of the idea that the arboreal decline was caused by economic intentionality on the part of human beings, although in the opinion of the author of the Los Tornos I study (Muñoz Sobrino 2001), in addition to this process there could have been some increase in the tree and shrub fractions that would colonize the mountain crests with heaths and pastures, along with the formation of bogs in the mountainous areas as a consequence of improved climatic conditions (moderate increase in temperature and high rainfall). In El Mirón we also detect this maximum of *Gramineae* during the Neolithic occupation, although, given the orography of the area, it seems fundamentally to have been a consequence of human activity.

The processes of anthropization of the landscape indicate to us that the type of economic activity conducted by the inhabitants of El Mirón beginning in the Neolithic generated necessities that interfered with the development of the natural vegetation in this habitat. Within the progress of the adoption of the food production economy, we know that livestock husbandry was a stable activity here. However, in the *pollen* sequence at El Mirón there are no indices of farming, at least in the area right around the cave. The cave itself is an excellent place of interest for human beings, as is demonstrated by its long human occupation, from the mid- to Upper Pleistocene up to modern days. Nevertheless, once again the morphology of the surrounding terrain played a conditioning role, since the rocky slopes did not provide favorable ground for farming. This would have to have been done in more favorable areas away from the cave, such as the valley floor of the Calera upstream of the site ("behind" Monte Pando) and that of the Asón at its confluence with the Calera-Gándara in Ramales. In reality, by the orography of the slope one can deduce that the pollen analysis is the reflection of the immediate vegetational surroundings of the cave and that it in turn could have acted as a "screen" blocking a more regional vision of the overall vegetation. For these reasons, the negative data on the presence of cereal pollen at El Mirón does not mean that the inhabitants of the cave did not practice agriculture (activities that in fact are attested from Neolithic times onward in places that are geographically very close by, such as Los Tornos and Ordunte). Rather, factors such as the limited properties for dispersal of this kind of pollen (i.e., cereal) and the location of the cave could have significantly conditioned the record of agricultural activities in the palynological sequence of El Mirón.

The beginning of the second period of reduction in the arboreal fraction happened in Level 5 of the Chalcolithic occupation (ZPL-4) of this cave. In the various records mentioned above, such as those of Ilso Betaio, Pico Ramos, and the megalithic complex of Ataun-Burunda, it was from this cultural period onward that the deforestation processes accelerated and then intensified even further in the Bronze and Iron Ages. When addressing these periods of woodland deterioration, it is important to mention two arboreal taxa: *Pinus sylvestris* tp. and *Betula*. Ever since the Tardiglacial the role played by these taxa was important, because it was they that initiated the process of arboreal recuperation, being as they are the two principal pioneer taxa (especially the pine). Once the deciduous forest reaches its maximum extent, these species descend to their minimum representation and can even disappear from the Holocene pollen record in some geographical areas, as is the case with birch. In El Mirón birch has only a limited presence and even this declines with time, as is also true in other deposits in the area (i.e., the bogs at the passes of Los Tornos and El Escudo). The broad ecological characteristics of this genus permit it to grow in various bioclimatic settings, forming part of riparian woods as well as high mountain ones or existing in the montane slope habitat type as an element in various stages of the degradation of the deciduous forest. For its part, the dynamic displayed by pine is clearly inversely correlated with that of the deciduous forest: pine's highest values (in ZPL-2 and ZPL-4) coincide with declines in the leafy woods, a circumstance that can be explained by pine's role as a colonizer of open spaces.

Acknowledgments

This study was carried out within the framework of the High Performance Research Group (IT-288-07) of the Prehistory Department of the University of the Basque Country, Vitoria/Gasteiz.

Appendix: Bibliographic References for Palynological Studies

ARCHAEOLOGICAL SITES

Open-Air Sites

1. **Herriko Barra (Zarautz, Guipúzcoa):** Mariezkurrena and Altuna 1995; Iriarte, Mujika, and Tarriño 2005
2. **Ilso Betaio (Arcentales-Sopuerta, Vizcaya):** Iriarte 1994b, Gorrotxategi et al. 1999
3. **Pareko Landa (Sollube, Vizcaya):** Iriarte, inédito
4. **Liencres (Liencres, Cantabria):** Clark, Menéndez-Amor 1979
5. **Ordunte (Valle de Mena, Burgos):** Iriarte, unpublished data

Caves and Rockshelters

6. **Amalda (Zestoa, Guipúzcoa):** Dupré 1988, 1990
7. **Ekain (Zestoa, Guipúzcoa):** Dupré 1984
8. **Urtiaga (Deba, Guipúzcoa):** Sánchez 1993
9. **Arenaza I (Galdames, Vizcaya):** Isturiz, Sánchez 1990
10. **Pico Ramos (Muskiz, Vizcaya):** Iriarte 1994a
11. **Covacho Arenillas (Islares, Cantabria):** Iriarte 2002
12. **El Salitre (Ajanedo-Miera, Cantabria):** López 1981, 1986
13. **El Otero (Secadura, Cantabria):** Leroi-Gourhan 1966
14. **El Pendo (Escobedo de Camargo, Cantabria):** Leroi-Gourhan 1980
15. **El Juyo (Camargo, Cantabria):** Boyer-Klein and Leroi-Gourhan 1985
16. **Morín (Villaescusa, Cantabria):** Leroi-Gourhan 1971
17. **Peña del Perro (Santoña, Cantabria):** López, López, and Uzquiano 1996
18. **Rascaño (Mirones, Cantabria):** Boyer-Klein 1981

Megalithic Monuments

19. **Estación megalítica de Atuna-Burunda (Guipúzcoa):** Iriarte 1994b, 1997, 1999
20. **Estación megalítica de Cotobasero-Basorogane (Vizcaya):** Iriarte 1995, Yarritu et al. 1999

NON-ANTHROPOGENIC DEPOSITS

21. **Playaundi (Estuario del Bidasoa, Guipúzcoa):** Sánchez 1996
22. **Urdaibai (Gernika, Vizcaya):** Iriarte et al. 2006
23. **Belate (Narvarte, Navarra):** Peñalba 1989
24. **Saldropo (Barazar, Vizcaya):** Ugarte, Garcia, and Ruiz 1986; García-Anton, Ruiz, and Ugarte 1989; Peñalba 1989
25. **Alsa (Barcena Pie de Concha, Cantabria):** Mariscal 1993
26. **Bederna (San Vicente de la Barquera, Cantabria):** Mary, Medus, and Delibrias 1975; Mary and Medus 1993
27. **Cueto de la Avellanosa (Polaciones, Cantabria):** Mariscal 1983
28. **El Escudo (Puerto de El Escudo, Cantabria):** Muñoz Sobrino 2001
29. **Estacas de Trueba (Puerto de Estacas de Trueba, Cantabria):** Mariscal 1995
30. **Jerra (San Vicente de la Barquera, Cantabria):** Mary, Medus, and Delibrias 1975; Mary and Medus 1993
31. **Los Tornos (Puerto de Los Tornos, Cantabria):** Peñalba 1989; Muñoz Sobrino 2001
32. **Pico Ano (Bárcena de Pie de Concha, Cantabria):** Salas 1993
33. **Pico Sertal (Rionansa, Cantabria):** Mariscal 1986
34. **Tina Mayor (Val de San Vicente, Cantabria):** Mary and Medus 1993

References Cited

Boyer-Klein, A.
1981 Análisis palinológico del Rascaño. In *El paleolítico superior de la Cueva de Rascaño (Santander)*, edited by J. González Echegaray and I. Barandiarán, pp. 217–220. Monografías 3. Centro de Investigación y Museo de Altamira, Santander, Spain.

Boyer-Klein, A., and Leroi-Gourhan, A.
1985 Análisis palinológico de la Cueva del Juyo. In *Excavaciones en la Cueva del Juyo*, edited by I. Barandiarán, L. Freeman, J. González Echegaray, and R. Klein, pp. 55–61. Monografías 14. Centro de Investigación y Museo de Altamira, Santander, Spain.

Castroviejo, S. E. A.
1986 *Flora Ibérica*. Vol. 1. *Plantas vasculares de la Península Ibérica e Islas Baleares (Lycopodiaceae–Papaveraceae)*. Real Jardín Botánico, CSIC, Madrid.
1990 *Flora Ibérica*. Vol. 2. *Plantas vasculares de la Península Ibérica e Islas Baleares (Platanaceae–Plumbaginaceae)*. Real Jardín Botánico, CSIC, Madrid.

Clark, G., and J. Menéndez-Amor
1979 Pollen samples from Liencres. *Quaternaria* 21:292–295.

Dupré, M.
1984 Palinología de los niveles VII a II del yacimiento del Ekain. In *El yacimiento prehistórico de la cueva de Ekain (Deba, Guipúzcoa)*, edited by J. Altuna and J. M. Merino, pp. 61–63. Eusko Ikaskuntza, Sociedad de Estudios Vascos, San Sebastián, Spain.
1988 *Palinología y peleoambiente. Nuevos datos Españoles.* Trabajos Varios 84. Servicio de Investigación Prehistórica, Valencia, Spain.
1990 Analisis palínico de la Cueva de Amalda. In *La Cueva de Amalda (Zestoa, Pais Vasco). Ocupaciones paleolíticas y postpaleolíticas*, edited by J. Altuna, A. Baldeón, and K. Mariezkurrena, pp. 49–51. Colección Barandiarán 4. Eusko Ikaskuntza, Sociedad de Estudios Vascos, San Sebastián, Spain.

Erdtman, G.
1971 *Pollen Morphology and Plant Taxonomy: Angiosperms.* Hafner, New York.

García-Anton, M., B. Ruiz, and F. M. Ugarte
1989 Análisis geomorfológico y palinológico de la turbera de Saldropo (Barazar, Zeanuri/Bizkaia). *Lurralde* 12:25–44.

Goeury, C., and J. Beaulieu
1979 Apropos de la concentration du pollen à l'oúde de la liqueur de Thoulet dans les sédiments minéraux. *Pollen et Spores* 21:239–251.

Gorrotxategi, X., M. J. Yarritu, M. Kandina, M. Sagarduy, M. J. Iriarte, and L. Zapata
1999 El poblado de montaña calcolítico al aire libre de Ilso Betaio (Bizkaia). *Isturitz* 10:3–204.

Herrera, M.
1995 *Estudio de la vegetación y flora vascular de la cuenca del Río Asón (Cantabria).* Serie Guineana 1. Universidad del País Vasco, Vitoria, Spain.

Iriarte, M. J.
1994a Estudio palinológico del nivel sepulcral del yacimiento arquelógico de Pico Ramos (Muskiz, Bizkaia). *Cuadernos de Sección. Prehistoria-Arqueología* 5:161–179.
1994b El paisaje vegetal de la Prehistoria reciente en el alto valle del Ebro y sus estribaciones atlánticas. Doctoral thesis, Universidad del País Vasco, Vitoria, Spain.
1995 Análisis palinológico del monumento megalítico de La Boheriza 2 (Karrantza, Bizkaia). Diputación Foral de Bizkaia, Bilbao. *Kobie* 22:158–161.
1997 El etorno arqueobotánico de la estación megalítica de Ataun-Burunda (Gupizkoa). I. Los dólmenes de Praalata y Aitxu (Ataun-Idiazabal). *Isturitz* 7:131–143.
1999 El etorno arqueobotánico de la estación megalítica de Ataun-Burunda (Gupizkoa). II. Los dólmenes de Unanabi y Napalatza (Idiazabal) y el túmulo de Txoritegui (Zerain). *Isturitz* 10:247–258.
2002 Cambios ambientales y adaptaciones humanas durante el inicio del Holoceno en el litoral cantábrica oriental. In *XV Congreso de Estudios Vascos*, vol. 1, pp. 139–151. Sociedad de Estudios Vascos, San Sebastián, Spain.

Iriarte, M. J., J. Mujika, and A. Tarriño
2005 Herriko Barra (Zarautz-Gipuzkoa). In *Mémoires de la Société Préhistorique Française*, 36, pp. 127–136. Paris.

Iriarte, M. J., C. Muñoz Sobrino, L. Gómez-Orellana, and P. Ramil-Rego
2006 Dinámica del paisaje en la Reserva de la Biosfera del Urdaibai durante el Holoceno. In *Actas del III Congreso Español de Biogeografía, Reserva de la Biosfera del Urdaibai (2004)*, edited by J. A. Cadiñanos, A. Ibabe, P. Lozano, G. Meaza, and M. Onaindia, pp. 113–117. Servicio de Publicaciones del Gobierno Vasco, Vitoria, Spain.

Isturiz, M. J., and M. R. Sánchez
1990 Investigaciones palinológicas en la prehistoria vasca. *Munibe* 42:277–286.

Leroi-Gourhan, A.
1966 Análisis polínico de la Cueva del Otero. *Excavaciones Arqueológicas en España* 53:83–85.
1971 Análisis polínico de Cueva Morin. In *Cueva Morin: excavaciones, 1966–1968*, edited by J. González Echegaray and L. Freeman, 359–369. Patronato de las Cuevas Prehistóricas, Santander, Spain.
1980 Analyse pollinique de la grotte du Pendo. In *El yacimiento de la cueva de El Pendo*, edited by J. González Echegaray, pp. 265–266. Bibliotheca Praehistorica Hispana, 17. Madrid.

López, P.
1981 Los pólenes de la Cueva de El Salitre. *Trabajos de Prehistoria* 38:93–96.
1986 Estudios palinológico del Holoceno español a través del análisis de yacimientos arqueológicos. *Trabajos de Prehistoria* 43:143–158.

López, P., J. López, and P. Uzquiano
1996 Paleoambiente y hábitat en las Marismas de Cantabria en las inicios del Holoceno: el caso del Abrigo de la Peña del Perro. In *Biogeografía Pleistocena-Holocena de la Península Ibérica*, edited by P. Ramil-Rego, C. Fernández Rodríguez, and M. Rodríguez Guitián, 333–348. Consellería de Cultura, Xunta de Galicia, Santiago de Compostela.

Mariezkurrena, K., and J. Altuna
1995 Fauna de mamíferos del yacimiento cosero de Herriko Barra (Zarautz, País Vasco). *Munibe* 47: 23–32.

Mariscal, B.
1983 Estudio polínico de la turbera del Cueto de la Avellanosa, Polaciones (Cantabria). *Cuadernos do laboratorio xeoloxico de laxe* 5:205–226.
1986 Análisis polínico de la turbera del Pico Sertal, de la Sierra de Peña Labra. In *Proceedings of the Symposium on Climatic Fluctuations during the Quaternary in the Western Mediterranean Regions*, edited by F. López Vera, pp. 205–220. Universidad Autónoma de Madrid, Madrid, Spain.
1993 Variación de la vegetación Holocena (4300–280 BP) de Cantabria a través del análisis polínico de la turbera del Alsa. *Estudios Geologicos* 49:63–68.
1995 La vegetación en Cantabria en los últimos 5000 años, según el análisis polínico de la turbera de Estacas de Trueba. In *Reconstrucción de los paleoambientes y cambios climáticos durante el Cuaternario*, edited by T. Aliexandre and A. Pérez-González, pp. 99–237. Centro de Ciencias Medioambientales, CSIC, Madrid, Spain.

Mary, G., and J. Medus
1993 El Holoceno de la región de San Vicente de la Barquera (Cantabria y Asturias). *El Cuaternario en España y Portugal* 2:961–963.

Mary, G., J. Medus, and G. Delibrias
1975 Le quaternaire de la côte asturienne (Espagne). *Bulletin de l'AFEQ* 1:13–23.

Moore, P. D., J. A. Webb, and M. E. Collinson
1991 *Pollen Analysis*. Blackwell Scientific, Oxford, UK.

Muñoz Sobrino, C.
2001 *Cambio climático y dinámica del paisaje en las montañas del noroeste de la Península Ibérica*. Universidad de Santiago de Compostela, Santiago, Spain.

Peñalba, M. C.
1989 Dynamique de végétation Tardiglaciaire et Holocène du centre-nord de l'Espagne d'après l'analyse pollinique. Doctoral thesis, Université d'Aix-Marseille III.

Punt, W.
1976 *The Northwest European Pollen Flora*. Vol. 4. Elsevier Scientific, Amsterdam.

Punt, W., and S. Blackmore
1991 *The Northwest European Pollen Flora*. Vol. 6. Elsevier Scientific, Amsterdam.

Punt, W., S. Blackmore, and C. G. S. Clarke
1988 *The Northwest European Pollen Flora*. Vol. 5. Elsevier Scientific, Amsterdam.

Punt, W., and C. G. S. Clarke
1980 *The Northwest European Pollen Flora*. Vol. 2. Elsevier Scientific, Amsterdam.
1981 *The Northwest European Pollen Flora*. Vol. 3. Elsevier Scientific, Amsterdam.
1984 *The Northwest European Pollen Flora*. Vol. 4. Elsevier Scientific, Amsterdam.

Ramil-Rego, P.
1992 *La vegetación Cuaternaria de las Sierras Septentrionales de Lugo a través del análisis polínico*. Universidade de Santiago de Compostela, Santiago, Spain.

Salas, L.
1993 Análisis palinológico de la turbera de Pico Ano, implicaciones en las variaciones climáticas del Holoceno Final. *Estudios sobre Cuaternario*, pp. 179–183. Universidad de Valencia, Valencia, Spain.

Sánchez, M. F.
1993 *De la taphonomie pollinique à la reconstitution de l'environnement. L'exemple de la région cantabrique*. British Archaeological Reports S-586. Oxford, UK.

Tutin, T. G. (ed.)
1980 *Flora Europaea*. Vol. 5. Cambridge University Press, Cambridge, UK.

Ugarte, F., M. Garcia, and B. Ruiz.
1986 La turbera de Saldropo: aspectos geomorfológicos y palinológicos. In *Gorbea mendikatearen antolaketarako azterketa*, edited by E. Gonzalez, pp. 110–199. Gobierno Vasco, Vitoria-Gasteiz, Spain.

Valdes, B., M. J. Diez, and I. Fernández (eds.)
1987 *Atlas polínico de Andalucía occidental*. Instituto de Desarrollo Regional de la Universidad de Sevilla and Diputación de Cádiz, Sevilla, Spain.

Yarritu, M. J., X. Gorrotxategi, L. Zapata, and M. J. Iriarte
1999 Investigación interdiciplinar del dólmen de La Cabaña 4 (Karrantza, Bizkaia). *Isturitz* 10:205–245.

Table 8.1. Stratigraphic position of palynological samples from the Cabin sondage (column H3)

Sample	Level	Cultural sequence
1	2	Bronze Age
2, 3	3	
4	4	Chalcolithic
5	5	
6, 7	6	
8	6.1	
9, 10	7	
11, 12	8	Neolithic
13–16	9	
17–19	10	
20	10.1	Mesolithic

Table 8.2. Stratigraphic position of palynological samples from the Mid-Vestibule Trench sondage (column L5, A and B)

Sample	Level	Cultural sequence
1a, 2a	5	Chalcolithic
3a–5a	301.0	Neolithic
6a	301.1	
7a–10a	303	
11a	303.1	
12a, 13a	303.2	
14a, 15b	303.3	
16b–18b	304	Mesolithic

Table 8.3. Statistically nonsignificant samples from the Cabin sondage

	Neolithic					Chalcolithic	Bronze
	Level 10			Level 9		Level 7	Level 2
	Square 19	Square 18	Square 17	Square 16	Square 15	Square 10	Square 1
Pinus sylvestris tp.	—	4	4	2	7	—	—
Quercus robur tp.	—	—	4	3	3	—	—
Corylus	11	30	20	16	20	2	1
Betula	—	—	1	2	—	—	—
Alnus	1	1	—	2	—	1	—
Juglans	—	—	—	—	—	—	—
Ulmus	—	—	—	—	1	—	—
IND	1	3	6	3	5	2	—
Poaceae	—	1	2	3	2	—	—
Ericaceae	—	—	1	—	—	—	—
Compositae liguliflora	—	2	—	—	—	—	—
Compositae tubuliflora	—	—	—	1	—	—	—
Centáurea	—	1	—	—	—	—	—
Cruciferae	—	—	1	—	2	—	—
Leguminosae	—	—	—	—	—	2	—
Caryophyllaceae	—	—	—	1	1	—	—
Labiatae	—	—	—	1	—	—	—
Dipsacaceae	—	2	—	1	—	1	—
Ranunculaceae	—	—	1	1	—	—	—
Cyperaceae	—	—	3	—	—	—	—
Polypodiaceae	—	1	6	12	11	1	—
Monolete	25	70	80	131	225	1	2
Trilete	3	11	20	17	81	—	—
TOTAL pollen + spores	**41**	**126**	**149**	**196**	**358**	**10**	**3**

Table 8.4. Statistically nonsignificant samples from the Mid-Vestibule Trench sondage (column A)

	Neolithic						
	Level 303.3	Level 303.2		Level 303.1	Level 303		
	Square 14	Square 13	Square 12	Square 11	Square 10	Square 9	Square 4
Pinus sylvestris tp.	—	—	—	—	1	2	—
Corylus	—	—	—	—	—	8	1
Betula	—	—	—	—	—	5	—
Alnus	—	—	—	—	—	—	1
IND	—	—	—	3	—	4	2
Poaceae	1	—	—	—	—	4	—
Compositae liguliflora	1	1	—	—	1	—	—
Compositae tubuliflora	1	—	—	—	—	—	—
Dipsacaceae	—	—	—	—	—	1	—
Polypodiaceae	1	—	—	—	1	2	—
Monolete	13	1	5	8	5	189	5
Trilete	—	—	—	—	—	4	—
TOTAL pollen + spores	**17**	**2**	**5**	**11**	**8**	**219**	**9**

Table 8.5. Statistically nonsignificant samples from the Mid-Vestibule Trench sondage (column B)

	Neolithic		Mesolithic	Magdalenian/Azilian
	Level 303.3		Level 304	Level 306
	Square 16	Square 17	Square 19	Square 20
Pinus sylvestris tp.	1	1	—	2
Quercus robur tp.	1	1	1	—
Corylus	5	2	—	4
Betula	—	1	—	—
Caprifoliaceae	—	—	—	1
Salix	1	—	—	—
IND	2	3	2	3
Poaceae	3	—	—	4
Compositae liguliflora	1	1		5
Compositae tubuliflora			1	1
Plantago	1	—	—	—
Rosaceae	—	—	—	1
Juncaceae	—	—	1	—
Monolete	36	37	60	34
Trilete	—	2	1	—
TOTAL pollen + spores	**51**	**48**	**66**	**55**

CHAPTER NINE

Analysis of Phytoliths from Holocene Levels of El Mirón Cave

Débora Zurro

TRANSLATED BY LAWRENCE GUY STRAUS

This chapter presents the preliminary results of phytolith analyses of three samples from two of the Holocene levels of El Mirón Cave. Due to the small size and stratigraphically incomplete nature of the overall Holocene sample, this study can only provide an idea about the potential of the post-Paleolithic strata for yielded evidence on recent vegetation and human gathering thereof in the vicinity of the site. Nevertheless, it does provide valuable information on the preservation of this kind of plant macrofossil, and it permits us to suggest specific hypotheses concerning their anthropic origin within the cave. Future comparative analyses of the phytoliths in the Upper Paleolithic and post-Paleolithic levels will allow us to explore the possibility of an intensification in the consumption of wild plant foods in the Mesolithic, as well as the beginnings of agriculture in the Cantabrian region (see Kubiak-Martens 1996, 2002; Zapata 2000; Zvelebil 1994).

The analysis of phytoliths, as explained in detail by Piperno (2006), provides archaeological research with many advantages in terms of the study of past environments and human activities (including construction, consumption, and subsistence) (Zurro 2006). First, it permits the detection of the presence of remains derived from the consumption of plants, independent of the action of fire on the original vegetal material. Second, it allows for the identification of plant parts beyond seeds, fruits, or wood, and it has demonstrated that phytoliths maintain a high level of survival in many different sedimentary contexts.

Materials and Methods

In the case of the present trial analysis conducted with sediments from El Mirón Cave, we wished to pursue several objectives, for which reason we took sediment samples from the two main excavation areas that were open during the summer excavation season of 2002. On the one hand, we sought to explore the possibilities of doing phytolith analysis in the humid, temperate but seasonal climate of the Atlantic Cantabrian region, in a cave site with a long record straddling the Pleistocene-Holocene boundary. This experiment might require modifications in the standard protocol of processing, slide preparation, and counting given the phytolith richness of the samples. On the other hand, with improvements in methodology and after overcoming possible technical obstacles, it became evident that the main interest of the study consisted of obtaining information on the consumption of plants in the interior of the cave vestibule by Mesolithic hunter-gatherer groups who utilized El Mirón as well as searching for evidence of the cultivation of cereal grasses beginning in the Neolithic. (The phytolith record from Upper Paleolithic levels will be described and discussed elsewhere.)

The three post-Paleolithic samples are all from Square 06, part of a sondage excavated in the center of the trench that connects the Cabin and Corral excavation areas in the El Mirón vestibule. Sample 06d/10–63 is from Level 304, which, although itself undated, is immediately bracketed by levels dating to 10,270 and 5790 BP (uncal.) (at ±1 sigma, 10,362–9818 and 4768–4540 cal BC) and is presumed to pertain to the Mesolithic, since among the few artifacts there are no ceramics, at least in the areas excavated. The other two samples, 06/9–59 and 06a/59–98, are both from Level 302 (pit 98a), uncertainly attributed to a late Neolithic or Chalcolithic and dated to 4910 BP (uncal.), or 3773–3641 cal BC.

The samples were processed following the methods of Madella, Power-Jones, and Jones (1998).[1] After drying a sample, the first step consisted of the extraction of carbonates by using a 10 percent solution of HCl. After this there was a deflocuation by means of $NaPO_3$, followed by extraction of organic matter with a 33 percent solution of H_2O_2. Finally, the amorphous silica was separated from the mineralogical fraction using a heavy liquid calibrated to a density of approximately 2.4 gm/cm^3 (SPT–polytungstate of sodium, Na_6 $(H_2W_{12}O_{40})$. H_2O.[2]

Once the silica fraction was procured, a rigid slide was made with a synthetic resin in order to study it by means of an optical microscope. Sample identification and counting were done with an Olympus BX-51 using a magnification of 400x. Counting proceeded up to a maximum of 400 phytoliths in the cases of slides that contained exceptional richness of sample materials. The final counts varied depending on corrections that were made after phytolithic identification.

Results

As could be expected in levels that are not very old, the phytoliths are well preserved. The percentages of physically damaged specimens (i.e., those that are physically degraded by breakage or surface erosion) or chemically damaged (i.e., partially dissolved) ones or those that are otherwise unidentifiable are all low in relation to those that could be identified. Damaged phytoliths are no more than 5–6 percent and unidentifiable ones range between 3 and 10 percent (figure 9.1).

The majority of the identified phytoliths pertain to the grasses (especially short cells) (table 9.1, figure 9.2). The samples are apparently relatively rich in vegetal material, with four silicate skeletons being identifiable in the first of the samples (06/302/9–59).[3] The presence of silicate skeletons has generally been interpreted as indicative of very stable preservation conditions. In all three analyzed samples there is a clear predominance of grass phytoliths, especially short cells, as noted above. These are followed by long cells (present in stems and leaves of grasses). Phytoliths of dicotyledons are very scarce, although one must keep in mind the low degree of silicification characteristic of these plants (Piperno 2006).

In the case of sample 06d/304/10–63, there is a noteworthy presence of five dendritic phytoliths (these being included in the final count of 108 long cells that correspond to the 26 percent given in table 9.1). Despite their small quantity relative to the total number of phytoliths in this group and even more so relative to the total of 400 counted phytoliths, the identification of dendritic morphotypes is of great importance because they correspond specifically to the cereal inflorescences (spikes, presumably of wild grasses).

It is also interesting to note that sample 06a/302/59–98 displays greater variability than the other two. This could indicate a greater original variety of plant materials in this particular spit within pit 98a than in the other Neolithic/Chalcolithic level or the Mesolithic one. The pit could have been used as a dumping place for residues of plant food consumption or of manufacturing activities involving plant matter.

Despite the scarcity of samples from the Holocene levels, the preliminary analyses reveal the presence of plant materials not visible without the use of the proper

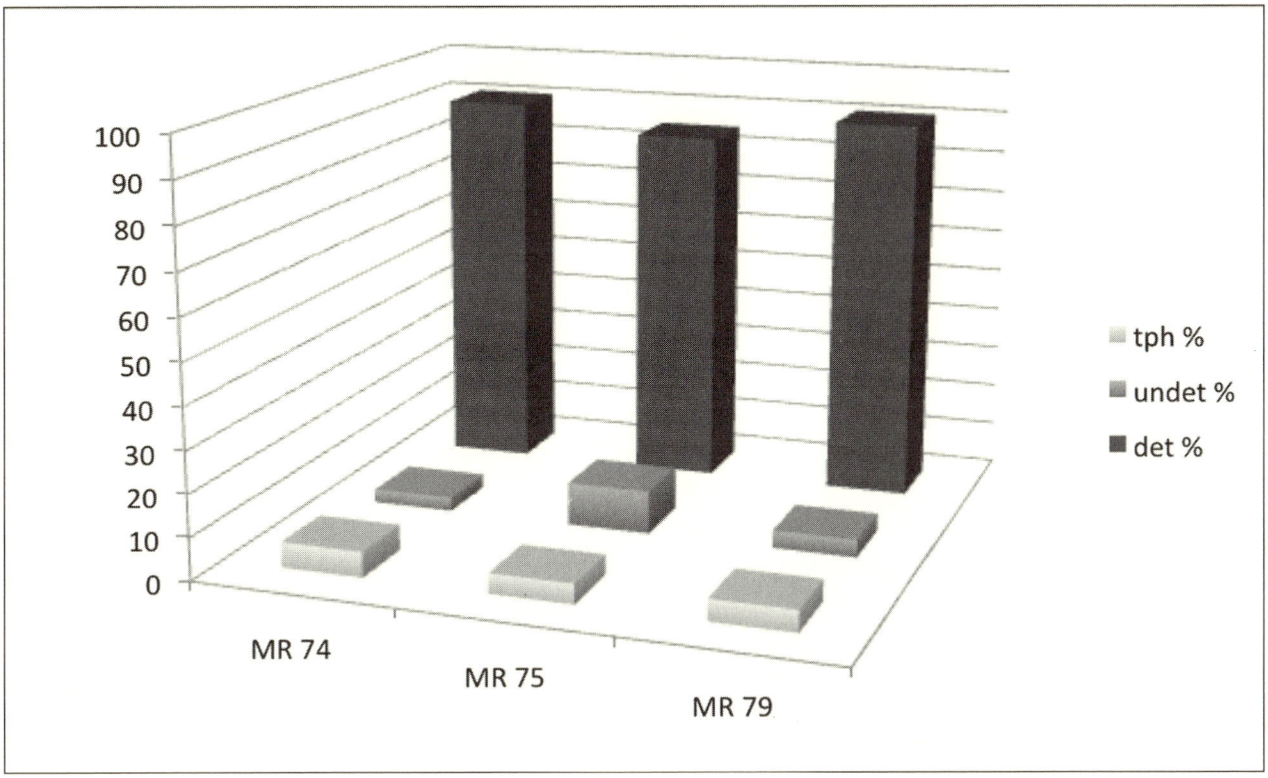

Fig. 9.1. Relationship between identified phytoliths and unidentified ones. Graph by Débora Zurro.

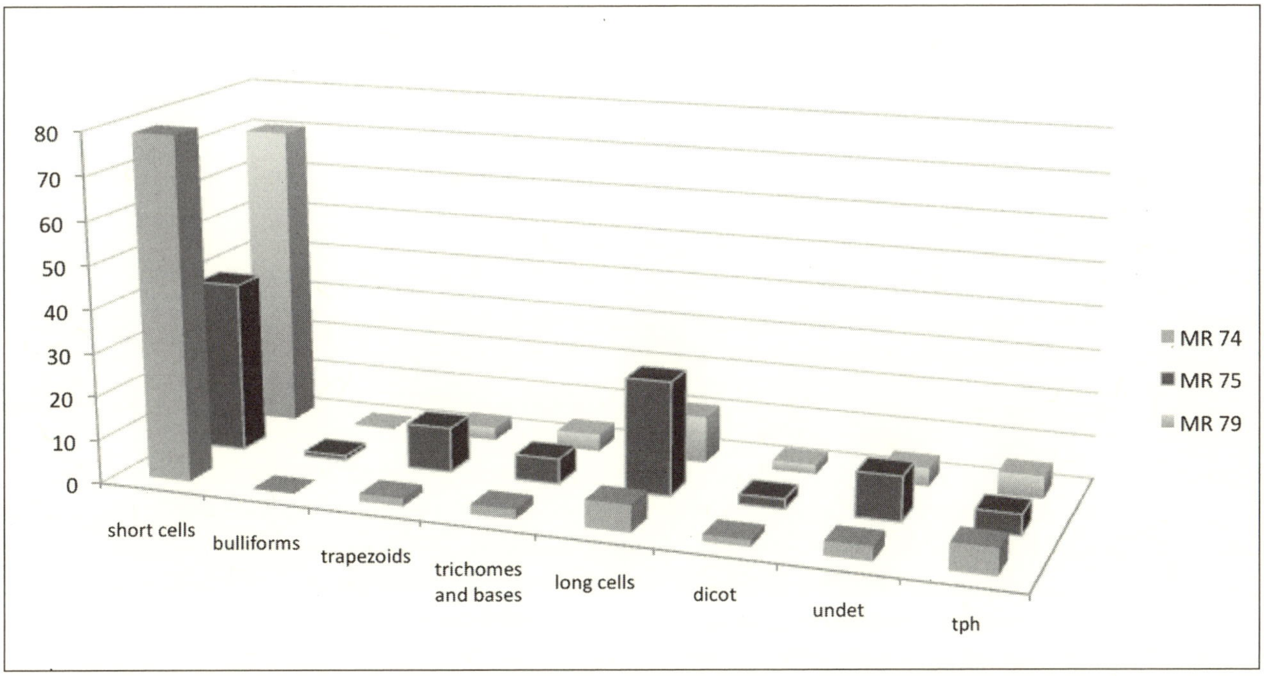

Fig. 9.2. Spectrum of phytolithic morphotypes among the analyzed samples (Levels 302 and 304). Sample 74 is from Square 06, Level 302, spit 9, no. 59; Sample 79 from Square 06, Level 302, spit 9, no. 98; Sample 75 from Square 06, Level 304, spit 10, no. 63. Graph by Débora Zurro.

specialized recovery methods. In addition, given the distance from the mouth of the cave at square 06 and given the archaeological context of the discovery of these phytoliths, it is very reasonable to argue that the plants were brought into the cave by humans.

Future analyses of sediments both in El Mirón and in other regional sites will no doubt shed more light on questions such as the beginnings of agriculture as well as on the general use of vegetal resources during the Mesolithic and Neolithic.

Notes

1. Processing of the samples was done in the Pitt-Rivers Bioarchaeology Laboratory at the McDonald Institute for Archaeological Research, University of Cambridge, to which I wish to express my profound thanks.
2. The amorphous silica fraction includes not only silicophytoliths but also other types of silicate particles of biological origin (e.g., diatoms, sponge spicules), as well as fragments of volcanic glass.
3. Standardized quantification of the amounts of phytoliths by AIF or per gram of sediment (Albert et al. 2000) is not presented here, while awaiting the mineralogical results by means of x-ray diffraction that are currently pending at the Instituto de Ciencias de la Tierra "Jaume Almera," CSIC.

References Cited

Albert, R., S. Weiner, O. Bar-Yosef, and L. Meignen
2000 Phytoliths in the Middle Palaeolithic deposits of Kebara Cave, Mt. Carmel, Israel: study of the plant materials used for fuel and other purposes. *Journal of Archaeological Science* 27:931–947.

Kubiak-Martens, L.
1996 Evidence for possible use of plant foods in Palaeolithic and Mesolithic diet from the site of Calowanie in the central part of the Polish Plain. *Vegetation History and Archaeobotany* 5:33–38.
2002 New evidence for the use of root foods in preagrarian subsistence recovered from the Late Mesolithic site at Halsakov, Denmark. *Vegetation History and Archaeobotany* 11:21–31.

Madella, M., A. Power-Jones, and M. Jones
1998 A simple method of extraction of opal phytoliths from sediments using a non-toxic heavy liquid. *Journal of Archaeological Science* 25:801–803.

Piperno, D.
2006 *Phytoliths: A Comprehensive Guide for Archaeologists and Paleoecologists*. AltaMira Press, Lanham, MD.

Zapata, L.
2000 La recolección de plantas silvestres en la subsistencia mesolítica y neolítica. Datos arqueobotánicos del País Vasco. *Complutum* 11:157–169.

Zurro, D.
2006 El análisis de fitolitos y su papel en el studio del consume de recursos vegetales en la prehistoria: bases para una propuesta metodólogica materialista. *Trabajos de Prehistoria* 63:35–54.

Zvelebil, M.
1994 Plant use in the Mesolithic and its role in the transition to farming. *Proceedings of the Prehistoric Society* 60:35–74.

Table 9.1. Percentages of phytolith morphotypes identified in samples from El Mirón Cave Levels 302 and 304

Sample-Square/ Level/ Spit-Item No.	n	Grass cells					Dicots	Indeterminate	Degraded
		Short cells	Trichome and bases	Trapezoid	Bulliform	Long cells			
74-O6/302/9-59	398	79	2	2	0	6	2	3	6
79-O6a/302/59-98	408	41	4	3	0	11	2	4	5
75-O6d/304/10-63	398	39	6	10	1	26	2	10	5

CHAPTER TEN

Neolithic, Chalcolithic, and Bronze Age Plant Remains from El Mirón Cave

Evidence on the First Farming Communities in the Cantabrian Region

Leonor Peña-Chocarro

This chapter reports on the analysis of the plant macroremains recovered from El Mirón Cave (Ramales de la Victoria, Cantabria) during three seasons of excavation (1997, 1998, and 1999).

El Mirón is a large cave situated in the eastern part of Cantabria province (northern Spain) on the edge of the Cantabrian Cordillera, at the interface between the Cantabrian coast and the Castilian plateau. It is located at 260 m above sea level, in a strategic position that dominates the Ruesga Valley and its tributary gorges: the Calera and Gándara.

El Mirón has a long stratigraphic sequence from the medieval period to the Middle Paleolithic, with levels assigned to Mousterian, Early Upper Paleolithic, Solutrean, Magdalenian, Azilian, Mesolithic, Neolithic, Chalcolithic, and Bronze Age periods. Excavations have been focused on two different areas within the vestibule: the outer vestibule, or Cabin, and the vestibule rear, or Corral (González Morales and Straus 1997, 2000a, b; Straus and González Morales 1996, 1998, 2001). The samples analyzed here cover the Neolithic, Chalcolithic, and Bronze Age levels.

Most discussions about the Neolithic in the Cantabrian area have failed to consider the role of arable farming and, more generally, the role of plants in human subsistence, denying their contribution to the economy of the first Neolithic communities (for a review of current theories on the adoption of agriculture in the Cantabrian area, see Zapata 2002b). It is only during the past few years that an increasing interest in agriculture has led to the application of systematic, on-site recovery techniques at several sites. This is starting to produce a new body of data based on archaeobotanical studies (e.g., Iriarte 1994a, b, 1998; Iriarte and Zapata 1996; Sánchez Goñi 1993; Zapata 2000, 2002a, b). Nevertheless, plant remains from Neolithic, Chalcolithic, and Bronze Age sites are still scarce, making the study of El Mirón of great interest for understanding the development of the first farming communities of the region.

The main objectives of this work are: (a) to identify the range of plants used by the inhabitants of the cave during the Holocene in order to improve our understanding of their use as part of human subsistence at the site; (b) to detect the appearance of the first cultivated plants, and (c) to analyze the variety and frequency of plant remains. The possibility of studying plant remains from different periods will also allow us to monitor diachronic change in their use.

Methodology

The interdisciplinary nature of the research carried out at El Mirón Cave has led to the application of an excellent strategy for recovery of organic material. Samples ranging from 37 to 0.5 l were taken from every level or feature excavated in the cave.

For the plant macro remains a flotation machine was set up at the field laboratory in the town of Ramales de la Victoria, where water was available. The machine was built especially for the purpose of recovering archaeobotanical material. To provide stronger water pressure, a pump was used. The working mechanism of the machine is very simple: water is pumped inside the tank, and when it is full, the soil sample is poured in over a 1 mm mesh situated inside the tank. The soil in contact with the water disaggregates, and the charred plant remains float and flow over the drain spout, being collected in a sieve lined with 250 μ mesh.

Information was recorded about each particular sample, including: context, volume, mesh size, and so on. Flotation samples were dried in the shade in a double 250 μ mesh to avoid possible loss of material after drying. Heavy residues were spread out and sun dried.

Due to the large amount of samples recovered, a selection was made in order to facilitate the study. In the lab, flotation samples were measured (they varied from 135 ml to less than 5 ml) and sieved to 2 mm, 1 mm, 500 μ, and 250 μ in order to facilitate sorting. All samples were fully sorted using a Wild M8 low-power microscope at magnifications of 6x to 50x. Seeds, chaff, and probable parenchymous tissue fragments were separated from the wood charcoal, which was the main component of the samples. All preserved plant remains from this site are charred.

Identifications have been carried out using the reference collection of the Laboratorio de Archeobiologia of the Museum of Como, Italy, and with the aid of several atlases currently used in archaeobotany.

Cereals have been ordered and named following the standard binomial classification, and they have been placed at the beginning of this chapter's tables, followed by the wild taxa in alphabetical order. Cereal fragments refer to parts of the grain, different from the embryo, which cannot be identified to genus or species level.

Due to conservation problems, some items could not be identified because the critical anatomical elements were absent. They were classified as "indeterminate." In some other cases we have used the term "type" (abbreviated "tp."; e.g., *Bromus* tp.), which indicates that the specimen is similar to the cited genus. As for parenchymatous fragments, we have only recorded their presence or absence within the samples.

Results

This section summarizes the plant remains only from the Holocene levels (Bronze Age, Chalcolithic, and Neolithic strata), from which a total of 368 samples have been analyzed (table 10.1).

Results of the analysis are presented below, including a description of the samples and the major plant categories present. In general, the number of remains is very low, and therefore it has not been possible to carry out analytical calculations.

THE NEOLITHIC LEVELS

Neolithic samples were collected from two different areas of the cave: the Cabin area, or outer vestibule (Strata 8, 9, 10), excavated in 1997 and 1998, and the Mid-Vestibule Trench (Strata 303, 303.1, 303.2, 303.3), excavated in 1999 to connect the outer vestibule with the vestibule rear, which itself presents no post-Paleolithic deposits.

The Cabin Area: Stratum 10

The excavation of this stratum during both the 1997 and 1998 seasons produced several samples for archaeobotanical study. Two AMS dates of 5570±50 BP and 5690±50 BP are available for Square I3. From the 1997 excavation, only one sample was analyzed (Sample 6, from Square I3B, Spit 18), while from the 1998 season 4 samples were sorted out (Samples 459, 460, 463, and 476 from Squares H2C and J4ABC). The only plant remnant is a seed of a sedge species (*Carex* sp.) from the 1997 season. In terms of seeds and fruits, the other samples were sterile.

Cabin Area: Stratum 9

The 1997 season of excavation in the Cabin area produced a total of 38 samples (all from target Square I3) assigned to the Neolithic period (Strata 9, 9.6, 9.7, 9.8, and 10). Three dates, ranging from 5250 BP (9.6) to 5170 BP (9), are available from these strata. A total of 25 samples have been analyzed (Sample 6 from Stratum 10 has been discussed above). In addition, six more samples from the 1998 excavation have also been studied. These come from Squares H3, H4, I4, and J4. From the whole group of samples, only eight have yielded plant remains (table 10.2). The assemblage is much reduced, consisting of two badly preserved seeds of the families Caryophyllaceae and Cyperaceae as well as some remains of parenchymatous tissues. So far, no evidence of cultivated plants has been identified.

Cabin Area: Stratum 8

A total of 19 samples from Stratum 8 (Squares H2, H3, H4, H5, I4, J4, and V7), excavated in 1998, have been investigated. Table 10.3 shows the few plant remains recovered from the three samples that contained seeds. The remains consist of a single seed of *Luzula* sp., a member of the Juncaceae family, a nonidentified seed and a possible, extremely distorted cereal fragment. The degree of preservation of this latter fragment prevents precise identification. As no other remains of cultivated plants have yet been found in this level, extreme caution should be taken when interpreting these remains.

Despite the number of samples analyzed (52) from Strata 8, 9, and 10, very few plant remains were recovered. The few samples containing seeds show an extremely low density of remains. Poor preservation has prevented precise identification of the specimens. Two large families (Caryophyllaceae and Cyperaceae) are represented in Stratum 9, whereas the genus *Luzula* is present in a sample of Stratum 8.

In addition to the vesicular fragments deriving most probably from the charring of sugar-rich tissues, different categories of parenchymous tissues were recovered. These are remains of the vegetative plant tissue of roots, rhizomes, tubers, bulbs, and so on, which, without any doubt, played an important role in the subsistence of prehistoric groups. Their identification is problematic, and only over the past 10 years have techniques been developed for their identification, using anatomical characters (Hather 1993, 1994).

THE MID-VESTIBULE TRENCH

Samples from this trench have produced an interesting assemblage of plant remains characterized by the presence of different cereal species. Four different strata have been studied: 303, dated to 5500±90 BP; 303.1, with a date of 5520±70 BP; 303.2; and 303.3, dated to 5790± 90 BP. All the samples come from Squares L5, M5, and N5. The number of samples analyzed is 79, of which 40 have yielded plant remains. Tables 10.4, 10.5, 10.6, and 10.7 show the various taxa identified. Samples marked with "*" are tiny samples that do not come from flotation but rather from wet sieving. No details on soil volumes are available for these.

Stratum 303

One hundred samples from Squares M5, N5, and L5 have been studied, of which only 23 yielded plant remains (table 10.4). The assemblage is composed of 65 remains (seeds and fruits). An important element among these samples are the cereals, which appear in Samples 39, 52, 54, 57, 139, 231, 233, and 237 in squares L5, M5, and N5. Apart from cereals, which represent about 25 percent of the total assemblage, several wild taxa, including acorns and hazelnuts, are also represented.

Stratum 303.1

A total of 23 samples (table 10.5) have been analyzed, but only 7 produced seeds. Very few plant remains have been retrieved (17), and no cereals have been identified at all. The wild taxa are represented by few specimens belonging large families such as Caryophyllaceae or Poaceae.

Stratum 303.2

From a total of 30 samples studied, only 4 produced plant remains (table 10.6). The number of seeds recovered is extremely low (only nine), with an indeterminate cereal and five different wild taxa. The only fruit species represented is hazelnut.

Stratum 303.3

This is the oldest Neolithic level in El Mirón. Twenty-six samples have been floated, but plant remains were recovered from only six of them. Seeds are extremely rare, represented by one single emmer grain, one fragment of hazelnut pericarp, and seven indeterminate seeds (table 10.7).

What follows is a description of the main cereal categories retrieved from the Neolithic levels of the Mid-Vestibule Trench. Identification of wheat species based on grain morphology is very difficult due to the enormous overlapping among species. However, there is a series of morphological traits that can be used to distinguish different species. Three wheat species have been so far identified, namely einkorn (*T. monococcum*), emmer wheat (*T. dicoccum*), and a naked wheat (*T. aestivum/durum*). The numbers are extremely reduced, and in most cases these categories are represented by just single grains. The caryopses are badly preserved and often greatly distorted, so in some cases identification remains uncertain between two different species, such as *T. monococcum/dicoccum*. In all cases, only grains have been recovered and there is no evidence of chaff remains.

***Triticum monococcum*:** The only example of this species showed a tapered end with a high back and rounded ventral surface with very narrow grooves.

***T. dicoccum*:** The grains show the characteristic pointed ends and marked dorsal ridge.

***T. durum/aestivum*:** Distinction between the tetraploid (*T. durum*) and the hexaploid (*T. aestivum*) on the basis of grain alone is not possible, so in the absence of chaff remains both species are considered together (Hillman et al. 1996).

***Triticum* sp.:** This designation was given to all grains or fragments that lacked features attributable to specific categories.

Cereal indeterminate: This category includes extremely badly preserved grains and fragments of domesticated cereal grains that we could not identify even to the genus level. These were distinguished from wild grasses by their size and general outline.

Wild taxa: These include some examples identified to genus level, such as *Artemisia* sp., *Galium* sp., and *Lolium* sp., and larger categories represented by the families Poaceae, Caryophyllaceae, or Leguminosae. Numbers are again very small. Acorns are represented by small cotyledon fragments, whereas from the hazelnuts only fragments of the hard pericarp tend to be preserved. Hazelnut shell fragments are characterized by a typical curved shape with a smooth external surface and the presence of vascular bundles visible in cross-section. In addition, several parenchymatous fragments have also been isolated

THE CHALCOLITHIC AND THE BRONZE AGE

Stratum 7 and Subdivisions

Twenty-eight samples from Squares H1, H2, H3, H4, I4, and J4 have been examined, but only six have yielded plant remains (table 10.8). The number of taxa represented is extremely low: three categories of cereals and one wild plant (acorn). A conventional radiocarbon date is available from Stratum 7, 3740±120 BP, but it is undoubtedly too young.

Stratum 6 and Subdivisions

Only 5 (table 10.9) of the 24 studied samples from Squares H1, H2, H3, H4, I4, and J4 have yielded plant remains. The total number of remains of this level is very low (19), with cereals present in a couple of samples and wild plants represented by only three taxa (acorns, hazelnuts, and the family of the Caryophyllaceae).

Stratum 5 and Subdivisions

Nineteen samples from Squares H1, H2, H3, H4, I4, and J4 have been analyzed, from which only 10 have yielded plant remains (table 10.10). Strata 5.1 and 5 have been dated to 4120±50 BP and 3820±240 BP. Free-threshing wheats (*Triticum aestivum/durum*) are present in almost all samples, whereas hulled wheats have not been identified. A single rachis of a free-threshing wheat has been found in Sample 322. This stratum has not produced any large legumes, but there are remains of edible fruits such as acorns and hazelnuts. A single naked wheat fragment has been recovered, so far representing the only chaff remain from the whole site. Although free-threshing or naked wheats can be separated from their chaff remains, the rachis from El Mirón is not well preserved, preventing its identification to species level. It is, however, clear that it belongs to a naked wheat form.

Stratum 4

A total of 25 samples from Squares H3, H4, I4, and J4 have been studied, producing 336 plant remains (table 10.11). Both free-threshing wheats (*Triticum aestivum/durum*) and hulled wheats (*T. dicoccum)* are present. Naked wheats are the most abundant and ubiquitous of the crop species, while emmer appears in only five samples. An interesting feature of this stratum is the presence of a single pea, which is the only example of a cultivated pulse from the site. In addition, a wide range of wild plants are represented, with more than 20 different taxa, including acorns (*Quercus* sp.), hazelnuts (*Corylus avellana*), and several other wild species.

Stratum 3 and Subdivisions

Twenty-three samples from Squares H1, H2, H3, and H4 have been analyzed, 15 of which did not produce seeds. Eight samples (table 10.12) contained seeds, and only two (221 and 348) yielded cereals, represented by few grains of a free-threshing wheat (*Triticum aestivum/durum*). No chaff remains have been identified. The few other taxa represented are wild plants. Due to poor preservation, most of the identifications have some degree of uncertainty, which is expressed with the name "type" (e.g., *Bromus* tp.). In some other cases, identification has been possible only at the family level (Chenopodiaceae or Caryophyllaceae). Fragments of parenchymatous tissues were found in three samples. Stratum 3 has an AMS date of 3700±40 BP.

In addition to these samples, several others have been examined from Stratum 2 (Sample 138 from Square J4B, Spit 2), Stratum 14 (Sample 92, Square I2C, Spit 24), Stratum 15 (Samples 93 and 100, Squares I2C and I3D, Spit 15) and Stratum 16 (Sample 101, Square I2D, Spit 16) from the 1997 season. Five samples from Stratum 10.1 (1999 season) have been also scanned: Sample 400 (Square J4A, Spit 34), Sample 401 (Square J4B, Spit 34), Sample 406 (Square J4A, Spit 34), Sample 417 (Square J4C, Spit 34), and Sample 426 (Square H2D, Spit 18). None of them produced plant remains.

Discussion

The excavation of El Mirón has produced an interesting assemblage of plant remains from a long sequence that spans the Upper Paleolithic and post-Paleolithic periods, unique for the region. While numbers of seeds and fruits from the Holocene strata are low, they represent an important source of information for our understanding of the first farming communities in the region.

THE FIRST AGRICULTURE IN THE CANTABRIAN REGION

The archaeobotanical data suggest that at least three different cereals (all wheat species) were used at El Mirón during the Neolithic: emmer (*Triticum dicoccum*), einkorn (*T. monococcum*), and a naked wheat (*T. aestivum/durum*). Strata 303.3, 303.2, and 303, from which the cereal grains were retrieved, have been radiocarbon dated between 5790±90 BP (4840–4410 cal BC) and 5500±90 BP (4520–4050 cal BC). The single grain of emmer from Stratum 303.3 (Square M5C, Spit 10, Specimen no. 374) has been directly dated by AMS to 5550±40 BP (4450–4340 cal BC with a ±1 sigma range; GX-30910). For now, at least, the El Mirón cereals are absolutely the earliest-known cultivated macroremains from the Cantabrian region of Spain.

The number of cereal grains is very low (table 10.13), and this limits their interpretation. However, even if small, the recovered amount is clear evidence of the presence of cultivated plants during the fifth millennium cal BC.

Despite the limited number of cereal remains (a total of 13 grains and 7 fragments), it is interesting to note that the variety of species represented is relatively large (einkorn, emmer, and free-threshing wheats). Such variety may suggest that a fully developed agriculture was available to the inhabitants of El Mirón. The only missing Mediterranean cereal is barley, which is otherwise the most common species for the Neolithic of the region (see below). No chaff has been identified from the Neolithic.

For comparison with the El Mirón evidence, it is possible to consider a few other sites along the Atlantic coast, particularly in the nearby Basque Country, from where most of the archaeobotanical data for this period come. The site of Kobaederra (Vizcaya) is particularly relevant, as it has a stratigraphic sequence that spans the period between the Bronze Age and the Neolithic. What is more, until now it had the earliest evidence of cultivated macroremains in the region with a single barley grain AMS dated to 5375±90 BP (4360–3990 cal BC) (Zapata 2002b). The complete assemblage of cultivated plants from the Neolithic period is composed of one emmer (*T. dicoccum*) grain and one barley (*Hordeum vulgare*) grain. In Pico Ramos (Vizcaya), Zapata (Zapata et al. 2005) has also identified another barley grain dated to 5370±40 BP (4330–4050 cal BC). A third Neolithic barley grain from a level dated to 5200–5000 BP (4220–3800 cal BC) comes from the site of Lumentxa (Vizcaya) (Zapata 2002b). In addition to the macroremains, cereal pollen from the site of Herriko Barra (Guipúzcoa) has been identified in a context dated to 6010±90 BP (5210–4690 cal BC) (Iriarte, Mujika, and Tarriño 2005). Within this context the assemblage from El Mirón is of paramount importance, as it confirms the existence of a well-developed agriculture already by the fifth millennium cal BC, perhaps slightly before the appearance of the first megaliths.

As evidenced above, and despite the enormous efforts put into recovery of plant remains, data from the Cantabrian region are very scarce and represented by just a few grains. This scarcity appears as one of the characteristics of the first archaeobotanical assemblages. Different causes may account for such paucity of remains:

1. The application of systematic recovery techniques has only recently become a common practice in research projects, providing the retrieval of the first assemblages of plant remains. In fact, the application of systematic flotation at several Neolithic sites in the region has resulted in the emergence of a scanty but most interesting body of data on Neolithic agriculture (see Zapata 2002b).
2. All sites where flotation has been applied are caves that, as has already been noted (Zapata 2002b), may not be the most representative examples of Neolithic occupation.
3. Several taphonomic issues (e.g., high humidity) may affect the conservation of plant remains.

In spite of such a paucity of remains, however, the assemblages show the presence of four different cereals: barley, emmer, einkorn, and a naked wheat in the earliest Neolithic of the region.

INFLUENCES ON CANTABRIAN AGRICULTURE: WHAT CROPS DID PEOPLE GROW DURING THE NEOLITHIC?

A further point relates to influences on the earliest Cantabrian agriculture. As has been discussed elsewhere (Zapata 2002b; Zapata and Peña-Chocarro in press[a]; Zapata et al. 2005), it was the northern part of the Iberian Peninsula that most likely had access to a wide range of crops, which arrived from different areas, such as the Pyrenean region, the Ebro Valley, and the Meseta.

Early sites are well known in the Pyrenean area, but the only evidence of plant remains comes from Balma Margineda (Andorra). It has yielded an assemblage that includes naked barley and naked wheats in a context dated to 6850±160 and 6670±120 BP (Marinval 1995). As for the Ebro Valley, sites are yet scarce but recent archaeobotanical research at several sites is producing an interesting botanical dataset. The site of Los Cascajos (Navarra) (García Gazólaz and Sesma 1999), currently under research by the author and L. Zapata, has already produced several barley grains in a context dated to 6185±75 BP (5300–4860 cal BC) (Zapata et al. 2003). In addition, recent archaeological investigations at the Neolithic complex of Ambrona (Soria), thought to have been colonized by Neolithic groups from the Ebro Valley (Kunst and Rojo 1999), indicate the presence of agriculture already by the sixth millennium BC. Hulled wheats, and particularly einkorn, seem to be the main cultivated species (Zapata et al. 2005; Stika 1999). Not far from Ambrona, the cave of La Vaquera has also yielded an assemblage comprising barley, einkorn, emmer, and free-threshing wheats (López et al. 2003).

Within this scenario and based on the archaeobotanical record, it appears that barley played a significant role as a crop for the first Neolithic communities of the area. It is present in all Neolithic sites of the Basque Country: Kobaederra, Lumentxa, and Pico Ramos (Zapata 2002b), as well as in Navarra at Los Cascajos (Peña-Chocarro and Zapata in press). It should be also remembered that both barley and free-threshing wheats (*Triticum aestivum/durum*) are the main elements of the Early Neolithic in southern France (Marinval 1991a, b, 1993). Their absence from El Mirón is difficult to interpret; they are absent not only from the Neolithic but also from the Chalcolithic, when one should expect a widening of cultivated species.

The scarcity of identifiable material preserved within the samples could be the reason for their absence here.

Another interesting crop is emmer, whose role within the first agriculture of the Cantabrian area appears quite important. Emmer is a tetraploid wheat belonging to the so-called hulled wheats, which are characterized by having the glumes adherent to the grain. These are not removed during threshing, as happens with naked cereals, and therefore they need to be dehusked prior to consumption. Emmer is a quite rustic and resistant cereal, well adapted to mountain areas. It has been identified in El Mirón, Kobaederra, Los Cascajos (from the early Neolithic), and Lumentxa (late Neolithic). In southern France it started to be present later on during the Epicardial. Emmer is also represented in sites in the northern Meseta (Vaquera and Ambrona). Its absence or little representation in the early Neolithic archaeobotanical record of southern France may suggest that emmer arrived into the Cantabrian region via the Ebro Valley from the Mediterranean region, where it is well known from the beginning of the Neolithic.

As for einkorn (only one, single grain from El Mirón), another hulled wheat, it is absent from both the Cantabrian region and southern France but appears well represented in the Meseta, particularly at Ambrona and with little importance at La Vaquera. There is also good evidence of its cultivation in the Mediterranean area.

Based on the available data, L. Zapata and I suggested (Zapata 2002b; Zapata and Peña-Chocarro in press[a]; Ibáñez et al. 2001) the idea of a specific crop pattern for Atlantic agriculture. Species such as emmer and barley would have been selected for their rusticity and adaptability to the specific Atlantic conditions (mountain and humid). As mentioned before, emmer is a rustic species well adapted to mountain areas, where it has survived until the present (Peña-Chocarro 1996, 1999; Peña-Chocarro and Zapata 1998), whereas barley does well in extreme climates. Central to these observations is the presence of naked wheats at El Mirón, which opens an interesting subject about the range of crops of the first Cantabrian Neolithic. Contrary to hulled wheats, free-threshing or naked wheats are characterized by the presence of loose glumes, which are easily removed during threshing. Two species are included into this group: the tetraploid, durum-type wheats, which have a more Mediterranean distribution, and the hexaploid bread-wheat. Free-threshing or naked wheats are, generally speaking, less adapted to these conditions, even though the range of ecological environments in which wheats may thrive is very large. One should also bear in mind that we know very little about the ecological adaptability of the wheat varieties that grew during the Neolithic, and therefore it is possible that particular varieties of free-threshing wheats, adapted to the Atlantic conditions, were under cultivation in the area. However, many more data are still necessary in order to get a better understanding of this aspect. In any case, free-threshing wheats (*Triticum aestivum/durum*) were the commonest cereals of the French Atlantic Neolithic, probably due to contacts with the Mediterranean (Marinval 1993). They are well represented in the Pyrenees, at Balma Margineda (Andorra) in contexts dated to 6850±160 BP and 6670±120 BP (Marinval 1995), and in the Meseta at La Vaquera (Segovia), where they were abundant.

To conclude, it is fair to assume that the agriculture of the Cantabrian region has a clear Mediterranean origin with inputs probably from different regions and perhaps individual peculiarities characteristic of the different source areas.

AGRARIAN PRACTICES

Our knowledge of Neolithic agriculture is generally limited to the range of possible crops grown in a particular area, but we know very little about the way crops were managed in the past (e.g., cropping systems, agrarian practices, techniques, implements, or crop-processing). Many authors (Peña-Chocarro 1999; Hillman 1981, 1984; Jones 1984; Ibáñez et al. 2001) have shown that ethnography is an appropriate source of information to gain a better understanding of these problems. It is true, however, that with the little information we currently have, tackling these questions is very problematic.

Many of the operations involved in growing cereals, from sowing to grinding (including weeding, threshing, winnowing, sieving, etc.) can be recognized in appropriate samples. At El Mirón the type of remains necessary to address such questions (chaff and weed seeds) are absent or extremely limited (see above), so very little can be said.

Current research combining archaeobotanical work, microwear analysis, and ethnographic investigations

(Ibáñez et al. 2001) has led to suggestions of alternative harvesting methods, different from the conventional sickles. Use-wear analysis of the lithic materials from several Neolithic sites in the Cantabrian region points to the absence of sickle elements, even in sites where agriculture is well documented, such as Kobaederra. If sickles were not used during the Neolithic, what alternative harvesting methods were practiced? Ethnographic research carried out over the past few years (Peña-Chocarro 1999; Peña-Chocarro et al. 2005; Peña-Chocarro et al. 2000; Ibáñez et al. 2001) proves the existence of a series of alternative methods used in cereal harvesting. Cereals can be uprooted or their ears plucked using particular tools such as the *mesorias* (a pair of about 50 cm long wooden sticks joined by a piece of rope), which actually snatches whole ears from their stems. These plucking clamps are still used in Asturias to harvest emmer and spelt, as well as in some valleys of Nepal (Toffin 1983) and in the Caucasus (Sigaut 1978; Steensberg 1943; Bregadze 1982). Different factors may favor this particular choice: ecogeographical conditions, field size, straw uses, height of the plant, and particularly the type of cereal involved (plucking clamps can only be used with hulled wheats, where rachises are semi-brittle, allowing the whole ear to be removed).

The lithic industry from El Mirón is still under study, and no microwear results are yet available. However, the occurrence of free-threshing wheats, which are characterized by the presence of tough rachises, would not allow the use of implements such as the mesorias. If sickles were not used, uprooting may be the alternative harvesting method. Techniques such as these are generally used in areas with humid environments and by groups whose economy does not heavily rely on agriculture. In the first case, harvesting is not a critical activity; it can be extended over a long period, as maturation takes place gradually. In the second, cereals are cultivated in small fields that can be easily harvested without sickles (Ibáñez et al. 2001).

Another interesting point refers to the processing of cereals. The two different types of cereals identified at El Mirón—hulled wheats (einkorn and emmer) and free-threshing wheats—require different processing techniques. In order to free the grain from the husks, hulled wheats need to be dehusked by pounding, whereas free-threshing wheats are dehusked during threshed. The management of these two cereal types entails a certain complexity of agriculture, which probably characterized the agricultural system of El Mirón.

COLLECTING FROM THE WOODS: THE ROLE OF FRUIT TREES AND WILD PLANTS

Plant gathering was certainly an important activity for the subsistence of the first European farmers. Research carried out during recent decades has emphasized the role of wild plant foods in the diet of many hunter-gatherers. During the Holocene wild plants were certainly abundant, easy to collect, and in many cases, likely to have been stored (Zapata 2000, 2002a). These characteristics made wild plants an important human resource. However, the archaeobotanical record is very limited and does not seem to reflect such an important role. Again, a major problem has been the lack of extensive programs of systematic sampling and recovery. In addition, taphonomic factors can prevent the recovery of evidence of the use of wild plants, as many of them may have been consumed raw or away from living sites.

Focusing on the Cantabrian area, palynological (e.g., Iriarte 1994b; Sánchez Goñi 1993) and anthracological data (Zapata 2002a) demonstrate that during the Holocene human groups had access to a wide spectrum of wild plant foods. The few archaeobotanical data suggest that wild fruits contributed to the diet of the Mesolithic and Neolithic population (see examples in Zapata 2002a). Acorns (*Quercus* sp.), hazelnuts (*Corylus avellana*), and Pomoideae fruits are the most common edible species. They are documented in Neolithic sites of the Basque Country, such as Kanpanoste Goikoa (Alava), Kobaederra, and Lumentxa (Vizcaya). Both acorns and hazelnuts are represented at El Mirón from the Neolithic onward.

According to the anthracological and palynological data from El Mirón (Zapata, chap. 11, and Iriarte, chap. 8, this volume), *Quercus* and *Corylus* were important elements of the landscape during the Neolithic. Hazelnuts are very energetic food; their high content of proteins and fats made them a valuable resource during prehistory. However, despite their significant role in the vegetation, the evidence offered by the preserved macroremains is very much limited. The Neolithic strata have produced only five remains of acorn cotyledons and eight fragments of hazelnut pericarps. Although some other authors have suggested that the dense structure of the hazelnut pericarps prevents fragments from floating and therefore they remain at the bottom of the residue, this does not seem the case in El Mirón. It is more likely that their scarcity could be explained as a result of consumption practices. In other words, hazelnuts may have been consumed raw and the shells discarded outside the site.

Hazelnut shells are probably one of the most ubiquitous types of remains in early Holocene Europe. One possible explanation is differential preservation and, consequently, visibility in the archaeological record (Zapata 2002a). The tissue of hazelnut pericarps is extremely hard and resistant, which increases the possibilities of conservation after charring. Besides, it should be borne in mind that hazelnut shells are the by-products of consumption, likely to be discarded or deliberately burnt as fuel or kindling (Jones 2000). This would explain their overrepresentation in the archaeological record. On the contrary, in the case of acorns, the cotyledons are the remains most commonly recovered, as they are denser than the shells. However, such differential preservation may also be related to their processing. In fact, acorns are generally processed after the shell has been removed; the nut is then roasted or boiled. In case of bitter populations, acorns are soaked to eliminate the tannins and then processed into flour (Mason 1992, 1995; Peña-Chocarro et al. 2000).

Acorns are also common archaeobotanical remains. They have been found throughout Europe from the Mesolithic onward (Mason 1992, 1995). Acorns are an excellent source of carbohydrates and fats and from the nutritional point of view are very close to cereals. In fact, in areas of North America and Japan acorns have played an important role as staple food (Mason 1992, 1995). Their presence in the archaeological record is not necessarily related to human consumption, though. Ethnographic data and historical accounts suggest that acorns can be used as tanning agents, medicines, or animal food (Mason 1992, 1995). However, further research is needed to evaluate their role in prehistoric subsistence (Zapata 2000).

OTHER WILD TAXA

The number of wild species identified at El Mirón is very reduced, with taxa generally represented by one or two badly preserved seeds, which makes it very difficult to assess their role within the plant assemblage. Eight genera have been recognized, although none has been identified to species level: *Artemisia* sp., *Bromus* sp., *Lolium* sp., *Luzula* sp., *Carex* sp., *Veronica* sp., and *Viola* sp. In addition, members of four different families have been also retrieved. If seeds can be identified to species level, the information provided by the wild taxa is of great interest. However, one should not forget that seed assemblages represent a tiny and highly selected sample of the original environment, and therefore landscape reconstructions beyond general observations are difficult to achieve (de Moulins 1994). On the other hand, identification has been achieved only to genus level—and then not even in all cases—so assigning large groups to particular ecological categories remains extremely difficult. In the case of El Mirón, the genera identified contain many species that belong to different ecological contexts. For example, the genus *Galium* includes species that can be weeds from arable fields, such as *G. aparine*; others are part of the woodland vegetation, such as *G. odoratum* or *G. mollugo*; and some grow well in pasture areas, such as *G. asturiocantabricum* or *G. saxatile*, among others. The same applies to other genera, such as *Viola*, with species characteristic of mixed forests (*V. alba* or *V. hirta*) or heathlands (*V. lactea* or *V. lancifolia*), and so on. Different *Luzula* species characterize pasturelands, while others are typical of mixed forests.

Regarding the families identified (Caryophyllaceae, Cyperaceae, Poaceae, and Leguminosae), the situation is even worse, as they include many different genera with different species adapted to different ecosystems.

The introduction of wild seeds into the archaeological layers may be due to different factors. The modes of arrival into the site are many and diverse: seeds may have entered the archaeological record as a consequence of various activities (e.g., as crop weeds), as parts of plants collected deliberately for diverse purposes (medicines, culinary uses, dyes, or symbolic or ritual uses), or they may simply have arrived accidentally into the site (transported by the wind or animals, collected with firewood, etc.).

As for their role as plant foods, certain families (Chenopodiaceae, Gramineae, Polygonaceae, and others) contain edible species, which may have been used at the site. However, data are limited and caution is important when interpreting them (Peña-Chocarro 1999; Peña-Chocarro et al. 2000; Colledge 1994).

The range of plant foods consumed at any one site is never complete when charring is the main form of conservation. As shown in the case of hazelnuts and acorns, only certain parts of the plant (those that are harder and denser) may survive charring. Consequently, species consumed raw or processed without fire are less likely to be represented in the archaeological record, even if they were important components of the vegetal diet. What is more, if preserved, they might be in the form of unidentifiable items. This is the case of parenchymatous tissues—the vegetative plant tissue of roots such as rhizomes, tubers, bulbs, and so on—which certainly played an important role in subsistence. Research carried out by Hather (1993, 1994) has allowed the identification of many of such remains, opening a new avenue of archaeobotanical research. Parenchymatous tissues have been already recovered from many European sites (Kubiak-Martens 1996, 1999; Mason, Hather, and Hillman 1994; Mithen et al. 2001; Holden, Hather, and Watson 1995). At El Mirón, small fragments of parenchymatous tissues from almost every stratum have been isolated and await future identification.

THE EXPANSION OF AGRICULTURE: THE CHALCOLITHIC AND THE BRONZE AGE

The Chalcolithic strata from El Mirón, dated about 4100–3800 BP, have produced a more consistent archaeobotanical dataset. Although still poor (81 cereal remains), the assemblage shows higher cereal frequencies, particularly in the more recent strata (4 and 5). The pattern observed shows a wider range of species, for the most part among wild plants. However, when focusing on plant foods, the spectrum is more or less the same as in the Neolithic. Cereals are represented by two wheat species—emmer (present only in Stratum 4) and free-threshing wheats, recovered from Stratum 7—which was the dominant crop. Compared to the previous period the only missing cereal is einkorn, which was, in any case, represented in the Neolithic by just one find. A single remain of chaff, a free-threshing rachis fragment (Stratum 5, dated to 3820±240 BP), has been identified. The Chalcolithic brings a new species to El Mirón: the pea (*Pisum sativum*). Only one seed has been retrieved from Stratum 4, and at present this represents the only record of cultivated legumes from the Cantabrian region.

Other Chalcolithic data come from Kobaederra (Vizcaya) (4405± 55 BP, 3310–2900 cal BC), where emmer has been identified. At the same time, a new species, the foxtail millet (*Setaria italica*), makes its first appearance (Zapata 2002b). More evidence on cereal cultivation is derived from pollen analysis from two sites in Vizcaya: Pico Ramos (ca. 4250 BP) and Ilso Betaio (Iriarte 1994a, b). This pattern continues during the Bronze Age. The site of Arenaza (3580±70 BP), also in Vizcaya, has produced an interesting assemblage of plant remains, with free-threshing wheats as the predominant species and emmer and millet represented by a few seeds (Zapata 2002b).

Examining the picture provided by these archaeobotanical analyses, and bearing in mind their limitations, two points are of relevance: (1) the increasing number of remains, and (2) the appearance of new crop species (millets, pea).

During the Chalcolithic data on cultivated plants start to be more consistent, with higher cereal frequencies. From the single finds of the Neolithic, now the assemblages include larger numbers of remains where cereals play a central role, representing between 40 and 50 percent of the total plant remains, although those percentages decline in the early Bronze Age (Stratum 3) (table 10.14).

This increasing number of remains probably reflects the expansion of agricultural activities in the area. There is evidence now of a more developed agriculture, with a certain degree of complexity revealed by the introduction of new crops. This may imply a change in the agricultural schedule, with species sown at different times of the year. Foxtail millet, for example, is a summer crop, sown during the spring, while both emmer and free-threshing wheats are winter-sown. The combination of two types of cereals (spring- and winter-sown) is certainly a contribution of risk buffering in case of crop failure. As for the legumes, the only find is the pea from El Mirón, so very little can be said in this regard. It is likely that legumes contributed to the diet, but more evidence is necessary to provide the full picture of their role in the prehistoric diet of the Cantabrian area.

Even if it seems likely—and data seem to support it—that agriculture began to diversify and transform, only further excavations of Chalcolithic and Bronze Age settlements and more systematic recovery of plant remains will help gain a better understanding of agricultural developments. One should remember, however, that the scanty Neolithic data are still inadequate to provide a detailed picture on the real scale of agriculture.

A further point regards the wild taxa that, at least at El Mirón, show an increase in terms of species represented. As in the Neolithic, identification to species level has been achieved in only a few cases. If, as the data seem to suggest, agriculture reached a greater degree of development, arable weeds would be expected to turn up in the archaeobotanical samples. Again, our information is very limited, but many of the families and genera represented at El Mirón include arable weed species (Poaceae, Caryophyllaceae, Chenopodiaceae, Cruciferae, *Asperula*, *Bromus*, *Medicago*, *Rumex*, etc.). *Solanum nigrum*, for example, is also field weed. Likewise, species belonging to other ecological settings are also included in these broad families. In essence, our knowledge of agricultural weeds is still very limited.

Exploitation of wild plant foods continued during the Chalcolithic and the Bronze Age. Remains of hazelnut shells and acorn cotyledons from El Mirón indicate their contribution to the diet of the human groups that inhabited the cave. Other plant foods certainly did play a role, but no clear evidence has been preserved. Most of the families identified contain edible plants. In addition, the presence of parenchymatous fragments points to the utilization of a much wider range of wild plants. However, taphonomic problems (see above) hinder a better understanding of their real contribution to human subsistence.

This study has brought together the evidence for Neolithic, Chalcolithic, and Bronze Age plant use that has emerged from the excellent systematic recovery of charred plant remains at El Mirón. This research was motivated by the desire to understand the role of farming in the first Neolithic communities of the Cantabrian region and to recognize the contribution of plants to the subsistence of prehistoric populations during the Holocene.

Although the plant record is limited, it is clear that cultivated plants are present from the fifth millennium BC in the form of three cereal species: einkorn, emmer, and free-threshing wheats. I have suggested that agriculture was probably a well-established practice, as demonstrated by the presence of both hulled and naked wheats with different processing requirements. The important presence of naked wheats offers new perspectives for research, enlarging the range of crops available to the Neolithic population. The absence of barley, common in other contemporaneous sites from the region, might be due to the paucity of remains and the quality of the data.

The Chalcolithic sees an increase in the frequencies of cultivated plants that is tentatively connected to a progressive agricultural development. The scale of such development is still poorly known, and deficiencies of the data impose caution when interpreting the plant assemblages. The presence of the pea in the late Chalcolithic is interesting, as it is the first record of a cultivated legume from the region, but only further data will help to improve our knowledge of pulse cultivation in Northern Atlantic Spain. There is an apparent decrease in the importance of cultivated plants in Bronze Age Stratum 3—as yet unexplained.

Gathering of wild plants was surely an important activity for the subsistence of human groups. Evidence is still limited to a few acorn cotyledons and hazelnut shells from both the Neolithic and Chalcolithic strata. The use of other plant foods is suggested, but reliable evidence is not yet available.

The question of the first Neolithic agriculture is still open to major debate. There are still many gaps in research, and our archaeobotanical record is still incomplete. Better archaeobotanical data are required and only future excavations where systematic sampling and recovery are applied will help to make new advances. I think, however, that the study of El Mirón is an important step for the comprehension of the first farming communities of the Cantabrian region.

Acknowledgments

The author's research is part of the ERC project Origins and Spread of Agriculture in the South-western Mediterranean Region, funded by the ERC (European Research Council) through an Advanced Grant (ERC-AdG 230561).

I would like to thank M. González Morales and L. G. Straus, directors of the excavation, for the opportunity to study the plant material from El Mirón. I am extremely grateful to Lydia Zapata for many inspiring discussions on prehistoric agriculture and for making comments on this paper. Thanks are also given to E. Castiglione for helping with some identifications and for her drawing and to the Laboratorio di Archeobiologia of the Museum of Como, where identifications were carried out. The original English version of the text has been improved by L. G. Straus.

References Cited

Bregadze, N.A.
1982 *Ocherki po agroetnograffii Gruzii* [Sketches on the agroethnography of Georgia]. Metsniereba, Tblisi, Georgia

Colledge, S.
1994 Plant exploitation in Epipaleolithic and Early Neolithic sites in the Levant. PhD thesis, University of Sheffield, UK.

de Moulins, D.
1994 Agricultural changes at Euphrates and steppe sites in the mid-eighth to the sixth millennium B.C. PhD thesis, University College London, UK.

Garcia Gazólaz, J., and J. Sesma
1999 Talleres de silex versus lugares de habitacion. Los Cascajos (Los Arcos, Mavarra), un ejemplo de neolitización en el Alto Valle del Ebro. *Saguntum* Extra-2:243–250.

González Morales, M., and L. Straus
1997 La prehistoria del Valle del Asón: excavaciones en La Cueva del Mirón. Campaña de 1996. In *II Congreso de Arqueología Peninsular. Paleolítico y epipaleolítico*, edited by R. de Balbín and P. Bueno, vol. 1, pp. 119–131. Fundación Rei Alfonso Henriques, Zamora, Spain.
2000a La Cueva del Mirón (Ramales de la Victoria, Cantabria): excavaciones 1996–1999. *Trabajos de Prehistoria* 57:121–133.
2000b El depósito arqueológico de la Prehistoria reciente de La Cueva del Mirón (Ramales, Cantabria): bases estratigráficas para una secuencia regional. In *Actas do 3º Congresso de Arqueologia Peninsular*, edited by V. Oliveira Jorge, vol. 4, pp. 49–62. ADECAP, Porto, Portugal.

Hather, J. G.
1993 *An Archaeobotanical Guide to Root and Tuber Identification*, vol. 1, *Europe and South West Asia*. Oxbow Books, Oxford, UK.
1994 A morphological classification of roots and tubers and its bearing on the origins of agriculture in Southwest Asia and Europe. *Journal of Archaeological Science* 21:719–724.

Hillman, G.
1981 Reconstructing crop husbandry practices from charred remains of crops. In *Farming Practice in British Prehistory*, edited by R. Mercer, pp. 123–164. Edinburgh University Press, Edinburgh, UK.
1984 Interpretation of archaeological plant remains: the application of ethnographic models from Turkey. In *Plants and Ancient Man: Studies in Palaeoethnobotany*, edited by W. van Zeist and W. Casparie, pp. 1–41. A. A. Balkema, Rotterdam.

Hillman, G., S. Mason, D. de Moulins, and M. Nesbitt
1996 Identification of archaeological remains of wheat: the 1992 London workshop. *Circaea* 12(2):195–209.

Holden, T. G., J. G. Hather, and J. P. N. Watson
1995 Mesolithic plant exploitation at the Roc del Migdia, Catalonia. *Journal of Archaeological Science* 22:769–778.

Ibáñez, J. J., J. E. Gonzalez Urquijo, L. Peña-Chocarro, L. Zapata, and V. Beugnier
2001 Harvesting without sickles: neolithic examples from humid mountain areas. In *Ethno-Archaeology and Its Transfers*, edited by S. Beyries and P. Petrequin, pp. 23–36. British Archaeological Reports S-983. Oxford, UK.

Iriarte, M. J.
1994a Estudio palinológico del nivel sepulcral del yacimiento arqueológico de Pico Ramos (Muskiz, Bizkaia). *Cuadernos de sección. Prehistoria-arqueología* 5:161–179.
1994b El paisaje vegetal de la prehistoria reciente en el alto valle del Ebro y sus estribaciones atlánticas. Doctoral thesis, Universidad del País Vasco, Spain.
1998 Análisis palinológico del depósito arqueológico de Kanpanoste Goikoa (Virgala, Álava). In *Kanpanoste Goikoa*, edited by A. Alday, pp. 85–91. Memorias de Yacimientos Alaveses, vol. 5. Diputación Foral de Álava, Vitoria-Gasteiz, Spain.

Iriarte, M. J., J. Mujika, and A. Tarriño
2005 Herriko Barra (Zarautz-Gipuzkoa): caractérisation industrielle et économique des premiers groupes de producteurs sur le littoral basque. In *Mémoires de la Société Préhistorique Française* 36 (Actes del colloque

unité et diversité des processus de néolithisation sur la façade atlantique de l'Europe), pp. 127–136. Paris.

Iriarte, M. J., and L. Zapata
1996 *El paisaje vegetal prehistórico en el País Vasco.* Diputación Foral de Álava, Vitoria-Gasteiz, Spain.

Jones, G. E. M.
1984 Interpretation of archaeological plant remains: ethnographic models from Greece. In *Plants and Ancient Man: Studies in Palaeoethnobotany*, edited by W. van Zeist and W. Casparie, pp. 43–61. A. A. Balkema, Rotterdam.
2000 Evaluating the importance of cultivation and collecting in Neolithic Britain. In *Plants in Neolithic Britain and Beyond*, edited by A. S. Fairbairn, pp. 79–84. Oxbow Books, Oxford, UK.

Kubiak-Martens, L.
1996 Evidence for possible use of plant foods in Palaeolithic and Mesolithic diet from the site of Calowanie in the central part of the Polish Plain. *Vegetation History and Archaeobotany* 5:33–38.
1999 The plant food component of the diet at the Late Mesolithic (Ertebolle) settlement at Tybrind Vig, Denmark. *Vegetation History and Archaeobotany* 8:117–127.

Kunst, M., and M. Rojo
1999 El Valle de Ambrona: un ejemplo de la primera colonización neolítica de las tierras del interior peninsular. *Saguntum* Extra 2:259–270.

López, P., A. Arnanz, P. Uzquiano, R. Macias, and P. Gil
2003 Informes técnicos: I. paleobotanica. In *Primeros agricultores y ganaderos en la Meseta Norte*, edited by M. Estemera, pp. 247–256. Junta de Castilla y Leon, Zamora, Spain.

Marinval, P.
1991a L'economia de subsistencia a Franca durant el Mesolític i el Neolític Antíc: les aportacions de la carpologia. *Cota Zero* 7:88–95.
1991b Recent developments in palaeocarpology in western and southern France. In *New Light on Early Farming. Recent Developments in Palaeoethnobotany*, edited by J. Renfrew, pp. 247–254. Edinburgh University Press, Edinburgh, UK
1993 Donnés récentes sur l'agriculture au Néolithique ancien et moyen dans le nord de la France. In *Le néolithique au quotidien. Documents d'archéologie Française* 39:12–18.
1995 Recollecció i agricultura de l'epipaleolític al neolític antic: anàlisi carpològica de la Balma de la Margineda. In *Las excavacions a la Balma de la Margineda*, edited by J. Guilaine and M. Martzluff, pp. 65–77. Edicions del Govern d'Andorra, Andorra.

Mason, S., J. G. Hather, and G. Hillman
1994 Preliminary investigation of the plant macro-remains from Dolní Vestonice II, and its implications for the role of plant foods in Palaeolithic and Mesolithic Europe. *Antiquity* 68:48–57.

Mason, S. L. R.
1992 Acorns in human subsistence. Doctoral thesis, University College London.
1995 Acornutopia? Determining the role of acorns in past human subsistence. In *Food in Antiquity*, edited by J. Wilkins, D. Harvey, and M. Dobson, pp. 12–24. Exeter University Press, Exeter, UK.

Mithen, S., N. Finlay, W. Carruthers, S. Carter, and P. Ashmore
2001 Plant use in the Mesolithic: evidence from Staosnaig, Isle of Colonsay, Scotland. *Journal of Archaeological Science* 28:223–234.

Peña-Chocarro, L.
1996 In-situ conservation of hulled-wheat species: the case of Spain. In *Hulled Wheats: Promoting the Conservation and Use of Underutilized and Neglected Crops*, edited by S. Padulosi, K. Hammer, and J. Heller, vol. 4, pp. 129–146. International Plant Genetic Resources Institute, Rome, Italy.
1999 *Prehistoric Agriculture in Southern Spain during the Neolithic and the Bronze Age. The Application of Ethnographic Models.* British Archaeological Reports S-818. Oxford, UK.

Peña-Chocarro, L., J. E. González Urquijo, J. J. Ibáñez Esterez, M. Moreno García, and L. Zapata Peña
2005 An ethnoarchaeological project in the Western Rif (Northern Morocco): first results. *Quaderni del Laboratorio di Archeobiologi* 1:21–32.

Peña-Chocarro, L., and L. Zapata
1998 Hulled wheats in Spain: history of minor cereals. In *Triticeae III*, edited by A. A. Jaradat, pp. 45–52. Science Publishers, Enfield, NH.
In press El cultivo del trigo en el s. XX en la Euskal Herria atlántica. *Cuadernos de Sección de la Sociedad de Estudios Vascos.*

Peña-Chocarro, L., L. Zapata, J. E. González Urquijo, and J. J. Ibáñez Estévez
2000 Agricultura, alimentación y uso del combustible: aplicación de modelos etnográficos en arqueobotánica. *Saguntum* Extra-3:403–420.

Sánchez Goñi, M. F.
1993 *De la taphonomie pollinique à la reconstitution de l'environnement: l'exemple de la région cantabrique.* British Archaeological Reports S-586. Oxford, UK.

Sigaut, F.
1978 Identification des techniques de récolte des graines alimentaires. *Journal d'Agriculture Traditionelle et de Botanique Appliquée* 25(3):145–162.

Steensberg, A.
1943 *Ancient Harvesting Implements: A Study in Archaeology and Human Geography.* Nordisk Forlag, Gyldendal, Copenhagen.

Stika, H. P.
1999 Erste archäobotanische Ergebnisse zu den neolithischen Ausgrabungen 1997 in Ambrona, prov. Soria. *Madrider Mitteilungen* 40:61–65.

Straus, L. G., and M. González Morales
1996 Preliminary excavations in El Mirón Cave (Ramales de la Victoria, Cantabria, Spain). *Old World Archaeology Newsletter* 20(1):14–18.
1998 Report on the initial excavations in El Mirón Cave (Ramales de la Victoria, Cantabria, Spain) with emphasis on the Magdalenian occupations. *Journal of Iberian Archaeology* 0:173–188.
2001 El Mirón Cave 2001: continuing excavations by the Universities of Cantabria and New Mexico. *Old World Archaeology Newsletter* 23(2):1–5.

Toffin, G.
1983 Moisson aux baguettes au Nepal central. *Objets et Mondes (Revue du Musée de l'Homme)* 23(3–4):173–176.

Zapata, L.
2000 La recolección de plantas silvestres en la subsistència mesolítica y neolítica. Datos arqueobotánicos del País Vasco. *Complutum* 11:157–169.
2002a Charcoal analyses from Basque archaeological sites: new data to understand the presence of *Quercus ilex* in a damp environment. In *Charcoal Analysis. Methodological Approaches, Palaeoecological Results and Wood Uses. Proceedings of the Second International Meeting of Anthracology (Paris, 2000)*, edited by S. Thiébault, pp. 121–126. British Archaeological Reports S-1063. Oxford, UK.
2002b *Origen de la agricultura en el País Vasco y transformaciones en el paisaje: análisis de restos vegetales arqueológicos.* Kobie Anejo 4. Diputación Foral de Bizkaia, Bilbao, Spain.

Zapata, L., and L. Peña-Chocarro
In press(a) L'agriculture néolithique de la façade Atlantique Européenne. *Bulletin de la Société Préhistorique Française.*
In press(b) Uso y gestión del bosque en la Euskal Herria atlántica. *Cuadernos de Sección de la Sociedad de Estudios Vascos.*

Zapata, L., L. Peña, G. Pérez, and H. P. Stika
2005 Difusión de la agricultura en la Península Ibérica. In *Actas del III Congreso del Neolítico de la Península Ibérica*, edited by P. Arias, R. Ontañón, and C. García-Moncó, pp. 103–113. Universidad de Cantabria, Santander, Spain.

Zapata, L., L. Peña-Chocarro, J. J. Ibáñez, and J. E. González Urquijo
2003 Ethnoarchaeology in the Moroccan Jebala (Western Rif): wood and dung as fuel. In *Food, Fuels and Fields—Progress in African Archaeobotany*, edited by K. Neumann, A. Butler, and S. Kahlheber, pp. 163–175. Africa Praehistorica, vol. 15. Heinrich-Barth-Institut, Cologne, Germany.

Table 10.1. Total samples analyzed

Level	Botanical number	Grid square	Water screening samples
2	138, 140	J4A, J4B	
3	221, 229, 244, 262, 347, 348, 355, 359, 360, 362, 364	H1A, H1B, H2A, H2B, H2D, H4A, H4B, H4D	
3.2	141, 142, 527, 529	H1C, H3D	
3.3	139, 363	H3C, H3D	
3.4	220	H4B	
3.5	249, 251, 257, 260, 365, 366, 369	H3B, H4A, H4B, H4D	
4	164, 165, 166, 167, 168, 169, 170, 171, 172, 173, 174, 175, 176, 177, 178, 179, 197, 198, 199, 200, 205, 210, 214, 215, 299	H3B, H4B, I4A, IAB, I4C, I4D, J4A, J4B, J4C, J4D	
5	294, 310, 314, 315, 316, 318, 322, 328, 329, 338, 341, 344, 345, 400, 515	H1D, H2A, H2B, H3B, H4B, H4D, I4B, I4C, J4A, J4C, J4D	
5.1	286, 291, 330, 399	H3B, H4B, H4D, H2C	

Table 10.1. *(continued)*

Level	Botanical number	Grid square	Water screening samples
6	29, 264, 269, 383, 386, 387, 415, 419, 420, 430, 431, 432, 433, 435, 437, 440, 493, 494, 500, 503, 531	H1A, H1D, H2A, H2B,H3B, H3D, H4B, H4C, H4D, I4B, I4C, J4A	
6.1	385, 388, 495, 499	J4A, H4B, H4D, I4B	
7	25, 27, 28, 447, 450, 453, 456, 459, 461, 463, 464, 467, 468, 471, 472, 473, 474, 478, 480, 484, 485, 491	H1A, H1D, H2B, H3B, H4A, H4B, H4D, I4B, I4C, I4D, J4A, J4C	
7.1	449, 454, 465	H2B, H3C	
7.2	455, 489	H3A, H3B, I4C	
7.5	490	J4C	
8	8, 11, 12, 13, 14, 15, 16, 17, 18, 19, 34, 36, 38, 39, 76, 91, 92, 93, 96	H2C, H3A, H3B, H3C, H3D, H4A, H4B, H4D, I4A, I4B, I4D, J4C	
9	2, 4, 7, 8, 11, 12, 13, 23, 95, 123, 124, 125, 186	I3A, I3B, I3C, I3D, I4D, H3B, H3D, H4D, J4B	
9.6	3, 9, 10, 14, 22	I3A, I3B, I3C, I3D	
9.7	1, 15, 18, 19, 20, 28	I3A, I3B, I3C, I3D	
9.8	5, 16, 21, 27	I3A, I3B, I3C, I3D	
10	6, 459, 460, 463, 476	I3B, J4A, J4B, J4C	
10.1	400, 401, 406, 417, 426	H2D, J4A, J4B, J4C	
14	92	2C	
15	93, 100	2C, I3D	
16	101	I2D	
301	12	M5C	
301.1	365, 366, 367	M5A, M5B, M5C	
303	9, 10, 17, 32, 33, 34, 35, 36, 37, 38, 50, 51, 52, 53, 54, 55, 56, 57, 58, 59, 60, 61, 62, 69, 102, 124, 127, 128, 129, 130, 131, 132, 133, 134, 135, 136, 226, 227, 228, 229, 230, 231, 232, 233, 234, 235, 236, 237, 424, 445, 451, 452, 453, 454, 455, 456, 469, 471	N5A, N5B, N5C, N5D, M5A, M5B, M5C, M5D, L5A, L5B, L5C, L5D	Square L5: 138, 139, 140, 141, 187, 188, 189, 190 Square M5: 4, 20, 30, 31, 33, 46, 47, 48, 49, 52, 53b, 54, 55, 92, 93, 94, 94b, 95, 95b, 95c, 138, 546 Square N5: 5, 6, 7, 8, 19, 20, 21, 22, 37, 39, 41, 43
303.1	63, 67, 68, 470, 472, 476, 479, 480, 481, 482, 493, 494, 498	N5A, N5D, L5A, L5B, L5C, L5D	Square L5: 224, 225, 226, 227 Square M5: 144, 145, 146 Square N5: 36, 38, 39
303.2	66, 71, 72, 81, 82, 364, 368, 369, 370, 371, 448, 495, 496, 497, 499, 502, 503, 504	N5A, N5B, N5C, N5D, M5A, M5B, M5C, M5D, L5A, L5B, L5C, L5D	Square L5: 384, 385, 386, 387 Square M5: 154, 155, 157, 158 Square N5: 47, 48, 49, 50
303.3	372, 373, 374, 375, 376, 377, 449, 500, 501, 514, 515, 516, 517, 524, 525, 526, 527	M5A, M5B, M5C, M5D, L5A, L5B, L5C, L5D	Square L5: 441, 442, 443, 444, 488 Square M5: 194, 195, 196, 197

Table 10.2. Plant remains from Stratum 9 and subdivisions

Botanical number	12	8	14	22	20	19	28	21
Stratum	9	9	9.6	9.6	9.7	9.7	9.7	9.8
Square and sector	I3C	I3A	I3D	I3C	I3C	I3C	I3D	I3D
Spit	?	14	15	15	16	16	16	17
Volume of soil sampled (l)		10	8	13	9	3	9	8
Volume of float (ml)	17	8	5	5	5	3	15	10
Wild plants								
Caryophyllaceae	1							
Cyperaceae								1
Indeterminate	2	1					1	
Parenchyma			*	*	*	*		

* = Tiny samples obtained from wet sieving rather than from flotation.

Table 10.3. Plant remains from Stratum 8

Botanical number	15	11	13
Square and sector	I4D	H4A	H3A
Spit	16	19	14
Volume of soil sampled (l)	2	12	15
Volume of float (ml)	5	10	10
Cultivated plants			
Cf. Cereal indeterminate	1		
Wild plants			
Luzula sp.			1
Indeterminate		1	

Table 10.4. Plant remains from Stratum 303

Botanical number	17	32	34	136	231	233	237	n55*	n33*	n20*
Square and sector	M5A	M5C	M5A	M5A/C	M5A	M5B	M5B	M5D	M5C	M5C
Spit	3	4	4	7	6	6	5	6	4	3
Volume of soil sampled (l)	8	16	35	16	27	37	12			
Volume of float (ml)	< 5	40	7	< 5	10	20	5			
Cultivated plants										
Triticum monococcum						1				
T. dicoccum										
T. monococcum/dicoccum										
T. aestivum/durum					1					
Triticum sp.					1					
Cereal indeterminate										
Cereal indeterminate (fragment)					1	1	1			
Wild plants										
Bromus tp.										
Caryophyllaceae				1						
Corylus avellana								1	1	4
Galium sp.			1							
Leguminosae	1									
Lolium sp.			2							
Luzula sp.	1									
Poaceae	1		2							
Polygonaceae				1						
Quercus sp.		2								
Veronica sp.			1							
Indeterminate	1		6							

* = Tiny samples obtained from wet sieving rather than from flotation.

61	50	57	69	52	54	55	53	n5*	n41*	n39*	n43*	n139*
N5D	N5C	N5A	N5C	N5B	N5B	N5A	N5B	N5A	N5A	N5D	N5C	L5B
4	3	3	4	3	3	2	2	2	4	4	4	12
17	19	32	21	19	8	12	36					
< 7	7	15	10	20	< 5	10	20					
				1								
		1		1								
		1			1					2		1
		1										
				1								
		2										
						1						
											1	
		1					1					
1	1											
2												
					1							
				3								
1	1		3					2	2			

Table 10.5. Plant remains from Stratum 303.1

Botanical number	58	63	68	498	482	366	n 225*
Sector	N5-B	N5-D	N5-A	L5-A	L5-B	M5-B	L5-B
Spit	4	5	5	14	14	8	14
Volume of soil sampled (l)	29	21	30	12	13	20	
Volume of float (ml).	22	5	8	5	5	< 5	
Wild plants							
Caryophyllaceae					1		
Corylus avellana							2
Leguminosae (small)				1			
Luzula sp.			1				
Poaceae	5						
Viola tp.							
Indeterminate	5				1	3	
Parenchyma		*					

* = Tiny samples obtained from wet sieving rather than from flotation.

Table 10.6. Plant remains from Stratum 303.2

Botanical number	66	81	n 385*	n 387*
Sector	N5-B	N5-C	L5-B	L5-D
Spit	6	6	15	15
Volume of soil sampled (l)	19	30		
Volume of float (ml)	< 5	15		
Cultivated plants				
Cereal indeterminate	1			
Wild plants				
Artemisia sp.	1			
Carex sp.	1			
Corylus avellana				1
Luzula sp.	1	1		
Viola tp.	1			
Indeterminate	1		1	
Parenchyma		*		

* = Tiny samples obtained from wet sieving rather than from flotation.

Table 10.7. Plant remains from Stratum 303.3

Botanical number	374	375	377	514	516	517
Sector	M5-C	M5-D	M5-B	L5-A	L5-B	L5-D
Spit	10	10	10	16	16	16
Volume of soil sampled (l)	16	19	17	0.5	16	15
Volume of float (ml)	7	10	5	< 5	7	< 5
Cultivated plants						
T. dicoccum	1					
Wild plants						
Corylus avellana				1		
Indeterminate		2	3		1	1

Table 10.8. Plant remains from Stratum 7 and subdivisions

Botanical number	472	480	467	474	455	490
Square and sector	7-I4C	7-I4C	7-H4A	7-H4D	7.2-H3B	7.5-J4C
Spit	16	11	18	17	12	16
Volume of soil sampled (l)	8	9	19	10	8	10
Volume of float (ml)	7	10	10	25	135	50
Cultivated plants						
Triticum aestivum/durum		1		1		1
Triticum sp.						1
Cereal indeterminate					1	
Wild plants						
Quercus sp.	2					
Indeterminate		3			2	
Parenchyma			*			*

* = Tiny samples obtained from wet sieving rather than from flotation.

Table 10.9. Plant remains from Stratum 6 and subdivisions

Botanical number	430	431	435	383	385
Square and sector	6-H3B	6-H4B	6-I4C	6-J4A	6.1-J4A
Spit	10	14	7	14	14
Volume of soil sampled (l)	10	14	20	16	4
Volume of flot (ml)	20	40	100	10	5
Cultivated plants					
Triticum aestivum/durum			3		
Triticum sp.			3		
Cereal indeterminate		1	2		
Wild plants					
Caryophyllaceae					1
Corylus avellana			1		
Quercus sp.				7	
Indeterminate	1				
Parenchyma	*				

* = Tiny samples obtained from wet sieving rather than from flotation.

Table 10.10. Plant remains from Stratum 5 and subdivisions

Botanical number	322	328	345	316	338	341	315	318	294	286
Square and sector	5-H2B	5-H2B	5-H3B	5-H4B	5-I4B	5-I4C	5-J4A	5-J4A	5-J4D	5.1-H4B
Spit	8	9	5	11	4	4	9	9	9	12
Volume of soil sampled (l)	12	11	11	10	12	5	12	12	10	6
Volume of float (ml)	35	15 l	40	5	50	40	11	15	20	5
Cultivated plants										
Triticum aestivum/durum	4			3					2	
T. aestivum/durum rachis	1									
Triticum sp.	5		1		1					
Cereal indeterminate					3					2
Cereal fragment	6		2				3			
Wild plants										
Compositae		1								
Corylus avellana						2				
Cyperaceae						1				
Luzula sp.										1
Medicago/Melitotus			1							
Quercus sp.								1		
Solanum sp.									1	
Indeterminate	5	2	4		2	6			1	
Parenchyma	*				*	*			*	

* = Tiny samples obtained from wet sieving rather than from flotation.

Table 10.11. Plant remains from Stratum 4

Botanical number	171	172	164	178	173	175	197	200	167	169
Square and sector	I4D	I4A	I4A	I4A	I4B	I4B	I4B	I4B	I4C	I4C
Spit	2	2	2upper	2	2	2	3	2	2	2
Volume of soil sampled (l)	20	14	11	18	10	11	10	8	11	11
Volume of float (ml)	20	25	25	30	50	30	15	20	15	80
Cultivated plants										
Triticum dicoccum									3	4
Triticum dicoccum/durum										
Triticum aestivum/durum		1	2	3		3			1	
Cf. *T. aestivum/durum*						1				
Triticum sp.										
Cereal indeterminate										
Cereal indeterminate (embryo)										
Cereal fragment	2	2	3					6	10	2
Pisum sativum							1			
Wild plants										
Asperula sp.							1			
Brassica tp.										
Bromus tp.		1								
Carex sp.						3	1	1		
Caryophyllaceae			2	2			1	2		
Chenopodium sp.				1						
Chenopodiaceae										
Corylus avellana										
Cruciferae										
Hypericum sp.										
Luzula sp.						1				1
Medicago/Melitotus										1
Mentha/Satureja sp.									1	
Poaceae (*Poa* tp.)									2	
Poaceae										1
Polygonaceae			1							
Quercus sp.		4	2				1			
Rumex sp.								1	1	

177	176	215	165	170	174	205	198	166	199	210	214	299
I4D	J4A	J4A	J4B	J4B	J4B	J4B	J4C	J4D	J4D	J4D	H3B	H4B
2	4	4	4	5	2	4	4	4	6	3	7	8
8	8	10	32	4	4	10	8	10	12	10	6	15
1	40	30	7	5	5	10	20	60	25	40	5	10
			2					2	6			
								2				
			2	1			1	5	14	5	1	
				3			1	7	9		1	
1												
			1									
			16	5			4	32	34			3
								1				
								1			1	
			1			1					1	1
		2	3			1		1				1
	1											
							19			1		
						1		2				
				1								
								1		1		
						1		1		1		
			1		1			2				
								1		1		
		1								4		

Table 10.11. *(continued)*

Botanical number	171	172	164	178	173	175	197	200	167	169
Rumex sanguineus						3				
Solanum nigrum										
Trifolium sp.									1	
Viola tp.										
Indeterminate		1	1	3	1		2		3	8
Parenchyma		*								

* = Tiny samples obtained from wet sieving rather than from flotation.

Table 10.12. Plant remains from Stratum 3 and subdivisions

Botanical number	347	348	221	139	363	220	257	369
Square and sector	3-H2A	3-H2D	3-H4D	3.3-H3C	3.3-H3D	3.4-H4B	3.5-H3B	3.5-H4A
Spit	1	1	3	3	3	6	4	7
Volume of soil sampled (l)	7		5	3	7	7	10	18
Volume of float (ml)	20	25	< 5	135	15	15	15	3
Cultivated plants								
Triticum aestivum/durum		1	2					
Cereal fragment			2					
Wild plants								
Bromus tp.							1	
Caryophyllaceae		1			1		1	
Chenopodiaceae					1			
Poaceae				1				1
Viola tp.					1			
Indeterminate	2				1			2
Parenchyma	*				*	*		

* = Tiny samples obtained from wet sieving rather than from flotation.

	177	176	215	165	170	174	205	198	166	199	210	214	299
								1					1
							1		2				
													2
				1	1	1	5		4	2	6	1	3
			*				*		*		*		

Table 10.13. Presence of cereals in the Neolithic samples

Botanical number	231	233	237	57	52	54	n 39*	n 139*	66	374
Stratum	303	303	303	303	303	303	303	303	303.2	303.3
Square and sector	M5-A6	M5-B6	M5-B5	N5-A3	N5-B3	N5-3	N5-D	L5-B	N5-B6	M5-C
Spit	6	6	5	3	3	3	4	12	6	10
Volume of soil sampled (l)	27	37	12	32	19	8			19	16
Volume of float (ml)	10	20	5	15	20	< 5	< 5	< 5	< 5	7
Cultivated plants										
Triticum monococcum		1								
Triticum monoc./dic.				1	1					
Triticum dicoccum					1					1
Triticum aestivum/durum	1			1		1	2	1		
Triticum sp.	1			1						
Cereal indeterminate (fragment)	1	1	1	2	1				1	

* = Tiny samples obtained from wet sieving rather than from flotation.

Table 10.14. Percentages of cereals within Chalcolithic and Bronze Age plant assemblages

Stratum	Cereal percentage
3	21
4	42
5	48
6	44
7	41

CHAPTER ELEVEN

Holocene Wood Charcoal from El Mirón Cave

Vegetation and Wood Use

Lydia Zapata

Wood charcoal is the most frequent macro-archaeobotanical material found at El Mirón. The wood was most likely brought to the cave as fuel for domestic hearths and also for other purposes, such as animal fodder, construction, or crafts. The subproducts of these activities could also end up in the fire. Unlike pollen, charcoal in an archaeological site is a direct product of human activities, but it also reflects, at least partially, the woody vegetation of the vicinity of the site.

The aims of the charcoal analysis at El Mirón are (1) to help reconstruct the past vegetation near the site during the Holocene, (2) to assess environmental change through time, (3) to study the diversity of exploited woods, and (4) in combination with other archaeobotanical remains, to develop the ethnoecological modeling of plant resources exploited by humans.

El Mirón is a large cave located on Monte Pando, at about 250 m above sea level in one of the coastward ranges of the Cantabrian Cordillera, at the eastern edge of the province of Cantabria. It is immediately surrounded by peaks that reach 600–900 m. It is about 100 m above the bed of the Calera river (a tributary of the Asón) and 20 km away from the coast of the Bay of Biscay as the crow flies.

Present climatic conditions are oceanic, with mild temperatures and very high rainfall (1,707 mm in Ramales, 80 m above sea level). Good syntheses of the present vegetation of the areas of the Asón basin and adjacent western Vizcaya can be found in Herrera (1995) and Onaindia (1986) respectively. The modern theoretical climax vegetation of the area is assumed to be dominated by forests, which have been decimated by human action. Two important climatic vegetation series that currently grow on well-drained soils are *Polysticho setiferi-Fraxineto excelsioris* S. and *Hyperico pulchri-Querceto roboris* S., both with *Quercus robus* as the main arboreal component of mature phases together with other trees, such as *Corylus avellana* and *Fraxinus excelsior*. In siliceous hilly areas with poor, dry soils *Melampyro-Querceto pyrenaicae* S. is also present, with *Quercus pyrenaica* as the main arboreal component. In rocky calcareous areas with dry soils *Lauro nobilis-Querceto ilicis* S.

grows. There can be an evergreen forest where *Quercus ilex* dominates, along other trees such as *Phillyrea latifolia*, *Rhamnus alaternus*, and *Arbutus unedo*. The series of *Hyperico androsaemi-Alneto glutinosae* S. spreads along the rivers in areas with permanent water. Here *Alnus glutinosa* is the main component, although *Quercus robur*, *Fraxinus excelsior*, and *Betula celtiberica* are also present (Herrera 1995).

The excavation of El Mirón Cave has been concentrated since 1996 mainly in two areas of the large vestibule (10 m wide × 30 m deep × 13 m high), the Cabin and the Corral. Each excavation block is at most about 10 m^2 in size, and they have been connected by a continuous stratigraphic trench. The cultural sequence includes contexts pertaining to the Mousterian, Early Upper Paleolithic, Solutrean, Magdalenian, Azilian, Mesolithic, Neolithic, Chalcolithic, and Bronze Age periods, with traces of medieval occupations (e.g., González Morales and Straus 2000a, b; Straus and González Morales 2000, 2003; Straus et al., 2001, 2002). In this work we will only deal with the Holocene wood charcoal record from the outer vestibule (Cabin) area (Levels 10.1–10.2) and the Mid-Vestibule Trench (Levels 303.1–303).

Materials, Methods and Sampling Strategy

Various types of archaeobotanical analyses have been undertaken at El Mirón, a fact that clearly improves possibilities for interpretation: besides charcoal, pollen and seeds have been studied by M. J. Iriarte and L. Peña-Chocarro, respectively.

Sampling for macroremains was carried out on-site during the excavation of El Mirón. In addition, hand-picked plant material was collected by the excavators. The sediment was processed with a Siraf-type flotation machine (de Moulins 1996) made by J. Quevedo, technician from the Laboratorio de Física Aplicada (Universidad de Cantabria). The floating and suspendable fractions were collected in a sieve with mesh size of 250 μm. The sunken residue was collected in a 1 mm sieve and has also been examined. The samples were sorted under a low-power, reflected-light microscope.

Samples were classified in the laboratory, grouping them by grid square and stratigraphic provenance units, here called *contexts*. They were many and varied; some had many fragments of wood, while others had few or pieces too small to be identified. The total number of fragments for each context that we give in the tables and diagrams are pieces that come from the summing of different samples. The grouping of samples has been done with the following criteria: (1) all samples combined come from the same stratigraphic context; (2) wood charcoal in these samples was scattered through the sediment—it was not concentrated charcoal or charcoal from hearths; (3) samples come from stratigraphically reliable areas or grid squares; (4) samples with only fragments smaller than 2 mm were rejected, since their identification by anatomical observation is difficult; and (5) when possible, a minimum of 100 fragments were identified for each context. In some contexts, all available charcoal has been identified (Levels 2, 9.6, 9.7, 9.8, 303.1, and 303). Following this criteria, table 11.1 shows the samples that have been combined for each context.

Identification of wood charcoal was done using epi-illuminated light microscopy, with reference to Schweingruber (1978, 1990) and to our own reference collection of modern woods from western Europe. Most of the wood taxa identified display characteristic anatomical features that clearly distinguish them and that are well defined in the bibliography, so we will comment only on those that pose particular problems. When a genus has several anatomically similar species but only one presently lives in the Iberian Peninsula, we assume we are dealing with this species (for example, *Cornus sanguinea*). The nomenclature used here corresponds to Castroviejo et al. (1980).

Regarding the identification and classification of the genus *Quercus*, we have followed Schweingruber (1990) and have recognized two groups: (1) evergreen oaks, *Quercus ilex* L./*Q. coccifera* L., species impossible to distinguish anatomically; and (2) *Quercus Subgenus Quercus* Oersted, which includes all deciduous and semi-deciduous oaks that now grow in the region, also impossible to distinguish anatomically: *Quercus faginea* Lam., *Quercus petraea* (Mattuschka) Liebl., *Quercus pubescens* Willd., *Quercus pyrenaica* Willd., and *Quercus robur* L.

Rosaceae is a big family with several woody species that display similar anatomy. At least two main groups—the Pomoideae group and the genus *Prunus*—may be distinguishable. Pomoideae includes different genera of trees and shrubs that are very difficult to distinguish

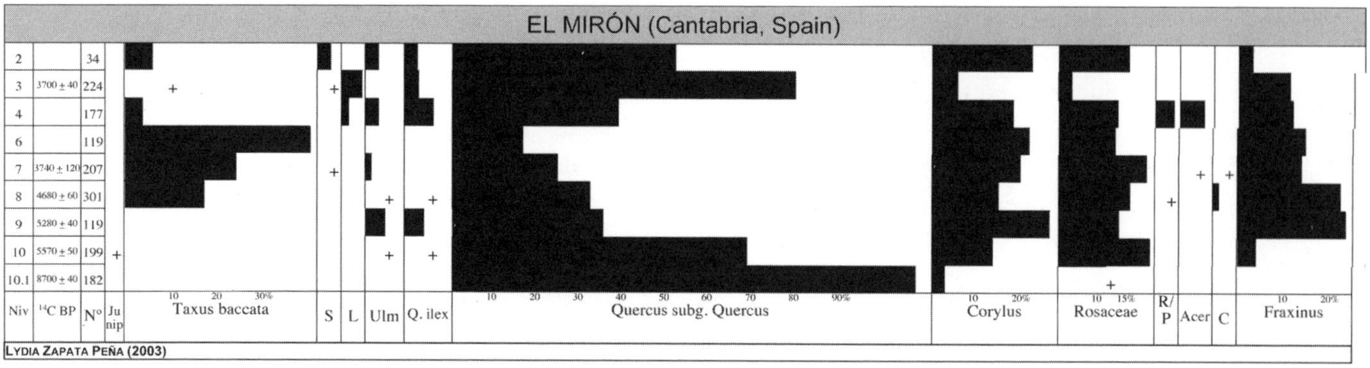

Fig. 11.1. Distribution of identified wood charcoal fragments from the Mesolithic, Neolithic, Chalcolithic, and Bronze Age levels in El Mirón Cabin excavation area. S: *Salix/Populus*; L: *Laurus*; Ulm: *Ulmus*; R/P: *Rhamnus/Phillyrea*; C: *Cornus*. Graph by Lydia Zapata.

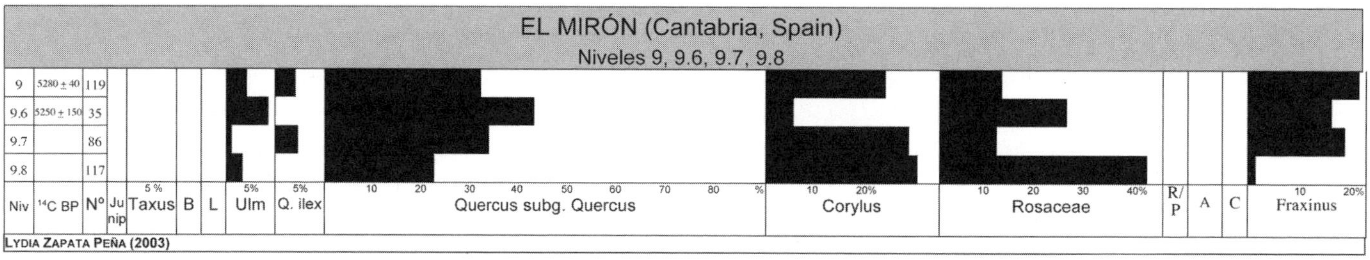

Fig. 11.2. Breakdown of identified wood charcoal samples from subdivisions of Neolithic Stratum 9. B: *Betula*; L: *Laurus*; Ulm: *Ulmus*; R/P: *Rhamnus/Phillyrea*; A: *Acer*; C: *Cornus*. Graph by Lydia Zapata.

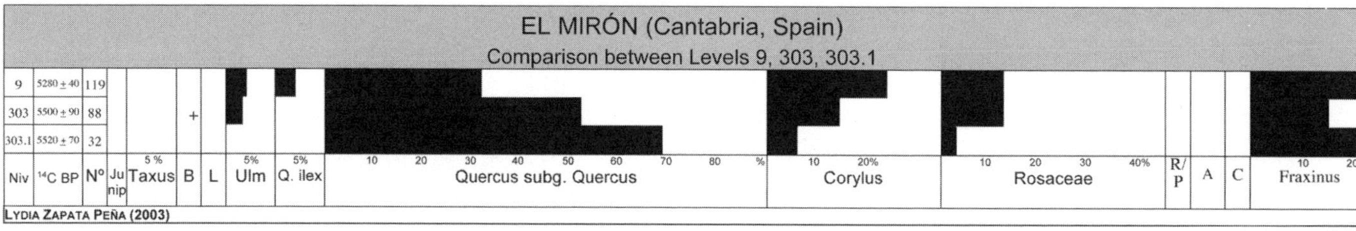

Fig. 11.3. Wood charcoal comparison of Neolithic Levels 9, 303, and 303.1. B: *Betula*; L: *Laurus*; Ulm: *Ulmus*; R/P: *Rhamnus/Phillyrea*; A: *Acer*; C: *Cornus*. Graph by Lydia Zapata.

anatomically: *Malus*, *Pyrus*, *Crataegus*, *Sorbus*, and so on (apple tree, pear tree, hawthorn, rowan/wild service/whitebeam). Thus all the fragments have been grouped into a single taxon. In the case of the genus *Prunus*, in spite of the difficulty in discrimination we have pointed out different types of anatomy following Schweingruber (1978): *Prunus* tp. *avium* (cherry), *Prunus* tp. *padus* (bird cherry) and *Prunus* tp. *spinosa* (blackthorn).

Rhamnus alaternus (buckthorn) and the genus *Phillyrea* show very similar anatomical features, so they have been joined in a single taxon. We note that the genus *Rhamnus* has been identified in the pollen analysis, so in fact it might be *Rhamnus alaternus* with which we are dealing.

Indeterminate fragments are those that do not show anatomical features in order to be identified. In most cases they are small, distorted pieces of knot wood or vitrified specimens. We also have a few cases of fragments that have been classified as "Indeterminate (*Ligustrum*?)." The reliability of this identification is very low, because anatomical features are unclear.

As conventions, in the botanical identifications "type" refers to an identification that is similar to the anatomy of a genus or species; "cf." means that almost certainly the item belongs to that genus or species.

In the tables we provide the raw data with the maximum level of identification that has been reached for each fragment. The diagrams (figures 11.1–11.3), on the other hand, are a summary of these data. Fragments identified as "cf." are summed with the ones that clearly belong to the taxon (for example, cf. *Fraxinus* is summed to *Fraxinus*). Also, all Rosaceae (Pomoideae and *Prunus*) are shown together and indeterminate specimens (1.7 percent of the total) are ignored.

Results

The charcoal analysis of El Mirón has been carried out by identifying 2,168 scattered wood fragments >4 mm. The results are shown in tables 11.2 and 11.3 and in figures 11.1–11.3. The figures correspond to fragment number, which is the most common unit for quantification (rather than volume or weight, which may also be used).

The taxa identified at the site correspond to a minimum of 17 species: the conifers include *Taxus baccata* (yew) and *Juniperus communis* (juniper); the angiosperms include *Acer* (maple), *Betula* (birch), *Cornus* (dogwood), *Corylus avellana* (hazel), *Fraxinus* (ash), *Laurus nobilis* (laurel), Pomoideae (apple tree, pear tree, hawthorn, rowan/wild service/whitebeam), *Prunus* tp. *avium* (cherry), *Prunus* tp. *padus* (bird cherry), *Prunus* tp. *spinosa* (blackthorn), *Quercus ilex*/*Q. coccifera* (holm oak/Kermes oak), *Quercus subg. Quercus* (deciduous and semi-deciduous oaks), *Rhamnus*/*Phillyrea* (buckthorn), *Salix* (willow), and *Ulmus* (elm).

Among the conifers, juniper has been identified only in Level 10 (two fragments). Yew, however, is a very important component of the assemblage from Level 8, around 4680±60 BP. From this level its values, in terms of fragment number, increase up to almost 40 percent at ca. 3800 BP in Level 6, but then it starts to decrease dramatically until it almost disappears, in a process that has also been seen in Aizpea in Navarra (Zapata 2001).

In terms of fragment number, the genus *Quercus* is the main contributor to the overall wood assemblage of El Mirón. However, its importance is variable within the sequence. Evergreen oak—most likely *Quercus ilex*—is infrequent but is present in Level 10. Deciduous oaks (*Quercus subg. Quercus*), however, constitute 97 percent of the fragments identified in Level 10.1, around 8700 BP. From this extremely high value the percentage starts to decrease in Level 10 (62 percent) and reaches its lowest value (15 percent) in Level 6, at around 3800 BP. Afterward it becomes more abundant again, with another peak (72 percent) in Level 3, at around 3700 BP.

Corylus avellana (hazel), Rosaceae, and *Fraxinus* (ash) are important components of the wood assemblage practically throughout the sequence. Values are particularly low in those samples where deciduous oaks dominate. Otherwise, they tend to form 10–20 percent of the samples. *Ulmus* is another minor taxon present throughout most of the sequence but with very low values (<5 percent). Other taxa, such as *Salix*/*Populus*, *Rhamnus*/*Phillyrea*, *Acer*, *Cornus*, and *Laurus nobilis* are all scarce (<5 percent) in the samples and are present only in particular parts of the sequence, although they tend to cluster in consecutive samples. Only one fragment of *Betula* has been identified in the entire sequence.

Contexts 9.6, 9.7, 9.8, 303, and 303.1 have been considered separately (table 11.3 and figures 11.2 and 11.3). These samples are consistent in terms of wood types, since the same taxa appear in all of them (*Ulmus*, *Quercus ilex*/*coccifera*, *Quercus subg. Quercus*, *Corylus avellana*, Rosaceae, and *Fraxinus*). The percentages are not uniform, however, something we will discuss later.

Discussion: Can We See the Forest for the Charcoal?

Charcoal analysis provides both paleoecological and paleoethnobotanical information. It shows the presence of certain woody species available in the vicinity of a site, but most likely it also reflects human use and selection of fuel in the past. Thus consideration has to be given to the cultural selection of wood and context-related variation within sites. How well charcoal assemblages reflect the paleoenvironment and the plant communities that grew in an area is a subject that has been much discussed in the literature (e.g., Asouti and Austin in press; Badal, Bernabeu, and Vernet 1994; Chabal 1997; Heinz 1990; Piqué 1999a; Shackleton and Prins 1992; Smart and Hoffman 1988; Théry-Parisot 2002a; Thompson 1994). There are different schools of thought according to the emphasis placed on ecological or ethnobotanical interpretation. The first group of specialists (e.g., Badal, Bernabeu, and Vernet 1994; Chabal 1997; Heinz 1990, 2002) carry out environmental reconstructions of vegetation dynamics from archaeological charcoal data, because they assume that charcoal is essentially a function of past vegetation. The conditions that would allow a palaeoecological interpretation would be scattered charcoal from domestic use that represents a long period of activity. Other archaeobotanists refuse to believe that charcoal can be interpreted strictly in this way, since fuelwood supply patterns may vary according to different factors, such as human selection, fuel properties, and changes in catchment areas (e.g., Piqué 1999a,b; Shackleton and Prins 1992; Smart and Hoffman 1988; Thompson 1994).

The presence or absence of taxa in our samples is conditioned by different factors, including, among others: (1) taphonomy, (2) wood availability in the vicinity of the site, (3) properties of the different woods as fuels, (4) other uses of the available trees—for crafts, construction, nuts or fodder, for example—and depositional habits, and (5) cultural choices. I will briefly discuss the ideas and the general assumptions behind these issues.

NON-ANTHROPIC TAPHONOMIC FACTORS

The following are poorly known factors that surely affect the representation of the taxa on a site: differential burning, fragmentation and conservation of the taxa, and differential production of charcoal. These are not minor questions, but unfortunately they often tend to be ignored when interpreting results. The conditions of burning and physical parameters of the wood can be crucial to preservation. Taxa with a soft anatomical structure (*Salix* or *Populus*, for example) reduce to ashes faster than denser woods such as *Buxus* or Rosaceae. Shrubs and small pieces of wood are often used as kindling and are very likely to be consumed entirely from the beginning. The structure of the hearths also influences the preservation of charcoal (March 1992) and pieces of wood located at the center of the fire can burn completely (Smart & Hoffman 1988).

WOOD AVAILABILITY AND THE PRINCIPLE OF LEAST EFFORT

In a woody environment like the one around El Mirón during the Holocene long-distance transport of fuelwood is unlikely, so we can assume that trees growing not far from the site would be preferentially collected. Also, dead wood would most likely be preferred, as we can see in other archaeological examples (e.g., Théry-Parisot 2001). Conifers may for this reason be overrepresented, because they tend to produce more dead wood; deciduous forests produce less and wood disintegration is faster (Peterken 1996). Shackleton and Prins (1992) discuss the "principle of least effort" model for the interpretation of charcoal datasets. It assumes that people collected what was easiest to get and that all species were collected in direct proportion to their occurrence in the environment. They predict conditions where this model would hold and others where it would not. For example, during the initial occupation of one area, dead and live wood availability and diversity would be high. People could afford maximum selectivity with low effort. These conditions would change through time. Dry wood and preferred species would tend to disappear in the vicinity of the settlement, and people would walk farther to collect the taxa they wanted. Eventually they would collect alternative species or whatever was available. Also, species that regenerate fast when cut and produce new shoots quickly might be collected more often.

FUELWOOD PROPERTIES, OR "IS IT A GOOD FUEL?"

The notion of good fuel is related to the purpose it is used for (heating, lighting, cooking, production of smoke, or manipulation of raw materials) (Théry-Parisot 2002b). The properties of the fuelwood (weight, how quickly it burns, smoke production, smell, etc.) condition its selection (Smart y Hoffman 1988). It is true that different species have different burning qualities. Conifers, rich in resin, have a higher calorific power than other trees, but *any* humid wood has much lower calorific power than a dry one. Healthy wood that has been dried under shelter for at least eight months after cutting would be the best option (Théry-Parisot 2002b). Other factors such as the provenance and the age of the tree also vary the properties of the wood when burning.

As for calorific output or how quickly combustion takes place, these depend partially on the species density, since light species will burn and liberate heat quickly. However, some charcoal analysts suggest that other conditions such as humidity, size morphology, and physiological state of the wood are more determinant than the species (Chabal 1997; Théry-Parisot 2002b). Calorific efficiency would be related more to the size of the fuel and its humidity than to the anatomical structure and properties of each species. Thus these authors minimize the importance of human selection of certain species. More than the species, aspects such as size, humidity, and condition of the wood might be the factors to control in order to get a "good fuel."

OTHER USES OF THE TREES, INDUSTRIAL ACTIVITIES, AND DEPOSITIONAL HABITS

The function of the site and the activities carried out on it are key factors in the composition of the charcoal assemblages (Piqué and Noguera 2002). Some woods may be underrepresented in our sites, because the trees might have been selected for other uses. For example, *Corylus avellana* (hazel) provides a good fuel and according to pollen records had to be an extremely frequent tree in Cantabrian Spain. However, it is not so common in charcoal datasets from Atlantic Iberia (Zapata 2002b), so we suggest that it was being used for other purposes, such as production of nuts and material for building and crafts like basketry. Other species may be overrepresented in the charcoal record because they are selected and transported to the site for other purposes but later, as a byproduct of other activities, end up in the fire. We have suggested this for the possible overrepresentation of *Pinus* tp. *sylvestris* in Mendandia, in Alava (Zapata and Peña-Chocarro in press), and other authors have suggested it in the case of tree-fodder species (Badal 1999), as ethnographic experience clearly shows (Zapata et al. 2003). This could also be the case for *Fraxinus* and *Taxus* in El Mirón, as we will discuss later.

Our ethnographic experience does show that farmers from different regions do not express clear preferences in the fuel they use for domestic hearths, intended basically for cooking and heating. Usually the most abundant trees tend to be highly valued (Zapata and Peña-Chocarro 2003; Zapata et al. 2003). However, industrial activities are subject to an intensive process of selection clearly linked to technical requirements. In the Moroccan Rif the firing of pottery in open bonfires is carried out with a combination of wood and cow dung cakes. When closed kilns are used, *Pinus* is selected because resins in conifers produce higher temperatures. In the case of open bonfires, lower temperatures are selected and therefore *Pinus* is avoided and slow-burning fuelwoods are chosen. The use of dung cakes to cover the bonfires also reflects technical properties, since potters regard dung as a slow-burning material that enhances the heat-enclosing, kiln-like effect.

The treatment, cleaning, and refuse-disposal activities linked to charcoal residues can be highly variable and depositional histories can be very complex. For example, in the Rif charcoal produced in the firing of pottery, which has been selected with great care, can be moved from one structure to another and reused for cooking in domestic hearths. Cases like this make context-related variation among archaeological samples a difficult question to address (Zapata et al. 2003).

CULTURAL CHOICES AND BELIEFS

Human decision-making may be important in the response different individuals give to particular needs for fuel. The use or avoidance of certain species for fuel may be linked not to their properties but rather to cultural perceptions and beliefs that may or may not correspond to objective, scientific facts (see examples in Smart and Hoffman 1988; Théry-Parisot 2002b). Cultural preference and avoidance of particular species may thus be embodied in a sociocultural context that goes beyond ecological determinism.

So we have to face controversial and poorly known issues when trying to interpret archaeological charred wood. However, we also have some advantages compared to other bioarchaeological information. Notwithstanding human selection of fuels, we cannot ignore the ecological information charcoal provides, at least about some of the arboreal taxa growing near the site. In a woody environment like Holocene Cantabria it is very unlikely that domestic fuel, which is needed in large amounts and on a daily basis, was transported long distances; thus it will give us a good picture of part of the vegetation near the site. Also, in the case of El Mirón we are in a good position because: (1) an important number of fragments and samples have been examined and a minimum number of 15 species has been recognized, and (2) a coherent internal dynamic can be seen when samples from different levels are compared (for example, the progressive decrease in *Quercus* wood in Levels 10.1 to 6, or the progressive increase and decrease of *Taxus*). It does not necessarily have to be like this, but the fact that we have these progressive diachronic trends in samples that have been independently analyzed shows that there is an internal logic—whether we interpret it in ecological or cultural terms. It is also obvious that, since we have other archaeological and environmental data from the same site, the final conclusions will have to integrate charcoal results with other types of evidence, something that, in the case of archaeobotany, we have done in another chapter.

The Woods Employed in El Mirón: History, Uses, and Selection

Many factors may result in one wood being used more than another. Its abundance in the landscape or its value as fuel or for other purposes all surely condition its representation in the archaeological record. The most used woods in El Mirón are considered good fuels by present farming communities from this region (Zapata and Peña-Chocarro 2003) and from other regions (Zapata et al. 2003); some of the poorly considered ones likewise are poorly represented—such as *Salix* and *Populus*—or even absent—such as *Alnus* and *Tilia*, which according to pollen data should be present in the vicinity of El Mirón. We shall briefly point out some data about the history and uses of the woods present in this cave (Aizpuru, Catalán, and Garin 1990; Iriarte 1994; López González 2002; Orme and Coles 1985; Peñalba 1994; Reille and Andrieu 1995; Rivera and Obón de Castro 1991; Vignote, Picos, and Zamora 2000; Zapata 2000, 2002b; Zapata and Peña-Chocarro, 1998, 2003).

JUNIPER (*Juniperus*)

Starting with the conifers, Juniper (*Juniperus*) was an important taxon during the Late Glacial and during the first part of the Holocene arboreal recolonization of the region. Afterward, parallel to the spread of deciduous forests, it became less common. This genus includes several species with different habitats and growing conditions. According to present distribution of these species in northern Spain, it is most likely that in El Mirón we are dealing with *Juniperus communis* (common juniper), whose wood is compact, resistant, aromatic, and easy to work. Its fruits are used in making beverages and as condiments. The distillation of the wood produces a type of tar. It is considered a good fuel that historically has been used for smoking foods.

YEW (*Taxus baccata* L.)

Yew (*Taxus baccata* L.) is a conifer that now is very scarce in the region. It can be found as a species in the chalk scrub community, scattered in rocky areas, often on slopes, though it does not form extensive populations. However, it is also present in deeper soils within deciduous forests.

When looking at its history, it is usually poorly represented in pollen diagrams although it produces abundant pollen and is wind-pollinating. It has been suggested that this is due to its difficult identification (Cortés, Vasco, and Blanco 2000). In northern Iberia it is present in pre-Holocene contexts like Lezetxiki (interior Guipúzcoa). In Urtiaga (coastal Guipúzcoa) it is present between 10,280 and 8700 BP (Sánchez Goñi 1993). It is present during the Azilian in Arenaza (Vizcaya) (Isturiz and Sánchez Goñi 1990). In Aizpea (Navarra) we have identified yew charcoal in a pattern that clearly resembles El Mirón, beginning around 7150 BP, but its expansion started around 6500 BP. At the top of the sequence of this rockshelter it disappears. In Mazaculos (eastern Asturias) it is present between 7000 and 5100 BP (Uzquiano 1992). In the dolmen of Collado del Mallo in La Rioja yew is the most abundant wood in the oldest contexts around 4700–4000 BP, including a hearth beneath the base of the megalith (Zapata 2002b). In the cave of Peña Parda (Sierra de Cantabria in Álava) during the Chalcolithic *Taxus* wood is also abundant (33 percent) (Ruiz and Zapata 2003). In the peat bog of Belate (Navarra) *Taxus* pollen is present at around 5900 BP, but it regressed sometime before 3000 BP (Peñalba 1994). The dramatic decrease that can be seen in most site and off-site deposits in the Late Holocene was most likely due to human action, since yew is a tree that regenerates very slowly. Yew provides a very good fuel. It is appreciated for its tough and compact wood and since prehistoric times has been used for spears and bows, due to its elasticity. Its bark has been used for braiding and weaving. Yew is an extremely toxic plant. Its alkaloid, taxine, acts upon the nervous system of humans and animals, producing convulsions, hypotension, cardiac depression, and ultimately death. In spite of this, its branches have occasionally been used as animal fodder—goats are most likely to be able to tolerate it—something that we will discuss later.

OAKS

Among the non-conifers, oaks, hazel, Rosaceae, and ash dominate in El Mirón. Oaks are very important, making up 62 percent and 97 percent of the fragments in the lower levels and 72 percent and 47 percent in the upper ones. Different species of *Quercus* and their hybrids are now potentially the most important arboreal taxa in Cantabrian Spain. The extension of mixed-oak forests (oaks, elm, lime, and ash) was possible because of the warmer Holocene climatic conditions, and its maximum expansion took place during the Climatic Optimum in the middle Holocene (i.e., the Atlantic pollen zone). All *Quercus* species, particularly the evergreen *Quercus ilex*, offer a highly valued fuel that traditionally has also been much used as timber, for building and for planking. Split roundwood has traditionally been used for hurdle work for sails, and it can be coppiced. The bark is used for tanning hides, and the acorns can be eaten by animals and humans, with nutritional values similar to those of cereals (Mason 1995, 2000).

HAZELWOOD (*Corylus avellana* L.)

Hazel wood (*Corylus avellana* L.) is also present in all the Holocene levels of El Mirón. This tree spread throughout northern Iberia around 9000 BP. Hazel grows in the understory and on the margins of deciduous forests, usually in shaded, cool areas. It tolerates different conditions, from dry to wet—but not waterlogged—and from basic to moderately acid soils. Cool, calcareous soils and oceanic conditions are optimum. It can grow as a tree and as a shrub, and it needs light to flower. It is easily managed by humans through coppicing and by the opening up of clearings. Hazel is a tree with multiple uses. It can be twisted and bent without snapping, so traditional uses include basketry and different types of building materials such as wattling and thatching spars. Of course, the nuts are edible, and they are indeed the most common fruit in archaeological sites from western Europe. In spite of its being a very good fuel that produces long-lasting embers, in nearby Basque sites it was apparently not as widely used as one could expect given its importance in the palynological record. That is why we have suggested that this tree may have been selected for purposes other than fuel. Compared to other contemporary sites from the region, in El Mirón the values of *Corylus* wood are quite high (ca. 10–25 percent) and may reflect its abundance in the vicinity of the site.

ROSACEAE

The wood from different species of the Rosaceae is also present throughout the sequence. This is a big family, with different types of shrubs and trees that usually grow best in light conditions, in woodland where the canopy has been opened, on thorny margins of deciduous forests, and in scrub. Frequently they are not well represented in the pollen record, where, in addition, they are identified only to family level. This is why the particular history of these trees is poorly known. Fortunately, in our case wood anatomy allows a higher-resolution identification to genus or species level. In El Mirón most fragments belong to the Pomoideae group, which includes hawthorn—particularly abundant in the area presently—as well as rowan/wild service/whitebeam, apple, and pear trees. The rest have been ascribed to the genus *Prunus*. An important group of fragments resembles the anatomy of *Prunus avium* (cherry), and a few are similar to *Prunus padus* (bird cherry) and *Prunus spinosa* (blackthorn). All these trees are highly valued for fuelwood. In some cases, their quality has been compared to beech wood. Blackthorn is considered to give very good embers for domestic hearths. All fruits of the rose family are edible and have been widely used in the area starting in Mesolithic times. Their use for fuel is also well represented in most of the Holocene sites from the region.

ASH (*Fraxinus excelsior* L.)

Ash (*Fraxinus excelsior* L.) seems to have been an important tree in El Mirón. It forms part of deciduous forests in areas of cool, rich, deep soils. It is a light-demanding species that grows well in chalk or limestone soils. Under favorable conditions it acts as a pioneer species. Its wood is very appreciated, elastic and hard, good for shafts and for spears, agricultural implements, and poles in general; it is highly valued as fuel. The leaves and the bark produce pigments and are slightly toxic. However, they are a highly valued fodder.

WILLOW (*Salix*)

The different species of willow (*Salix*) cannot be distinguished by their wood anatomy. They live in varied habitats, particularly along rivers and streams and in wet areas, although also in clearings and margins of evergreen forests. They can be pioneer species on wet soils. Willow wood is not usually considered a good fuel. The flexible shoots lend themselves to coppicing or pollarding to produce raw materials for basketry, ties, hurdling, and so on. The bark is used for hide tanning. In El Mirón only one *Salix* fragment has been clearly identified in Level 2. The other two fragments, in Levels 7 and 3, have been classified as *Salix/Populus*.

MEDITERRANEAN BUCKTHORN (*Rhamnus alaternus*)

Rhamnus alaternus (Mediterranean buckthorn) is a small tree that lives on all types of soils, in hedges, forest margins, and bushy communities. The wood of the different *Rhamnus* species is appreciated and is usually considered a good fuel. The fruits, wood, and root provide pigments and are poisonous. In the case of *R. alaternus* the bark has purgative properties and the leaves and shoots are astringent because of their high tannin content. Because of this, they have been used in gargling for mouth inflammation and sore throat. In El Mirón *Rhamnus alaternus/Phillyrea* has been identified in only three levels (8, 5, and 4).

ELM (*Ulmus*)

Elms (*Ulmus*) live in deep and fertile riverbank soils or as part of mixed-broadleaf forests, on their own or forming small groups. They are a component of the mixed-oak forest that spreads from the early and middle Holocene. According to Reille and Andrieu (1995), from 4000 BP on *Ulmus* decreases in the western Pyrenees. In El Mirón it is present in small percentages throughout the whole sequence, with the exception of Level 3. Elm wood is considered a good fuel, and it can be used in basketry and cord making. The bark of some elms has also been used for making flour.

MAPLE (*Acer* L.)

The various species of the genus *Acer* L. (maple) that currently grow in northern Spain live in different plant communities. They are usually components of deciduous forests—alder, oak or beech—and they are also found in hedges and on wood margins. In northern Iberia the use of maple wood as fuel is well documented from the Mesolithic—for example, in the rockshelters of Aizpea and Mendandia from about 7600 BP—but never in large numbers, probably because maples never were extremely abundant in the landscape, due to its resistance to shade and poor colonizing ability. *Acer platanoides* and *Acer pseudoplatanus* provide sugary sap, and maple wood is very appreciated for working. All maples are very good fuels, and they are considered to have calorific values close to that of beech.

DOGWOOD (*Cornus sanguinea*)

Cornus sanguinea (dogwood) is a calcicolous shrub that, like most shrubs, grows best on forest margins, in clearings, and in hedges. Its wood is very hard and horny and has been used for hafting and dyeing. It is a good fuel. The fruits are astringent and may be toxic. The seeds of dogwood berries have oil that has been used in lighting and for making soap. The small branches have been used in basketry. Historically, magic and wicked properties have been attributed to dogwood, linked to the red sap ("sanguinea") it produces when cut. It is an extremely uncommon taxon in wood archaeological samples from northern Iberia, so its history in the area is poorly known. In El Mirón, only seven fragments have been identified, from Levels 8, 7, and 5.

LAUREL (*Laurus nobilis*)

Laurus nobilis (laurel) is a thermophilous plant that grows in wet and shady gorges of the Cantabrian region, usually in alder communities, mixed-oak forests, and evergreen oak forests. Its history is poorly known, but it might be linked to the development of evergreen oak communities. In Basque archaeological sites its wood has been identified from the Neolithic (ca. 5500 BP). The wood is hard, strong, and aromatic. It has been used for starting fires and for giving a peculiar aroma to smoked and other foods. Its leaves and fruits have various food, flavoring, and psychotropic uses.

BIRCH (*Betula*)

Betula (birch) has been identified in only one fragment in our samples. *Betula alba*, which could be the species we are dealing with, grows on siliceous and wet soils. It is a light-demanding species and a highly efficient colonizer, favoring abandoned clearings. For this reason it flourished early in the Late Glacial and Early Postglacial, along with *Pinus* and *Juniperus*. Afterward its importance decreased, but it still remained an important component of Holocene forests in northern Iberia. It is also part of the plant communities that substitute for various deciduous forests. Human action, namely through fire and pastoralism, can also favor it. Birch can be coppiced; its wood is soft and rots quickly. It is one of the more difficult trees to fell and process with an ax. However, it is traditionally considered a good fuel, and it gives off a nice smell when burning. A sugary sap can be extracted from incisions in the trunks. The bark has been used in basketry as well as for making containers that can hold liquids and can be exposed to fire.

Charcoal Phases in El Mirón

The sequence from El Mirón covers an important part of the Holocene (ca. 8700–3700 BP). Four wood charcoal phases may be defined according to the presence and proportion of the taxa identified, without considering other archaeological or chronological criteria:

Charcoal phase	Levels	Diagnostic features
4	4, 3, 2	*Taxus* decreases while *Quercus* recovers higher percentages
3	8, 7, 6, 5	Appearance and increase of *Taxus* while *Quercus* decreases
2	10, 9, 303	Medium values for deciduous oaks. *Corylus* and *Fraxinus* increase. Absence of *Taxus*
1	10.1	Deciduous oaks dominate (96.7 percent)

Environment and Human Selection of Wood

The general trends of vegetation dynamics during the Holocene in the western Pyrenees, Basque Mountains, and Cantabrian Mountains are well known thanks to studies carried out by several authors (e.g., Iriarte 1994; Jalut et al. 1992; Peñalba 1992b, 1994; Reille and Andrieu 1995; Sánchez Goñi 1993, 1994, 1996). The Holocene arboreal recolonization started around 10,000 BP (uncal.). The gradual arrival of more temperate and humid conditions permitted the existence of a pioneer vegetation of *Pinus*, *Betula*, and a few deciduous *Quercus*. This first stage was followed by the development of a deciduous forest of *Quercus* and *Corylus*. The colonization by *Quercus* preceded that by *Corylus* (Sánchez Goñi 1993), which took place around 8290 BP (Peñalba 1992a). The Atlantic climatic period (ca. 7500 BP) saw the arrival of the mixed-oak forest (*Quercus*, *Tilia*, *Alnus*, *Ulmus*, *Fraxinus*, and *Corylus*). The first indications of farming are documented about 6000 BP (Zapata 2002b), and soon thereafter anthropogenic disturbances became clear in the vegetation through an intensification of forest clearance, the transformation of some of the existing arboreal communities, and the spread of new ones, such as beech forests.

The use of woodland resources in the area during the Holocene also is starting to be known, thanks to analyses of archaeological charcoal (e.g., Uzquiano 1992; Zapata 2002b). In many Holocene archaeological sites from Cantabria and the Basque region, *Quercus* wood was the most-used fuel, probably due to its excellent properties when burning and to its abundance in the landscape, since most likely different types of *Quercus* have been the natural vegetation of the most favorable soils over most of this region for long periods.

Based on fragment numbers, yew, oaks, hazel, Rosaceae, and ash seem to have been the most-used fuels during the Holocene occupations at El Mirón. In order to evaluate the importance of these trees in the landscape, we need to look at other types of evidence, particularly the pollen record from the same site, something that we will consider in another chapter (Iriarte, chap. 8, this volume). Here we will simply start with some questions:

LOW DIVERSITY IN LEVEL 10.1

The oldest context studied here, Level 10.1—around 8700 BP—has a very low diversity of taxa, the lowest in the sequence. *Quercus* wood represents 94.6 percent of the fragments, *Corylus avellana* 4.5 percent, and there is a single fragment of Pomoideae wood (0.5 percent). These results come from the identification of 182 fragments, a higher number than in other contexts with more taxa, which come from 14 samples from different grid squares (table 11.1), so we can assume they are a good sample for the context. When one taxon is practically the only one present in one context, we must consider the possibility that we are dealing with hearths or concentrated charcoal, since archaeological and ethnographic experience show that these contexts tend to offer very low diversity (Chabal 1997; Ntinou 2002). We might also be dealing with a very short occupation that did not allow for the collection of a diversity of woods. If we could exclude this possibility due to other archaeological information, we could suggest that *Quercus* formations might have been particularly important in this period, but it would be necessary to check other types of evidence in order to discard the hypothesis of human selection bias. Micromammalian fauna from the site indicate a wooded environment for this period (Cuenca Bescós and García Pimienta, chap. 12, this volume).

RESULTS FROM LEVELS 9, 9.6, 9.7, 9.8, 303, AND 303.1

In these cases we are dealing with chronologically very similar early Neolithic Cantabrian contexts, which span the period from around 5520 to 5280 BP. Two of them (9.6 and 9.8) are specifically bonfire lenses, but they are not the contents of concentrated charcoal from hearths. The results from these contexts are very consistent in terms of presence and absence of taxa. However, there is a variability in percentages that at least in part might be explained by the low number of fragments available in some levels (9.6 and 303.1 particularly).

The Use of Hazel (*Corylus avellana*)

The high values of *Corylus* in both the pollen and charcoal records suggest that hazel is a very important component of the landscape around the cave. In other Holocene sites that we have analyzed from Vasco-Cantabrian Spain, hazel wood percentages are much lower in spite of very high values in pollen, something we have interpreted as a possible selection of hazel for purposes other than fuelwood. In this case, both types of evidence match, since higher values in pollen might be due to the fact that *Corylus* produces a lot of pollen rain.

The Use of Ash (*Fraxinus*) and Oak as Tree Fodder

The use of tree leaves as fodder for domestic animals is a well-documented practice during recent prehistory throughout Europe (Akeret and Jacomet 1997; Austad 1988; Badal 1999; Haas, Karg, and Rasmussen 1998; Rasmussen, 1990a, 1993) and in the region close to El Mirón at present (Zapata & Peña-Chocarro 2003). Trees can be cut fresh for stall-feeding or can be directly browsed in the forest by domestic animals. Stock that is stall-fed for long periods can have leafy hay as a major contribution to diet when overwintering. The woody remnants, a byproduct, can be used as kindling and fuel in fires.

In El Mirón ash and oak could be good candidates for this practice. Compared to what we can see in the pollen diagram, *Fraxinus* is a very common fuel in El Mirón and might be overrepresented. It reaches the highest values during the late Neolithic in Levels 9 and 8, around 5280–4680 BP, where, in terms of fragment numbers, it sums to over 20 percent of the wood we have identified. Since *Fraxinus* is traditionally considered an excellent tree fodder, we might suggest its use for this purpose. This has been suggested in other cases (e.g., Uzquiano 2002:111), although it is not easy to prove.

Another good candidate to be used as tree fodder is deciduous oak. Oaks are usually preferred for leafy hay cut for use over winter to feed to stalled ovicaprids, because the leaves are durable in storage. The leaves of other deciduous trees, like ash, would most likely disintegrate, so they would be cut as fresh leaf fodder for immediate consumption (Halstead and Tierney 1998).

If animals were being fed with tree fodder, it is possible that collection was systematic, thereby involving some kind of forest management, although opportunistic exploitation is also possible. Leafy hay should have a high ratio of leaf to wood, and ethnographic observations point to an ideal cutting cycle for oak of 3–5 years. This yields bushy trees with a high proportion of leaves (Halstead and Tierney 1998:72–73). Since the trees are not destroyed, use as tree fodder results in a sustained exploitation of woodland resources.

The Use of Rosaceae

Rosaceae trees and shrubs seem to be an important component of the fuel used in the cave during the Holocene. Pomoideae such as hawthorn and *Prunus* similar to wild cherry are very good fuels that provide long-lived embers for domestic hearths.

Taxus* vs. *Quercus

As figure 11.1 shows, while other taxa do not show as much variation throughout the sequence, *Taxus* and *Quercus* show a negative correlation; *Taxus* increases at the expense of *Quercus* from Level 8 to Level 6, and the opposite dynamic can be seen after Level 5: regression of *Taxus* and increase of *Quercus*. Was there competition between the two species in the vicinity of the cave? Did they occupy independent niches? Is this alternation the result of a change in activities on the site? Different explanations, both ecological and cultural, can be suggested for these dynamics:

1. If we followed a strictly ecological explanation, the percentages of *Quercus* and *Taxus* wood in figure 11.1 would relate to their relative importance in the vegetation around the site. According to Chabal (1997:59–60), the following hypotheses could theoretically explain a situation like this, although not all are equally possible in terms of forest ecology:

Fact: Increase of *Taxus* and regression of *Quercus* (El Mirón, Chalcolithic-Bronze Age, Levels 8–6) Possible explanations in ecological terms:
1. *Taxus* expands, *Quercus* decreases in the landscape
2. *Taxus* is constant, *Quercus* decreases sharply
3. *Taxus* decreases very little, *Quercus* decreases a lot
4. *Taxus* expands sharply, *Quercus* is constant
5. *Taxus* expands a lot, *Quercus* expands very little

2. The values of *Quercus* and *Taxus* might also reflect a change in the catchment area, if they were indeed growing in different areas, something that unfortunately we cannot know. At present yew grows in marginal areas, mostly on rocky chalk. However, it is a tree that is also present on better soils in deciduous mixed forests.
3. The dynamics among different fuels might be strongly biased by activities carried out in the cave related to fuel gathering. The subproduct of woods that are carefully selected for particular crafts or as tree fodder will most likely end up as fuel.

In the last two cases, the spread in the use of *Taxus* wood and its decrease in the most recent samples could be unrelated to the dynamics of oak forests. However, a combination of these explanations might be possible.

According to what we observe in Levels 10.1–9, it seems plausible that *Quercus* forests were retreating around the site during the Neolithic due to human action. Deforestation processes in the vicinity of El Mirón have been suggested for this period from the micromammalian fauna from the site (Cuenca Bescós and García Pimienta, chap. 12, this volume).

At around 4680 BP we record the use of *Taxus* wood for the first time at El Mirón, and it seems very likely that it expanded in the region starting in the late Neolithic. Not much pollen information is available for this taxon in other deposits, and it is not present in El Mirón pollen either, but available paleobotanical information points to a maximum spread of yew between 6000 and 3000 BP (Cortés, Vasco, and Blanco 2000; Peñalba 1994; Zapata 2002b) in the context of favorable climatic conditions. Ecologically, the fact that *Taxus* expands during the Atlantic—a warm and humid period—might be linked to its bioclimatic requirements. First, yew does not tolerate frost and is very sensitive to low temperatures and continental conditions; particularly in the first months of the vegetative period, in February and March, it needs mild temperatures and high rainfall. Second, it needs high edaphic or environmental humidity; it grows in areas with at least 600 mm of rainfall, but its optimum is between 850 and 900 mm.

In terms of present forest ecology, yew grows in different communities of beech and oak forests. In Cantabrian Spain it is also common on the fringes of humid, deciduous forests next to other trees that have similar requirements and that are also present in the Holocene wood from El Mirón: *Betula*, *Corylus avellana*, *Sorbus*, *Ulmus glabra*, *Prunus avium/padus*, *Fraxinus excelsior*, and so on. It is a tree that may appear alone or in small pure and mixed groups, commonly under other dominant trees. It may be favored by human action, since this tree thrives on the fringes of broadleaved forest formations after clearings. Besides, although yew prefers good edaphic conditions, it can grow successfully on very poor soils, and it can colonize extremely thin soils, playing a pioneer role (Cortés, Vasco, and Blanco 2000), so it is very likely that it also grew in the rocky areas around the cave.

We must also take into consideration the possibility that yew was being carefully selected as a fuel during the Late Neolithic and the Bronze Age. We can suggest different explanations for overrepresentation of this taxon:

1. Its proximity to the cave, if it was growing on the rocky slopes immediately surrounding the site.
2. Its excellent quality as a fuel. It burns slowly, does not extinguish easily, and has very good calorific value.
3. Excellent quality of the wood for different crafts. Ethnographically well-documented uses of yew as raw material include the making of spears and bows, braiding, and weaving. The byproducts of these activities would end up in the fire.
4. Its use as tree fodder, which would leave branches as a byproduct. After the leaves were eaten, the branches would most likely be used as fuel. However, the use of yew as tree fodder is controversial (Bevan-Jones 2002; Cortés, Vasco, and Blanco 2000; López González 2002). Much evidence points to the fact that it is highly toxic for humans and for some animals. Many classical authors mention the use of yew in different ways as a poison for euthanasia and suicide, since ingestion of the diterpenoid alkaloids contained in the leaves, bark, roots, and wood is highly cardiotoxic. It causes heart stress and ultimately heart failure, with many historical examples of fatal poisoning. Strabo (Peralta 2000:153) mentions that people from northern Iberia used arrows poisoned with yew and that they committed suicide with yew leaves before being captured by the Romans. Dogs die after eating approximately 30 g of yew leaves. It produces intoxication and abortion in sheep and cows, although a cow has to eat large amounts for it to be fatal. In spite of this, toxicity is variable and partly depends on tolerance. There is evidence of its use as tree fodder, particularly for goats, although it is not a good practice. Domestic and wild goats eat the leaves, as do red deer and roe deer. These animals, in fact, can have a strong impact on the tree, since they stop its growth and regeneration. Allué (2002:258) mentions its presence in different archaeological "corral" contexts related to the exploitation of ovicaprids. Other authors (Thiébault 1983 and Ros 1985, cited in Allué 2002:259) establish a relationship between the presence of *Taxus* and the abundance of ovicaprids in "corral" levels. *Taxus* decreases are linked to the increase of bovids and other possible tree fodders like *Fraxinus*.
5. Intentional cutting and burning so that animals in the area do not feed on it. Yew is not a serious hazard in the field if properly cut back to keep stock from access to the foliage (Thiébault 1988:49; Bevan-Jones 2002:9). If it were cut for this reason, it is very likely that twigs and branches were used as fuel.
6. The burning of the wood for hygenic reasons. The poisonous wood of the yew contains many diterpenoid alkaloids and an insecticide substance, ecdisone, which discourage many parasites and act as insect and rodent repellent. The smoke produced by the burning of leaves seems to keep mice away. The wood might have been selected for burning for such reasons.

Taking into consideration the available paleobotanical data, it seems clear that after an optimum around 6500–3500 BP, *Taxus* decreased sharply in northern Iberia. The dating of this process may not have been the same in the whole region, but it seems to span the period between 3500 and 2000 BP. This was most likely induced by human action, due to overexploitation for the possible reasons we have pointed out. Human impact is obviously reinforced by the poor reproductive success of yew, with its high bioclimatic requirements, very low growing rate, and long germination process. This makes it a very poor competitor with the other trees that form Holocene mixed-oak forests.

In the case of El Mirón, it seems very plausible that yew became scarce during the Bronze Age and people returned to the use of other available fuels in the area, particularly *Quercus*, which could sustain heavier exploitation. These might not be growing on the limestone slopes close to the cave, but they also provide excellent fuel and raw material. Thus the evolution of *Quercus* forests in the area is difficult to evaluate from our charcoal results alone, since it might be strongly associated with the evolution and use of *Taxus*.

Some Absences: *Pinus, Alnus, Tilia, Frangula, Erica*

In spite of similarities, pollen and charcoal analyses obviously provide different sets of archaeobotanical data (table 11.4). Differential pollen rain, human selection of fuels, and the different scales of the data (since pollen have a regional component and charcoal tends to come from a closer area) are only some the issues that may explain differences. In our case, there is a group of trees present in the pollen record (Iriarte, chap. 8, this volume) that are not present in the wood charcoal from the site: pine, alder, lime, alder buckthorn, and heather.

I can suggest some explains for the absence of these taxa:

1. *Pinus*: The pollen percentages of pine are not very high in El Mirón. Since pines are very good producers of pollen, with very good dispersal, it is possible that *Pinus* was not growing in the immediate vicinity of the site or it grew in a location that made it difficult to collect.
2. *Frangula* and *Erica* were not abundant in the area according to pollen data.
3. *Tilia* and *Alnus* might have been avoided as fuelwoods, since they are often considered poor fuels (Zapata and Peña-Chocarro 2003). They might also be underrepresented for taphonomic reasons related to their physical parameters, since woods with a soft structure, such as lime, tend to reduce to ashes easier than dense ones.

Quercus ilex: A Mediterranean Tree in a Damp Environment

The presence of *Quercus ilex* in Cantabrian Spain has long been assumed to represent the existence of relict forests, remnants of ancient natural formations that in the past were more extensive. Our work in the area rejects this idea and shows that extensive *Quercus ilex* formations are quite recent and are probably linked to human action.

HUMAN IMPACT

Definitive conclusions about human impact on the landscape cannot be made based on wood charcoal alone. We do not have clear indicators of the opening of the landscape or clear substitutions of some plant communities by others, but we do have some indirect data. The presence of light-demanding species such as members of the Rosaceae family, *Corylus*, and some shrubs points to the existence of clearings and wood margins throughout the sequence. Some of these species are also very common as hedges, which could have begun to be abundant among the first farming societies. As I have already pointed out, I also interpret the decrease of *Taxus* wood use during the late Holocene in terms of human overexploitation of the tree.

The presence of *Quercus ilex* and associated species like *Laurus*, particularly at the top of the sequence, might also result from human action, as we have pointed out in other works (Zapata 2002a, b; Zapata and Meaza 1998). In these works we have suggested that the spread of *Lauro nobilis-Querceto ilicis* S. ("encinar cantábrico"), as the extensive formations we know today in Cantabrian Iberia, might have been a recent process. Although it has traditionally been taken as a relict natural forest in the area, it is not documented in paleobotanical data from sites where it thrives today. Most likely, after the Bronze Age human action on rocky slopes might have favored the extension of these evergreen forests through different processes: intensification of agropastoral practices; slash and burn agriculture; grazing by domestic animals; periodic pruning for fuel, tree fodder, or industrial production of charcoal, and so on. As we can see in Mediterranean areas (e.g., Chabal 1997; Vernet 1999; Fabre 1996, cited in Chabal 1997), these activities, on their own or in interaction, interfere with the dynamics of *Quercus* forests and may favor the expansion of *Quercus ilex* formations. The degradation and erosion of soils (Martínez Cortizas and Moares 1995) would also help the spread of this less-demanding species.

Conclusions: Out of the Woods

First, we have to make some reflections and evaluate our sampling strategy. El Mirón is a big site that has provided abundant flotation samples, from which in turn a variety of bioarchaeological materials have been recovered. As we have noted previously, our results come from the summing of many samples of scattered charcoal from two contiguous areas of the vestibule.

We think our results provide a good general picture of the wood contents of the site's Holocene deposits. The results are coherent intrasite because the sequence has an internal logic and the results are consistent with other environmental evidence from the site and with the vegetation dynamics of Holocene northern Spain. However, we must point out that, due to the excellent recovery techniques used at the site, many more charcoal samples remain unidentified and might be of interest for future analyses with different objectives. For example, the spatial variability within a level has not been fully explored.

Assuming that in a well-wooded environment long-distance transport of wood for fuel is unlikely, notwithstanding numerous taphonomic reasons that can affect the representation of fuel on archaeological sites, the sequence shows the importance of mixed-oak forest species in the vicinity of the cave. During the Mesolithic, in Level 10.1, the charcoal analysis shows an almost exclusive use of oak as fuel, along with some hazel and Rosaceae (hawthorns, apple, cherry). During the Neolithic (Levels 10 and 9), around 5570–5280 BP, the use of oak decreases, but it continues to be an extremely important fuel, while other taxa such as hazel, Rosaceae, and ash expand. Thus, as oak decreases, the representation of mixed-oak forest species increases.

From around 4700 to 3800 BP the main feature of the charcoal sequence is the significant use of yew as fuel, while oaks decrease. The dynamics between *Quercus* and *Taxus* might be interpreted in different ways, including a change in the catchment area. However, considering other paleobotanical information, we think it is reasonable to suggest that yew expanded in northern Spain in the context of favorable climatic conditions. It was likely present in deciduous forest formations, but it would also probably colonize rocky areas with thin soils. Its presence points to mild temperatures and high rainfall from the Late Neolithic to the Bronze Age. It is possible that its wood was selected by the people in El Mirón for the various reasons we have discussed, including fodder for certain domestic animals, namely goats, the animals that could better cope with this toxic plant.

The significant presence of ash, which is not so important in the pollen record, and oaks might also be related to the practice of feeding animals with tree leaves. In the case of oaks, the cutting of the branches can be carried out systematically in a sustained way without producing strong deforestation processes, although it generates managed stands of oak, very different visually from natural ones (Rasmussen 1990b).

From Level 4 on the most important features of the El Mirón charcoal assemblage are the decrease in the use of yew wood, while oak percentages increase again, something that might be linked to human impact on yew stands but might also indicate to a change in the activities carried out in the cave.

The small representation of species related to the *Lauro nobilis-Querceto ilicis* S. community throughout the sequence reinforces our hypothesis that the "encinar cantábrico" is a recent formation, most likely related to human disturbance of the natural dynamics of *Quercus* forests.

Acknowledgments

L. Zapata's study was done within the UPV/EHU IT-288–07 Research Group of the University of the Basque Country, the AGRIWESTMED project funded by ERC grant no. AdG-230561, and project HAR2011–23716 of the Ministerio de Ciencia e Inovación. Part of this work was also done at the Department of Archaeology, University of Cambridge. My thanks are extended to the directors of the El Mirón excavation, L. G. Straus and M. González Morales, (and their crews) for carrying out good recovery of plant macroremains, for suggestions, and for discussions about the site. I also thank M. J. Iriarte and L. Peña-Chocarro for sharing and discussing their archaeobotanical data from the site. G. Renobales and J. C. García Codrón also provided interesting information on tree ecology and the "*Taxus* issue." The original English version of the text has been edited and improved by L. G. Straus.

References Cited

Aizpuru, I., P. Catalán, and F. Garin

1990 *Guía de los árboles y arbustos de Euskal Herria.* Gobierno Vasco, Vitoria-Gasteiz, Spain.

Akeret, O., and S. Jacomet

1997 Analysis of plant macrofossils in goat/sheep faeces from the Neolithic lake shore settlement of Horgen Scheller—an indication of prehistoric transhumance? *Vegetation History and Archaeobotany* 6:235–239.

Allué, E.

2002 Dinámica de la vegetación y explotación del combustible leñoso durante el Pleistoceno Superior y el Holoceno del noreste de la Península Ibérica a partir del análisis antracológico. PhD diss., Universitat Rovira i Virgili, Tarragona, Spain.

Asouti, E., and P. Austin

In press Reconstructing woodland vegetation and its exploitation by past societies, based on the analysis and interpretation of archaeological wood charcoal macro-remains. *Environmental Archaeology.*

Austad, I.

1988 Tree pollarding in western Norway. In *The Cultural Landscape: Past, Present and Future,* edited by H. H. Birks, H. J. B. Birks, P. E. Kaland, et al., pp. 11–29. Cambridge University Press, Cambridge, UK.

Badal, E.

1999 El potencial pecuario de la vegetación mediterránea: las Cuevas Redil. *Saguntum* Extra-2:69–75.

Badal, E., J. Bernabeu, and J. L. Vernet

1994 Vegetation changes and human action from the Neolithic to the Bronze Age (7000–4000 B.P.) in Alicante, Spain, based on charcoal analysis. *Vegetation History and Archaeobotany* 3:155–166.

Bevan-Jones, R.

2002 *The Ancient Yew. A History of* Taxus baccata. Windgather Press, Macclesfield, UK.

Castroviejo, S., M. Laínz, G. López González, P. Montserrat, F. Muñoz Garmendia, J. Paiva, and L. Villar

1980 (and ongoing) *Flora Iberica.* C.S.I.C., Madrid. www.floraiberica.es.

Chabal, L.

1997 *Forêts et sociétés en Languedoc (Néolithique Final, Antiquité Tardive). L'anthracologie, méthode et paléoécologie.* Éditions de la Maison des Sciences de l'Homme, París.

Cortés, S., F. Vasco, and E. Blanco

2000 *El Libro del tejo (*Taxus baccata L.*). Un proyecto para su conservación.* ARBA, Madrid.

de Moulins, D.

1996 Sieving experiment: the controlled recovery of charred plant remains from modern and archaeological samples. *Vegetation History and Archaeobotany* 5:153–156.

González Morales, M., and L. G. Straus

2000a La Cueva del Mirón (Ramales de la Victoria, Cantabria): excavaciones 1996–1999. *Trabajos de Prehistoria* 57(1):121–133.

2000b El depósito arqueológico de la Prehistoria reciente de La Cueva del Mirón (Ramales, Cantabria): bases estratigráficas para una secuencia regional. In *Actas do 3º Congresso de Arqueologia Peninsular,* edited by V. Oliveira Jorge, pp. 49–65. Vol. 4. ADECAP, Porto, Portugal.

Haas, J. N., S. Karg, and P. Rasmussen

1998 Beech leaves and twigs used as winter fodder: examples from historic and prehistoric times. *Environmental Archaeology* 1:81–86.

Halstead, P., and J. Tierney

1998 Leafy hay: an ethnoarchaeological study in NW Greece. *Environmental Archaeology* 1:71–80.

Heinz, C.

1990 Dynamique des végétations holocènes en Méditerranée Nord-occidentale d'après l'anthracoanalyse de sites préhistoriques: méthodologie et paléoécologie. *Paléobiologie Continentale* 16(2):1–212.

2002 Evidence from charcoal analysis for palaeoenvironmental change during the Late Glacial and Post-Glacial in the central Pyrenees (France). In *Charcoal Analysis. Methodological Approaches, Palaeoecological Results and Wood Uses. Proceedings of the Second International Meeting of Anthracology (Paris, 2000),* edited by S. Thiébault, pp. 95–101. British Archaeological Reports S-1063. Oxford, UK.

Herrera, M.

1995 *Estudio de la vegetación y flora vascular de la Cuenca del Río Asón (Cantabria).* Guineana, vol. 1. UPV/EHU, Bilbao, Spain.

Iriarte, M. J.

1994 El paisaje vegetal de la prehistoria reciente en el Alto Valle del Ebro y sus estribaciones atlánticas. Datos polínicos. Antropización del paisaje y primeros estadios de la economía de producción. PhD diss., Departamento de Geografía, Prehistoria y Arqueología, UPV/EHU, Vitoria-Gasteiz, Spain.

Isturiz, M. J. de, and M. F. Sánchez Goñi

1990 Investigaciones palinológicas en la prehistoria vasca. *Munibe (Antropología-Arkeología)* 42:277–285.

Jalut, G., J. M. Marti, M. Fontugne, G. Delibrias, J. M. Vilaplana, and R. Julia

1992 Glacial to interglacial vegetation changes in the northern and southern Pyrénées: deglaciation, vegetation cover and chronology. *Quaternary Science Reviews* 11:449–480.

López González, G.
2002 *Guía de los árboles y arbustos de la Península Ibérica y Baleares*. Ediciones Mundi-Prensa, Madrid.

March, R. J.
1992 L'utilisation du bois dans les foyers préhistoriques: une approche expérimentale. *Bulletin de la Société Botanique de France* 139(Actual. bot. 2/3/4):245–253.

Martínez Cortizas, A., and C. Moares
1995 *Edafología y arqueología: Estudios de yacimientos arqueológicos al aire libre en Galicia*. Xunta de Galicia, Santiago de Compostela, Spain.

Mason, S. L. R.
1995 Acornutopia? Determining the role of acorns in past human subsistence. In *Food in Antiquity*, edited by J. Wilkins, D. Harvey, and M. Dobson, pp. 12–24. Exeter University Press, Exeter, UK.
2000 Fire and Mesolithic subsistence. Managing oaks for acorns? *Palaeogeography, Palaeoclimatology, Palaeoecology* 164:139–150.

Ntinou, M.
2002 *La paleovegetación en el norte de Grecia desde el Tardiglaciar hasta el Atlántico. Formaciones vegetales, recursos y usos*. British Archaeological Reports S-1038. Oxford, UK.

Onaindia, M.
1986 *Ecología vegetal de las Encartaciones y Macizo del Gorbea -Vizcaya*. Servicio Editorial Universidad del País Vasco, Bilbao, Spain.

Orme, B. J., and J. M. Coles
1985 Prehistoric woodworking from the Somerset levels: 2. Species selection and prehistoric woodlands. *Somerset Levels Papers* 11:6–14.

Peñalba, C.
1992a Biogeografía holocena de las principales especies forestales del norte de la Península Ibérica. *Cuadernos de Sección Historia. SEV* 20:391–409.
1992b La vegetación y el clima en los Montes Vascos durante el Pleistoceno Superior y el Holoceno según los análisis polínicos. In *The Late Quaternary in the Western Pyrenean Region*, edited by A. Cearreta and F. M. Ugarte, pp. 171–182. Universidad del País Vasco, Bilbao, Spain.
1994 The history of the Holocene vegetation in northern Spain from poll analysis. *Journal of Ecology* 82(4):815–832.

Peralta, E.
2000 *Los Cántabros antes de Roma*. Real Academia de la Historia, Madrid.

Peterken, G. F.
1996 *Natural Woodland. Ecology and Conservation in Northern Temperate Regions*. Cambridge University Press, Cambridge, UK.

Piqué, R.
1999a *Producción y uso del combustible vegetal: Una evaluación arqueológica*. Universidad Autónoma de Barcelona. CSIC, Madrid.
1999b Quantification in archaeobotany: charcoal analysis and fire-wood management. In *Computer Applications and Quantitative Methods in Archaeology*, edited by J. A. Barceló, I. Briz, and A. Vila, pp. 189–200. British Archaeological Reports S-757. Oxford, UK.

Piqué, R., and M. Noguera
2002 Landscape and management of forest resources in the Balearic Islands during the II-I millenium B.C. In *World Islands in Prehistory*, edited by W. H. Waldren and J. A. Ensenyat, pp. 292–300. International Insular Investigations. V Deia International Conference of Prehistory. British Archaeological Reports S-1095. Oxford, UK.

Rasmussen, P.
1990a Leaf foddering in the earliest neolithic agriculture. Evidence from Switzerland and Denmark. *Acta Archaeologica* 60:71–86.
1990b Pollarding of trees in the Neolithic: often presumed—difficult to prove. In *Experimentation and Reconstruction in Environmental Archaeology*, edited by D. E. Robinson, pp. 77–99. Oxbow Books, Oxford, UK.
1993 Analysis of goat/sheep faeces from Egolzwil 3, Switzerland: evidence for branch and twig foddering of livestock in the Neolithic. *Journal of Archaeological Science* 20:479–502.

Reille, M., and V. Andrieu
1995 The late Pleistocene and Holocene in the Lourdes Basin, western Pyrénées, France: new pollen analytical and chronological data. *Vegetation History and Archaeobotany* 4(1):1–21.

Rivera, D., and C. Obón de Castro
1991 *La guía INCAFO de las plantas útiles y venenosas de la Península Ibérica y Baleares (excluídas medicinales)*. INCAFO, Madrid.

Ruiz, M., and L. Zapata
2003 Análisis antracológico del yacimiento arqueológico de Peña Parda. *Cuadernos de arqueología de la Universidad de Navarra* 11:217–251.

Sánchez Goñi, M. F.
1993 *De la taphonomie pollinique à la reconstitution de l'environnement. L'exemple de la région cantabrique*. Tempus Reparatum. Archaeological and Historical Associates Limited, Oxford, UK.
1994 L'environnement de l'homme préhistorique dans la région cantabrique d'après la taphonomie pollinique des grottes. *L'Anthropologie* 98(2–3): 379–417.
1996 Vegetation and sea level changes during the Holocene in the estuary of the Bidasoa (southern part of the Bay of Biscay). *Quaternaire* 7(4):207–219.

Schweingruber, F. H.
1978 *Anatomie europäischer Hölzer.* Paul Haupt Publishers, Bern, Switzerland.
1990 *Mikroskopische Holzanatomie.* Eidgenössische Forschungsanstalt für Wald, Schnee und Landschaft, Birmensdorf, Switzerland.

Shackleton, C. M., and F. Prins
1992 Charcoal analysis and the "principle of least effort"—a conceptual model. *Journal of Archaeological Science* 19:631–637.

Smart, T. L., and E. S. Hoffman
1988 Environmental interpretation of archaeological charcoal. In *Current Paleoethnobotany. Analytical Methods and Cultural Interpretations of Archaeological Plant Remains*, edited by C. A. Hastorf and V. S. Popper, pp. 167–205. University of Chicago Press, Chicago.

Straus, L., and M. González Morales
2000 The El Mirón Cave/Asón River Valley Prehistoric Project (Cantabria, Spain): the Upper Palaeolithic components. In *Actas do 3° Congresso de Arqueologia Peninsular*, vol. 2, edited by V. Oliveira Jorge, pp. 351–362. ADECAP, Porto, Portugal.
2003 El Mirón Cave and the 14C chronology of Cantabrian Spain. *Radiocarbon* 45(1):41–58.

Straus, L., M. González Morales, W. R. Farrand, and W. J. Hubbard
2001 Sedimentological and stratigraphic observations in El Mirón, a Late Quaternary cave site in the Cantabrian Cordillera, northern Spain. *Geoarchaeology* 16(5):603–630.

Straus, L., M. R. González Morales, M. P. García-Gelabert, and M. A. Fano
2002 The Late Quaternary human uses of a natural territory: the case of the Río Asón drainage (eastern Cantabria Province, Spain). *Journal of Iberian Archaeology* 4:21–61.

Théry-Parisot, I.
2001 *Économie des combustibles au Paléolithique. Expérimentation, taphonomie, anthracologie.* CNRS Éditions, París.
2002a Fuel management (bone and wood) during the Lower Aurignacian in the Pataud Rock Shelter (Lower Palaeolithic, Les Eyzies de Tayac, Dordogne, France). Contribution of experimentation. *Journal of Archaeological Science* 29:1415–1421.
2002b Gathering of firewood during the Palaeolithic. In *Charcoal Analysis. Methodological Approaches, Palaeoecological Results and Wood Uses. Proceedings of the Second International Meeting of Anthracology (Paris, 2000)*, edited by S. Thiébault, pp. 243–249. British Archaeological Reports S-1063. Oxford, UK.

Thiébault, S.
1988 *L'Homme et le milieu végétal. Analyses anthracologiques de six gisements des Préalpes au Tardi et au Postglaciaire. Documents d'archéologie française.* Maison des Sciences de l'Homme, Paris.

Thompson, G. B.
1994 Wood charcoals from tropical sites: a contribution to methodology and interpretation. In *Tropical Archaeobotany. Applications and New Developments*, edited by Jon G. Hather, pp. 9–34. Routledge, London.

Uzquiano, P.
1992 Recherches anthracologiques dans le secteur pyrénéo-cantabrique (Pays Basque, Cantabria et Asturias): environnements et rélations homme-milieu au Pléistocène Supérieur et debut de l'Holocène. PhD diss., Université Montpellier II—Sciences et techniques du Languedoc, Montpellier, France.
2002 Vegetation and firewood management at Cueva de la Vaquera (Segovia, Spain) between 6 and 3.7 kyr. BP: anthracological contribution to the landscape archaeology of the Spanish central mountains. In *Charcoal Analysis. Methodological Approaches, Palaeoecological Results and Wood Uses. Proceedings of the Second International Meeting of Anthracology (Paris, 2000)*, edited by S. Thiébault, pp. 109–112. British Archaeological Reports S-1063. Oxford, UK.

Vernet, J. L.
1999 Reconstructing vegetation and landscapes in the Mediterranean: the contribution of anthracology. In *Environmental Reconstruction in Mediterranean Landscape Archaeology*, edited by P. Leveau, F. Trément, K. Walsh, et al., pp. 25–36. Oxbow Books, Oxford, UK.

Vignote, S., J. Picos, and R. Zamora
2000 *Características de las principales maderas utilizadas en Bizkaia: Tecnología y aplicaciones.* Diputación Foral de Bizkaia, Bilbao, Spain.

Zapata, L.
2000 La recolección de plantas silvestres en la subsistencia Mesolítica y Neolítica. Datos arqueobotánicos del País Vasco. *Complutum* 11:157–169.
2001 El uso de los recursos vegetales en Aizpea (Navarra, Pirineo occidental): la alimentación, el combustible y el bosque. In *Cazadores-recolectores en el Pirineo navarro. El sitio de Aizpea entre 8000 y 6000 años antes de ahora*, edited by I. Barandiarán & A. Cava (eds.), pp. 325–359. Veleia. Anejos Series Maior 10. UPV/EHU, Vitoria-Gasteiz, Spain.
2002a Charcoal analyses from Basque archaeological sites: new data to understand the presence of *Quercus ilex* in a damp environment. In *Charcoal Analysis. Methodological Approaches, Palaeoecological Results and Wood Uses. Proceedings of the Second International Meeting of Anthracology (Paris, 2000)*, edited by S. Thiébault, pp. 121–126. British Archaeological Reports S-1063. Oxford, UK.
2002b *Origen de la agricultura en el País Vasco y transformaciones en el paisaje: análisis de restos vegetales arqueológicos.* Kobie. Anejo 4. Diputación Foral de Bizkaia, Bilbao, Spain.

Zapata, L., and G. Meaza

1998 Procesos de antropización y cambios en el paisaje vegetal del País Vasco atlántico en la prehistoria reciente: su incidencia en la expansión de hayedos y encinares. *Munibe (Ciencias Naturales)* 50:21–35.

Zapata, L., and L. Peña-Chocarro

1998 La historia del bosque y su explotación en el pasado: evidencia arqueológica y etnográfica. *Zainak* 17:87–99.

2003 Uso y gestión del bosque en la Euskal Herria atlántica: aprovechamiento tradicional de los recursos forestales en Encartaciones y Gorbea. *Zainak* 22:201–215.

2005 Los macrorrestos vegetales del yacimiento de Mendandia. In El campamento prehistórico de Mendandia: ocupaciones mesolíticas y neolíticas entre el 8500 y el 6500 B.P., edited by A. Alday, pp. 411–425. Fundación J.M. Barandiarán, Ataun, Spain.

Zapata, L., L. Peña-Chocarro, J. J. Ibáñez, and J. E. González Urquijo

2003 Ethnoarchaeology in the Moroccan Jebala (Western Rif): wood and dung as fuel. In *Food, Fuels and Fields—Progress in African Archaeobotany*, edited by K. Neumann, A. Butler, and S. Kahlheber, pp. 163–175. Africa Praehistorica 15. Heinrich-Barth-Institut, Cologne, Germany.

Table 11.1. Provenance of samples analyzed, by level

Level	Flotation sample no.	Grid squares
2	21, 26, 24, 27	K5-D, K5-B, K5-A, K5-C
3	390, 389, 233	H1-B, H1-B, H2-D
4	210, 175, 166	J4-D, I4-B, J4-D
5	339, 295, 321, 301, 300, 293, 343	J4-B, J4-D, H3-D
6	264, 430, 431	H2-A, H3-B, H4-B
7	478, 5, 6, 20, 21, 22, 65, 58	H4-D, H1-A, H1-B, H2-C, H2-D
8	30, 31, 32, 37, 41, 43, 45, 48, 49, 70, 72, 73, 74	J4-A, J4-B, J4-C, J4-D, H2-A, H2-B, H2-C, H2-D
9	2, 4, 7, 8, 11, 12, 13	I3-A, I3-B, I3-C, I3-D
9.6	3, 9, 14, 22	I3-A, I3-B, I3-C, I3-D
9.7	1, 15, 18, 19, 20, 28	I3-A, I3-B, I3-C, I3-D
9.8	5, 16, 21, 27, 29	I3-A, I3-B, I3-C, I3-D
10	6, 279, 280, 281, 282, 283, 286, 296, 307, 309, 314, 315, 320, 325, 327, 395, 459, 460, 463, 476	H1-A, H1-B, H2-A, H2-B, H2-C, H2-D, I3-B, J4-A, J4-B, J4-C, J4-D
10.1	289, 293, 400, 401, 406, 414, 416, 417, 418, 419, 426, 454, 455, 456	H2-A, H2-B, H2-C, H2-D, J4-A, J4-B, J4-C, J4-D
303	38, 50, 51, 52, 54, 56, 57, 58, 59, 61, 62, 69	N5-A, N5-B, N5-C, N5-D
303.1	63, 67, 68	N5-A, N5-C, N5-D

Table 11.2. Absolute results (fragment counts) of the charcoal analysis of the main Holocene levels (n = 1810)

	2	3	4	5	6	7	8	9	10	10.1
		3700±40		3820±240		3740±120	4680±60	5280±40	5570±50	8700±40
Gymnosperms										
Juniperus				47					2	
Taxus baccata	2	2	7		46	48	50			
Angiosperms										
Acer			9	1		1				
Betula										
Cornus						2	4			
Cf. *Cornus*				1						
Corylus avellana	7	12	30	51	24	40	42	29	25	5
Fraxinus	1	23	20	22	17	28	65	27	8	
Cf. *Fraxinus*		1								
Laurus nobilis		10	3	5						
Cf. *Laurus nobilis*				1						
Pomoideae		2	22		1	7	7	7	13	1
Cf. *Pomoideae*							1			
Prunus tp. *avium*	5			29	13	25	37	7	24	
Prunus tp. *padus*						1		1		
Prunus tp. *spinosa*									1	
Prunus sp.		2				3				
Cf. *Prunus* sp.										
Rosaceae		2				1				
Cf. Rosaceae						1				
Quercus subg. Quercus	16	161	62	42	18	46	88	38	123	176
Cf. *Quercus subg. Quercus*										
Quercus ilex/Q. coccifera	1	1	11	2			2	5		
Cf. *Quercus ilex/Q. coccifera*		6							1	
Quercus sp.		1								
Cf. *Quercus* sp.										
Rhamnus alaternus / Phillyrea			7	4			2			
Salix	1									
Salix / Populus		1				1				
Cf. *Salix / Populus*										
Ulmus	1		5	4		3	3	5	2	
Indeterminate (Angiosperm inmature)			1							
Indeterminate (*Ligustrum*?)				4						
Indeterminable		2	10			10	8	1	4	
TOTAL (IDENTIFIED)	**34**	**224**	**177**	**213**	**119**	**207**	**301**	**119**	**199**	**182**

Table 11.3. Absolute results (fragment counts) of the charcoal analysis of levels 9.6, 9.7, 9.8, 303, and 303.1 (n = 358)

	9.6	9.7	9.8	303	303.1
	5250±150			5500±90	5520±70
Gymnosperms					
Juniperus					
Taxus baccata					
Angiosperms					
Acer					
Betula				1	
Cornus					
Cf. *Cornus*					
Corylus avellana	2	25	36	13	2
Fraxinus	6	17	2	14	7
Cf. *Fraxinus*					
Laurus nobilis					
Cf. *Laurus nobilis*					
Pomoideae	4	4	4	2	
Cf. *Pomoideae*					
Prunus tp. *avium*	5	5	45	9	1
Prunus tp. *padus*					
Prunus tp. *spinosa*					
Prunus sp.					
Cf. *Prunus* sp.		1			
Rosaceae					
Cf. Rosaceae		1			
Quercus subg. Quercus	15	29	26	45	22
Cf. *Quercus subg. Quercus*					
Quercus ilex/Q. coccifera		1			
Cf. *Quercus ilex/Q. coccifera*		3			
Quercus sp.				1	
Cf. *Quercus* sp.					
Rhamnus alaternus / Phillyrea					
Salix					
Salix / Populus					
Cf. *Salix / Populus*					
Ulmus	3		4	3	
Indeterminable				4	
TOTAL (IDENTIFIED)	**35**	**86**	**117**	**88**	**32**

Table 11.4. Presence (shaded area) and absence of arboreal taxa in pollen and charcoal analyses

Pollen	Common name	Charcoal
Pinus sylvestris tp.	pine	
	juniper	*Juniperus*
	yew	*Taxus baccata*
	laurel	*Laurus nobilis*
	elm	*Ulmus*
Quercus ilex tp.	holm/prickly oak	*Quercus ilex/coccifera*
Quercus robur tp.	deciduous oak	*Quercus subg. Quercus*
Corylus avellana	hazel	*Corylus avellana*
Salix	willow/poplar	Salix/Populus
Rosaceae	hawthorn/apple/ pear tree cherry tree bird cherry blackthorn	Rosaceae: Pomoideae *Prunus* tp. *avium* *Prunus* tp. *padus* *Prunus* tp. *spinosa*
Rhamnus	buckthorn	*Rhamnus alaternus / Phillyrea*
Betula	birch	*Betula*
Alnus	alder	
Tilia	lime	
Frangula	alder buckthorn	
Erica	heather	
	dogwood	*Cornus sanguinea*
	maple	*Acer*
Fraxinus	ash	*Fraxinus*

Plate I.1. San Vicente Peak from inside El Mirón Cave. Photo by Lawrence G. Straus.

Plate 1.2. "Dissection" of an initial Magdalenian occupation surface in the Corral area. Photo by Lawrence G. Straus.

Plate 1.1. Excavation of post-Paleolithic levels in the outer vestibule (Cabin) area. Photo by Lawrence G. Straus.

Plate 1.3. Chalcolithic burning lenses (Levels 7–4) in the Cabin area. Photo by Lawrence G. Straus.

Plate 1.4.
Chalcolithic rock-filled pit features (Level 5, pits G and H). Photo by Lawrence G. Straus.

Plate 1.6. East stratigraphic section (J–K/2–4) of the Cabin area. Photo by Lawrence G. Straus.

Plate 1.5. Sondage P6 stratigraphic section, Mid-Vestibule Trench. Photo by Lawrence G. Straus.

Plate 1.7. Excavation of a Magdalenian occupation surface in the Cabin area. Photo by Lawrence G. Straus.

Plate 2.1.
The Río Carranza gorge,
an eastern tributary of the middle
Asón. Photo by Lawrence G. Straus.

Plate 2.2. The Sierra del Hornijo from the north.
Photo by Lawrence G. Straus.

Plate 2.3. Monte Pando (location of El Mirón)
from San Vicente Peak. Photo by Lawrence G. Straus.

Plate 2.4. West face of Monte Pando, showing mouths of El Mirón, Covalanas, and La Haza caves.
Photo by Manuel R. González Morales.

Plate 3.1. El Mirón Cave mouth (close-up). Photo by Lawrence G. Straus.

Plate 3.3. The Ruesga Valley from El Mirón. Photo by Lawrence G. Straus.

Plate 3.2. Cuevamur. Photo by Lawrence G. Straus.

Plate 3.4. San Vicente Peak from El Mirón. Photo by Lawrence G. Straus.

Plate 3.5. El Mirón Cave vestibule at the start of the excavation in 1996. Photo by Lawrence G. Straus.

Plate 3.6. El Mirón Cave vestibule looking toward the cave mouth from the rear slope. Photo by Lawrence G. Straus.

Plate 3.7. The colluvial-alluvial slope between the El Mirón Cave vestibule and inner cave. Note $CaC0_3$ stain of rivulet after winter rains. Photo by Lawrence G. Straus.

Plate 3.8. Breccia remnant with Azilian material at the top of the colluvial-alluvial slope. Photo by Lawrence G. Straus.

Plate 3.9. Niche A in the southern cave wall midway up the colluvial-alluvial slope. Photo by Lawrence G. Straus.

Plate 3.10. The stone cabin foundations near the front of the cave vestibule. (Cabin area excavation of Bronze Age levels.) Photo by Lawrence G. Straus.

Plate 5.1. Geological trench cut into the midslope of the colluvial ramp, exposing poorly stratified, sandy, cobble-rich sediment; the cobbles in the pile at the top were extracted from the matrix. 10 cm units on scale. Photo by William R. Farrand.

A.

Plate 5.2.
Mid-Vestibule section photos:

A. South wall, O–P/6, including Strata 301 through 306. Note especially two pale mondmilch Layers, 304 and 305, separated by a light brown oxidized zone at the top of 305, marking a hiatus. Photo by William R. Farrand.

B. West end, O–N/6, from Stratum 303.1 through 312 at floor of trench. Stratum 305 is absent here. Note color contrasts from 306 to 308 and 311 to 312. Ca. 20 cm segments on folding scale. Photo by William R. Farrand.

B.

Plate 5.3. Lower west wall of Cabin area excavation from Stratum 14 (*top*) through 17 (*at floor*), contrasting the smooth, silty clay of Stratum 14 and the rocky stratum complex of Strata 15–17. 10 cm units on scale. Photo by William R. Farrand.

Plate 5.4. East wall of Square X10 exposing coarsely stratified sequence of Strata 121 through 127, dipping distinctly to the south, overlying the massive rocky load of Stratum 128 with large limestone blocks, with a wavy erosional surface between 127 and 128. 10 cm scale. Photo by William R. Farrand.

A.

B.

Plate 5.5.
Cuevamur entrance and close-up of sediments:

A. View into Cuevamur showing a steep, barren slope of slits and gravels overlain by several large limestone blocks. Entrance is 15 m wide. Photo by William R. Farrand.

B. Detail of coarse cobbles and pebbles overlying gravels and pale-colored silts (at bottom of scale), as well as large, angular blocks near the top of the photo. Ca. 20 cm segments on folding scale. Photo by William R. Farrand.

A.

B.

Plate 5. 6. Cuevamur travertine and gravels:

A Mid-slope sediments in Cuevamur showing poorly sorted, rounded cobbles over silt (behind notebook) and small gravel. Notebook is 18 cm high. Photo by William R. Farrand.

B. Gray travertine hanging from ceiling of Cuevamur and sealing the top of reddish-brown sediments in Cuevamur, estimated to be at least 200,000 years old per uranium-series dating by J. Bischoff. 10 cm scale. Photo by William R. Farrand.

Plate 8.1. Western cliff face of Monte Pando, showing modern vegetation near El Mirón. Photo by María Jose Iriarte Chiapusso.

CHAPTER TWELVE

Holocene Biostratigraphy and Climatic Change in Cantabria

The Micromammalian Faunas of El Mirón Cave

Gloria Cuenca Bescós and Juan Carlos García Pimienta

TRANSLATED BY LAWRENCE GUY STRAUS

El Mirón Cave, near Ramales de la Victoria on the northern edge of the Cantabrian Cordillera in eastern Cantabria, has a large mouth and moderate overall length. It is at an elevation of about 260 m above present sea level, 3°27' West of the Greenwich Meridian and 43°15' North (figures 12.1 and 12.2). Its favored location at a crossroads of valleys leading into Vizcaya and over the cordillera to Burgos but only about 20 km from the Holocene coastline, with its mouth facing due west, led to its having been utilized by humans ever since at least Mousterian times (Straus and González Morales 2001b). Until just a few years ago shepherds used the vestibule as a livestock stable. Although El Mirón was discovered scientifically at the beginning of the twentieth century, systematic excavations did not begin until 1996. Ever since then, under the direction of L.G. Straus and M. González Morales (2001a, b), a team of five to twenty-five students has been excavating in the cave for

Fig. 12.1. Location of El Mirón Cave on the Iberian Peninsula. Map by Gloria Cuenca.

two months every summer. The main excavation areas have been near the front of the vestibule (the Cabin) and at its rear (the Corral).

The Holocene sediments from the Cabin area were water-screened for the recovery of small archaeological and paleontological remains. Sorting of the screen residues involved separation of artifacts, vertebrates, and botanical remains, which were then prepared for study and curation (Straus and González Morales 2001a, b).

This study presents the first results of our analyses of the micromammalian remains from the Holocene levels of El Mirón, namely from the Cabin area of the outer vestibule. At the time of writing this excavation area had reached a total depth of about 3 m, down to the early to mid-Magdalenian. During the course of the excavation some 14 Holocene-age levels and lenses were defined (Levels 1–10.1). Some of the lenses are in-fillings of pit features. For the effects of this study several of the minor lenses were grouped with their corresponding levels, but Level 3.5 (a channel fill posterior to deposition of Level 3) and Level 10.1 (a major, distinct stratum of Mesolithic age) were considered separately, as significant stratigraphic units in their own right.

The wet-screening operation conducted on the deposits of Levels 1–10.1 by the El Mirón team provided an interesting collection of micromammals, which is the basis of the present study. In our laboratory at the University of Zaragoza we carried out the analyses of these fossils and prepared them for their long-term curation. We have identified remains of 21 species of micromammals from the Holocene assemblages of El Mirón (table 12.1). They are distributed among three orders: Rodentia, Eulipotyphla, and Chiroptera. Pragmatically, we group them together under the category of micromammals because of their small size. We can also add to this list some small carnivores pertaining to the Mustelid family, which, because of their small size, are also found in the microvertebrate samples from El Mirón, for which reason we also studied them in this context. Summaries of the whole El Mirón micromammalian sequence (including the Upper Paleolithic and terminal Middle Paleolithic levels) are presented in Cuenca Bescós et al. (2008, 2009).

The complete faunal collection of micromammals from the Holocene levels of El Mirón is described as follows (see also tables 12.2 and 12.3):

Fig. 12.2. Location of El Mirón Cave and other Paleolithic sites in the Río Asón valley. Map by L. G. Straus, redrafted by Ronald Stauber.

Rodentia (12 species): *Arvicola terrestris*, *A. sapidus*, *Microtus agrestis*, *M. arvalis*, *M. oeconomus*, *Chionomys nivalis*, *Clethrionomys glareolus*, *Terricola lusitanicus*, *Pliomys lenki*, *Apodemus sylvaticus-flavicollis*, *Glis glis*, *Eliomys quercinus*

Eulipotyphla (6 species): *Talpa europaea*, *Sorex coronatus-araneus*, *Sorex* "robust sp.," *S. minutus*, *Neomys* sp., *Crocidura russula*

Chiroptera: one large species, *Myotis* sp.; and one small species, *Miniopterus* sp.

Carnivora: *Mustela nivalis*

Chronology of El Mirón Cave

Most of the micromammalian samples have been indirectly dated by means of radiocarbon dating of charcoal from the levels (Straus and González Morales 2001a, b). The Holocene levels can be grouped into five periods:

Modern: Level 1
Protohistoric (?): Level 3.5 (pit fill)
Early Bronze Age: Levels 2 and 3 (+ Lenses 3.0–3.4)
Chalcolithic: Levels 4–7
Neolithic: Levels 8–10
Mesolithic: Level 10.1

The radiocarbon dates for the Holocene levels are summarized in table 12.4.

Summary of the Micromammalian Faunas of El Mirón Cave in Stratigraphic Context

The micromammalian remains analyzed here come from the Cabin excavation area in the outer vestibule, squares H–J/1–4; their vertical provenance context consists of levels and spits and their horizontal provenance consists of meter squares and quarter-meter subsquares (figures 12.3 and 12.4).

In this section we describe the stratigraphic units with their micromammalian contents from top to bottom within the Holocene deposit. Preliminary publications as well as chapters in the present volume by the project directors (Straus et al. 2001) and geologists give sedimentological and stratigraphic details that we will not repeat here. In contrast, we concentrate on the micromammalian faunas that characterize each level (and the relative importance of the different taxa) (see also tables 12.2 and 12.3). From top to bottom:

Level 1: This is a gray-brown deposit topped by recent dung, straw, and modern refuse. It contains abundant organic matter and rubble fill from the collapsed cabin walls. It is essentially a modern deposit relating to use of the cave as a livestock stable. We have two micromammal samples from Level 1; they included only two taxa: *Arvicola terrestris* and *Talpa*.

Level 3.5: This depositional unit is the in-filling of a broad channel or pit in the western side of the excavation area. This feature manifestly cut through at least Levels 3, 3.2, 4, and 5 and into the tops of Level 6 and even Level 7. Its relationship to the discontinuous patches that constitute Level 2 is unclear, and neither is it radiometrically dated. It does predate Level 1 and did not yield modern artifacts (see figure 12.4). This dug-out or erosive (?) feature could be filled with sediments pertaining to the so-called Iron Age Cold Phase (2900–2300 BP) that has been observed in northeastern Spain in open-air contexts and described by the authors as a cold, humid episode that coincided with a minor advance of montane glaciers and higher precipitation than in the preceding warm, subatlantic pollen zone—hence an erosive period (Gutierrez Elorza and Peña-Monné 1998). However, no Iron Age artifacts were found in it. The microfaunal assemblage from Level 3.5 consists of the following taxa: *Arvicola terrestris, Microtus agrestis, M. arvalis, M. oeconomus, Chionomys nivalis, Clethrionomys glareolus, Terricola lusitanicus, Pliomys lenki, Apodemus sylvaticus-flavicollis, Glis glis, Talpa* sp., *Sorex coronatus-araneus, S. minutus, Neomys* sp., *Crocidura* sp., *Myotis*, and *Mustela* sp. The presence of *Pliomys* in such a recent deposit is highly unusual, as will be commented upon below.

Level 2: Discontinuous patches of beige or yellowish, sandy clay below the cabin floor, essentially at the eastern side of the excavation area. Domesticated animal bones are abundant, but the only micromammals are again *Arvicola terrestris* and *Talpa*.

Level 3 (+ Lenses 3.0–3.4): A complex of the main level (3), which is a brown, silty loam and various multicolored lenses formed by such diverse processes as calcium carbonate precipitation and burning. This complex presents a much higher diversity of micromammalian taxa than overlying Levels 1 and 2: *Arvicola terrestris, Microtus agrestis, M. arvalis, M. oeconomus, Chionomys nivalis, Clethrionomys glareolus, Terricola lusitanicus, Pliomys lenki, Apodemus sylvaticus-flavicollis, Glis glis, Talpa* sp., and *Mustela*. The apparent presence of *Pliomys* is also notable but probably misleading in this 3700 BP (uncal.) Bronze Age context, as will be discussed at length below.

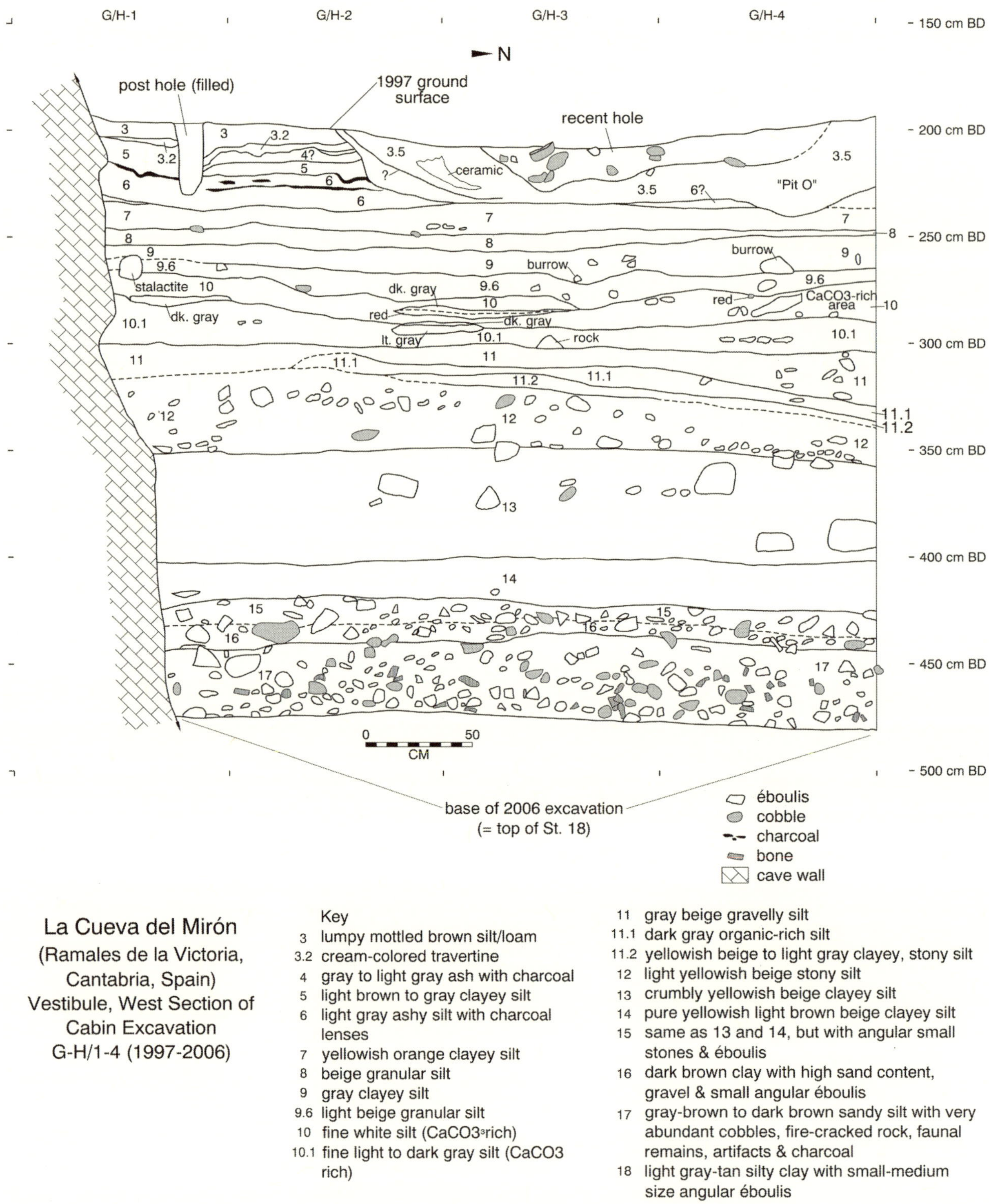

Fig. 12.3. West section of the Cabin area excavation (G–H/1–4). Section by L. G. Straus, redrafted by Ronald Stauber.

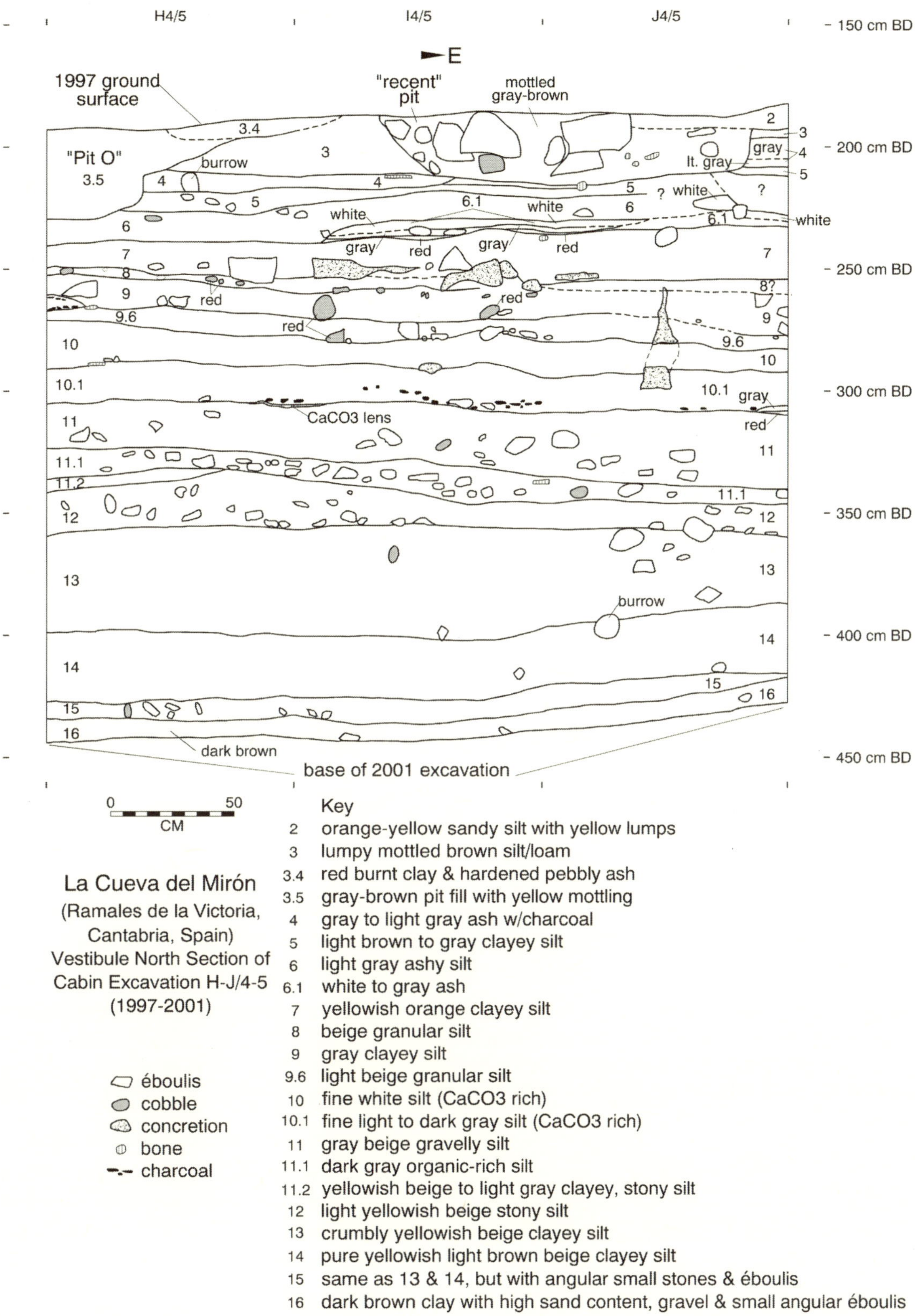

Fig. 12.4. North section of the Cabin area excavation (H–J/4–5). Note the boundary between Levels 10.1 and 10 and the position of Level 3.5 relative to the underlying levels. Level 3.5 has the geometry of a trench or channel that had cut into Level 3 and lower levels and thus is later than them. Section by L. G. Straus, redrafted by Ronald Stauber.

Level 4: This is a fire deposit formed of alternating white and light-gray ash lenses that are locally cemented and charcoal patches, with fire-cracked rocks. At least two fire foci were found in 1996. Ceramic sherds are concentrated around these fires, and some of them are burned. The micromammals are less diverse than in the Level 3 complex above: *Arvicola terrestris*, *Microtus arvalis*, *Clethrionomys glareolus*, *Apodemus sylvaticus-flavicollis*, *Glis glis*, *Talpa* sp., *Sorex coronatus-araneus*, and *Mustela* sp.

Level 5 (+ Lenses 5.1–5.4): This is a complex of gray-brown, clayey silt with abundant organic matter, including charcoal. There are several pit fills and lenses. Level 5 *sensu stricto* dates to 3820 BP and Lens 5.1 to 4120 BP. Most of the micromammalian remains are from five *sensu stricto*. The combined assemblage includes the following taxa: *Arvicola terrestris*, *Microtus agrestis*, *M. oeconomus*, *Chionomys nivalis*, *Clethrionomys glareolus*, *Terricola lusitanicus*, *Apodemus sylvaticus-flavicollis*, *Glis glis*, *Talpa* sp., *Sorex coronatus-araneus*, and *Mustela* sp.

Level 6: This is a layer of white ash with fragments of charcoal. Although there is a Lens 6.1, all the micromammalian remains came from six *sensu stricto* and include: *Arvicola terrestris*, *Microtus agrestis*, *Chionomys nivalis*, *Clethrionomys glareolus*, *Apodemus sylvaticus-flavicollis*, *Glis glis*, *Eliomys quercinus*, *Talpa* sp., *Sorex coronatus-araneus*, and *Neomys* sp.

Level 7 (+ Lenses 7.1–7.5): This is a complex composed of Level 7 *sensu stricto* (a silty clay) and a series of lenses rich in ash and charcoal. Micromammals are rare but display a difference vis-à-vis the overlying strata, which will be discussed below. Specifically, there are many fewer arvicolines than in the later levels, with this being the only level in which *A. terrestris* is absent. The taxa that are present include: *Microtus agrestis*, *Chionomys nivalis*, *Apodemus sylvaticus-flavicollis*, *Glis glis*, *Eliomys quercinus*, *Talpa* sp., *Sorex coronatus-araneus*, and *Crocidura* sp. (A date of 3740 BP from Level 7 is out of stratigraphic order but would make more sense at + two standard deviations [i.e., 3980 BP].—LGS)

Level 8: Level 8 *sensu stricto* is a beige-gray clay with charcoal fragments; it is underlain by hearth-fill Lens 8.1, ash with charcoal and fire-cracked rock. Lens 8.1 is dated to 4680 BP, but all the micromammalian remains come from Level 8 *sensu stricto*. They are: *Arvicola terrestris*, *Microtus agrestis*, *M. arvalis*, *Chionomys nivalis*, *Clethrionomys glareolus*, *Terricola lusitanicus*, *Apodemus sylvaticus-flavicollis*, *Glis glis*, *Talpa* sp., *Sorex coronatus-araneus*, *S. minutus*, *Neomys* sp., and Chiroptera.

Level 9 (+9.1–9.8): This is a complex composed of Level 9 *sensu stricto* and a series of lenses; in general the whole complex is dominated by clay with ashes and charcoal. A charcoal layer seems to define the base of the complex in the northern edge of the excavation area, without any more charcoal lenses below Lens 9.8 (see figure 12.4). The micromammals from this complex include: *Arvicola terrestris*, *Microtus agrestis*, *M. arvalis*, *Chionomys nivalis*, *Clethrionomys glareolus*, *Terricola lusitanicus*, *Apodemus sylvaticus-flavicollis*, *Glis glis*, *Eliomys quercinus*, *Talpa* sp., *Sorex coronatus-araneus*, and *Sorex* "robust species."

Level 10: This is a whitish, calcium carbonate–rich silt. It is more extensive, continuous, and thicker than most of the overlying units. Its microfauna include *Arvicola terrestris*, *Microtus agrestis*, *M. arvalis*, *Chionomys nivalis*, *Clethrionomys glareolus*, *Terricola lusitanicus*, *Apodemus sylvaticus-flavicollis*, *Glis glis*, *Eliomys quercinus*, *Talpa* sp., *Sorex coronatus-araneus*, *Sorex* "robust species," *Mustela* sp., and the Chiroptera *Myotis* and *Miniopterus*. Levels 10 and 10.1 are faunistically relatively similar, being distinguished by the presence in Level 10 of *Sorex minutus* and Chiroptera, which are absent in Level 10.1. The presence of the small cave bat *Miniopterus* could indicate that either the cave was partially abandoned by humans between 5690 and 5570 BP or the sampling methods were insufficient to reliably control for the presence of this small Chiroptera at all times. Also notable is the presence of *Arvicola sapidus* in Level 10.1, which represents the first reference to the water rat during the Holocene in the northern part of the Iberian Peninsula. There are two other cases that show the presence of *Arvicola sapidus* during the Upper Pleistocene in the Cantabrian region: one in the

Solutrean of La Riera (Altuna 1986) and the other—also in the later Upper Paleolithic—at Las Caldas, both in Asturias (Corchón 1981).

Level 10.1: This is a grayish, silty clay, also very enriched in $CaCO_3$. Like Level 10, it is thick and continuous. According to Straus and González Morales (2001a), there seem to have been depositional hiati between Levels 10 and 10.1 and between Level 10.1 and underlying Level 11. Level 10.1 is the last level to lack ceramic sherds and remains of domesticated animals. It has small numbers of lithic artifacts and only wild ungulate remains. Together with these characteristics, its dates, which range between 9550 and 8380 BP, imply a Mesolithic age. The Level 10.1 microfaunal assemblage is composed of *Arvicola terrestris*, *A. sapidus*, *Microtus agrestis*, *Chionomys nivalis*, *Clethrionomys quercinus*, *Talpa* sp., *Sorex coronatus-araneus*, and *Mustela* sp. Notable are the absences of *M. arvalis*, both species of *Sorex*, *Crocidura*, and the Chiroptera *Myotis* and *Miniopterus*.

Paleontological Study of the Micromammals from the Holocene of El Mirón Cave

Rarely are microvertebrates the subject of study in Spanish archaeological sites of Holocene age; indeed, they are hardly ever mentioned as having been found in the deposits—as if they did not exist. Exceptions can be noted the studies of Arturo Morales Muñiz and associates (Morales Muñiz 1986; Morales Muñiz et al. 1995; Morales and Rodríguez 1997), as well as those of Pemán (1987, 1988, 1990a, b) and Laplana Conesa and Cuenca Bescós (1995). The small size of the remains (2.0–0.5 mm in most cases) has probably been the main reason for micromammals being ignored in Holocene-age excavations. Methods for the recovery of the remains of small mammals, birds, amphibians, reptiles, and fish are rather complex, as large amounts of deposit have to be water-screened through fine mesh—either in their entirety or in several column samples of at least 1 m^2 in area. This costs a great deal of time, energy, and money but is worth it in terms of the long-term paleoenvironmental detail that can be obtained with systematic micromammalian analysis.

Although the study of micromammals from Quaternary sites basically began as a result of their biostratigraphic relevance for dating and correlation, late Upper Pleistocene and Holocene sites are now dated by radiometric methods, especially C-14. However, in older sites (Pliocene and Lower, Middle, and early Upper Pleistocene), when other dating methods are difficult or impossible to apply, the micromammalian faunas can be used to provide information on geological age of the deposits under study. In addition, as shown in the latest studies, Quaternary micromammals can provide a great deal of paleoecological and paleoclimatic information and thus must be taken into consideration in any Quaternary paleoenvironmental reconstructions. In the most recent periods they can also shed light on possible commensalism and other aspects of the interactions of humans with their environment, such as sedentism or seasonality of site occupation (e.g., Morales Muñiz et al. 1995; Tchernov 1992). Recent studies (e.g., Pokines 1998; Pemán 1985, 1990b) in fact demonstrate that aspects of the late Upper Pleistocene and Holocene can be reconstructed from the study of successions or stratigraphic distributions of microvertebrates. The undoubted value of certain species as environmental indicators must be emphasized in the reconstruction of local climatic variations: for example, the proximity of water bodies is indicated by *Neomys*, *Arvicola*, and *Microtus oeconomus*; the presence of woodland by *Apodemus*; humid prairie by *Talpa*; sunlit, rocky ground by *Chionomys nivalis*; and humid climate by *Sorex*.

To obtain fossils of small size from the Holocene deposits of El Mirón Cave, nested screens ranging in mesh size from 2 cm to 2 mm were used. The microvertebrate remains were picked from the screen residues by hand. This means that the smallest remains (e.g., those of the majority of the shrews and bats in the orders Eulipotyphla and Chiroptera, respectively, as well as the tinier rodents) are frequently underrepresented, thus biasing the composition of the overall collection. In order to determine the extent of the possible bias, we have compared the Mirón Holocene collection results with those of other authors working in the Cantabrian region, notably Pokines (1998), whose principal study was of micromammals from the Magdalenian deposit of El Juyo Cave (coastal Cantabria), which was subjected to systematic flotation recovery methods (Freeman et al. 1998). Later, for the excavation of the Upper Pleistocene deposits in El Mirón, a flotation machine designed by González Morales was used to recover the finest fractions and minimum mesh size for screening was reduced (Straus, pers. comm.).

MATERIALS, METHODS, AND NOMENCLATURE

Despite its imperfections in the first season of excavations (which is when most of the Holocene levels reported on here were dug), the system of water- and dry-screening and residue hand-picking at El Mirón did result in the collection of a large quantity of micromammalian remains. There are 309 sorted lots of material from Level 1 to Level 10.1. These yielded a total MNI (minimum number of individuals) of 1,542 individuals (tables 12.2 and 12.3), distributed among 20 species based on counts of the first lower molars (both as loose teeth and as teeth still set in mandibles), except in the cases of *Talpa* (mole)—for which the count is based on the humerus—and the sorids—for which, in the absence of whole mandibles or crania, the articular condyle of the mandible is used in classification. The samples are labeled "El Mirón Microfauna Levels 1–10.1," and in the figures the different species are grouped by square and excavation spit as well as by sample number (figures 12.5–12.21). All the samples contain variable numbers of fossils, among which we have prepared the classification elements (teeth and other diagnostic skeletal items). The whole collection has been curated in small boxes labeled with complete provenance information (El Mirón, year, sample number, square, subsquare, level, and spit). The collection will be stored in the Cantabrian Regional Museum of Prehistory in Santander.

Our analysis, as indicated above, was generally limited to teeth (either loose or in mandibles) of rodent, insectivore, and chiropteran species, except in a few insectivore cases for which identifications and counts of minimum numbers of individuals can be made with other elements. The MNI per species is usually calculated on the basis of the number of lower first molars in the case of the arvicolines, a diagnostic molar for the rest of the micromammals, or some equally identifiable post-cranial element, in all cases divided by two, since most of the diagnostic elements in vertebrate skeletons are paired.

The methodology applied is biostratigraphic, consisting of both qualitative and quantitative study of the distribution of species by levels (in this case, Levels 1–10.1). For this study we have assumed that the principal agents of accumulation of microvertebrates were nocturnal birds of prey, since, in fact, most of the remains display a moderate degree of weathering by digestion but are whole—signatures of having been caught and swallowed by raptorial birds. The absence of remains in a few levels or lenses could be an indication that raptors were less common in the cave at those times, which could have been due to a variety of factors, such as climatic conditions or a sustained human presence. Moles (*Talpa*) and the *Arvicola* species could have lived in El Mirón Cave occasionally; thus the presence of their remains in the deposits could have two causes: raptorial transport and natural in situ death. We believe that their persistence in the most recent levels suggests that they still live in the cave today. (The excavators occasionally have found live voles in the Corral excavation area.—LGS)

Each of the 309 samples was studied separately (see table 12.3), but for the purposes of the biostratigraphic analysis we grouped all the samples corresponding to each level and its related lenses. Levels 3.5 and 10.1 are considered as separate stratigraphic units, as explained above, since Level 3.5 has nothing to do with Level 3 (and is more recent) and Level 10.1 is older and culturally distinct from Level 10.

Study of the Micromammals from the Post-Paleolithic Levels of El Mirón Cave

As noted earlier, there are 21 species of small mammals distributed among the orders Eulipotyphla, Chiroptera, Rodentia, and Carnivora in the Holocene deposit from El Mirón. The leporids (rabbits and hares), considered to be micromammals in paleontological research, are studied separately in archaeological sites since most of their anatomical elements are large enough to be piece-plotted.

In this section we present a brief description of each species found in the Holocene levels of El Mirón as well as a brief summary of the habitat and present and possible Holocene geographic distribution of the species throughout the Iberian Peninsula. The classifications are based mainly on the dentition, since in most cases only loose teeth were preserved. For the description of the teeth of the various small mammal families we adopt the nomenclature that has been established by the standard authors, and we use the following abbreviations: I = incisors, C = canines, P = premolars, and M = molars. A superscript indicates that the dental element is an upper tooth and a subscript means a lower tooth. In arvicolines the letter *t* followed by a number refers to a dental cusp triangle with its standard position designation.

ORDER RODENTIA

Family Muridae

Subfamily Arvicolinae

The arvicolines are generally burrowing rodents that live in open settings with deep soils that permit the digging of extensive tunnels. They almost exclusively inhabit the Northern Hemisphere. The systematics of arvicolines is based on the jugal teeth (molars), which are above all characterized by hypsodonty, that is, a high crown that among most present species reaches the condition of hypselodontia, or the loss of roots and continuous growth. Roots are not even found in the final stages of development. The morphometric parameters of the dentition are the principal distinguishing characteristics of the arvicolines; they include size (length, width, development of labial and lingual triangles, and development of the anterior complex), crown height, presence/absence of roots, cementum in the recesses, enamel thickness, number and morphology of recesses, and projections (triangles [t]) on the occlusal surface. The phylogenetic hypotheses and thus classification of the arvicolines are based on the evolution of these characteristics. The most important dental element is the lower first molar (M_1). Most of the morphometric analysis is done on the occlusal surface of this tooth.

Arvicola terrestris (Linnaeus, 1758)

DESCRIPTION AND DISCUSSION

The species *A. terrestris*, or mole rat (also known as the northern water rat and ground vole), is a large arvicoline characterized by the absence of tooth roots, cementum in the recesses, M_1 with three more-or-less closed triangles, and an anterior complex formed by two triangles that join together with the anterior lobe. *A. terrestris* has differentiated enamel; among modern Spanish populations it is slightly thicker on the medial part of the triangles of the occlusal surface than on the distal part. This discrepancy in the differentiation of the enamel of *A. terrestris* in Spain with respect to other populations in Europe is reflected in the taxonomy of the species.

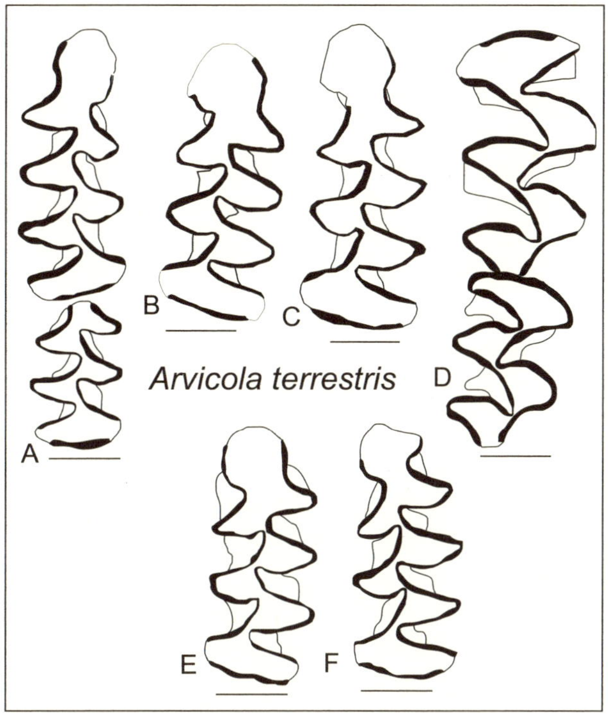

Fig. 12.5. *Arvicola terrestris* (Linnaeus, 1758). Level 3.5: A (no. 123), B (no. 122); Level 6: C (no. 103); Level 3: D,E,F (no. 2). Drawing by Gloria Cuenca.

HABITAT AND GEOGRAPHIC DISTRIBUTION

At present the northern water rat has a wide distribution across Europe, although in Spain it is found only in the north. In the rest of Europe, where it has no competition, *A. terrestris* lives in fluvial environments. However, where *A. sapidus* is present, it retreats to fields and ground near rivers.

Arvicola sapidus (Miller, 1908)

DESCRIPTION AND DISCUSSION

Morphologically, *A. sapidus*—the southern water rat or water vole—is similar to *A. terrestris*. At the present time it is the largest arvicoline among the Spanish fauna. It is distinguished from *A. terrestris* by having

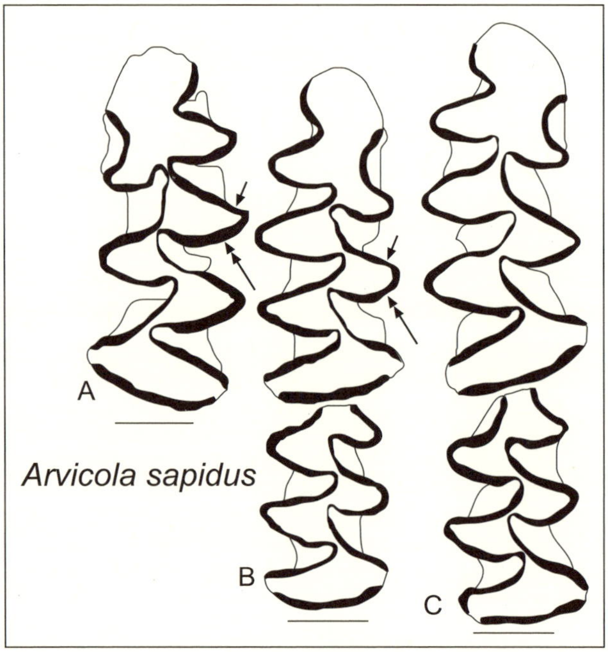

Fig. 12.6. *Arvicola sapidus* (Miller, 1908). Level 10: A,B (no. 249), C (no. 262 b). Drawing by Gloria Cuenca.

differentiated dental enamel in plesiomorphic state, of the type of *Mimomys*—that is to say, thicker on the distal tooth face than on the mesial (Maul 2001).

HABITAT AND GEOGRAPHIC DISTRIBUTION

This a species typical of the banks of rivers and irrigation ditches in the Iberian Peninsula and the south-center of France. *A. sapidus* prefers stagnant waters with dense bank vegetation; it disappears from polluted areas, where it is replaced by the common rat (Arrizabalaga i Blanch, Montagud i Blas, and Gosálbez i Noguera 1986). It is not a common species in northern Spain either today or in the earlier Holocene (Zabala 1983; Ventura and Sans-Fuentes 1997; Pokines 1998). Nor do we know when it made its first entry into the Cantabrian region, since it is also rare in the Upper Pleistocene, having been identified only in the Solutrean levels at Las Caldas (Corchón 1981) and La Riera (Altuna 1986), both in Asturias, and probably in the Lower Magdalenian of El Juyo Level 8 in Cantabria (Pokines 1998). In the Middle Pleistocene, however, there are several references to *A. sapidus* cited as *confer* ("cf.") or *affinis*, which suggests that this species is in need of systematic revision (Cuenca-Bescós, Canudo, and Laplana 1999).

Microtus agrestis (Linnaeus, 1761)

El Mirón Level 3.5, no. 122

DESCRIPTION AND DISCUSSION

All the species of the genera *Microtus*, *Terricola*, and *Chionomys* have differentiated, *Microtus*-type tooth enamel, that is, thicker on the mesial face than on the distal face of the occlusal triangles.

M. agrestis is of medium size; its teeth have cementum but no roots. The M_1 has three closed triangles in the posterior complex and four closed or minimally converging ones in the anterior complex. It is characterized by asymmetry and alternation of triangles t4, t5, and t6–t7, as well as marked alternation of the inward-oriented angles. The M^2 has three lingual projections, which is the same thing as four closed triangles.

Fig. 12.7. *Microtus agrestis* (Linnaeus, 1761). Level 3.5 (no. 122). Drawing by Gloria Cuenca.

HABITAT AND GEOGRAPHIC DISTRIBUTION

This species is commonly known as the mountain or short-tailed vole. At present *M. agrestis* lives in northern Europe and Asia, from Scandinavia to the Lena River, although its distribution is discontinuous in Siberia and it is also found in Great Britain. In the northern Iberian Peninsula it is distributed from the Pyrenees to Galicia and northern Portugal. It is absent from the Mediterranean regions. Its favorite habitats are wooded prairies, forest edges, and even upland moors and dunes in Europe (Musser and Carleton 1993; Pokines 1998). In Spain it is an indicator of Atlantic climate. In Montseny (Catalonia), although a Mediterranean montane area, because of its altitude there are vegetal and animal communities that are more "northern" in character—including *M. agrestis*, which is found from the oak zone to the beech zone but in deforested patches with grass cover, even occupying stone walls and fields (Arrizabalaga i Blanch, Montagud i Blas, and Gosálbez i Noguera 1986).

Microtus arvalis (Pallas, 1779)

DESCRIPTION AND DISCUSSION

The M_1 is similar to that of *M. agrestis*, being different because of the nearly symmetrical and parallel disposition of t4–t5 and especially t6–t7, which creates a more rounded perimeter than in *M. agrestis*. M^2 has two lingual projections, which are equivalent to three closed triangles.

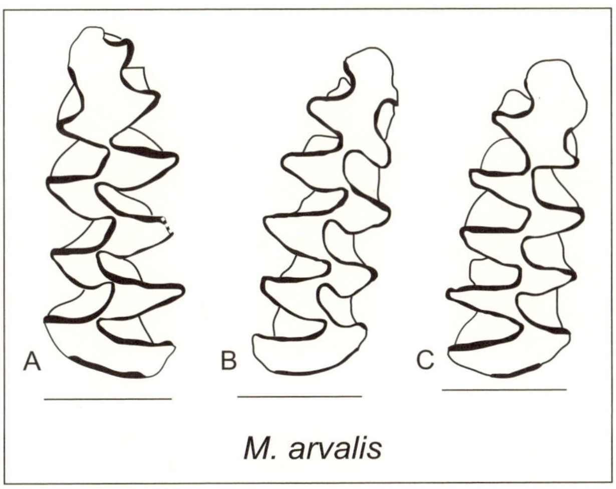

Fig. 12.8. *Microtus arvalis* (Pallas, 1779). Level 3.5: A,B,C no. 25). Drawing by Gloria Cuenca.

HABITAT AND GEOGRAPHIC DISTRIBUTION

The distribution of *M. arvalis* is similar to that of *M. agrestis*, although it is a more opportunistic species and is distinguished by having a more extensive and continuous spread than its relative. In the Iberian Peninsula it extends farther south and it can be found in a wide variety of habitats, from pastures to deciduous and coniferous woods (Pokines 1998). However, under present conditions of aridity in the Mediterranean region it has taken refuge in mountainous massifs like Javalambre and Peñagolosa (in the provinces of Teruel and Castellón, respectively), having disappeared from coastal areas where it had lived during the Upper Pleistocene (Guillem 1995).

Microtus oeconomus (Pallas, 1776)

El Mirón Level 5, no. 220a

DESCRIPTION AND DISCUSSION

The marsh, Nordic, or root vole prefers cold, humid habitats and is the most rare of the *Microtus* species in the Holocene of El Mirón, having been found only in Levels 5 and 3 (see table 12.2). Its M_1 has three closed triangles, with alternating t4 and t5, the latter converging

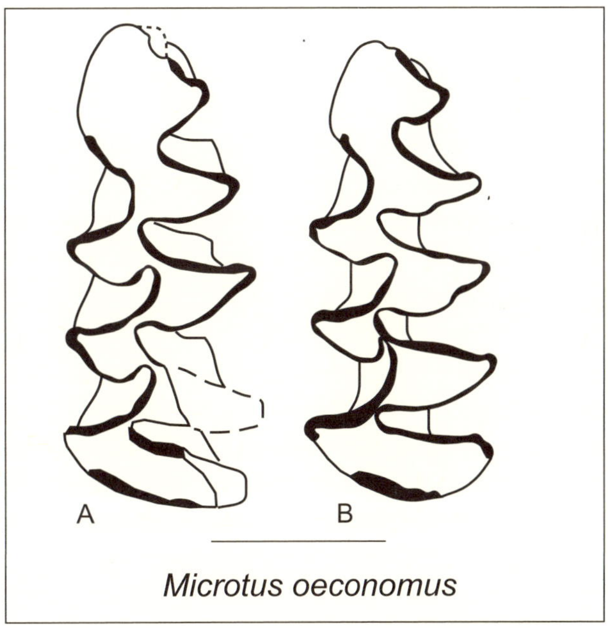

Fig. 12.9. *Microtus oeconomus* (Pallas, 1776). Level 3.5: A (no. 122d); Level 5: B (no. 220). Drawing by Gloria Cuenca.

significantly with t7 and the anterior lobe. The anterior complex is asymmetrical, since t6 is a barely marked projection. In *M. oeconomus* there are five lingual recesses with cementum, although the most mesial of them is reduced. The fourth lingual recess is almost parallel to the third labial recess, in contrast to *M. agrestis*, in which both recesses are alternating.

HABITAT AND GEOGRAPHIC DISTRIBUTION

M. oeconomus is a typical inhabitant of the tundra and northern taiga zones of the Holarctic, from Canada to Holland and Scandinavia (Musser and Carleton 1993). Although today it is well represented in northern and central Europe (Pokines 1998; Nadachowski 1982), it disappeared from the British Isles (and probably from the midlatitudes of western Europe north of the Pyrenees) at the beginning of the Holocene (Sutcliffe and Kowalski 1976). Its extinction in Spain is very recent, however, since it has been found in Roman-age levels at Amalda Cave in Guipúzcoa (Pemán 1990a) and now in Levels 5 (Chalcolithic) and 3 (early Bronze Age) in El Mirón (see table 12.2). It competes with *M. arvalis*, so that when this species is present *M. oeconomus* inhabits colder, more humid habitats than pastures, namely bogs and marches, wet grasslands, tundra, and wooded steppe (Pokines 1998). According to Pokines, this is a solitary animal that cannot compete directly with *M. arvalis*, which, in contrast, lives in groups—a factor that has to be taken into consideration when analyzing its stratigraphic distribution. Although there are few data on this species in the Iberian Peninsula, one of the latest citations of *M. oeconomus* is from Tardiglacial deposits in Els Ermitons Cave in Gerona, Catalonia (Alcalde Gurt 1982). Its presence in El Mirón is one of the latest known for the Cantabrian region. It is found in large quantities in the Magdalenian levels (V, VI, and VI) of Erralla Cave (Guipúzcoa), along with *M. arvalis-agrestis* (Pemán 1985). It must have been very humid in the environs of the site during the Magdalenian. Erralla is close to Amalda, and this area of Guipúzcoa must have had humid, cool conditions with open grasslands not only during the Late Glacial but also in the Holocene, at least up to Roman times. In El Mirón *M. oeconomus* is scarce, but its presence coincides with that of *M. arvalis* as well as with episodes with the greatest abundance of *Arvicola* (figure 12.22), for which reason we believe that these late Chalcolithic/early Bronze Age times were cool and humid.

Chionomys nivalis (Martins, 1842)

DESCRIPTION AND DISCUSSION

Chionomys nivalis is the largest *Microtus*-type vole on the peninsula (although *Iberomys* has molars that are transversally wider). (In fact, it used to be called *Microtus nivalis*.—LGS) Its M_1 is characterized by an anterior lobe that is mushroom-shaped and tilted toward the labial side. On the lingual side there are only four recesses, one fewer than among most species of *Microtus*. Among young individuals this trait is less marked, leading to possible misidentification as *M. agrestis*, since the fourth lingual recess can range from slightly provergent (forward-oriented) in juveniles to very provergent in adults.

This vole is an inhabitant of high mountain areas above the tree line in alpine regions, up to 2,000 m in elevation—whence its name, snow vole. However, its habitat seems to be controlled more by the presence of rocky, unvegetated ground and the presence

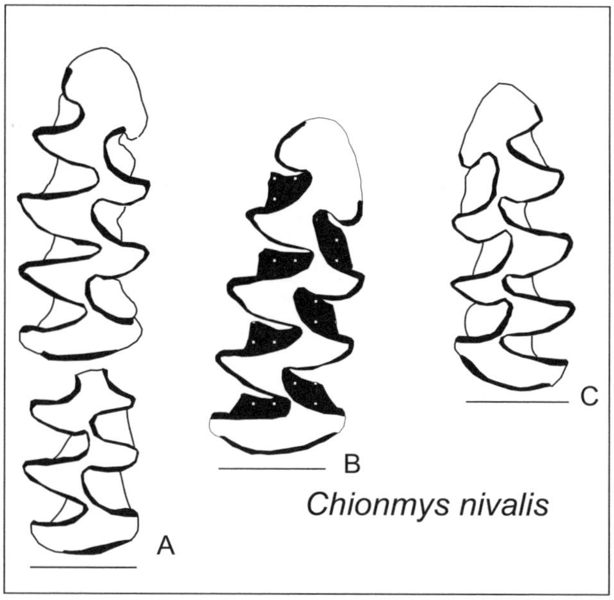

Fig. 12.10. *Chionomys nivalis* (Martins, 1842). Level 3.5: A (no. 25); Level 3.5: B (no. 122); Level 6: C (no. 103). Drawing by Gloria Cuenca.

of cracks and fissures than by altitude, since today in the Cantabrian region *C. nivalis* can be found in areas near the seacoast (Junco 1987; Pokines 1998). During the Upper Pleistocene and Holocene (up to the Chalcolithic) it has also been found in sites at low elevations: in the Mousterian through Chalcolithic of Amalda (205 m above present sea level), in the Aurignacian through Azilian of Aitzbitarte (220 m), in the Magdalenian of Ekain (90 m) (all in Guipúzcoa), and in the Aurignacian of Cueto de la Mina adjacent to La Riera (35 m) in Asturias, among others. It is of course possible that the increase in elevation with respect to glacial-age sea level may also have had some effect on the distribution of this species in the region. In the rest of Spain and the south of France the Mousterian may have been a time that was favorable to the extension of the snow vole, since at that time it was found in Mediterranean areas where it is now absent, such as at Hortus in French Languedoc (Chaline 1972; Cabrera-Millet, Britton-Davidian, and Orsini 1982) and in Spain at Carigüela (Granada) (Ruiz Bustos 2000); Los Toros (Teruel) (Gil and Sesé 1985); and Gabasa (Huesca) (Gil and Lanchares 1988) (although this last case is in need of confirmation). It is probable that during the coldest phases of the Upper Pleistocene the increase in rocky areas without vegetation may have been favorable to the spread of this species into lower elevations or into upland areas of the Iberian interior. The climatic improvement of the Holocene was apparently not so uniform as to completely all populations of *C. nivalis*, some of which have survived in isolated pockets up to the present.

HABITAT AND GEOGRAPHIC DISTRIBUTION

At present *C. nivalis* is found in the mountainous areas of the Mediterranean regions of western Europe, from the Pyrenees to the Alps and even in Slovakia's Tatra Mountains, as well as in the eastern Mediterranean from Mount Olympus in Greece to Lebanon and Israel in the Levant and as far east as the Zagros Mountains in Iran (Musser and Carleton 1993). In Spain it is found in the Pyrenees, the Cantabrian Cordillera as far west as Pontevedra, and the Sierras de Gredos, Guadarrama, and Nevada, normally above 800 m, except in Cantabria and regions of the former Yugoslavia, where it can be found below 300 m (Junco 1987; Pokines 1998).

Clethrionomys glareolus (Schreber, 1780)

El Mirón Level 10, no. 62, and El Mirón Level 5, no. 220a

DESCRIPTION AND DISCUSSION

The species *C. glareolus* is the only present-day arvicoline on the Iberian Peninsula that has both roots and cementum in its dental recesses. Another peculiarity of the red or bank vole is its continuous enamel, which lacks open zones and has the same thickness across the whole occlusal perimeter. The molars have convergent triangles.

HABITAT AND GEOGRAPHIC DISTRIBUTION

The bank vole is one of the few modern arvicolines that live in deciduous, more-or-less open woodlands or in areas vegetated with bushes or high grasses. It is a good indicator of a well-vegetated landscape conditioned by a temperate, humid climate (Pemán 1985, 1990a, b). It is found in the temperate latitudes of Eurasia, from Lake Baikal to the British Isles. In Spain

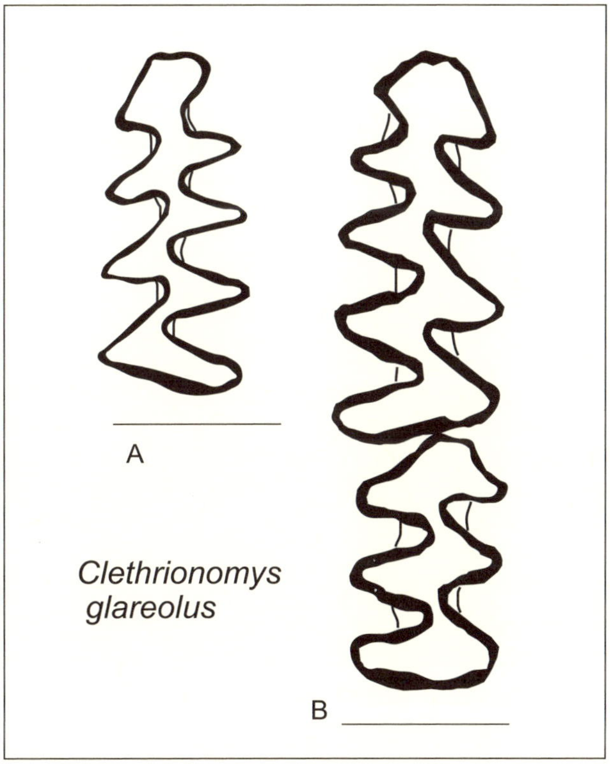

Fig. 12.11. *Clethrionomys glareolus* (Schreiber, 1780). Level 5: A (no. 25); Level 10: B (no. 162). Drawing by Gloria Cuenca.

its present distribution is confined mainly to the northernmost regions (Ventura, López-Fuster, and Gosálbez 1993; Pokines 1998), although it can also be found in the mountainous massifs of Levantine Spain (e.g., Montseny in Catalonia), which have a Eurosiberian climate despite their proximity to the Mediterranean (Arrizabalaga i Blanch, Montagud i Blas, and Gosálbez i Noguera 1986).

Terricola lusitanicus (Gerbe, 1879)

El Mirón Level 10, no. 249

DESCRIPTION AND DISCUSSION

The degree of separation of t6 and t7, which make up the second pitymyan rhombus, as well as the inclination of the labial triangles are the keys to classification of a species of *Terricola* as *T. lusitanicus*.

On the Iberian Peninsula there are presently three species of *Terricola*: *T. duodecimcostatus*, *T. pyrenaicus gerbei*, and *T. lusitanicus* (Brunet-Lecomte and Chaline 1993) (or four species, if *T. pyrenaicus* and *T. gerbei* are considered separate species, as argued by Niethammer and Krapp [1982] and Giannoni, Borghi, and Martínez Rica [1993]).

The systematics of this group is complicated, and for a taxonomic discussion of the genus *Terricola*, one should consult Brunet-Lecomte (1988), Brunet-Lecomte and Chaline (1990, 1993), and Brunet-Lecomte et al. (1987). The species of *Terricola* present at El Mirón has an M_1 with the morphology of *T. lusitanicus*: the "second pitymyan rhombus" is fairly isolated in its anterior part, as in *T. lusitanicus* and *T. pyrenaicus*, while that of *T. duodecimcostatus* converges significantly. When *T. lusitanicus* and *T. duodecimcostatus* are allopatric this trait is more variable, and some isolated individuals are difficult to distinguish on its basis alone (Brunet-Lecomte et al. 1987). The distinction between *T. lusitanicus* and *T. pyrenaicus* is based on the greater distal inclination of the labial triangles in the former relative to the latter. Also noteworthy is the protuberance of the labial apices of these triangles (figure 12.8), which constitutes an evolutionary novelty of the species *T. lusitanicus*.

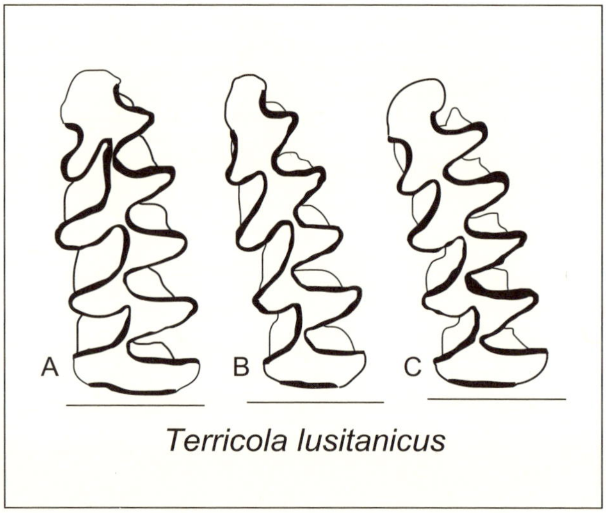

Fig. 12.12. *Terricola lusitanicus* (Gerbe, 1879). Level 10: A,B,C (no. 249). Drawing by Gloria Cuenca.

HABITAT AND GEOGRAPHIC DISTRIBUTION

The three species of *Terricola* on the Iberian Peninsula are burrowing animals that are similar in the kinds of tunnels they dig. However, *T. duodecimcostatus* digs and pushes the dirt forward with the help of its incisors, while *T. lusitanicus* and *T. pyrenaicus*—both usually digging in deeper soils than *T. duodecimcostatus*—push the dirt back using their hind limbs (Giannoni, Borghi, and Martínez Rica 1993). The use of the incisors (as in the case of *Arvicola terrestris*) is related to the fact that *T. duodecimcostatus* digs in sediments that are harder than those in which that the species using their hind limbs dig (Giannoni, Borghi, and Martínez Rica 1992). *T. lusitanicus* needs fairly deep soils to dig its burrows, for which reason it is linked to a more oceanic (Atlantic) climate than *T. duodecimcostatus*. Both species are sympatric in Castile and Leon, while in Cantabria one finds *T. lusitanicus* in sympatry with *T. pyrenaicus* at the present time (Brunet-Lecomte 1988, 1991; Brunet-Lecomte et al. 1987). The Pyrenean vole prefers deep soils, while the Lusitanian vole prefers somewhat shallower soils with some rocky outcrops (Borghi, Giannoni, and Martínez Rica 1990).

Subfamily Murinae

The murines are rodents whose teeth have roots, low crowns (brachydont), and occlusal surfaces formed by four to six principal cusps and a variable number of secondary cusps on the labial side of the molars. Their dental formula is 1I, 3M, as among the arvicolines.

Apodemus sylvaticus or *A. flavicollis*

El Mirón Level 4

DESCRIPTION AND DISCUSSION

Apodemus is characterized by having a low occlusal surface with six principal cusps on the first upper and lower molars and four cusps each on the second and third molars. On the labial faces of the M1s there are three secondary cusps while on the M2s there are one or two, and the M3s are reduced, with the posterior part being tapered. The present-day species of *Apodemus* on the Iberian Peninsula are *A. sylvaticus* (wood mouse) and *A. flavicollis* (yellow-necked field mouse). Morphologically they are very similar, sharing several dental traits. Although *A. flavicollis* is on average larger than *A. sylvaticus*, when the two are sympatric their dimensions overlap, which makes it very hard to differentiate them without a detailed morphological analysis. The size difference, which is more evident in central and northern Europe, diminishes toward the south such that the distinction between the two species on the Iberian Peninsula (and in general on all the Mediterranean peninsulas) is problematic (Nores 1988).

HABITAT AND GEOGRAPHIC DISTRIBUTION

Although both species have a broad distribution, *A. flavicollis* has a discontinuous one throughout Europe and in Spain is found only in the north, from the Pyrenees to Asturias, while *A. sylvaticus* is more Mediterranean and is found throughout the whole Iberian Peninsula.

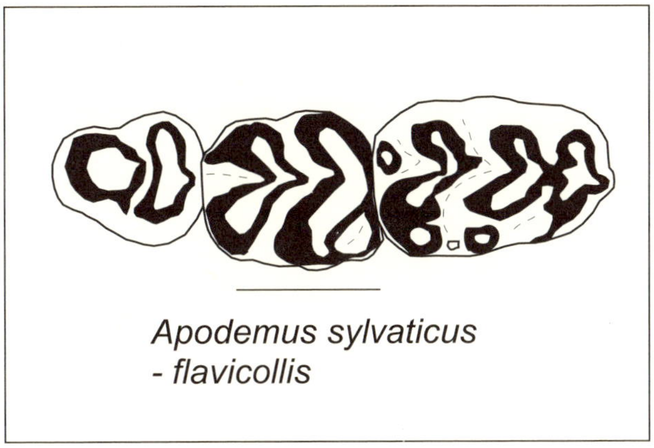

Fig. 12.13. *Apodemus sylvaticus* or *A. flavicollis*. Level 4. Drawing by Gloria Cuenca.

Family Myoxidae

The rodents of the glirid family are characterized by teeth with roots, low cusps (brachydont), and occlusal surfaces furrowed by transversal crests separated by valleys. The taxonomic nomenclature of the glirids (dormice) is reviewed by Holden (1993).

Myoxus (Glis) glis (Linnaeus, 1766)

El Mirón Level 10

DESCRIPTION AND DISCUSSION

This is the largest of the European dormice. It is distinguished by having teeth with rectangular to square outlines and flat occlusal planes, with transversal crests

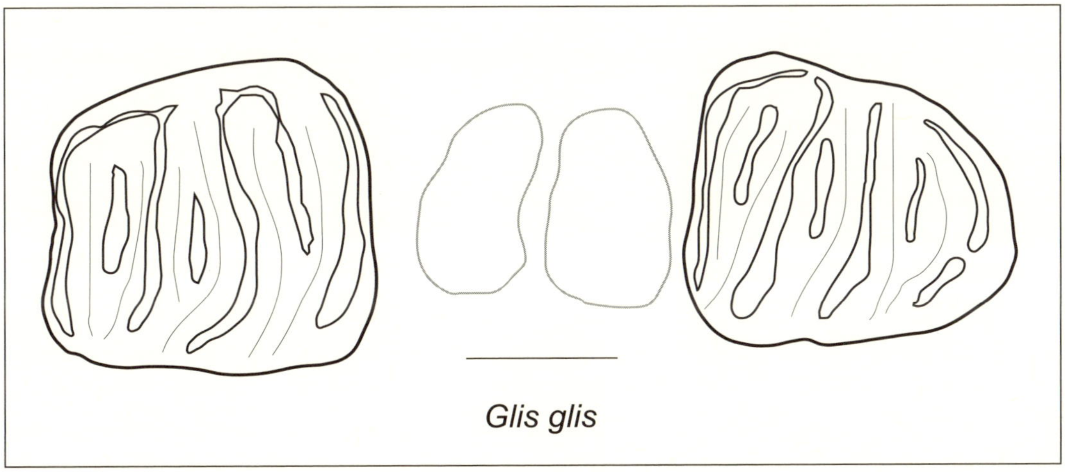

Fig. 12.14. *Myoxus (Glis) glis* (Linnaeus, 1766). Left mandible with M1 and M2. Level 10. Drawing by Gloria Cuenca.

that are clearly separated by relatively wide valleys. These crests are developed labio-lingually. On the lower molars the crests turn toward the anterior part of the teeth, and on the upper molars they curve toward the posterior part. Between the principal crests there are secondary crests, which are shorter and do not reach either the labial or lingual faces.

HABITAT AND GEOGRAPHIC DISTRIBUTION

The gray or edible dormouse is a warm woodland species that at present lives in central and southern Europe, from the Caucasus to northern Spain. It is also found in southern England, but there it may have been introduced in Roman times and may be an occasional commensal with humans (Pokines 1998). In the site of Erralla (Guipúzcoa), it was found in warm-climate levels of the sequence (II, III, V), associated with heliophile species such as *Chionomys nivalis* (Pemán 1985).

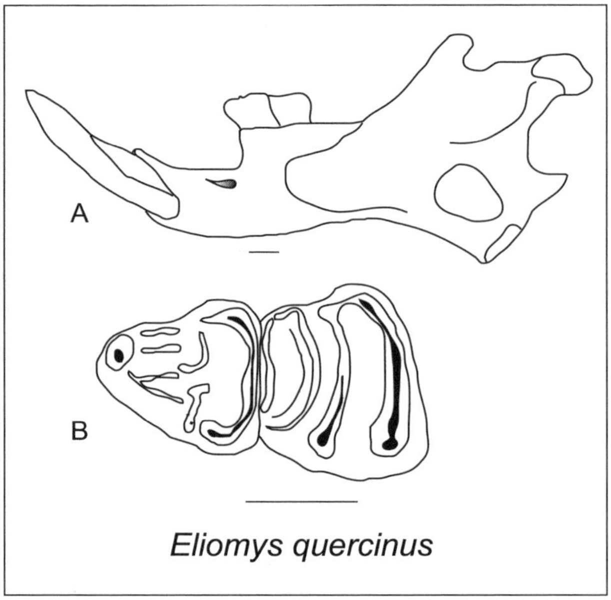

Fig. 12.15. *Eliomys quercinus* (Linnaeus, 1766). A: left mandible with M1 and M2; B: occlusal of one of the teeth. Level 10. Drawing by Gloria Cuenca.

Eliomys quercinus (Linnaeus, 1766)

DESCRIPTION AND DISCUSSION

E. quercinus is glirid of medium size, smaller than *M. glis* but with molars of almost the same size. It is the only dormouse that has a distal mandibular foramen that is completely perforated. The molars are characterized by concave occlusal surfaces on which the main crests run in the labial-lingual direction, but at their labial and lingual ends the principal cusps are especially notable for their large size.

HABITAT AND GEOGRAPHIC DISTRIBUTION

The garden dormouse is a species that is broadly distributed throughout temperate Europe and the Mediterranean regions, including North Africa and the Levant. It is found in forests and other wooded zones all over the Iberian Peninsula in areas where there are also stony clearings. Like the edible dormouse, it is occasionally a commensal with humans. Notable is its near-disappearance from the Cantabrian region during the Upper Pleistocene, since it has been cited only in the Aurignacian of Cueto de la Mina (Asturias) (Pokines 1998). Its presence in sites dating to the Bronze Age (such as La Encantada in Ciudad Real, La Mancha) could be related in some way to humans, albeit indirectly, since they create artificial environments that are favorable to certain micromammalian species while eliminating others (Morales Muñiz 1986). In this sense the presence of *E. quercinus* seems to be linked to human activity, although the exact nature of the relationship is difficult to specify. It could have to do with actual human occupation or condition of the site afterward, due to the creation of artificial habitation places (hollows formed by burials, storage pits, or chambers), which can favor the expansion of this species within its natural habitat of rocky areas near the edge of woods. The man-made hollows may make it easier for them to dig their burrows. Another example of Bronze Age burials and garbage pits that have favored the presence of *E. quercinus* is the site of La Balsa la Tamariz in Tauste, Zaragoza (Laplana Conesa and Cuenca Bescós 1995).

Talpa europaea (Linnaeus, 1758)

DESCRIPTION AND DISCUSSION

These moles are characterized by having elongated mandibles and robust angular and coronoid processes that contrast with the delicate articular condyle, which is simple and cylindrical. This structure is very distinct from that of the insectivores of the soricid family. The

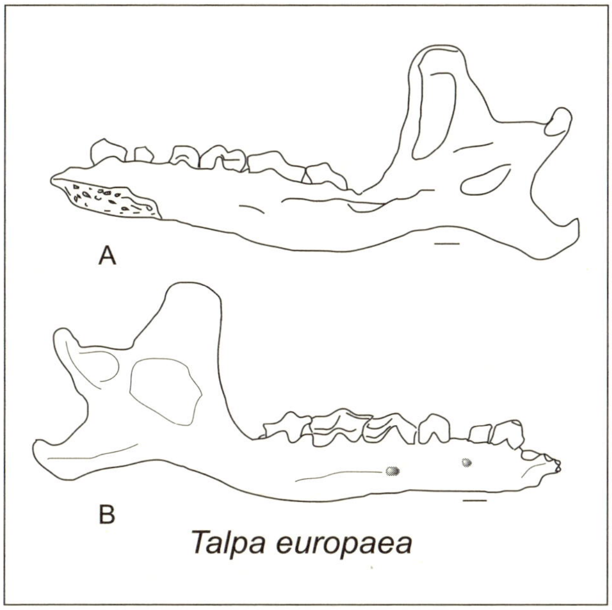

Fig. 12.16. *Talpa europaea* (Linnaeus, 1758). Right mandible, A: lingual view; B: labial view. Level 10, (no. 261). Drawing by Gloria Cuenca.

Fig. 12.17. *Sorex coronatus-araneus*. A and B: left and right articular condyles. Level 10 (no. 260). Drawing by Gloria Cuenca.

mole remains from El Mirón belong to a large species and, although teeth and other cranial remains are scarce, because of their great size we think that they can be assigned to the species *T. europaea*.

HABITAT AND GEOGRAPHIC DISTRIBUTION

Moles are burrowing insectivores that require deep soils to dig the galleries in which they live. They are the most underground of all the mammals found in El Mirón, and the accumulation of their remains here could be due to their occasional habitation of the cave, probably during the most humid periods.

Sorex coronatus-araneus

DESCRIPTION AND DISCUSSION

At the present time the three species of *Sorex* on the Iberian Peninsula (*S. araneus*, *S. granarius*, *S. coronatus*) are difficult to distinguish on the basis of cranial characteristics or especially on isolated teeth, except if one does an extensive morphometric analysis with large collections (López-Fuster and Ventura 1996). The *Sorex* remains at El Mirón are scarce and fragmentary, so it is not possible to do such an analysis. However, we can distinguish a more robust form and a more gracile one among this material.

Based on known geographic distributions, it seems probable that the species present at El Mirón is like the present-day *S. coronatus* that is found in the area of Ramales de la Victoria. This, however, is something that would have to be verified in the future, since we know that the present distributions of other species of micromammals do not necessarily have to coincide with those of the earlier Holocene.

HABITAT AND GEOGRAPHIC DISTRIBUTION

The three *Sorex* species are found in the north of the Iberian Peninsula: the common shrew (*S. araneus*) is distributed in a band in northern Catalonia from the eastern Pyrenees to the Catalan pre-Pyrenees; *S. coronatus* extends in the north from the Pyrenees to Galicia and to the Sistema Ibérico mountains in the south. The Iberian shrew (*S. granarius*) is found from the Sistema Central mountains to the mouth of the Tagus River and from there northward to Galicia (López-Fuster and Ventura 1996). All are species that inhabit humid places with dense grass cover (wet pastures for *S. coronatus*) or shrubs and even trees (Pokines 1998).

Sorex minutus (Linnaeus, 1766)

DESCRIPTION AND DISCUSSION

This is the smallest of the red-toothed shrews of the genus *Sorex*. *S. minutus* (the pygmy shrew) is so small that it is easily distinguishable from other species.

HABITAT AND GEOGRAPHIC DISTRIBUTION

It occupies all of temperate Europe, and in northern Spain it extends via the Duero Basin to Portugal (Blanco 1998). Its presence is linked to dense vegetal cover, preferably grassy, in humid, temperate habitats.

Neomys sp.

DESCRIPTION AND DISCUSSION

Neomys is a large soricid that lives near watercourses. It is clearly distinguished from the other soricids by its articular conyle (figures 12.14 and 12.15) and by its robust teeth. There are two species of water shrew in Spain: *N. fodiens* (whitefooted water shrew) and *N. anomalus* (Cabrera's or Miller's water shrew). They are easily distinguishable by their external characteristics: long neck hair in *N. fodiens*, which is not found in *N. anomalus*, as well as the large size of the former relative to the latter (Pemán 1983). However, differentiation of the two species is more problematic when one has only a few isolated remains with which to work.

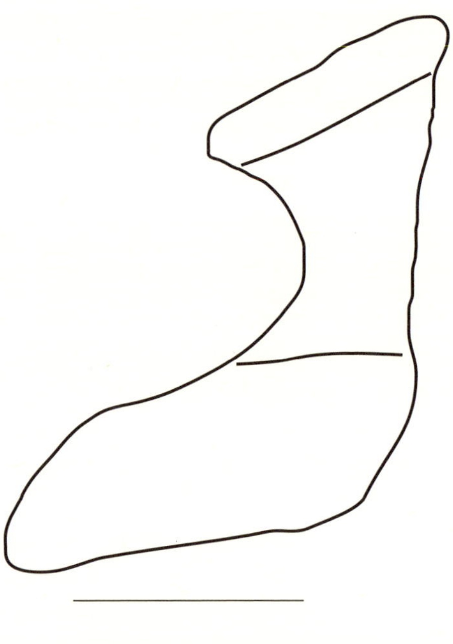

Fig. 12.18. *Neomys* sp. Left articular condyle. Level 10 (no. 260). Drawing by Gloria Cuenca.

HABITAT AND GEOGRAPHIC DISTRIBUTION

At the present time *N. fodiens* is broadly distributed across Europe, including northern Spain, while *N. anomalus* has a distribution linked to mountain massifs in the Iberian Peninsula and central-eastern Europe (Blanco 1998). The water shrews (as indicated by their name) are always connected to water bodies, including arroyos and lakes, since they feed on aquatic invertebrates. However, *N. anomalus* seems to be less closely linked to aquatic habitats than *N. fodiens*, since it can also be found in humid fields and woods.

Crocidura russula (Hermann, 1780)

DESCRIPTION AND DISCUSSION

Remains of white-toothed shrew are very scarce at El Mirón; there are mandible fragments (including some well-preserved condyles) and a fragment of cranium. In the latter (El Mirón Level 7, no. 128) one can see that the P^3 protocone forms the antero-lingual edge of the tooth. In addition, the base of the crown, when viewed labially, has an angular anterior edge. These characteristics are really not sufficient to separate the two species of *Crocidura* that presently live on the Iberian Peninsula, since we would need to have much more material to do a detailed biometric analysis. However, the P^3 morphology combined with the geographical location of El Mirón lead us to conclude that this is probably the species *C. russula*.

HABITAT AND GEOGRAPHIC DISTRIBUTION

The species of genus *Crocidura* on the Iberian Peninsula are *C. suaveolens*—northern and Atlantic in distribution—and *C. russula*—more frequent throughout the whole peninsula, even in humanized environments. With its Atlantic environmental preferences, *C. suaveolus* is found in areas along the coast of northern Spain with moderate, humid climate. It even penetrates humid areas of the southern Mediterranean coast of

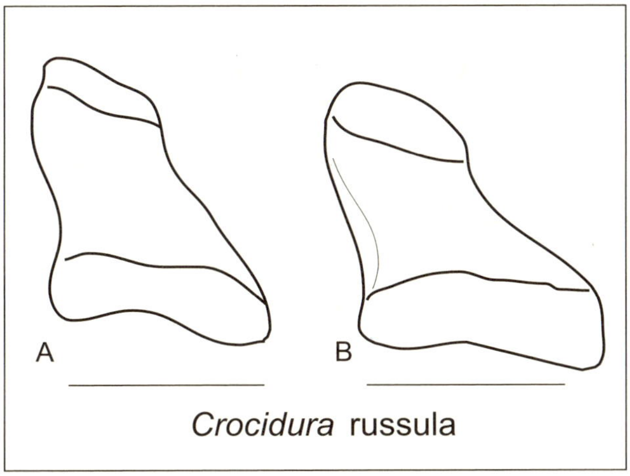

Fig. 12.19. *Crocidura russula* (Hermann, 1780). Right articular condyles. Level 10 (no. 260). Drawing by Gloria Cuenca.

Spain, such as the marshes and other habitats around the Guadalquivir (Rey and Landin 1973). These same authors also mention that *C. suaveolens* is linked to the narrow coastal strip of the Cantabrian region, where the mountains rise practically from the shore, strictly limiting the non-montane zone favorable to this species in this region. From the stratigraphic distribution of *Crocidura* in El Mirón it is clear that it was not abundant in this site, lacking as it is from most of the Holocene levels except Level 7, which was apparently one of the driest periods during the postglacial in the area of this site.

ORDER CHIROPTERA

Myotis sp.

DESCRIPTION AND DISCUSSION

The remains of chiropterans are very scarce in the Holocene deposits of El Mirón, although this scarcity could be due more to the recovery methods (screen mesh size) used for these levels than to a real paucity of bat remains in the sediments. (Nonetheless, bats are in fact very rare in the cave today.—LGS) Only in Levels 3, 8, and 10 did we find remains of a large bat, probably belonging to a species of the genus *Myotis*, such as the large, cave-dwelling "mouse-eared bat," which is a current resident of all of southern and central Europe (Sevilla García 1988).

Miniopterus sp.

DESCRIPTION AND DISCUSSION

This form is one of the smallest of the European chiropterans and the smallest known cave-dwelling bat. In El Mirón what we are dealing with is probably *Miniopterus schreibersii*, the only species of bat cave of such small size to be found in Spain (Sevilla García 1988; Cuenca Bescós et al. 2008).

ORDER CARNIVORA

Mustela nivalis (Linnaeus, 1766)

DESCRIPTION AND DISCUSSION

The small carnivores are rare in the Holocene levels at El Mirón. We only found remains of a mustelid in Levels 3, 4, 5, and 10. By their small size, these can all be attributed to *Mustela nivalis*, the weasel, which, although it lives in dens, is not much of a burrower (Van den Brink 1971). (The larger mammalian faunal remains analyzed

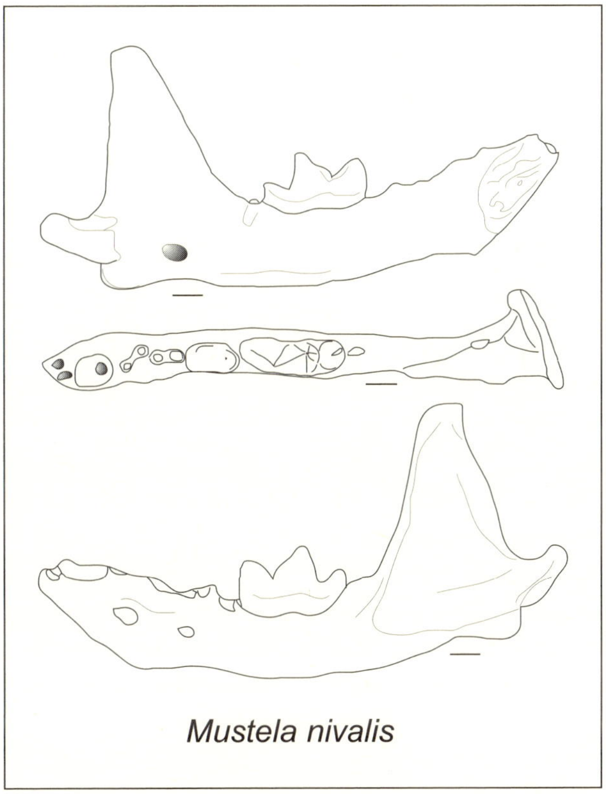

Fig. 12.20. *Mustela nivalis* (Linnaeus, 1766). Left mandible. Level 10 (no. 261). Drawing by Gloria Cuenca.

by Altuna and Mariezkurrena [chap. 16, this volume] did not include much else by way of small carnivores either: a very few marten remains in Levels 8 and 10, wildcat in Level 8, fox in Level 7, and dog in Levels 7, 8, and 9.6. There is also one hedgehog bone in Level 9.)

Results and Conclusions

TAPHONOMY AND DIVERSITY OF THE MICROMAMMALS OF THE HOLOCENE OF EL MIRÓN CAVE

For methodological reasons, our analysis has been limited to the samples from the Holocene deposits of the Cabin excavation area in the outer vestibule. These samples constitute a single stratigraphic column that corresponds to a relatively uniform area. Ultimately the Mesolithic- and early Neolithic-age materials from Levels 10.1 and 10 could be compared from equivalent levels exposed in the Mid-Vestibule Trench and (for the Mesolithic only) in the Corral excavation area at the rear of the vestibule (Levels 304–303 and Levels 102–101, respectively). Such a comparison could tell us whether potential differences might be due more to location within the cave than to stratigraphic position (i.e., real temporal changes in assemblage composition). The 309 microfaunal samples that we did analyze are grouped by Holocene age level within the Cabin area, as described in the section on stratigraphy. The contribution of each level to the ensemble is more or less uniform across the sequence, with the exception of Levels 1 (mixed, recent fill, only partially screened in a systematic way—LGS) and 2 (a thin, discontinuous, yellow-beige clay lens that might represent some specific human earth-moving activity—LGS). These samples are so small and poor that they were left out of any statistical analyses (see table 12.3, which gives the list of all microfaunal remains by sample and level).

In tables 12.1 and 12.2 we observe that the number of species declines dramatically in those most recent levels (Levels 1 and 2), probably mostly for reasons of sampling but perhaps also as a consequence of the permanent use of the cave by people and their livestock.

The high species diversity in Levels 3 and 10 is notable. Level 3, with 17 species (tables 12.1 and 12.2), has the same species as the other levels, except the garden dormouse (*Eliomys quercinus*) and the robust form of *Sorex*. However, the most notable fact—as commented upon both above and below—is the persistence of *Microtus oeconomus*. The presence of *Pliomys lenki* in Level 3.5 needs to be interpreted with extreme caution here, however, as admixture by erosion from Magdalenian levels upslope of the El Mirón Cave vestibule is indicated by the presence of typical artifacts and wild ungulate bones more heavily fossilized than those normally found in the Chalcolithic and Bronze Age deposits of that area.

In table 12.2 we can see four first appearances of species: Levels 10, 9, 5, and 3, with *A. sapidus*, a robust form of *Sorex*, and *M. oeconomus*. Leaving aside the indeterminate form of *Sorex*, it is noteworthy that *M. oeconomus*, which appears in the Holocene of El Mirón, has since disappeared from the region because of migration to more favorable regions. The species *Arvicola sapidus*, although present, is very rare in northern Spain (Zabala 1983). It is also relevant that, like *M. oeconomus*, *Chionomys nivalis* is also absent today in Cantabria, although in this case the species has not been completely extirpated from the Iberian Peninsula; it can be found in relict populations in mountain massifs, where there are habitats that are favorable for it, such as sunny, unvegetated rock outcrops.

It is also worth noting the (at least local) extinction of the garden dormouse (*Eliomys quercinus*), a warm woodland taxon.

The rest of the species have a more or less continuous distribution throughout the Holocene sequence at El Mirón, with increases and decreases in representation (as quantified in terms of MNI) or in species diversity explainable in relation to various non–mutually exclusive factors:

1. Differential sampling could skew results. The number of analyzed samples per level is variable (see table 12.1), but this includes essentially all the micromammalian material collected by the excavation team during several campaigns. Thus the total number of samples from Levels 1–10.1 studied here is 309 individual units, each with its square, subsquare, level, and spit provenance and a unique sample number. The number of samples per level is virtually the same in all cases except for Levels 1 and 2, so the paucity of finds in these latter two levels is clearly related to this factor. Except for Levels 1 and 2, the number

of samples is homogeneous, so differential sampling can be discounted as an explanatory factor for inter-level variations in species representations.

2. Lesser human occupation (or lesser intensity thereof) could be correlated with greater use of the cave by the predators responsible for accumulating at least most of the micromammals. Such conditions could translate to the formation of assemblages more fully representative of the species that lived in the area around the site.
3. Changes in climatic conditions could have favored or suppressed greater diversity among the local fauna.
4. Changes in the conditions of sedimentation within the cave could favor preservation or destruction of the remains of vertebrates, as in the singular case of Level 3.5, which is clearly a feature excavated into Level 3. It is in both these unit that remains of *Pliomys lenki* were found—an arvicoline that had arguably gone extinct around the end of the Last Glacial (figure 12.21). The geometry of Level 3.5 (figure 12.4) indicates that it is the infilling of a pit or channel into which materials from older levels had eroded. However, the good preservation of the remains as well as the faunal association with the *Pliomys* finds might not seem at first glance to support this hypothesis, although other evidence for the admixture of Magdalenian-age materials into Cabin area layers dating to the Metal Ages clearly supports an age no later than the Last Glacial for *P. lenki*.
5. The alteration of the levels by processes of bioturbation posterior to stratification of the sediments could mix the microfaunal contents of the layers. For example, it is possible that the use of the vestibule for human residence and as a stable in both the recent and less recent past caused some degree of bioturbation, particularly resulting from pit digging, as well as trampling.

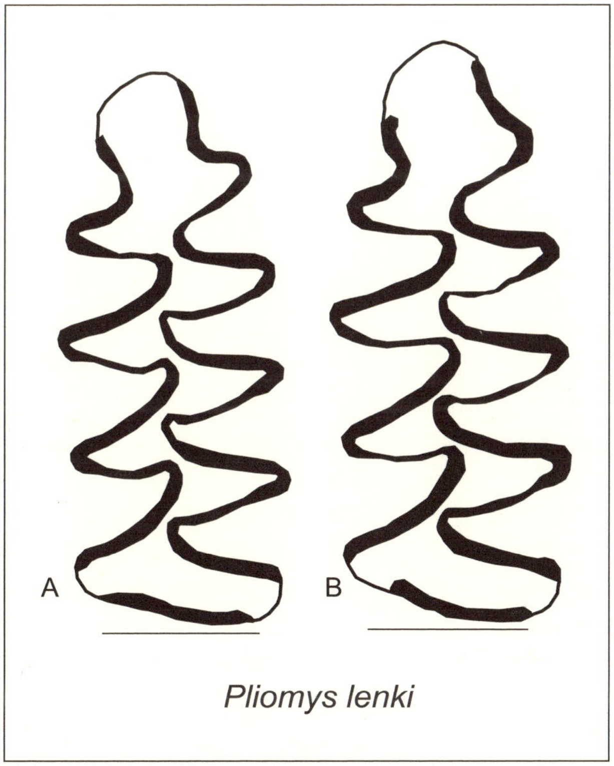

Fig. 12.21. *Pliomys lenki* (Heller, 1930). Level 3.5: A and B (no. 122). Drawing by Gloria Cuenca.

THE EXTINCTION OF *PLIOMYS LENKI*

Independently of the stratigraphic position and age of Level 3.5, the extinction of *Pliomys lenki*—a very long-lived species that according to different authors was present since Middle or even Lower Pleistocene times—must have been caused by various factors:

1. A climatic change that affected the environment in which this arvicoline lived. Most authors assert that *P. lenki* was fundamentally a cold steppe rodent (e.g., Bartolomei et al. 1975; Pemán 1990a).
2. Alterations in the species's habitation areas. In general, the different species of this genus existed under a diversity of conditions throughout the complex climatic changes that have taken place since the Lower Pleistocene. However, *Pliomys* is generally found in karstic sites (Cuenca Bescós et al. 2010; Cuenca Bescós, Canudo, and Laplana 1999), which makes us think that it is in reality a hypogeous (subterranean) species that is controlled less by climate than by the substrate. The continuous or even occasional use of caves by humans during the Upper Pleistocene could have affected *P. lenki* to the extent of gradually causing its extinction, with the Cantabrian region becoming a refugium enclave because of its wealth of caves and rockshelters with a more moderate climate than the center of the peninsula.

3. An unrepresentative picture of the faunal composition during the Upper Pleistocene. The frequently invoked discontinuity of the fossil record has provided us with an imperfect picture of the faunal composition during this period. One would also have to admit that all the systematic studies of the modern-day fauna of northern Spain are incomplete, since there is no explanation for considering *P. lenki* to have gone extinct in the Bronze Age without also admitting that it could in fact be a living species—albeit one that has not yet been scientifically observed.
4. A stratigraphic inversion. Level 3.5 could actually be a deposit composed of sediments eroded by water (or shoveled together by humans) from a mixture of recent and older layers. If the presence of *P. lenki* is to be explained by its having been mixed in from older layers, this could be supported by its presence in the Upper Magdalenian, as is the case in Corral Level 106. But there is no evidence that Level 3.5 could anywhere have cut into such old strata, which lie very far below it in the outer vestibule (Cabin) area. However, part of its fill could have come from Magdalenian deposits eroded down from the steep slope at the rear of the vestibule in Holocene times (attested by fossilized bones and Upper Paleolithic artifacts in post-Mesolithic layers and features of the Cabin area of the cave vestibule). More detailed studies of the taphonomy would be needed to determine the mode(s) of deposition of these "late-occurring" *P. lenki* remains. It would ideally also be useful to radiometrically date Level 3.5—preferably with multiple determinations to ascertain the existence of bones of mixed ages versus homogeneous contents.

In the absence of new data, we would like to note that the extant studies of the modern microfauna of the Cantabrian region and northern Spain in general as well as the comparative collections that have been loaned to us for study (by the CSIC Instituto Pirenaico de Ecología and by the Museo Nacional de Ciencias Naturales in Madrid) lead us to think that this fauna is sufficiently well known that it is unlikely an arvicoline as characteristic as *Pliomys* would have been overlooked (Pemán 1990a; Pokines 1998; Brunet-Lecomte and Delibes 1984). For this reason, we are certain that it is indeed convincingly extinct, although the exact time of its disappearance remains somewhat uncertain. As noted above, in El Mirón Cave its last unquestioned appearance was in Upper Magdalenian Level 106 (12,460 BP uncal.).

On the other hand, there does not seem to have been any important change in the composition of the microfauna of Level 3.5 vis-à-vis the rest of the Level 3 complex, as can be seen in tables 12.1, 12.2, and 12.3. Except for *Pliomys*, the micromammals of Level 3.5 are also found in the rest of the levels labeled "3" (3, 3.0–3.4), and they are found in the same approximate proportions in terms of numbers of individual animals.

As a working hypothesis, we could propose that the final disappearance of *Pliomys* is linked to a deterioration of its preferred climate, first in the rest of Europe and only later in Spain, with the eventual complete loss of its last refuges, most likely in the Tardiglacial of the Cantabrian region.

CLIMATIC INFERENCES FOR THE HOLOCENE SEQUENCE OF EL MIRÓN BASED ON MICROFAUNAL ANALYSIS

Quantitative and qualitative data from stratigraphic distributions provide classic tools for the detection of faunal composition changes that may or may not be related to climatic changes.

To learn about the stratigraphic evolution of this fauna, we analyze both the nature of the composition and the proportions of the different microfaunal species for each level, focusing specially on the arvicolines (rodents) and soricids (insectivores) and the ratio of arvicolines to murids (tables 12.1, 12.2, and 12.3).

A very interesting fact is that the murid species *Mus* (house mouse) and *Rattus* (black rat), which presently live in Cantabria as well as in the rest of Spain, are absent from the Holocene levels in El Mirón Cave. The only murid in the Holocene deposit at the site is *Apodemus*, which is well represented in all the levels (table 12.2). The lack of "modern" murids (*Mus*, *Rattus*) gives us some information about the Holocene dispersal of these taxa. In fact, these murids didn't colonize northern Spain until very recently (Pokines 1998; Pemán 1985, 1990b)—a conclusion supported by this study. They probably appeared as a consequence of the (relatively late) spread of agriculture into this Atlantic region, as well as the generalized extension of

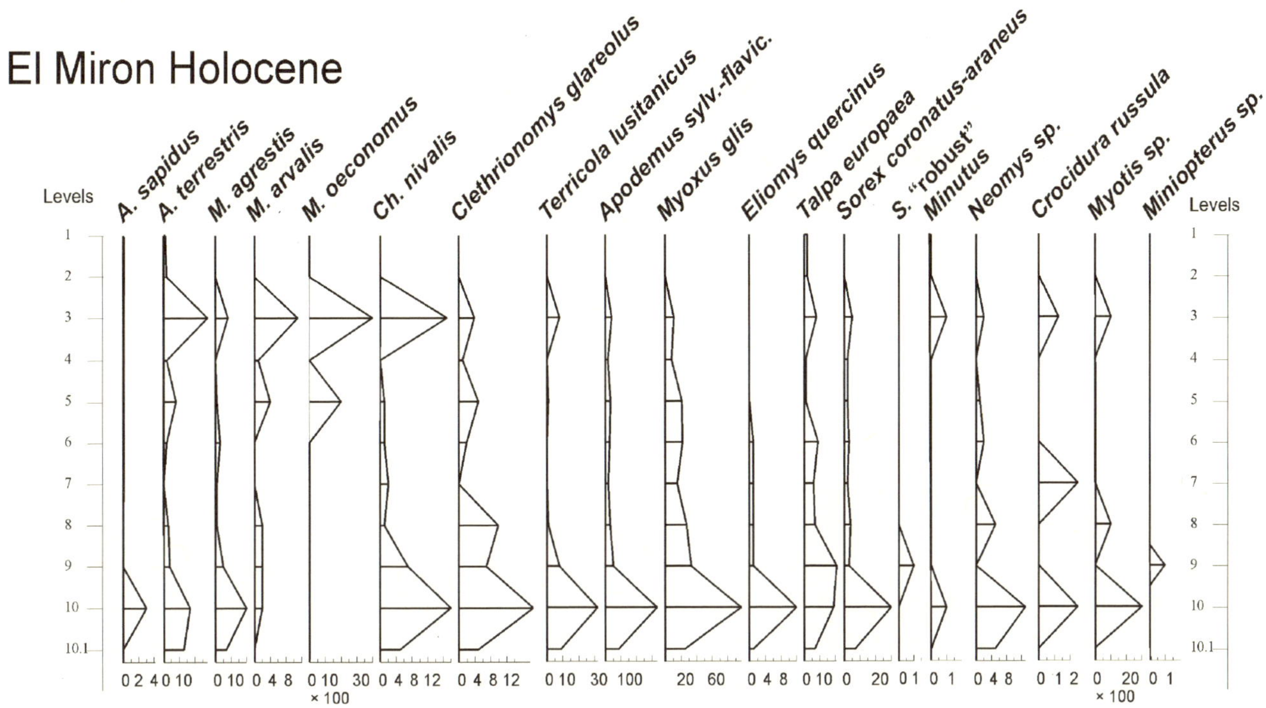

Fig. 12.22. Stratigraphic distribution of micromammal species in the Holocene levels of El Mirón Cave. Each column represents the minimum number of individuals calculated from a significant skeletal element (see text). The horizontal scale is different for each species. Drawing by Gloria Cuenca.

commerce via trade routes that were created to connect the center of the peninsula with the north coast. Trade routes such the so-called Silver Road were major axes of dispersion for commensal animals such as the house mouse and the house sparrow (Morales Muñiz et al. 1995).

In tables 12.1, 12.2 and 12.3 and figure 12.22 we can see that the distribution of rodents in the Holocene of El Mirón is characterized by the following:

1. The sporadic presence of a few species, such as *Arvicola sapidus* in Level 10, that constitute unique occurrences. Also, there is a species of *Sorex* that is more robust than the rest and is found only in Level 9.
2. The practically continuous presence of *Arvicola terrestris* and *Talpa europaea* throughout the sequence up to the present.
3. A faunal discontinuity in Levels 6 and 7 that culminates with the arrival (in Level 5) of colder-climate species such as *Microtus oeconomus* and the disappearance of species typical of deciduous forest such as *Eliomys quercinus* (whose last appearance is in Level 6).
4. The apparent recovery of the microfauna in Level 3.5, which has the highest species diversity in the whole sequence and the presence of *Pliomys lenki*, this latter probably the result of mixing with older deposits by natural and/or anthropic erosion.
5. The decrease in species numbers in Levels 7 and 4.
6. The elevated number of arvicoline species relative to the very poorly represented murids at this site in comparison to other Cantabrian Holocene sites.

As noted earlier, some rodents are occasionally commensals with humans, as in the cases of the edible dormouse (*Myoxus glis*) and the garden dormouse (*Eliomys quercinus*). If there had been a fully continuous occupation of El Mirón Cave during the Holocene, these species would have proliferated in all the levels.

Also as noted above, most of the micromammals in the El Mirón Holocene deposits have living representatives in the Cantabrian region. In general, these are taxa typical of the Eurosiberian biome within Spain, a zone characterized by its Atlantic climate, generally more humid than "Mediterranean" Spain. The habitat preferences of the various species represented at El Mirón are summarized below.

CLIMATIC AND ECOLOGICAL ASPECTS OF THE HOLOCENE MICROFAUNAS FROM EL MIRÓN CAVE

Micromammal species are linked to particular habitats; the "trick" is determining which aspect(s) of each habitat is (are) the critical controlling factor(s) in each species' presence, absence, or relative abundance. Without entering into a very detailed analysis, we observe that most authors agree in grouping particular species according to their ecological preferences for woodlands, humid grasslands, open pastures, water bodies, and unvegetated rocky areas (also known as alpine habitats). Merging the classifications of various authors (Pokines 1998; Marquet 1989; Chaline 1975), we have developed a matrix of relationships between species and environments (not taking variations in latitude and longitude into account, since this is a very small geographic region). In general terms (and subject to future refinements), we can suggest that there are seven groups of micromammals as a function of habitat characteristics:

1. **Woodland species:** *Eliomys quercinus*, *Myoxus glis*, *Apodemus*, *Clethrionomys glareolus*
2. **Humid meadow species:** the three species of *Sorex*, *Arvicola terrestris*, *Terricola lusitanicus*, *Talpa europaea*
3. **Dry grassland species:** *Crocidura russula*, *Microtus agrestis*, *M. arvalis*, *Pliomys lenki*
4. **Species from water bodies and ponded fields:** *Microtus oeconomus*, *Neomys fodiens*, *Arvicola sapidus*
5. **Alpine species:** *Chionomys nivalis*
6. **Generalists:** *Mustela nivalis*
7. **Cave dwellers:** *Miniopterus*, *Myotis*

This classification lacks sharp boundaries between the groups because, for instance, woodland species can also be found on the edges of dry grasslands with abundant shrubs (*Eliomys*, *Apodemus*) and the species that prefer water bodies can also adapt to wet meadowlands (*Arvicola*). One has to keep in mind that fossil species did not necessarily have the same ecological characteristics as do the modern representatives. Modern distribution and habitat relationships can be the consequences of biogeographic factors. Nonetheless, if we apply this classification to attempt a climatic interpretation of the micromammalian faunas in the Holocene sequence in El Mirón, we can propose the following:

1. There is an overall trend for woodland cover to decrease gradually through time from Level 10.1 (Mesolithic-age) to Level 3 (early Bronze Age), when there is a drastic reduction in woods in favor of humid meadows (probably in the nearby Calera, Gándara, and Asón valley floors—LGS) and alpine habitats almost completely devoid of trees (presumably the rocky slopes immediately surrounding the cave, which were being intensively grazed by domesticated ovicaprines—LGS).
2. The humid meadows are always present in the environment around the site, but they increase significantly (perhaps almost twofold) relative to some of the earlier levels (possibly signifying agricultural intensification by increased forest clearance on the valley floors—LGS).
3. The dry grasslands are also best represented in Level 3, although this biotope is always scarcely manifested by the micromammals in the El Mirón Holocene sequence (not surprising, given the wet, Atlantic climate of the Cantabrian region—LGS).
4. Water is present permanently in almost all the levels, except 9, 7, and 4. Level 9 could in fact correspond to a relatively dry climatic episode (Subboreal?—LGS), since there is also a marked drop in woodland species at this time. In Levels 7 and 4 we observe a general decrease in faunal diversity, which could be related more to greater human activity in the cave than to any climatic change.
5. The rocky areas without vegetation increase notably in Level 3, although they are represented throughout the whole sequence except Level 4. This habitat is more extensive in Level 9 than in the rest of the Level 10.1–4 sequence. Level 9 could mark a

short period of relative aridity, since there is also a decrease in woodland species at this time.

From what we argued earlier, we deduce that there were four climatic phases during the times represented by Holocene sedimentary deposits in the El Mirón vestibule, a period of major burning in the cave that affected species diversity, and a final period of generalized and intensified anthropization of the environment. (Keep in mind that there were significant hiati at least before the beginning of Level 10.1 deposition and between Levels 10.1 and 10, and post–Bronze Age deposits are essentially missing, so the whole Holocene record is not represented in the site.—LGS) The hypothesized phases are:

Phase 1: The beginning of the Holocene at El Mirón is warm and humid, with a wooded landscape interrupted by some meadows and open rocky areas. This phase is represented by Levels 10.1 and 10.

Phase 2: At the start of deposition of the Level 9 complex, the woods undergo a regression in favor of wet meadows. These conditions are maintained, albeit with some fluctuations, until Level 5 times, corresponding therefore to Levels 9–6.

Phase 3: There is a slight increase in both species diversity and numbers of individuals. *Microtus oeconomus* makes its appearance in Level 5; this is a species that today is found in northern Europe under cold, marshy conditions.

Phase 4: There is a drastic decrease in micromammalian diversity. The bonfires of Level 4 could indicate intensive human use of the cave (especially for stabling livestock—LGS), and this could be the cause of the decline in species diversity.

Phase 5: Woodlands decrease markedly and are replaced by fields and unvegetated rocky areas; the climate must have been colder and humid. Levels 3–3.5.

Phase 6: There is total humanization of the environment with near-total loss of woodlands and drastic decline in species diversity. Levels 2 and 1, modern times.

CORRELATION WITH GENERAL CLIMATIC SCHEMES BASED ON PALYNOLOGY AND GEOMORPHOLOGY

The climate of the past 10,000 years in Iberia is known mainly from pollen analyses (e.g., Sánchez Goñi 1993; Sánchez Goñi et al. 2000) and from studies of geomorphological developments on the peninsula (e.g., Gutierrez Elorza and Peña Monné 1998; Gutierrez Elorza and Sesé Martínez 2001), backed up by studies of ice cores in Greenland and Antarctica (e.g., Lowe and Walker 1997) and even evidence from icebergs (Heinrich 1988). The vegetation during the Holocene in the north-central Spain has been characterized by Peñalba (cited in Sánchez Goñi 1993).

In the Holocene of the southeast of the Iberian Peninsula, Sánchez Goñi et al. (2000) detect three climatic phases (which then served as a model for interpreting the Eem Interglacial—stage 5e—in the region). These phases are of interest here to determine whether the Holocene micromammalian record from El Mirón has some wider (perhaps global) significance. The phases for southeast Iberia are as follow:

Phase a: Beginning of the Holocene: warm, Mediterranean, wooded (*Quercus*-dominated)

Phase b: Humid phase characterized by an increase in ferns

Phase c: Major decrease of woods; increase in ericaceous heathlands; fluctuations in humidity, which continues, although rather less than in the previous phase. The disappearance of olive (*Olea*) indicates colder conditions.

In terms of Phases 1–5 at El Mirón, we believe that we can tentatively correlate Phases 1–2 with southeast Pollen Phase a, Phase 3 with Pollen Phase b, and Phases 4–5 with Pollen Phase c. Finally, El Mirón Phase 6 represents the near-continuous recent occupation of the cave by humans and their livestock.

Summary of Results and Conclusions

- The study of the biostratigraphic distribution, composition, and state of preservation of the micromammals in the upper levels of the El Mirón Cave vestibule allow us to deduce that the faunal associations in the Holocene deposits do reflect the contemporaneous composition of the faunas in the area around the cave, mainly as "sampled" by cave-roosting, raptorial birds (notably owls). For this reason the micromammal remains are useful in the paleoclimatic and paleoenvironmental reconstruction of the cave's surroundings.

- Human occupation does not seem to have been continuous in the El Mirón vestibule during the Holocene. Had human occupation been continuous, the commensal micromammalian species would have proliferated in all the levels, but that is not the case. In support of this observation, one can also point to the high species diversity in all the levels except the topmost (essentially modern) one, a fact that shows the cave to have often been occupied by raptors, which are incompatible with humans.
- The appearance of the modern murids, which are human commensals, came later in the Cantabrian region in northern Spain than in the south and center of the country—probably in relation to the late arrival of agriculture in the Atlantic environments of the north.

The biostratigraphic distribution of rodents, insectivores, chiropterans, and murids in El Mirón demonstrates that there were at least five climatic/environmental phases during those parts of the Holocene represented by Level 10.1 (9000–8000 BP) through Level 3 (3700 BP) and up to the present: the first was wooded; the second saw regression of woodlands; the third had variable conditions, probably under significant human influence; the fourth had drier, more steppic conditions; and the fifth is not so much a climatic phase, as it represents intensive human occupation and use of the Cantabrian landscape that continues at the present time.

- The Level 3 complex (3, 3.0–3.4—and possibly including later-formed feature fill Level 3.5) is the coldest unit in the Holocene sequence; it corresponds to the known period of cold conditions that coincided with the Bronze Age throughout Europe.
- El Mirón contains a late (but problematic) record of the species *Pliomys lenki*—in a Holocene level (Level 3.5) of undermined specific age and origin, probably the result of mixture.

Acknowledgments

First of all, we would like to thank the El Mirón excavation team members who, since 1996, have been digging, screening, separating, and curating the archaeological and paleontological finds from this site under the direction of Professors Lawrence Straus and Manuel González Morales. We acknowledge help from the Instituto Pirenaico de Ecología, Museo Nacional de Ciencias Naturales, Fundación Atapuerca, Atapuerca Project (CGL2006–13532-C03–02) and El Mirón Project. Our thanks also go to María Fernanda Sánchez Goñi and Cristina Roc Gargallo for their help with the pollen graphs. The early excavation campaigns at El Mirón (whence the Cabin area Holocene deposits) were funded by the Fundación Marcelino Botín, the L. S. B. Leakey Foundation, and the National Geographic Society. The town of Ramales de la Victoria (Cantabria) has provided material support every year. Lawrence Straus and Manuel González Morales have provided essential information for and constructively reviewed this chapter, which was translated into English and edited by Straus.

References

Alcalde Gurt, G.
1982 Presènsia interessant de *Pliomys lenki* i de *Microtus oeconomus* en el reompliment de la cova dels Ermitons (La Garrotxa, Girona). *Acta Geologica Hispanica* 17:281–282.

Altuna, J.
1972 Fauna de mamíferos de los yacimientos prehistóricos de Guipúzcoa. *Munibe* 24:1–464.
1986 The mammalian faunas from the prehistoric site of La Riera. In *La Riera Cave: Stone Age Hunter-Gatherer Adaptations in Northern Spain*, edited by L. G. Straus and G. A. Clark, pp. 237–274. Anthropological Research Papers 36. Arizona State University, Tempe.

Arrizabalaga i Blanch, A., E. Montagud i Blas, and J. Gosálbez i Noguera
1986 *Introducció a la biologia i zoogeografia dels petits maífers (insectivors i rosegadors) del Montseny (Catalunya).* Papers de Treball. Generalitat de Catalunya, CIRIT, Barcelona, Spain.

Bartolomei, G., J. Chaline, O. Fejfar, D. Jánossy, M. Jeannet, W. Koenigswald, and K. Kowalski
1975 *Pliomys lenki* (Heller 1930) (Rodentia, Mammalia) in Europe. *Acta Zoologica Cracoviensia* 20(10):393–467.

Blanco, J. C.
1998 *Mamíferos de España.* Vols. 1 and 2. Guías de Campo. Planeta, Barcelona.

Borghi, C. E., S. M. Giannoni, and J. P. Martínez Rica
1990 Soil removed by voles of the genus pitymys in the Spanish Pyrenees. *Pirineos* 136:3–18.

Brunet-Lecomte, P.
1988 Les campagnols souterrains (Terricola, Arvicolidae, Rodentia) actuels et fossiles d'Europe occidentale. Doctoral thesis, Université de Bourgogne.
1991 Répartition géographique des campagnols du genre *Microtus* (Arvicolidae, Rodentia) dans le nord-ouest ibérique. *Arquivos do Museu Bocage* 2(2):11–29

Brunet-Lecomte, P., G. Brochet, J. Chaline, and M. Delibes
1987 Morphologie dentaire comparée de *Pitymys lusitanicus* et *P. duodecimcostatus* (Arvicolidae, Rodentia) dans le nord-ouest de l'Espagne. *Mammalia* 51:145–158.

Brunet-Lecomte, P., and J. Chaline
1990 Relations phylogénétiques et évolution des campagnols souterrains d'Europe (Terricola, Arvicolidae, Rodentia). *Comptes Rendus de l'Académie des Sciences* 311, series 2:745–750.
1993 Mise au point sur *Microtus (Terricola) pyrenaicus gerbei* (Gerbe, 1879) (Rodentia, Arvicolidae). *Mammalia* 57(1):139–142

Brunet-Lecomte, P., and M. Delibes
1984 Alimentación de la lechuza común *Tyto alba* en la cuenca del Duero, España. *Doñana, Acta Vertebrata* 11:213–229.

Cabrera-Millet, M., J. Britton-Davidian, and P. Orsini
1982 Génétique biochimique comparée de *Microtus cabrerae* (Thomas 1906) et de trois autres espèces d'Arvicolidae méditerranéens. *Mammalia* 46(3):381–388.

Chaline, J.
1972 Les rongeurs du Würmien II de la grotte de l'Hortus (Valflaunès, Hérault). *Etudes quaternaires* 1:233–240.
1975 Evolution et rapports phyletiques des campagnols (Arvicolidae, Rodentia) apparentés à *Dolomys* et *Pliomys* dans l'hemisphère nord. *Comptes Rendus de l'Académie des Sciences D* 281:33–36

Corchón, M. S.
1981 *Cueva de las Caldas.* Excavaciones arqueológicas en España 115. Madrid, Spain.

Cuenca Bescós, G., J. I. Canudo, and C. Laplana
1999 Análisis bioestratigráfico de los roedores del Pleistoceno Medio del yacimiento de Galería (Sierra de Atapuerca, Burgos). In *Memorias. Arqueología en Castilla y León*, vol. 7, pp. 189–210. Junta de Castilla y León, Valladolid, Spain.
2001 La séquence des rongeurs (Mammalia) des sites du Pléistocène inférieur et moyen d'Atapuerca (Burgos, Espagne). *L'Anthropologie* 105:115–130.

Cuenca Bescós, G., L. G. Straus, J. García-Pimienta, M. González Morales, and J. López-García
2010 Late Quaternary small mammal turnover in the Cantabrian region: the extinction of *Pliomys lenki* (Rodentia, Mammalia). *Quaternary International* 212:129–136.

Cuenca Bescós, G., L. G. Straus, M. R. González Morales, and J. C. García Pimienta
2008 Paleoclima y paisaje del final del cuaternario en Cantabria: los pequeños mamíferos del Mirón (Ramales de la Victoria). *Revista Española de Paleontología* 23:91–126.
2009 The reconstruction of past environments through small mammals: from the Mousterian to the Bronze Age in El Mirón Cave (Cantabria, Spain). *Journal of Archaeological Science* 36: 947–955.

Ferraris, M., B. Sala, and V. Scola
1990 The Late Pleistocene fauna with *Pliomys lenki* from the Ghiacciaia Cave loess (northern Italy). *Quaternary International* 5:71–79.

Freeman, L. G., J. González Echegaray, J. Pokines, H. Stettler, and M. Krupa
1998 Tamisage ultra fin et récupération de l'outillage: observations réalisées a El Juyo (Espagne cantabrique). *L'Anthropologie* 102:35–44.

Giannoni, S. M., C. E. Borghi, and J. P. Martínez Rica
1992 New data on the burrowing behaviour of *Microtus (Pitymys) duodecimcostatus. Zeitschrift für Säugetierkunde* 57:23–28.
1993 Comparing the burrowing behaviour of the Iberian mole voles (Microtus (Terricola) lusitanicus, M. (T,) pyrenaicus and M. (T.) duodecimcostatus). *Mammalia* 57(4):483–490.

Gil, E., and E. Lanchares
1988 Los roedores del yacimiento musteriense de la Cueva de Gabasa (Pirineo Aragonés). Interés paleoecológico. *Geogaceta* 3:5–7.

Gil, E., and C. Sesé
1985 Los roedores (Mammalia) del sitio de ocupación Musteriense de la Cueva de los Toros (Teruel). *Col-Pa* 40:41–49.

Guillem, P.
1995 Paleontología continental: microfauna. In *El Cuaternario del País Valenciano*, edited by V. M. Rosselló i Verger et al., pp. 227–233. Asociación Española para el Estudio del Cueternario y Universitat de Valencia, Valencia, Spain.

Gutierrez Elorza, M., and J. L. Peña Monné
1998 Geomorphology and Late Holocene climatic change in northeastern Spain. *Geomorphology* 23:205–217.

Gutierrez Elorza, M., and V. H. Sesé Martínez
2001 Multiple talus flatirons, variations of scarp retreat rates and the evolution of slopes in Almazán Basin (semi-arid central Spain). *Geomorphology* 38:19–29.

Heinrich, H.
1988 Origin and consequences of cycle ice rafting in the northeast Atlantic Ocean during the past 130,000 years. *Quaternary Research* 9:142–152.

Holden, M. E.
1993 Family Myoxidae. In *Mammal Species of the World*, 2nd ed., edited by D. E. Wilson and D. M. Reeder, pp. 763–770. Washington, D.C.: Smithsonian Institution Press.

Junco, E.
1987 *Mamíferos salvajes de Asturias.* Caja de Ahorros de Asturias, Oviedo, Spain.

Laplana Conesa, C., and G. Cuenca Bescós
1995 Los microvertebrados (anfibios, reptiles y mamíferos) asociados al yacimiento de la Edad del Bronce de la Balsa la Tamariz (Tauste, Zaragoza). *Col-Pa* 47:55–69.

López Antoñanzas, R., and G. Cuenca Bescós
2002 The Gran Dolina Site (Lower to Middle Pleistocene, Atapuerca, Burgos, Spain): new palaeoenvironmental data based on the distribution of small mammals. *Palaeogeography, Palaeoclimatology, Palaeoecology* 186:311–334.

López-Fuster, M. J., and J. Ventura
1996 A morphometrical review of the Sorex araneus-arcticus species group from the Iberian Peninsula (Insectivora, Soricidae). *Bonner Zoologische Beiträge* 46:327–337.

Lowe, J. J., and M. J. Walker
1997 *Reconstructing Quaternary Environments.* Addison Wesley Longman, London.

Marquet, J. C.
1989 Paleoenvironnement et chronologie des sites du domaine Atlantique français d'age Pleistocene moyen et supèrieur d'apres l'étude des rongeurs. Doctoral thesis, Université de Bourgogne.

Maul, L.
2001 The transition from hypsodonty to hypselodonty in the *Mimomys savini-Arvicola* lineage. *Lynx* (Prague) 32:247–253.

Morales, A., and J. Rodríguez
1997 Black rats (*Rattus rattus*) from medieval Mertola (Baixo Alentejo, Portugal). *Journal of Zoology* 241:623–642.

Morales Muñiz, A.
1986 Análisis de la fauna de vertebrados recuperada en las sepulturas del poblado del Bronce del Cerro de la Encantada (Provincia de Ciudad Real). *Oretvm* 2:159–196.

Morales Muñiz, A., M. A. Cereijo Pecharroman, F. Hernández Carrasquilla, and C. Liesau von Lettow-Vorbeck
1995 Of mice and sparrows: commensal faunas from the Iberian Iron Age in the Duero Valley (central Spain). *International Journal of Osteoarchaeology* 5:127–138.

Musser, G. G., and M. D. Carleton
1993 Family Muridae. In *Mammal Species of the World*, 2nd ed., edited by D. E. Wilson and D. M. Reeder, pp. 501–756. Washington, D.C.: Smithsonian Institution Press.

Nadachowski, A.
1982 Late Quaternary rodents of Poland with special reference to morphotype dentition analysis of voles. *Pánstwowe Wydawnictwo Naukowe* (Warsaw, Cracow):1–108.

Niethammer, J., and F. Krapp (eds.)
1982 *Handbuch der Säugertiere Europas*, vol. 1, *Nagetiere I. (Muridae).* Akademische Verlagsgesellschaft, Wiesbaden, Germany.

Nores, C.
1988 Diferenciación biométrica de *Apodemus sylvaticus* y *Apodemus flavicollis* en la Cordillera Cantábrica. Primeros resultados. *Revista de Biología de la Universidad de Oviedo* 6:109–116.

Pemán, E.
1983 Biometría y sistemática del género *Neomys* (Kaup, 1771) (Mammalia, Insectivora) del País Vasco. *Munibe* 35:115–148.
1985 Aspectos climáticos y ecológicos de los Micromamíferos del yacimiento de Erralla. *Munibe* 37:49–57.
1987 Los micromamíferos del yacimiento de Iruaxpe I (Guipúzcoa). *Munibe* 39:85–86.
1988 Los micromamíferos de Urtao II (Oñati, Guipúzcoa). *Munibe* 41:77–78.
1990a Los micromamíferos de la cueva de Amalda y su significado. Comentarios sobre *Pliomys lenki* (Heller, 1930) (Rodentia, Mammalia). In *La Cueva de Amalda (Zestoa, Pais Vasco) Ocupaciones paleolíticas y postpaleolíticas*, edited by J. Altuna, A. Baldeón, and K. Mariezkurrena, pp. 225–238. Colección Barandiarán 4. Eusko Ikaskuntza, Sociedad de Estudios Vascos, San Sebastián, Spain.
1990b Los micromamíferos en el Pleistoceno Superior del País Vasco. *Munibe* 42:259–262.

Pokines, J. T.
1998 *The Paleoecology of Lower Magdalenian Cantabrian Spain*. British Archaeological Reports S-713. Oxford, UK.

Rey, J. M., and A. Landin
1973 Sobre la presencia de *Crocidura suaveolens* en el sur de Andalucía (Mammalia, Insectivora). *Boletín de la Real Sociedad Española de Historia Natural (Sección Biológia)* 71:9–16.

Ruiz Bustos, A.
2000 *Estudio paleoecológico de los sedimentos con presencia del hombre de Neandertal en la cueva de la Carihuela (Piñar, Granada). Síntesis ambiental del Würm mediterráneo en la Cordillera Bética*. Edita Ayuntamiento de Piñar, Granada, Spain.

Sánchez Goñi, M. F.
1993 *De la taphonomie pollinique à la reconstitution de l'environnement*. British Archaeological Reports S-586. Oxford, UK.

Sánchez Goñi, M. F., J. L. Turon, F. Eynaud, N. J. Shackleton, and O. Cayre
2000 Direct land/sea correlation of the Eemian and its comparison with the Holocene: a high-resolution palynological record off the Iberian margin. *Geologie en Mijnbouw* (Netherlands Journal of Geosciences, NJG) 79:345–354.

Sevilla García, P.
1988 Estudio paleontológico de los Quirópteros del Cuaternario español. *Paleontologia i Evolució* 22:113–233.

Straus, L., and M. González Morales
2001a El Mirón Cave 2001: continuing excavations by the Universities of Cantabria and New Mexico. *Old World Archaeology Newsletter* 23(2):1–5.
2001b The upper paleolithic in El Mirón Cave (Ramales, Cantabria, Spain). In *Le Paléolithique Supérieur européen. Bilan quinquennal 1996–2001, Commission 8, XIVe Congrès UISPP*, pp. 135–139. ERAUL 97. Liège, Université de Liège.

Straus, L .G., M. González Morales, W. R. Farrand, and W. J. Hubbard
2001 Sedimentological and stratigraphic observations in El Mirón, a Late Quaternary cave site in the Cantabrian Cordillera, northern Spain. *Geoarchaeology* 16:603–630.

Sutcliffe, A. J., and K. Kowalski
1976 Pleistocene rodents of the British Isles. *Bulletin of the British Museum (Natural History, Geology)* 27:147.

Tchernov, E.
1992 Biological evidence for human sedentism in Southwest Asia during the Natufian. In *The Natufian Culture and the Origins of the Levantine Neolithic*, edited by O. Bar-Yosef and F. Valla, pp. 315–340. International Monographs in Prehistory, no. 1. Ann Arbor, MI.

Van den Brink, F. H.
1971 *Guía de campo de los mamíferos salvajes de Europa Occidental*. Omega, Barcelona, Spain.

Ventura, J., M. J. López-Fuster, and J. Gosàlbez
1993 A morphometric analysis of cranial variation in the bank vole, *Clethrionomys glareolus* (Schreber, 1780) (Rodentia, Arvicolidae) from the Iberian peninsula. *Zoologischer Anzeiger* 231:183–193.

Ventura, J., and M. A. Sans-Fuentes
1997 Geographic variation and divergence in nonmetric cranial traits of Arvicola (Mammalia, Rodentia) in southwestern Europe. *Zeitschrift für Säugetierkunde*, 62:99–107

Zabala, J.
1983 Primera cita de *Arvicola sapidus* (Miller 1908, Mammalia, Rodentia) para la provincia de Guipúzcoa. *Munibe* 35:109–114.

Table 12.1. Number of micromammal samples per level and species diversity in the Holocene levels of El Mirón Cave

Level	Number of samples	Number of species per level
1	2	2
2	2	2
3	65	17
4	20	8
5	44	13
6	31	10
7	33	8
8	17	13
9	51	12
10	31	17
10.1	13	12
TOTAL	**309**	**21**

Table 12.2. Distribution of micromammal minimum number of individuals (MNI) in levels 1–10 at El Mirón Cave

Level	*P. lenki*	*Arvicola sapidus*	*Arvicola terrestris*	*Microtus agrestis*	*Microtus arvalis*	*Chionomys nivalis*	*Microtus oeconomus*	*Clethrionomys glareolus*	*Terricola pyrenaicus*	*Apodemus sylvaticus flavicollis*	*Glis glis*	*Eliomys quercinus*	*Talpa europaea*	*Sorex coronatus -araneus*	*Sorex "robust sp."*	*Sorex minutus*	*Neomys,*	*Crocidura*	*Quiropteros*	*Mustela*	MNI[a]
1			1										2								3
2			2										2								4
3	2		44	18	23	32	5	6	14	67	18		19	13		2	3	1	1	1	269
4			2		1			1		17	8		1	2						1	33
5			8	1	4	1	2	5	1	33	21		1	2			1			1	81
6			2	3		1		2		28	22	1	9	3			2				73
7				1		2				19	15	1	6	2				2			48
8			3	1	2	1		10	1	31	27	1	7	4			5		1		94
9			4	5	2	7		7	8	52	33	1	21	3	1						144
10		3	17	20	2	18		19	32	332	97	12	19	30		1	13	2	4	1	622
10.1			13	7		5		5	9	86	25	1	7	7			5			1	171
TOTAL	**2**	**3**	**96**	**56**	**34**	**67**	**7**	**55**	**65**	**665**	**266**	**17**	**94**	**66**	**1**	**3**	**29**	**5**	**6**	**5**	**1,542**

[a]Based on M_1 or, if lacking, on another classifiable postcraneal element.

Table 12.3. Distribution of micromammal species in the samples from the Holocene levels of El Mirón Cave

Sample No.	Level	Label	*Pliomys*	*A. sapidus*	*A. terrestris*	*M. agrestis*	*M. arvalis*	*M. oeconomus*	*Ch. nivalis*	*Clethrionmys glareolus*	*Terricola lusitanicus*	*Apodemus sylvaticus*	*Glis glis*	*Eliomys*	*Talpa*	*S. coronatus-araneus*	*S. "robust sp."*	*S. minutus*	*Neomys*	*Crocidura*	*Myotis*	*Miniopterus*	*Mustela*
1	1	97-1/2-1-J4-A-1													1								
2	1	97-1-1-J4-B-2			1										1								
3	2	97-2-1-H4-C-?			1										1								
4	2	97-2-2-J4-A-11			1										1								
5	3	97-2+3-1-I4-A-1			1																		
6	3	97-2+3-1-I4-B-2						1															
7	3	97-2y3-1-I4-A-1													1								
8	3	96-3-1-J1-A-1											1										
9	3	96-3-20-I2-D-89					1																
10	3	97-3.2-3-H1-A-102										1											
11	3	97-3.2-3-H1-D-104										1											
12	3	97-3.2-5-H2-A-118			1																		
13	3	97-3.2-5-H2-D-119								1		1											
14	3	97-3.3-6-H2-A-179			1							1											
15	3	97-3.3-6-H2-A-181B										1											
16	3	97-3.4-6-H4-B-108													1								
17	3	97-3??-3-H4-C-49			1	1			3	1	2	3				1		1					
18	3	97-3?-3-H4-C-49B						1															
19	3	97-3-1-H1-A-1			1				1														
20	3	97-3-1-H1-A-1B			1	1	1				1	3											
21	3	97-3-1-H1-B-2			2		1		1		1	1			1	1							
22	3	97-3-1-H1-B-2C										2											
23	3	97-3-1-H2-A-27			1							1							1	1			
24	3	97-3-1-H2-B-1							1				1		1								
25	3	97-3-1-H2-B-2			2	3	1		1		1												
26	3	97-3-1-H2-B-3											1										
27	3	97-3-1-H2-C-3			1								1										
28	3	97-3-1-H3-B-2			1	1			1														
29	3	97-3-1-H3-D-4										1				1							
30	3	97-3-1-H3-D-96			1							1											
31	3	97-3-1-H4-A-1B										1	1										
32	3	97-3-1-H4-B-2			1																		
33	3	97-3-1-H4-B-2B			1				1						1								
34	3	97-3-1-H4-C-3			1																		
35	3	97-3-1-H4-D-4										1				1							
36	3	97-3-2-H1-A-22			1				2			1							1				
37	3	97-3-2-H1-A-22B										1											
38	3	97-3-2-H1-B-23			1							3											
39	3	97-3-2-H1-B-236						1	1		1	2											
40	3	97-3-2-H2-A-27			1		2					2											
41	3	97-3-2-H2-B-28										1											
42	3	97-3-2-H2-C-29			1		2		1		1				1	1							
43	3	97-3-2-H2-D-30			2	1	2		2			1			1								
44	3	97-3-2-H4-A-23			1							2	1										
45	3	97-33-6-H2-D-182											2										

Table 12.3. *(continued)*

Sample No.	Level	Label	*Pliomys*	*A. sapidus*	*A. terrestris*	*M. agrestis*	*M. arvalis*	*M. oeconomus*	*Ch. nivalis*	*Clethrionmys glareolus*	*Terricola lusitanicus*	*Apodemus sylvaticus*	*Glis glis*	*Eliomys*	*Talpa*	*S. coronatus-araneus*	*S. "robust sp."*	*S. minutus*	*Neomys*	*Crocidura*	*Myotis*	*Miniopterus*	*Mustela*
46	3	97-3-3-H4-A-47			1	1	1			1		4	1								1		
47	3	97-3-3-H4-D-50							1	1		3	1										
48	3	97-3-3-J4-A-19			1																		
49	3	97-3-3-J4-B-20			1			1															
50	3	97-3-4-H2-B-73							1			1			1								
51	3	97-3-4-H2-C-74										1	1										
52	3	97-3-4-H2-D-75			1						1												
53	3.5	97-3.5-7-H4-D-123			1				1			2											
54	3.5	97-3.5-4-H3-B(+A)-25B								1			1		1								1
55	3.5	97-3.5-4-H3-B-25			2	2	4		4			7				2			1				
56	3.5	97-3.5-4-H3-B-25C			1							1			1								
57	3.5	97-3.5-4-H3-C-51			1		1		1	1	1	3	1		1	1		1					
58	3.5	97-3.5-4-H4-C-92											1										
59	3.5	97-3.5-4-H4-D-93			1	1			2			1			1	1							
60	3.5	97-3.5-7-H4-A-122	2		2	3	1		1			4	2		2								
61	3.5	97-3.5-7-H4-A-122B			1	1			1			1			1								
62	3.5	97-3.5-7-H4-A-122B(?)			1						1				1								
63	3.5	97-3.5-7-H4-A-122C			2	2	4		2			3	1		1	1							
64	3.5	97-3.5-7-H4-A-122D			1			1							1	1							
65	3.5	97-3.5-7-H4-B-217										2											
66	3.5	97-3.5-7-H4-C/D-123			1																		
67	3.5	97-3.5-7-H4-C/D-123B				1	2		1		3	1				2							
68	3.5	97-3.5-7-H4-C-221			1				2				1		1								
69	3.5	97-3.5-7-H4-D+C-123			1						1	1											
70	4	97-4-8-H4-B-236										4	1										
71	4	97-4-1-I4-D-4					1																
72	4	97-4-1-I4-C-3			1																		
73	4	97-4upper-2-I4-A-24											1										
74	4	97-4-2-I4-C-26											1										
75	4	97-4-2-I4-D-27											1										
76	4	97-4-2-I4-B-25B										1	1										
77	4	97-4-2-I4-B-25D										1											
78	4	97-4-6-J4-A-140											1										
79	4	97-4-40-54(J)-B-40B								1													
80	4	97-4-5-J4-B-130										2											
81	4	97-4-4-J4-B-40											1										
82	4	97-4-3-J4-D-22																					1
83	4	97-4-4-J4-B-40C										1											
84	4	97-4-4-J4-C-9			1																		
85	4	97-4upper-2-I4-A-24b										5											
86	4	97-4-2-I4-A-24B										1											
87	4	97-4upper-2-I4-A-24D										1	1			1							
88	4	97-4-2-I4-C-26B													1								

Table 12.3. *(continued)*

Sample No.	Level	Label	*Pliomys*	*A. sapidus*	*A. terrestris*	*M. agrestis*	*M. arvalis*	*M. oeconomus*	*Ch. nivalis*	*Clethrionmys glareolus*	*Terricola lusitanicus*	*Apodemus sylvaticus*	*Glis glis*	*Eliomys*	*Talpa*	*S. coronatus-araneus*	*S. "robust sp."*	*S. minutus*	*Neomys*	*Crocidura*	*Myotis*	*Miniopterus*	*Mustela*
89	4	97-4-4-J4-A-40A										1				1							
90	5	97-5-4-H1-D-167										1											
91	5	97-5-4-H1-B-157										1	1										
92	5	97-5-5-H1-A-175								1		1	1										
93	5	97-5.2-4-H1-B-158											1										
94	5	97-5-4-H1-A-156B(?)										1	1		1								
95	5	97-5-4-H1-A-156																					1
96	5	97-5-5-H1-B-176										1											
97	5	97-5.1-12-H4-B-270										1											
98	5	97-5-4-J4-D-135										1	1										
99	5	97-5-9-I4-D-156										1											
100	5	97-5-4-I4-A-132			1																		
101	5	97-5-9-J4-D-156											1										
102	5	97-5-9-H4-D-244																	1				
103	5	97-5.4-18-J4-D-228											1										
104	5	97-5-3-J2-B-Bag8			1		1																
105	5	97-5-8-H2-C-222											1										
106	5	97-5.1-12-H4-D-271										1											
107	5	97-5-4-I4-B-133											1										
108	5	97-5-5-J1-C-12										1											
109	5	97-5.1-8-H3-D-76										1											
110	5	97-5-9-J4-D-167										1											
111	5	97-5.4-11-J4-D-19										1											
112	5	97-5-7-J4-A-144					1	1															
113	5	97-5-9-J4-A-161B											1										
114	5	97-5-9-H4-B-243								1		2	1			1							
115	5	97-5-9-J4-A-161										1	1										
116	5	97-5-9-J4-A-161A								1													
117	5	97-5.1-9-H2-B-251										1											
118	5	97-5-7-J4-B-141			1							1											
119	5	97-5.4-17-J4-B-224											1										
120	5	97-5.1-9-H2-D-253										1											
121	5	97-5-8-H2-D-223										3	1										
122	5	97-5.2-11-H2-D-289										1											
123	5	97-5-6-H3-D-70								1		1	1										
124	5	97-5.4-18-J4-D-225B										1											
125	5	97-5.4-17-J4-A-218							1			1											
126	5	97-5-9-J4-B-159														1							
127	5	97-5-5-H3-C-39										1	1										
128	5	97-5-5-H3-B-38			1							1											
129	5	97-5.1?-9-H2-C-252										1	2										
130	5	97-5-5-H3-D-40			1							1											
131	5	97-5-8-H2-A-220A			1			1		1		1											
132	5	97-5-6-H3-C-69			1								1										
133	5	97-5.1?-9-H2-A-250			1	1	2				1	2	2										

Table 12.3. *(continued)*

Sample No.	Level	Label	*Pliomys*	*A. sapidus*	*A. terrestris*	*M. agrestis*	*M. arvalis*	*M. oeconomus*	*Ch. nivalis*	*Clethrionmys glareolus*	*Terricola lusitanicus*	*Apodemus sylvaticus*	*Glis glis*	*Eliomys*	*Talpa*	*S. coronatus-araneus*	*S. "robust sp."*	*S. minutus*	*Neomys*	*Crocidura*	*Myotis*	*Miniopterus*	*Mustela*
134	6	97-6-7-H1-B-202										1											
135	6	97-6-6-H1-D-190													1								
136	6	97-6-6-H1-B-189											2										
137	6	97-6-6-H1-A-188										1	1		1								
138	6	97-6-7-H1-A-201											1										
139	6	97-6-9-H3-C-79										1	1		1								
140	6	97-6-14-J4-A-196													1								
141	6	97-6-10-H2-D-287										1	2										
142	6	97-6-10-H2-C-286				1						1											
143	6	97-6-10-H3-A-103			1				1			1											
144	6	97-6-11-I4-B-227										1	1										
145	6	97-6-12-H2-A-336											2										
146	6	97-6-12-H2-B-337											1										
147	6	97-6.1-16-H4-B-291											1										
148	6	97-6-9-I4-B-186										1											
149	6	97-6-10-H2-A-284				1						4	1		1								
150	6	97-6-14-H4-C-288				1									1								
151	6	97-6.1-111-I4-A-239										1	1										
152	6	97-6-9-I4-A-185										3				1							
153	6	97-6-10-H3-B-8611											1										
154	6	97-6-10-H3-C-96											1										
155	6	97-6-9-I4-B-186C										1											
156	6	97-6-9-I4-B-186B										1											
157	6	97-6-9-H3-B-77C										1											
158	6	97-6-10-H3-D-87										2	1						1				
159	6	97-6-10-H3-C-96A										1	1		1								
160	6	97-6-11-I4-A-226								2		2	2		1	1			1				
161	6	97-6-14-J4-A-203													1								
162	6	97-6-7-H1-D-203											1										
163	6	97-6-14-H4-B-276										1		1		1							
164	6	97-6-10-H2-B-285			1							3	1										
165	7	97-7.1-13-H3-A-128																		1			
166	7	97-7.1-13-H3-A-128A										1	1										
167	7	97-7.2-12-H3-B-124										1			1								
168	7	97-7.4-13-J4-C-198											1										
169	7	97-7.5-16-J4-C-216B													1								
170	7	97-7-11-H3-A-115											1										
171	7	97-7-11-H3-C-116										1				1							
172	7	97-7-11-I4-C-280											1										
173	7	97-7-15-H4-D-289										1	1										
174	7	97-7-16-I4-A-244										1											
175	7	97-7-16-I4-C-246													1					1			
176	7	97-7-17-H4-A-298										1	1										
177	7	97-7-17-H4-B-293B										1											
178	7	97-7-17-H4-D-294B										1	1										

Table 12.3. *(continued)*

Sample No.	Level	Label	*Pliomys*	*A. sapidus*	*A. terrestris*	*M. agrestis*	*M. arvalis*	*M. oeconomus*	*Ch. nivalis*	*Clethrionmys glareolus*	*Terricola lusitanicus*	*Apodemus sylvaticus*	*Glis glis*	*Eliomys*	*Talpa*	*S. coronatus-araneus*	*S. "robust sp."*	*S. minutus*	*Neomys*	*Crocidura*	*Myotis*	*Miniopterus*	*Mustela*
179	7	97-7-17-I4-A-25											1		1								
180	7	97-7-17-I4-B-252										1											
181	7	97-7-17-I4-B-254B													1								
182	7	97-7-17-I4-D-254										1	1										
183	7	97-7-18-H4-A-306										1											
184	7	97-7-18-H4-B-307							1			2											
185	7	97-7-18-H4-B-307B														1							
186	7	97-7-18-H4-C-308										1	1										
187	7	97-7-18-H4-D-309										1											
188	7	97-7-18-H4-D-309B							1				1										
189	7	97-7-18-J4-B-227				1																	
190	7	97-7-19-J4-A-240											1	1									
191	7	97-7-19-J4-B-250										2											
192	7	97-7-19-J4-B-250A											1										
193	7	97-7-19-J4-B-250D										1											
194	7	97-7-20-J4-A-249													1								
195	7	97-7-20-J4-C-267										1											
196	7	97-7-20-J4-D-267											1										
197	7	97-7-5-J2-B-18											1										
198	8	96-8-7-J3-B-30											1										
199	8	96-8-12-I3-C-42												1									
200	8	96-8-12-I3-D-43											2										
201	8	96-8-12-I3-B-41											1										
202	8	98-8-19-I4-A-298										1											
203	8	98-8-19-I4-C-300											1										
204	8	98-8-20-H4-D-352										1											
205	8	98-8-20-H4-A-349										1	1		1								
206	8	98-8-19-H4-D-323								1		3	2										
207	8	98-8-18-I4-A-268				1				1		2	3		1				1				
208	8	98-8-19-H4-B-32								2		2	2		1				1		1		
209	8	98-8-19-H4-A-320								1		2	1		1								
210	8	98-8-19-H4-C-322			1				1	2		6	4		1	3							
211	8	98-8-18-I4-D-271			1		1					3	2										
212	8	98-8-18-I4-B-269					1					2	2						1				
213	8	98-8-19-I4-D-30			1					1		4	3		1				1				
214	8	98-8-18-I4-C-270								2	1	4	2		1	1			1				
215	9	98-9.6-23-H4-D-400													1								
216	9	98-9.7-24-I4-B-553													1								
217	9	97-9-14-I3-C-64B													1								
218	9	98-9.8-25-I4-C-666													1								
219	9	96-9-7-J1-D-24													1								
220	9	97-9.7-16-I3-D-168B											1										
221	9	98-9.7-24-I4-C-554										1			1								
222	9	98-9.6-23-I4-A-484										1			1								
223	9	96-9-13-I3-D-48			1																		

Table 12.3. *(continued)*

Sample No.	Level	Label	*Pliomys*	*A. sapidus*	*A. terrestris*	*M. agrestis*	*M. arvalis*	*M. oeconomus*	*Ch. nivalis*	*Clethrionmys glareolus*	*Terricola lusitanicus*	*Apodemus sylvaticus*	*Glis glis*	*Eliomys*	*Talpa*	*S. coronatus-araneus*	*S. "robust sp."*	*S. minutus*	*Neomys*	*Crocidura*	*Myotis*	*Miniopterus*	*Mustela*
224	9	98-9-21-I4-B-371										1	1										
225	9	97-9-14-I3-B-63A							1														
226	9	98-9.8-25-I4-B-665							1			1											
227	9	97-9.6-15-I3-A-102B				1					1												
228	9	98-9-18-H4-C-470											1	1									
229	9	98-9-24-J4-B-382											1										
230	9	97-9.7-16-I3-D-168								1		1											
231	9	97-9.8-17-I3-B-215							1			2				1							
232	9	98-9.6-23-I4-D-487										1											
233	9	98-9-22-H4-D-389													1								
234	9	98-9.6-25-H4-A-542										1											
235	9	98-9.7-24-I4-D-555				1																	
236	9	98-9-21-I4-C-372													1								
237	9	97-9.7-16-I3-C-167B			1																		
238	9	98-9.6-22-I4-D-413				1			1			1											
239	9	98-9.6-22-I4-A-410										1											
240	9	98-9.8-25-I4-D-667											3										
241	9	97-9.8-17-I3-A-214							1	2													
242	9	98-9.6-25-H4-B-543					1			1		1			1								
243	9	98-9.6-24-H4-D-439										2											
244	9	98-9.6-24-H4-A-436										1	1		1								
245	9	98-9-21-I4-A-370											2		1								
246	9	98-9.6-23-I4-C-486							1			1											
247	9	97-9-?-I3-D-65B										2	1		1								
248	9	98-9-21-I4-D-373			1							1											
249	9	98-9-20-I4-C-320										1	1		1								
250	9	97-9-14-I3-A-62B										3	1			1							
251	9	97-9-14-I3-B-63B									1	1	1										
252	9	97-9-14-I3-D-65A				1							1			1							
253	9	98-9.6-25-H4-D-545											3										
254	9	97-9-14-I3-A-62A								1	1	2	1		1								
255	9	97-9.8-17-I3-C-215							1			4	1				1						
256	9	98-9-20-I4-B-319										1	2		1								
257	9	97-9.6-15-I3-D-105										1	1										
258	9	97-9.6-15-I3-B-103											1										
259	9	98-9.6-25-H4-C-544				1				1		10			1								
260	9	98-9.6-22-I4-B-411									1	1	1		1								
261	9	98-9.6-24-H4-C-438										1	2		1								
262	9	97-9-14-I3-C-64A										2	1		1								
263	9	97-9.7-16-I3-C-167									2	1	2		1								
264	9	97-9.8-17-I3-D-215					1			1	1	2	1										
265	9	97-9.7-16-I3-D-168B			1						1	3	2										
266	10	97-10-19-I3-A-259B			1	1					1	5				1			1		1		
267	10	97-10-18-I3-C-249		3		3			2	2	4	45	20	2	1	4			1		1		
268	10	97-10-18-I3-B-248			1	2			1		2	21	1	1	1	1							

Table 12.3. *(continued)*

Sample No.	Level	Label	*Pliomys*	*A. sapidus*	*A. terrestris*	*M. agrestis*	*M. arvalis*	*M. oeconomus*	*Ch. nivalis*	*Clethrionmys glareolus*	*Terricola lusitanicus*	*Apodemus sylvaticus*	*Glis glis*	*Eliomys*	*Talpa*	*S. coronatus-araneus*	*S. "robust sp."*	*S. minutus*	*Neomys*	*Crocidura*	*Myotis*	*Miniopterus*	*Mustela*
269	10	97-10-19-I3-A-259			1	1					2	11	2	1	1	1							
270	10	97-10-19-I3-C-261			2	2					2	40	10	2	2	4			2			1	1
271	10	97-10-18-I3-D-250				2	1		4		2	54	15	1	1	3			1				
272	10	97-10-19-I3-D-262A			1		1		2	5	3	22	9	1	1	1			1				
273	10	96-10-14-J3-C-51			1							1	3										
274	10	98-10-20-H4-C-705								1		3	1		1				1				
275	10	98-10-27-I4-D-686			1	2			1			5							1				
276	10	98-10-26-I4-D-677				1			1	1	1	2	1		1								
277	10	98-10-27-I4-B-684			1	1						5				1			1				
278	10	98-10-26-I4-A-674				2			1	2	1	9	1		1								
279	10	98-10-26-I4-C-676										2	3								1		
280	10	98-10-26-H4-D-706								1		7	2		1	1			1				
281	10	97-10-19-I3-B-260B			1						1	13	2			1			1				
282	10	97-10-18-I3-A-241							1			15	3			3		1	1				
283	10	97-10-18-I3-A-241B				1			1	1	1	12	4	1		2							
284	10	97-10-19-I3-D-262B				1			1	3	3	4			1								
285	10	97-10-19-I3-B-260			1	1			1		5	16	3	1	1	3			1	2			
286	10	96-10-14-J3-D-52			1								1										
287	10	96-10-14-J3-B-50											2										
288	10	98-10-26-H4-B-704							1	1		2	1		1								
289	10	97-10-18-I3-A-247											1										
290	10	98-10-27-H4-D-718												1									
291	10	98-10-26-H4-A-703								1	1	4				1							
292	10	98-10-27-H4-B-716										1											
293	10	98-10-27-H4-C-717											1										
294	10	96-10-14-J3-A-49			2								2		1								
295	10	97-10-18-I3-C-249B			1					1	2	6	1		1	2							
296	10	97-10-19-I3-D-262B			3	1			1		2	32	8	1	3	2			1		1		
297	10.1	97-10.1-20-I3-A-266A										3											
298	10.1	97-10.1-21-I3-A-272			2							2			1								
299	10.1	97-10.1-20-I3-D-269							1			1	1		1								
300	10.1	97-10.1-20-I3-A-266B			1	2			2		2	8	1		1	1			1				
301	10.1	97-10.1-2-I3-C-274			3	1				1	1	20	10										
302	10.1	97-10.1-20-I3-C-268			1					2	1	34	5	1		1			2				
303	10.1	97-10.1-21-I3-D-275			3	1			1		1	5				1							1
304	10.1	97-10.1-20-I3-B-267			1	1				1	2	13	1		1	1			1				
305	10.1	97-10.1-21-I3-B-273			1											1							
306	10.1	98-10.1-28-H4-A-725				1			1				1		1								
307	10.1	97-10.1-28-I4-C-699													1								
308	10.1	98-10.1-28-H4-C-725											1										
309	10.1	97-10.1-20-I3-D-269			1	1				1	2		5		1	2			1				

Columns are in taxonomic order: Rodentia: Pliomys–Eliomys; Insectivora: Talpa–Crocidura; Chiroptera: Myotis–Miniopterus; Carnivora: Mustela.

Table 12.4. El Mirón Cave Holocene radiocarbon dates (1996–2002)

Square	Level	Spit	Period[a]	Date BP	1 SD	Lab no.	Method[b]	Calibrated Date* BC
H2	3	4	BA	3700	40	GX-25851	AMS	2140–2030
J2	5	3	Chal	3820	240	GX-22127	Conv.	2575–1931
I3	5.1	4	Chal	4120	50	GX-22130	AMS	2858–2586
H2	7	14	Chal	3740	120	GX-24460	Cxcnt	2305–1963
I3	8.1	13	Neo?	4680	60	GX-22131	AMS	3612–3371
J2	9	8	Neo	5170	170	GX-22128	Conv.	4221–3789
H4	9	22	Neo	5280	40	GX-24461	AMS	4217–4001
I4	9.6	22	Neo	5250	150	GX-24462	Cxcnt	4318–3945
I3	10	19	Neo	5570	50	GX-23414	AMS	4449–4359
I3	10	19	Neo	5690	50	GX-23413	AMS	4582–4458
J4	10.1	34	Mes	8380	175	GX-24463	Cxcnt	7586–7182
I4	10.1	28	Mes	8700	40	GX-25852	AMS	7745–7609
I4	10.1	29	Mes	9550	50	GX-24464	AMS	9119–8792

Note: All items are from the outer vestibule (Cabin) excavation zone; material is charcoal.

[a]BA = Bronze Age; Chal = Chalcolithic; Neo = Neolithic; Mes = Mesolithic

[b]AMS = accelerator; Conv. = conventional; Cxcnt. = extended count

*CALIB 4.1.2

CHAPTER THIRTEEN

Bird Remains from the Post-Azilian Levels of El Mirón Cave

Mikelo Elorza
TRANSLATED BY LAWRENCE GUY STRAUS

This chapter analyzes the bird remains from Mesolithic, Neolithic, Chalcolithic, and Bronze Age levels in the Cantabrian site of El Mirón Cave. The sample is not very representative but indicates environmental surroundings very similar to those of the present on the edge of the Cantabrian Cordillera. The part of the site sequence studied here suggests a development toward denser deciduous forests in the region.

El Mirón Cave is located at about 260 m above sea level at a distance of about 20 km from the present shore of the Bay of Biscay, near the town of Ramales de la Victoria. The cave, with a mouth measuring about 18 m wide and at least as high, faces the west from a vertical cliff-face. It is about 100 m above the valley floor and is surrounded by mountain summits at or around 1,000 m above sea level (González Morales and Straus 2000a, b). It is surrounded by a landscape that is presently dominated by pastureland and small thickets in the lower elevations and matorral on the higher, rocky slopes. The modern environment is characterized by a humid, temperate climate and vegetation defined by the presence of acidophilous deciduous woods that feature *Quercus robur* in the phytosociological series known as "*Tamo communis-Quercetum roboris*" within the Eurosiberian ecological province (Rivas Martínez 1987).

The excavations conducted in El Mirón since 1996 have so far revealed a cultural sequence from the Mousterian to the Bronze Age, and even traces of medieval visits. One of the most important aspects of this site is indeed the length and relative continuity of its stratigraphy, with one of the most complete series of radiocarbon dates for the prehistory of the Cantabrian region (González Morales and Straus 2000a, b).

The main excavations have been carried out in three areas of the cave vestibule: the Cabin (outer vestibule), Mid-Vestibule Trench, and Corral (vestibule rear). The latter area has no substantive post-Paleolithic deposits, so it is not dealt with here. The first two areas yielded culturally poor levels of Mesolithic age (Levels 10.1 and 304, respectively), rich early Neolithic levels (Levels 10–9 and 303.3–303, respectively), late Neolithic (Cabin Level 8), Chalcolithic (Cabin Levels 7–5 and possibly some of partially mixed Mid-Vestibule Level 302), late Chalcolithic–initial Bronze Age (Cabin Level 4), and early Bronze Age (Cabin Level 3). These levels have been dated by 17 radiocarbon determinations (González Morales and Straus 2000a, b) that can be summarized as follows by giving only the extreme values for each major Holocene cultural phase: Mesolithic, 9550±50–8380±175 BP; early Neolithic, 5790±90–5250±150 BP; Chalcolithic, 4120±50–3740±120 BP; and Bronze Age, 3700±40 BP. Given the limited number of prehistoric Cantabrian sites with ornithological studies (see Hernández Garrasquilla 1993), the long, well-dated El Mirón sequence obviously takes on considerable interest.

Materials

The avian sample studied here consists of 172 vertebrae and 476 other skeletal elements, of which only 40 could be determined to the genus or species level (table 13.1). Identification was rendered difficult due to the highly fragmented state of preservation, with only 30 bones having been found whole. Among the indeterminable material, which is mainly composed of non-corvid passeriforms, there are four bones of galliforms of medium size (from the Mesolithic, Neolithic, and Bronze Age) and an anatine (from the Neolithic). The small numbers of remains from 12 individual stratigraphic levels have been grouped by cultural phases in order to the increase the size of the various nonhomogeneous subsamples. The Bronze Age horizon produced only one identifiable bone and the recent Neolithic only two. Despite these limitations, the overall sample does provide some indications about the environments in which the inhabitants of El Mirón lived during the course of the Holocene, which are discussed below.

Systematics

ORDER FALCONIFORMES (SHARPE, 1874)

Family Accipitridae (Vieillot, 1816)

Accipiter gentilis (Linnaeus, 1758)

This species is represented in Level 7 by a proximal portion of a tibiotarsal. The northern goshawk is a raptor of the Holarctic faunal province that is linked exclusively to wooded areas. It nests in trees from sea level up to the tree line in subalpine zones, both in conifers and in oaks, with a preference for mature woods broken by clearings (Cramp and Simmons 1980–1983). It is distributed widely across temperate and boreal regions of Eurasia and North America. Its possible existence in the present surroundings of the site has been suggested by Purroy (1997).

Family Falconidae (Leach, 1819)

Falco peregrinus (Tunstall, 1771)

The peregrine falcon is represented by two cranial portions of coracoids and scapula in Levels 9 and 303. It is a poor environmental indicator as it is presently distributed across much of the world at all elevational ranges, although always associated with rocky cliffs and gorges. It does always avoid broad, dense forests (Cramp and Simmons 1980–1983). The surroundings of the site are presently populated by *F. peregrinus* (Purroy 1997).

ORDER GALLIFORMES (TEMMINCK, 1820)

Family Tetraonidae (Vigors, 1825)

Tetrao tetrix (Linnaeus, 1758)

The morphology and biometry of a distal portion of femur (Erbersdobler 1968) reveal the presence of the black grouse in Level 304 of the Mid-Vestibule Trench. This mondmilch level is culturally virtually sterile but almost certainly coincides with Level 10.1 in the Cabin, which has traces of Mesolithic visits to the cave (González Morales and Straus 2000a, b). The black grouse is a sedentary bird that currently occupies boreal, arctic, and alpine zones of the Palearctic province. *Tetrao tetrix* acts like a typical inhabitant of transitional zones between small open woods on the one hand and a variety of open habitats such as fallow fields, heaths, humid pastures, or even steppe on the other hand (Cramp and Simmons 1980–1983).

The present-day population of black grouse closest to El Mirón is located in the Alps, where it is considered a glacial relict. During the Pleistocene *T. tetrix* has been cited in the Iberian Peninsula in the Magdalenian of Lamiñak II in Vizcaya (Hernández

Carrasquilla 1994), with Eastham's determination at Urtiaga in Guipúzcoa in reality being a *Gallus gallus* (personal observation). Citations of this bird in southwest France are more numerous (Tyrberg 1998). It is very probable that *T. tetrix* did reach the north of the Iberian Peninsula during the Pleistocene and some of its populations managed to persist in the foothills of the Cantabrian Cordillera into the Holocene, at least until the early Mesolithic (Preboreal). This also occurred with its relative, the Capercaille (*Tetrao urogallus*), which still maintains two isolated populations in the Cantabrian Cordillera and in the Pyrenees (Cramp and Simmons 1980–1983).

ORDER GRUIFORMES (BONAPARTE, 1854)

Family Rallidae (Vigors, 1825)

Porzana porzana (Linnaeus, 1766)

An incomplete distal portion of a tarsometatarsal from Level 8 seems to be assignable to the spotted crake. This is a very timid rallid whose distribution on the Iberian Peninsula is little known (Purroy 1997). From its winter headquarters in Africa, the Near East, and India (Hoyo, Elliot, and Sargatal 1996), *P. porzana* nests in the western Palearctic near a variety of shallow, non-stagnant, usually fresh watercourses with abundant marsh vegetation (Cramp and Simmons 1980–1983). *P. porzana* would find an adequate habitat in the humid valley floor along the Río Asón and its tributaries that flow below the site.

ORDER CHARADRIIFORMES (HUXLEY, 1877)

Family Scolopacidae (Vigors, 1825)

Scolopax rusticola (Linnaeus, 1758)

The Eurasian woodcock, which left behind a coracoid in Level 10, is an inhabitant of temperate and boreal forest environments of the Palearctic. *S. rusticola* lives in mosaic zones among deciduous, coniferous, and mixed woods with shady, humid conditions and undergrowth (Cramp and Simmons 1980–1983). In the Iberian Peninsula its distribution includes Portugal, the Cantabrian region, and the Pyrenees, with a resident population that is notably swelled in fall and spring by winter visitors and birds migrating toward northern Europe. At times of peak seasonal population the spectrum of wooded habitats that this woodcock uses is expanded substantially (Purroy 1997).

ORDER COLUMBIFORMES (LATHAM, 1790)

Family Columbidae (Illiger, 1811)

Columba livia (Gmelin, 1789)/*Columba oenas* (Linnaeus, 1758)

El Mirón yielded a distal portion of an ulna from Cabin Level 10.1, a distal portion of a coracoid from Mid-Vestibule Level 303.2, a proximal portion of a carpometacarpal from Cabin Level 5, and an anterior phalange from Cabin Level 3.5. The morphologies of these bones do not permit distinction between these two species of dove (Fick 1974). *C. livia*, the rock dove, is a rock-dweller that (in its nondomesticated form) occupies the temperate and warm zones of the Palearctic province. *C. oenas*, the stock dove, is associated with wooded environments of the Palearctic, from the boreal to the Mediterranean regions (Cramp 1985–1988). The rock dove lives on coastal cliffs and rock walls, where it nests and rests, and it feeds in open areas, avoiding high, dense vegetation, either arboreal or herbaceous. The stock dove prefers to live in the ecotonal zone between clearings and wooded areas, with mature trees in which to nest. Both species currently live in the Iberian Peninsula, although they are relatively rare in the Cantabrian region (Purroy 1997).

Columba palumbus (Linnaeus, 1758)

A humerus from Level 10.1, a distal portion of a radius from Level 10, and a proximal portion of a carpometacarpal from Level 9.6 (all in the Cabin area) are assigned to the common wood pigeon, which nests at the present time in the vicinity of the site (Purroy 1997). This species is found in the western Palearctic in significantly wooded areas but also along forest edges and in mosaic zones with clearings. *C. palumbus* is widely distributed throughout Europe, having adapted well to cultivated fields. It only avoids rocky mountains, densely vegetated humid zones, treeless uplands, and open coasts (Cramp 1985–1988). Thus, in the El Mirón area it would presumably favor the lower valley habitats and not the rocky slopes and cliffs above the site.

ORDER STRIGIFORMES (WAGLER, 1816)

Family Strigidae (Vigors, 1816)

Otus scops (Linnaeus, 1758)

The morphology of a proximal portion of tibiotarsal from Level 10 permits its attribution to the Eurasian scops owl. This nocturnal raptor is a trans-Saharan migrant that in summer breeds in Europe in both temperate Mediterranean and continental areas with alternating wooded and open vegetation (Cramp 1985–1988). It mainly feeds on insects, which means that it avoids dense woods and mountainous areas. In the Iberian Peninsula its densest populations are currently in the Mediterranean region, while in the Cantabrian region it is far rarer and citations of its nesting are very scarce (Purroy 1997).

Strix aluco (Linnaeus, 1758)

The tawny owl is represented by a distal portion of a humerus from culturally sterile Level 304 in the Mid-Vestibule Trench. This nocturnal raptor lives in all types of Palearctic biotopes, from sea level to the subalpine zones, so long as there are at least minimal areas of woods, either deciduous or perennial. Thus it can be found in forests, fields, gardens, or riparian gallery woods. *S. aluco* ranges from boreal to Mediterranean zones, avoiding humid and unvegetated rocky areas (Cramp 1985–1988). It presently nests in the vicinity of the site (Purroy 1997).

Asio sp.

A distal tibiotarsal fragment from Level 303.3 (early Neolithic) does not allow for classification among the three species of this genus that are found in the western Palearctic: *A. otus*, *A. flammeus*, and *A. capensis*. The first of these has greater arboreal preferences than the latter two. Among these midsize owls, the third is only rarely found today in the Iberian Peninsula.

ORDER PASSERIFORMES (LINNAEUS, 1766)

Family Hirundinidae (Rafinesque, 1815)

Delichon urbica (Linnaeus, 1758)

A humerus from Level 10 (HUM-GL: 13.97 mm, as measured according to Driesch 1976) has been assigned to the northern house martin, a Eurasian summer trans-Saharan guest, which currently breeds in the vicinity of El Mirón (Purroy 1997). *D. urbica* originally occupied rocky cliff faces but today fundamentally lives in human constructions. It is not a good environmental indicator since it breeds in both oceanic and continental zones, from the subarctic to Mediterranean steppes, and avoids only very dense wooded areas or very arid places (Cramp 1985–1988).

Family Paridae (Vigors, 1825)

Parus major (Linnaeus, 1758)

Level 10 in El Mirón produced a humerus (HUM-GL: 17.68 mm) of the great tit, a sedentary insect eater that is abundant in all wooded environments of Eurasia from subarctic to Mediterranean zones, preferring low elevations but capable of ascending in montane regions up to the tree line. With this broad range of habitats, it lives in all kinds of settings so long as there are some trees for nesting (Cramp and Perrins 1993–1994). It is presently found in the area of the site (Purroy 1997). In the Cantabrian region the highest densities of *P. major* are found in oak stands, followed (in decreasing order of importance) by riparian thickets, beech groves, conifer plantations, and fields (Alvarez et al. 1998).

Family Turdidae (Rafinesque, 1815)

Turdus sp.

The material from this genus consists of 17 humeri and a coracoid from Levels 8, 9, 9.8, and 10. These bones could not be securely attributed to any of the six species of *Turdus* that live in the Palearctic. The thrushes in general inhabit a wide variety of wooded areas that usually also have clearings.

Family Corvidae (Vigors, 1825)

Pyrrhocorax graculus (Linnaeus, 1766)

Represented by a distal portion of humerus from Level 10, the yellow-billed chough is a corvid that lives and breeds in fissures in cliffs and in avens in the upper montane zones of Eurasia and North Africa. In terms of elevation, *P. graculus* reaches as far up as the snow line, well above the tree line. It is a sedentary species that does not descend to lowland plains or coast. It feeds in alpine pastures and in the mountains (Cramp and Perrins 1993–1994). Today it is found in the Cantabrian Cordillera and has been cited as possibly nesting near the site (Purroy 1997).

Corvus corax (Linnaeus, 1758)

The common raven left behind a distal portion of a tibiotarsal in Level 303.3 of the Mid-Vestibule Trench and a distal portion of a tarsometatarsal in Cabin Level 10.1. This corvid, cited as possibly nesting in the environs of the site at the present time (Purroy 1997), is sedentary and is amply distributed throughout the Holarctic region of the world. In terms of latitude, *C. corax* occupies a variety of environments from the tundra to the savanna, and in terms of altitude, from sea level to 2,400 m in the Alps and south Siberian Altai Mountains (Cramp and Perrins 1993–1994). It is thus a poor environmental indicator, since it lives in such a wide range of habitats. It requires only sufficient open areas in which to feed, avoiding the interior of areas of woods and shrubs, dense vegetation in general, extensive humidity, or zones of intensive agriculture.

Family Fringillidae (Leach, 1820)

Coccothraustes coccothraustes (Linnaeus, 1758)

Level 6 yielded a complete humerus of a hawfinch (HUM-GL: 23.26 mm). *C. coccothraustes* habitually breeds in the western Palearctic in both low and montane zones of temperate regions as well as more discontinuously in boreal, Mediterranean, and steppic zones. This grain eater is irregularly distributed at present across the Iberian Peninsula, with its densest populations being in the southwest (Purroy 1997). The hawfinch is a typical inhabitant of the canopy of forests with large trees, preferably deciduous taxa and habitually of the genera *Quercus* and *Carpinus*. It also visits trees of the genera *Fagus*, *Fraxinus*, *Ulmus*, and *Acer* if it finds sufficient nuts in them. It only occasionally lives in conifers, this being the case with some of the overwintering populations in the Iberian Peninsula as well as some of the populations at the eastern end of the bird's present range (Cramp and Perrins 1993–1994).

Discussion

The absence of dismemberment, butchery, cut marks, or burning marks on the studied remains prevents us from confirming any avian contribution to the human diet during the successive early- and middle-Holocene occupations of El Mirón Cave. If we add the fact that there are several species likely to have simply died in situ in a cave whose walls they could have used as living or resting places (i.e., *F. peregrinus*, *C. livia*, *S. aluco*, *D. urbica*, *P. graculus*, *C. corax*), we can conclude that the human contribution to the accumulation of the taphocenesis studied here was at best very slight. The human groups that visited or lived in El Mirón in recent prehistoric periods paid little attention to the birds as they exploited other resources of the upper Asón valley. The abundance of micromammals (see Cuenca Bescós and García, chap. 12, this volume) and the range of sizes of birds studied here indicate that it was a raptor of medium size that was the main agent of accumulation of the other, non-rock-dwelling birds whose remains were left in the cave (Cramp 1985–1988). However, it is likely that a mosaic of predators was responsible for the accumulation of remains of micromammals and small birds in the cave. It is worthy of note that *Bubo bubo* (the eagle owl) is absent from the Mirón Holocene avifaunal samples. This raptor tends to prey on larger animals (including partridges and lagomorphs, which are also absent from the samples). (The relative abundance of bird remains in the Mesolithic and Neolithic levels as opposed to the Chalcolithic and Bronze Age ones might suggest that the human presence was less than permanent or even sporadic in the earlier periods, while with massive animal stabling during the later periods the cave became more permanently inhospitable to birds of all sorts, both raptors and others.—LGS)

As a whole, the post-Azilian birds of El Mirón Cave present a picture of climatic and overall environmental conditions similar to those of the present. We have observed no birds associated with environments unlike those of today, and in fact, with the exception of *Tetrao tetrix*, all the identified taxa, including representatives of the genera *Asio* sp. and *Turdus* sp., can be found at present in the environs of the site (Purroy 1997).

During the Mesolithic of El Mirón, the presence of *T. tetrix*, *C. palumbus*, and *S. aluco* indicate the existence of localized woods with a low density of trees and the presence of clearings. The birds of the Neolithic levels suggest the existence of alpine or montane meadows (*P. graculus*) and humid woodlands (*S. rusticola*), combined with open spaces (*C. palumbus, O. scops*). Finally, the birds of the Chalcolithic levels show the presence of forested areas—probably deciduous in composition as well as extensive, dense, and mature in nature (*A. gentilis, C. coccothraustes*).

With all due caution, given the small size and problematic representativeness of the sample studied, we propose that the ensemble of levels has provided the following tentative information: continued presence of both wooded and open environments in the area around the cave during the Mesolithic and Neolithic, with higher arboreal density at least in some areas near the cave in the Chalcolithic period.

Acknowledgments

Lawrence Straus and Manuel González Morales invited me to conduct the avian faunal analysis from the El Mirón excavations, which they direct. Jesús Altuna and Gloria Cuenca Bescós separated out bird bones found among the macro- and micromammalian remains that they respectively studied. Francisco Hernández, Antonio Sánchez, and Lawrence Straus improved the text with their suggestions on an earlier draft. Arturo Morales facilitated my consultation of the comparative osteological collection of the Laboratory of Archeozoology at the Universidad Autónoma de Madrid. Lawrence Straus translated the original Castilian text into English.

References Cited

Alvarez, J., J. Aihartza, J. T. Alcalde, A. Bea, L. F. Campos, L. M. Carrascal, E. Castién, T. Crespo Pérez, J. A. Gainzarain, A. Galarza, E. García Tejedor, I. Mendiola, G. Ocio Andrés, and I. Zuberogoitia
1998 *Vertebrados continentales: Situación actual en la Comunidad Autónoma del País Vasco.* Servicio Central de Publicaciones del Gobierno Vasco, Vitoria-Gasteiz, Spain.

Cramp, S.
1985–1988 *The Birds of the Western Palearctic.* Vols. 4 and 5. Oxford University Press, Oxford, UK.

Cramp, S., and C. M. Perrins
1993–1994 *The Birds of the Western Palearctic.* Vols. 7 and 8. Oxford University Press, Oxford, UK.

Cramp, S., and K. E. L. Simmons
1980–1983 *The Birds of the Western Palearctic.* Vols. 2 and 3. Oxford University Press, Oxford, UK.

Driesch, A. von den
1976 A guide to the measurement of animal bones from archaeological sites. *Bulletin of the Peabody Museum of Archaeology and Ethnology* 1. Cambridge, MA.

Erbersdobler, K.
1968 *Vergleichend morphologische Untersuchungen an Einzelknochen des postcranialen Skeletts in Mitteleuropa vorkommender mittelgroßer Hühnervögel.* Ludwig-Maximilians-Universität, Munich.

Fick, O. K. W.
1974 *Vergleichend morphologische Unte rsuchungen an Einzelknochen europäischer Taubenarten.* Ludwig-Maximilians-Universität, Munich.

González Morales, M., and L. Straus
2000a La Cueva del Mirón (Ramales de la Victoria, Cantabria): excavaciones 1996–1999. *Trabajos de Prehistoria* 57:121–133.
2000b El depósito arqueológico de la Prehistoria reciente de La Cueva del Mirón (Ramales, Cantabria): bases estratigráficas para una secuencia regional. In *Actas do 3° Congresso de Arqueologia Peninsular*, edited by V. Oliveira Jorge, vol. 4, pp. 49–62. ADECAP, Porto, Portugal.

Hernández Carrasquilla, F.
1993 Catálogo provisional de los yacimientos con aves del Cuaternario de la Península Ibérica. *Archaeofauna* 2:231–275.
1994 La avifauna no passeriforme de Lamiñak II. *Kobie. Paleoantropología* 21:189–202.

Hoyo, J., A. Elliot, and J. Sargatal
1996 *Handbook of the Birds of the World.* Vol. 3. Lynx Edicions, Barcelona.

Purroy, F. J.
1997 *Atlas de la aves de España (1975–1995).* Lynx Edicions, Barcelona.

Rivas Martínez, S.
1987 *Memoria del mapa de series de vegetación de España.* ICONA, Madrid.

Tyrberg, T.
1998 *Pleistocene Birds of the Palearctic: A Catalogue* Publications of the Nuttall Ornithological Club 27. Nuttall Ornithological Club, Cambridge, MA.

Table 13.1. Post-Azilian bird remains from El Mirón Cave (number of remains by cultural period)

	Mesolithic	Early Neolithic	Late Neolithic	Chalcolithic	Bronze
Accipter gentilis				1	
Falco peregrinus		2			
Tetrao tetrix	1				
Cf. *Porzana porzana*			1		
Scolopax rusticola		1			
Columba livia/oenas	1	1		1	1
Columba palumbus	1	2			
Otus scops		1			
Strix aluco	1				
Asio sp.		1			
Delichon urbica		1			
Parus major		1			
Turdus sp.		17	1		
Pyrrhocorax graculus		1			
Corvus corax	1	1			
Coccothraustes coccothraustes				1	
Aves indeterminate	63	380	86	44	34
TOTAL	68	409	88	47	35

CHAPTER FOURTEEN

The Holocene Herpetofauna of El Mirón Cave

Borja Sanchiz, Jorge M. Lobo, and Salvador Bailon

TRANSLATED BY LAWRENCE GUY STRAUS

The excavations directed by M. González Morales and L.G. Straus in El Mirón Cave yielded relatively few remains of amphibians and reptiles from the Holocene deposits. These are the subject of the present zooarchaeological analysis, complementing the studies of other zoological groups reported upon in this monograph. González Morales and Straus (2000a, b; Straus et al. 2001) provide the relevant details about the site and its stratigraphy, archaeology, and chronology.

Methods

TAXONOMIC IDENTIFICATION

The taxonomic nomenclature used here follows Salvador (1998) and Frost (2010). The morphological traits used for identification are based on the criteria developed by Bailon (1999) and Böhme (1977) for the anurans and Szyndlar (1984) for the ophidians, as well as on direct comparisons with material in the collections of the Museo Nacional de Ciencias Naturales (CSIC) in Madrid.

ARCHAEOZOOLOGICAL CRITERIA

The calculation of minimum numbers of individuals (MNI) is based on the most numerous single skeletal element per species from each stratigraphic level, taking into account side as well as sex in cases of clear sexual dimorphism.

BIOLOGICAL INFORMATION

The information that we provide on ecological aspects of the species (periods of activity, predators, etc.) is taken from works by Barbadillo et al. (1999), Salvador (1998), Salvador and García-París (2001), and Salvador and Pleguezuelos (2002). All the species represented in the early to mid-Holocene levels of El Mirón are found today in Europe. For their continental distributions, one can consult Gasc et al. (1997). The syntheses of Holman (1998) and Martín and Sanchiz (2010) provide data on the presence of these taxa in other European Quaternary sites.

BIOGEOGRAPHICAL CRITERIA

Information on modern species's geographical ranges is taken from the recent herpetological atlas of Spain (Pleguezuelos, Márquez, and Lizana 2002), which was done on a UTM projection with a resolution of 10 × 10 km cells. We selected as our study area those 100 km² cells within the Iberian northern Atlantic climatic zone (with the exclusion of small portions of northern Portugal, for which adequate information is lacking).

The geographic analysis was done using the IDRISI 32 GIS software (Eastman 2001). For each 100 km² grid we included the following information: (1) climatic data (average annual temperature, total annual precipitation, and annual insolation) as provided by the Spanish Instituto Nacional de Meteorología, (2) altitudinal data (average elevation) according to a digital terrain model (Clark Labs 2000) with a resolution of 1 km, (3) data on groundcover and land use (woods, brush or heathlands, cultivated fields, pastures) according to the CORINE Land Cover 2000 Project database (freely available at http://terrestrial.eionet.europa.eu/CLC2000) and (4) lithological data (siliceous or calcareous soils) derived from the Spanish national soils atlas (ITGE 1988). The data on lithology and groundcover/land use were transformed into qualitative variables that represent the presence or absence of high percentages (>75% of land area) of each category.

Following the methodology detailed in Sanchiz and Lobo (2006), we develop probabilistic models for species distributions using the Biomapper 2.1 program (Hirzel et al. 2002). To do this, we calculate a series of uncorrelated environmental factors that summarize the main environmental gradients that exist in the region. These factors allow us to generate a map (or maps) of environmental suitability for each species by means of a comparison of the general values from the region with those that occur in the cells in which each species or group of species is actually present. One can thus estimate how probable it is that any particular place belongs to the environmental domain of those places in which the presence of a particular species (or group of species) has been ascertained. This methodology is an alternative to the climatic inferences exclusively based on the observed occurrences of a species, as recently done for the Pleistocene (e.g., Blain, Bailon, and Cuenca-Bescós 2008; Blain et al. 2009) and tries to skip the nonrandom prospection in the making of herpetological distribution atlases.

The variables selected for making paleoclimatic inferences are: (1) annual average temperature (°C), (2) annual precipitation (mm), (3) annual insolation (hours of sunlight), and (4) altitude (m). The last variable, whose value obviously has not changed for the site during the Holocene, acts in fact in zoogeography as a resultant factor that incorporates several occult populational variables, such as longevity or growth rate (see, e.g., Esteban and Sanchiz 2000).

The species selected for biogeographic analysis were *Anguis fragilis*, *Bufo bufo*, and *Rana temporaria*, in two different combinations representative of the herpetofaunas from most of the levels. The remains of other species are sporadic, and their absences from some of the levels are not statistically significant. The environmental conditions of the grids showing the joint presence of a selected group of species are compared with those for which these species have not been cited. We also compare the environmental values for those zones with high and low probability of presence of the abovementioned species groups, according to their predictive models of habitat suitability. These calculations were done with Statistica 6.1 software (StatSoft Inc. 2003).

Material

The Holocene herpetological material is stored in the Museo de Prehistoria y Arqueología de Cantabria, Santander. It consists of 356 bone fragments, of which 331 could be identified to species or genus level. The distribution of the finds by taxa and levels is presented in table 14.1. All remains seem to belong to adult individuals.

Results

PALEONTOLOGICAL ANALYSIS AND ECOLOGICAL AND TAPHONOMIC OBSERVATIONS

UTM grid cell VN69, where El Mirón Cave is located, has been at present relatively little surveyed for herpetofauna (only 6–10 citations, according to Pérez-Mellado and Cortázar [2002]). As a result, the national atlas (Pleguezuelos, Márquez, and Lizana 2002) lists only the documented presence of common wall lizard (*Podarcis muralis*) and Iberian adder (*Vipera seoanei*) in this grid. However, many of the (apparent) absences are presumably due to sampling deficiencies, and in fact all the species found in the archaeological site are present today in nearby grids. The species identified in the site are as follow:

Family Salamandridae
Salamandra (Garsault, 1764)

Salamandra salamandra (Linnaeus, 1758)

A single vertebra from Level 8 attests to the presence of the fire salamander at the site in late Neolithic times. On the Iberian Peninsula this is a polytypic species, with several subspecies that are very diverse in terms of morphology, reproduction, and autoecology. In general, these are animals characteristic of areas with high humidity, abundant in the Cantabrian region in deciduous forests. It is a nocturnal species that is inactive at temperatures <3°C. Its annual period of activity is highly variable; depending on local conditions, it can be active from February to March and October to November, with a period of hibernation, or even all year except for a period of aestivation in summer. Thus this species can provide no archaeozoological information on seasonality.

Several reptiles (*Natrix natrix*, *Vipera seoanei*), birds (*Buteo buteo*, *Strix alauco*, and *Turdus torquatus* but never *Tyto alba*) and mammals (*Lutra lutra*, *Meles meles*, *Sus scrofa*) are known predators of adult *Salamandra salamandra*. Such predation is, however, rare in nature, since the fire salamander has a conspicuous aposematic coloration and secretion glands that produce a strong neurotoxic alkaloid.

Family Alytidae
Alytes (Wagler, 1830)

Alytes obstetricans (Laurenti, 1768)

The common midwife toad is currently diversified into several subspecies on the Iberian Peninsula. The European form, *A. o. obstetricans*, is present in Asturias, Cantabria, Basque Country, and Navarra. However, it is not possible to identify the available archaeofaunal material to the subspecific level. This species is found in a wide variety of habitats. It is active at dusk and during part of the night, although it can also be seen during the day under cloudy, humid conditions. Its annual period of activity is restricted to the months between February and November, with a short period of hibernation. Adults are predated by snakes (*Natrix natrix*, *Natrix maura*, *Vipera seoanei*), birds (*Lanius excubitor*, *Strix alauco*, *Tyto alba*), and carnivores. In the case of *Tyto alba*, the owl pellet collection of J.M. Rey (Universidad de Santiago de Compostela) contains 16 places where *A. obstetricans* was hunted by the barn owl, making up about 1.5 percent of the prey among the lots in which this amphibian is found. These findspots are in Alava, northern Burgos, Cantabria, and Vizcaya. Judging by the periods in which the barn own hunted this toad, the annual maxima clearly correspond to the periods of March–April and July–August, which is an indirect indicator of the times during which the midwife toad is more active and conspicuous (and thus available for predation) in this region.

Family Bufonidae
Bufo (Garsault, 1764)

Bufo bufo (Linnaeus, 1758)

The common toad is one of the most frequently represented amphibian species in European archaeological sites. Unfortunately the lack of osteometric studies of living populations makes it impossible to identify the subspecies found in El Mirón, there being currently at least three reasonably differentiated subspecies on the Iberian Peninsula (Lizana 2002).

The wide variety of environmental conditions under which *Bufo bufo* lives in its very broad distribution throughout Europe means that as a species it is a poor environmental indicator (figure 14.1). In the study region the common toad always has a period of hibernation, being active mostly between March and October, although sporadically it can also be seen outside of this period, depending on local circumstances. It is occasionally the prey of a great number of species, from snakes (*Natrix natrix*), diurnal birds, and nocturnal raptorial birds (*Bubo bubo*, and *Strix alauco*, but not *Tyto alba*) to mammals (e.g., *Lutra lutra*, *Meles meles*, *Mustela putorius*, *Sus scrofa*, *Vulpes vulpes*). In the case of El Mirón Cave, the predators could have belonged to the last two

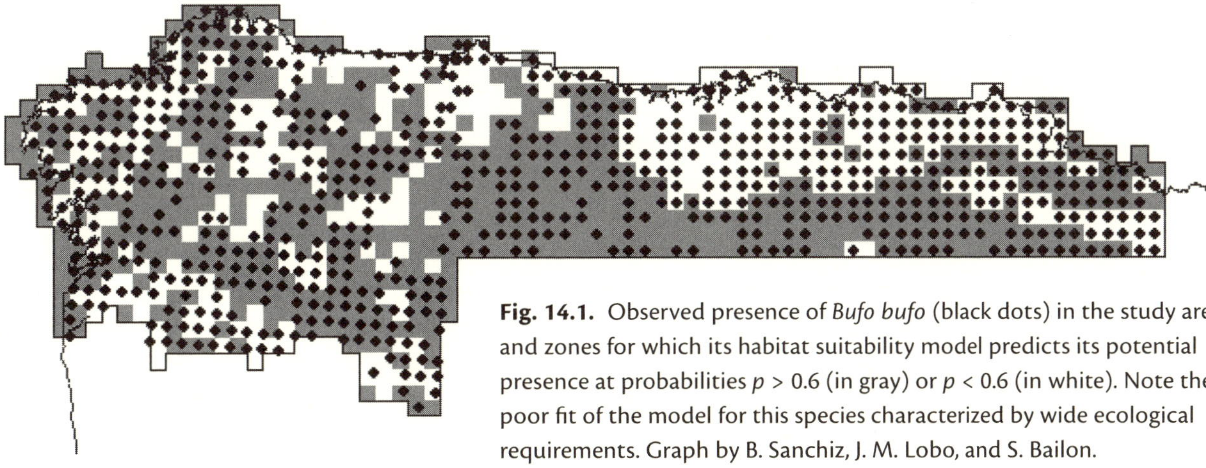

Fig. 14.1. Observed presence of *Bufo bufo* (black dots) in the study area and zones for which its habitat suitability model predicts its potential presence at probabilities $p > 0.6$ (in gray) or $p < 0.6$ (in white). Note the poor fit of the model for this species characterized by wide ecological requirements. Graph by B. Sanchiz, J. M. Lobo, and S. Bailon.

groups, judging from the fragmentation of the toad bones and traces of digestion, as described by Pinto Llona and Andrews (1999).

Family Ranidae

Rana (Linnaeus, 1758)

Rana temporaria (Linnaeus, 1758)

The grass frog is without doubt the amphibian species most commonly found in Quaternary sites in Atlantic Europe. Today it is widely distributed throughout all of non-Mediterranean Europe, even reaching northern Scandinavia (Gasc et al. 1997). On the Iberian Peninsula, the southern limit of its distribution lies precisely in the study region, for which reason we could expect that at the El Mirón site it should behave as a fairly sensitive environmental indicator at the extreme edge of its ecological range (figure 14.2).

The period of activity of *R. temporaria* throughout the year is highly variable. Hibernation is a requirement that is both ecological and physiological for this species. It has been demonstrated experimentally that gametogenesis requires a temperature below 6–7°C during two or more weeks for maturation. Thus the distribution of the species, according to Balcells (1975), approximately fits the 5°C isotherm for average January temperature. Like other anurans, the grass frog is generally nocturnal in its habits. Its period of reproduction can even include some days in December or January, but the time when adults are at greatest risk of being killed by predators of archaeozoological interest (such as the barn owl) seems to be between April and September. Thus attests J. M.

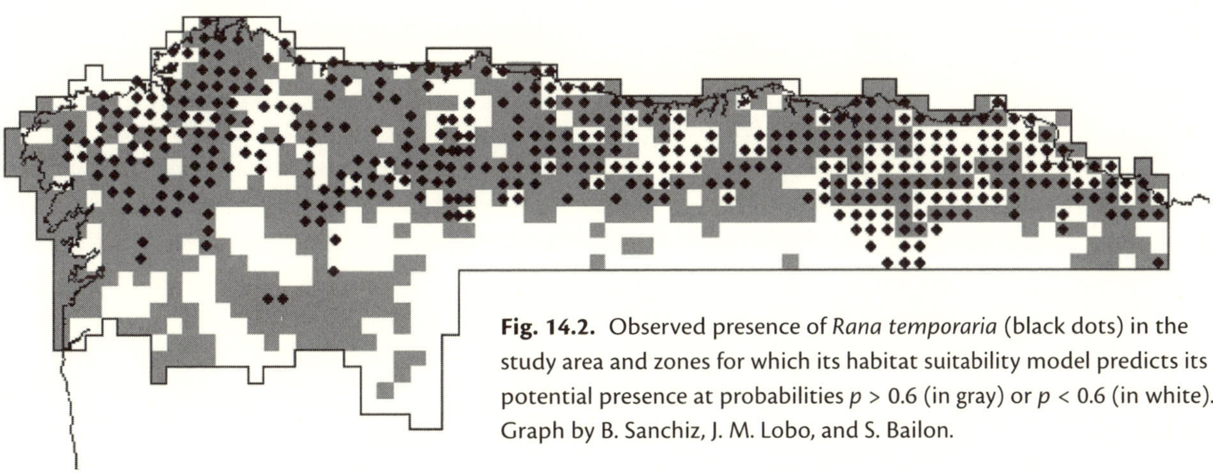

Fig. 14.2. Observed presence of *Rana temporaria* (black dots) in the study area and zones for which its habitat suitability model predicts its potential presence at probabilities $p > 0.6$ (in gray) or $p < 0.6$ (in white). Graph by B. Sanchiz, J. M. Lobo, and S. Bailon.

Rey's barn owl pellet collection at the Universidad de Santiago, in which there are 14 findspots in Cantabria with remains of grass frog kills that occurred during this period, with a maximum in July.

Known predators of *R. temporaria* include some snakes (particularly during mating season), occasionally diurnal birds and mammals, and especially nocturnal raptorial birds. In the El Mirón case, and following the criteria given by Pinto Llona and Andrews (1999), a large part of the material seems to have come from barn owl pellets, but there are also clearly some bones that, because of their types of breakage and digestion traces, must be attributed to predation by other owls and perhaps even by mammals. Some of the remains present a characteristic kind of alteration in the form of deep arabesque etchings that might be attributed to small rootlets (Y. Fernández-Jalvo, pers. comm., 2003).

Anura indet

This group includes poorly preserved fragments that most likely belong to either *Bufo bufo* or *Rana temporaria*. In the samples that we have studied there is no evidence of any amphibian species different from the ones already mentioned above.

Family Anguidae

Anguis (Linnaeus, 1758)

Anguis fragilis (Linnaeus, 1758)

The slow worm usually buries among dead leaves or under stones, superficially digging in loose soils, but it is not considered able to contaminate deep strata. This species is active between March and October and is nocturnal (or crepuscular) in its habits. Among its known predators are (sporadically) certain snakes (*Vipera seoanei*), birds of prey (e.g., *Buteo buteo*, *Strix alauco*), and mammals (*Lutra lutra*, *Martes martes*, *Meles meles*, *Sus scrofa*, *Vulpes vulpes*). We cannot determine the taphonomic origin of the slow worm remains in El Mirón (figure 14.3).

Family Colubridae

Coronella (Laurenti, 1768)

Coronella sp.

The only two remains of this genus (found in Levels 3.5 and 10.1) cannot be securely identified to the species level. However, they might very possibly pertain to the smooth snake (*Coronella austriaca*), a species that is common in Europe and is present in the Eurosiberian region of the Iberian Peninsula, where it inhabits humid areas with annual precipitation over 600–800 mm (Pleguezuelos, Márquez, and Lizana 2002). The two European species of this genus (*C. austriaca* and *C. girondica*) are active between March and October.

Zanemis (Wagler, 1830)

cf. *Zanemis longissimus* (Laurenti, 1768)

The only fragment of this genus (from Level 4) possibly belongs to the Aesculapian snake (*Z. longissimus*), since the only other similar Iberian species (*Rhinechis scalaris*) is a Mediterranean form that does not currently live in the Cantabrian region (Pleguezuelos and

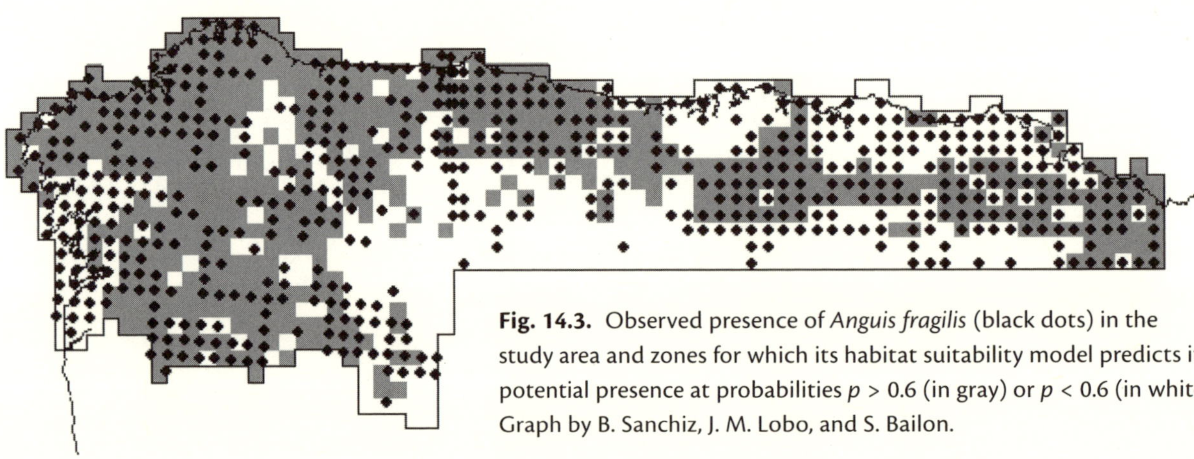

Fig. 14.3. Observed presence of *Anguis fragilis* (black dots) in the study area and zones for which its habitat suitability model predicts its potential presence at probabilities $p > 0.6$ (in gray) or $p < 0.6$ (in white). Graph by B. Sanchiz, J. M. Lobo, and S. Bailon.

Honrubia 2002). However, this identification cannot be confirmed morphologically. These two species were previously included in the genera *Elaphe* and exhibit a close vertebral morphology.

Natrix (Laurenti, 1768)

Natrix natrix (Linnaeus, 1758)

The grass snake is a basically diurnal species that is active mainly between March and October (Braña 1998). It is prey to diurnal raptorial birds (e.g., *Circaetus gallicus*, *Milvus milvus*, *M. migrans*) and mammals (e.g., *Lutra lutra*, *Martes foina*, *Meles meles*).

Natrix sp.

It is very possible that all this material may also pertain to the grass snake (*N. natrix*), although it cannot be unequivocally identified to the specific level.

Family Viperidae

Vipera (Laurenti, 1768)

Vipera sp.

There is only one bone (a vertebral fragment from Level 9.6) that attests to the presence of the viper at this site, although it is not clearly identifiable to the species level. At the present time there are two species that live together in part of Cantabria: *Vipera aspis* and *V. seoanei* (although they may not be strictly syntopic). This region also borders to the south with the distribution of *V. latasti*; thus a fairly short radius around the site may include the range of all three of the Peninsular vipers (Pleguezuelos, Márquez, and Lizana 2002). All the Iberian species of *Vipera* are diurnal and active between April and October.

ABSENT SPECIES

Negative evidence is rarely admissible in paleontological studies. In this case, however, it is worthwhile to note that we have found no remains of *Discoglossus galganoi*. Among the anurans, this species is consumed by *Tyto alba* relatively often (similarly to *Rana temporaria*). *D. galganoi* has been found at present in at least 11 barn owl pellet spots in Cantabria (J. M. Rey, pers. comm., 2003). Given that this owl was probably the agent of accumulation for most of the remains of *Rana temporaria* in the Holocene deposits of El Mirón Cave, it would be logical to detect the presence of *D. galganoi* in these archaeofaunal assemblages *if* it had been living in the region at the time. It can thus be supposed that this species has been expanding eastward along the Cantabrian region during the Holocene and that it had not reached the Río Asón drainage basin by Bronze Age times. Similar reasoning was used (Sanchiz 1980) in the case of La Paloma Cave in central Asturias, 200 km west of El Mirón. The barn owl today preys in situ at La Paloma on this species, which, however, does not appear in the most recent levels studied in the archaeological site, namely the Magdalenian and Azilian.

TAPHONOMIC NOTES AND INTER-LEVEL COMPARISONS

The above-mentioned annual periods of species activity would seem to indicate that death and subsequent burial of the herpetofaunal remains took place essentially between the months of March and October. All the remains seem to have been the result of animal predation, there being no evidence of human interaction with the herpetofauna.

The small sample size makes it difficult to be sure whether the inter-level differences among identified herpetofaunas are real or not. In the case of the sporadic trace appearances of *Salamandra salamandra*, *Alytes obstetricans*, and the ophidians (table 14.1), it is impossible to establish faunal comparisons. The grass frog, on the contrary, appears in all the Holocene levels. Assuming theoretically (but perhaps not very realistically) that the taphonomic processes acting on the herpetofauna had remained the same among the levels, one could determine whether the absences of *Bufo bufo* and *Anguis fragilis* in any given level are genuine and not simply accidents of sampling. The results of such a statistical determination are, however, not clear in our case, since a chi-square analysis on *Bufo bufo* shows that the expected proportion of minimum numbers of individuals does not differ significantly from the observed MNI. This could point to some possibly true absences. However, in terms of number of identified bones, there is a significant difference ($p < 0.001$). For *Anguis fragilis* the statistical results are the reverse, although in this species the tiny size of the remains could have affected collection, thereby causing a biased impression of (false) species absences. Given all this, we have opted in the ecogeographic section of this chapter to separately analyze two faunal assemblages that cover almost all the excavated levels. These are, on the one hand, the group make up of *Bufo bufo* + *Rana temporaria* (B+R group) and, on the other hand, the group of *Anguis fragilis* + *Bufo bufo* + *Rana temporaria* (A+B+R group).

ZOOGEOGRAPHIC AND PALEOCLIMATIC ANALYSES

The geographic distribution of amphibians and reptiles, given their ectothermic (cold-blooded) metabolism, is heavily conditioned by environmental factors. For this reason, quantitative biogeographic analysis is of great potential archaeozoological interest. In the present case there are limitations deriving mainly from the information provided by distributional atlases. First, the best resolution (Pleguezuelos, Márquez, and Lizana 2002) is the 10 × 10 km UTM grid. This means that each spatial unit is defined by the *average* of the different local values present, which might be quite diverse in a mountainous area. Second, the atlas provides indications about species "presence" or "absence," but it is a virtually certain fact that there are grids with false absences due to incomplete prospection.

The species that we have selected as environmental indicators (*Anguis fragilis*, *Bufo bufo*, and *Rana temporaria*) are easy to find in a survey campaign, a fact that minimizes the "false absence" problem. However, we preferred the strategy of making predictive maps of habitat suitability, based exclusively on verified presences and on global conditions within the study zone—that is, on positive evidence. Figures 14.1–14.3 display the individual cases of *Bufo bufo*, *Rana temporaria*, and *Anguis fragilis*, respectively.

These probabilistic maps can also be made for groups of species, such as our B+R group (places where *Bufo bufo* and *Rana temporaria* potentially coexist) and our A+B+R group (where the former two species can live together with *Anguis fragilis*). These maps can give us a first estimate of the most probable paleoenvironmental conditions. Next we statistically compare the environmental conditions among the places with high and low probabilities of being ideal for these herpetological groups and, in the event that the results are significant, we establish their confidence intervals.

The results of the analysis of the potential joint distribution model for group B+R (from Sanchiz and Lobo, 2006) and group A+B+R, created as if they were virtual single taxa, are presented in table 14.1. The values inferred for B+R indicate that the insolation factor does not differ between places with low and high habitat suitability for this group. However, both the precipitation and altitude factors are significantly greater in the more suitable grids for the joint presence of these two amphibians, while, to the contrary, the temperature factor is lower. In the case of the A+B+R group altitude does not yield significant differences, but precipitation (much greater in the more suitable zones) and temperature and insolation (both smaller in the most suitable zones) do show significant differences.

Table 14.2 shows the confidence intervals of these variables for the habitat suitability analysis as well as their current values in the 10 × 10 km UTM cell corresponding to the El Mirón location.

The provided results show that modern values for precipitation and temperature in the UTM grid cell VN69 lie outside the 95 percent confidence interval of a normal distribution for these variables in the zones whose potential habitat suitability is high, with respect to both faunal groups, B+R and A+B+R (table 14.3). This situation also obtains—albeit at a more marginal level—in the case of the insolation factor for group A+B+R.

If the habitat suitability values inferred for the El Mirón area may be extrapolated to past conditions, these results would allow us to deduce that, for practically all the levels with studied herpetofauna, early to mid-Holocene climatic conditions at the cave were significantly colder and drier than those of the present. The method of numeric biogeography also permits quantification of this paleoenvironmental inference. In this case it assigns annual precipitation values that are never above 1,192 mm for levels with the B+R group or never above 1,303 mm for levels with the A+B+R group and average annual temperatures never above 10.8°C for levels with the B+R group or 11°C for levels with the A+B+R group.

These results are congruent with information derived from the other vertebrates (birds and mammals) from the site (as reported by Elorza [chap. 13], Cuenca Bescós and García Pimienta [chap. 12], and Altuna and Mariezkurrena [chap. 16], this volume). They amount to one more confirmation of the prevailing paleoenvironmental scheme for northern Spain, according to which this area of Cantabria has never been outside of the Eurosiberian biogeographic zone during the course of the Holocene. In fact, it is for the present time that the herpetofaunal data show the highest levels of temperature and precipitation for this area—the closest ever during isotope stage 1 to the Mediterranean biogeographical province.

The scarcity of remains and the low species diversity in the sample that we have been able to identify from the Holocene deposits excavated in El Mirón Cave do not allow for more precise ecogeographic analyses, but their comparison with the Pleistocene levels—whose herpetofaunas have not yet been studied—may provide very interesting results in terms of environmental changes in the Cantabrian mountain zone.

Acknowledgments

We thank Professors M. González Morales and L. G. Straus for having provided the materials that are the subject of this study. Professor G. Cuenca Bescós prepared the material, separating it from the micromammalian remains that she studied; she also provided critical help and information. Dr. Y. Fernández-Jalvo identified the rootlet markings on some of the bones. J. E. González efficiently facilitated our frequent consultations of the comparative collections in the Museo Nacional de Ciencias Naturales (CSIC). J. M. Rey provided critical information from his extensive modern owl pellet collections from northern Spain. The text was translated into English by Professor Straus, who also provided information on the site and feedback on earlier drafts. This study was partially supported by a Spanish Government grant, no. REN 2001–0466 and CGL 2008–03881.

References Cited

Bailon, S.
1999 Diférentiation ostéologique des anoures (Amphibia, Anura) de France. *Fiches d´Ostéologie Animale, Archéologie* 1:41.

Balcells, E.
1975 Observaciones en el ciclo biológico de anfibios de alta montaña y su interés en la detección del inicio de la estación vegetativa. *Publicaciones Centro Pirenaico de Biología Experimental* 7(2):55–153.

Barbadillo, L. J., J. I. Lacomba, V. Pérez-Mellado, V. Sancho, and L. F. López-Jurado
1999 *Anfibios y reptiles de la Península Ibérica, Baleares y Canarias.* Planeta, Barcelona.

Blain, H. A., S. Bailon, and G. Cuenca-Bescós
2008 The Early-Middle Pleistocene palaeoenvironmental change based on the squamate reptile and amphibian proxies at the Gran Dolina site, Atapuerca, Spain. *Palaeogeography, Palaeoclimatology, Palaeoecology* 261:177–192.

Blain, H. A., S. Bailon, G. Cuenca-Bescós, J. L. Arsuaga, J. M. Bermúdez de Castro, and E. Carbonell
2009 Long-term climate record inferred from Early-Middle Pleistocene amphibians and squamate reptile assemblages at the Gran Dolina cave, Atapuerca, Spain. *Journal of Human Evolution* 56:55–65.

Böhme, G.
1977 Zur Bestimmung quartärer Anuren Europas an Hand von Skelettelementen. *Wissenschaftliche Zeitschrift Humbold-Universität Berlin (Math.-Nat. Rehie)* 36(3):283–300.

Braña, F.
1998 *Natrix natrix* (Linnaeus, 1758). In *Fauna Ibérica*, vol. 10, *Reptiles*, edited by A. Salvador, pp. 454–466. Museo Nacional de Ciencias Naturales, CSIC, Madrid.

Clark Labs
2000 Global Change Data Archive. Vol. 3. 1 km Global Elevation Model. CD-Rom, Clark University, Worcester, MA.

Eastman, J. R.
2001 Idrisi 32 release 2. Manual version 32.20. Clark Labs, Clark University, Worcester, MA.

Esteban, M. L., and B. Sanchiz
2000 Differential growth and longevity in low and high altitude *Rana iberica* (Anura, Ranidae). *Herpetological Journal* 10:19–26.

Frost, D. R.
2010. Amphibian species of the world: an online reference. Version 5.4 (8 April 2010). Electronic database accessible at http://research.amnh.org/vz/herpetology/amphibia/. American Museum of Natural History, New York.

Gasc, J. P., A. Cabela, J. Crnobrnja-Isailovic, D. Dolmen, K. Grossenbacher, P. Haffner, J. Lescure, H. Martens, J. P. Martínez Rica, H. Maurin, M. E. Oliveira, T. S. Sofianidou, M. Veith, and A. Zuiderwijk

1997 *Atlas of Amphibians and Reptiles in Europe.* Societas Europaea Herpetologica and Muséum National d'Histoire Naturelle, Paris.

González Morales, M. R., and L. G. Straus

2000a La Cueva del Mirón (Ramales de la Victoria, Cantabria): excavaciones 1996–1999. *Trabajos de Prehistoria* 57(1):121–133.

2000b El depósito arqueológico de la Prehistoria reciente de La Cueva del Mirón (Ramales, Cantabria): bases estratigráficas para una secuencia regional. In *Actas do 3° Congresso de Arqueologia Peninsular*, edited by V. Oliveira Jorge, pp. 49–62. Vol. 4. ADECAP, Porto, Portugal.

Hirzel, A. H., J. Hausser, D. Chessel, and N. Perrin

2002 Ecological-niche factor analysis: how to compute habitat-suitability maps without absence data? *Ecology* 83:2027–2036.

Holman, J. A.

1998 *Pleistocene Amphibians and Reptiles in Britain and Europe.* Oxford University Press, Oxford, UK.

ITGE

1988 *Atlas geocientífico y del medio natural de la comunidad de Madrid.* Medio Ambiente. Instituto Tecnológico GeoMinero de España, Madrid.

Lizana, M.

2002 *Bufo bufo.* In *Atlas y libro rojo de los anfibios y reptiles de España*, edited by J. Pleguezuelos, R. Márquez, and M. Lizana, pp. 103–106. Dirección General de Conservación de la Naturaleza and Asociación Herpetológica Española, Madrid.

Martín, C., and B. Sanchiz

2010 Lisanfos KMS. Version 1.2. Online reference accessible at http://www.lisanfos.mncn.csic.es/. Museo Nacional de Ciencias Naturales, CSIC. Madrid, Spain.

Pérez-Mellado, V., and G. Cortázar

2002 Bases metodológicas del atlas de los anfibios y reptiles de España. In *Atlas y libro rojo de los anfibios y reptiles de España*, edited by J. Pleguezuelos, R. Márquez, and M. Lizana, pp. 21–32. Dirección General de Conservación de la Naturaleza and Asociación Herpetológica Española, Madrid.

Pinto Llona, A. C., and P. J. Andrews

1999 Amphibian taphonomy and its application to the fossil record of Dolina (middle Pleistocene, Atapuerca, Spain). *Palaeogeography, Palaeoclimatology, Palaeoecology* 149:411–429.

Pleguezuelos, J. M., and S. Honrubia

2002 *Elaphe scalaris* (Schinz, 1822). Culebra de escalera. In *Atlas y libro rojo de los anfibios y reptiles de España*, edited by J. Pleguezuelos, R. Márquez, and M. Lizana, pp. 201–283. Dirección General de Conservación de la Naturaleza and Asociación Herpetológica Española, Madrid.

Pleguezuelos, J. M., R. Márquez, and M. Lizana (eds.)

2002 *Atlas y libro rojo de los anfibios y reptiles de España.* Dirección General de Conservación de la Naturaleza and Asociación Herpetológica Española, Madrid.

Salvador, A.

1998 *Reptiles.* Vol. 10 of *Fauna Ibérica*, edited by M. A. Ramos et al. Museo Nacional de Ciencias Naturales (CSIC), Madrid.

Salvador, A., and M. García-París

2001 *Anfibios españoles. Identificación, historia natural y distribución.* Canseco Editores, Talavera de la Reina, Spain.

Salvador, A., and J. M. Pleguezuelos

2002 *Reptiles españoles. Identificación, historia natural y distribución.* Canseco Editores, Talavera de la Reina, Spain.

Sanchiz, B.

1980 Anfibios fósiles de la Cueva de la Paloma (Pleistoceno terminal, Asturias). In *La Cueva de la Paloma (Soto de las Regueras, Asturias)*, edited by M. Hoyos, pp. 101–108. Excavaciones Arqueológicas en España. Ministerio de Cultura, Madrid.

Sanchiz, B., and J. M. Lobo

2006 Restos herpetológicos de niveles auriñacienses de la Cueva de El Castillo (Puente Viesgo, Cantabria). In *En el centenario de la Cueva de El Castillo: El ocaso de los Neandertales*, edited by V. Cabrera, F. Bernaldo de Quirós, and J. M. Maíllo, pp. 403–416. Centro Asociado a la Universidad Nacional de Educación a Distancia en Cantabria, Santander, Spain.

StatSoft Inc.

2003 STATISTICA, data analysis software system and user's manual. Version 6.1. StatSoft Inc., Tulsa, OK.

Straus, L., M. González Morales, W. Farrand, and W. Hubbard

2001 Sedimentological and stratigraphic observations in El Mirón, a Late Quaternary cave site in the Cantabrian Cordillera, northern Spain. *Geoarchaeology* 16:603–630.

Szyndlar, Z.

1984 Fossil snakes from Poland. *Acta Zoologica Cracoviensia* 28(1):1–156.

Table 14.1. Number of fragments (FF) and minimum number of individuals (MNI) for each taxon and Holocene level in El Mirón Cave

Culture[a]	Level	*Salamandra salamandra*		*Alytes obstetricans*		*Bufo bufo*		*Rana temporaria*		Anura indeterminate	*Anguis fragilis*		*Natrix natrix*		*Natrix* sp.		*Coronella* sp.		*Elaphe* sp.		*Vipera* sp.	
		FF	MNI	FF	MNI	FF	MNI	FF	MNI	FF	FF	MNI	FF	MNI	FF	MNI	FF	MNI	FF	MNI	FF	MNI
BRN	03			1	1	8	1	9	3	5	5	1			1	1						
BRN	03.2							2	1		1	1										
BRN	03.4							1	1													
BRN/Fe	03.5					2	1	4	4				1	1			1	1				
BRN	03?							1	1													
CALC	04					2	1	3	1										1	1		
CALC	04-up					1	1	1	1													
CALC	05							6	2		1	1										
CALC	05.1?							1	1	1	7	1										
CALC	06					1	1	24	3		3	1										
CALC	07					1	1	7	2	2	6	1										
NEOL	08	1	1	2	1	10	3	65	7	8	8	1	1	1	2	1						
NEOL	09							42	7	4	1	1										
NEOL	09.6					1	1	45	6	2	2	1									1	1
NEOL	09.7							4	3		1	1										
NEOL	09.8					1	1	8	3	1												
NEOL	10			1	1			27	3	1	4	1										
MESO	10.1									1					1	1	1	1				
FF	**356**	**1**		**4**		**27**		**250**		**25**	**39**		**2**		**4**		**2**		**1**		**1**	
MNI	**84**		**1**		**3**		**11**		**49**			**11**		**2**		**3**		**2**		**1**		**1**

[a]BRN: Bronze Age; Fe: Iron Age; CALC: Calcolithic; NEOL: Neolithic; MESO: Mesolithic

Table 14.2. Environmental values for areas having high ($p > 50$) or low ($p < 50$) probabilities of joint presence of *Bufo bufo* and *Rana temporaria* (group B+R) or *Anguis fragilis, Bufo bufo,* and *Rana temporaria* (group A+B+R), according to the habitat suitability models for each group

Group	Variables	N[a]	Mean	Minimum	Maximum	SD[b]	*T* p[c]
B+R	**Probabilistic suitability ($p > 50$)**						
	precipitation (mm)	640	1,164.42	419.44	2,090.30	349.67	
	temperature (°C)	640	10.68	5.00	14.50	1.86	
	altitude (m)	640	757.65	5.00	1,994.00	416.39	
	Insolation (sunlight hours)	640	2,001.13	1,583.92	2,313.58	98.16	
B+R	**Probabilistic suitability ($p < 50$)**						
	precipitation (mm)	355	1,071.20	399.88	2,336.73	400.72	≤0.001
	temperature (°C)	355	11.54	5.80	14.80	1.49	≤0.001
	altitude (m)	355	590.23	4.00	1,730.00	344.23	≤0.001
	insolation (sunlight hours)	355	1,998.08	1,606.25	2,381.17	82.61	>0.05
A+B+R	**Probabilistic suitability ($p > 50$)**						
	precipitation (mm)	532	1,274.50	604.85	2,090.26	311.82	
	temperature (°C)	532	10.88	5.00	14.80	2.02	
	altitude (m)	532	699.17	4.00	1,994.00	443.82	
	insolation (sunlight hours)	532	1,991.00	1,583.92	2,381.17	103.11	
A+B+R	**Probabilistic suitability ($p< 50$)**						
	precipitation (mm)	463	972.83	399.88	2,336.73	368.00	≤0.001
	temperature (°C)	463	11.13	5.80	14.80	1.46	=0.03
	altitude (m)	463	692.17	6.00	1,916.00	344.60	>0.05
	insolation (sunlight hours)	463	2,010.44	1,606.25	2,307.25	78.34	≤0.001

[a]Number of meshes
[b]Standard deviation
[c]Probability of difference between the means of the two suitability areas (Student *T* test)

Table 14.3. Means and confidence intervals (95 percent) in the high habitat suitability models ($p > 50$) for the joint presence of *Bufo bufo* and *Rana temporaria* (group B+R) or *Anguis fragilis, Bufo bufo*, and *Rana temporaria* (group A+B+R) for some relevant paleoclimatic variables

Group	Variables	Mean	Interval		Present[a]
B+R	**Probabilistic suitability ($p > 50$)**				
	precipitation (mm)	1,164	1,135	1,192	1,450
	temperature (°C)	10.7	10.5	10.8	13.0
A+B+R	**Probabilistic suitability ($p > 50$)**				
	precipitation (mm)	1,274	1,246	1,303	1,450
	temperature (°C)	10.9	10.7	11.1	13.0
	insolation (sunlight hours)	1,991	1,982	2,000	2,003

[a]Average values on the UTM 10 x 10 km square where El Mirón Cave is located

CHAPTER FIFTEEN

Taphonomic Study of the Large and Medium Mammals from the Post-Paleolithic Occupations in El Mirón Cave

Manuel Pérez Ripoll and María Dolores López Gila

TRANSLATED BY LAWRENCE GUY STRAUS

Taphonomic study of the faunal sample from the Mesolithic, Neolithic, Chalcolithic, and Bronze Age levels in El Mirón Cave (Ramales de la Victoria, Cantabria, Spain) has yielded valuable information and served to highlight the differential destruction of the various skeletal elements and their respective parts, among bones of both adult animals and juveniles. Three principal causes that seem to have led to the degradation of the original bone assemblages: human breakage for marrow extraction, human (and livestock) trampling, and breakage by carnivores, notably dog. The study of marks on the bones permits us to characterize the butchering processes that were carried out through time at the site. It also suggests the possibility of intrusions of earlier faunal materials into Bronze Age and late Chalcolithic levels. Finally, this analysis has resulted in a greater, more detailed understanding of the economic transition between the Mesolithic and Neolithic in the Cantabrian region.

The sample of bones studied totals 47,234 items, of which 43,024 are taxonomically unidentifiable (table 15.1) and 4,220 are identifiable (table 15.2). Loose teeth are excluded. We inspected all the bones for marks, which can be separated into those produced by humans during marrow-cracking; tool marks; carnivore gnaw marks, presumably made by dogs; rodent gnaw marks; and burning traces.

We measured the lengths of both identifiable and unidentifiable bone fragments. We did not include whole bones (such as carpals and tarsals, unbroken phalanges, etc.), since they are not fragments, which are the objects of interest in this chapter. The sample of identified and measured bones reached a total of 2,813 items (table 15.2).

Degree of Fragmentation in the Sample

The lengths of bone fragments have been grouped by 1 cm intervals. Among the taxonomically unidentified bones, maximum length attains 8 cm (table 15.3); however, the majority (71 percent, or 30,540 out of a total of 43,024) of these fragments fall into the first length interval, that is, <1 cm. In all the levels these tiny fragments exceed 70 percent of the total, except in Level 10, where they make up only 50 percent of the unidentified items. Starting at 2 cm, the number of bone fragments drops off sharply in all levels (figure 15.1). The overwhelming predominance of these small splinters is mainly the result of trampling, given the repeated concentration of activities within the enclosed space defined by the cave walls. The intensity of human activities is obviously another fundamental factor in the high degree of bone fragmentation.

The distribution of fragments by size is different among the taxonomically identified remains. The

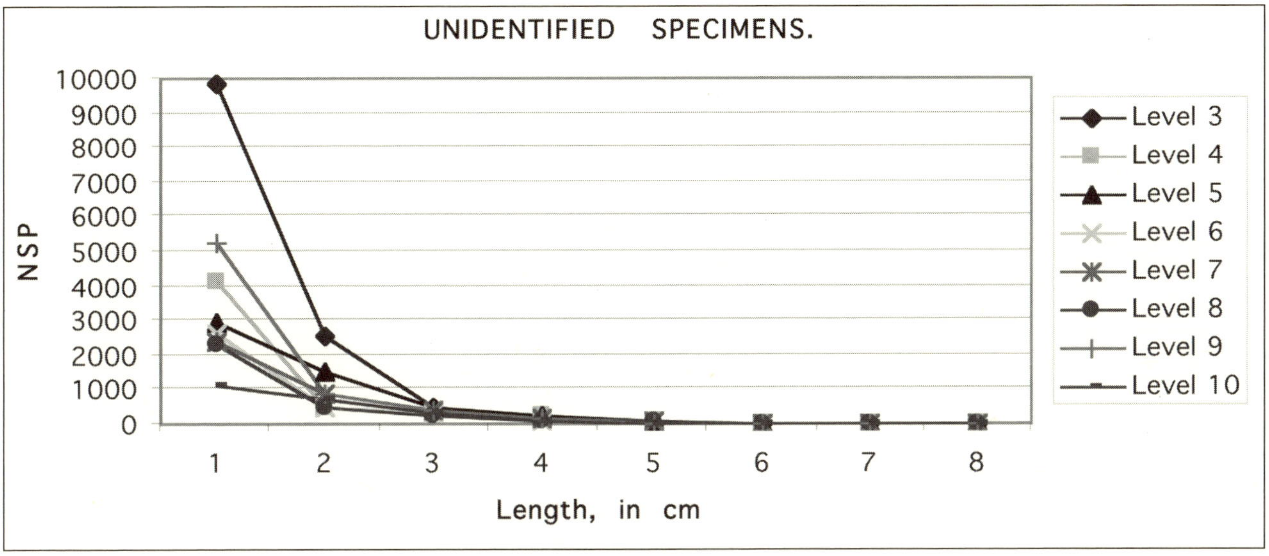

Fig. 15.1. Unidentifiable bone fragments by length. Note that majority are < 1 cm. Graph by M. Pérez Ripoll and M. D. Gila.

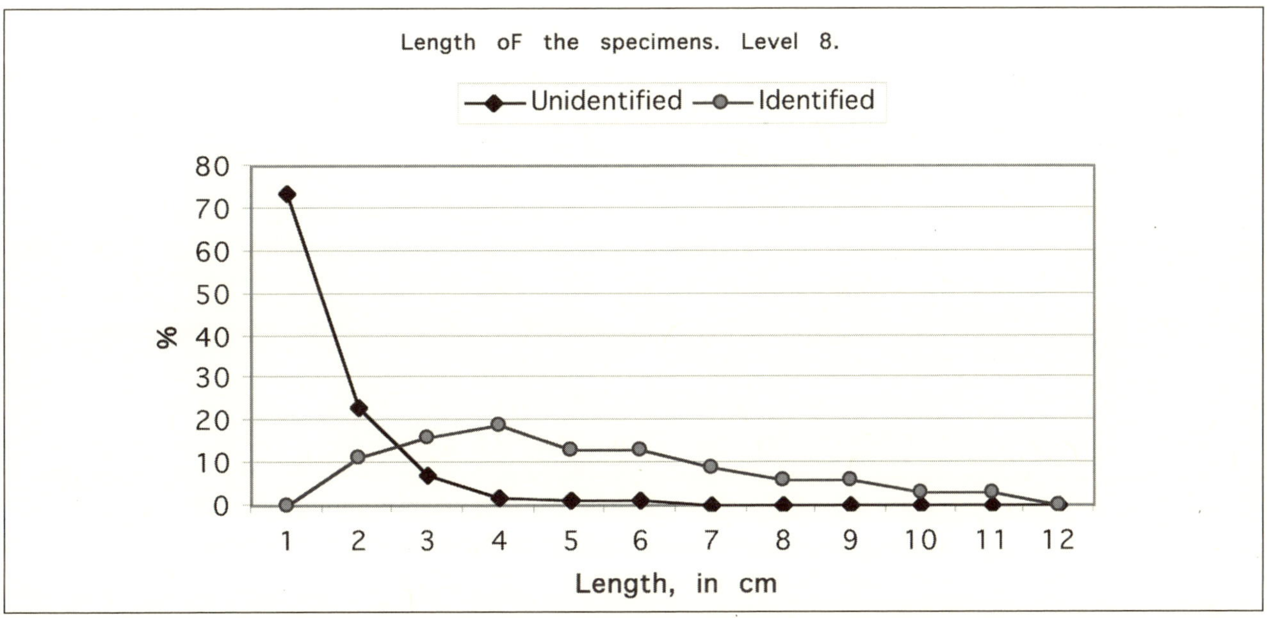

Fig. 15.2. Lengths of identified vs. unidentified bones from Level 8. Note the extreme size difference between the two groups: identifiable fragments generally 3–7 cm long; unidentifiable fragments usually splinters < 1 cm. Graph by M. Pérez Ripoll and M. D. Gila.

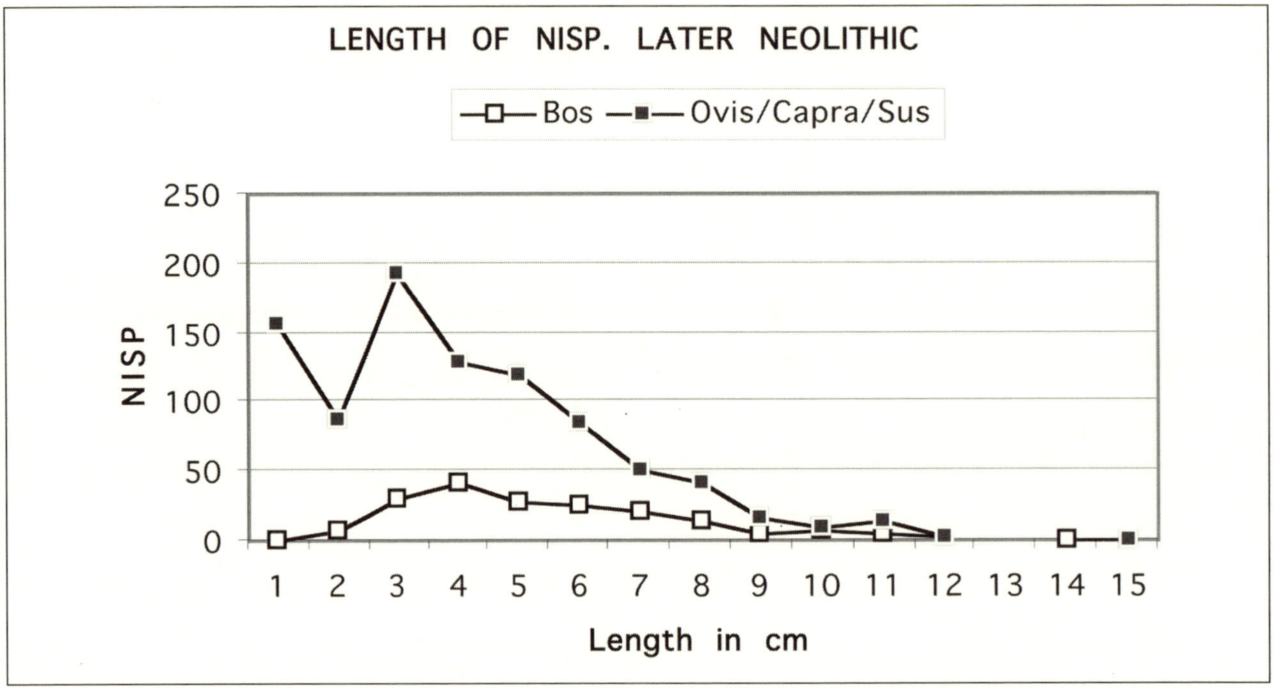

Fig. 15.3. Lengths of unidentifiable *Bos* bones compared to those of *Ovis, Capra,* and *Sus*: the former are 1–6 cm, the latter are 3–8 cm long. Graph by M. Pérez Ripoll and M. D. Gila.

maximum lengths attain 18 cm, but we have simplified the tables by grouping all values of 15 cm or above (table 15.4). On average, the lengths of the identified bones are clearly greater than those of the indeterminate ones (figure 15.2). The bone fragments of cattle (*Bos*) are concentrated in the interval of 3–7 cm, but there are also many that measure 7–14 cm. The bones of goat, sheep, and pig are concentrated between 2 and 6 cm and decrease in number in the interval between 7 and 11 cm (figure 15.3). The size of the identified fragments can be explained by two main causes: first, human breakage for carcass dismemberment and, in the case of *Bos*, for marrow extraction; and second, breakage done by dogs to the bone residues, both from cattle and from the medium-size ungulates (goat, sheep, and pig). These conclusions are corroborated by study of the marks on the bones.

Study of the Marks on the Bones

The distribution of marks on the unidentified bones is quite variable. Cut marks from tools are rare, as are dog bite marks on the bones: the latter range from 0.7 to 12 percent in Levels 3, 4, 5, 7, 9, and 10 and are rather more frequent (24–30 percent) in Levels 6 and 8 (table 15.5). This suggests that dog agency in bone breakage was relatively rare compared with trampling, which was much more significant. On the other hand, bones with traces of burning are variably represented: there are few in Levels 3, 7, 8 , 9, and 10 (5–10 percent) and more in Levels 4, 5, and 6 (20–40 percent). The scarce Mesolithic bone fragments include few or no burned items: there are four (13 percent) from Cabin Level 10.1 but none from Mid-Vestibule Level 304. The scarcity of bones and of burning in the Mesolithic layers is in sharp contrast with the situation in the underlying Magdalenian sequence.

DOG AND RODENT GNAW MARKS

Not only are the identifiable bones larger than the unidentifiable ones, but they also include more bones with gnaw marks. The percentage of cattle bones with gnaw marks, relative to the total number of identifiable bones, ranges from 33 to 46 percent (early Neolithic: 38 percent; later Neolithic: 46 percent; Chalcolithic: 33 percent; Bronze Age: 46 percent). The bones of goat and sheep display higher relative frequencies of gnaw marks, between 38 and 53 percent (early Neolithic: 53 percent; later Neolithic: 53 percent; Chalcolithic: 50 percent; Bronze Age: 38 percent). Pig bones, in contrast, generally have lower values (early Neolithic: 14 percent; later Neolithic: 20 percent; Chalcolithic: 26 percent; Bronze Age: 46 percent) (tables 15.6a–d). The explanation for this difference is so far elusive.

In general, the high number of gnaw marks indicates that dogs were important agents of bone breakage. Such marks are found on all major anatomical elements (tables 15.7a–d).

Among the bones altered by dogs some have traces of corrosion that was probably produced by gastric juices. These are bones that habitually correspond to regurgitated food remains. However, they are quantitatively few (tables 15.6a–d).

There are also a few bones with rodent gnaw marks (tables 15.6a–d). It is frequent to find commensal rodents living alongside humans within food-producing economies. They use bones to sharpen their teeth, often leaving marks.

ANTHROPIC MARKS ON BONES

There are three types of man-made marks on bones: tool cuts, fractures for marrow extraction, and burning traces.

Butchery Marks on Bones of Domesticated Animals

The tool cut marks can be metallic or lithic. The first are found in the Bronze Age and late Chalcolithic levels (4 and 5). The second are found in the early Chalcolithic and Neolithic levels (6 and 7).

The morphology of the cut marks made with metal tools is very different from that of marks made by stone tools. Metal cut marks are deep, long, and rectilinear, with angular edges and very marked, narrow *V*-shaped cross-sections (figure 15.4). There are three types of such marks: (1) those found at joints that are transversal, similar to lithic cut marks, having to do with carcass disarticulation; (2) those located on diaphyses, pelves, ribs, and vertebral spines, which are elongated, oblique, or longitudinal, having to do with defleshing; and (3) those that result from a combination of cutting and chopping, caused by blows dealt with a cutting edge to section long bones at the articular area or to split certain bones, such as vertebrae or ribs to expose their spongy inner structure.

In the bones of bovines, defleshing marks are more abundant than those related to disarticulation, indicating that filleting was a common practice in the butchering of cows (table 15.8). On the contrary, among the bones of goats, sheep, and pigs disarticulation marks outnumber defleshing marks. Many of their anatomical parts—with the exception of the axial skeleton—were not usually defleshed. It is the bones of the axial skeleton that display the highest concentration of cut marks related to meat removal (table 15.9).

The stone tool cut marks are morphologically different from the metal ones. They are not as deep or as straight, but they are wider, their *V* profile is not as marked, and along the edges of the cuts one can see micro-striations made by micro-wear irregularities on the stone cutting edges (Pérez Ripoll 1992). The process of butchery was not much different from that conducted

Fig. 15.4. Photo of Bronze Age *Bos* rib with cut marks from the use of a metal tool. Photo by M. Pérez Ripoll and M. D. Gila.

with metal tools. Defleshing marks are the only ones that have been seen on the cattle bones (table 15.8), while on the bones of goat, sheep, and pig disarticulation marks again predominate (table 15.9). The latter are localized at and near joints, are usually transversal, and are sometimes quite deep, no doubt because of the need to sever tendons and articular ligaments. The defleshing marks are oblique and elongated, being more longitudinal on long bones; their morphology is usually in the form of slight cuts and not in the form of scrape marks. This characteristic differentiates them from bones of the Paleolithic or Mesolithic, since the pre-agricultural (hunted) animal bones have abundant defleshing marks that are often in the form of scraping traces.

We deduce from the study of the butchery cut marks that there were two different processes involved. Defleshing was done to *Bos*, while only the axial skeletons of goat, sheep, and pig were systematically defleshed. Other skeletal elements of the medium-size ungulates were only occasionally defleshed, notably some humeri, radii, pelves, and femora. Possibly the very different body sizes of these animals can explain the differences in butchery practices. The availability of ceramic vessels allowed for cooking the various parts of the different anatomical elements of goat, sheep, and pig—especially young and very young animals, as is the case at El Mirón—along with cereal grains and legumes. It would have been necessary only to section and disarticulate each skeleton element into a size that could be accommodated by the pot and therein cooked. Once the meat was eaten, the bones could be fed to the dogs, and dogs then become the most important agent for bone breakage and alternation, hence the large number of gnaw marks. The anatomical elements of cattle, because of this animal's size, were filleted and the meat then used in stews or preserved by drying or smoking.

With respect to bone breakage, the process of defleshing is connected to the consumption of marrow. The bones of goat, sheep, and pig have few breaks; in addition, the majority of these are concentrated on metapodials and the tibia, especially among remains from the Neolithic, since these bones were chosen to make awls and their breakage does not have to do with marrow or grease extraction (table 15.10). The extraction of marrow is part of a more general process that starts with the dismemberment of carcasses by anatomical units, followed by the defleshing of each of these, the disarticulation and cleaning of the bones, and finally

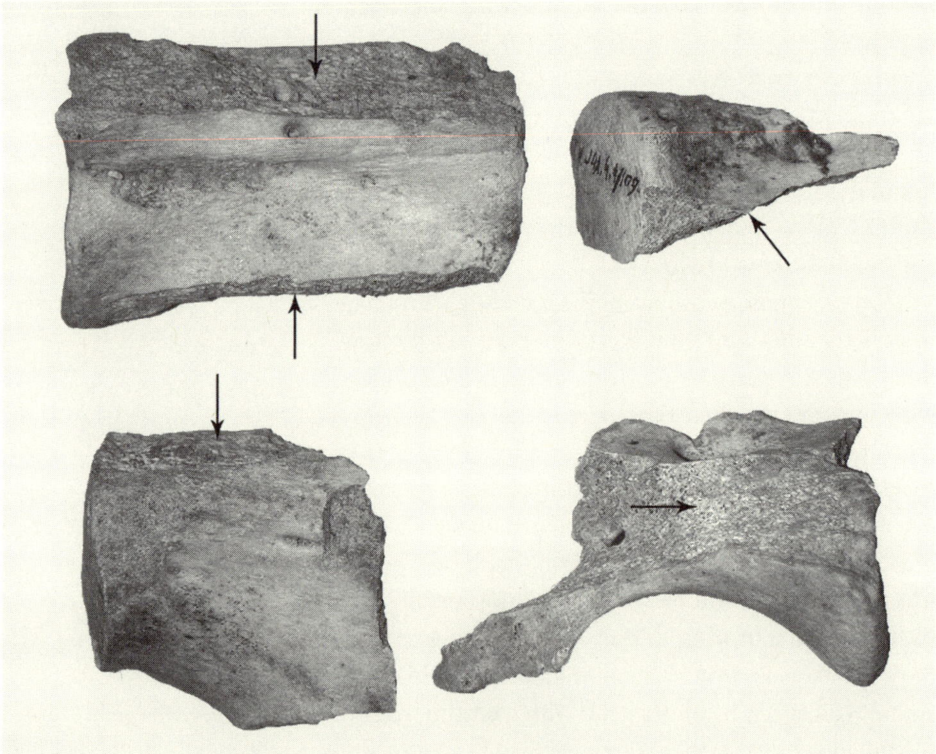

Fig. 15.5.
Photo of *Bos* vertebrae split open with metal tools, possibly axes, from Bronze Age Level 3 (with lenses) and Chalcolithic Levels 4 and 5. Photo by M. Pérez Ripoll and M. D. Gila.

the breaking of the bones. Consequently, the scarcity of breaks on goat, sheep, and pig bones makes sense in terms of the low number of defleshing marks, and both kinds of traces relate to the same pattern of butchering for these three midsize, domesticated ungulates.

Among pig bones from the Bronze Age there are two ribs that had been broken by blows with a metal tool, which caused cuts and fractures. These marks indicate that the ribs were split into portions.

Among cattle bones, fractures for marrow extraction are numerous. They are located both at the articular ends and on the diaphyses (table 15.11). Their existence is related to the high number of defleshing marks, which indicate the use of both meat and marrow, even among bones containing relatively little marrow, such as the first and second phalanges. In the Bronze Age and late Chalcolithic levels (4 and 5) we studied vertebrae, ribs, and the distal end of a humerus that had been split by blows from a hard metal tool, probably an axe; the objective of these fractures is the extraction of marrow from the spongy, cancellous core of these bones (figure 15.5).

The unequal representation of butchery marks between the large cattle on the one hand and the medium-size ungulates on the other is not unique to El Mirón. In the Neolithic site of Cova de l'Or (Valencia, Mediterranean Spain) we have observed that the bones of goats, sheep, and pigs have few defleshing marks but many from disarticulation; the fractures are few, but animal gnaw marks numerous (Pérez Ripoll 1992). In the Neolithic and Chalcolithic levels of Nerja Cave (Málaga, Andalucía) we observe the same distribution (Bernabeu, Barton, and Pérez Ripoll 2001; Bernabeu, Pérez Ripoll, and Valle 1999). This is true even in the Neolithic levels of the settlement at Knossos in Crete (Pérez Ripoll 2002).

Bone Deformation

Some deformations of bones are not the direct consequence of human intervention. Protection measures taken by humans for their livestock to avoid losses to predators such as wolves have as a consequence the survival of old animals and those with deformities (i.e., handicapped animals), which under natural conditions would have been easy prey for carnivores (Pérez Ripoll 2001). In Level 3.5 (Bronze Age or later) there is a first phalange of *Bos* with an exostosis on the proximal articulation, something that is normally produced with old age and by continual application of force throughout the animal's long life (figure 15.6). This pathology indicates that the individual was used as a draft animal. Such deformations have been identified at various sites throughout the Iberian Peninsula beginning in the Chalcolithic, linked to the introduction of the plow (Pérez Ripoll 1999). In ethnoarchaeological studies we conducted in the village of Pomakos, Greece, we observed this type of deformity among bones of cattle that during their lives had been used to pull a plow (Pérez Ripoll, unpublished data).

Fig. 15.6. Photo of Bronze Age and Chalcolithic first phalanges of *Bos*. The phalanx on the left has an osseous deformation in the proximal, articular zone, while the rest were split for marrow extraction. Photo by M. Pérez Ripoll and M. D. Gila.

Butchery Marks on the Bones of Wild Animals

Bones of wild animals are analyzed here separately because of the special information that they can yield. The cut marks on bones of red deer and ibex from Bronze Age levels are more numerous than the marks on the bovine and sheep/goat bones (7 percent for red deer and 9 percent for ibex versus 2 percent for cattle and 5 percent for sheep/goat) (table 15.12). To these differences of frequency we can add information on the morphology of the marks. The cut marks on the red deer and ibex were made by lithic tools—not metal tools, as is the case among the bones of domesticated animals from the same levels. In addition, most of the marks on the wild animals are longitudinal scrapes. The percentage of marked bones in Chalcolithic Level 5 is high (40 percent of the red deer bones), in contrast to the low percentages for Levels 6 and 7. In the later Neolithic the cut marks are few, but in the early Neolithic they are relatively numerous, especially on bones of wild animals (5 percent of the red deer bones and 18 percent of the ibex bones, versus 2 percent of the sheep/goat bones and 1.5 percent of the cattle bones). Defleshing marks—particularly from scraping—are predominant, especially in the early Neolithic (table 15.12).

Something similar is found among bone fractures. In the Bronze Age and combined late Chalcolithic Levels 4 and 5, fractures are common among ibex bones (21 percent and 35 percent, respectively) and for red deer (24 percent and 38 percent, respectively). In contrast to these high values for wild animals, fractured cattle bones are relatively few (9 percent in the Bronze Age and 7.6 percent in the late Chalcolithic) and even rarer among the combined group of sheep, goats, and pigs, for which they never exceed 0.5 percent (table 15.13). The majority of the fractures occur on foot bones: metapodials, phalanges, and even a calcaneus. The latter elements contain very little marrow, yet they were split. This kind of exhaustive use of bones for marrow does not seem to make sense in the heavily pastoral/agricultural context of these levels. Because of this disconnect, we wonder whether the original provenance of the wild and domesticated animal bones was the same or whether the wild animal bones might somehow be intrusive. In this regard, we note the discovery among the bone splinters of an antler *sagaie* base in Level 3.5 (a large trench that cuts into—and thus postdates—Bronze Age Level 3). In Level 3 we found part of a red deer metatarsal with deep grooves no doubt gouged for the extraction of bone tool blanks ("groove-and-splinter" technique). The color and texture of the sedimentary matrix still adhering to the wild animal bones with scrape marks and fractures are the same as the dirt adhering to the sagaie, which also suggests that *some* of the red deer and ibex bones may have been intrusive in these late levels. (A few other sagaie fragments and a straight-backed, stiletto-like stone point, classifiable as a Gravette, were also found in situ in the Chalcolithic levels. The presumption is that Magdalenian levels, once existing near the surface of the slope at the rear of the vestibule, had been disturbed and some of their contents eroded downward and westward

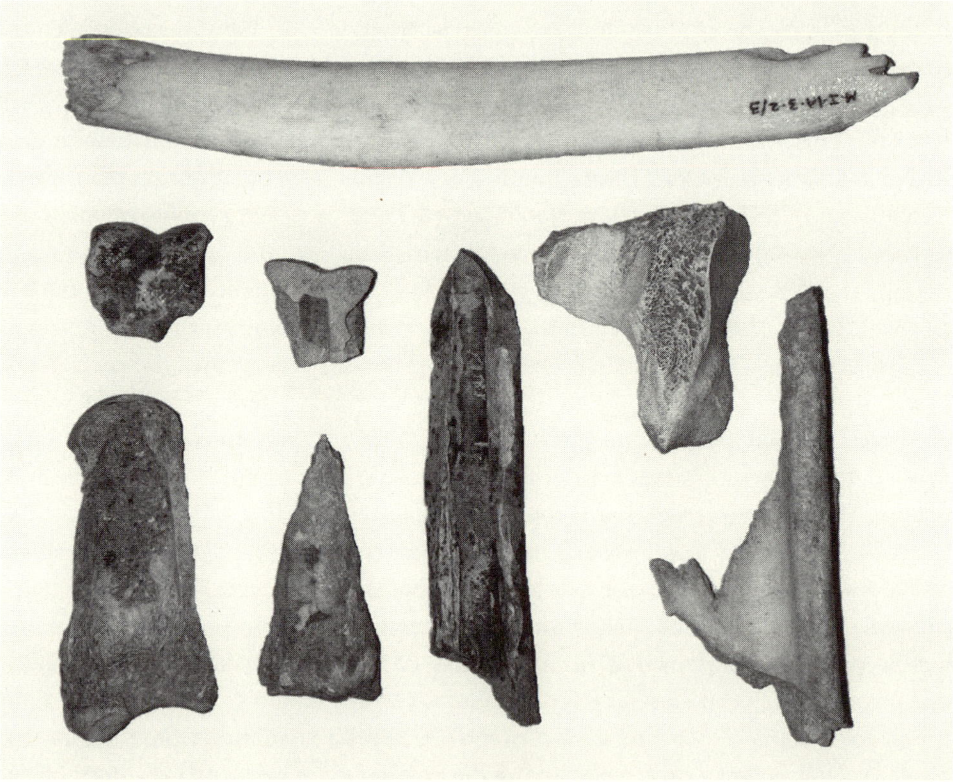

Fig. 15.7. Red deer bones from Bronze Age levels: rib (*top*), fragments of a vertebra and scapula (*right*), all bitten by dogs. Red deer phalanx fragments and ibex metapodial fragments that were fractured for marrow extraction (*center* and *left*); the structural condition and color of these bones differ from those of the gnawed bones. Photo by M. Pérez Ripoll and M. D. Gila.

by people and livestock during the Metal Ages. Some of these old bones and artifacts then became incorporated in the recent deposits and features.—LGS and MGM) The remainder of the remains of red deer and ibex really do seem to have come from the Bronze Age and Chalcolithic levels, since they also have high numbers of (presumed) dog gnaw marks and their coloration is the same as that of the domesticated animal bones (figure 15.7).

Levels 6, 7, 8, and 9 (early Chalcolithic and late Neolithic) are characterized by high numbers of gnawed bones of red deer and ibex. These have either no or very few cut marks or fractures. Their coloration is homogeneous with that of the domesticated animal bones.

However, in the early Neolithic levels (9.6, 9.7, and 9.8) of the Cabin area one observes a slight increase in fractured and cut-marked bones. Levels 303.1, 303.2, and 303.3 in the adjacent Mid-Vestibule Trench have few wild animal bones. Level 303 has two red deer bones: a medial proximal fragment of humerus with several scrape marks from defleshing and a lateral proximal fragment of a tibia with many cut and scrape marks from defleshing and a fracture on its lateral/caudal edge. These bones fit in well with those from the earliest Neolithic level in the Cabin (Level 10), all of which are from red deer, with many fractures, especially of metapodials and phalanges (figure 15.8). There are also many (21) red deer bones with gnaw marks. All the broken wild animal bones—both those with cut and defleshing scrape marks and those with gnaw marks—have the same coloration as the bones of domesticated animals from the early Neolithic levels. In addition, it is noteworthy that the bones from Level 10.1, which lacks domesticated animals and ceramics and is Mesolithic, have fractures and cut marks that are very similar to those found in the early Neolithic (table 15.14). This could suggest two alternate hypotheses to explain the marks on the wild animal bones from Neolithic Levels 10 and 303.3–303:

1. That part of the red deer bones actually came from underlying Levels 10.1 and 304–305 (although the latter are virtually devoid of finds in the Mid-Vestibule Trench); or
2. That in the early Neolithic levels we are observing the interrelationship of different cultural and economic influences: some "foreign" (or allochthonous) customs relating to the treatment of domesticated animals, including the dog, which is a significant agent of bone alteration, and others "native" (or autochthonous), linked to a hunting-based economy focused on red deer and accompanied by traditional practices for the full utilization of wild food resources, as based on the defleshing of anatomical units by ancestral methods (such as longitudinal scraping and filleting) and also marrow cracking of both large bones (humerus, tibia, metapodials) and even small ones (first and second phalanges). Perhaps the existence and survival of these patterns

Fig. 15.8. Fragments of first and second phalanges from early Neolithic–age red deer, all of which were fractured for marrow extraction. Photo by M. Pérez Ripoll and M. D. Gila.

of meat and marrow exploitation are proof of the persistence of the activities and behaviors of foraging peoples undergoing a process of acculturation into the Neolithic way of life (albeit tardily in Cantabrian Spain vis-à-vis the Mediterranean zone and even the nearby upper Ebro Basin).

In purely foraging contexts in Spanish Mediterranean sites, defleshing marks are very numerous on bones of ibex and red deer. Almost all the bones are broken for marrow extraction, even the phalanges (Aura et al. 2002; Nadal 1998; Pérez Ripoll and Martinez Valle 2001). The types of stone tool marks (especially scraping) and the types of fractures (Pérez Ripoll 1992) are the same as those that we observe on the bones from the Bronze Age (intrusive?) and from the early Neolithic levels (non-intrusive?) at El Mirón, indicative of butchering practices different from those that are characteristic of food production economies in Spain.

Traces of Burning

Traces of burning on the bones from the Holocene levels at El Mirón are generally rare. They are very scarce in the Bronze Age levels (1–2 percent of the taxonomically identified bones), more numerous in the Chalcolithic (11–25 percent), again very scarce in the late Neolithic (1–2 percent), and relatively scarce in the earlier Neolithic (4–10 percent). The explanation must lie in the mode of food preparation. In Spanish Mediterranean sites the direct action of fire is quite manifest in the remains from foraging societies: meat roasting on coals, placement of bones directly on the coals to facilitate breakage, smoking of meat, and so on. However, in the context of food production economies the cooking of foods is more often done in ceramic receptacles, which eliminates direct contact of meat and bones with the fire. This could explain the low levels of burning on the bones from the post-Mesolithic occupations at El Mirón. Unfortunately, the number of bones from the Mesolithic levels in this cave is so small as to obviate any meaningful direct comparisons.

Representation of Long Bone Parts

The study of the preserved bone parts is of great importance in trying to understand the differential destruction of bones.

We group the long bones (humerus, radius, femur, tibia, and metapodials) of goat, sheep, and pig on the one hand and those of cattle on the other, in order to study the preservation of each body part in relation to animal age, expressed simply as juvenile versus adult (or indeterminate). We distinguish among whole bones, proximal ends, fragments of proximal parts, diaphyses, fragments of diaphysis (splinters), distal ends, and fragments of distal parts (table 15.14).

Whole bones of the three medium-size ungulates are relatively abundant: in all the levels they are of young sheep, goats, and pigs—mostly fetal or neonatal and not chewed by dogs. In the Chalcolithic levels there are three whole adult bones, all pig metapodials. The whole long bones of cattle are all from young animals.

The proximal ends of long bones from the three medium-size ungulates are less abundant than are fragments of proximal parts of the same elements, except in the Chalcolithic levels. For cattle they are less abundant in all the levels. This disproportion is explained by the low bone density of those proximal parts whose articular area is spongy (as is true of the proximal humerus, femur, and tibia) relative to the denser, more compact structure in other bones (proximal radius and metapodials). In the former articular areas one finds red marrow, which is an attractive food for dogs; because of their low density, these areas are not difficult to bite and break, thereby producing proximal fragments. In addition, in all these cases bones of young animals predominate, except for the Chalcolithic, where bones of adult, midsize ungulates dominate and where complete proximal heads outnumber proximal fragments.

The presence of whole, "headless" diaphyses (bone tubes) is a peculiarly characteristic result of the agency of carnivores, notably dogs (figure 15.9). These develop when carnivores chew and destroy both articular ends (epiphyses) of a bone, leaving only the diaphysis. The majority of the bone tubes came from young animals in the early Neolithic (74 percent of the total number of diaphyses) and in the later Neolithic (63 percent). The number of whole diaphyses is more balanced in the Chalcolithic (50 percent), and there are fewer in the Bronze Age (44 percent). The abundance of diaphyses of young animals in the Neolithic is indicative of a pattern of selection of young animals for slaughter for meat in

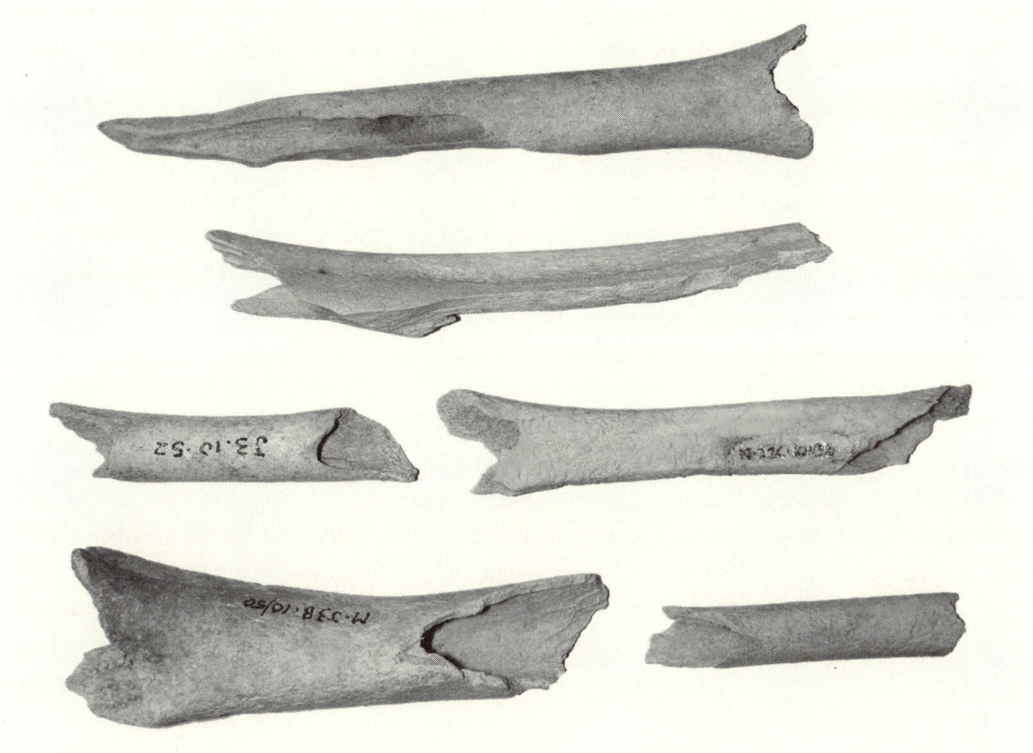

Fig. 15.9.
Sheep and goat long bone diaphyses gnawed by dogs. Photo by M. Pérez Ripoll and M. D. Gila.

this period. The decrease in numbers of young animal bones in the Chalcolithic and the Bronze Age is indicative of the so-called secondary products revolution—an increase in the relative importance of products besides meat, namely milk, wool, and in the case of cattle, traction power, as has also been demonstrated at other sites in Mediterranean Spain as early as the late Neolithic as well as in the Chalcolithic (Pérez Ripoll 1999). All the cattle remains at El Mirón are of young animals, except in the Chalcolithic levels.

Whole distal ends of the three medium-size ungulate taxa always outnumber fragments of distal parts. This is because of the importance of the distal epiphyses of humerus and tibia, which are very dense and resistant to fragmentation. Dogs generally do not break these parts or, for that matter, distal radii. Consequently, once again it is the agency of the dog that explains the differential preservation of distal epiphyses. Among the cattle bones, however, there is greater equilibrium between proximal and distal epiphyses, a fact explained in part by human fracture, which tends to be done near the articular ends but without actually breaking the ends per se.

The taxonomically unidentified bone fragments are very abundant in El Mirón. Their small sizes are evidence of the intensity of repeated human occupations in the cave. This is different from the situation in villages, where activities were much more dispersed through space.

The marks we have observed demonstrate the important role of dogs in breakage and alteration of the identifiable bones. Nonetheless, human responsibility for the final state of the faunal assemblages is also clearly attested, especially in the case of the bovine bones. The cut marks have revealed the existence of different butchering processes: one for *Bos* and the other for goat, sheep, and pig. Among the cattle bones, defleshing marks predominate and marrow-cracking evidence is also abundant. This means that fileting was a common practice in meat consumption, as was marrow exploitation—including both white marrow, which is found within the diaphyses, and red marrow, which is located in the spongy structure of certain articulations and other areas of bone. In contrast, the processing of goats, sheep, and pigs is based on the segmenting of anatomical units by means of disarticulation, preliminary to the cooking of joints (i.e., meat still on the bones). The dogs ate bone remains both of the three medium-size ungulates and of cattle.

The high number of gnawed bones underlines the important role that dog had throughout the whole post-Mesolithic sequence, even though actual remains of this carnivore are quite few in the complete assemblages analyzed by Altuna and Mariezkurrena (chap. 16, this volume). The probable role of dogs in these societies was to guard and control livestock. Gnawed bones are constants within the debris of food-producing economies and are found, for example, both in Near Eastern sites (e.g., Davis 1985, 1991; Helmer 1991; Sana 1999) and in Spanish Mediterranean-zone sites—both in caves (such as Santa Maira, Cendres, Or) and in open-air villages (such as Arenal de la Costa, Jovades) (Bernabeu, Barton, and Pérez Ripoll 2001; Bernabeu, Pérez Ripoll, and Martínez Valle 1999).

The preservation of the various bone parts is the result of differential destruction by dogs and by human actions. In this regard, the distribution of diaphyses is peculiarly characteristic of the agency of dogs when they chew articular parts, leaving behind bone tubes. The survival of bones of young animals is quite high, both among whole bones and among skeletal parts, a fact that serves to evaluate representativeness and differential destruction by age class among the bones of the overall sample.

The butchery marks on red deer and ibex suggest the intrusive character of some of the wild animal bones in the Bronze Age and late Chalcolithic levels (4 and 5)—a possibility that is supported by the discovery of a few antler points (not characteristic artifacts in those periods, but definitely suggestive of a late Upper Paleolithic origin) in these heavily pitted levels within the El Mirón Cabin area. In Levels 6–9 of the Cabin area (early Chalcolithic and late Neolithic), in contrast, wild animal bones are few, but gnaw marks on them are abundant, a fact that is coherent with the domesticated animal bones from the same levels. In early Neolithic Levels 10 and 303.3–303 (in the Cabin and Mid-Vestibule Trench) there are major series of red deer bones with cut and scrape marks and fractures that are not typical of the butchering practices of fully agricultural societies. These bones could be items that were derived from contact with underlying Mesolithic Level 10.1, or they could be due to survival of practices characteristic of forager butchery during a time of contact or acculturation with food-producing systems. It is also possible that the early Neolithic sample from Level 10 was contaminated by a few Mesolithic bones, since, in the first test excavation of 1996, the distinction between Levels 10 and 10.1 was not recognized. On the other hand, the absolute scarcity of bones in Level 10.1 and especially Levels 304–305 makes these Mesolithic-age layers unlikely sources for massive "contamination" of the overlying and faunally richer early Neolithic levels.

References Cited

Aura, J. E., V. Villaverde, M. Pérez Ripoll, R. Martinez Valle, and P. Guillem
2002 Big game and small prey. Paleolithic and Epipaleolithic economy from Valencia (Spain). *Journal of Archaeological Method and Theory* 9:215–268.

Bernabeu, J., C. M. Barton, and M. Pérez Ripoll
2001 A taphonomic perspective on Neolithic beginnings: theory, interpretation, and empirical data in the western Mediterranean. *Journal of Archaeological Science* 28:597–612.

Bernabeu, J., M. Pérez Ripoll, and R. Martínez Valle
1999 Huesos, neolitización y contextos arqueológicos aparentes. In *Actes del II Congrés del Neolític a la Península Ibérica*, edited by J. Bernabeu and T. Orozco, pp. 589–596. Departament d'Arqueología i Prehistoria, Universitat de Valencia, Valencia, Spain.

Davis, S. J. M.
1985 A preliminary report of the fauna from Hatoula: a Natufian-Khiamian (PPNA) site near Latroun, Israel. In *Le site Natufian-Khiamien de Hatoula, près de Latroun, Israel*, edited by M. Lechevalier and A. Ronen, pp. 71–98. Centre de Recherche Français de Jerusalem, Jerusalem.
1991 When and why did prehistoric people domesticate animals? Some evidence from Israel and Cyprus. In *The Natufian Culture in the Levant*, edited by O. Bar-Yosef and F. R. Valla, pp. 101–390. Archaeological series 1. International Monographs in Prehistory, Ann Arbor, MI.

Helmer, D.
1991 Étude de la faune de la phase IA (Natoufien final) de Tell Mureybet (Syrie), fouilles Cauvin. In *The Natufian Culture in the Levant*, edited by O. Bar-Yosef and F. R. Valla, pp. 359–370. Archaeological series 1. International Monographs in Prehistory, Ann Arbor, MI.

Nadal, J.
1998 Les faunes del Pleistocè Final-Holocène a la Catalunya meridional i de ponent. interpretacions tafonòmiques i paleoculturals. Doctoral thesis, Universitat de Barcelona, Barcelona.

Pérez Ripoll, M.
1992 *Marcas de carnicería, fracturas intencionadas y mordeduras de carnívoros en huesos Prehistóricos del Mediterráneo español.* Instituto de Cultura Juan Gil-Albert, Alicante, Spain.
1999 La explotación ganadera durante el III milenio a.C. en la Península Ibérica. In *Actes del II Congrés del Neolític a la Península Ibérica*, edited by J. Bernabeu and T. Orozco, pp. 95–103. Departament d'Arqueología i Prehistoria, Universitat de Valencia, Valencia, Spain.
2001 El proceso de domesticación animal en el Próximo Oriente. Planteamiento y evolución. *Archivo de Prehistoria Levantina* 24:65–96.
2002 Extinción de la fauna endémica y colonización humana de las grandes islas del Mediterráneo. *Saguntum PLAV* Extra 5:147–163.

Pérez Ripoll, M., and R. Martinez Valle
2001 La caza, el aprovechamiento de las presas y el comportamiento de las comunidades cazadoras prehistóricas. In *De Neandertales a Cromañones. El inicio del poblamiento humano en las tierras valencianas*, edited by V. Villaverde, pp. 73–98. Departament d'Arqueología i Prehistoria, Universitat de Valencia., Valencia, Spain.

Sana, M.
1999 *Arqueología de la domesticación animal. La gestión de los recursos animales en Tell Halula (Valle del Eufrates-Siria) del 8.800 al 7.000 BP.* Universitat Autònoma de Barcelona, Barcelona.

Table 15.1. Unidentified specimens by level

	Level 3	Level 4	Level 5	Level 6	Level 7	Level 8	Level 9	Level 10	Total
Burns	839	1,596	2,093	730	292	260	342	246	**6,398**
Burns + cuts	2	0	1	0	1	0	0	0	**4**
Burns + gnawing	6	2	12	0	12	0	0	2	**34**
Gnawing	98	716	120	1,076	494	776	508	103	**3,891**
Gnawing + cuts	3	0	1	0	3	0	0	0	**7**
Cuts	46	25	4	2	2	2	1	0	**82**
No marks	12,022	2,904	2,958	1,751	3,148	2,085	5,843	1,897	**32,608**
TOTAL	**13,016**	**5,243**	**5,189**	**3,559**	**3,952**	**3,123**	**6,694**	**2,248**	**43,024**

Table 15.2. Identified number of individual species (NISP) (teeth not included)

	Bronze Age				Chalcolithic					Later Neolithic				Early Neolithic			
Levels	3	3.5	301	Total	4	5	6	7	Total	8	9	9.6–9.8	Total	10	303	303.1–303.3	Total
Bos taurus	134	49	3	186	165	173	91	57	486	69	58	117	244	7	44	15	66
Ovis/Capra	90	20	62	172	76	162	60	242	540	245	319	488	1,052	123	199	285	607
Capra hircus	30	2	6	38	8	2	0	1	11				0				0
Ovis aries	12	0	3	15	15	3	5	20	30	9	33	69	111	8	16	24	48
Sus domesticus	28	14	8	50	18	74	29	43	164	24	26	49	99	9	9	11	29
Canis familiaris	0	0	0	0	0	0	0	0	0	1	0	1	2				
Total domesticated	294	85	82	461	282	414	185	363	1,231	348	436	724	1,508	147	268	335	750
Capra pyrenaica	29	17	14	60	6	11			17					3	7	1	11
Rupicapra ru.						1			1	1			1				
Cervus elaphus	8	7	18	33	3	10			13	9	46	59	114	85	2	9	94
Capreolus capreolus			1	1													
Sus scrofa						1	1		2	4			4				
Equus sp		1	2	3													
Ursus arctos									1								
Felis silvestris			1	1						3			3				
Martes sp.										1			1				
Total wild	37	25	36	98	9	23	1		34	18	46	59	123	88	9	10	105
TOTAL																	4,220

Table 15.3. Unidentified specimens by length

Length (cm)	1	2	3	4	5	6	7	8	Total
Level 3	9,823	2,553	459	122	37	16	5	1	**13,016**
Level 4	4,134	495	309	226	50	22	4	3	**5,243**
Level 5	2,917	1,484	477	206	65	26	8	6	**5,189**
Level 6	2,583	494	285	112	58	21	6	0	**3,559**
Level 7	2,416	869	392	160	79	27	8	1	**3,952**
Level 8	2,303	464	217	80	46	11	2	0	**3,123**
Level 9	5,227	858	434	135	26	13	1	0	**6,694**
Level 10	1,137	713	278	89	22	9	0	0	**2,248**
TOTAL	**30,540**	**7,930**	**2,851**	**1,130**	**383**	**145**	**34**	**11**	**43,024**

Table 15.4. Identified specimens, number of individual species (NISP) by length

Length (cm)	1	2	3	4	5	6	7	8	9	10	11	12	13	14	15	Total
Bronze Age, Levels 3, 3.5, 301																
Bos taurus	4	4	17	20	22	15	11	13	8	4	5	7	1	1	8	**140**
Ovis, Capra, Sus	32	38	21	12	10	8	8	8	4	1	1					**143**
Cervus, Capra pyrenaica	1	10	6	5	4											**309**
Total	**36**	**42**	**38**	**32**	**32**	**23**	**19**	**21**	**12**	**5**	**6**	**7**	**1**	**1**	**8**	**283**
Chalcolithic, Levels 4, 5, 6, 7																
Bos taurus	0	23	53	41	46	31	30	22	17	19	7	9	6	9	19	**332**
Ovis, Capra, Sus	31	67	61	89	56	55	43	31	17	13	12	4	2	1	1	**483**
Cervus, Capra pyrenaica	2	3	3	1	4	4	1									**18**
Total	**31**	**90**	**114**	**130**	**102**	**86**	**73**	**53**	**34**	**32**	**19**	**13**	**8**	**10**	**20**	**833**
Late Neolithic, Levels 8, 9, 9.6, 9.7, 9.8																
Bos taurus	1	7	29	41	27	26	21	14	4	6	4	3		1	1	**185**
Ovis, Capra, Sus	155	88	192	128	119	84	50	42	16	9	14	2			1	**900**
Cervus, Capra pyrenaica	0	13	18	16	10	8	5	7	6	1		1				**85**
Total	**156**	**95**	**221**	**169**	**146**	**110**	**71**	**56**	**20**	**15**	**18**	**5**	**0**	**1**	**2**	**1,170**
Early Neolithic, Levels 10, 303, 303.1, 303.2, 303.3																
Bos taurus	1	1	2	3	7	5	8	1	3	0	1					**32**
Ovis, Capra, Sus	52	80	92	73	60	28	17	13	6	6	1	1				**429**
Cervus, Capra pyrenaica	3	12	17	6	7	7	2	3	6		1	1			1	**66**
Total	**53**	**81**	**94**	**76**	**67**	**33**	**25**	**14**	**9**	**6**	**2**	**1**			**1**	**527**
TOTAL IDENTIFIED SPECIMENS																**2,813**

Table 15.5. Unidentified specimens and marks by length intervals

Length (cm)	1	2	3	4	5	6	7	8	Total
Bronze Age, Level 3									
Burns	662	144	27	5	1				839
Burns + cuts	0	0	2	0					2
Burns + gnawing	0	3	0	3					6
Gnawing	6	38	27	14	7	5	1		98
Gnawing + cuts	0	1	1	0	1				3
Cuts	1	18	12	12	3				46
No marks	9,154	2,349	390	88	25	11	4	1	12,022
TOTAL	9,823	2,553	459	122	37	16	5	1	13,016
Chalcolithic, Levels 4, 5, 6, 7									
Burns	3,403	883	267	106	33	15	9	2	4,718
Burns + cuts			1		1				2
Burns + gnawing		4	6	6	6	4			26
Gnawing	1,345	435	306	198	72	42	6	1	2,405
Gnawing + cuts			2	2					4
Cuts	4	6	9	9	3	0	1	1	33
No marks	7,298	2,014	872	383	137	35	17	5	10,761
TOTAL	12,050	3,342	1,463	704	252	96	33	9	17,949
Later Neolithic, Levels 8, 9									
Burns	567	29	5	1					602
Burns + cuts									0
Burns + gnawing									0
Gnawing	756	277	161	60	24	6	1		1,285
Gnawing + cuts									0
Cuts		1		1	1				3
No marks	6,207	1,015	485	153	47	19	2		7,928
TOTAL	7,530	1,322	651	215	72	25	3		9,818
Early Neolithic, Level 10									
Burns	112	98	26	4	4	2			246
Burns + cuts									0
Burns + gnawing		1	1						2
Gnawing	28	49	20	5	1				103
Gnawing + cuts									0
Cuts									0
No marks	997	565	231	80	17	7			1,897
TOTAL	1,137	713	278	89	22	9			2,248

Table 15.6a. Bronze Age number of individual species (NISP) and number of marks

Level	3						3.5					301					
	Total	Fr	Cu	Gn	Gr	Bu	Total	Fr	Cu	Gn	Bu	Total	Fr	Cu	Gn	Gr	Bu
Bos taurus	134	5	4	75	1	3	49	8		12	1	3			1		
Ovis/Capra	90		2	48			20		1	5		62		3	23		2
Capra hircus	30		2	2			2					6		1	3		
Ovis aries	12		2	4		3	0					3					1
Sus domesticus	28		2	14			14		2	6	1	8			1		
Total domesticated	294	5	12	143	1	6	85	8	3	23	2	82		4	28		3
Equus							1					2					
Capra pyrenaica	29	3	1				17	7		1		14	3	4	2		
Cervus elaphus	8	2	1	3			7	2	1	1		18	4	3	7	1	1
Capreolus capreolus												1		1			
Felis sp.												1					
TOTAL WILD	37	5	2	3	0	0	25	9	1	2	0	36	7	8	9	1	1

Note: Types of marks: Fr = produced by humans during marrow-cracking; Cu = tool marks; Gn = carnivore gnaw marks, presumably made by dogs; Gr = rodent gnaw marks; Bu = burning traces.

Table 15.6b. Chalcolithic number of individual species (NISP) and number of marks

Level	4					5					6						
	Total	Fr	Cu	Gn	Bu	Total	Fr	Cu	Gn	Bu	Total	Fr	Cu	Gn	Dg	Gr	Bu
Bos taurus	165	12	13	67	9	173	12	10	32	32	91	5	26	30		1	16
Ovis/Capra	76		7	12	7	162	3	3	44	23	60		5	21	4		9
Capra hircus	8			3	1	2			1	1	0						
Ovis aries	2		1			3		2	1	2	5		1	4			1
Sus domesticus	18		1	3	5	74	2	1	13	10	29		1	8			
Canis familiaris																	
Total Domesticated	269	12	22	85	22	414	17	16	91	68	185	5	33	63	4	1	29
Equus						1					1						1
Capra pyrenaica	6	2		1		11	4										
Rupicapra ru.						1											
Cervus elaphus	3			1	1	10	4	4	1								
Ursus arctos	1			1													
Vulpes vulpes																	
TOTAL WILD	10	2		3	1	24	8	4	1	1	1						

Note: Types of marks: Fr = produced by humans during marrow-cracking; Cu = tool marks; Gn = carnivore gnaw marks, presumably made by dogs; Gr = rodent gnaw marks; Bu = burning traces; Dg = digestion traces.

Table 15.6c. Chalcolithic (Level 7) number of individual species (NISP) and number of marks

	Total	Fr	Cu	Gn	Dg	Gr	Bu
Bos taurus	57	8	5	34		1	9
Ovis/Capra	242		4	195	3		18
Capra hircus	1			1			
Ovis aries	20		1	17	2	1	2
Sus domest.	43		3	27			1
TOTAL DOMESTICATED	364	8	13	274	5	2	30

Note: Types of marks: Fr = produced by humans during marrow-cracking; Cu = tool marks; Gn = carnivore gnaw marks, presumably made by dogs; Gr = rodent gnaw marks; Bu = burning traces; Dg = digestion traces.

Table 15.6d. Early Neolithic number of individual species (NISP) and number of marks

Level	10							303						303.1–303.3					
	Total	Fr	Cu	Gn	Dg	Gr	Bu	Total	Fr	Cu	Gn	Gr	Bu	Total	Fr	Cu	Gn	Gr	Bu
Bos taurus	7							44			17		2	15		1	7		1
Ovis/Capra	123	1		82	1		2	199		3	138		9	284		5	104	1	22
Ovis aries	8			6				16	1	3	3	1	3	24	1		12		2
Sus domesticus	9			1			2	9		1	3	1	1	11					
Total domesticated	147	1		89	1		4	268	1	7	161	2	15	334	1	6	123	1	25
Capra pyrenaica	3			1			4	7		2	3			1					
Cervus elaphus	85	13		21	1	1		2	1	2				9	1		3		3
TOTAL WILD	88	13		22	1	1	4	7	1	4	3			10	1		3		3

Note: Types of marks: Fr = produced by humans during marrow-cracking; Cu = tool marks; Gn = carnivore gnaw marks, presumably made by dogs; Gr = rodent gnaw marks; Bu = burning traces; Dg = digestion traces.

Table 15.7a. Bronze Age anatomical parts (number of gnawed specimens)

	Skull	Md-Mx	Axial	Front legs	Rear legs	Phalanges	Total
Level 3							
Bos taurus	31 (12)		33 (25)	33 (18)	22 (11)	15 (9)	134 (75)
Ovis/Capra	10	9 (4)	15 (12)	19 (10)	26 (18)	11 (4)	90 (48)
Capra hircus				1 (1)	1 (1) + 12	16	30 (2)
Ovis aries				9 (4)	2	1	12 (4)
Sus domesticus	2 (1)	3 (1)	3	7 (5)	9 (5)	4 (2)	28 (14)
Capra pyrenaica	1		1	9	7	11	29
Cervus elaphus		1	1 (1)	2 (1)	2 (1)	2	8 (3)
TOTAL LEVEL 3	**44 (13)**	**13 (5)**	**53 (38)**	**80 (39)**	**82 (36)**	**60 (15)**	**331 (146)**
Level 3.5							
Bos taurus	4 (1)	3 (1)	15 (4)	9 (3)	7 (3)	11	49 (12)
Ovis/Capra				9 (1)	3 (2)	8 (2)	20 (5)
Capra hircus				1		1	2
Sus domesticus	2	5	6 (3)	4 (3)	1		18 (6)
Equus sp.		1					1
Capra pyrenaica				5	4 (1)	8	17 (1)
Cervus elaphus		1	1 (1)		2	3	7 (1)
TOTAL LEVEL 3.5	**6 (1)**	**10 (1)**	**22 (8)**	**28 (7)**	**17 (6)**	**31 (2)**	**114**
Level 300.1							
Bos taurus			1			2 (1)	3 (1)
Ovis/Capra	13 (6)	3 (2)	5 (5)	13 (4)	9 (4)	19 (2)	62 (23)
Ovis aries				2	1		3
Capra hircus	1			2	2	1	6
Sus domesticus			2	2 (1)		4	8 (1)
Equus sp.				1		1	2
Capra pyrenaica				4	5 (1)	5 (1)	14 (2)
Cervus elaphus		1 (1)	1 (1)	3 (1)	6	7 (4)	18 (7)
Capreolus capreolus				1			1
Felis sp.				1			1
TOTAL LEVEL 300.1	**14 (6)**	**4 (3)**	**9 (6)**	**29 (6)**	**23 (5)**	**39 (8)**	**118 (34)**

Table 15.7b. Chalcolithic anatomical parts (number of gnawed specimens)

	Skull	Md-Mx	Axial	Front legs	Rear legs	Phalanges	Total
Level 4							
Bos taurus	37 (8)	14 (9)	62 (38)	22 (5)	10 (5)	20 (2)	165 (67)
Ovis/Capra	3	11	30 (3)	11 (5)	16 (4)	5	76 (12)
Capra hircus	3			3 (1)	2 (2)		8 (3)
Ovis aries				1	1		2
Sus domesticus	3	1	6 (3)	2	3	3	18 (3)
Capra pyrenaica			1 (1)	1	1	3	6 (1)
Cervus elaphus	1		1 (1)		1		3 (1)
Ursus arctos					1 (1)		1 (1)
TOTAL LEVEL 4	**47 (8)**	**26 (9)**	**100 (46)**	**40 (11)**	**35 (12)**	**31 (2)**	**279 (88)**
Level 5							
Bos taurus	35	7	26 (11)	30 (9)	35 (11)	14 (11)	173 (42)
Ovis/Capra	14	12 (7)	31 (17)	19 (11)	23 (7)	7 (2)	162 (44)
Ovis ariess				3 (1)			3 (1)
Capra hircus				1 (1)		1	2 (1)
Sus domesticus	5 (1)	7 (3)	1	15 (6)	17 (3)	8	74 (13)
Capra pyrenaica			1	2	4	3	11
Rupicapra rup							1
Cervus elaphu.		1	1 (1)	1	2	2	10 (1)
Equus sp.							1
TOTAL LEVEL 5	**54 (1)**	**136 (10)**	**60 (29)**	**70 (28)**	**81 (21)**	**35 (13)**	**437 (102)**
Level 6							
Bos taurus	22	9	29 (19)	12 (6)	14 (3)	5 (2)	91 (30)
Ovis/Capra	2	4 (1)	16 (1)	17 (9)	11 (6)	10 (4)	60 (21)
Ovis aries				2 (2)	2 (2)	1	5 (4)
Sus domesticus	5	3	4	8 (4)	5 (1)	4 (3)	29 (8)
Capra pyrenaica					1		1
TOTAL LEVEL 6	**29**	**16 (1)**	**49 (20)**	**39 (21)**	**33 (12)**	**20 (9)**	**186 (63)**
Level 7							
Bos taurus	8 (5)	4 (4)	11 (6)	10 (7)	9 (7)	15 (5)	57 (34)
Ovis/Capra	32 (28)	31 (23)	39 (35)	54 (42)	72 (58)	14 (9)	242 (195)
Ovis aries				11 (9)	5 (4)	4 (4)	20 (17)
Capra hircus				1 (1)			1 (1)
Sus domesticus	6	12 (8)	3 (3)	6 (5)	10 (8)	6 (3)	43 (27)
Canis familiaris			1				1
Cervus elaphus					1		1
Vulpes vulpes			1				1
TOTAL LEVEL 7	**46 (33)**	**47 (35)**	**55 (44)**	**82 (64)**	**97 (77)**	**39 (21)**	**366 (274)**

Table 15.7c. Later Neolithic anatomical parts (number of gnawed specimens)

	Skull	Md-Mx	Axial	Front legs	Rear legs	Phalanges	Total
Level 8							
Bos taurus	12 (7)	2 (1)	17 (9)	12 (3)	15 (10)	11 (7)	69 (37)
Ovis/Capra	36	22 (14)	47 (31)	55 (35)	62 (50)	15 (8)	237 (138)
Ovis aries				3 (2)	6 (2)		9 (4)
Capra hircus				3 (3)	1 (1)		4 (4)
Sus domesticus	2 (2)		1 (1)	7 (4)	8 (6)	6	24 (13)
Canis familiaris			1				1
Sus Scrofa				2 (1)	1	1 (1)	4 (3)
Cervus elaphus	2 (1)		5 (3)		2 (1)		9 (5)
Felis silvestris					3		3
Meles meles		1					1
Rupicapra ru.					1		1
TOTAL LEVEL 8	**52 (10)**	**25 (15)**	**71 (44)**	**82 (48)**	**99 (70)**	**33 (16)**	**362 (204)**
Level 9							
Bos taurus	6 (2)	6 (2)	10 (4)	12 (7)	11 (2)	13 (6)	58 (23)
Ovis/Capra	26 (10)	31 (8)	64 (44)	92 (64)	90 (57)	16 (7)	319 (190)
Ovis aries	1 (1)		1	16 (9)	11 (2)	4	33 (12)
Sus domesticus	3	9		2 (2)	4 (2)	8 (2)	26 (6)
Cervus elaphus	3 (1)		12 (10)	9 (1)	5 (3)	17 (9)	46 (24)
TOTAL LEVEL 9	**39 (14)**	**46 (10)**	**87 (58)**	**131 (83)**	**121 (66)**	**58 (24)**	**482 (255)**
Level 9.6–9.8							
Bos taurus	16 (2)	31 (5)	16 (16)	27 (14)	19 (10)	8 (5)	117 (52)
Ovis/Capra	2 (2)	38 (3)	101 (37)	168 (113)	156 (118)	23 (17)	488 (290)
Ovis aries				22 (12)	46 (6)	1 (1)	69 (19)
Sus domesticus	27	9	2	3 (1)	3	5 (1)	49 (2)
Canis familiaris		1					1
Cervus elaphus		2	13 (6)	19 (5)	13 (1)	12	59 (12)
TOTAL LEVEL 9.6–9.8	**45 (4)**	**81 (8)**	**132 (59)**	**239 (145)**	**237 (135)**	**49 (24)**	**783 (375)**

Table 15.7d. Early Neolithic anatomical parts (number of gnawed specimens)

	Skull	Md-Mx	Axial	Front legs	Rear legs	Phalanges	Total
Level 10							
Bos taurus	2		3	1		1	7
Ovis/Capra	8 (3)	10 (1)	22 (10)	43 (38)	33 (27)	7 (3)	123 (82)
Ovis aries	1			3 (2)	2 (2)	2 (2)	8 (6)
Capra hircus						1 (1)	1 (1)
Sus domesticus	1		2	3 (1)	1	2	9 (1)
Cervus elaphus	5	3 (1)	22 (9)	17 (4)	14 (3)	24 (4)	85 (21)
Capra pyrenaica					1	2 (1)	3 (1)
TOTAL LEVEL 10	**15 (3)**	**13 (2)**	**46 (19)**	**66 (45)**	**51 (32)**	**37 (11)**	**236 (115)**
Level 303							
Bos taurus	7 (2)	2	9 (6)	7 (3)	13 (6)	6	44 (17)
Ovis/Capra	18 (16)	15 (8)	53 (37)	41 (26)	53 (39)	19 (12)	199 (138)
Ovis aries				2	8 (1)	6 (2)	16 (3)
Sus domesticus		2		2 (2)	2 (1)	3	9 (3)
Capra pyrenaica			4 (2)		3 (1)		7 (3)
TOTAL LEVEL 303	**25 (18)**	**19 (8)**	**66 (45)**	**52 (31)**	**79 (48)**	**34 (14)**	**275 (164)**
Level 303.1							
Bos taurus	2		1 (1)	1	3 (2)	1 (1)	8 (4)
Ovis/Capra	12 (2)	7 (2)	40 (14)	34 (14)	48 (15)	14 (5)	155 (52)
Ovis aries	1 (1)			7 (5)	3	5	16 (6)
Capra hircus				1			1
Sus domesticus.	1		1	1	1	3	7
TOTAL LEVEL 303.1	**16 (3)**	**7 (2)**	**42 (15)**	**44 (19)**	**55 (17)**	**23 (6)**	**187 (62)**
Level 303.2							
Bos taurus						3	3
Ovis/Capra	13 (4)	3 (1)	19 (8)	16 (8)	18 (11)	6 (3)	75 (35)
Ovis aries				4 (4)	3 (2)	1	8 (6)
Sus domesticus				1		3	4
Cervus elaphus	1 (1)	1		1	3 (2)	1	7 (3)
Capra pyrenaica						1	1
TOTAL LEVEL 303.2	**14 (5)**	**4 (1)**	**19 (8)**	**22 (12)**	**24 (15)**	**15 (3)**	**98 (44)**
Level 303.3							
Bos taurus			2 (2)		2 (1)		4 (3)
Ovis/Capra	5	2	13 (10)	18 (2)	13 (4)	3 (1)	54 (17)
Cervus elaphus				1		1	2
TOTAL LEVEL 303.3	**5**	**2**	**15 (12)**	**19 (2)**	**15 (5)**	**4 (1)**	**60 (20)**

Table 15.8. *Bos taurus,* distribution of cut marks

	Bronze Age		Chalcolithic		Later Neolithic	Early Neolithic
	Dismemberment	Filleting	Dismemberment	Filleting	Filleting	Filleting
Mandible			2	1		
Hyoides					1	
Vertebrae				2	1	
Ribs		3		28		1
Scapula			1	1		
Shaft fg. humerus				3	1	
Distal humerus			1			
Cubitus			1	2		
Proximal radius			1			
Fg. shaft radius			1			
Pelvis				1		
Shaft femur				1		
Fg. shaft femur		1		2	1	
Distal tibia			1			
Astragalus			1			
Frag. Px. Metatarsal			1			
Distal metatarsal			2			
Phalanx II			1			
TOTAL	0	4	13	41	4	1

Table 15.9. *Ovis, Capra,* and *Sus,* distribution of cut marks

	Bronze Age		Chalcolithic		Later Neolithic		Early Neolithic	
	Dismem- berment	Filleting	Dismem- berment	Filleting	Dismem- berment	Filleting	Dismem- berment	Filleting
Maxilar			1 (*Sus*)					
Mandible			2 (1 *Sus*)	12 (1 *Sus*)			1	
Vertebrae		1						
Ribs		4 (*Sus*)			4	1	4	1
Scapula			2	1		1 (*Sus*)		
Distal humerus			1 (*Sus*)					
Distal fg. humerus								
Proximal radius			1 (*Sus*)		2			
Proximal fg. radius					1			
Shaft fg. radius		1						
Cubitus			1					
Proximal metacarpus	1		2 (1 *Sus*)					
Proximal fg. metacarpal								
Distal metacarpal			1				1	
Pelvis		1	1			1		1
Femur proximal								1
Shaft femur								1
Astragalus	1		1		3			
Proximal metatarsal			3 (1 *Sus*)					
Proximal fg. metatarsal	1							
Distal phalanx I	1							
Phalanx III							1	
TOTAL	4	7	16	13	10	3	7	4

Table 15.10. *Ovis, Capra,* and *Sus,* distribution of fracture marks

	Early Bronze	Chalcolithic	Later Neolithic	Early Neolithic
Proximal metacarpus	1			
Prox. fg. metacarpus				1
Shaft fr. tibia		1 (*Sus*)		
Distal tibia				1
Prox. fg. metatarsus			1	1
Shaft fg. metatarsus			1	
TOTAL	1	1	2	3

Table 15.11. *Bos taurus*, distribution of fracture marks

	Bronze Age	Chalcolithic	Later Neolithic
Skull		1	
Mandible		5	
Vertebrae		4	
Ribs	1		
Shaft fg. humerus		3	1
Distal humerus	1	1	1
Distal fg. humerus		1	
Proximal radius		5	
Shaft radius		1	
Shaft fg. radius	2	4	
Proximal metacarpus			1
Proximal fg. metacarpal		1	1
Distal metacarpal		1	1
Shaft femur		1	
Shaft fg. femur	1	1	1
Distal tibia		2	
Astragalus			
Proximal metatarsal	1	1	
Proximal fg. metatarsal	1	1	
Distal metatarsal		1	1
Lateral phalanx 1	1		
Distal fg. phalanx I		3	1
Lateral phalanx II	1		
TOTAL	9	37	8

Table 15.12. *Cervus elaphus* and *Capra pyrenaica*, distribution of cut marks

	Bronze Age		Chalcolithic		Later Neolithic		Early Neolithic	
	Cervus	*Capra*	*Cervus*	*Capra*	*Cervus*	*Capra*	*Cervus*	*Capra*
Mandible	1		1					
Hyoides					1			
Ribs	1						3	1
Shaft fg. humerus			1		1		1	
Proximal radius					1			
Carpals	1							
Proximal fg. metacarpal		1						
Pelvis								1
Shaft femur								
Shaft fg. femur			2				1	
Shaft fg. tibia			1					
Proximal metatarsal	1							
Shaft fg. metatarsal		1						
Centrotarsal	1							
TOTAL	4	3	5	0	3	0	5	2

Table 15.13. *Cervus elaphus* and *Capra pyrenaica*, distribution of fracture marks

	Bronze Age		Chalcolithic		Later Neolithic		Early Neolithic	
	Cervus	*Capra*	*Cervus*	*Capra*	*Cervus*	*Capra*	*Cervus*	*Capra*
Shaft fg. humerus			1					
Proximal fg. radius					3			
Proximal fg. metacarpal		1						
Shaft fg. metacarpal							2	
Distal metacarpal	1							
Distal fg. metacarpal		1						
Shaft fg. femur			2					
Shaft fg. tibia							1	
Calcaneus				1				
Proximal fg. metatarsal		1		1				
Shaft fg. metatarsal		1					1	
Distal fg. metatarsal							1	
Sagittal phalanx I	1				2		2	
Proximal phalanx I		2		2				
Proximal fg. phalanx I		1					1	
Lateral phalanx 1								
Distal phalanx I	2	1					3	
Distal fg. phalanx I			2				3	
Sagittal phalanx II		1						
Proximal phalanx II	1	1						
Lateral phalanx II							1	
Distal phalanx II	1			1	1			
Distal fg. phalanx II	2	3		1			3	
TOTAL	**8**	**13**	**5**	**6**	**6**	**0**	**18**	**0**

Table 15.14. Epipaleolithic number of individual species (NISP)

	Level 10.1					Level 304
	Total	Fr	Cu	Gn	Bu	Total
Bos sp.	1					
Capra pyrenaica		3	3			2
Cervus elaphus	29					6
Capreolus capreolus	6					
Sus sp.	3					

Note: Types of marks: Fr = produced by humans during marrow-cracking; Cu = tool marks; Gn = carnivore gnaw marks, presumably made by dogs; Bu = burning traces.

CHAPTER SIXTEEN

Macromammalian Remains from the Holocene Levels of El Mirón Cave

Jesús Altuna and Koro Mariezkurrena

TRANSLATED BY LAWRENCE GUY STRAUS

The prehistoric site of El Mirón Cave is located in Ramales de la Victoria (southeastern Cantabria) and had been known archaeologically since the beginning of the twentieth century, although its importance had not been determined until recently (e.g., González Morales and Straus 1997, 2000). It strategically dominates valleys that connect eastern Cantabria with Vizcaya in the Basque Country to the east and with the Castilian Meseta of Burgos to the south.

The faunal remains of large mammals that are studied here derive from excavations of Holocene deposits conducted in the front (western) part of the El Mirón Cave vestibule by M. González Morales and L. G. Straus between 1996 and 2002. Specifically, they come from a culturally poor Mesolithic stratum and a series of much richer Neolithic, Chalcolithic, and Bronze Age levels, mainly in the outer vestibule Cabin block excavation area but also from the western end of a 1 m wide × 14 m long trench that connects that area with the Corral block excavation area at the rear of the vestibule. (The Pleistocene faunal assemblages from both the front and rear of the vestibule will be published later.) The levels analyzed here are the following (González Morales and Straus 2000):

10.1: Mesolithic, with three uncalibrated ^{14}C dates ranging between 9550 and 8380 BP
10: Early Neolithic, with dates of 5690 and 5570 BP
9: Neolithic, with dates of 5250 and 5280 BP
9.6, 9.7, and 9.8: Neolithic lenses, with a date of 5250 BP from 9.6
303, 303.1, 303.2, 303.3: Neolithic levels in the Mid-Vestibule Trench, equivalent to Level 10, with three coherent dates ranging between 5500 and 5790 BP
8, 8.1: Late Neolithic, with a date of 4680 BP from 8.1
5, 5.1, 6, and 7: Chalcolithic, with dates of 3740, 4120, 3820 BP from 7, 5.1, and 5
3, 4: Final Chalcolithic and early Bronze Age, with a date of 3700 BP from 3 (Level 3 also has several lenses)

302: Mixture of late Neolithic, Chalcolithic, and Bronze Age, with a date of 4910 BP from the base of an associated pit

301: Mixed deposit containing possible Bronze Age and more recent materials

The eastward continuation of the Mesolithic-age deposit (Mid-Vestibule Trench Stratum 304 and Corral Strata 101–102) is essentially devoid of identifiable macrofaunal remains.

Materials and Methods

The total macromammalian assemblage is large, with identifiable remains alone totaling 6,008 items. The remains are highly fragmented, however, thus significantly limiting both biometric analysis and species-level determinations of many of the bones, in particular the differentiation between *Ovis aries* and *Capra hircus* on the one hand and between *Sus domesticus* and *Sus scrofa* on the other. To illustrate all one has to do is to indicate that the total weight of identifiable remains from all these levels is no more than 30,448 grams, which works out to an average weight of 5 g per fragment. And this is true despite the fact that 1,139 of the identifiable remains pertain to a large animal, the domesticated cow.

The compositional distribution of the assemblages by species and levels is displayed as absolute numbers in tables 16.1–16.3. Tables 16.4–16.6 present the percentages of remains by species and levels for the same data included in the previous tables. Tables 16.7–16.9 display the raw weights of the remains by levels, while tables 16.10–16.12 give the percentages for weights. Figures 16.1–16.6 graphically represent the same results as are given in the tables.

These tables and figures clearly illustrate the following conclusions concerning the development of the roles of both domestic livestock and wild-game hunting in the subsistence of the prehistoric inhabitants of El Mirón during the early and middle Holocene.

First of all and as is usual in the Vasco-Cantabrian region, Mesolithic Level 10.1 (whose assemblage of identifiable remains is very small) yielded only wild animal bones and teeth: red deer, roe deer, and boar. These three species constitute the trilogy of the most

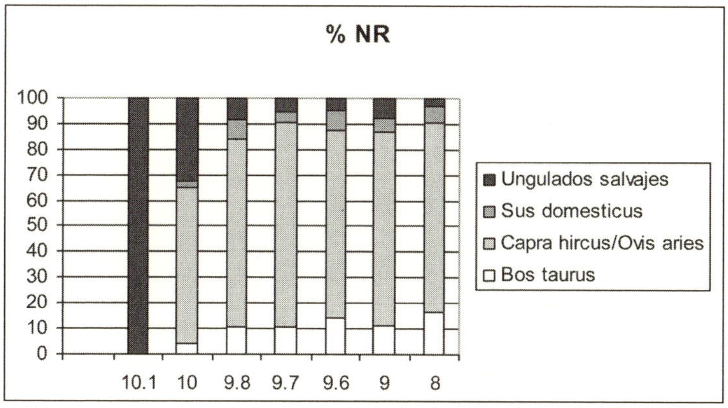

Fig. 16.1. Ungulate faunal spectra (percentages) for the Mesolithic and Neolithic levels in the outer vestibule (Cabin) area. "NR" = number of remains (NISP); "ongulados salvajes" = wild ungulates. Graph by J. Altuna and K. Mariezkurrena.

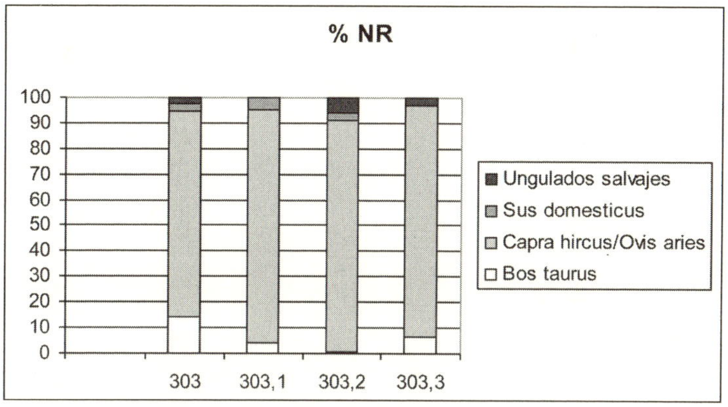

Fig. 16.2. Ungulate faunal spectra (percentages) for the Neolithic levels in the Mid-Vestibule Trench. Graph by J. Altuna and K. Mariezkurrena.

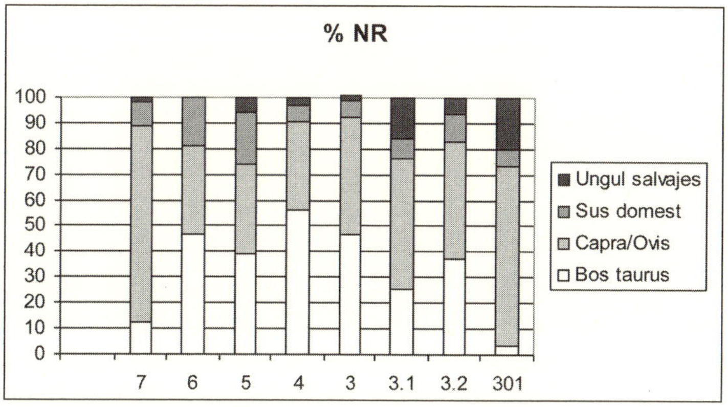

Fig. 16.3. Ungulate faunal spectra (percentages) for the Chalcolithic and early Bronze Age levels. (Level 301 is mixed.) Graph by J. Altuna and K. Mariezkurrena.

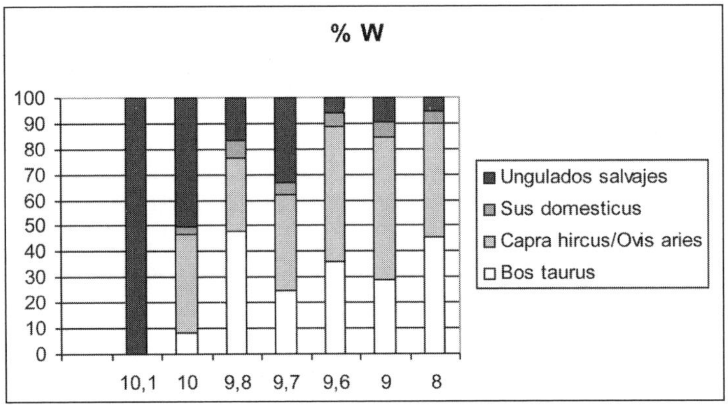

Fig. 16.4. Weights of bones of ungulates (percentages) in the Mesolithic and Neolithic levels in the outer vestibule (Cabin) area. Graph by J. Altuna and K. Mariezkurrena.

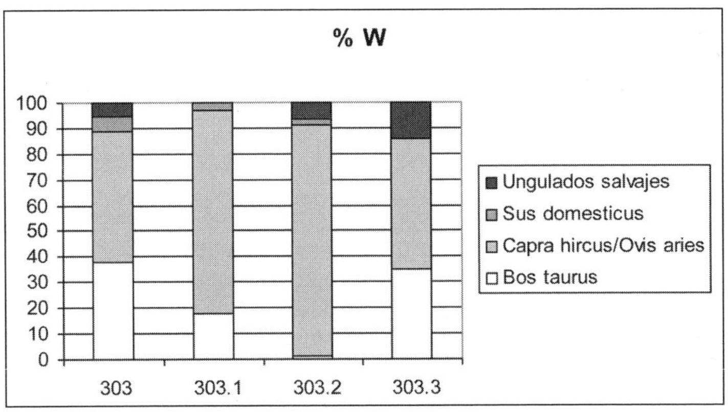

Fig. 16.5. Weights of bones of ungulates (percentages) in the Neolithic levels in the Mid-Vestibule Trench. Graph by J. Altuna and K. Mariezkurrena.

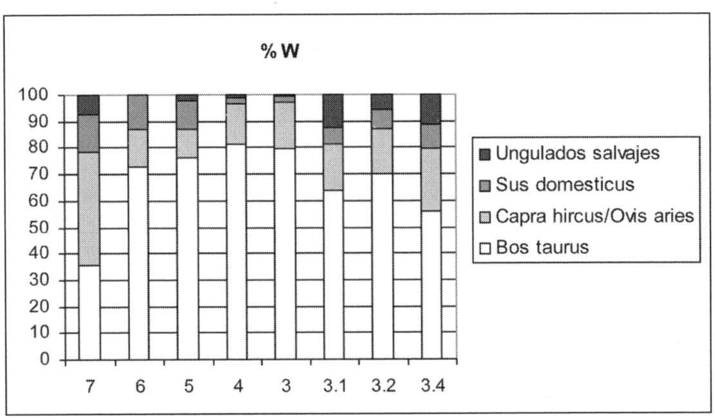

Fig. 16.6. Weights of bones of ungulates in the Chalcolithic and early Bronze Age levels. Graph by J. Altuna and K. Mariezkurrena.

frequent ungulates in Cantabrian Mesolithic sites. In contrast, in Solutrean, Magdalenian, and Azilian levels at Cantabrian sites situated in steep, rocky habitats, ibex generally dominates the fauna (e.g., Bolinkoba, El Rascaño, Collubil). Thus it is striking that at El Mirón, which is in a montane setting on a steep, rocky cliff, ibex is absent from the Mesolithic assemblage.

In the oldest Neolithic at El Mirón (Level 10 + Levels 303–303.3), hunting continued to be significant, but it decreased over time throughout the period. The dominant wild species is red deer, but boar, roe deer, ibex, chamois, and aurochs are also present (the latter two being represented by only one remain each, however). (A potential problem involves that fact that Level 10.1 was not observed as being different from Level 10 in the initial 1996 test excavation in three meter squares [I2, J2, J3], and hence its faunal remains were not separated. This might tend to inflate the relative frequency of wild animals in Level 10. However, in Level 10 *sensu stricto* [i.e., in the other seven "unmixed" squares], 30.6 percent of the ungulate remains are of wild animals [n = 49], while in the three "mixed" squares [Level 10 + 10.1] there are 32.8 percent wild ungulate remains [n = 39]. The statistical insignificance of these results shows that the overall "contamination" of the early Neolithic Level 10 assemblage by wild animal remains from underlying Mesolithic Level 10.1 is not an important distorting factor.) The domesticated livestock of the early Neolithic are dominated by ovicaprines, followed by cattle and finally small numbers of pig. Among the ovicaprine remains that can be distinguished as to species, sheep far outnumber goats.

In terms of potentially available meat, which is proportional to bone weight (tables 16.7–16.12; figures 16.4–16.6), we can see that cattle could have had a dietary significance that was greater than might be implied by the simple bone counts, given their much greater size than sheep and goats.

The later Neolithic levels (9.8–8) show substantial decreases in wild animals (and almost all of these are red deer). While ovicaprines are dominant, there is somewhat of a trend (also visible within the 303 series of levels in the Mid-Vestibule Trench) for cattle to gain in importance in terms of numbers of identifiable specimens, although their relative weight contribution was always substantial within these assemblages.

The oldest Chalcolithic level (7) continues to show the same overall spectrum of fauna as in the late Neolithic, with apparently even less emphasis on hunting. But there are drastic changes in the later Chalcolithic levels (6–4), with cattle becoming the numerically dominant species, followed by ovicaprines and finally pig. The dominance of cattle is even more notable in terms of potentially available meat weight.

Cattle decline somewhat in numeric importance in the Bronze Age levels (3 + associated lenses). Pig also declines, while ovicaprine remains increase somewhat in terms of both number of identified species (NISP) and weight. As one can see in table 16.3, the number of goat remains in Lens 3.1 is 41 items, but this is because there are simply two complete hind feet of this species in this unit. They pertain to one kid of somewhat less than six months of age. On the other hand, there is evidence of a slight increase in hunting, not only of red deer but also of ibex. (Some of the bones of wild animals in the metal ages levels and pits, more fossilized than the rest, were probably introduced into these mid-Holocene deposits from Magdalenian levels that had once been banked up against the colluvial-alluvial ramp that rises from the rear of the vestibule and that were destroyed and eroded downslope by natural and/or anthropogenic processes during these periods of often high humidity and intensive human and livestock occupation.—LGS)

The methods we used are the same that we have described earlier in similar archaeozoological analyses. For species identifications we basically employed the large comparative collection we have assembled over more than thirty years in the Department of Prehistory of the Sociedad de Ciencias Aranzadi in San Sebastián. For age and sex determinations, the distribution of remains of each species according to their anatomical position, the calculation of the minimum number of individuals (MNI), and biometric measurements we have used the criteria set forth in detail in our earlier publications (e.g., Altuna 1980; Altuna and Mariezkurrena 1984; Mariezkurrena 1986). However, here we would like to specify more clearly the determination of age among domesticated livestock:

Neonates: individuals whose milk dentition had not yet begun to wear (i.e., animals that were still suckling).

Infantiles: individuals whose milk dentition had begun to wear or whose first molar (only) had erupted and just started to wear. These also include individuals whose pelvic acetabulum was still unfused and whose long bones were both unfused and of size and structure indicative of early age.

Juveniles: individuals that, having passed the previous stage, nevertheless still did not have their complete permanent dentition or had premolars or third molars that were only slightly worn. They also include individuals whose distal humerus, distal tibia, and proximal radius had unfused epiphyses.

Pre-adults: individuals whose epiphyses were fused on the above-named long bones but still unfused for the proximal humerus, ulna, femur, and tibia and for the distal radius and femur. They also include individuals whose vertebrae were of adult size but whose heads had not yet fused to the centra.

Adults: individuals who had passed the previous stage and whose premolars and third molars had slight wear on all cusps. Authors usually indicate this degree of wear with a + sign.

Mature individuals: those whose premolars and third molars had moderate wear, indicated by ++.

Seniles: individuals for whom the above-cited teeth were heavily worn, indicated as +++.

The ages at which all these stages (except the neonatal one) occur are different among the different species represented at El Mirón:

Cattle:

Infantile: from around birth to less than about 8–10 months
Juvenile: between 8–10 months and 3 years
Pre-adult: 3–4 years
Adult: 4–6 years
Mature: 6–9 years
Senile: over 9 years

Ovicaprines:

Infantile: from around birth to less than 4–5 months
Juvenile: between 4–5 months and 2 years

Pre-adult: between 2 and 3–3.5 years
Adult: between 3–3.5 and 5 years
Mature: 5–8 years
Senile: over 8 years

Pig:

Infantile: from around birth to less than 6 months
Juvenile: between 6 months and 2.5 years
Pre-adult: between 2.5 and 3.5 years
Adult: 3.5–5 years
Mature: 5–8 years
Senile: over 8 years

For the specific determination of ovicaprine livestock (i.e., sheep versus goat), in addition to the classic work of Boessneck, Muller, and Teichert (1964) we have used the criteria of Payne (1969, 1985) and Prummel and Frisch (1986).

Analyses of the Remains by Level

MESOLITHIC LEVEL (10.1)

This level yielded only 42 identifiable remains, which are distributed in the following fashion:

Cervus elaphus: NISP = 28: a cranial fragment, a mandibular fragment, a molar, eight vertebral fragments, four rib fragments, a pelvis fragment, two metapodial fragments, two tarsals, seven phalanges, and a sesamoid. MNI = 2: an adult and an old red deer.

Capreolus capreolus: NISP = 11: a maxillary fragment, two mandibular fragments, five loose teeth, a calcaneum, and two phalanges. MNI = 2: an adult and an old roe deer.

Sus scrofa: NISP = 3: a molar fragment, a carpal, and a phalange. MNI = 1 juvenile.

Since there are no infantile individuals, one cannot macroscopically estimate the age in months of these animals and thus the season(s) of kills cannot be determined. In sum, Mesolithic-age Level 10 has three adult wild ungulates and two juveniles. This is a very small assemblage (despite the fairly large area excavated), but it does contrast sharply with what is found in the overlying Neolithic levels. The measurements that can be taken on these remains are reported in the biometric tables.

NEOLITHIC LEVELS

Domesticated Animals

As indicated above, ovicaprines left the most remains in the Neolithic levels (figures 16.1 and 16.2), but if we focus on the amount of meat that could have been provided by livestock (figures 16.4 and 16.5) cattle has increased importance, since in some levels or lenses (8 and 9.8) they equal or surpass sheep and goats in meat weight.

The distributions of livestock remains by MNI and age are displayed in table 16.13. One can see that young (i.e., non-adult) animals largely outnumber adults. However, the presence of livestock—especially ovicaprines—of all ages suggests that the cave was occupied throughout the whole year. The presence of heavily worn molars (especially of ovicaprines and cattle), signifying animals that had been kept alive into old age, would suggest their use for secondary products such as milk and (in the case of sheep) wool. On the other hand, there are very few heavily worn pig teeth.

The distribution of anatomical elements by species and by period are detailed as absolute numbers in table 16.15 and summarized by major anatomical groups in terms of percentages in table 16.16. These tables show that the treatment of carcasses of cattle and ovicaprines is similar among the various levels, except in the case of the terminal Chalcolithic/early Bronze Age, when remains of the trunk are much more abundant for both cattle and sheep/goats (ca. 26 percent vs. 12–18 percent for the other periods). On the contrary, pig trunk remains (never abundant) decrease through time, while head remains increase (although the grouped Neolithic and Chalcolithic samples are limited and the Bronze Age one too small to meaningfully analyze in this regard). In any event, it is clear that butchery took place in the cave, since all major anatomical parts are represented among the remains of the domesticated animals.

A separate analysis of the anatomical elements present in the Neolithic levels in the Cabin area (Levels 10–8) and in the Mid-Vestibule Trench (303.3–303) in tables 16.15–16.17 shows that there is no significant difference in terms of cattle and sheep/goat butchering operations between the front and middle of the vestibule. The representation of major anatomical regions is virtually the same between the two excavation areas. (This is not surprising, since the areas are contiguous insofar as concern the Neolithic levels [10 and 303–303.3] in the

H–O rows, which are interrupted only by unexcavated square K5.—LGS) We did not do this analysis for pig remains due to their small number, especially in the 303 series of levels, where calculation of percentages would be meaningless.

We wish at this point to return to a subject that we have dealt with in two of our earlier studies of Neolithic fauna from the Cantabrian region (Altuna 1980; Altuna and Mariezkurrena 2001), in which we suggested the possible *local* domestication of cattle from aurochs. Among the materials that we have analyzed over the years are some remains of small-size bovines imported as domesticates from other regions of Europe. But there are other remains that are larger than all the cattle we have studied in the Chalcolithic and Bronze and Iron Ages of the Cantabrian region. We have compared these remains with those of the aurochs that in fact survived in the region until the Roman era as well as with data from the abundant literature on aurochs in the Mesolithic and Neolithic, and their measurements in fact fall below those of aurochs. The sources consulted include those of Bökönyi (1962) on aurochs in Hungary, Stampfli (1963) on those of Burgäschisee Sud in Switzerland, Degerböl and Fredskilde (1970) on those of Denmark, Driesch, and Boessneck (1976) on those of Portugal, Breu (1986) on those of Eggolsheim in Baveria, and our own publications on the aurochs of the sites of Gibijo (Altuna 1974) and La Renke (Altuna and Mariezkurrena 2001), both in the interior Basque province of Alava.

In our first specific study of domestication in the Cantabrian region (Altuna 1980), we took note of a few bovine specimens in Neolithic levels in Arenaza (ca. 30 km east of El Mirón in Vizcaya) and in Los Husos (Alava) that were larger than all the cattle of the Chalcolithic, Bronze Age, and Iron Age known up to that time in the region. However, the Neolithic material available over 20 years ago was so scarce that we could do no more than mention the possible local domestication of aurochs as a modest working hypothesis. In principle this was a plausible scenario, given that the "agritype" or potential wild progenitor species from which cattle derived—the aurochs—had existed in the region, being well documented still in Mesolithic and even Neolithic times in Vasco-Cantabria.

Later, when we studied the Neolithic materials from La Renke, we again found a few similar remains—larger

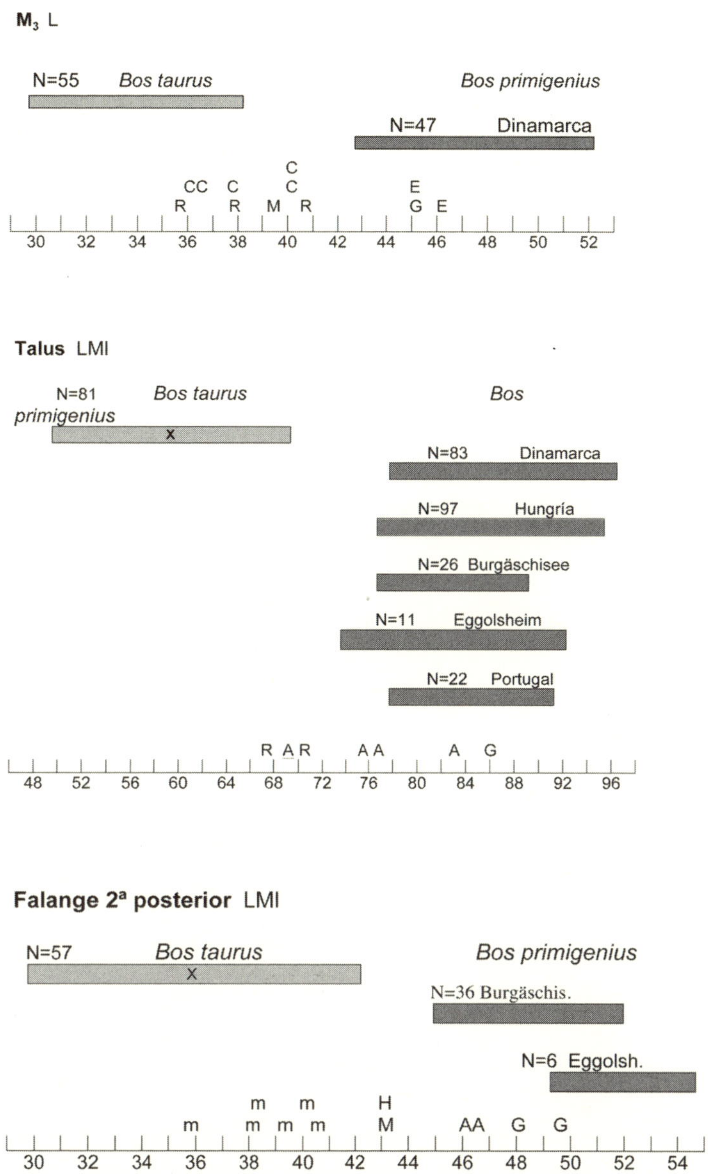

Fig. 16.7. Ranges and average lengths of lower M3, astragalus, and second hind phalanx for domesticated cattle at Basque sites of the Iron and Bronze Ages and ranges for Holocene aurochs from other European sites compared to individual measurements from Vasco-Cantabrian sites. M: El Mirón Neolithic; m: El Mirón Chalcolithic; R: La Renke Neolithic; C: Los Cascajos; A: Arenaza Neolithic; A: Arenaza Mesolithic; H: Los Husos Neolithic; G: Gibijo; "Dinamarca" = range for Danish sites; "Hungría" = range for Hungarian sites; Eggolsheim ("E") and Burgäschisee are sites in Baveria and Switzerland, respectively. Graph by J. Altuna and K. Mariezkurrena.

than those of contemporaneous cattle but smaller than aurochs. At that time we were also analyzing faunal remains from the site of Los Cascajos in Navarra, where there are two lower third molars, a metacarpal, a pelvis, a metatarsal, and a first phalange that fall in between the measurements for domesticated cattle and those for aurochs. In the present study of El Mirón more such remains have been found: a third lower molar and a second hind phalange. The total number of all these finds is still small, but they seem to still point in the direction of the original hypothesis of a local domestication of cattle from aurochs in northern Spain (figure 16.7).

Among the other domesticated animals were also found two remains of dog (*Canis familiaris*): one mandibular fragment without teeth from Lens 9.6 and a cervical vertebra from Level 8, both of later Neolithic age. The latter is from a pre-adult individual, since the vertebral disk (or "head") had not yet fused to the centrum. Dog has not been identified in the earlier Neolithic levels at El Mirón.

Wild Animals

As indicated earlier in our discussion of materials and methods, hunting was still relatively important in the early phase of the Neolithic. The significant role of wild game is principally evident in the outer (Cabin) excavation area (Level 10) but far less so in the Mid-Vestibule Trench (Levels 303–303.3). (The former observation is partly inflated since, as noted above, Level 10.1 was not identified until after the initial excavation of the test pit in squares I2, J2, and J3, when it was seen in the sections. Thus materials labeled as Level 10 from those squares include some wild fauna from Level 10.1.—LGS)

Tables 16.1 and 16.2 show the wild species that are represented. By far the most abundant one is red deer, with 219 remains from all the Neolithic levels combined. The distribution of *Cervus* anatomical elements is totally different from that which characterizes the domesticated animals. In the case of the red deer there are, in fact, 26 head elements, 43 of the trunk, and fully 150 of the limbs. This translates to 68.5 percent of the remains belonging to the appendicular skeleton and only 11.9 percent to the cranial skeleton. In contrast, among the cattle and ovicaprines from the Neolithic, the percentages are roughly equal for the two skeletal regions (tables 16.15a and 16.15b), while among the pigs the distribution of recovered anatomical parts is exactly opposite of the red deer distribution. Thus, pig cranial remains far outnumber appendicular ones. The other wild ungulates (boar, roe deer, chamois, and ibex) are represented by very few remains in the Neolithic.

Finally, there are three remains of *Martes* of an indeterminate species (two mandible fragments and one of femur), three others of *Felis sylvestris* (wild cat: femur, calcaneum, and metatarsal) and a mandible of *Erinaceus europaeus* (hedgehog)—an animal that is very rare in prehistoric sites.

CHALCOLITHIC LEVELS

Domesticated Animals

The livestock composition changed in the later Chalcolithic levels relative to the Neolithic occupations, although the oldest of them (Level 7) has a faunal composition similar to that of the late Neolithic (figure 16.3). In the rest of the Chalcolithic assemblages, cattle clearly numerically dominate, a fact that is even clearer in the potential amounts of meat that could have been provided by the various species (figure 16.6).

The distribution of the materials in terms of MNI and age can be seen in table 16.14. Once again, young animals outnumber adults. Among the juveniles there are infantile and very young individuals of all ages, indicating that they were killed at all seasons of the year and suggesting that humans were occupying the cave continuously.

As in the Neolithic levels, tooth wear among the cattle and ovicaprines shows that some of these animals were kept alive until they were very old, probably to use their milk (and, in the case of sheep, their wool).

The distributions of remains by anatomical elements for the different species are given in tables 16.15 and 16.16. These are similar to the distributions found among the Neolithic levels. This includes the fact that remains of pig trunk are less numerous, while head parts are more abundant than among either the cattle or the sheep and goats.

There is just one dog bone: a cervical vertebra in Level 7.

Wild Animals

Among the wild animals are the same species as before except for roe deer, which is absent from the Chalcolithic levels. Red deer and ibex are the most frequent game species, but the relative frequency of *Cervus* is much less than in the Neolithic levels.

Of note is the presence of a horse bone in Level 5. We have designated it as *Equus ferus* in the tables, not to raise any phylogenetic issue but simply to indicate that it is a wild horse, not a domesticated one, which Linnaeus called *Equus caballus*. As we have indicated elsewhere (Altuna 1998), it had been thought that the horse had gone extinct in northern Iberia during the Mesolithic and then later been reintroduced in a domesticated form during the Bronze Age. But in several recent faunal analyses we have been able to show the horse persisted in the region during all times between the Azilian and the Iron Age. Thus at Zatoya in Navarra (Altuna and Mariezkurrena 1989) it is present in the Azilian, the post-Azilian Epipaleolithic, and the Neolithic. We have also found it in the Neolithic of La Renke (Altuna and Mariezkurrena 2001). So the horses of the Chalcolithic were relicts of a small wild population that had survived in the region despite the environmental changes of the early and middle Holocene. It is very difficult to know exactly when the domesticated horse was introduced, since this taxon did not undergo the kinds of changes that happened to the other species—notably cattle and pigs—as a consequence of domestication. Whether the horses of the Iberian Bronze Age are (1) relicts of ones that had remained wild, (2) domesticated, or (3) both wild survivors and domesticated is something that cannot be ascertained at this time, given the scarcity of finds. Only if and when the available sample becomes much larger, so that studies of ages at death, anatomical part representation, and so on can be conducted as has been the case in eastern Europe, will we be able to solve this problem definitively.

BRONZE AGE LEVELS

Faunal remains are much less abundant in the early Bronze Age levels than in the Chalcolithic ones. Only Level 4 has a substantial amount of remains, and it may correspond to the time of transition between the Chalcolithic and Bronze Age. The faunal spectum in Levels 4–3 is similar to that of the underlying Chalcolithic levels, but after that (admittedly in small assemblages) the representation of cattle appears to decline, especially in Level 3.1 (tables 16.3 and 16.6). At the same time, evidence of hunting (mostly of red deer and ibex) increases—but the absolute number of remains and MNI are very small (table 16.17).

We must recall that in Lens 3.1 there are two complete hind feet of a goat kid that died at less than six months of age, since all the ephiphyses are unfused—even the proximal heads of the second phalanges. In the centrotarsal one can clearly see the fusion groove of the original tarsals. The bones of these two feet had not been scattered across the site, however, as is the case with the other body parts. This fact thus inflates the representation of goat in this lens (and consequently decreases the relative frequency of cattle). Because the phalanges, sesamoids, and other foot elements were all collected together, this individual could be specifically identified as *Capra hircus*; otherwise the remains would have had to be assigned to the category of undifferentiated ovicaprine.

BIOMETRICS

In tables 16.18–16.27 we present the measurements that could be taken for the record. We have already commented upon the bovine measurements for the Neolithic. In regard to the biometrics of Chalcolithic and Bronze Age cattle, we can affirm that they are comparable with those of cattle that have been analyzed for these periods throughout the Cantabrian region (table 16.18). There are still a few remains that fall within the range of variability for aurochs (*Bos primigenius*), and we thus attribute them to that wild species. Thus there is a pelvis in Level 5 and two third phalanges (one each from Lens 9.6, Neolithic, and mixed Level 302) whose measurements are well above those of contemporaneous domesticated cattle, as shown below:

Level 5		
Pelvis:	LA:	71 mm

Levels:		9.6	302
3rd Phalanx:	LdS:	74 mm	82.2 mm
	Ldo:	55 mm	62.9 mm

Aurochs appears sporadically in Vasco-Cantabrian Chalcolithic sites. There is even a problematic aurochs bone in an early Iron Age level at the site of La Hoya (Alava). It is a centrotarsal whose maximum width is 61.5 mm, which is clearly greater than the measurements for all centrotarsals of domesticated cattle that we know of in the region.

On the other hand, there exists a Roman text of *Servius Grammaticus* (fourth century AD) that says, "Siluestres uri qui in Pyreneo monte nascuntur inter Gallias. Uri autem boues siluestres, quos uulgo bobalos appellant" (Thilo and Hagen 1878–1902:para 2374) (Loosely translated as, "Wild aurochs originate in the Gallic Pyrenees. Aurochs, or wild oxen, are called buffalos by the people."—LGS) There are frequent graphic representations of bovines on Ibero-Roman stellae, on terra sigillata vessels, on mosaics and coins struck in Hispania (by Greeks, Hispano-Phoenicians, Ibero-Tartessians, Hispano-Latinos, and others), but one cannot deduce from them whether the animals are wild or domesticated.

Measurements for the ovicaprine livestock have no unusual characteristics. They fall within the ranges of variation that are well known for the region (tables 16.19–16.21). Some of the remains identifiable as *Capra* display measurements that are typical of Iberian ibex (*C. pyrenaica*) as opposed to domesticated goat (*C. hircus*) of Near Eastern origin. These are presented with their measurements in table 16.22.

The measurements for pig also fall within the known ranges of variation for this species during these periods of early domestication in the region (table 16.23). There are no intermediate values in measurements between wild boar and domesticated pig, such as we have observed in the Basque Country sites of Arenaza and Los Husos, although to a lesser extent than in the case of the bovines. Boar is indeed still (albeit spottily) represented in all the Mesolithic and post-Mesolithic phases at El Mirón, as can be seen in the measurement values given in table 16.24.

Dog, which is represented by so few remains (as shown in tables 16.1 and 16.3), yielded none that could provide a meaningful measurement, although it is clearly not wolf.

Among the wild ungulates there is a measurable distal end of a first phalanx (width = 41.2 mm), unfortunately from mixed surficial Level 301 in the Mid-Vestibule Trench. Its presence suggests that at least part of the materials that make up Level 301 are older than the Iron Age (which is itself a period not archaeologically represented among the artifacts found at El Mirón—LGS)—for example, Chalcolithic and/or Bronze Age. Tables 16.25–16.27 present the measurements for red and roe deer and chamois.

The ensemble of Neolithic and Chalcolithic (a.k.a. "Eneolithic") levels in El Mirón Cave add substantially to the relatively few known occupation sites for these periods in the Cantabrian region and adjacent zones of northern Spain. The ability to distinguish different phases within these periods, the existence of multiple radiocarbon dates, and the high number of archaeozoological remains makes this site one of the most important to be excavated in this part of the Iberian Peninsula.

However, the highly fragmented state of the material limits the importance of the assemblages from this site, particularly in two aspects:

1. Reduction in the ability to provide species-level identifications within the ovicaprine group, thus limiting the amount of detail that could potentially be derived concerning the composition of the small livestock component in the economy as it may have changed through time; and
2. Reduction in the ability to do biometric studies, which are so important in the detailed definition of the processes of domestication within the Neolithic. Since wild agritypes (potential ancestor species) of cattle and pigs existed in the region, it would have been very interesting to follow leads concerning the possibility of in situ domestication of these animals on a stronger empirical basis. In any event, this site did provide additional evidence on this subject, adding to what we already have from the nearby site of Arenaza (western Vizcaya) and from the more distant, trans-cordilleran sites of Los Husos and La Renke in Alava and Los Cascajos in Navarra.

All these data reinforce—albeit weakly—the hypothesis of in situ domestication of aurochs, in addition to the importation of domesticated cattle into the Vasco-

Cantabrian region. Certainly there are several other known cases of local cattle domestication around the Mediterranean Basin in the Neolithic.

As we have observed in other sites in the region, at El Mirón ovicaprine livestock were very important in the Neolithic occupations. However, even in the later Neolithic one can see a slight increase in the importance of cattle, which becomes all the more significant and ultimately dominant in the Chalcolithic occupations. This is especially true when one considers the amount of meat and milk that cattle could have provided. The use of milk is suggested by the fact that adult, mature, and even old cattle, as well as ovicaprines, were kept alive, in sharp contrast to the situation with pigs, which were mostly slaughtered at infantile, juvenile, and pre-adult ages. Pig also increases in importance in the Chalcolithic, a trend that is inconsistently observed among other sites in the region.

Hunting, a major source of subsistence during the minor Mesolithic visits to El Mirón, continued in the early Neolithic, but it was clearly a secondary source of meat and—in sharp contrast to the apparently ibex-dominated Magdalenian assemblages—seems to have focused on red deer. Scatterings of wild ungulate remains are found among the later Neolithic, Chalcolithic, and Bronze Age faunal assemblages. They include a few wild aurochs, horse, boar, and ibex, as well as red deer, roe deer, and chamois. Whether the humans had any role in the presence of trace remains of fox, wild cat, bear, mustelid, or hedgehog among these levels is unknown.

References Cited

Altuna, J.

1974 Hallazgo de un uro (*Bos primigenius* Boj.) en la sierra de Gibijo (Alava). Estudio de su esqueleto y de la fauna asociada al mismo. *Munibe* 26:27–51.

1980 Historia de la domesticación animal en el País Vasco desde sus orígenes hasta la romanización. *Munibe* 32:1–163.

1998 The horse at the end of the Würm glaciation and the post-glacial period in the north of Spain. *Archaeolingua* 8:31–40.

Altuna, J., and K. Mariezkurrena

1984 Bases de subsistencia de origen animal en el yacimiento de Ekain. In *El yacimiento prehistórico de la cueva de Ekain (Deba, Guipúzcoa)*, edited by J. Altuna and J. M. Merino, pp. 211–280. Eusko Ikaskuntza, Sociedad de Estudios Vascos, San Sebastián, Spain.

1989 Análisis arqueozoológico de los macromamíferos del yacimiento de Zatoya. In *El yacimiento prehistórico de Zatoya (Navarra)*, edited by I. Barandiarán and A. Cava, pp. 237–266. Vol. 8. Trabajos de Arqueología Navarra, Pamplona, Spain.

2001 La cabaña ganadera del yacimiento de La Renke (Alava, País Vasco). *Munibe (Antropología-Arkeología)* 53:75–86.

Boessneck, J., H. H. Muller, and M. Teichert

1964 Osteologische Unterscheidungsmerkmale zwischen Schaft (Ovis aries Linne) und Ziege (Capra hircus Linne). *Kuhn-Archiv* 78(1–2):1–129.

Bökönyi, S.

1962 Zur Naturgeschichte des Ures in Ungarn und das Problem der Domestikation des Hausrindes. *Acta Archaeologica Academiae Scientiarum Hungaricae* 14:175–214.

Breu, W.

1986 *Tierknochenfunde aus einer germanischen Siedlung bei Eggolsheim in Oberfranken.* Institut für Palaeoanatomie, Domestikationsforschung und Geschichte der Tiermedizin, Munich.

Degerböl, M., and B. Fredskilde

1970 The Urus (*Bos primigenius* Bojanus) and Neolithic domesticated cattle (*Bos taurus domesticus* Linné) in Denmark. *Det Kongelige Danske Videnskabernes Selskab. Biologiske Skrifter* 5:1–117.

Driesch, A. von den, and J. Boessneck

1976 Zur Grösse des Ures, Bos primigenius Bojanus, 1827, auf der Iberischen Halbinsel. *Säugetierkundliche Mitteilungen* 24:66–77.

González Morales, M., and L. Straus

1997 La prehistoria del Valle del Asón: excavaciones en La Cueva del Mirón. La campaña de 1996. In *Actas del II Congreso de Arqueología Peninsular*, edited by R. de Balbín and P. Bueno, pp. 119–131. Vol. 1. Fundación Rei Afonso Henriques, Zamora, Spain.

2000 El depósito arqueológico de la Prehistoria reciente de La Cueva del Mirón (Ramales, Cantabria): bases estratigráficas para una secuencia regional. In *Actas do 3º Congresso de Arqueologia Peninsular*, edited by V. Oliveira Jorge, pp. 49–62. Vol. 4. ADECAP, Porto, Portugal.

Mariezkurrena, K.
1986 La cabaña ganadera del Castillar de Mendavia (Navarra). *Munibe (Antropología-Arqueología)* 38:119–169.

Payne, S.
1969 A metrical distinction between sheep and goat metacarpals. In *The Domestication and Exploitation of Plants and Animals*, edited by P. Ucko and G. Dimbleby, pp. 295–305. Gerald Duckworth, London.
1985 Morphological distinctions between the mandibular teeth of young sheep, *Ovis*, and goats, *Capra*. *Journal of Archaeological Science* 12:139–147.

Prummel, W., and H. J. Frisch
1986 Guide for the distinction of species, sex and body side in bones of sheep and goat. *Journal of Archaeological Science* 13:567–677.

Stampfli, H. R.
1963 Wisent, *Bison bonasus* (Linne,1758) Ur, Bos primigenius (Bojanus, 1827) und Hausrind, Bos taurus (Linne, 1758). In *Seeberg. Burgäschisee-Süd: die Tierreste; Acta Bernensia II Teil* 3, edited by J. Boessneck, J. P. Jequire, and H. R. Stampfli, pp.1–215. Stämpfli, Bern, Switzerland.

Thilo, G., and H. Hagen (eds.)
1878–1902 *Servius Grammaticus*. In *Vergilii carmina commentarii*. 3 vols. Teubner, Leipzig, Germany.

Table 16.1. Number of remains in Mesolithic (10.1) and Neolithic (10–8) levels

Level	10.1	10	9.8	9.7	9.6	9	8
Bos taurus		12	14	26	109	77	89
Capra hircus/Ovis aries		171	96	200	573	504	399
Ovis aries[a]		7	2	12	46	15	21
Capra hircus[a]		1		1	1	1	
Sus domesticus		7	10	10	59	34	36
Canis familiaris					1		1
Total domesticated		**199**	**120**	**236**	**742**	**615**	**525**
Sus scrofa	3					3	4
Cervus elaphus	28	86	11	13	36	48	10
Capreolus capreolus	11	8					
Bos primigenius					1		
Rupicapra rupicapra							1
Capra pyrenaica		3					
Martes sp.		2					1
Felis silvestris							3
Erinaceus europaeus						1	
Total wild	**42**	**99**	**11**	**13**	**37**	**52**	**19**
TOTAL	**42**	**298**	**131**	**249**	**779**	**667**	**544**

[a]These identifiable remains of goat and sheep are also included in the *Capra hircus/Ovis aries* category.

Table 16.2. Number of remains in Neolithic levels

Level	303	303.1	303.2	303.3
Bos taurus	62	11	1	5
Capra hircus/Ovis aries	354	237	171	66
Ovis aries[a]	13	8	1	
Capra hircus[a]	1	1		
Sus domesticus	15	12	6	
Total domesticated	**431**	**260**	**178**	**71**
Sus scrofa	2			
Cervus elaphus	2		11	2
Capra pyrenaica	6			
Total wild	**10**		**11**	**2**
TOTAL	**441**	**260**	**189**	**73**

[a]These goat and sheep remains are also included in the *Capra hircus/Ovis aries* category.

Table 16.3. Number of remains in Chalcolithic (7–4) and Bronze Age (3, 3.1, 3.2, 3.4) levels

Level	7	6	5	4	3	3.1	3.2	3.4	302
Bos taurus	76	124	138	208	65	44	56	12	1
Capra hircus/Ovis aires	474	91	125	129	64	88	68	9	18
Ovis aries[a]	33		2	4		3	4		
Capra hircus[a]				7	1	41	2	1	
Sus domesticus	59	51	70	24	9	13	16	1	2
Canis familiaris	1								
Total domesticated	**610**	**266**	**333**	**361**	**138**	**145**	**140**	**22**	**21**
Equus ferus			1						
Sus scrofa	8				3				
Cevus elaphus	1		10	4		10	5	1	
Bos primigenius			1	1					1
Rupicapra rupicapra			2			1			
Capra pyren.			8	6		17	5	1	1
Vulpes vulpes	1								
Ursus arctos				1					
Total wild	**10**		**21**	**12**	**3**	**28**	**10**	**2**	**2**
TOTAL	**620**	**266**	**354**	**373**	**141**	**173**	**150**	**24**	**23**

Note: Level 302 is probably a mixture of late Neolithic, Chalcolithic, and/or early Bronze Age.

[a]These goat and sheep remains are also included in the *Capra hircus/Ovis aries* category.

Table 16.4. Percentage of number of identified species (NISP) of domesticated and wild ungulates in Mesolithic (10.1) and Neolithic (10–8) levels

Level	10.1	10	9.8	9.7	9.6	9	8
Bos taurus		4.0	10.7	10.4	14.0	11.6	16.5
Capra hircus/Ovis aries		57.4	73.3	80.4	73.7	75.7	74.0
Sus domesticus		2.3	7.6	4.0	7.6	5.1	6.7
Domesticated	**0**	**66.8**	**91.6**	**94.8**	**95.4**	**92.4**	**97.2**
Sus scrofa	7.1					0.4	0.7
Cervus elaphus	66.7	28.9	8.4	5.2	4.5	7.2	1.9
Capreolus capreolus	26.2	2.7					
Bos primigenius					0.1		
Rupicapra rupicapra							0.2
Capra pyrenaica		1.0					
Wild	**100.0**	**33.2**	**8.4**	**5.2**	**4.6**	**7.6**	**2.8**
TOTAL NISP	**42**	**296**	**131**	**249**	**779**	**667**	**539**

Table 16.5. Percentage of the number of remains in Neolithic levels

Level	303	303.1	303.2	303.3
Bos taurus	14.1	4.2	0.5	6.8
Capra hircus/Ovis aries	80.3	91.2	90.5	90.5
Sus domesticus	3.4	4.6	3.2	
Domesticated	**97.8**	**100.0**	**94.2**	**97.3**
Sus scrofa	0.4			
Cervus elaphus	0.4		5.8	2.7
Capra pyrenaica	1.4			
Wild	**2.2**	**0.0**	**5.8**	**2.7**
TOTAL	**441**	**260**	**189**	**73**

Table 16.6. Percentage of number of individual species (NISP) of domesticated and wild ungulates in Chalcolithic (7–4) and Bronze Age (3, 3.1, 3.2) levels

Levels	7	6	5	4	3	3.1	3.2
Bos taurus	12.3	46.6	39.6	56.1	46.9	25.4	37.3
Capra hircus/Ovis aries	76.7	34.2	35.3	34.7	45.5	50.9	45.4
Sus domesticus	9.5	19.2	19.8	6.5	6.5	7.5	10.7
Domesticated	**98.5**	**100.0**	**94.7**	**97.3**	**97.9**	**83.8**	**93.4**
Equus ferus			0.3				
Sus scrofa	1.3				2.1		
Cervus elaphus	0.2		2.8	1.1		5.8	3.3
Bos primigenius			0.3				
Rupicapra rupicapra			0.6			0.6	
Capra pyrenaica			2.3	1.6		9.8	3.3
Wild	**1.5**	**0.0**	**6.3**	**2.7**	**2.1**	**16.2**	**6.6**
TOTAL NISP	**619**	**266**	**354**	**372**	**141**	**173**	**150**

Table 16.7. Weight (g) of remains of various species in Mesolithic and Neolithic levels

Level	10.1	10	9.8	9.7	9.6	9	8
Bos taurus		85	493	97	922	717	1,126
Capra hircus/Ovis aries		412	293	149	1,415	1,361	1,100
Sus domesticus		27	77	17	137	139	116
Canis familiaris					3		2
Total domesticated	**0**	**524**	**863**	**263**	**2,477**	**2,217**	**2,344**
Sus scrofa	2					7	41
Cervus elaphus	142	522	169	132	158	228	67
Capreolus capreolus	30	12					
Bos primigenius					38		
Rupicapra rupicapra							24
Martes sp.		3					1
Felis silvestris							3
Erinaceus europaeus						1	
Total wild	**174**	**537**	**169**	**132**	**196**	**236**	**136**
TOTAL	**174**	**1,061**	**1,032**	**395**	**2,673**	**2,453**	**2,480**

Table 16.8. Weight (g) of remains in Neolithic levels

Level	303	303.1	303.2	303.3
Bos taurus	518	106	3	51
Capra hircus/Ovis aries	704	482	305	75
Sus domesticus	30	18	6	
Total domesticated	**1,252**	**606**	**314**	**126**
Bos primigenius				
Sus scrofa	82			
Cervus elaphus	1		23	21
Capra pyrenaica	20			
Total wild	**103**	**0**	**23**	**21**
TOTAL	**1,355**	**606**	**337**	**147**

Table 16.9. Weight (g) of remains in Chalcolithic and Bronze Age levels

Level	7	6	5	4	3	3.1	3.2	3.4	302
Bos taurus	1,117	1,329	3,301	2,696	1,411	678	855	64	5
Capra hircus/Ovis aries	1,342	259	472	520	310	188	206	27	39
Sus domesticus	439	233	488	67	36	61	85	10	5
Canis familiaris	6								
Total domesticated	**2,904**	**1,821**	**4,261**	**3,283**	**1,757**	**927**	**1,146**	**101**	**49**
Equus ferus			5						
Sus scrofa	203				15				
Cervus elaphus	24		55	27		62	60	10	
Bos primigenius			64						27
Rupicapra rupicapra			1			4			
Capra pyrenaica			41	18		69	12	3	8
Vulpes vulpes	1								
Ursus arctos				14					
Total wild	**228**	**0**	**166**	**59**	**15**	**135**	**72**	**13**	**35**
TOTAL	**3,132**	**1,821**	**4,427**	**3,342**	**1,772**	**1,062**	**1,218**	**114**	**84**

Table 16.10. Percentage of weights of ungulate remains in Neolithic levels

Level	10.1	10	9.8	9.7	9.6	9	8
Bos taurus		8.0	47.8	24.6	34.5	29.2	45.5
Capra hircus/Ovis aries		39.0	28.4	37.7	52.9	55.5	44.4
Sus domesticus		2.6	7.5	4.3	5.1	5.7	4.7
Total domesticated		**49.6**	**83.7**	**6.6**	**92.5**	**90.4**	**94.6**
Sus scrofa	1.1					0.3	1.7
Cervus elaphus	81.7	49.3	16.4	33.4	5.9	9.3	2.7
Capreolus capreolus	17.2	1.1					
Bos primigenius					1.4		
Rupicapra rupicapra							1.0
Total wild	**100.0**	**50.4**	**16.4**	**33.4**	**7.3**	**9.6**	**5.4**
TOTAL WEIGHT (g)	**174**	**1,058**	**1,032**	**395**	**2,673**	**2,452**	**2,476**

Table 16.11. Percentage of weights of remains in Neolithic levels

Level	303	303.1	303.2	303.3
Bos taurus	37.6	17.5	0.9	34.7
Capra hircus/Ovis aries	51.1	79.5	90.5	51.0
Sus domesticus	6.0	3.0	1.8	
Total domesticated	**94.7**	**100.0**	**93.2**	**85.7**
Bos primigenius				
Sus scrofa	3.8			
Cervus elaphus			6.8	14.3
Capra pyrenaica	1.5			
Total wild	**5.3**	**0.0**	**6.8**	**14.3**
TOTAL WEIGHT (g)	**1,377**	**606**	**337**	**147**

Table 16.12. Percentage of weights of remains in Chalcolithic and Bronze Age levels

Level	7	6	5	4	3	3.1	3.2	3.4	302
Bos taurus	35.7	73.0	74.6	81.1	79.7	63.9	70.1	56.1	6.0
Capra hircus/Ovis aries	42.9	14.2	10.7	15.6	17.5	17.7	17.0	23.7	46.4
Sus domesticus	14.1	12.8	11.0	2.0	2.0	5.7	7.0	8.8	6.0
Total domesticated	**92.7**	**100.0**	**96.3**	**98.7**	**99.2**	**87.3**	**94.1**	**88.6**	**58.4**
Equus ferus			0.1						
Sus scrofa	6.5				0.8				
Cervus elaphus	0.8		1.2	0.8		5.8	4.9	8.8	
Bos primigenius			1.4						32.1
Rupicapra rupicapra						0.4			
Capra pyrenaica			0.9	0.5		6.5	1.0	2.6	9.5
Total wild	**7.3**	**0.0**	**3.6**	**1.3**	**0.8**	**12.7**	**5.9**	**11.4**	**41.6**
TOTAL WEIGHT (g)	**3,125**	**1,821**	**4,427**	**3,328**	**1,772**	**1,062**	**1,218**	**114**	**84**

Table 16.13 Minimum number of individuals (MNI) and summary of individual ages in Neolithic levels

Level 10	MNI	Age		
		A+M+S	P+J	I+N
Bos taurus	2		1	1
Capra hircus/Ovis aries	8	3	2	3
Sus domesticus	2	1	1	
Cervus elaphus	3	2	1	
Capreolus capreolus	1	1		
Martes sp.	1	1		
TOTAL	**17**	**8**	**5**	**4**

Levels 9.8+9.7+9.6	MNI	Age		
		A+M+S	P+J	I+N
Bos taurus	6	2	3	1
Capra hircus/Ovis aires	27	7	9	11
Sus domesticus	5	1	3	1
Canis familiaris	1	1		
Cervus elaphus	4	2	1	1
TOTAL	**43**	**13**	**16**	**14**

Level 9	MNI	Age		
		A+M+S	P+J	I+N
Bos taurus	4	1	2	1
Capra hircus/Ovis aries	12	5	4	3
Sus domesticus	8	1	6	1
Sus scrofa	1	1		
Cervus elaphus	3	1	1	1
Erinaceus europaeus	1	1		
TOTAL	**29**	**10**	**13**	**6**

Level 8	MNI	Age		
		A+M+S	P+J	I+N
Bos taurus	4	2	1	1
Capra hircus/Ovis aries	13	6	3	4
Sus domesticus	2		2	
Canis familiaris	1		1	
Sus scrofa	1	1		
Cervus elaphus	2	1	1	
Rupicapra rupicapra	1	1		
Martes sp.	1	1		
Felis silvestris	1	1		
TOTAL	**26**	**13**	**8**	**5**

Table 16.13 *(continued)*

Level 303	MNI	Age		
		A+M+S	P+J	I+N
Bos taurus	4	1	3	
Capra hircus/Ovis aries	13	3	5	5
Sus domesticus	3	1	2	
Sus scrofa	1	1		
Cervus elaphus	1	1		
Capra pyrenaica	1	1		
TOTAL	**23**	**8**	**10**	**5**

Level 303.1	MNI	Age		
		A+M+S	P+J	I+N
Bos taurus	3		2	1
Capra hircus/Ovis aries	9	2	4	3
Sus domesticus	2		1	1
TOTAL	**14**	**2**	**7**	**5**

Level 303.2	MNI	Age		
		A+M+S	P+J	I+N
Bos taurus	1			1
Capra hircus/Ovis aries	11	4	2	5
Sus domesticus	3		1	2
Cervus elaphus	1	1		
Vulpes vulpes	1	1		
TOTAL	**17**	**6**	**3**	**8**

Level 303.3	MNI	Age		
		A+M+S	P+J	I+N
Bos taurus	1	1		
Capra hircus/Ovis aries	4	2	1	1
Cerus elaphus	1	1		
TOTAL	**6**	**4**	**1**	**1**

Note: A+M+S: Adults+Mature+Old; P+J: Preadults+Juveniles; I+N: Young+Neonates

Table 16.14. Minimum number of individuals (MNI) and summary of individual ages in Chalcolithic levels

Level 7	MNI	Ages		
		A+M+S	P+J	I+N
Bos taurus	4	1	2	1
Capra hircus/Ovis aries	20	8	7	5
Sus domesticus	3	1	2	
Canis familiaris	1	1		
Sus scrofa	6	1	5	
Cervus elaphus	1	1		
Vulpes vulpes	1	1		
TOTAL	36	14	16	6

Level 6	MNI	Ages		
		A+M+S	P+J	I+N
Bos taurus	5	1	2	2
Capra hircus/Ovis aries	6	2	1	3
Sus domesticus	5	2	2	1
TOTAL	16	5	5	6

Level 5	MNI	Ages		
		A+M+S	P+J	I+N
Bos taurus	7	3	1	3
Capra hircus/Ovis aries	9	3	3	3
Sus domesticus	6	3	2	1
Equus ferus	1	1		
Cervus elaphus	2	1		1
Rupicapra rupicapra	2			2
Capra pyrenaica	1	1		
TOTAL	28	12	6	10

Note: A+M+S: Adults+Mature+Old; P+J: Preadults+Juveniles; I+N: Young+Neonates

Table 16.15a. Distribution of remains of the most abundant species by skeletal element

	Bos taurus				*Ovis aries/Capra hircus*				*Sus domesticus*		
	NEO-LITHIC	CHALCO-LITHIC	CHALCOLITHIC-BRONZE	BRONZE	NEO-LITHIC	CHALCO-LITHIC	CHALCOLITHIC-BRONZE	BRONZE	NEO-LITHIC	CHALCOLITHIC	BRONZE
Horn	1	5	2		2		4	6			
Cranium	41	53	32	34	119	40	3	21	39	18	
Maxilla		5	7	2	17	11	5	3	13	7	3
Upper teeth	34	27	7	9	406	107	14	26	26	24	12
Mandible	32	15	12	19	130	37	10	11	9	9	2
Lower teeth	60	43	18	22	526	170	11	58	40	42	12
Loose teeth	8	5	3	3	27	20	9	1	6	3	5
Hyoid					7			4			
Vertebra	40	16	30	12	157	21	11	19	3		
Sacrum			2		2						
Sternum					1						
Ribs	16	46	24	24	199	61	19	16	9	9	3
Scapula	2	11	5	8	64	16	6	3	4	4	
Humerus	16	21	5	6	113	28	3	6		4	
Radius	7	4	3	5	126	18	2	10	3	5	
Ulna	11	16	8	3	58	10	3	5	2	1	
Carpal	10	9	6	5	53	11		4		4	
Metacarpal	21	8	5	5	96	22	3	3	1	2	1
Pelvis	7	11		6	96	19	4	3			
Femur	10	16	4	5	102	20	2	8	4	11	
Patella					4	1		1			
Tibia	13	14	4	9	150	29	4	8	4	13	1
Fibula				2	11	1		3	4	3	
Tarsal	20	5	4	2	86	15	5	18	3	6	2
Metatarsal	13	14	1	4	101	15	3	6	1	7	3
Metapodials	8	1	4		31	5		6	6	5	6
Phalanges	36	27	18	11	107	35	5	30	16	3	2
Sesamoid	1	7	5	5	12	4		7			
TOTAL	**407**	**379**	**209**	**201**	**2,803**	**716**	**118**	**286**	**193**	**185**	**52**

Table 16.15b. Distribution of anatomical parts of cattle and ovicaprines among the combined Neolithic levels of the front and middle zones of the cave vestibule

	Level	Head	Trunk	Limbs	Total
Bos taurus	10-9-8	148	43	138	**329**
	%	45.0	13.1	41.9	
	303	28	13	39	**80**
	%	35.0	16.3	48.7	
Ovicaprines	10-9-8	883	227	949	**2059**
	%	42.9	11.0	46.1	
	303	366	136	344	**783**
	%	46.7	17.4	43.9	

Table 16.16. Summary of tables 16.15a and 16.15b with percentage of remains by anatomical part

	Horizons	Head	Trunk	Limbs	Total
Bos taurus	Neolithic	176	56	175	**407**
	%	43.2	13.8	43.0	
	Chalcolithic	153	62	164	**379**
	%	40.4	16.4	43.2	
	Chalcolithic-Bronze	81	56	72	**209**
	%	38.8	26.8	34.4	
	Bronze	89	36	76	**201**
	%	44.3	17.9	37.8	
Ovicaprines	Neolithic	1,234	359	361	**2,803**
	%	45.0	12.8	43.2	
	Chalcolithic	385	82	249	**716**
	%	53.8	11.5	34.8	
	Chalcolithic-Bronze	54	30	34	**118**
	%	45.8	25.4	28.8	
	Bronze	120	35	74	**229**
	%	52.4	15.3	32.3	
Sus domesticus	Neolithic	133	12	40	**193**
	%	68.9	6.2	20.7	
	Chalcolithic	103	9	73	**185**
	%	55.7	4.9	39.5	

Table 16.17. Minimum number of individuals (MNI) and distribution by age of domesticated and wild animals in Bronze Age/Late Chalcolithic levels

Level 4	MNI	Ages		
		A+M+S	P+J	I+N
Bos taurus	6	2	3	1
Capra hircus/Ovis aries	8	3	4	1
Sus domesticus	2	1	1	
Cervus elaphus	1	1		
Capra pyrenaica	1	1		
Ursus arctos	1	1		
TOTAL	**19**	**9**	**8**	**2**

Level 3	MNI	Ages		
		A+M+S	P+J	I+N
Bos taurus	4	2	2	
Capra hircus/Ovis aries	8	4	1	3
Sus domesticus	5	2	1	2
Sus scrofa	1	1		
Cervus elaphus	2	1	1	
Rupicapra rupicapra	1	1		
Capra pyrenaica	3	2	1	
TOTAL	**24**	**13**	**6**	**5**

Note: A+M+S: Adults+Mature+Old; P+J: Preadults+Juveniles; I+N: Young+Neonates

Abbreviations used in tables 16.18–16.27

A	Width	El	Lateral thickness
AA	Acetabulum width	EmO	Minimum olecranon thickness
Ad	Distal width	EPA	Thickness above Anconian process
AF	Articular facet width	Extr	Extremity
AFT	Talus facet width	H	Height
AM	Maximum width	Hax	Axial height
AmD	Minimum diaphysis width	HCaM1	Corpus height in front of M1
AmitD	Mid-diaphysis width	Hml	Minimum illium height
AMT	Maximum trochlea width	HP	Extensorial process height
AMV	Maximum vertebra width	HScr	Cranial surface height
ant	Front Limb	L	Length
Ap	Proximal width	LA	Acetabulum length
APC	Coronoid process width	Ldo	Dorsal length
AS	Articular surface width	LdS	Diagonal floor or base length
AScd	Caudal articular surface width	LF	Articular facet length
AScr	Cranial articular surface width	LM	Maximum length
ASd	Distal articular surface width	LMC	Maximum length from head
ASp	Proximal articular surface width	LmC	Minimum neck length
AT	Troclea width	LMl	Maximum lateral length
Desg	Wear	LMm	Maximum medial length
DMB	Maximum base diameter	LMP	Maximum process length
DmB	Minimum base diameter	LMS	Maximum inter-surface length
EC	Caput width or thickness	LS	Articular surface length
Ed	Distal thickness	post	Hind limb

+	Slight wear on all cusps of premolars and third molars (adult)
++	Moderate wear on premolars and third molars (mature)
+++	Heavily worn premolars and third molars (senile)

Table 16.18. Dimensions of measurable skeletal elements of *Bos taurus*

Maxilla		
LM^3	30.5	30
AM^3		24
Level	4	8

Mandible				
L P_2–M_3	142			
L P_2–P_4	48.5			
L M–M_3	90			
LM_3	36	37	37	39.7
A M_3	15.5	16.6	16.7	16.9
Wear	++	+	++	++
Level	3	5	9	9

Scapula			
LmC		43.3	49.5
LS	55		56.5
AS			45
Level	3	5	7

Humerus		
AT	71	75.5
Level	4	8

Radius		
Ap	80	85
Asp	75.5	72
Level	5	5

Table 16.18. *(continued)*

Ulna		
EmO	46.1	
EPa	61	
APc		48.5
Level	4	5

Metacarpal			
Ap	61		
AmD	30.7		
Ad		57.5	64.5
Ed		30	34.6
Level	9	8	4

Femur	
Ap	115
EC	46
Level	5

Tibia				
AmD	35.4			
Ad		59	59.2	61.2
Ed		42	44	46
Level	3	5	5	8

Centrotarsal	
AM	51.7
Level	4

Astragalus									
LMl	59.5	62	63	65.5	66	62.5	65	54.6	61.3
LMm	55.2	57.4	58	60.7	61.5	58	60		
El	32.4	35.3	35.5	37	37.5	36.5		40	
Ad	35.3	41	42	44	46.7	40			
Level	4	5	5	6	6	8	8	9	9

Metatarsal	
Ap	42.5
Level	4

Phalange 1					
LM	60	56	63.4	56.5	
Ap	31.2	31.5	32	27.7	26.7
AmD	27	26.8	25.5		
Ad	29.5	29	31		
Extr	ant	ant	post		
Level	6	7	7	3	3

Phalange 2																	
LM	35.2	35.5	38.7	33.5	37.5	38.5	38	39	38.5	38.1	39.1	40.5	38	40	35.5	40	43
Ap	26.1	28.6	31	26.5		33.5	30	29	30.2	29.4	26.1	30.1	28.5	30.1	27.6	28.5	32.6
AmD	23.4	22.2	24.9	20.7	24.1	26	24.5	24.5	25.1	22.3	21.9	23.9	22.2	24.1	21.1	22.5	25
Ad	26	25.1	27.9	22.7	25.6	28.5	28	23.5	26.5	25.9	22.2	25.6	25.3	25.5	24	23.5	26.8
Ed	31.8	30	34.6		32	33.5			34.1	27.5	27.4	30.5	31	31.2	26.1		32.2
Extr	ant	ant	ant	ant	ant	ant	ant	ant	ant	post	post	post	post	post	post	post	post
Level	4	4	4	5	5	5	7	7	9	4	4	4	5	5	6	7	8

Phalange 3				
LdS	61.6	61.3	60.5	67.5
Ldo	50.6	50.3	49.1	54.2
Level	4	4	4	6

Table 16.19. Dimensions of measurable skeletal elements of *Ovis aries*

Scapula								
LmC	15.7	20.5	18.2	19.2	18.6	18.4	18.1	19.3
LMP	30		32		33.5		33	
LS	25		26		27		25	
AS	20		21.5		20		20.5	
Level	4	5	7	7	10	300	9	9

Humerus								
AmD					12.4	15.5		
Ad	28.6	32	31	27.6	26.3		30.5	30.5
AT	27.6	30.5	29.1	27	25.3		31.3	29.2
Ed		27.5	25.5	24	23.5	26.7	27	27
Level	3	3	8	8	300	9	9	9

Metacarpal						
Ap	24	23.8	24.8	22	24	23.5
AmD	13.4	13		12.4	13	
Level	7	7	7	9	9	9

Ulna		
EmO	21.2	
EPA		23.9
Level	9	**9**

Radius													
Ap	31.1	32	29.2	32.2	29.9	30.5	31.7	29.1	29.5	29.7	30.5	30.6	29.6
Asp	29.3	30.1	28	30.2	27.6	27.5	30	26.9	27.8	28.2	28.9	28.8	28.2
AmD							16.7	15	15.3				
Level	3	3	7	7	9	10	300	300	9	9	9	9	

Tibia							
AmD		14	13.3				
Ad	25.6	25.8	25.1	24.1	27.1	27.5	27.5
Ed	20	19.3	19	18.3	21.5	20.5	20.2
Level	300	9	9	9	9	9	**9**

Metatarsal		
Ap	19.1	19.3
LEVEL	7	300

Astragalus									
LMl	28.5	29.6	28.6	24.8	28.4	28.6	29.8	28.8	32.2
LMm	27.3	27.2	17.7	24	27.1	27	27.6	26.7	30.5
El	16.1	16.8	15.5	14.2	15.5	16.5	15.9	15.2	18.8
Ad	18.2	19	18.5	15.9	17.4	17.8	18.6	18.2	20.5
Level	3	4	300	300	300	300	301	301	301

Calcaneus	
AM	53.3
Level	10

Table 16.20 Dimensions of measurable skeletal elements of *Capra hircus*

Horn Core			
DMB	29.2	26.1	25.5
DmB	19	17.5	
Level	4	4	4

Humerus		
AmD	12.8	
Ad	28.7	28.6
AT	28	27.6
Ed	25.4	
Level	4	3

Femur	
Ap	41.3
EC	19.2
Level	303

Table 16.21. Dimensions of measurable skeletal elements of *Ovis aries/Capra hircus*

Maxilla			
$LP^2–P^4$	24.6	22.4	23.2
Wear	++	++	++
Level	4	4	8

M3											
L	19.8	20.4	19.1	20	19	19.9	18.7	17.7	20.5	19.1	19.9
A	12.1	12.5	11	11.4	11.9	11.6	11.9	11.1	11	11.1	12.1
Wear	+++	++	++	+	++	+++	+++	++	+	+	+
Level	8	8	8	8	8	9	9	9	9	9	9

Mandible							
LP2–M3						72	
$LP_2–P_4$	25.2	22	22.5	23	24.5	22	20.3
$LM_1–M_3$						49.5	
LM_3						22.2	
AM_3						8.3	
Wear	++	++	++	++	+	++	++
Level	3	3	3	4	4	7	7

M3											
L	23.5	21.8	19.4	24	22.6	23.9	23.6	22.9	22.2	23	23.8
A	8.8	8.2	7.5	8.6	8	8.5	8.6	8.4	8.5	8.8	7.4
Wear	++	++	+++	+++	+	+	++	++	+++	+	++
Level	7	7	7	7	7	7	7	7	7	8	8
M3 *(continued)*											
L	22.4	22.5	22	22	22	21.5	22	21.4	21.8	22.6	21.6
A	8	7.9	7.8	8.3		7.7	7.6	8.2	7.9	7.9	8.1
Wear	+++					+	++	+++	++		+
Level		9	9	9	9	9	9	9	9	9	9

	M3											
	21.3	18.2	18.2	17.9	18.4	19.5	20.5	21.2	20.6	19.5	18.7	
	12	12	10.8	10.5	10.6	11.3	10.9	12	13.2	12	11	10.2
	++	++	++	++	++	++	++		++	+++	+	+
	9	9	9	9	9	9	9	9	9	9	9	10

	Mandible								
			73.5	78	71	79			
	22.3		22.6	24.8	20.7	24.5	23.5	21	19.5
		52	49.5	52.6	49	53.5			
		22.5	22.1	25	22.6				
		8.2	8.7	9.2	8.5				
	++	+	++	+++	++		++	++	++
	7	7	8	303	303	9	9	9	9

	M3										
	19.7	24.2	23	22.6	22	21.1	21.4	21	20.7	21.2	
	7.8	8.1	8.6	8.1	7.8	8.1	7.7	8.3	7.7		
	++	++	++	+++		++	++	++	++	+++	
	8	8	8	8	9	9	9	9	9	9	
	M3 *(continued)*										
	22.2	24	23	21.1	20.5	23	23	23.3	23.7	21.6	
	7.6	7.9	8.8	8.3	8.1	7.7	8.5	8.9	8.1	8.3	7.7
	+	+	++	+++	+	++	++	++	++		
	9	9	9	9	9	9	9	9	303	303	303

Table 16.21. *(continued)*

Axis		
ASCr	41	45.7
Level	8	301

Sacrum	
ASCr	22
HSCr	11.4
Level	9

Scapula					
LmC	20.5	17.6	18.2	17.2	19.1
LS		25.5			
AS		19.2			
Level	5	9	9	9	9

Humerus														
AmD			14.7	14.8		13		14.8		14.2				
Ad	32	29			28.7	28.5	28.2				30		31	35
AT	30.8	27.6				27	27		30		28.4	28	30.3	30.5
Ed						25.3					27.7	25.3	28.1	28
Level	5	6	7	7	7	9	9	9	9	10	303	300	301	301

Radius							
Ap			29.9			30.3	28
Asp			27.6	26.4	26.4	29	27
AmD	16.7	16.4				16.9	
Ad	32	28.8					
ASd	25.5						
Level	3	8	9	9	9	9	9

Ulna				
EmO			24.8	20.5
EPA	23.3	22.5	28.4	24.1
APC	17.7	17.3	19.5	17.4
Level	7	9	303	303

Metacarpal						
Ap	23.4	22	23.7			
AmD		12.8				
Ad				25	24.9	23.2
Level	8	9	300	303	303	303

Pelvis				
Hml	15.9			16.1
LA		29.5	27	29
AA		26.5		
Level	3	6	9	303

Femur		
EC	21	
Ad		36
Level	303	303

Tibia		
AmD		14.4
Ad	27	25.2
Ed	20.2	20
Level	7	8

Calcaneus	
AM	21.5
Level	6

Centrotarsal			
AM	26.1	24.5	22.6
Level	8	10	303

Astragalus												
LMl	29	29.5	27	26.8	27.3	27.6	27.4	27.1	26.7	29.2	26.4	28.8
LMm	27.3	28.1	26.2	27.3	25.8	26.1	26.5	26.2	25.3	28		
El	6.6	16	15.6	15.3	15.8	15.2	15	15.5	14.6	16.7	14.8	
Ad	18	18.5	18.6	17.1	18.3		16.8	17.6	16.8	18.2		17.1
Level	8	8	8	8	8	9	9	9	9	9	9	9

Table 16.21. *(continued)*

Metatarsal						
Ap	19		19.7	21.5		
AmD					11.1	
Ad		23.8			24	23
Level	4	5	9	9	9	300

Phalange 1				
LM	35.7	41	35.4	36
Ap	11.4	12.4	11.6	11.8
AmD	8.9	9.6	9	9.8
Ad	11.1	11.2	10.8	
Extr	post	post	post	
Level	9	303	303	10

Phalange 2								
LM	21.8	19.6	23	19.2	18.5	21	21	20.5
AM	12.1	11.7	11.5	10.1	10.6	11.5	12	10.7
AmD	8.9	8.2	8.5	6.9	7.7	8.2	8.6	7.7
Ad	10.1	9.6		7.9	8.7	9.1	9.3	8.3
Extr	ant	post	post	post	post	post	post	post
Level	6	3	6	10	303	303	9	9

Phalange 3	
LdS	26
Ldo	21.5
Extr	post
Level	8

Table 16.22. Dimensions of measurable skeletal elements of *Capra pyrenaica*

M3		
L	19.8	21.6
Level	301	301

Carpal 2+3	
AM	21.1
Level	3

Carpal 4	
AM	15.6
Level	5

Metacarpal	
Ed	20.3
Level	301

Patella	
AM	26.6
Level	5

Astragalus			
LMl	38.8	37.9	35.5
LMm	36.3	36.3	33.5
El	21.5	21	18.9
Ad	25	25.9	22.6
Level	3	301	10

Centrotarsal	
AM	28.2
Level	3

Metatarsal		
Ed	21.8	20.7
Level	301	301

Phalange 1				
LM	42.4	43.5		
Ap	13.4	14	15.2	
AmD	11.1	10.3		
Ad	13.2	13.2		13.8
Extr	post	post		
Level	303	5	5	10

Phalange 2						
LM	27.2	31	28	26.5	27.5	28.7
Ap	19	18.4	14.2	14.9	13.4	13.2
AmD	13.1	13.1	10.7	11	10.2	10.3
Ad	14.5	13.7	11.8	10.8	11.1	
Extr	ant	ant	post	post	post	
Level	3	3	3	3	4	10

Phalange 3	
LdS	42
Ldo	39.5M
Level	301

Table 16.23. Dimensions of measurable skeletal elements of *Sus domesticus*

Maxilla	
LP_1–P_4	46.5
Wear	+
Level	5

Mandible		
LP_2–P_4	34.2	
LM_1–M_3		62
LM_3		32.6
AM_3		14.5
Wear	++	++
Level	5	9

Loose Teeth

M^1			
L	14	15.7	17.5
A	12.9	14.6	13.3
Wear	++	+	++
Level	5	8	9

M^2			
L	18	17.6	20.4
A	16	16.8	18.6
Wear	++	o	+
Level	5	5	6

M^3				
L	31.1	32.4	32.6	32.7
A	19.4	19.8	17.5	
Wear	+	+	+	++
Level	3	4	6	6

M_1					
L	17.3	16.2	17.3	17.9	17.8
A	11.4	11	11.4	13.5	11.1
Wear	+	+	+	+++	
Level	7	8	9	9*	10

M_2		
L	20.2	23
A	14.4	14.7
Wear	+	+
Level	8	9

M_3		
L	36	35.8
A	16.5	16.4
Wear	++	++
Level	6	7

Scapula			
LmC	22.3	23	24.9
LMP		35.5	39.2
LS		31.6	32
AS		24.8	25.3
Level	5	9	9

Humerus	
AmD	18.5
Level	7

Radius	
Ap	27.4
Level	7

Ulna	
APC	21.4
Level	303

Metacarpal 5	
LM	56.5
Level	6

Femur	
AmD	16
Level	6

Tibia		
AmD	18.3	20.5
Level	7	9

Astragalus	
LMl	40.5
LMm	38.56
Level	6

Calcaneus	
AM	21
Level	7

Metatarsal 5	
LM	59.6
Level	4

Phalange 1			
LM	33.5	33.8	25.8
Ap	15.5	16.9	12.4
AmD	13.1	12.9	9.7
Ad	15	14.8	11.3
Extr	ant		
Level	5	4	10

Table 16.23. *(continued)*

Phalange 2			
LM	22	21	21.5
Ap	16.1	14.9	12.8
AmD	12.5	12.1	
Ad	14.6	13.1	
Extr	ant	ant	
Level	5	303	9

Phalange 3					
LdS	33.7	30	29.7	29.5	27
Ldo	32.2	29	26.5	26.5	26
Level	4	5	7	8	8

*Possibly *Sus scrofa*

Table 16.24 Dimensions of measurable skeletal elements of *Sus scrofa*

Mandible	
LP_2–M_3	114
LP_2–P_4	39.7
LM_1–M_3	76
Level	7

M_3	
L	40
A	20.2
Level	9

Radius	
Ap	34.1
Level	8

Astragalus	
LMl	46.4
LMm	42.7
Level	303

Phalange 1	
Ad	19.5
Level	8

Table 16.25. Dimensions of measurable skeletal elements of *Cervus elaphus*

Mandible	
LP_2–P_4	37
Level	9

Radius		
Ap	53.2	
Asp	51	41
AmD		23.5
Level	9	9

Metacarpal	
Ed	28.7
Level	301

Astragalus			
LMl	57.8	48	
LMm	49.8	45	
El	25.6	25	
Ad	30.5	30.2	36
Level	10	9	301

Centrotarsal				
AM	47.8	39.6	38	46.2
Level	301	10	10	10.1

Phalange 1				
LM	55			
Ap				19.1
Ad		19.2	21	
Level	10.1	10.1	10	9

Phalange 2		
LM	35	
Ap	17	18.9
AmD	12.8	
Ad	13.8	
Level	10	9.8

Table 16.26. Dimensions of measurable skeletal elements of *Capreolus capreolus*

Maxilla	
LP^2–P^4	28.5
Level	10.1

M3		
L	16.2	16.3
A	7.2	7.2
Level	10	10

Phalange 1		
LM	36	
Ap		11.4
Level	10.1	10

Phalange 2	
Ap	10.2
Level	10.1

Table 16.27. Dimensions of measurable skeletal elements of *Rupicapra rupicapra*

M3	
L	20.5
A	7
Level	3

Radius	
LM	185.5
Ap	28.6
AS	25.8
AmD	16.8
Ad	26.6
ASd	25
Level	301

Metatarsal	
LM	159.5
Ap	22.3
AmD	11.9
Ad	15.1
Ed	17.4
Level	8

CHAPTER SEVENTEEN

Descriptions of the Post-Paleolithic Pit and Hearth Features of El Mirón Cave

Lawrence Guy Straus and Manuel R. González Morales

The objective of this chapter is to present the basic descriptions of man-made features encountered in the Neolithic, Chalcolithic, and Bronze Age deposits in El Mirón Cave. The sources of information vary (field notes on excavation square "spit forms," the general excavation journal, drawings, photos, special rock fill count lists for pits, and general artifact inventories coded by level, spit, square, and subsquare). Together they constitute a significant corpus of data on a relatively under-studied aspect of the archaeological record for post-Mesolithic/pre-Roman Cantabria, namely non-burial structures in residential contexts for early food-producing societies.

Almost all the Holocene-age features (or structures) were found in the Cabin excavation area, and coincidentally, most of these (including almost all the largest, most spectacular pits) were uncovered in the first days of the initial (1996) campaign of test excavations, whose goal was to ascertain whether El Mirón Cave contained intact Paleolithic deposits. Thus, although excavated carefully by trowel, the (at first unexpected) Bronze Age, Chalcolithic, and late Neolithic features were relatively quickly dug and documented. Because parts of some features were dug in one year and other parts in later campaigns (mainly in 1997 and 1998), drawings and photos were done piecemeal and sometimes at different depths, leading to discordances in the documentary record. The figures provided here are thus often composites whose different parts do not always seamlessly fit together. In 1996 only six meter-squares were dug (two of which are partial squares), a fact that caused some large pits to be only partially excavated in that year. Sometimes a feature was not initially recognized as such, if only a slight amount of it entered into a particular excavation square, so the composite drawings may have missing areas. The speed with which some of the features had to be cleared in order to attain underlying deposits sometimes precluded full documentation. Nonetheless, these structures are among the better documented ones in the current record for the pre–Iron Age ceramic periods of Cantabrian prehistory, since many of the known sites (particularly of the Chalcolithic and Bronze Age) in this region are funerary (ossuary) caves, not living sites.

Another unfortunate circumstance stemming from the discovery of most of these features during the initial (and somewhat shorter than usual) testing season at El Mirón is that systematic sediment flotation in a field laboratory had not yet been arranged for and instituted (this began in 1997 and has continued ever since). Water screening was done, but on-site, with the recycling of water supplied by local firemen. This arrangement was less than ideal, as the water quickly became very muddy and opaque due to the clayey nature of the sediments. Late, in-lab water screening and systematic flotation did yield better recovery results, but not in time for many of the largest and most spectacular pits of the Chalcolithic and Bronze Age uncovered in 1996.

What follows are simple descriptions of the structures based on the drawings, content lists (both piece-plotted finds and those recovered in screening—all computerized—plus less formal on-site counts of pit rock fills, which generally were not saved except in the case of possible grinding stones), and notes made by the excavators on the spit forms and by Straus in his excavation journals. Provenance treatment of pit contents varied on a case-by-case basis. Any finds from provenance units specifically identified as feature fill (i.e., exclusive decimalized level numbers for features—see below) and those corresponding to spits and subsquares occupied by particular features without an exclusive stratigraphic designation are included in our listing of contents. Once recognized as features (and such recognition sometimes came quickly but in other cases more slowly, depending on whether the feature was situated within a single square or among excavation units and whether or not the latter were being dug simultaneously or sequentially), many of their contents were given separate stratigraphic unit (lens) designations as decimals, generally after the number assigned to the major level (apparently) most closely associated with the feature—often the level into which the features were dug, meaning that the feature (completely or partially) postdated the level in question. Our best estimate of the stratigraphic association of each feature is given in the individual descriptions that follow, but there is sometimes unavoidable ambiguity in specific attribution. In some cases where pits cut into other pits, a sequence of events can be developed. In some cases, based on attributes observed in situ, we can suggest whether a structure was a primary fire feature (hearth) or simply a pit that may have become backfilled with materials including ash, charcoal, and burned items presumably derived from nearby, (pene-)contemporaneous hearths. A few other phenomena classified in the field as "feature" may have been simply accumulations of stones or masses of burned sediments lying on surfaces (i.e., not dug-out structures), while a few others (mostly not in post-Paleolithic deposits but rather in late Magdalenian ones in the vestibule rear, or Corral, area) turned out upon excavation most likely to be rodent burrows, not man-made pits. The post-Paleolithic features are described here in approximate stratigraphic order, from oldest to youngest (lowest to highest).

As described in the chapters on taphonomy and lithic artifacts (chaps. 15 and 18, this volume), it is apparent that small quantities of Upper Paleolithic faunal remains and artifacts (including fragments of rather classic antler *sagaies*) were introduced into post-Paleolithic deposits (especially pits), probably as a consequence of erosion from a wedge of Magdalenian- and Azilian-age materials that had originally continued far up the ramp leading east toward the inner cave and that later was moved downslope by a combination of water flow and the activity of humans and livestock in the cave during the mid-Holocene.

Neolithic Features

FEATURE L

This is a semicircular pit located in square I2b and much of I2d, extending very slightly into I1b at the edge of the cave wall. It measured about 90 cm long × 80 cm wide × 20 cm deep. However it may have also extended into J2a+c, although this area was not included in the drawing (figure 17.1). Feature L was dug from Stratum 9 into Stratum 10. It contained charcoal overlain by ash. There were some fire-cracked rocks. A total of 69 stones measuring >3 cm each were recovered from the pit fill in I2; 14 of these measure >10 cm; 4 are waterworn cobbles (2 of which have use wear). Only three artifacts—all debitage items—were recovered from this feature. Feature L was probably an in situ hearth pit. It was excavated in 1996.

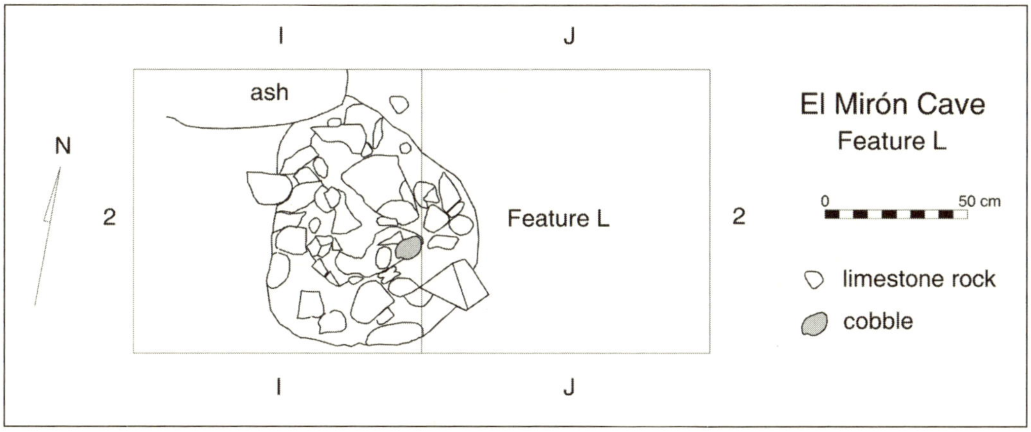

Fig. 17.1. Feature L. Drawing by L. G. Straus, redrafted by Ronald Stauber.

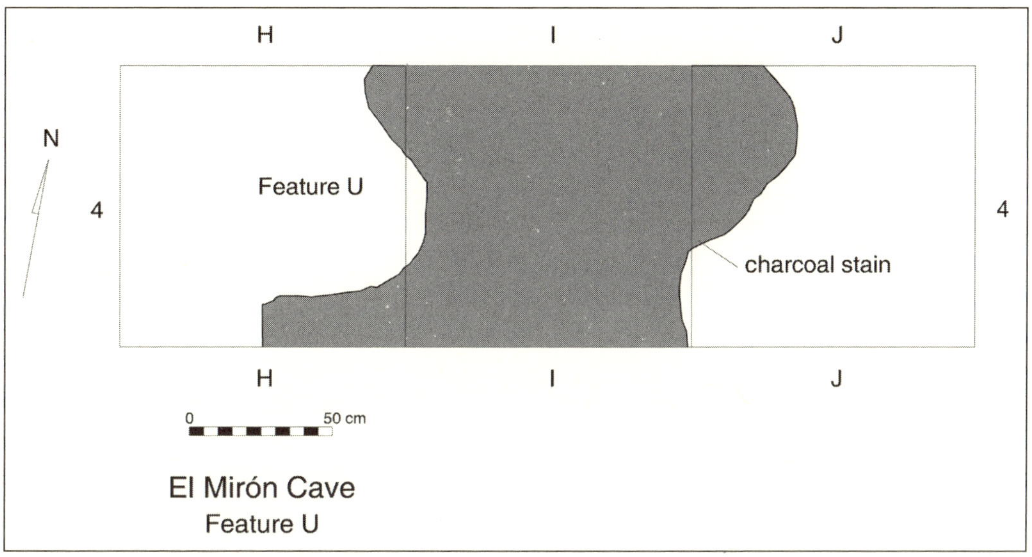

Fig. 17.2. Feature U. Drawing by L. G. Straus, redrafted by Ronald Stauber.

FEATURE U

This is a problematic feature that may be a natural or seminatural depression (i.e., probably not an artificially dug-out pit) in squares I4b–d and J4a–c. Irregular in outline, it (vaguely) measured about 1.5–1 m long × 1 m wide × 1–5 cm deep (figure 17.2). This shallow double basin was filled with a lens of sediments called Stratum 9.8. These were charcoal-stained and lay directly atop Stratum 10. Contents were few and include cow bones and a mandible. Apparently this depression was used by humans in which to build a fire. There is a small hole closely associated with this depression. It was excavated in 1998.

FEATURE T

No plan was drawn of this small pit located in I4c+d. It measured about 50 cm × 50 cm × 10 cm deep. It was dug entirely within Stratum 9.6 and was covered by the top of that layer. Pit T contained red-pink sediments but only two small pieces of fire-cracked rock. It was probably a small hearth. It was excavated in 1998.

FEATURE S

This is a small pit mainly in Subsquare J4c, measuring about 50 cm × 50 cm × 17 cm deep (figure 17.3). It was dug within Stratum 9 and lay above Stratum 9.6. The pit was filled with dark ash, although the top and the bottom had lenses of white ash with charcoal flecks. Contents also included several angular (fire-cracked) rocks, including four large ones. The various ash lenses mainly lay atop the stones, although the stones did lie atop a basal charcoal lens that was in direct contact with the top of Stratum 9.6. This manifestly was an in situ hearth. It was excavated in 1998.

Fig. 17.3. Feature S. Drawing by L. G. Straus, redrafted by Ronald Stauber.

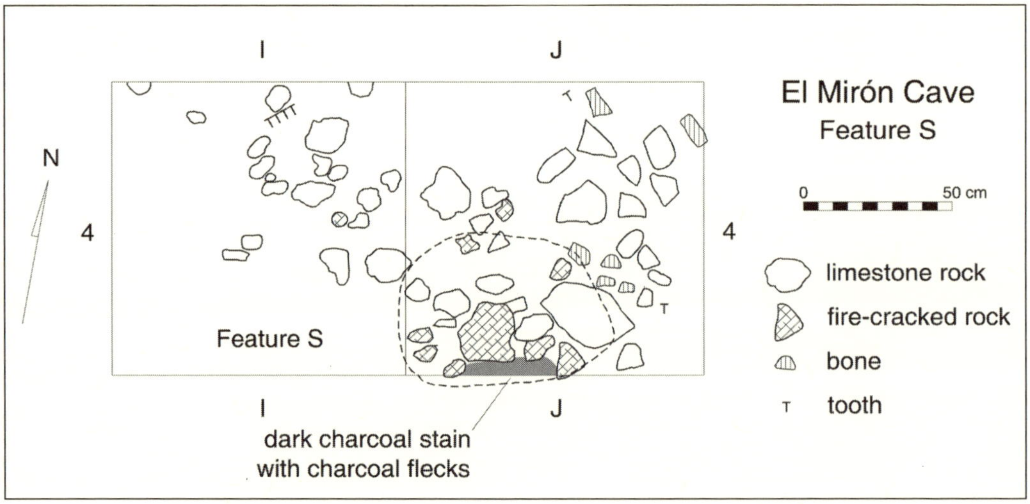

Fig. 17.4. Feature M. Drawing by L. G. Straus, redrafted by Ronald Stauber.

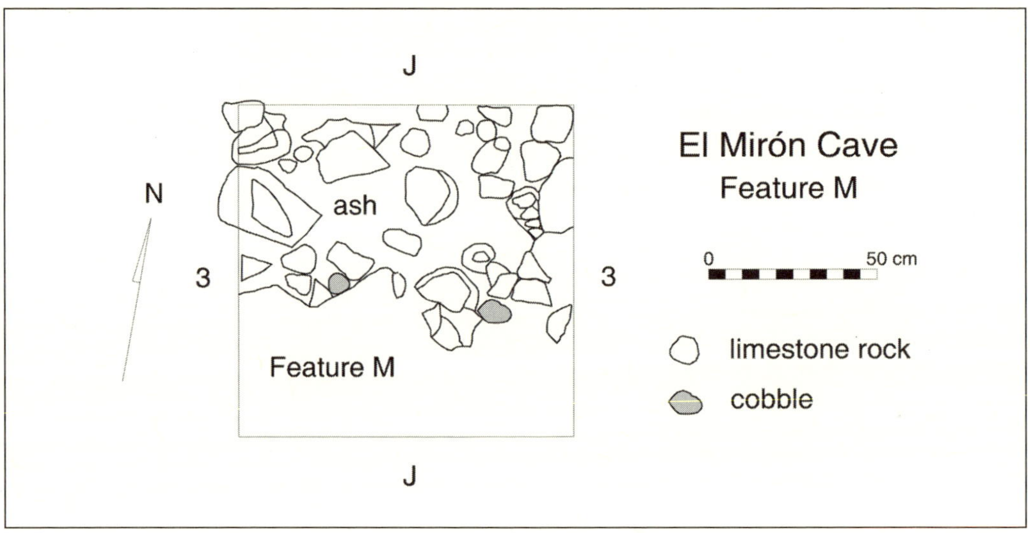

FEATURE M

This is a "rock pile" constituted by angular stones around an area filled with ash above a lens of reddish-brown, (apparently) burnt silt in Square J3a, b, c and a small eastern part of I3 (figure 17.4). It is not clear that these rocks and burnt sediments were actually in a pit. Feature M measured a bit more than 1 m square and was at least 5 cm thick. The feature lay within Stratum 9, and the contents were designated as a separate lens, Stratum 9.2. Some of the rocks underlay the ash lens. A total of 140 rocks measuring >3 cm each were recovered. Of these, 31 measured >10 cm and 12 were waterworn cobbles (of which 5 had traces of use wear). There were virtually no other finds (two flakes). This feature may have been a surface bonfire. It was excavated in 1996.

Chalcolithic Features

FEATURE I

This is a large, semicircular pit located in square I3a+c and H3b+d, with an associated ash lens that continued into I2a. The pit measured about 100 cm long × 80 cm wide × 15 cm deep (figure 17.5). It had been dug entirely within Stratum 7 into the base of that layer and was covered by the top of Stratum 7 as well. The contents are called Stratum 7.2. This is a case of a structure dug during an unambiguous period, namely Stratum 7—early Chalcolithic. Its eastern edge was defined by two large *éboulis* blocks in I3a+c. It contained charcoal and ash and a few bones but no sherds. The fill included about 100 stones measuring >3 cm each, of which 12 were

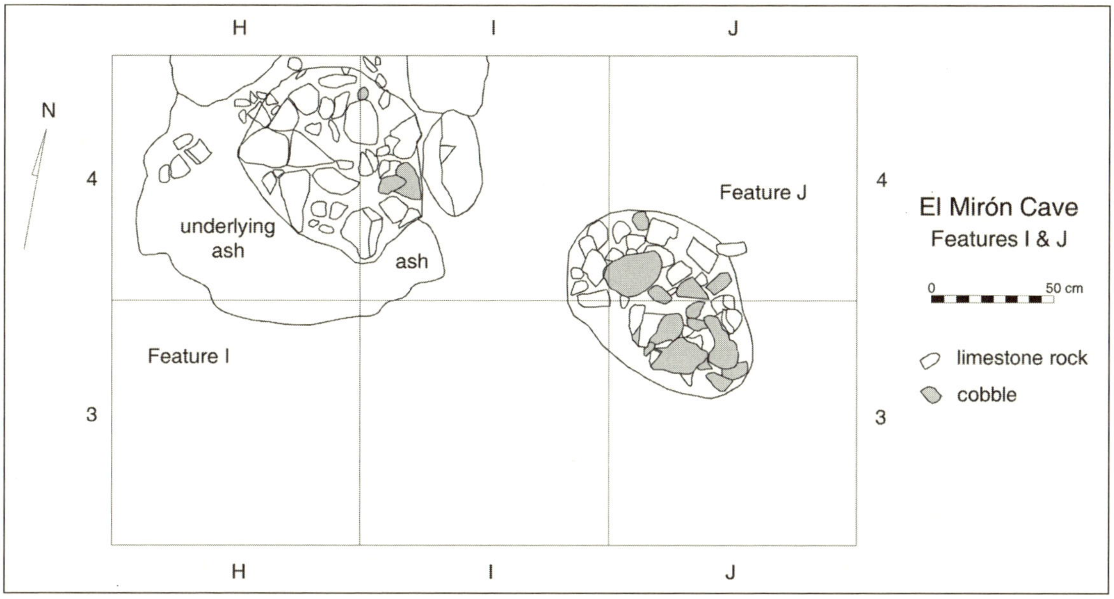

Fig. 17.5. Features I and J. Drawing by L. G. Straus, redrafted by Ronald Stauber.

waterworn cobbles (4 of which had traces of use wear). Pit I was excavated in 1996 and 1997.

FEATURE J

Feature J is another large pit, similar to Feature I, located in Subsquares J3a+b, J4c+d, and a small area of I3b. It had been dug into Stratum 7 and was capped by a localized layer of beige clay with lenses of ash and charcoal called Stratum 6.1—perhaps representing continued fire episodes in much the same spot as pit J. Feature J might be a little later in construction than Feature I (but this is unsure), and it does seem to predate pits G and D. The top of pit J was actually cut by the base of pit D in I3b. Pit J measured about 95 cm long × 70 cm wide × 15 cm deep (figure 17.5). Its upper, dark, organic-rich fill was labeled Stratum 7.4, with the basal lens being called Stratum 7.5. The whole fill was very rich in charcoal. Stratum 7.5 clearly represents the first generation of fires in this hearth pit, with abundant ash, charcoal, a carbonized branch, and numerous cobbles about 2 cm thick. Overall, the contents included at least 98 rocks measuring >3 cm each, of which 29 were waterworn cobbles and 1 was a large cobble anvil. Broken down into subdivisions, Stratum 7.5 specifically included five flakes, a backed bladelet, and two bones, while Stratum 7.4 yielded a burnt and pecked cobble, a second pecked cobble, a couple of flakes, and a retouched blade. Pit J was excavated in 1996 and 1997.

FEATURE K

This is simply an ash patch originally measuring a bit more than 40 cm × 50 cm × 15 cm thick (figure 17.6). The western part of the feature was excavated in J3d and J2b, but it continued into K2a and K3c, where it was

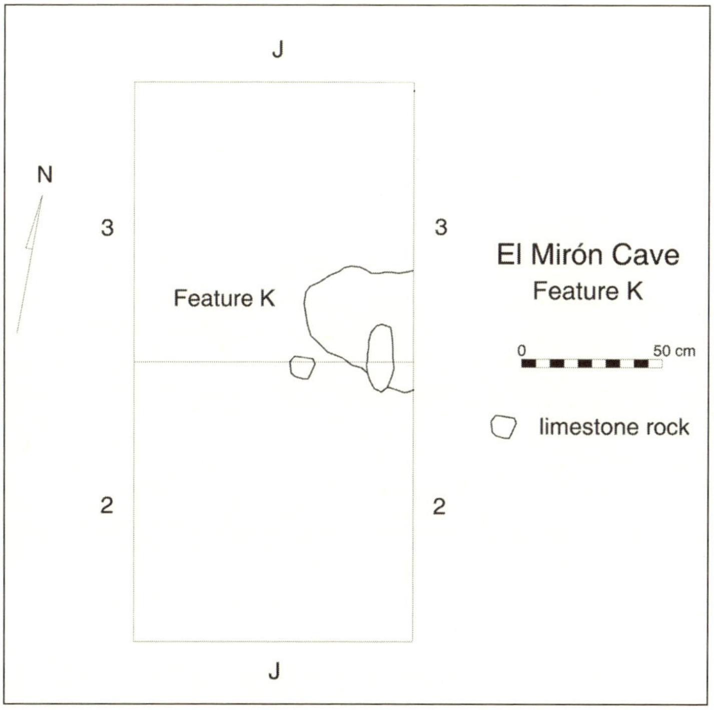

Fig. 17.6. Feature K. Drawing by L. G. Straus, redrafted by Ronald Stauber.

not excavated. It was within Stratum 6.1 and lay atop Stratum 7. Aside for one large éboulis rock, contents of the excavated portion of this ash lens were nil. It was excavated in 1996.

FEATURE Q

Feature Q is a small, sub-circular basin dug into the top of the white ash Stratum 6. It was located in I4 and measured about 25 cm × 22 cm × 3 cm deep. It contained dark grey ash with a black charcoal ring around its circumference. It was filled with fire-cracked rocks and one piece-plotted sherd. It must have served as a hearth/roasting pit, though clearly functionally distinct from the very large rock-filled pits such as pits I, J, H, D, and C. It was excavated in 1997.

FEATURE H

Feature H is a very large pit, measuring about 130 cm long × 105 cm wide × 25 cm deep and located in squares H2b+d and I2, plus a small northern strip of I1a+b adjacent to the cave wall (figure 17.7). It is regularly circular in outline and bowl-shaped in section. It was dug through Stratum 5 into the top of Stratum 7. The fill of Feature H is called Stratum 5.2. It is similar to and probably a bit older than adjacent pit C, which seems to cut into Feature H. It is perhaps contemporaneous with Feature D. It was filled with fire-cracked rocks and cobbles: 168 stones measuring >3 cm each (of which 51 measure >10 cm each); 15 of these rocks were cobbles, of which 7 had use wear. There were many bones of cattle, pigs, and ovicaprines—some of which were burnt. There is also a burnt segment of stalagmite. Sherds and charcoal pieces were abundant. Lithic artifacts, on the other hand, were rare: only two pieces of shatter (<1 cm) were recorded. This structure probably functioned (at least in its last use) as a roasting pit. Feature H was excavated in 1996 and 1997.

FEATURE D

Pit D is the largest of the features excavated: at least 150 cm long × at least 75 cm wide × 20 cm deep (figure 17.8). It is located in J4a, b, d and the north edge of J3. It cut from Stratum 5 through or into Strata 6, 6.1, 7 and—locally—even the top of 8. Its fill was called Stratum 5.4. It is probably (pene-)contemporaneous with pit C; it predates pit R which cuts into it; it may be predated by pit J, into which Feature D may have cut. Feature J underlies Stratum 6.1, which Feature D cuts through in one place. Irregularly oval (definitely not round) in shape, "D" was filled with stones: a total of about 530

Fig. 17.7. Feature H. Drawing by L. G. Straus, redrafted by Ronald Stauber.

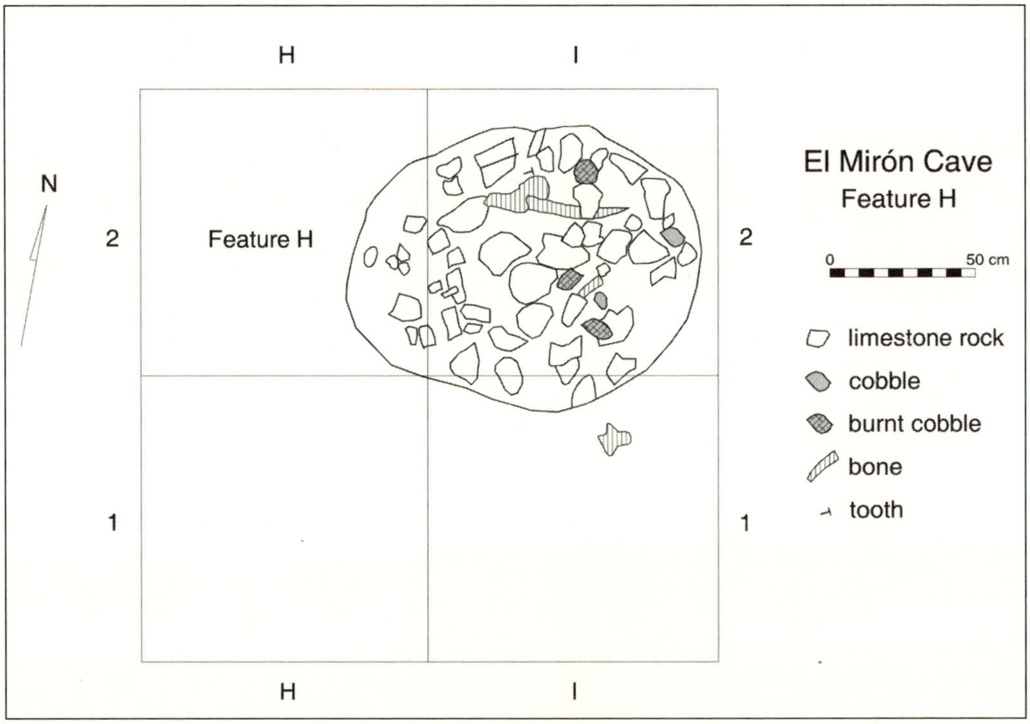

Fig. 17.8.
Features C and D. Drawing by L. G. Straus, redrafted by Ronald Stauber.

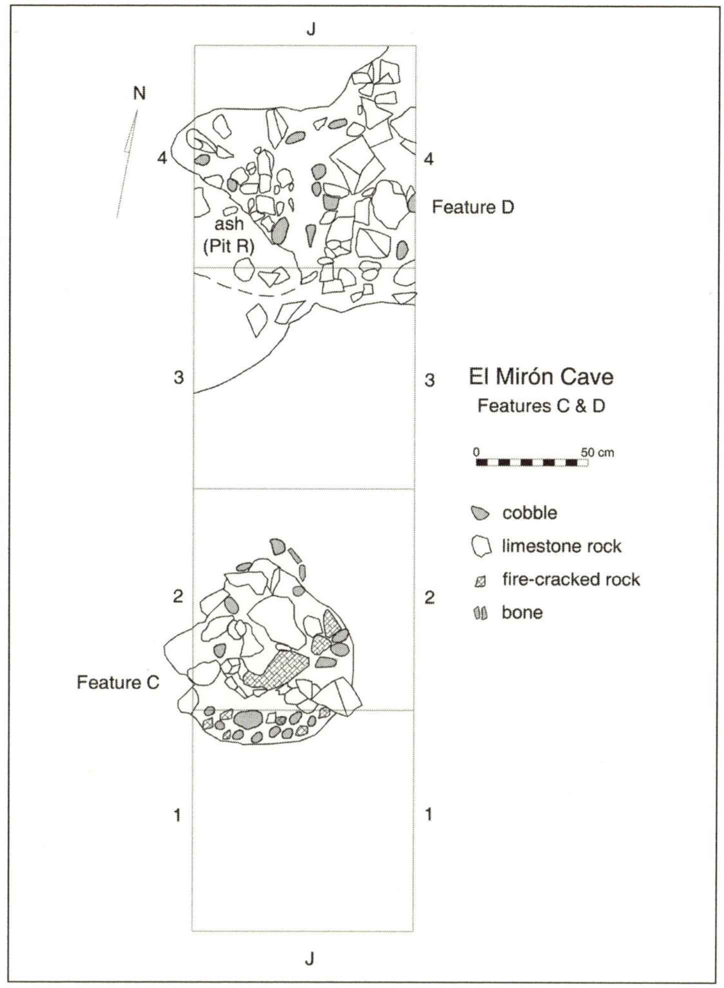

stones of >3 cm in length. Of these, 66 measured >6 cm; 15 were waterworn cobbles (4 with use wear traces). Many of the stones were fire-cracked. Objects recovered from the fill include four flakes, a backed, pointed blade (typologically a Gravette point—possibly intrusive or collected as a curiosity by Chalcolithic people) from the base of the pit, four bones, and four large chunks of charcoal. It is not clear whether this was originally a hearth, but at some point it may have been used as a roasting pit as well as a trash pit. It was excavated in 1996 and 1997.

FEATURE C

This is another circular, basin-shaped, rock-filled pit located in J2c, parts of J2a+d and J1a, and the eastern edge of I2. It measured about 90 cm long × 75 cm wide × 25 cm deep (figure 17.8). Feature C was dug from Stratum 5 through Stratum 6 and into Stratum 7. Two possible postholes were physically associated with it. The base of a rhomboidal arrowhead of Chalcolithic type was found nearby in J2a but not in pit C. The rock fill included 180 stones measuring >3 cm each, of which 60 measured >10 cm; 36 of the stones were waterworn cobbles (13 of which had use wear). Many of the rocks were apparently fire-cracked. The fill contained ash and abundant charcoal (including three large chunks), a few sherds, a few bones, some mussel shells, and a perforated shell of a univalve mollusk. This structure may also have been used as a roasting pit. It was excavated in 1996.

FEATURE G

This is an oval pit with a pile of stones atop it in Subsquares I3a+b and I4d. It is associated with a large, square "posthole" in I3b that was filled with loose, blackish sediment. Feature G measured about 110 cm long × 65 cm wide × 25 cm deep (figure 17.9). It was

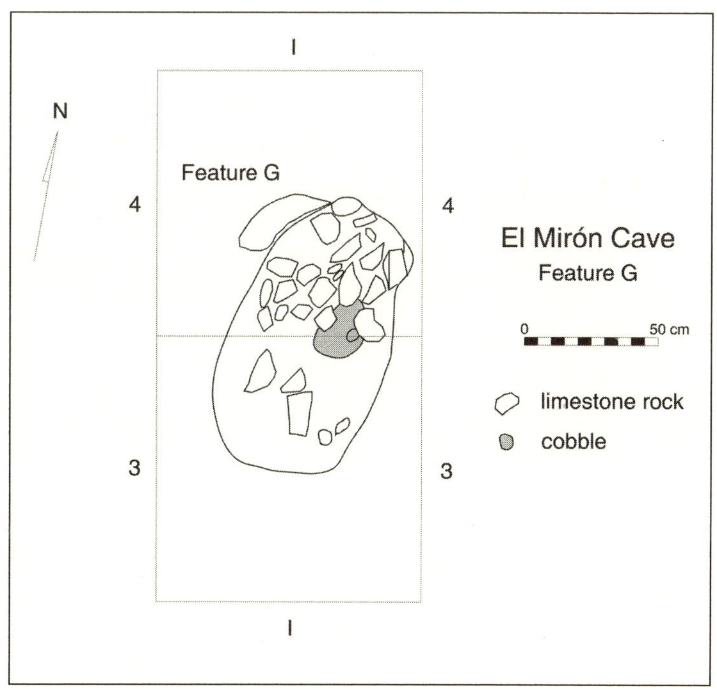

Fig. 17.9.
Feature G. Drawing by L. G. Straus, redrafted by Ronald Stauber.

dug from the base of Stratum 5.0 through Strata 5 and 6 and into the top of Stratum 7. It predated pit R and postdated pit J. Its fill was called Stratum 5.3. The base was regularly rounded and was covered with ash and charcoal. There were a few sherds and flint artifacts at the pit base, along with one stone mortar (there was also a second, less clear mortar). The pit fill was otherwise archaeologically nearly sterile except for a large number of limestone rocks and waterworn cobbles, especially those of medium size (5–15 cm). In total there were about 72 stones of >3 cm in length (28 of which equaled or exceeded 10 cm), of which 9 were cobbles. Three of the latter had use wear traces. This pit may also have had multiple functions during its lifespan: storage (?), hearth/roasting pit, trash pit (?). It was excavated in 1996 and 1997.

FEATURE P

This is a small hearth pit filled with fire-cracked rocks. It was circular and shallow, measuring about 40 cm × 35 × 5 cm deep (figure 17.10). It was located in H4b+d and a small part of I4a+c. It was dug into Stratum 5.0 from Stratum 5. Some lithics, bones, or ceramics were found in this feature. It was excavated in 1997.

FEATURE R

This is a small pit (50 cm × 35 cm × 15 cm deep) located in Subsquare J4c (figure 17.11). It was dug into pit G. The fill was called Stratum 5.5 and contained 34 rocks >6 cm in length in white-ash matrix. There was a large piece of wood in the fill and burnt cobbles at the base. It was probably a hearth pit. Feature R was excavated in 1997.

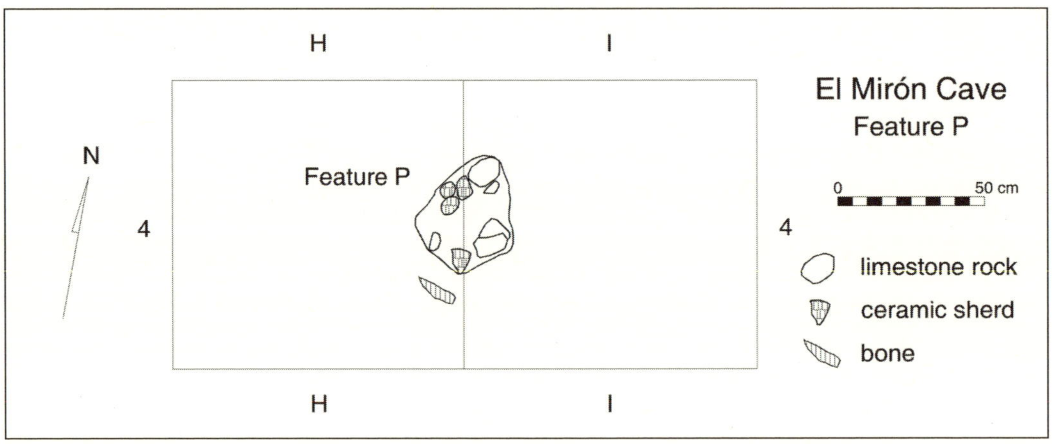

Fig. 17.10. Feature P. Drawing by L. G. Straus, redrafted by Ronald Stauber.

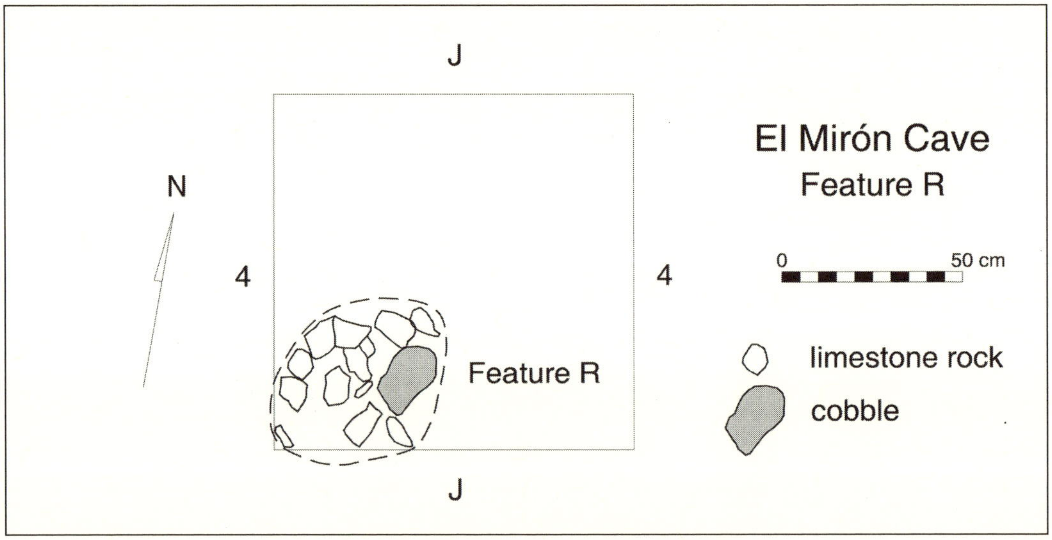

Fig. 17.11. Feature R. Drawing by L. G. Straus, redrafted by Ronald Stauber.

FEATURE F

Feature F is a small (50 × 50 × 5–10 cm deep) pit centered in Subsquare I3c and continuing at least into I2a+b (and perhaps parts of I2d and H3d) (figure 17.12). It was semicircular but very irregular in form, probably due to rodent burrows. Indeed, it is possible that Feature F itself is not (at all?) a man-made pit but rather (wholly or partially) a large burrow, although this is not proven. It was dug through Stratum 5 and into Stratum 6, probably from the base of Stratum 4. It was filled with gray, ashy sediments and a relatively large number of stones: 51 measuring >3 cm in length (16 of which were >10 cm), of which 7 were waterworn cobbles (1 with use wear). It was excavated in 1996.

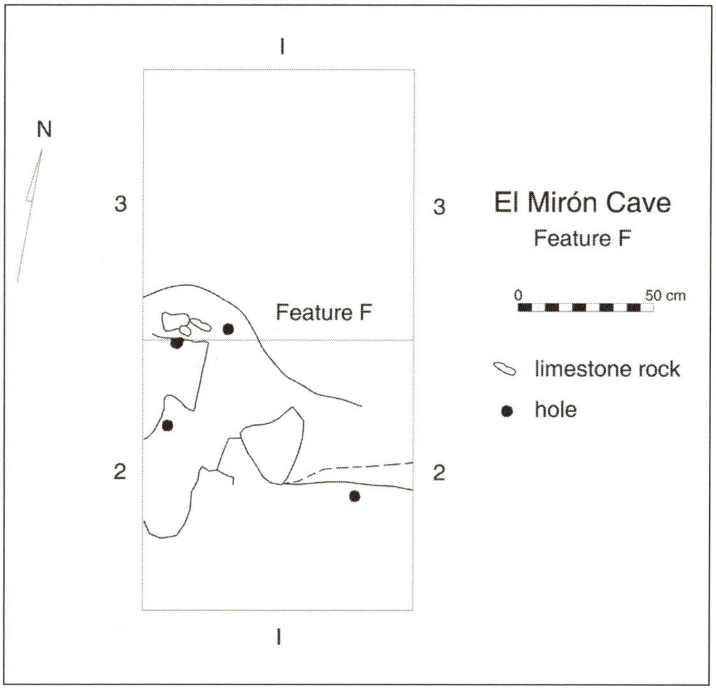

Fig. 17.12. Feature F. Drawing by L. G. Straus, redrafted by Ronald Stauber.

FEATURE E

Feature E is a small pit filled with stones dug from the top of Stratum 4 *sensu stricto* into its lower part in Subsquare I3d. It measured about 40 cm × 35 cm × 20 cm deep (figure 17.13). It is apparently associated with two small (post- or burrow) holes in I3c. The fill of pit E was rich in charcoal. It lacked finds and was probably a small hearth pit. It was excavated in 1996.

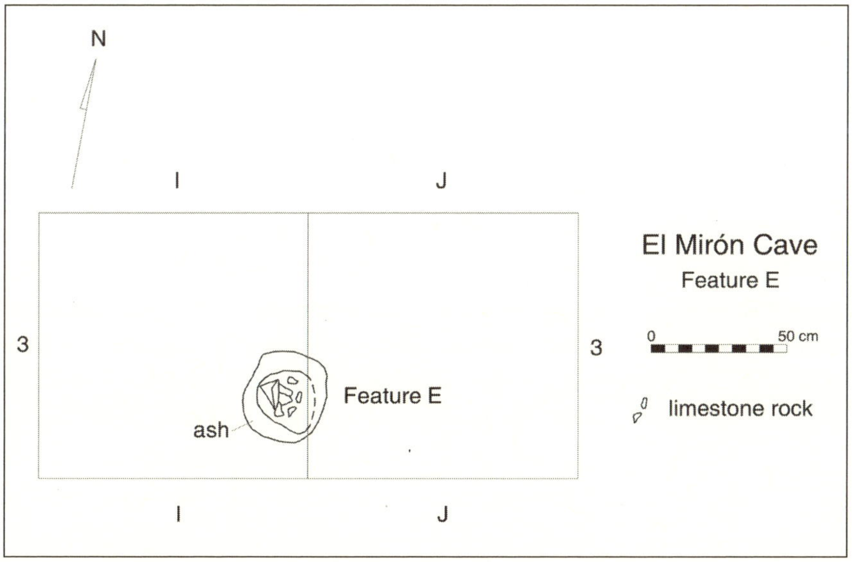

Fig. 17.13. Feature E. Drawing by L. G. Straus, redrafted by Ronald Stauber.

Late Chalcolithic or Early Bronze Age Features

FEATURE A

Feature A is a hearth pit dug into Stratum 4 and was filled with material from either later Stratum 4 or early Stratum 3 occupations. It did not reach the base of Level 4. It was located in Squares J2 and J3 and continues into unexcavated Squares K2 and K3. It measured at least about 75 cm × 65 cm × 20 cm deep (figure 17.14). It is associated with nine possible postholes (some of which, however, were more problematic [i.e., possibly rodent burrows] than others). The matrix of the pit A fill was ash. That fill included 180 stones measuring >3 cm in length (of which 38 were >10 cm long). Of those stones, 23 were waterworn cobbles, 8 of which had wear traces. The top of the fill yielded a very large, battered cobble (a possible anvil, measuring about 40 × 20 cm); two pieces of a large, broken, fire-reddened slab (a possible grinding stone or metate, also measuring about 40 cm long by about 20 cm wide when reassembled); a cluster of six to seven small pebbles (water- or milk-boiling stones?); a circularly pecked stone (a jar stopper?); several sherds; a few flint artifacts; and various cow bones (not all burnt). There were two fire foci and several lenses of ash and charcoal within the pit, which seems to have had at least two burning episodes. The upper of these episodes had a fire-reddened zone in its center with a mass of fire-cracked limestone rocks and cobbles in a basin scooped into the ash of the earlier fire. A small pit dug from the base of Feature A cut through Stratum 5 and even into Stratum 6. It too was filled with fire-cracked rocks and cobbles. Pit A, along with pit B, was archaeologically excavated during the very first days of the project in 1996.

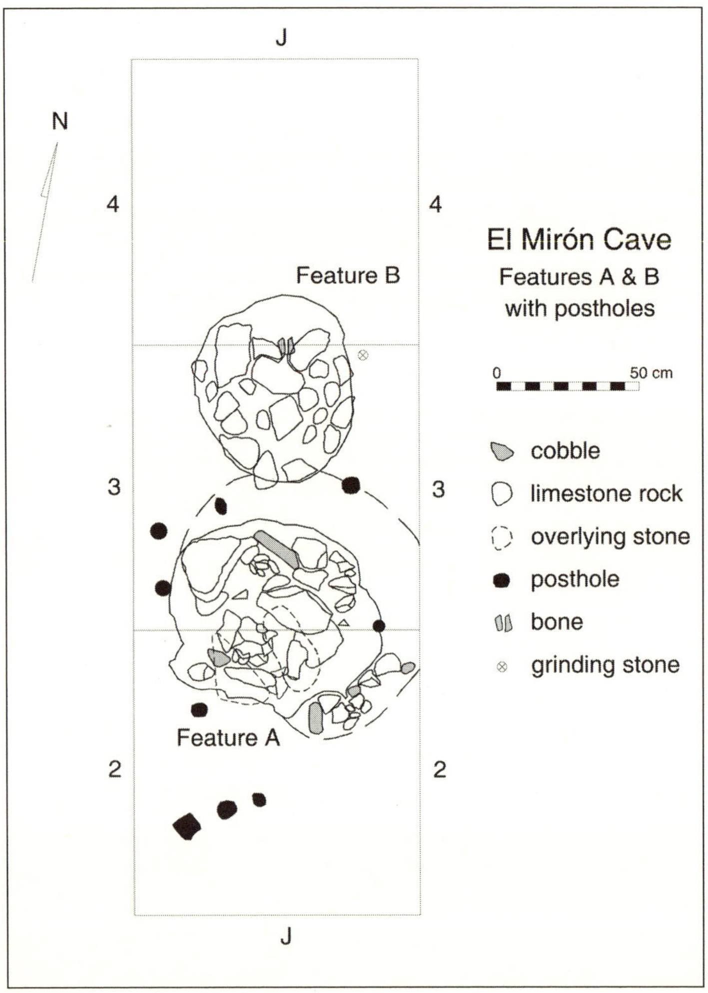

Fig. 17.14.
Features A and B. Drawing by L. G. Straus, redrafted by Ronald Stauber.

FEATURE B

Pit B is another large hearth in J3 extending into J4. It measured about 65 cm × 55 cm × 15 cm deep (figure 17.14). It was dug into Stratum 4 and was filled with either late Stratum 4 or early Stratum 3 sediments, consisting of ash and stones. It was associated with the same nine possible postholes as pit A. Feature B is possibly slightly younger than Feature A, which is deeper relative to the surrounding levels. Feature B may cut partially into Feature A. The rocks in the fill were angular and some were waterworn cobbles. There were 30 stones measuring >3 cm in length, 8 of which were >10 cm. There were some sherds. They were not intensely burned. There was a very large block in essentially all of Subsquare J4d that might also be part of the fill of pit B. Near pit B there was an apparent upper grinding stone (a handstone, or *mano*). The possible postholes associated with pits A and B have depth ranges between 3 and 23 cm (averaging 12.4 cm). One is large and square in shape; two are small and regularly circular in outline. The rest are less regular. These holes seem to have been dug through Stratum 4 either from the top of that level or from the base of Stratum 3.

GENERAL OBSERVATIONS ON THE LARGE CHALCOLITHIC/EARLY BRONZE AGE PITS

A distinct type of round or oval pit of ±1 m in average diameter by only 15–20 cm in depth emerges from these descriptions of features dug in Levels 4, 5, and 7—the principal Chalcolithic occupation horizons (a couple of them—Features A and B—might have been dug through Level 4 from the base of early Bronze Age Level 3, but this is not entirely clear). These strikingly large, similar pits include Features A, B, C, D, G, H, I, and J. A few precedents for these large features do seem to have existed in late Neolithic times (pits L and 98a, the latter having been enlarged and reused, probably in Chalcolithic times). All these big pits are characterized by the presence of large numbers of fire-cracked rocks, ash, and charcoal. Several contained probable grinding stones, but ceramic sherds, lithic artifacts, and bones were generally fairly scarce.

Bronze Age or Iron Age Feature

FEATURE O

This is actually a complex consisting of Strata 3.4 and 3.5, the former of which is a localized lens and the latter the fill of a trench or channel that measured at least 160 cm long × at least 70 cm wide × 20–35 cm deep. It continues westward into unexcavated Squares G3–4 (figure 17.15). The complex may have been some sort of combustion feature and residue pile adjacent to a trench. The Stratum 3.4 patch in Subsquare H4b only was clayey silt that was reddened (from burning) and white indurated ash, packed with small pebbles in a hard matrix that underlies Stratum 1 or 3. It is unclear whether this material lay under Stratum 1 or Stratum 3, and it could be contemporaneous to the scattered yellow, clayey patches that we labeled Stratum 2, in intimate contact with both Strata 1 and 3 in other areas of the Cabin excavation area. The hypothesis that Stratum 3.4 lay between Strata 1 and 3 would make it late Bronze or Iron Age in date. This is the more likely hypothesis. The Stratum 3.5 fill was a mottled, locally variable, yellow-orangish, gritty (decomposed sandstone?) or clayey sediment that was apparently in a basin, trench, or channel, possibly equivalent to Stratum 2 as defined at the eastern end of the Cabin area. Stratum 3.5 was mainly excavated in Subsquare H4b+c as well as in part of H3a+b. This whole complex lies atop the ash and charcoal lenses of Stratum 4. Stratum 3.5 is a wedge overlying Stratum 4 in the west, but it abuts Stratum 4 to the east, where Stratum 3.5 ends.

Stratum 3.5 contained a medium-size, sub-angular cobble. Artifacts in Stratum 3.5 included 11 large cores and items of debitage, 714 small items of debitage, a partially backed blade, a backed bladelet, a notch, a notch + bec, a burin on transverse truncation, and a denticulate, as well as 15 sherds. There was also a double-bevel base antler sagaie of Magdalenian aspect. How these

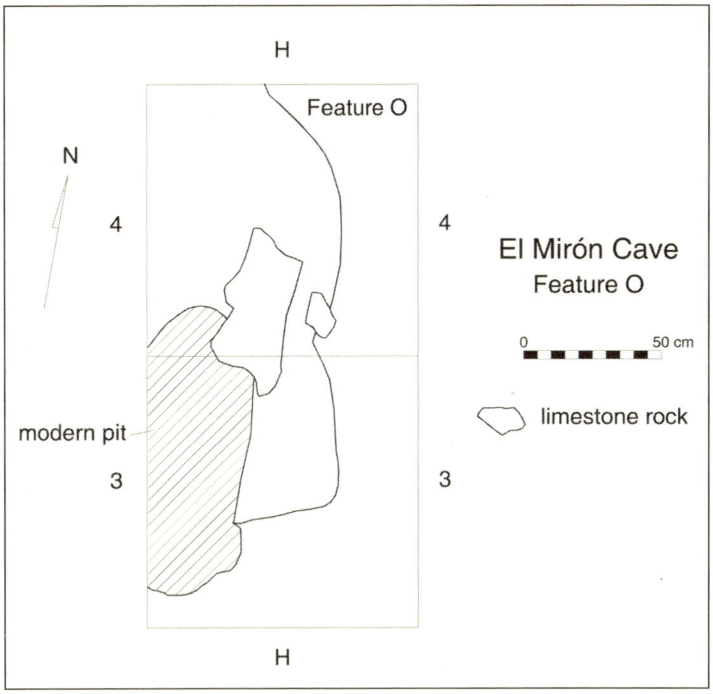

Fig. 17.15. Feature O (Levels 3.4 and 3.5). Drawing by L. G. Straus, redrafted by Ronald Stauber.

Upper Paleolithic–type artifacts got into this level is unresolved, although it probably involved Bronze Age disturbance and/or erosion of Magdalenian deposits that may have once continued in the form of a wedge atop the now-exposed colluvial-alluvial slope that leads back and up into the inner cave. This is probably also how remains of the extinct vole *Pliomys lenki* got into this surficial unit—through redeposition from late Last Glacial sediments. However, there were no modern artifacts in Stratum 3.5, despite the closeness of this feature to the 1996 ground surface. Faunal remains included 65 piece-plotted bones and 11 teeth. The Stratum 3.4 red and white combustion "patch" yielded a backed bladelet, 1 continuously retouched piece, 3 large flakes, 103 pieces of microdebitage, 4 ceramic sherds, 14 bone fragments, and a tooth.

The Stratum 3.5 trench or channel does cut through Stratum 3 but does not cut through Stratum 1, which is of mixed origin (including some modern artifacts). In turn, there is a "modern" basin-like hole dug into trench O (Stratum 3.5); that recent hole does have modern artifacts. Stratum 3.5 was excavated in 1997. (Feature N turned out to be a rodent burrow.)

Late Neolithic through Chalcolithic (or Bronze Age) Pit in the Mid-Vestibule Trench

FEATURE 98A

Excavation of the Mid-Vestibule Trench revealed the existence of a huge, man-made pit that seems to have existed across a substantial span of time during the middle Holocene. This pit—labeled Feature 98a—was excavated only in Square 06, mainly in Subsquare a, plus small areas of b and c. Because the surface deposits in this area of the vestibule have been intensively disturbed both in recent and, apparently, in more remote times, it was very difficult to distinguish between intact late post-Paleolithic layers and churned materials. The latter clearly characterized the uppermost sediments (Levels 300–301), which included small amounts of recent and old bones and ancient and modern artifacts, sometimes including glass shards. The integrity of Level 302 is more problematic, and distinguishing it from Level 301 was made very difficult in this area by the dark, blackish-reddish-brown color of all these surface deposits, between which there are no clearly visible stratigraphic breaks. Modern artifacts were not found in Level 302 in this area, but there were flint artifacts of "Magdalenian" aspect and an antler sagaie in the Level 302 fill above the top of pit 98a. This feature is drawn in cross-section in the stratigraphic section O–P/6 (figure 4.9).

Insofar as can be determined, pit 98a cuts through the upper part of Level 306 and all of Levels 305, 304, 303.1, and 303. The base of the pit is markedly smaller than the top, as if a relatively small hole (radius of approximately 75 cm) had been dug in Level 303 times and then, apparently in Level 302 times (i.e., post-Neolithic), a much wider basin (radius of approximately 130 cm), centered or extending farther to the east, was dug. After it was first dug, the pit seems to have been used as a hearth, as the base was full of alternating ash and charcoal lenses. The lower fill had no sherds, bones, or slag.

These basal charcoal and ash lenses were overlain by a mass of large rocks, some clearly fire-cracked. They were surrounded by a charcoal and ash matrix. This matrix among (and just below the rocks) yielded a few pieces of burned bone and 36 lithic artifacts (all unretouched debris: 19 trimming flakes, 5 shatters, 6 flakes, 3 chunks, and 2 bladelets). No sherds were found, despite water screening. Given that we excavated only about one-quarter of the lower (first-generation) pit, the fact that we exposed about a dozen large rocks—one measuring about 30 × 30 cm—is quite impressive. This rock layer is in the lower-middle part of the pit in-fill. Among the rocks were several chunks of what appeared to be slag.

Above the rock layer, the broader, more recent part of the pit (or the re-excavated pit) was filled with organic-rich, loamy sediments loosely attributable to Level 302. A large sample of charcoal from one of the basal lenses (i.e., below the rock-and-slag layer) yielded a radiocarbon date of 4910±80 BP, suggesting that the first stage of this hearth pit had been dug earlier and then partially filled with combustion material in the late Neolithic (after early Neolithic Level 303, which dates to 5500±90 BP) and that the later pit (perhaps used in metallurgy, based on the presence of possible slag) was (re-)dug and used in the Chalcolithic and/or Bronze Age. The uppermost deposit was a yellowish-beige, clayey slit—material with which the pit had been backfilled after it had ceased to be used as a combustion structure. The actual contents of the whole pit (as opposed to the mixed, capping sediments of Level 301) were devoid of modern artifacts. Finds described as specifically coming from the upper part of pit 98a fill (i.e., above the rock and slab layer) include three trimming flakes, one bladelet, and a prismatic blade core, along with a few burnt and even calcined bone fragments and possible slag. As with the large, round, late Neolithic and Chalcolithic pits in the nearby Cabin area, this feature is notable for the scarcity of artifacts—including ceramics.

The total depth of the pit as we can perceive it at present is at least about 80 cm. With a top diameter of over 2.5 m, this was an enormous crater. The rock fill of its lower-middle part must have been used to store a considerable amount of heat. This vast hole must have been a central focal point for human activities in this—middle—part of the cave vestibule for a long time. It was a major feature in the living space and was clearly reused and frequently reconfigured by filling and re-excavation. There were several generations of pits here, and even after the huge, final pit was mostly filled it must have left a shallow, basin-like depression in the ground surface used in post-Chalcolithic times.

The first-generation pit 98a is reminiscent of (though larger than) the rock-filled pits of the Chalcolithic levels in the nearby Cabin area. The upper part (later generation) of the pit has no analogy in terms of scale among any of the features in the Cabin, with the possible exception of the large trench or pit called Feature O/Level 3.5, which could be late Bronze or Iron Age. Both of these holes seem to have finished their lives as trash dumps, although Feature O is far richer in artifactual and faunal contents than Feature 98a.

CHAPTER EIGHTEEN

The Post-Paleolithic Stone Industries of El Mirón Cave

Lawrence Guy Straus, Manuel R. González Morales, and John D. Rissetto

This chapter summarizes and describes the relatively small lithic industries of the post-Paleolithic levels in El Mirón Cave. It also describes the one metal object found, an awl from the early Bronze Age. The Holocene-age cultural deposits in El Mirón Cave are (insofar as we know from our trenches) concentrated in the outer and middle sectors of the vestibule, although there is evidence of Bronze Age and medieval explorations of the dark inner cave (i.e., pieces of radiocarbon-dated charcoal). The archaeological materials from the vestibule are a disparate lot. From bottom to top, they represent the following cultural manifestations:

1. rather slight traces of visits to the cave during the Mesolithic (Cabin Stratum 10.1, Mid-Vestibule Stratum 304, and Corral Strata 101–102: all highly enriched in calcium carbonate, attesting to very wet conditions, possibly including ponding in the front part of the vestibule and some gentle runoff at the rear);
2. denser remains of cultural activity dating to the early and later Cantabrian Neolithic with moderately abundant (domesticated and wild) faunal remains and ceramics, some grains of wheat, and limited lithic artifacts, but without many structural features (Cabin Strata 10–8, Mid-Vestibule Strata 303.3–303, plus possibly the basal fill of pit 98a [lower Stratum 302]);
3. layers that are extremely rich in pits; fire-cracked, rock-filled hearths; bonfire residue; domesticated livestock remains; and ceramics but with only modest lithic artifact assemblages and no clear evidence of metallurgy, though dating to the Chalcolithic (Cabin Strata 7–4 and possibly—in part—Mid-Vestibule Stratum 302);
4. a complex of deposits (Cabin Strata 2 and 3, with associated lenses) dating to the Bronze Age and containing very abundant ceramics and domesticated livestock remains as well as evidence of in situ metallurgical activity (slag, combustion structures, a copper awl) but again relatively few lithics; and

5. a large intrusive pit or trench containing *non-modern* ceramics, domesticated livestock remains, and some lithic artifacts (Stratum 3.5, undated but possibly late Bronze or Iron Age or mixed). (The charcoal fragments yielding medieval-age radiocarbon determinations in both the Corral and inner cave excavation areas were not associated with any artifacts, either lithic or ceramic.)

The present chapter therefore presents and analyzes the lithic industries from Mesolithic, Neolithic, Chalcolithic, and Bronze Age levels from the 1996–2000 excavations in the El Mirón Cave vestibule. The complete collection of Holocene-age lithic artifacts (not including manuports, such as fire-cracked rocks and cobbles) sums to 8,664 items (including burin spalls and splintered pieces), all but 151 of which are unretouched knapping debris (cores and debitage) (tables 18.2 and 18.3). Total weight of the lithics is about 3.75 kg. These totals do not include materials from manifestly mixed deposits, in which "old" lithics, ceramics, and bones are found in false association with modern or sub-modern artifacts (including glass, porcelain, and iron objects such as horseshoes) and unmineralized faunal remains.

However, the absolute integrity of "intact" but surficial Stratum 1 (in the Cabin) and Stratum 302/pit 98a (in the Mid-Vestibule Trench) is not guaranteed, and for that matter—given the amount of prehistoric pit digging manifested especially in the Chalcolithic and Bronze Age levels—it is very likely that lithic artifacts (and sherds) have moved around within the stratigraphy. Indeed, the materials include a backed blade very reminiscent of a Gravette point in a pit within Stratum 6, antler points from Strata 4 and 3.5 (the latter with a double-bevel base) that the excavators called "Magdalenian-like," and a few fragments of worked antler items very similar to Upper Paleolithic *sagaies*, plus some remains of wild animals that (according to Pérez Ripoll López Gila, chap. 15, this volume), in terms of degree of fossilization, color, and butchering damage, look like bones from pre-Neolithic assemblages (but *not* resembling the bones of domesticated animals from the same levels)—all in various post-Mesolithic levels (notably Strata 5 and 3). All these finds might suggest that some Holocene-age pits reached Upper Paleolithic levels in other (as yet unexcavated) parts of the cave vestibule and/or that Holocene-age people collected such "antiquities" as curiosities from elsewhere within or outside the cave.

The most likely hypothesis, however, is that Upper Paleolithic layers once extended back toward the inner cave, banked up as a wedge-shape deposit atop the ancient, sterile colluvial-alluvial ramp. These layers would have been destroyed by erosion (including the activities of people and movements of their livestock), and their contents—including lithic and osseous artifacts and faunal remains—would have been transported downslope, to be worked into the post-Mesolithic layers by trampling and (especially) extensive pit digging. This hypothesis is supported by a radiocarbon date of 16,600±90 BP on a bone from a dark brown, organic, loamy silt deposit (typical of the early Magdalenian levels of the vestibule) containing faunal remains and lithic artifacts that filled a small concavity (Niche A) within the southern cave wall about a third of the distance (10 m) up (east) the ramp and 25–40 cm above its now–culturally bare surface. The hypothesis of significant Holocene erosion of late Last Glacial cultural deposits down the ramp is also supported by two dates of 10,740±40 and 10,390±50 BP on a tooth from a breccia also containing bones and lithic artifacts and that adheres to the base of a remnant of a flowstone surface along the northern cave wall, above the top of the ramp. The once-capping flowstone might have been dissolved by dripping water in which the acidity had increased as a consequence of the reforestation of the mountain slopes above the cave in the mid-Holocene (M. A. Courty, pers. comm., July 2008). The ramp area thus once did contain deposits of Magdalenian and Azilian age, and these may well have been the sources of artifacts of late Paleolithic aspect that became incorporated within post-Mesolithic layers downslope in the El Mirón vestibule because of water flow from the inner cave during the humid mid-Holocene as well as the extensive pit digging and livestock trampling of the Neolithic, Chalcolithic, and Bronze Age occupations. Thus at least a few of the lithic artifacts described below (numbers unknown) *might* have originated in late Paleolithic assemblages that were redeposited downslope, toward the front of the vestibule.

Analytical Methodology

The post-Paleolithic lithic artifacts are classified in the same ways we classify the much larger Upper Paleolithic assemblages. All items are classified by debris type (i.e., core or debitage) according to a typology developed by Straus and associates (notably G. A. Clark) over the course of many excavations conducted in Spain, Portugal, France, and Belgium since the mid-1970s (see table 18.1). This system emphasizes technological attributes useful in giving a first approximation of manufacturing stages within the reduction sequence. In addition, all debris longer than 1 cm are classified according to an ad hoc typology of lithic raw materials that was developed empirically during the course of the excavation campaigns, in which many of the more recently defined (and generally very rare) raw material types are not represented in the Holocene collections. (Most such recent materials were found in the first few years of the project and come from the Cabin area only.) Because so many of the artifacts are very small trimming flakes and shatter (mostly found in the water-screening process), most items are listed as being made on "unidentified stone" and are thus not counted in the percentages of raw material types. All piece-plotted debris are measured (length, width, thickness). All lithics are weighed, individually for piece-plotted items and globally by type class for non-tools; per subsquare and spit for (usually very small) non-piece-plotted and screen-found items, which are usually microdebitage such as trimming flakes and shatter that by definition are less than 1 cm in maximum dimension. Weights are measured by digital scale to the nearest tenth of a gram and dimensions are measured by dial or digital calipers to the nearest millimeter.

Objects that are identified as formal (i.e., retouched) tools are also classified according to the classic descriptive Upper Paleolithic typology of D. de Sonneville-Bordes and J. Perrot (1954, 1955, 1956a, b), which is most closely applicable, of course, to collections from southwestern France. This scheme was used because it is well known and is appropriate to most of the tools from the El Mirón post-Paleolithic collections due to the near-absence of esoteric geometric microlith types. (Had such types been abundant, a more appropriate typology might have been that of the G.E.E.M.—the Epipaleolithic/Mesolithic Study Group—also developed in France.) In fact, most of the El Mirón post-Paleolithic tools are fairly simple endscrapers, retouched pieces, and backed bladelets. Thus, tools are classified in three-fold fashion:

1. according to our debris/blank typology,
2. by the El Mirón raw material typology, and
3. by the de Sonneville-Bordes and Perrot tool typology.

All tools are measured (length, width, thickness), and all are individually weighed (regardless of size), even if they were not piece-plotted and even if they were found in the screen. Edge angles are measured by goniometer to the nearest 5° for all retouched edges (including retouched notch edges); where there is variability within a single edge, these angles represent "averages" determined empirically but always measured right along the actual edge (i.e., these are not spine plane angles). When there is more than one retouched (or burinated) edge on a single blank and when the resulting tool cannot be classified by one of the existing "compound" tool types in the de Sonneville-Bordes and Perrot typology (e.g., endscraper-burin, perforator-endscraper, perforator-burin, perforator-truncation, endscraper-truncation, burin-truncation, or piece continuously retouched on two edges; types 18–22 and type 66), we classified the item more than once in terms of tool type. Thus there can be more tools than tool blanks, although the dimensions and weight are taken and counted only once per blank. The multiple-edged tool phenomenon is relatively frequent among the Magdalenian assemblages, but it is essentially nonexistent in the post-Paleolithic ones (there is only one case: a doubly truncated blade with a retouched notch in Stratum 10). "Manuports" (fire-cracked rocks and cobbles that sometimes have evidence of use, such as concentrated pecking [anvils, hammers] or ground surfaces) are not counted among the retouched lithics and are not dealt with here.

Known Flint Sources

There are some poor-quality types of flint not far from El Mirón, notably in the Carranza Valley immediately northeast of the Pando-Moro mountain massif, including thin veins in the Aptian (Lower Cretaceous) limestone cliff face in which the cave site of Venta de la Perra is located, 5.5 linear km from El Mirón. However, the closest presently identified outcrops of excellent-quality flints are located east of the mouth of the Asón valley along the modern-day Atlantic coast. Some of these outcrops are in the Upper Cretaceous Plentzia Formation. They are found along the sea cliffs east of the mouth of the Abra (the outer port of Bilbao) at Barrika Beach and on both sides of the inlet of Plentzia (Elorza and Bustillo 1989; Elorza et al. 1991). This area of the Bilbao shore includes the well-known multi-component quarry-workshop site of Kurtzio at Sopelana. The Plentzia Formation is Cenomanian (early Upper Cretaceous) to Coniacian/Santonian (middle Upper Cretaceous) in age (Elorza et al. 1991). Plentzia flint is found in both bedded and nodular forms that range in maximum length from around 2 to 20 cm. Flint deposits were formed by the replacement of carbonate host rock by silica derived from the dissolution of Opal A siliceous sponge spicules and radiolarians. Microscopic analyses of bedded and nodular flint samples indicate the presence of detrital quartz particles as well as patches and piloid forms of unsilicified carbonate (Rissetto 2009; Elorza and Bustillo 1989). Material color varies among black, blue, gray, and white. Flint quality also varies from fine to coarse grain, depending on location of layering. Higher-quality flints are situated in the initial (oldest) layers of the Plentzia Formation. We have sampled the Barrika outcrop, which is located 5 km east of Punta de la Galea at the mouth of the Abra (Bilbao) estuary and 45 km east of the mouth of the Asón. The impressive, 60–70 m high cliffs at this vast beach expose vertically tilted limestone and marl strata with bedded and nodular flints of excellent (homogeneous) quality. The cobbles generally range in size from 5 to 20 cm or more in length. So far, at least, only our El Mirón Type A flint group has been tentatively identified at Barrika. (For petrographic and geochemical descriptions of Barrika samples, see Rissetto [2009] and Elorza and Bustillo [1989]. For a quartz crystallinity index of the bedded flint from Barrika, see Elorza. Arriortua, and Amigo [1985].)

Another known source of excellent flint is also along the present shore, at and near Sonabia Point, at the entrance of the Oriñón inlet, 10 km east of the mouth of the Asón in eastern Cantabria. The outcrop here pertains to the end of the Lower Cretaceous (Upper-Middle Albian) and hence is somewhat older than the Plentzia Formation. The flint-bearing bedrock here (and farther east along the shore at Islares, Castro Urdiales, and Mioño) is a clayey limestone with spicules. Thin veins of dark gray/black flint can be seen as veins in the limestone, and small cobbles (2–15 cm in maximum size) of this material can be found along the small beaches of Sonabia Point at the foot of Monte Candina. Given the mapped east-southeast to west-northwest trend of this formation, such material might be found on the now-inundated continental shelf just off Monte Buciero (Santoña) at the mouth of the Asón. The Sonabia flints are macroscopically quite similar to those of Plentzia, but this remains to be demonstrated petrographically and/or geochemically. Subjectively, the flints from Barrika seem to be of better quality, and they certainly occur in larger cobble or vein sizes. Sonabia has yielded samples that seem very similar to both El Mirón flint groups A and B. In the descriptions that follow we will refer to El Mirón flint groups A and B as "coastal," on the (as yet unproven) assumption that they came from Sonabia and/or Barrika-Plentzia or from presently unknown (or now unavailable) outcrops of the same Lower or Upper Cretaceous formations along the Holocene shore. (Another possible coastal source is at Llaranza, between Langre Beach and the mouth of Santander Bay, about 45 km from El Mirón via the Asón and the coastal plain of eastern Cantabria.)

Mesolithic Stone Artifacts

Lithic artifacts from Mesolithic-age deposits are truly scarce, totaling only slightly more than 260 g in weight (tables 18.4–18.8). Cabin Stratum 10.1 (9550–8380 BP) yielded only 151 lithic artifacts, of which only 5 are formal tools (figure 18.1). The physically correlated (but undated) Mid-Vestibule Stratum 304 is even poorer: 101 lithics, of which only 1 is a tool. The Stratum 304 tool is an obliquely truncated flake; the Stratum 10.1 tools are an atypical endscraper, a thick-nosed scraper on a pyramidal bladelet core, a bladelet continuously retouched on one edge, a lightly retouched bladelet (pseudo-Dufour), and a perforator. Those Stratum 10.1 artifacts for which identification could be made (i.e., non-microdebitage) are distributed as follows in terms of lithic raw materials by count: Type A coastal-type flints: 23.2 percent; Type B coastal-type flints: 16.1 percent; other flint types combined: 46.4 percent; "lutites"/mudstones: 10.7 percent; and other (non-flint) stones: 3.6 percent. By far the heaviest artifact (40 g) is the single piece of lutite—probably of local origin (this material is presently found in the bed of the Río Calera below the cave). In contrast, all of the few tools from this small assemblage are flint: 40 percent each are Type A and "other" flints and 20 percent are Type B flints. Level 101 (a surficial, calcium carbonate–enriched, granular, sandy silt in the vestibule rear/Corral area) yielded 95 lithic artifacts, 2 of which are tools (a thumbnail endscraper on a flake and a notched blade). Of the 23 identified stones, 47.8 percent are Type A by count, 17.4 percent are Type B, 34.8 percent are other flint types, and only 4.3 percent (1 item each) are lutite and another class of stone (quartz), respectively. The proportions of tools relative to unworked debris in these two Mesolithic assemblages (Strata 10.1 and 101) are 3.4 percent and 2.2 percent, respectively. In Strata 304 and 102 the proportions are 0.1 percent and 2.6 percent, respectively. All told, the proportions of tools to debris are very low. Noteworthy among the (mere) 11 tools from the four Mesolithic-age stratigraphic units is the absence of geometric microliths. There is a manuport from Stratum 10.1 that is classified as groundstone. However, it is very small (4.2 × 2.5 × 1.8 cm; 15.5 g), so its function is unclear.

For purposes of comparison with Stratum 10.1, the small assemblages of spatially contiguous Strata 304 (with a total of only 16.4 g of lithics; 86 percent of the identifiable stones by weight are coastal-type flints) and 101 were combined to sum 196 total lithic artifacts, versus 151 for Stratum 10.1. There are some notable differences between the two collections. Although both have a high percentage of cortical debitage (12.3 percent for Stratum 10.1 and 9.9 percent for Strata 101+304, versus 12.2 percent in Stratum 102), Stratum 10.1 has the highest percentage of cores of any level (0.7 percent), while Strata 101+304 have none. There is only one formal bladelet core (a pyramidal one retouched into a scraper in Stratum 10.1) in these Mesolithic-age collections. Chunks (large angular debris, which can include exhausted cores) are only moderately abundant in both collections (2.7–2.6 percent). Crested blades and platform renewal flakes are barely represented (0.7 percent—one of the latter) in Stratum 10.1 but are absent in Strata 101+304. An argument could be made for some direct access to flint nodules and their transport back to the cave for occasional knapping during the Mesolithic. This would make sense, given the known outcrops of excellent-quality flint-bearing limestone formations along the present-day shore of eastern Cantabria and western Vizcaya and the abundance of Mesolithic sites around the estuaries of the Asón and Agüera rivers. However, clearly the amount of flint that could be transported was very limited and many of the flints are not (yet?) demonstrated to belong to the gamut

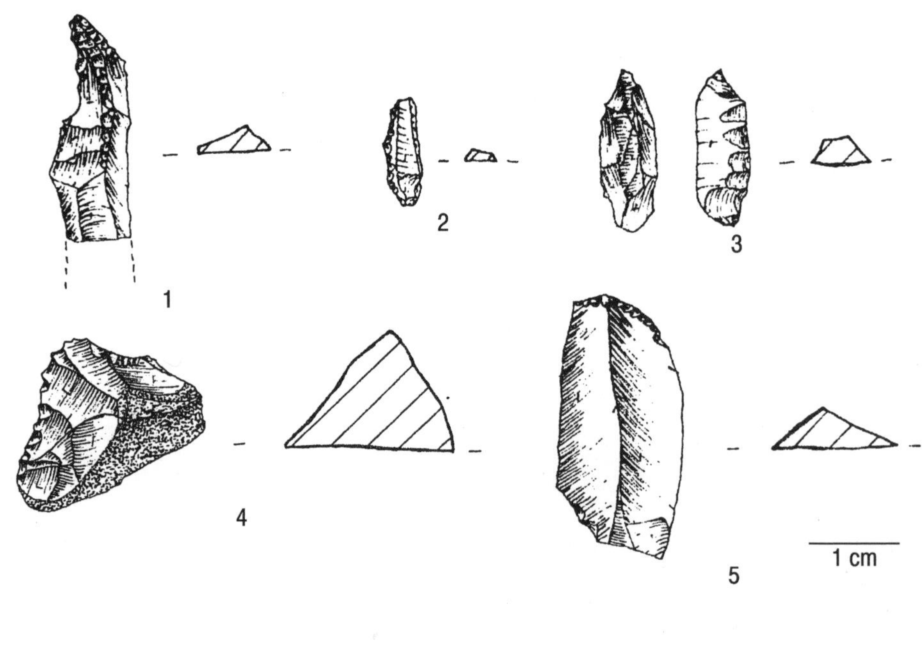

Fig. 18.1.
Mesolithic stone tools. Level 10.1. 1. Perforator (23): I4c, no. 729; 2. Retouched bladelet (90): I3c, no. 274.2; 3. Continuously retouched piece, 1 edge (65): H2b, no. 580.3; 4. Nucleiform endscraper (15): H4c, no. 725.6; 5. Atypical endscraper (2) + piquant trièdre: I4c, no. 727.2. Drawing by Susana Ruiz Pedraza and Natalia Suárez.

of known coastal flint types. Both Strata 10.1 and 101+304 have about 20 percent large flakes (23.3 percent and 14.0 percent, respectively), but Stratum 10.1 has a relatively low percentage (52.1 percent) of microdebitage products (trimming flakes and shatter, both <1 cm), while Strata 101+304 have a relatively high percentage (67.9 percent). Stratum 10.1 has the highest percentage (17.8 percent) of blades of all the Holocene assemblages, but Strata 101+304 has a low 3.6 percent. Bladelets (<2 cm) are not abundant in Stratum 10.1 (4.1 percent) but are more so in Strata 101+304 (11.9 percent).

Stratum 102 lies directly below the calcium carbonate–enriched Stratum 101, but both levels are essentially the same granular, sandy silt. Neither level is radiocarbon dated, as there are virtually no organic materials in either. However, Stratum 102 is locally underlain by a fire lens, 102.1, which is dated to 11,950±70 BP—final Magdalenian or initial Azilian. The stratigraphic equivalences and age of Stratum 102 are somewhat uncertain, but—given geomorphological evidence of high humidity (namely sheetwash)—the level is likely to pertain to the very beginning of the Holocene, and so its scanty lithic artifact assemblage is included in this monograph. It contains 116 chipped stone artifacts, of which only 3 are tools: an angle burin on break made on a flake, a microgravette point on a bladelet, and a blade continuously retouched on two edges. The first two tools are made on coastal Type A flints and the third on another type of flint. Tools represent 2.7 percent of the unretouched debris. There are no cores, but chunks make up 3.4 percent. Microdebitage makes up a high 62.1 percent of the assemblage and cortical material a high 13.8 percent. Blades and bladelets are both relatively plentiful at 6.0 percent and 10.3 percent, respectively. In contrast, large (>1 cm) flakes are not abundant (18.1 percent). The debris/blanks are dominated by lutites and other flint types (2.3 and 51.2 percent by count, respectively), versus 48.8 percent coastal-type flints (39.5 percent Type A and 9.3 percent Type B, by count). Only 4.7 percent by count are other stones. This composition differs somewhat from the other relatively large Mesolithic collection (Stratum 10.1), which has much fewer Type A items, somewhat more Type B, and more other flint types.

The grand total of tools from all the definite and possible Mesolithic horizons comes to only 11 items: 3 endscrapers, 1 burin, 1 perforator, 1 microgravette point, 1 truncation, 1 notch, 2 continuously retouched pieces, and 1 nibbled bladelet. Notable is the near absence of possible weapon tips and absence of geometric microliths. The atypical endscraper from Stratum 10.1 is on a blade with a *piquant trièdre* at the opposing end (figure 18.1, no. 5). The total number of debris items (mainly debitage and, in reality, in the majority microdebitage at that) comes to only 455, with retouched tools equaling 2.4 percent of that total. Faunal remains are scanty and man-made features absent. All in all the evidence points to short, ephemeral visits to the cave (at least to those areas of the vestibule that we have excavated) during early Holocene times prior to the arrival of Neolithic livestock and ceramic technology in the Cantabrian region. This contrasts strikingly with the density of Mesolithic sites along the coast, especially around the new (Holocene) mouth of the Asón—although formal tools are very rare there as well (see González Morales 1999:25), as is generally true of the coastal Mesolithic, such as the Asturian farther west.

Neolithic Stone Artifacts

The oldest levels with ceramic sherds are Cabin Stratum 10 (5690±50 BP) and Mid-Vestibule Stratum 303.3 (5790±90 BP). Although the transition between late Neolithic and Chalcolithic is gradual and definitional differences rather arbitrary in practice, we consider Stratum 8 to be the top of the Neolithic sequence in the Cabin (outer vestibule) area, with Lens 8.1 dating to 4680±60 BP. The attribution of Mid-Vestibule Stratum 302 is ambiguous, as it shows possible traces of being partly mixed (possibly Neolithic + Chalcolithic?). However, a very large, deep pit (pit 98a, only one quarter of which was excavated—in square 06) with several episodes of apparent combustion use, filling, and redigging yielded a date of 4910±80 BP from its *lower* fill, which could have derived from surrounding Stratum 302 (while the upper fill of this crater-like feature, which contained apparent slag, could be Chalcolithic or Bronze Age). So the Neolithic at El Mirón seems to represent a period of some 1,200 radiocarbon years but in terms of calibrated ages the period between about 4700 and 3400 BC

All the Neolithic levels of the Cabin and Mid-Vestibule Trench combined (but not including Stratum

302) yielded a grand total of slightly less than 1 kg of lithic artifacts (931 g). Stratum 302 yielded 454 g of lithics, but nearly a quarter of that total is from just one item.

The individual early Neolithic stone artifact assemblages are small—like the Mesolithic ones (tables 18.9–18.13). Several are too small to include at all meaningfully in any collections (Strata 303.3, 9.2–9.8, 8.1). Only Stratum 303 (5500±90 BP) has a fairly large assemblage (787 items), the rest of the larger assemblages (Strata 10, 9, 8, 303.2, 303.1) ranging from 143 to 222 items. Chronoculturally ambiguous Stratum 302 yielded 612 lithic artifacts. Likewise, only Stratum 303 has more than 10 tools (12), while the rest range from only 1 to 8 (and 9 in Stratum 302). The total number of lithic artifacts from all these Neolithic levels (including Stratum 302) is 2,546. But the making (and/or abandonment) of formal tools seems to have been rare at El Mirón.

Among the seven large assemblages, there is a striking difference between Cabin Stratum 10 and all the rest in terms of the representation of coastal-type flints: in Stratum 10 Type A flints make up only 26 percent of the lithic-identifiable artifacts by count and Type B only 9.6 percent, while in the rest of the assemblages the percentages range between 22.8 and 45.6 percent and 7.0 and 26.1 percent, respectively. In contrast, other flint types are much better represented in Stratum 10 (48.0 percent) than in the rest of the strata (18.3–36.1 percent). The lutites make up 6.6 percent in Stratum 9, 8.2 percent by count in Stratum 10, and 25.6 percent in Stratum 8. In the latter level the heaviest raw material represented is in fact lutite (153 g, a truly unusual amount). With the exception of Stratum 303.1 (18.8 percent), other stones (e.g., quartz, quartzite, limestone) are fairly scarce among the Neolithic knapped stone artifacts. Strata 9 and 10 stand out for the great diversity of identifiable raw materials present (18 and 19 types, respectively)—many of which are other flint types (totaling 44.3 and 36.4 g, respectively), although coastal-type flints are also relatively well represented (34.2 and 30.4 g, respectively).

Artifacts large enough to be identifiable as to raw material in the Mid-Vestibule Trench total 401 g for Stratum 302, 190 g for Stratum 303, 47 g for Stratum 303.1, 28 g for Stratum 303.2, and only 11 g for Stratum 303.3. The overall proportions of coastal-type flints by weight among these assemblages range from 27 to 71 percent, but the assemblage at the low extreme is the culturally ambiguous Stratum 302, half of whose identifiable lithics are "flint-like" mudstones, which are called lutite in Spain and which can be found in the bedload of the streams immediately below the site. The unambiguously Neolithic assemblages sampled in the Mid-Vestibule Trench are, in terms of weight, made up of one-third to two-thirds flints that were apparently transported from outcrops along the Holocene shore. (Stratum 303, the level with the lowest percentage by weight of coastal-type flints, has its weight distribution distorted by the presence of one item of quartzitic sandstone—probably local—weighing 41 g.) However, the absolute amounts of flint that were carried are very small in all these levels, although they do attest to probable human contact with the coastal zone, where there are Neolithic megaliths dated to approximately the same time as the early Neolithic occupations of El Mirón. In the Cabin excavation area, Stratum 8 yielded 212 g of chipped stone artifacts identifiable as to raw material; Stratum 9, 91 g; and Stratum 10, 115 g. The other Cabin levels have very little identifiable lithic raw materials. Lutite (mudstone) overwhelmingly dominates the Stratum 8 assemblage (72 percent of the total weight of identifiable stones), but coastal-type flints are very significant in Strata 9 (37.4 percent) and 10 (26.4 percent). Almost all the Neolithic tools are made on flint—half being provisionally identified as coastal-types of flint. Although lutites and other non-flint materials are common among the Neolithic stone debris, they were hardly used at all to make retouched implements. Perhaps, however, at least the larger ones were in fact employed as informal or expedient tools.

In terms of debris/blank characteristics, items with full or some dorsal cortex are scarce (4.8–8.3 percent) in the larger assemblages, except in Stratum 10 (9.8 percent). The same is true for chunks, Stratum 10 having a very high 8.4 percent, while the other strata range between 0 and 3.3 percent, but cores are absent (or extremely rare) in all the Neolithic assemblages. Crested blades and platform renewal flakes are essentially absent. Microdebitage ranges from 61 to 73 percent in all strata except 10, which yielded only 42 percent trimming flakes + shatter. On the other hand, large (>1 cm) flakes make up 34.5 percent of the Stratum 10 assemblage and 31 percent of Stratum 8, while the range is only 13–17 percent among the rest of the larger assemblages. Blades are relatively common in

Strata 10 and 303.1 (10 percent and 8 percent, respectively), with the rest having only 2.4–5.9 percent. Bladelets are not very common in most assemblages (1.8–4.9 percent), except in Strata 303–303.2 (9.6–10.4 percent)—similar to the range of variability among the Mesolithic assemblages. Excluding Lens 8.1 (which has one retouched piece but only three items of debris) and the several levels with no tools at all, formal tools represent 1.4–4.0 percent of the unretouched debris, with the high outlier percentage corresponding to Stratum 303.1.

The Neolithic tools (total including Stratum 302 is 45 items, 9 being from Stratum 302) include only one atypical endscraper but no sidescrapers or perforators. There are five burins and three points: one reminiscent of a Font-Yves (figure 18.2, no. 9), one microgravette, and one Gravette (figure 18.3, no. 6). There is one endscraper/burin. Most notable among these sparse collections are truncated and, especially, backed pieces: these include five backed bladelets and three backed blades (the latter all in Stratum 302; one is a doubly backed "rod" [figure 18.5, no. 4]). There are also a couple of circle segments: one with bifacial beveled (Helwan) retouch from Stratum 9.6 (figure 18.4, no. 1), and the other with abrupt backed retouch from Stratum 9 (figure 18.4, no. 2); plus

Fig. 18.2. Early Neolithic stone tools. 1. Continuously retouched piece, 1 edge (65) + notch (74): H1b, Level 10, no. 345.2; 2. Straight dihedral burin (27): I2b, Level 10, no. 54.1; 3. Continuously retouched piece, 1 edge (65): H2a, Level 10, no. 524; 4. (Atypical) segment (83): I4b, Level 10, no. 675.1; 5. Continuously retouched piece, 1 edge (65): L5c, Level 303.3, no. 443.5; 6. Multiple burin on retouched truncation (40): L5c, Level 303.3, no. 443.6; 7. Atypical endscraper (2): L5d, Level 303.1, no. 238; 8. Continuously retouched piece, 1 edge (65): 06c, Level 303.1, no. 40; 9. "Font-Yves" point (52): M5a, Level 303.1, no. 144.1; 10. Backed bladelet (85): L5a, Level 303.1, no. 224.9; 11. Backed bladelet (85): L5a, Level 303.1, no. 224.10. Drawing by Susana Ruiz Pedraza and Natalia Suárez.

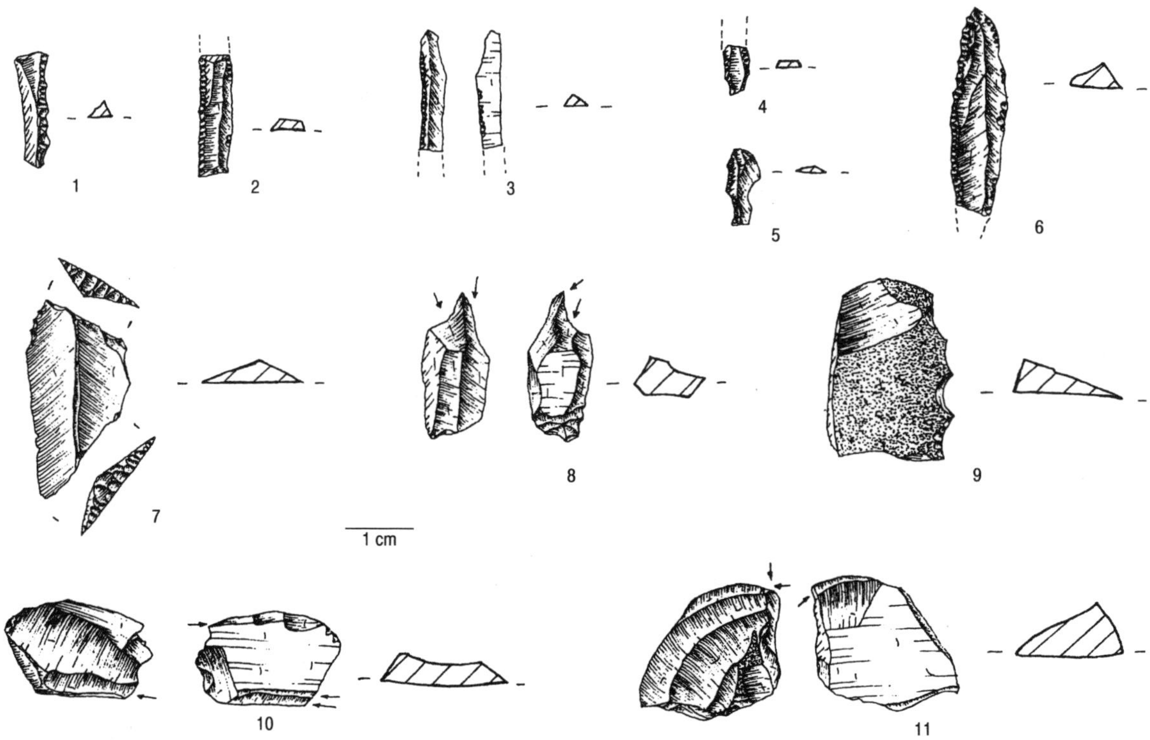

Fig. 18.3. Early Neolithic stone tools. Level 303. 1. Backed bladelet (85): 06c, no. 27.7; 2. Backed bladelet (85): M5c, no. 33.1; 3. "Dufour" bladelet (85): N5b, no. 37.2; 4. Backed bladelet (85): 06c, no. 27.3; 5. Backed bladelet (85): 06c, no. 27.4; 6. "Gravette" point (51): N5b, no. 37.12; 7. Trapeze (81): M5d, no. 55.1; 8. Splintered piece (76): L5b, no. 188.1; 9. Denticulate (75): M5c, no. 33.1; 10. Flake with prior removals: L5b, no. 153; 11. Bladelet core: L5c, no. 189.1. Drawing by Susana Ruiz Pedraza and Natalia Suárez.

one other less typical segment from Stratum 10. Stratum 303 yielded one trapeze (figure 18.3, no. 7). And there is a nibbled (Dufour-like) bladelet. There are a total of 8 denticulates and notches and 12 continuously retouched pieces, plus 2 splintered pieces (which could be bipolar core remnants). As noted above, although rare, geometric microliths are found in the El Mirón Neolithic (one item each in Strata 9, 9.6, 10, and 303). Most of the items (n = 10) that may have been weapon tips/barbs (armatures) are from contiguous Strata 303–303.1, suggesting some concentration of activities involving hunting equipment in the Mid-Vestibule area. The presence of a circle segment with invasive double-bevel (Helwan) retouch is typical of the early Neolithic in the Vasco-Cantabrian region and links it with the even earlier Neolithic of the adjacent upper Ebro Basin in La Rioja, Alava, Navarra, and Aragón (see Arias 1999:419; Alday 1998; Gorrochategui, Yarritu, and Zapata 1995:112). Such tools are also found in the aceramic coastal site of Herriko Barra (Guipúzcoa), where cereal pollen has been identified in association with only wild fauna at around 5900 BP (Iriarte, Mujika, and Tarriño 2005). As a group, the tools represent only 1.7 percent of the debris total from the Neolithic levels (figures 18.2–18.5).

Slightly over a quarter (27.3 percent) of the Neolithic tools are made on flakes, but 40.9 percent are made on blades and 25.0 percent on bladelets, making this a rather laminar industry. Since there are only two cores in the Neolithic assemblages (both from Stratum 302), however, the blades and bladelets may have been struck elsewhere: either in other parts of the site (such as the terrace in front of the cave mouth) or at some off-site locations (such as the flint-source spots themselves). The blanks were probably brought to the site, with little decortication done on-site. There are a total of 57 chunks (mostly non-cortical—potentially exhausted core remnants), but again it is Stratum 302 that dominates this category, with a total of 20.

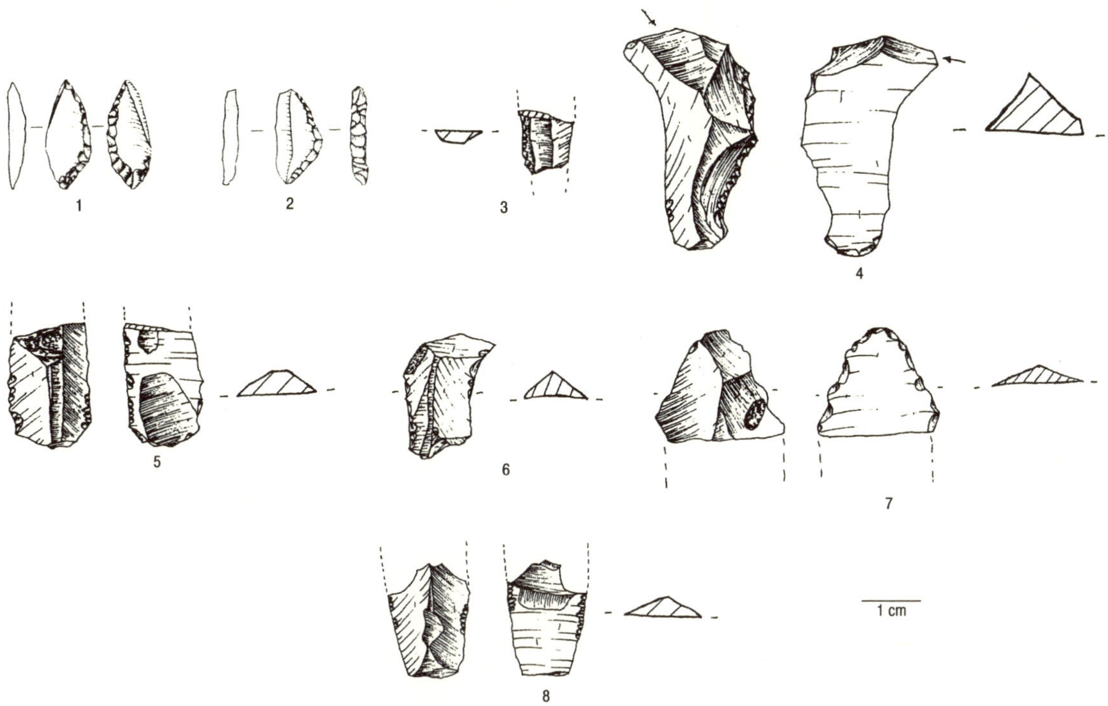

Fig. 18.4. Neolithic stone tools. 1. Segment (with beveled retouched) (83): H3a, Level 9.6, no. 261.4; 2. Segment (backed) (83): I3b, Level 9, no. 63.3; 3. Continuously retouched piece, 1 edge (65): I3d, Level 9.6, no. 105.2; 4. Notch (74): J2a, Level 9, no. 26.1; 5. Continuously retouched piece, 2 edges (66): I3b, Level 8.1, no. 46.1; 6. Continuously retouched piece, 1 edge (65): H2d, Level 8, no. 448.9; 7. Denticulate (75): I2a, Level 8, no. 40.1; 8. Continuously retouched piece, 2 edges (66): H2d, Level 8, no. 448.1. Drawing 1 by Christina Sinkovec, redrawn by Luis Teira; drawing 2 by Alejandra Silva, redrawn by Luis Teira; drawings 3–7 by Susana Ruiz Pedraza and Natalia Suárez.

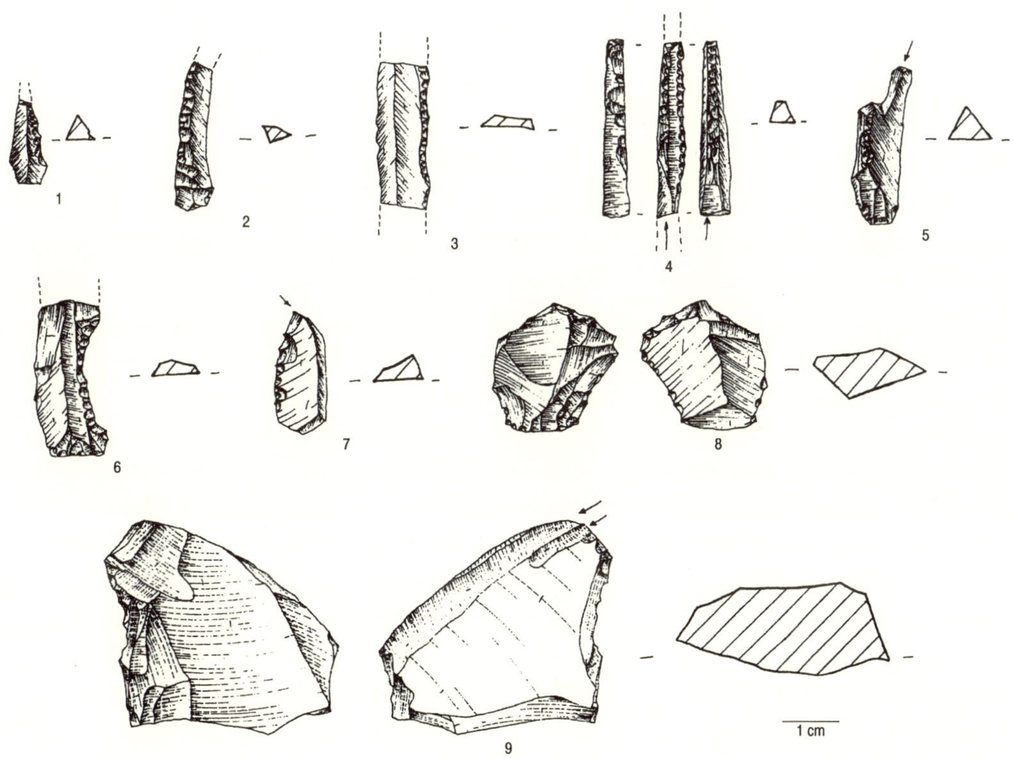

Fig. 18.5. Late Neolithic/Chalcolithic stone tools. Level 302. 1. Platform renewal flake: 06d, no. 18.33; 2. Burin spall: 06c, no. 17.3; 3. Completely backed blade (58): 06b, no. 16.8; 4. Double-backed blade ("rod"): P6b, no. 16.1; 5. Burin spall: 06b, no. 16.7; 6. Notch (74): 06c, no. 17.3; 7. Continuously retouched piece (65): 06d, no. 18.4; 8. Splintered piece (76): 06c, no. 17.1; 9. Notch (74) on large flake with prior bladelet removals: P6a, no. 11.1. Drawing by Susana Ruiz Pedraza and Natalia Suárez.

The 4 burins and the 11 continuously retouched pieces are made on flakes, blades, and bladelets, while the 9 notches and denticulates are made on flakes, blades, and chunks. Microliths and retouched/backed bladelets are mainly on bladelets, but there are several that are on small blades, this being a question of the arbitrary (and, in our case, low) metric threshold definition for bladelets (length <2 cm). The two items classified as points are made on small blades. In addition to the stone tools, there is a bone awl from Stratum 10. Interestingly, there are no groundstones or hammerstones/anvils in the Neolithic level, either in the Cabin excavation area or Mid-Vestibule Trench.

Chalcolithic Stone Artifacts

Cabin Strata 7–4 are assigned to the Chalcolithic, although we are cognizant that the distinctions between it and late Neolithic on the one hand and early Bronze Age on the other are tenuous to say the least. The distinctions are based on ceramic types, but the material in Cantabrian Spain is not always as diagnostic as in Mediterranean regions of the country. It is fortunately also the case that at El Mirón we found a couple of bifacially worked arrowhead fragments: the first is a basal portion of a rhomboidal point (figure 18.6, no. 5) and the other is corner notched (stemmed and winged) (figure 18.6, no. 1) and quite diagnostic of the Chalcolithic period. These points (which were listed as types 70 [bifacial foliate point] and 55 [stemmed point] in the de Sonneville-Bordes and Perrot type list, for lack of better pigeonholes in which to place them) are from Strata 5.1 and 5, respectively. The three radiocarbon determinations from the Chalcolithic levels are not stratigraphically coherent (perhaps due to the mixture of charcoal fragments caused by the abundant pit digging that was going on during this period in the cave vestibule). However, these rapidly accumulated deposits seem to have centered on about 4100–3800 BP (2800–2000 cal. BC). Unfortunately, the faunally, structurally, and artifactually rich Stratum 4 is not dated. However, the levels yielding the two invasively retouched, bifacial arrowhead fragments (Strata 5.1 and 5) are specifically and coherently dated to 4120±40 and 3820±240 BP. These dates (albeit perhaps a bit young by comparison) relate fairly well to a series of three uncalibrated dates (4790–4100 BP, all ±110—the calibrated age seems to be centered around 3000 BC but with a 2-sigma range from about 3900 to 2300 BC for the series) from a Chalcolithic ossuary deposit (Level 3) in recently excavated Pico Ramos Cave (western coastal Vizcaya), which yielded rhomboidal and corner-notched (stemmed), bifacial, invasively retouched points, as well as to other dated sites with such arrowheads in the Basque Country, Cantabria, northern Castile, and La Rioja (Zapata 1995; Gorrochategui, Yarritu, and Zapata 1995; Ontañón 1996).

In terms of lithic artifacts, the total counts (not counting minor lenses, most of which have only a few items) range from 205 in Stratum 6 to 361 in Stratum 5 (tables 18.14–18.18). All the Chalcolithic levels combined produced only 647 g of lithic artifacts. In the large assemblages (Strata 4, 5, 6, and 7), the representation of coastal-type flints among the lithic artifacts large enough to reasonably identify the raw material is substantial: by weight, Type A flints range from 25 to 58 percent and Type B from 0 to 6 percent, with totals ranging from about 27 to 62 percent of all the lithics, very similar to the range for the Neolithic assemblages (except Stratum 10). Other flint (and flint-like) types combined range between 17 and 43.5 percent of the larger Chalcolithic knapped stone assemblages, and other stones make up between 4 and 26 percent. The flint-like mudstones (lutites) per se are usually rare but do make up 5.2 percent of the identifiable lithics by count in Stratum 7, 1.5 percent in Stratum 5, and 3.0 percent in Stratum 4. In terms of weight, the only fairly substantial amount of lutite among the Chalcolithic levels comes from Stratum 7: 53 g. This same level also yielded 56 g of quartzites—also locally available stones. However, Stratum 7 also has substantial amounts of coastal-type (almost all Type A) and other flints: 59 and 51 g, respectively. The other major Chalcolithic assemblages are dominated by coastal-type and other true flints: Stratum 6, 50 and 14 g; Stratum 5, 57 and 49 g; and Stratum 4, 38 and 40 g, respectively. By whatever means (trade or travel), high-quality, non-local flints were still being procured by Chalcolithic-age people, presumably from sources along the seashore. But in any event, the absolute amounts of lithic raw materials—local or not—are small, so transport problems were not significant.

Among the larger assemblages (Strata 4, 5, 6, 7), microdebitage makes up 57–68 percent of the totals,

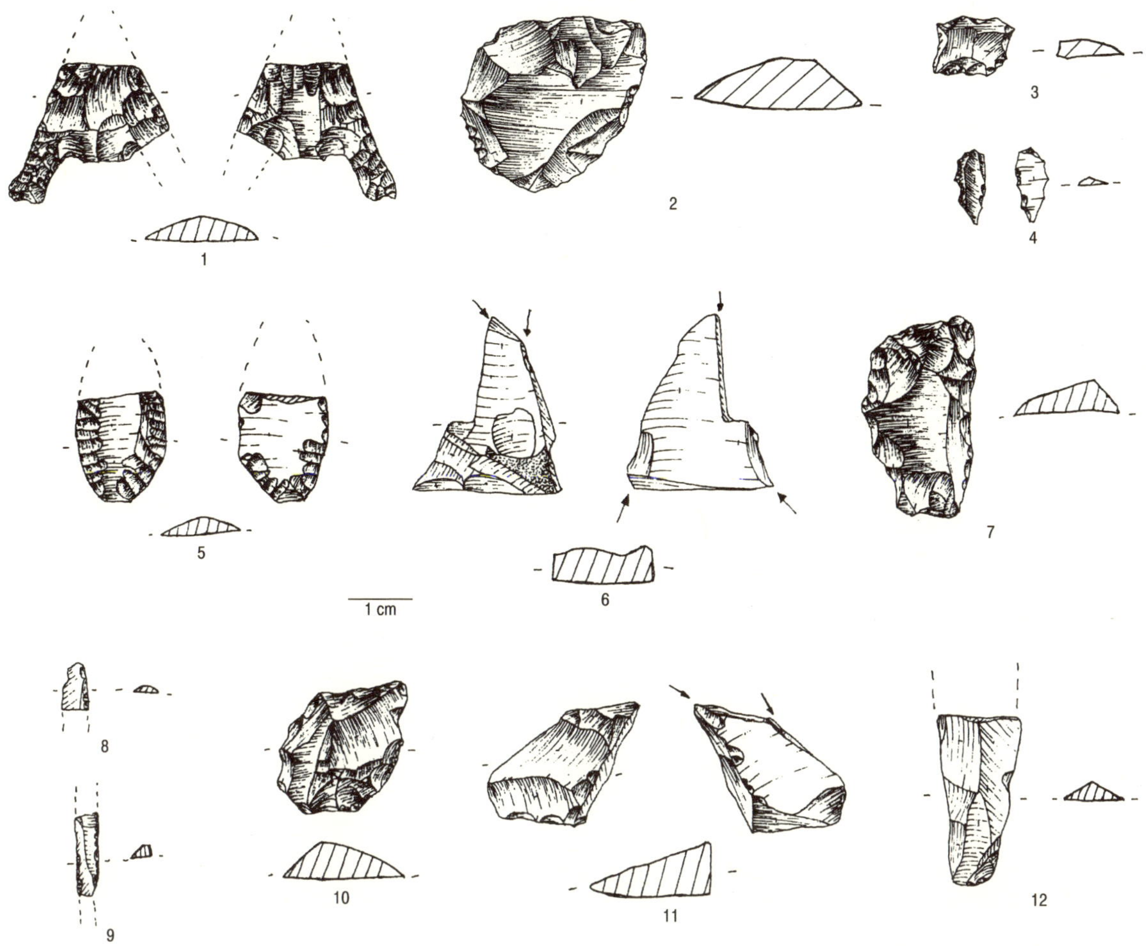

Fig. 18.6. Chalcolithic stone tools. 1. Stemmed arrowhead (55): I3d, Level 5.1, no. 17; 2. Continuously retouched piece, 1 edge (65): I3d, Level 5.0, no. 12.1; 3. Bec (24): J4c, Level 5.0, no. 155.10; 4. Denticulated backed bladelet (87): J4c, Level 5.0, no. 155.16; 5. Foliate point base (70): J2a, Level 5, no. 7.1; 6. Multiple dihedral burin (31): I2b, Level 5, no. 13.1; 7. Splintered piece (76): J4d, Level 5, no. 265; 8. Backed bladelet (85): L5a, Level 5, no. 47.24; 9. Backed bladelet (85): L5b, Level 5, no. 48.22; 10. Bec (24): L5b, Level 5, no. 48.10; 11. Mixed core: L5c, Level 5, no. 99; 12. Blade fragment: L5c, Level 5, no. 49.1. Drawing by Susana Ruiz Pedraza and Natalia Suárez.

cortical items between 7 and 14 percent, and chunks between 1 and 2. Cores (and platform renewal flakes) are completely absent. There are a few crested blades in Strata 4, 5, and 7, but this is only scant indication of some in situ blank production. Larger flakes (>1 cm) account for 18–23 percent of the debris + blanks, bladelets (including only a couple of possible burin spalls) 6–13.5 percent, and blades a similar range of 3–7 percent.

The tools represent only 1–2 percent of the unretouched debris (figures 18.6–18.8). As a group, more than half (52.3 percent) of the Chalcolithic tools/weapon tips are made on flakes, only 14.3 percent on blades, and 23.8 percent on bladelets. There is one tool on a core (an endscraper) and a splintered piece or bipolar core. All told, there are only three endscrapers (the other two being on flakes), two perforators (both on flakes), two burins (one on a flake and the other on a blade), a Gravette-like point (figure 18.7, no. 8), and a microgravette-like point (figure 18.8, no. 3)—the Gravette-like point on a blade, the microgravette-like point on a bladelet. In addition, there is a concave truncation, only two continuously retouched pieces, only two notches, one sidescraper (all but one notch made on flakes), and only four backed bladelets (one denticulated and all made on bladelets) but no geometric microliths. Formal, retouched lithic tools were obviously of minimal significance in the Chalcolithic levels, although the two (broken) arrowheads and two Gravette points (plus perhaps the backed bladelets) might be significant as possible indicators of inter-human violence, particularly in light of the scarcity of wild game animals in the Chalcolithic levels. By contrast, the sepulchral deposit in Pico Ramos yielded 14 stone arrowheads (both rhomboidal and corner-notched) and a so-called stone dagger (also bifacial and invasively retouched), but very few other lithic implements: an endscraper, two geometric microliths, a bladelet, four blades, three retouched flakes, and an unspecified number of debitage items (presumably flakes and chips) (Gorrochategui, Yarritu, and Zapata 1995).

The Chalcolithic levels yielded several apparent grinding stones: a mortar and a stone ball (handstone?) from Stratum 7; a manuport, which could be a grinding stone, from Stratum 6.0; a possible handstone (*mano*) from Stratum 6; a mortar from Stratum 5.3; and a handstone and two lower grinding stones (metates) from Stratum 4 (the pit A complex). In addition, there are two bone awls from Stratum 7, one from Stratum 5.1, and one from Stratum 5. These are polished to extremely sharp tips. Finally, there is one small piece of possible copper (not positively identified) in both Strata 7 and 6.1.

Since almost all the (relatively few) excavated Chalcolithic sites in the Cantabrian region are burial

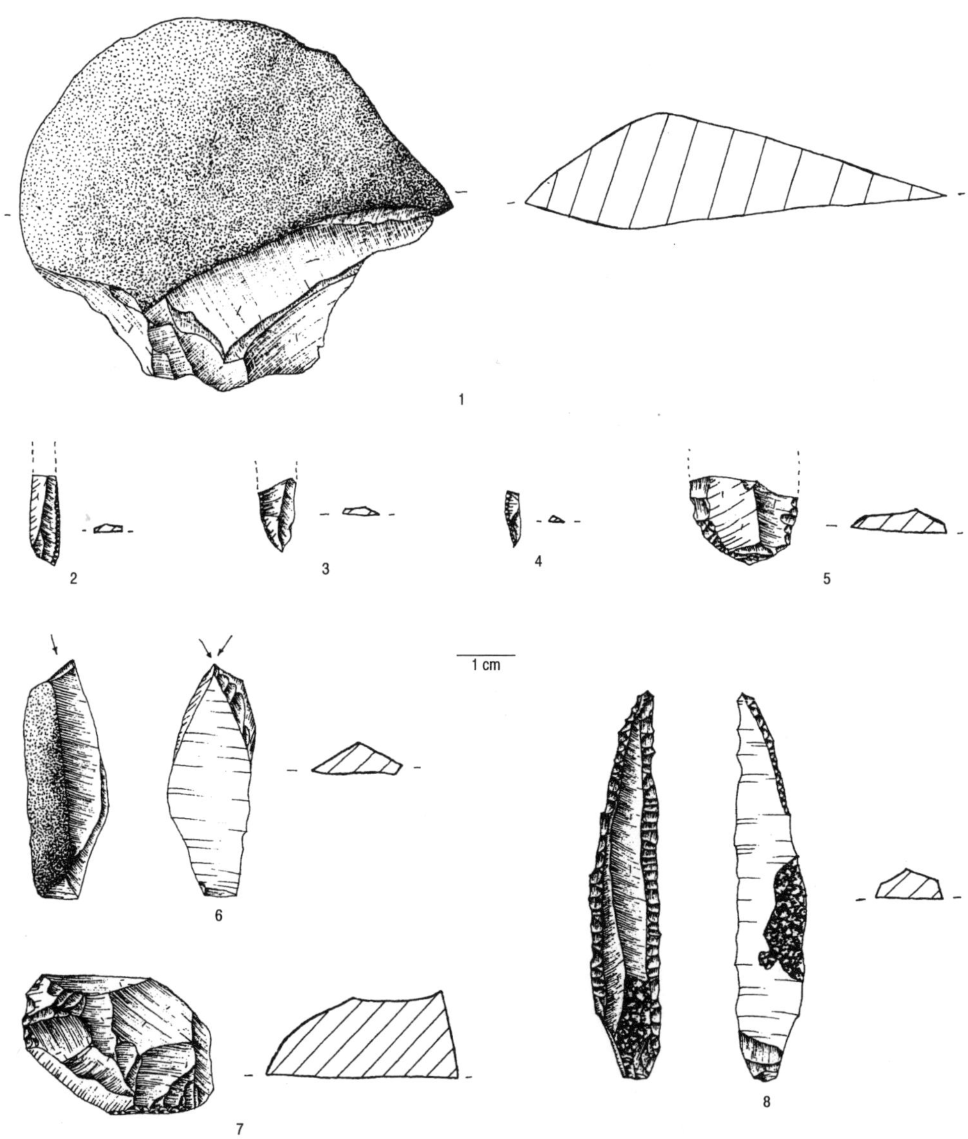

Fig. 18.7. Chalcolithic stone tools. 1. Secondary decortication flake. L5b, Level 301.2, no. 133; 2. Backed bladelet (85): J4c, Level 7.5, no. 216.1; 3. Backed bladelet (85): J4b, Level 7, no. 250.13; 4. Backed bladelet (85): J4b, Level 7, no. 250.12; 5. Continuously retouched piece, 2 edges (66): H2c, Level 7, no. 382; 6. Canted dihedral burin (28): J3c, Level 7, no. 22.1; 7. Nucleiform endscraper (15): J2a, Level 7, no. 17; 8. "Gravette" point (48): J3b, Level 6, pit D, no. 25. Drawing by Susana Ruiz Pedraza and Natalia Suárez.

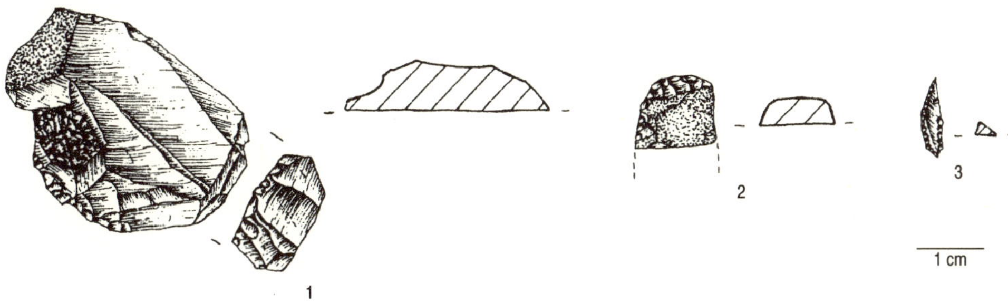

Fig. 18.8.
Late Chalcolithic stone tools. Level 4. 1. Sidescraper (77): H3d, no. 40.1; 2. Thumbnail endscraper (10): I4b, no. 25.9; 3. "Microgravette" point (51): J4a, no. 39.3. Drawing by Susana Ruiz Pedraza and Natalia Suárez.

caves and most Chalcolithic finds are from disturbed contexts (clandestine pothunter or spelunker finds, etc.), there is at present nothing comparable to the El Mirón assemblages, which are clearly from residential-cum-stabling contexts, not a human ossuary.

Bronze Age Stone Artifacts

The only date for an occupation deposit pertaining to the Bronze Age time range is 3700±40 BP (2140–2030 cal. BC) from Stratum 3, but there is a charcoal (torch?) fragment unassociated with any cultural finds from the Inner Cave Trench Stratum VII that is dated to 3230±40 BP (1525–1443 cal. BC). While Stratum 2 is probably also attributable to the Bronze Age, Stratum 1 (though lacking modern artifacts) may be of more mixed origin, lying as it did immediately beneath the loose debris and dung filling the interior of the Cabin when we started our excavations in 1996. In addition, there were considerable sediments that seemed to represent a combination of the yellow-beige patches of Stratum 2 and the gray-brown Stratum 1; artifacts from this mixture are labeled Stratum 1–2. Their "pure" attribution to the Bronze Age is problematic.

Various lenses associated with Stratum 3 (3.2, 3.3, 3.4) and Stratum 2 (2.3), as well as Stratum 1 itself, yielded very few lithic artifacts. In contrast, Stratum 3 itself has 1,052 items, Stratum 2 has 716, and Stratum 1–2 has 1,086. All the Bronze Age levels together produced just about exactly 1 kg of lithic artifacts. Among the artifacts large enough to be identified as to lithic raw material from these three large assemblages, coastal-type flints continue to be dominant: Type A, 37.3–44.55 percent and Type B, 9.2–26.1 percent (total coastal-type flints, 49.0–63.4 percent). Other flints and flint-like mudstones make up between 29.8 and 46.6 percent, and other stones combined constitute only 4.5–5.6 percent of these assemblages. The mudstones (lutites) are significant only in Stratum 3 (5.3 percent—18 items by count). In Stratum 3 there are 129 g of coastal-type flints (mostly Type A), 183 g of other true flints, 29 g of mudstones, and 28 g of other stones (limestones, quartzites, and quartzes). It is interesting to note that the vast majority (at least 312 g) of the knapped lithics are apparently nonlocal, which implies direct and/or indirect contacts with the coastal flint sources. However this still amounts to less than a third of a kilogram of transported stone, highlighting the relatively minor importance of chipped lithic technology in the main, otherwise culturally rich Bronze Age occupation of the El Mirón vestibule. The lithic materials from Strata 2, 1–2, and 1 are similarly dominated by true, nonlocal flints, but the weights are far smaller. The local stones are quite scarce among these assemblages, except for some quartzites in Strata 2 and 1–2 (total other stones: 26 and 21 g, respectively) (see tables 18.19–18.23).

Microdebitage makes up between 60.3 percent (Stratum 3) and an overall high of 75.1 percent (mixed Stratum 1–2) of all the lithic artifacts. Cortical material constitutes 3.9–7.5 percent, cores 0.1–0.3 percent, chunks 0.5–1.3 percent, and crested blades and platform renewal flakes 0.1–0.6 percent. The percentages of large flakes varies widely, from 9.7 percent (Stratum 1–2) to 24.3 percent (Stratum 3); the relative frequencies of bladelets are more even and rather high: 9.5 percent (Stratum 3) to 15.5 percent (Stratum 2) and 13.2 percent (Stratum 1–2). Blades, in contrast, are rare: 1.1–4.1 percent. In all of these levels there are only seven cores: almost all are flake (or mixed) cores; only one is a prismatic blade core (but, curiously, no bladelet cores—perhaps bladelets were imported in bulk from coastal flint quarry-workshop localities?).

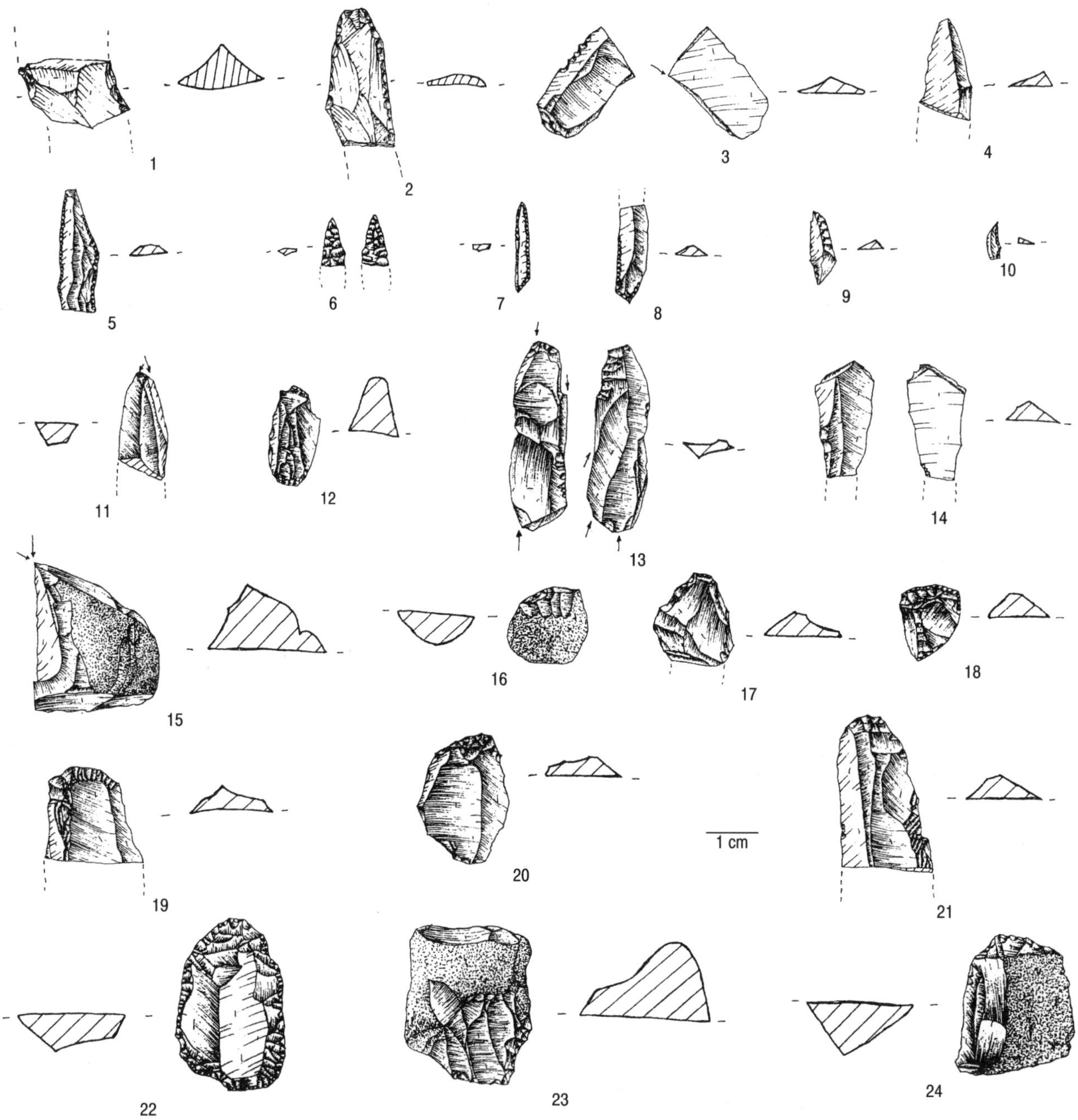

Fig. 18.9. Bronze Age stone tools. 1. Continuously retouched piece, 1 edge (65): H4b, Level 3.4, no. 108.11; 2. Completely backed blade (58): H4b, Level 3.4, no. 107; 3. Denticulate (75): H1b, Level 3.2, no. 103.1; 4. Blade fragment: H3b, Level 3, no. 2.2; 5. "Font-Yves" point (52): J4b, Level 3, no. 20.2; 6. Foliate point tip (70): H1a, Level 3, no. 1.21; 7. Truncated backed bladelet (86): H2d, Level 3, no. 30.7; 8. Backed bladelet (85): J4b, Level 3, no. 20.19; 9. Triangle (79): H2d, Level 3, no. 30.11; 10. Backed bladelet (85): H2b, Level 3, no. 28.4; 11. Straight dihedral burin (27): J4b, Level 3, no. 20.14; 12. Angle burin on break (30): H3c, Level 3, no. 3.1; 13. Multiple dihedral burin (31): H1a, Level 3, no. 1.22; 14. Truncated bladelet (84): H1a, Level 3, no. 22.8; 15. Angle dihedral burin (29): H1a, Level 3, no. 22.8; 16. Thumbnail endscraper (10): H1a, Level 3, no. 1.9; 17. Endscraper on retouched flake (5): H4b, Level 3, no. 2.1; 18. Thumbnail endscraper (10): H1b, Level 3, no. 23.13; 19. Endscraper on retouched blade (5): H4b, Level 3, no. 2.2; 20. Sidescraper (77): H4d, Level 3, no. 4.11; 21. Endscraper on blade (1): H1a, Level 3, no. 1.20; 22. Endscraper on retouched blade (5): H4b, Level 3, no. 5; 23. Keeled endscraper (11): H4d, Level 3, no. 4.11; 24. Endscraper on flake (8): H2b, Level 3, no. 28.6. Drawing by Susana Ruiz Pedraza and Natalia Suárez.

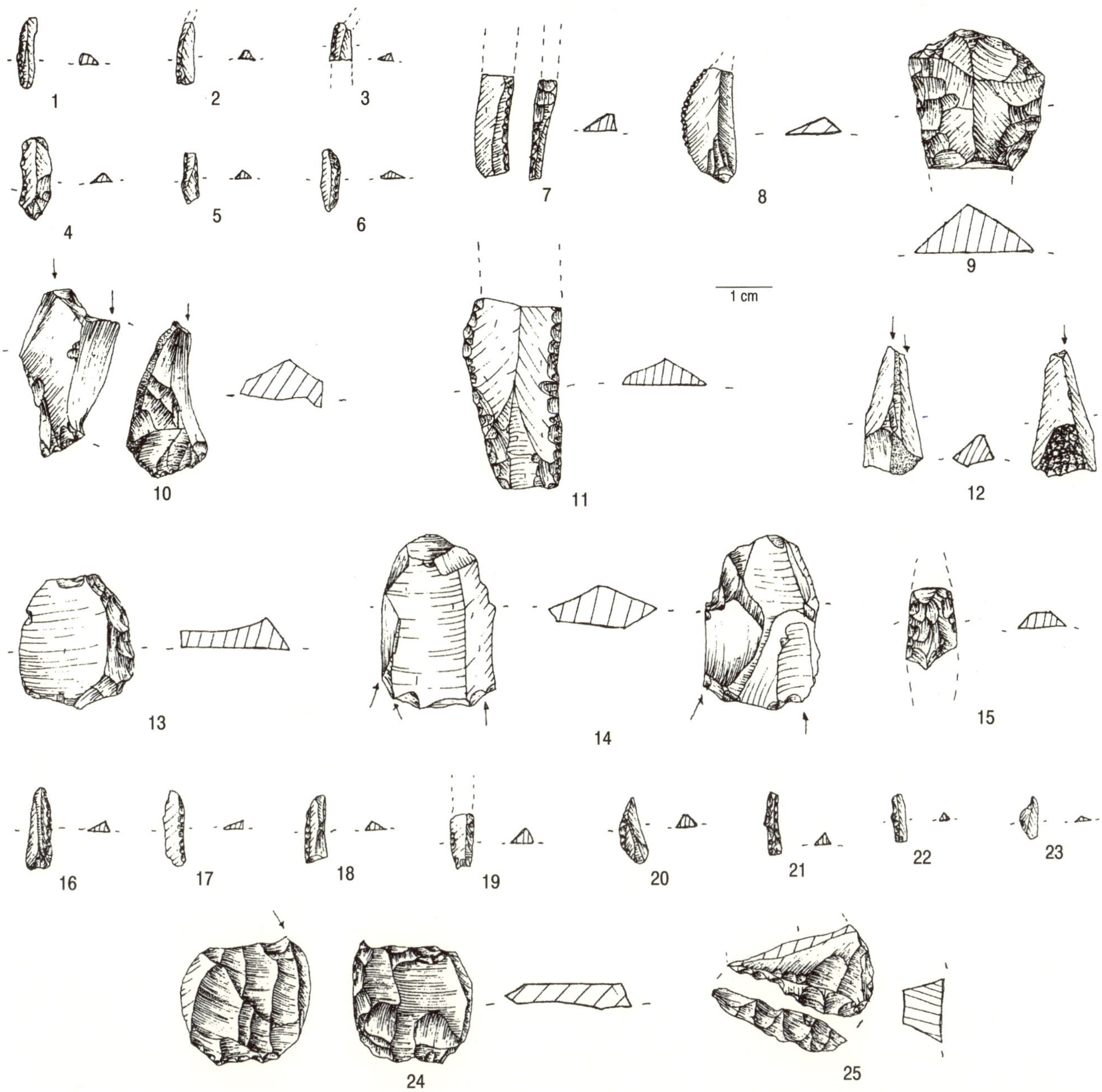

Fig. 18.10. Bronze Age stone tools. 1. Backed bladelet (85): K5c, Level 2, no. 36.28; 2. Backed bladelet (85): K5c, Level 2, no. 36.29; 3. Backed bladelet (85): K5c, Level 2, no. 3.37; 4. Backed bladelet (85): K5c, Level 2, no. 34.14; 5. Backed bladelet (85): K5b, Level 2, no. 35.28; 6. Backed bladelet (85): K5c, Level 2, no. 36.30; 7. Backed bladelet (85): K5c, Level 2, no. 36.26; 8. Retouched bladelet (90): J4a, Level 2, no. 11.15; 9. Endscraper on retouched blade (5): K5b, Level 2, no. 35.31; 10. Nucleiform endscraper (15): K5a, Level 2, no. 34.13; 11. Continuously retouched piece, 2 edges (66): K5b, Level 2, no. 67; 12. Bladelet core remnant: K5c, Level 1/2, no. 3.26; 13. Continuously retouched piece, 1 edge (65): K5d, Level 1/2, no. 4.24; 14. Splintered piece (76): K5d, Level 1/2, no. 22; 15. Continuously retouched piece, 2 edges (66) or unifacial point fragment: K5b, Level 1/2, no. 2.29; 16. Truncated backed bladelet (86): K5a, Level 1/2, no. 36.27; 17. Backed bladelet (85): K5d, Level 1/2,n0.4.25; 18. Backed bladelet (85): K5c, Level 1/2, no. 3.34; 19. Truncated backed bladelet (86): K5b, Level 1/2, no. 2.28; 20. Triangle (79): K5a, Level 1/2, no. 1.6; 21. Backed bladelet (85): K5c, Level 1/2, no. 332; 22. Backed bladelet (85): K5d, Level 1/2, no. 4.26; 23. Trimming flake: K5a, Level 1/2, no. 1.5; 24. Splintered piece (76): J4b, Level 1, no. 2.13; 25. Bec (24): K5d, Level 1, no. 27. Drawing by Susana Ruiz Pedraza and Natalia Suárez.

There is one tool in Stratum 3.2 (a denticulate) and three in Stratum 1 (a micro-perforator, a splintered piece, and a truncated bladelet). Stratum 3.4 has eight tools: six backed bladelets, a backed blade, and a piece retouched on one edge. The main Strata (1–2, 2, and 3) have 50 tools among them (figures 18.9, 18.10). They include 14 end-scrapers, 1 perforator, 5 burins, 1 backed blade, 3 continuously retouched pieces, 2 splintered pieces, 1 sidescraper, 2 triangles, 1 truncated bladelet, 17 backed bladelets (3 of which are truncated), and 1 nibbled (Dufour-like) bladelet. There are two points in Stratum 3: one classifiable as similar to an Upper Paleolithic Font-Yves point (figure 18.9, no. 5) and the other a small tip fragment of a bifacial, invasively retouched point (figure 18.9, no. 6). There is a possible mesial fragment of an invasively retouched unifacial point in Stratum 1–2 (figure 18.10, no. 15). In terms of blanks, 33.3 percent of the Bronze Age tools are made on flakes, 0.9 percent on blades, 37.8 percent on bladelets, 4.4 percent on cores, and 11.1 percent on chunks, while the 4.4 percent splintered pieces are just that (perhaps exhausted bipolar cores). Tools represent 1.2–2.6 percent of the unretouched debris total. Stratum 3 also yielded a pair of pecked stones that could be either anvils or hammerstones.

Most of the tools are made on coastal-type flints in Strata 1–2 and 2 (61.6 and 80.0 percent, respectively), but in Stratum 3 the tools are roughly evenly divided between good-quality coastal types and other type flints (40.9 and 50.0 percent). Other stones were not used for tools in the former two levels and constitute only 4.5 percent in Stratum 3. Debris is likewise heavily dominated by coastal and other type (apparently strictly nonlocal) flints. Only in Stratum 3 do both lutite and other non-flints have any relative importance (7.9 percent and 7.6 percent of the total weight of the identifiable chipped stones, respectively). El Mirón in the Bronze Age continued to be tied to the coastal zone of Vasco-Cantabria. Indeed, we found a few marine shells in these levels.

Including the 13 tools from Stratum 3.5 (pit O), 8 of the Bronze Age endscrapers are on flakes, 4 on cores/chunks, and 3 on blades. Burins are equally divided (three each) between flakes and cores/chunks. The one perforator is on a flake. Backed/truncated, retouched, and notched/denticulated pieces are mostly on blades. Both microliths (triangles) are made on bladelet segments, but backed/retouched bladelets—though mainly (28) on bladelets—can also include small blades (2). Points are made on a blade and a bladelet. In general, the Bronze Age assemblages are quite lamellar (i.e., bladelet-rich) in terms of both unretouched debris and tools/armatures.

Given the chronostratigraphic ambiguity of pit O fill/Stratum 3.5, it is worthwhile detailing its artifact contents separately from the more clearly Bronze Age assemblages of Strata 2–3. Stratum 3 yielded a copper pin as well as a couple of pecked stones that could be either anvils or hammerstones.

Bronze Age Awl

Stratum 3 yielded a copper awl (*punzón*). This is a bi-pointed object measuring 71 × 4 × 4 mm (figure 18.11). Its elemental composition was determined by Dr. Salvador Rovira of the Museo Arqueológico Nacional in Madrid and Dr. Jesus Setien of the Universidad de Cantabria, Departamento de Ciencia e Ingeniería del Terreno y los Materiales, to whom we extent our thanks. The method used was energy dispersive spectrometry by means of X-ray fluorescence. The composition is as follows: 98.328 percent copper, 1.440 percent arsenic, 0.178 percent nickel, and 0.055 percent iron.

As stated above, Stratum 3 has one radiocarbon date: 3700±40 BP (ca. 2100 cal. BC). A copper awl very similar to the one from El Mirón has recently been found in the funerary cave of Pico Ramos in western-most coast Vizcaya, about 35 km from El Mirón. The awl in question came from Level 3 in Pico Ramos, dated by three C14 assays between 4790 and 4100 BP (ca. 3900–2300 cal. BC). Like the El Mirón piece, it is

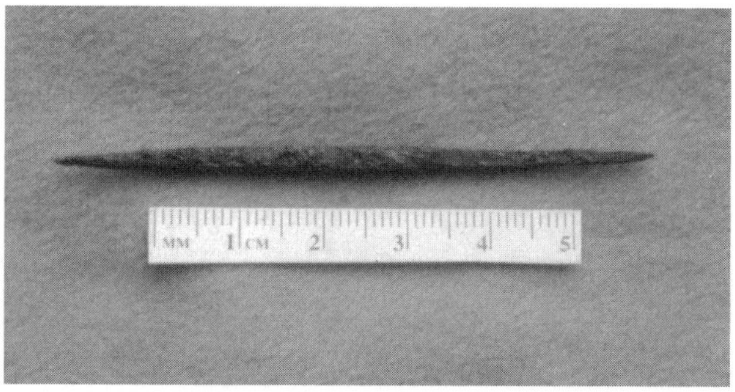

Fig. 18.11. Copper awl from Level 3 (Bronze Age).

quadrangular in cross-section, with one pointed end a bit more gradually tapered than the other. Its mineralogical composition is reported as 96.03 percent copper, 3.83 percent arsenic, 0.09 percent nickel, 0.04 percent iron, plus very small traces of two other elements (Salgado and Zapata 1995).

Bi-pointed awls have been found at a few other regional Chalcolithic and Bronze Age sites, notably in Gobaederra Cave (Alava), with 11 such pieces (along with 7 other copper points) in a context dated to about 3700 BP. There is also a bi-pointed awl from Amalda Cave (Guipúzcoa) and a pair of them from two neighboring dolmens at Ithé in the French Basque province of Soule (with one date of 4000 BP) (Salgado and Zapata 1995). The Amalda awl is from Level III, an undated Chalcolithic funerary deposit that also yielded four arrowheads, a few decorated ceramics, and a dozen perforated beads (Armendáriz 1990). Longer than the El Mirón awl but also quadrangular in section and more tapered at one end than at the other, the Amalda piece is also arsenical copper. Its composition is as follows: 96.15 percent copper, 1.348 percent nickel, 1.132 percent arsenic, 0.22 percent iron, and mere traces of two other elements. The El Mirón awl thus links this residential site to the world of burial caves and megaliths in the Chalcolithic and early Bronze Age. El Mirón Stratum 3 (especially Lens 3.4) had abundant evidence of intense (but uncontained) fire and possible pieces of slag, opening the possibility of in situ metallurgy.

Post–Bronze Age Stone Artifacts

The fill of pit O in the Cabin area, Stratum 3.5, is posterior to the deposition of Stratum 3 and the other 3-numbered lenses (except 3.4). It *might* be roughly equivalent in age to the patchy yellow lenses that as a group are called Stratum 2, but this is not clear. Although Stratum 3.5 is undated, it is not modern (it is itself cut into by a recent pit), containing no modern artifacts. It is conceivable that the contents of pit O/Stratum 3.5 could date to the later Bronze or Iron Age, but those contents might be mixed.

There are 1,027 lithic artifacts (total weight: 295 g) from Stratum 3.5, of which 14 are tools (figure 18.12). The tools represent 1.4 percent of the unretouched debris, which is quite standard among all the Holocene levels at El Mirón. Microdebitage also represents a quite standard 62.4 percent of the lithics, cortical items a quite standard 7.7 percent, cores 0.3 percent, chunks 2.4 percent, blades 4.2 percent, bladelets 6.6 percent, and flakes 24.0 percent. Only the flake percentage is on the somewhat high side relative to most of the other Holocene assemblages. There are no crested blades or platform renewal flakes in Stratum 3.5. The tools include an endscraper, a perforator/endscraper, a burin, a microgravette, a backed blade, a blade with scalariform ("Aurignacian"-like) retouch, two notches, a truncated bladelet, two backed bladelets, and two truncated backed bladelets. In addition there is a "bec" (stubby or atypical perforator) on the same blank as one of the notches. The rather Upper Paleolithic–like composition of this assemblage is striking (including the absence of "Mousteroid" types). The burin on a large blade with a straight retouched truncation blade (figure 18.12, no. 13) is a very Upper Paleolithic–looking tool, suggestive of possible mixture in late to middle Holocene times (due to human pit digging and/or erosion from the ramp leading up to the inner cave?). Five of the tools are on flakes, three on blades, and five on bladelets. In addition, there is one splintered piece. Part of the mix (as noted earlier) is the presence of a double-bevel base antler *sagaie* of Magdalenian aspect. However, again it is worth noting that there were *no modern intrusive artifacts* in this large pit (or channel?) fill.

By count, the Stratum 3.5 lithic assemblage contains 35.5 percent coastal-type flints, 31.7 percent other flints (including almost no mudstone), and 32.8 percent other stones. By weight, 100 g are coastal-type flints, 76 m are other true flints, 1 m is mudstone, and 28 m are other local stones (limestones, quartzes, quartzites). This profile is very similar to that of the Bronze Age and much of the Chalcolithic: mostly nonlocal and probably transported from the shore but with really insignificant amounts of material in absolute terms.

The ground surface and the mixed surficial deposits immediately below modern dung and debris in the mid- and rear (Corral) vestibule area—the former called Stratum 301—are rich in lithic artifacts. These are probably derived from a variety of post-Paleolithic and even late Paleolithic levels, brought to the surface by recent and late prehistoric pitting that seems to have been common in this area. Based on what

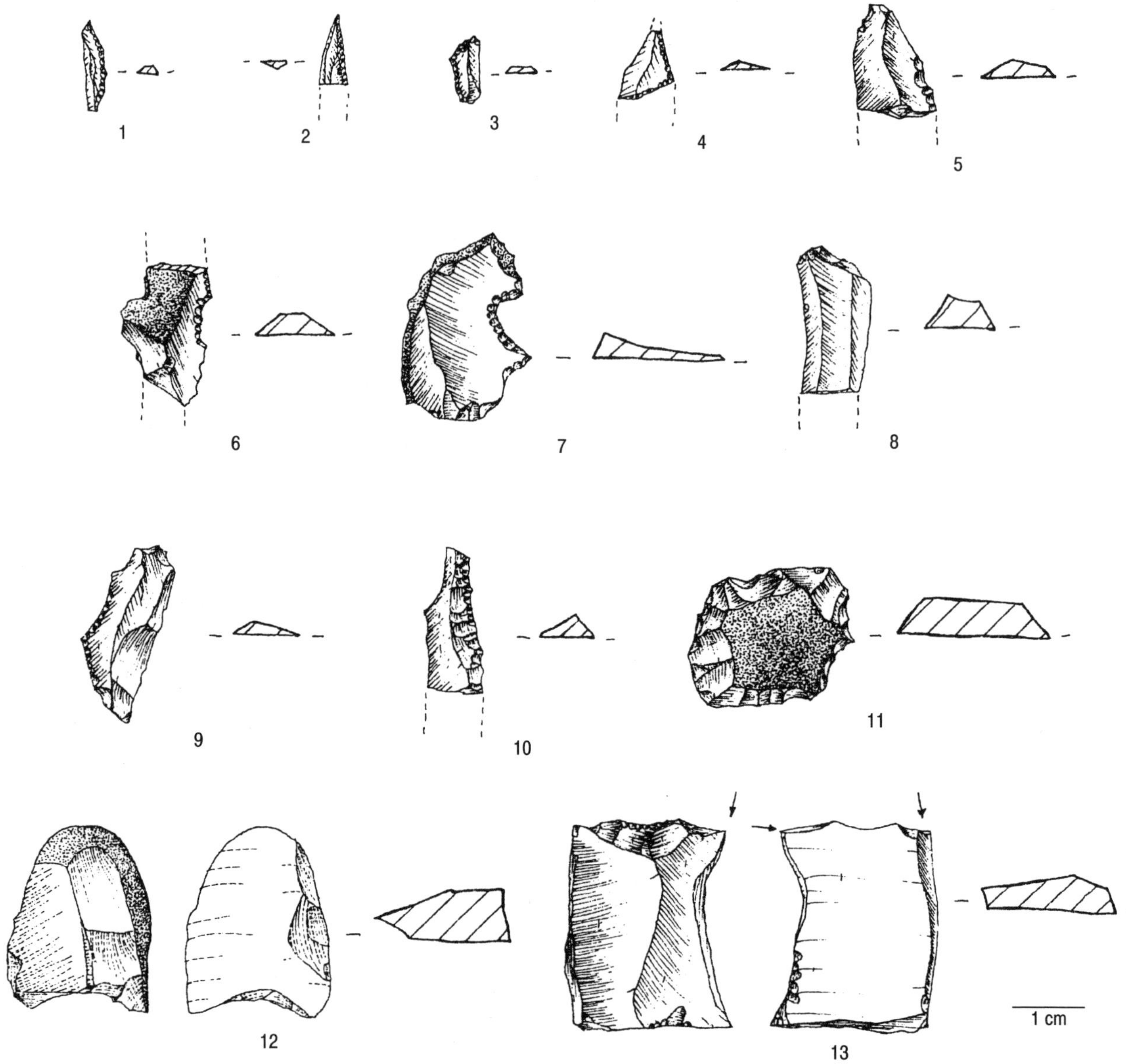

Fig. 18.12. Late Bronze Age (or Iron Age?) stone tools. Pit O, Level 3.5. 1. "Microgravette" point (51): H3a–b, no. 25.27; 2. Backed bladelet (85) or micro-point: H3a–b, no. 25.28; 3. Truncated backed bladelet (86): H3a–b, no. 25.26; 4. Backed bladelet (85): H3a–b, no. 25.29; 5. Partially backed blade (59): H4a, no. 122.55; 6. Notch (74): H3b, no. 25c; 7. Notch (74) + Bec (24): H4b, no. 124; 8. Atypical (canted) endscraper (2): H3b, no. 25.18; 9. Blade with oblique retouched truncation (61): H3a–b, no. 25.23; 10. Backed bladelet (85): H4a, no. 122.56; 11. Endscraper-Perforator (21): H3a–b, no. 25.13; 12. Notch (24): H3a–b, no. 25.15; 13. Burin on straight retouched truncation (34): H4c, no. 221.1. Drawing by Susana Ruiz Pedraza and Natalia Suárez.

we can see in the Mid-Vestibule Trench, mixing of at least Bronze Age, Chalcolithic, and Neolithic materials seems to have occurred, but late Upper Paleolithic materials were also probably brought down from once-existing cultural deposits atop the sterile colluvial-alluvial ramp at the vestibule rear by erosion and by the movements of humans and livestock. A few of the many stone tools found in these surface contexts are illustrated in figure 18.13.

There are 8,664 knapped lithic artifacts (3.75 kg) from 35 intact Holocene levels and lenses in the outer and

Fig. 18.13. Mixed surface materials (stone tools). Level 301. 1. Backed bladelet (85): L5c, no. 6,24; 2. Truncated bladelet (84): 06c, no. 7.2; 3. Bec (24): 06c, no. 7.1; 4. Flat nosed endscraper (14): 06a, no. 5.2; 5. Continuously retouched piece, 2 edges (66) (or unifacial point fragment): P6c, no. 3.1; 6. Continuously retouched piece, 1 edge (65): 06a, no. 1; 7. Continuously retouched piece, 1 edge (65): 06d, no. 4.2; 8. Multiple dihedral burin (31): M–N/7–9 (unnumbered); 9. Continuously retouched piece, 1 edge (65): 06b, no. 2.3.

mid-vestibule excavations of El Mirón. This is a very small quantity in comparison to the huge lithic assemblages typical of most of the Magdalenian levels that underlie them. In addition, the numbers of cores and tools are absolutely and relatively small. Decortication blades and flakes and other materials with cortex are also rare, as are crested blades and platform renewal flakes. It does not look as if much primary knapping was going on, in at least in those areas of the cave vestibule that we have excavated. Flint was favored for knapping in general and for tools in particular. As it does *not* seem to be geologically or fluvially available very near the cave, and as at least the identifiable types of flint are known to outcrop along the Holocene shore of eastern Cantabria and western Vizcaya, human visits to and from or contact with the coast during the Mesolithic, Neolithic, Chalcolithic, and Bronze Age seem assured. The nature of the relationships between human occupations of El Mirón and the coastal flint source areas may have been different among these periods. One can imagine that in Mesolithic times foraging humans may have visited the interior from their bases along the shore, bringing flint materials (mainly in the form of portable equipment, little of which was abandoned at El Mirón), while agro-pastoral humans in the later periods may have been located more permanently in the upper Asón valley and may have had to obtain their flint by trade with coast-based neighbors or through occasional visits to the shore (a few marine mollusk shells having been found in post-Paleolithic deposits at El Mirón). Such visits could have been related to ritual and/or social activities (witness the abundance of megalithic monuments on ridge-crests dominating the lower Asón), as well as undertaken for economic reasons (livestock sale or purchase, sale of secondary products?). In all cases, the amounts of stone being transported (mainly in the form of blanks or tools—not cores or nodules) were modest. Blanks were retouched into tools and tools were resharpened in the cave, as attested by the presence of relatively abundant microdebitage.

The 151 tools from all the post-Paleolithic levels are distributed altogether among 41 types, but never in large numbers (most types are represented by only one item per level, if any). Tools usually make up no more than 1–2 percent of the lithic artifact total among the larger assemblages in any of the post-Paleolithic periods. The only individual retouched type that is fairly consistently represented by more than one item per assemblage is type 85: backed bladelets (maximum $n = 6$). While the whole class of backed bladelets is generally represented among the level assemblages except in the Mesolithic ones, burins—though rare—are also fairly generally distributed, as are continuously retouched pieces (types 65–67). Endscrapers are mainly found in the Chalcolithic and Bronze Age levels (with only a few in the Mesolithic and none in the Neolithic); denticulates and notches are found mainly in the Neolithic and Chalcolithic. The absence or scarcity of geometric microliths (and, indeed, of evidence of the microburin technique) in the Mesolithic and Neolithic levels is notable (although this is not an uncommon phenomenon in Cantabrian Spain). Temporally diagnostic types do exist (a trapeze and the Helwan or double-bevel-retouched circle segments in the Neolithic, the arrowheads in the Chalcolithic), but they are extremely rare. Clearly it is ceramic technology that carries the weight of cultural and temporal significance in the period after about 6000 BP, along with the copper pin in early Bronze Age Stratum 3. Unretouched blades, bladelets, and flakes may have become more important (as sickle elements, knives) than formal tools as we define them in the post-Paleolithic world of Cantabria. Certainly the relative (and sometimes absolute) abundance of bladelets in a few of the Neolithic (Stratum 303), Chalcolithic (Strata 5 and 7), and Bronze Age (Strata 1–2, 2, and 3) assemblages is suggestive of the existence of hafted composite tools and/or weapons. They do contrast with the paucity of bladelets (and the total absence of blades) in the small Mesolithic assemblages.

References Cited

Alday, A.

1998 *Kanpanoste Goikoa*. Memorias de Yacimientos Alaveses 5. Diputación Foral de Álava, Vitoria-Gasteiz.

Arias, P.

1999 The origins of the Neolithic along the Atlantic coast of continental Europe: a survey. *Journal of World Prehistory* 13:403–464.

Armendáriz, A.

1990 Los niveles postpaleolíticas de la Cueva de Amalda. Estudio de las industrias. In *La Cueva de Amalda (Zestoa, Pais Vasco). Ocupaciones paleolíticas y post-paleolíticas*, edited by J. Altuna, A. Baldeón, and K. Mariezkurrena, pp. 117–174. Colección Barandiarán 4. Eusko Ikaskuntza, Sociedad de Estudios Vascos, San Sebastián, Spain.

Elorza, J. J., M. I. Arriortua, and J. M. Amigo

1985 Indices de cristalinidad en los silex de carácter turbiditíco de Barrika (NE de Bilabo). *Boletín Geológico y Minero* 94:44–81.

Elorza, J. J., and M. A. Bustillo

1989 Early and late diagenetic chert in carbonate turbites of the Senonian flysch, northeast Bilbao, Spain. In *Silicious Deposits of the Tethys and Pacific Regions*, edited by J. R. Hein and J. Obradović, pp. 93–105. Springer-Verlag, New York.

Elorza, J. J., F. García-Garmilla, M. I. Arriortua, and M. A. Bustillo

1991 Chert in marine environments. In *Excursion Guidebook: Marine and Continental Facies with Silicious Sedimentary Rocks. VI International Flint Symposium*, edited by M. A. Bustillo, J. J. Elorza, I. Armenteros, F. García-Garmilla, J. A. Blanco, and M. I. Arriortua, pp. 1–92. Instituto Tecnológico GeoMinero de España, Madrid.

González Morales, M. R.

1999 La perhistoria de Santoña. *Monte Buciero* 2:17–28.

Gorrochategui, X., M. J. Yarritu, and L. Zapata

1995 La industria lítica del depósito sepulcral de Pico Ramos (Muskiz. Bizkaia). *Munibe* 47:101–113.

Iriarte, M. J., J. Mujika, and A. Tarriño

2005 Herriko Barra (Zarautz-Gipuzkoa): caractérisation industrielle et économique des premiers groupes de producteurs sur le littoral basque. In *Unité et diversité des processus de néolithisation sur la façade atlantique de l'Europe*, edited by G. Marchand and A. Tresset, pp. 127–136. Mémoires de la Société Préhistorique Française, vol. 36. Paris.

Ontañón, R.

1996 Las industrias líticas del Neolítico final—Calcolítico en Cantabria. *Munibe* 48:15–51.

Rissetto, J. D.

2009 Late Pleistocene Hunter-Gatherer Mobility Patterns and Lithic Exploitation in Eastern Cantabria (Spain). PhD diss., University of New Mexico.

Salgado, J., and L. Zapata

1995 La industria metálica del depósito sepulcral de Pico Ramos (Muskiz, Bizkaia). *Munibe* 47:115–126.

Sonneville-Bordes, D. de, and J. Perrot

1954 Lexique typologique du Paléolithique Supérior. *Bulletin de la Société Préhistorique Française* 51:327–333.

1955 Lexique typologique du Paléolithique Supérior. *Bulletin de la Société Préhistorique Française* 52:76–79.

1956a Lexique typologique du Paléolithique Supérior. *Bulletin de la Société Préhistorique Française* 53:406–412.

1956b Lexique typologique du Paléolithique Supérior. *Bulletin de la Société Préhistorique Française* 53:547–559.

Zapata, L.

1995 La excavación del depósito sepulcral calcolítico de la cueva Pico Ramos (Muskiz, Bizkaia). La industria ósea y los elementos de adorno. *Munibe* 47:35–90.

Table 18.1. Lithic debris, tool blank, and manuport typology

Type no.	Type name	Attributes
Chipped stone		
0	Microburin	Notch + snap at end of blade(let)
1	Non-cortical trimming flake	L ≤1 cm with Hertzian morphology, no cortex
2	Cortical trimming flake	Like no. 1, but with some cortex
3	Non-cortical shatter	L ≤1 cm angular debris without Hertzian morphology or cortex
4	Cortical shatter	Like no. 3, but with some cortex
5	Plain flake	L >1 cm, no cortex
6	Primary decortication flake	Cortex covers whole dorsal surface
7	Secondary decortication flake	Some dorsal cortex
8	Plain, whole, or proximal blade	L >2 cm; at least twice as long as wide (when whole) with ±parallel edges, a definite butt, but no cortex
9	Broken plain blade	Mesial or distal blade fragment; no cortex
10	Primary, whole, or proximal decortication blade	L ≥ W and >2 cm; cortex covers dorsal surface; butt present
11	Secondary, whole, or proximal decortication blade	Like no. 10, but with dorsal surface only partly cortical
12	Mesial/distal cortical blade	Like no. 9, but with some cortex
13	Plain, whole, or proximal bladelet	L ≤2 cm; parallel-sided, narrow, thin; whole or proximal fragment with butt present; no cortex
14	Broken, plain bladelet	Like no. 13, but mesial or distal fragment, with no cortex or butt
15	Whole/proximal cortical bladelet	Like no. 13, but with some cortex
16	Mesial/distal cortical bladelet	Like no. 14, but with some cortex
17	Burin spall	Thick blade(let) with tri- or quadrangular section
18	Unidirectional crested blade	Crest formed by flake scars perpendicular to blade axis in one direction; longitudinal blade scar on adjacent crest face
19	Bidirectional crested blade	Same as no. 18, but flake scars form both faces of crest
20	Flake core	Core with only flake removals (usually globular)
21	Prismatic blade core	Cylindrical core with only blade (L >2 cm) removals
22	Pyramidal blade core	Pyramidal core with only blade removals
23	Prismatic bladelet core	Cylindrical core with only bladelet (L ≤2 cm) removals (can include bladelet cores on flakes)
24	Pyramidal bladelet Core	Pyramidal core with only bladelet removals (includes many "nucleiform endscrapers")
25	Mixed core	Core with both flake and blade(let) removals
26	Non-cortical chunk	Large angular debris, L >1cm, without Hertzian morphology or cortex; includes core remnants/fragments of exhausted cores
27	Cortical chunk	Like no. 26, but with some cortex
28	Platform renewal flake/blade	Has lip of a core platform with nibbling from core preparation before perpendicular removal to correct flaking platform angle (includes "sausage slices")
29	Splintered piece ("pièce esquillée")	Bipolar blade or core remnant; bifacial splintering at both ends (also classified as tool type no. 76)

Table 18.1. *(continued)*

Type no.	Type name	Attributes
"Manuports"/cobble artifacts		
30	Fire-cracked rock	Cobble (usually sandstone) with angular fractures; reddened and/or blackened
31	Pebble or cobble	Unmodified, smooth, water-worn stone
32	Hammerstone/anvil	Cobble with concentrated peck marks at end(s) or on surface(s) (some later used as fire-cracked rock)
33	Slab ("plaquette")	Thin, flat stone
34	Groundstone ("mano")	Cobble with evidence of grinding
35	Structural element	Stone (non-fire-cracked) recovered from a feature such as a wall or pit lining

Table 18.2. Major blank and debris category percentages, post-Paleolithic levels

Level	*n*	Blank / debris categories (%)							
		Microdebitage[a]	Flakes[b]	Blades[c]	Bladelets[d]	Crested blades + platform renewal flakes[e]	Cores[f]	Chunks[g]	Cortical material[h]
1	36	27.7	44.4	8.3	8.3	2.7	0.0	8.3	10.8
1–2	1,083	75.3	9.7	1.1	13.0	0.1	0.3	0.5	3.9
2	711	69.5	12.0	1.3	15.5	0.3	0.1	1.3	7.0
2.3	106	44.3	31.1	6.7	6.7	0.0	0.9	10.3	21.7
3	1,048	60.5	24.3	4.1	9.2	0.6	0.2	1.1	8.0
3.2	43	79.0	7.0	4.6	9.4	0.0	0.0	0.0	2.3
3.3	9	44.4	22.2	0.0	22.2	11.1	0.0	0.0	0.0
3.4	114	55.3	22.7	7.1	12.3	0.0	0.0	2.6	14.0
3.5	1,026	62.4	24.1	4.2	6.6	0.0	0.3	2.4	9.5
4	337	67.4	18.4	3.9	7.7	0.3	0.0	2.4	10.4
5	359	63.5	23.1	1.4	9.2	1.4	0.0	1.4	6.6
5.1	48	68.6	23.1	0.0	6.2	0.0	0.0	2.1	2.1
5.2	5	40.0	40.0	0.0	20.0	0.0	0.0	0.0	40.0
5.4	18	55.5	39.0	0.0	5.5	0.0	0.0	0.0	17.0
6	206	67.5	17.5	7.3	5.8	0.0	0.0	1.9	14.1
6.1	14	28.6	35.7	7.1	28.6	0.0	0.0	0.0	28.6
7	295	57.3	21.4	5.4	13.2	0.7	0.0	2.0	11.5
7.1	25	56.0	8.0	8.0	0.0	0.0	0.0	28.0	12.0
7.4	8	37.5	37.5	25.0	0.0	0.0	0.0	0.0	12.5
7.5	5	40.0	20.0	0.0	40.0	0.0	0.0	0.0	20.0
8	216	61.1	31.0	4.2	2.8	0.0	0.0	0.9	8.3
8.1	3	0.0	0.0	66.7	0.0	0.0	33.3	0.0	33.3
9	221	72.9	16.7	5.9	1.8	0.0	0.0	2.7	8.1
9.2	2	50.0	50.0	0.0	0.0	0.0	0.0	0.0	50.0

Table 18.2. *(continued)*

Level	*n*	Blank / debris categories (%)							
		Microdebitage[a]	Flakes[b]	Blades[c]	Bladelets[d]	Crested blades + platform renewal flakes[e]	Cores[f]	Chunks[g]	Cortical material[h]
9.6	96	81.3	4.2	11.5	1.0	0.0	0.0	2.0	8.2
9.7	77	83.1	5.2	1.3	10.4	0.0	0.0	0.0	3.9
9.8	5	100.0	0.0	0.0	0.0	0.0	0.0	0.0	20.0
10	142	42.3	34.5	9.9	4.9	0.0	0.0	8.4	9.8
10.1	146	52.1	22.6	17.8	4.1	0.7	0.0	2.7	12.3
302	606	73.3	16.8	3.0	2.8	0.3	0.5	3.3	7.5
303	779	73.1	13.1	2.4	9.6	0.3	0.1	1.4	5.1
303.1	176	62.5	18.2	8.0	10.2	0.0	0.0	1.1	5.7
303.2	144	69.5	14.6	5.5	10.4	0.0	0.0	0.0	4.8
303.3	59	61.0	18.6	5.1	13.6	0.0	0.0	1.7	3.4
304	100	73.0	12.0	2.0	12.0	0.0	0.0	1.0	8.0
101	93	62.3	16.1	5.4	11.8	0.0	0.0	4.3	11.8
102	116	62.1	18.1	6.0	10.3	0.0	0.0	3.4	13.8

[a]Cortical and non-cortical trimming flakes and shatter (types 1–4)

[b]Plain and decortication flakes (types 5–7)

[c]Plain and cortical, whole, proximal, mesial, and distal blades (types 8–12)

[d]Plain and cortical, whole, proximal, mesial, and distal bladelets (types 13–16)

[e]Unidirectional and bidirectional crested blades; platform renewal flakes (types 18, 19, 28)

[f]Flake cores; prismatic and pyramidal blade cores; prismatic and pyramidal bladelet cores; mixed cores (types 20–25)

[g]Cortical and non-cortical chunks (types 26–27)

[h]Cortical trimming flakes; cortical shatter; primary and secondary decortication flakes; primary and secondary, whole, or proximal decortication blades; medial/distal cortical blades; primary and secondary, whole, or proximal decortication bladelets; medial/distal cortical bladelets; cortical chunks (types 2, 4, 6, 7, 10–12, 15, 16, 27): sum divided by total stone tool blanks and knapping debris

Table 18.3. Number and type of lithic blanks for tools, post-Paleolithic levels

	Blank Types									
LEVEL	UK	5	6	7	8	9	11	12	13	14
1	1	.	.	.	.	.	.	.	.	1
1–2	.	2	.	1	1	.	.	.	7	.
2	.	2	.	.	1	.	.	.	4	1
3	2	6	1	1	1	1	.	.	6	2
3.2	.	.	.	.	1	.	.	.	.	.
3.4	.	.	.	.	1	.	.	.	4	2
3.5	.	1	1	3	1	2	.	.	4	1
4	.	1	.	2	.	.	.	.	2	.
5	.	3	.	1	.	.	.	.	1	.
5.1	1	.	.	.	.	.	.	.	.	.
5.4	.	1	.	.	.	.	.	.	.	.
6	.	1	.	.	1	.	.	.	.	.
7	1	2	.	.	.	.	1	.	2	.
7.5	.	.	.	.	.	.	.	.	1	.
8	.	1	.	1	1	.	.	.	.	.
8.1	.	.	.	.	1	.	.	.	.	.
9	.	1	.	1	.	.	.	1	.	1
9.6	.	.	.	.	1	1	.	.	.	.
10	.	.	.	.	1	3	.	.	.	.
10.1	.	.	.	.	1	.	.	.	2	.
302	.	1	1	.	.	3	1	.	.	.
303	.	2	1	1	1	1	.	.	2	3
303.1	.	1	.	.	.	1	1	.	1	3
303.3	.	1	.	.	.	.	.	.	.	.
304	.	1	.	.	.	.	.	.	.	.
101	.	.	.	1	1	.	.	.	.	.
102	.	1	.	.	1	.	.	.	1	.
TOTAL	5	28	4	12	15	12	3	1	37	14

	Blank Types *(continued)*									
LEVEL	15	17	18	21	24	25	26	27	28	29
1	.	.	.	.	.	.	.	.	.	1
1–2	.	.	.	.	.	.	1	.	.	1
2	.	.	.	.	.	.	.	1	.	1
3	.	.	.	1	.	1	1	2	1	.
3.2	.	.	.	.	.	.	.	.	.	.
3.4	.	.	.	.	.	.	1	.	.	.
3.5	.	.	.	.	.	.	.	.	.	.
4	.	.	.	.	.	.	.	.	.	.

Table 18.3. *(continued)*

	Blank Types *(continued)*									
LEVEL	15	17	18	21	24	25	26	27	28	29
5	.	.	.	.	.	.	.	.	.	1
5.1	.	.	.	.	.	.	.	.	.	.
5.4	.	.	.	.	.	.	.	.	.	.
6	.	.	.	.	.	.	.	.	.	.
7	.	.	.	.	.	.	.	.	.	.
7.5	.	.	.	.	.	.	.	.	.	.
8	.	.	.	.	.	.	.	.	.	.
8.1	.	.	.		.	.	.	.	.	.
9	.	.	.	.	.	.	.	.	.	.
9.6	.	.	.	.	.	.	.	.	.	.
10	.	.	.	.	.	.	1	.	.	.
10.1	.	.	1	.	1	.	.	.	.	.
302	1	1	.	.	.	.	.	.	.	1
303	.	.	1	.	.	.	.	.	.	.
303.1	.	.	.	.	.	.	.	.	.	.
303.3	.	.	.	.	.	.	.	.	.	.
304	.	.	.	.	.	.	.	.	.	.
101	.	.	.	.	.	.	.	.	.	.
102	.	.	.	.	.	.	.	.	.	.
TOTAL	1	1	2	1	1	1	4	3	1	5

Blank Types: UK = unknown blank; 5 = plain flake; 6 = primary decortication flake; 7 = secondary decortication flake; 8 = plain, whole, or proximal blade; 9 = broken plain blade; 10 = primary, whole, or proximal decortication blade; 11 = secondary, whole, or proximal decortication blade; 12 = medial/distal cortical blade; 13 = plain, whole, or proximal bladelet; 14 = broken plain bladelet; 15 = whole or proximal decortication bladelet; 16 = medial/distal cortical bladelet; 18 = unidirectional crested blade; 19 = bidirectional crested blade; 20 = flake core; 21 = prismatic blade core; 22 = pyramidal blade core; 23 = prismatic bladelet core; 24 = pyramidal bladelet core; 25 = mixed core; 26 = non-cortical chunk; 27 = cortical chunk; 28 = platform renewal flake; 29 = pièce esquillée (splintered piece)

Table 18.4. Mesolithic stone tool blanks and knapping debris (1996–2001)

	Level			
	10.1	304	101	102
0-Microburin	.	.	.	.
1-Non-cortical trimming flake	43	56	44	49
2-Cortical trimming flake	1	1	.	4
3-Non-cortical shatter	29	14	14	19
4-Cortical shatter	3	2	.	.
5-Plain flake	25	10	9	14
6-Primary decortication flake	1	.	3	2
7-Secondary decortication flake	7	2	3	5
8-Plain, whole or proximal blade	1	.	2	5
9-Broken plain blade	22	1	1	1
10-Primary, whole, or proximal decortication blade	1	.	.	.
11-Secondary, whole or proximal decortication blade	.	1	1	1
12-Medial/distal cortical blade	2	.	1	.
13-Plain, whole or proximal bladelet	5	4	10	9
14-Broken, plain bladelet	1	7	1	1
15-Whole/proximal cortical bladelet	.	1	.	1
16-Medial/distal cortical bladelet	.	.	.	1
17-Burin spall	.	.	.	.
18-Unidirectional crested blade	.	.	.	.
19-Bidirectional crested blade	.	.	.	.
20-Flake core	.	.	.	.
21-Prismatic blade core	.	.	.	.
22-Pyramidal blade core	.	.	.	.
23-Prismatic bladelet core	.	.	.	.
24-Pyramidal bladelet core	.	.	.	.
25-Mixed core	.	.	.	.
26-Non-cortical chunk	1	.	1	2
27-Cortical chunk	3	1	3	2
28-Platform renewal flake	1	.	.	.
29-Pièce esquillée	.	.	.	.
TOTAL LITHIC DEBRIS (1–29)	**146**	**100**	**93**	**116**

Table 18.5. Mesolithic tools (1996–2001)

	Level			
TOOLS	10.1	304	101	102
2-Atypical endscraper	1	.	.	.
10-Thumbnail endscraper	.	.	1	.
15-Nucleiform endscraper	1	.	.	.
23-Perforator	1	.	.	.
30-Burin on break	.	.	.	1
51-Microgravette	.	.	.	1
61-Oblique truncation	.	1	.	.
65-Continuously retouched piece, 1 edge	1	.	.	.
66-Continuously retouched piece, 2 edges	.	.	.	1
74-Notch	.	.	1	.
90-Retouched bladelet	1	.	.	.
TOTAL	**5**	**1**	**2**	**3**

Table 18.6. Blank types on which tools were made, Mesolithic levels (10.1, 304, 101, 102)

	Flakes	Blades	Bladelets	Cores/chunks	Splintered pieces
Endscrapers	1	1	.	1	.
Perforators	.	1	.	.	.
Burins	1	.	.	.	.
Backed and truncated pieces	1	.	.	.	.
Continuously retouched pieces and sidescrapers	.	1	1	.	.
Notches and denticulates	.	1	.	.	.
Microliths	.	.	.	.	.
Backed and retouched bladelets	.	.	1	.	.
Splintered pieces	.	.	.	.	.
Points	.	.	1	.	.
Others	.	.	.	.	.
TOTAL (n = 11)	**3**	**4**	**3**	**1**	**0**

Table 18.7. Mesolithic lithic raw materials assemblages counts by level for all lithic tools and debris

	Lithic blank and debris counts			
LEVEL	10.1	304	101	102
A flints[a]	11	8	10	17
B flints [a]	8	5	4	4
Other flints	24	3	7	22
Lutites[b]	6	1	1	1
Other non-flints[c]	2	.	1	2
Unidentified stones[d]	95	83	70	70
TOTAL	**146**	**100**	**93**	**116**

	Lithic tool counts			
LEVEL	10.1	304	101	102
A flints[a]	2	.	1	2
B flints[a]	1		1	.
Other flints	2	.	1	1
Lutites[b]	.	.	.	.
Other non-flints[c]	.	.	.	.
Unidentified stones[d]	.	.	.	.
TOTAL	**5**	**1**	**2**	**3**

[a]Groups of flints that (based on color, texture, inclusions) are apparently varieties of materials that can all be found at present in the Cretaceous levels outcropping along the Holocene shore of eastern Cantabria and western Vizcaya

[b]Mudstone

[c]Limestones, quartzes, quartzites, calcites, and sandstones

[d]Lithic chips or shatter too small (<1 cm) to accurately identify by raw material type

Table 18.8. Weight (g) of all artifacts (debris and tools) by major lithic group, Mesolithic assemblages

Level	10.1	304	101	102
A flint[a]	11.6	4.9	9.9	29.7
B flint[a]	7.5	2.6	12.3	3.0
Other flint	17.3	0.9	14.0	22.9
Lutites[b]	39.9	0.3	7.0	12.0
Other non-flints[c]	2.0	.	3.4	2.0
Unidentified stones[d]	11.6	7.7	8.3	31.2
TOTAL	**89.9**	**16.4**	**54.9**	**100.8**

[a]Groups of flints that (based on color, texture, inclusions) are apparently varieties of materials that can all be found at present in the Cretaceous levels outcropping along the Holocene shore of eastern Cantabria and western Vizcaya

[b]Mudstone

[c]Limestones, quartzes, quartzites, calcites, and sandstones

[d]Lithic chips or shatter too small (<1 cm) to accurately identify by raw material type

Table 18.9. Neolithic stone tool blanks and knapping debris (1996–2001)

	Level													Type
	8	8.1	9	9.2	9.6	9.7	9.8	10	302	303	303.1	303.2	303.3	
0-Microburin	.	.	.	.	.	.	.	.	.	.	.	.	.	0
1-Non-cortical trimming flake	73	.	68	.	33	6	3	39	372	473	89	77	34	1
2-Cortical trimming flake	1	.	2	.	2	1	2	1	8	3	1	.	.	2
3-Non-cortical shatter	55	.	89	.	43	57	.	16	56	89	20	23	2	3
4-Cortical shatter	3	.	2	1	.	.	.	4	8	4	.	.	.	4
5-Plain flake	56	.	30	1	3	2	.	42	79	75	25	15	10	5
6-Primary decortication flake	.	.	1	.	.	.	.	.	6	5	3	1	.	6
7-Secondary decortication flake	11	.	6	.	1	2	.	7	17	22	4	5	1	7
8-Plain, whole, or proximal blade	4	1	7	.	6	1	.	8	6	11	7	4	1	8
9-Broken plain blade	4	1	3	.	2	.	.	4	9	8	6	3	2	9
10-Primary, whole, or proximal decortication blade	1	.	.	.	.	.	.	1	.	.	.	.	.	10
11-Secondary, whole, or proximal decortication blade	.	.	.	.	2	.	.	1	2	.	1	1	.	11
12-Medial/distal cortical blade	.	.	3	.	1	.	.	.	1	.	.	.	.	12
13-Plain, whole, or proximal bladelet	1	.	1	.	1	.	.	3	11	54	6	9	4	13
14-Broken, plain cladelet	3	.	3	.	.	8	.	3	5	17	12	6	3	14
15-Whole/proximal cortical bladelet	.	.	.	.	.	.	.	1	1	4	.	.	1	15
16-Medial/distal cortical bladelet	2	.	.	.	.	.	.	.	.	.	.	.	.	16
17-Burin spall	1	.	1	.	1	.	1	1	5	8	.	1	.	17
18-Unidirectional crested blade	.	.	.	.	.	.	.	.	2	2	.	.	.	18
19-Bidirectional crested blade	.	.	.	.	.	.	.	.	.	.	.	.	.	19
20-Flake core	.	.	.	.	.	.	.	.	1	1	.	.	.	20
21-Prismatic blade core	.	.	.	.	.	.	.	.	.	.	.	.	.	21
22-Pyramidal blade core	.	.	.	.	.	.	.	.	.	.	.	.	.	22
23-Prismatic bladelet core	.	.	.	.	.	.	.	.	.	.	.	.	.	23
24-Pyramidal bladelet core	.	.	.	.	.	.	.	.	1	.	.	.	.	24

Table 18.9. *(continued)*

	Level													Type
	8	8.1	9	9.2	9.6	9.7	9.8	10	302	303	303.1	303.2	303.3	
25-Mixed core	.	.	.	.	.	.	.	.	1	.	.	.	.	25
26-Non-cortical chunk	2	.	2	.	.	.	.	6	17	9	1	.	1	26
27-Cortical chunk	.	1	4	.	2	.	.	6	3	2	1	.	.	27
28-Platform renewal flake	.	.	.	.	.	.	.	.	.	.	.	.	.	28
29-Pièce esquillée	.	.	.	.	.	.	.	.	1	.	.	.	.	29
TOTAL LITHIC DEBRIS (1–29)	217	3	222	2	97	77	6	143	612	787	176	145	59	

Table 18.10. Neolithic lithic raw materials assemblages counts by level for all lithic tools and debris

Lithic blank and debris counts													
LEVEL	8	8.1	9	9.2	9.6	9.7	9.8	10	302	303	303.1	303.2	303.3
A flints[a]	32	.	26	1	5	3	.	19	65	41	19	22	3
B flints[a]	10	.	4	.	2	1	1	7	23	47	12	5	3
Other flints	15	2	22	.	7	1	.	35	37	69	18	13	6
Lutites[b]	21	.	4	.	1	.	.	6	15	8	3	3	.
Other non-flints[c]	4	.	5	1	1	.	.	6	13	12	12	2	3
Unidentified stones[d]	135	1	161	.	81	72	5	70	459	610	112	100	44
TOTAL	217	3	222	2	97	77	6	143	612	787	176	145	59

Lithic tool counts													
LEVEL	8	8.1	9	9.2	9.6	9.7	9.8	10	302	303	303.1	303.2	303.3
A flints[a]	2	.	1	.	.	.	.	2	4	4	5	.	1
B flints[a]	.	.	.	.	.	.	.	.	.	3	.	.	.
Other flints	1	1	2	.	1	.	.	3	3	4	1	.	.
Lutites[b]	.	.	.	.	.	.	.	.	1	1	.	.	.
Other non-flints[c]	.	.	.	.	.	.	.	.	1	.	.	.	.
Unidentified stones[d]	.	.	1	.	1	.	.	.	.	.	1	.	.
TOTAL	3	1	4	.	2	.	.	5	9	12	7	.	1

[a]Groups of flints that (based on color, texture, inclusions) are apparently varieties of materials that can all be found at present in the Cretaceous levels outcropping along the Holocene shore of eastern Cantabria and western Vizcaya

[b]Mudstone

[c]Limestones, quartzes, quartzites, calcites, and sandstones

[d]Lithic chips or shatter too small (<1 cm) to accurately identify by raw material type

Table 18.11. Weight (g) of all artifacts (debris and tools) by major lithic group, Neolithic assemblages

Level	8	8.1	9	9.2	9.6	9.7	9.8	10	302	303	303.1	303.2	303.3
A flint[a]	32.0	.	32.1	1.0	6.1	6.7	.	22.6	89.0	25.3	17.2	18.5	2.3
B flint[a]	8.8	.	2.1	.	0.6	.	0.3	7.8	17.8	33.4	4.0	1.1	1.4
Other flint	15.7	8.0	44.3	.	7.0	4.0	.	37.1	38.5	57.3	18.6	3.5	2.8
Lutites[b]	153.1	.	2.6	.	0.8	.	.	36.4	198.1	17.1	1.9	0.8	.
Other non-flint[c]	2.9	.	9.7	44.0	0.1	.	.	10.9	57.6	57.0	5.5	4.4	4.7
Unidentified stone[d]	11.1	1.0	11.8	.	12.6	3.9	0.4	42.6	53.4	33.5	30.9	6.3	3.3
TOTAL	223.6	9.0	102.6	45.0	27.2	14.6	0.7	157.4	454.4	223.6	78.1	34.6	14.5

[a]Groups of flints that (based on color, texture, inclusions) are apparently varieties of materials that can all be found at present in the Cretaceous levels outcropping along the Holocene shore of eastern Cantabria and western Vizcaya

[b]Mudstone

[c]Limestones, quartzes, quartzites, calcites, and sandstones

[d]Lithic chips or shatter too small (<1 cm) to accurately identify by raw material type

Table 18.12. Neolithic tools (1996–2001)

	Level								
TOOLS	8	8.1	9	9.6	10	302	303	303.1	303.3
2-Atypical endscraper								2	
17-Endscraper on burin	.	.	1	.	.	.	.	.	.
27-Straight dihedral burin	.	.	.	.	1	.	.	.	.
30-Burin on break	.	.	.	.	.	.	1	1	.
34-Burin on straight truncation	.	.	.	.	.	.	.	1	.
40-Multiple burin on truncation	.	.	.	.	.	.	1	.	.
48-Gravette point	.	.	.	.	.	.	1	.	.
51-Microgravette	.	.	.	.	.	.	1	.	.
52-Font-Yves point	.	.	.	.	.	.	.	1	.
58-Completely backed blade	.	.	.	.	.	3	.	.	.
65-Continuously retouched piece, 1 edge	1	.	1	1	2	2	1	2	1
66-Continuously retouched piece, 2 edges	.	1	.	.	.	.	.	.	.
74-Notch	1	.	1	.	1	3	.	.	.
75-Denticulate	1	.	.	.	.	.	1	.	.
76-Pièce esquillée	.	.	.	.	.	1	1	.	.
81-Trapeze	.	.	.	.	.	.	1	.	.
83-Segment	.	.	1	1	1	.	.	.	.
85-Backed bladelet	.	.	.	.	.	.	3	2	.
90-Retouched bladelet	.	.	.	.	.	.	1	.	
TOTAL	3	1	4	2	5	9	12	9	1

Table 18.13. Blank types on which tools were made, Neolithic levels

	Flakes	Blades	Bladelets	Cores/chunks	Splintered pieces
Endscrapers	.	.	.	.	.
Perforators	.	.	.	.	.
Burins	1	2	1	.	.
Backed and truncated pieces	2	4	1	.	1
Continuously retouched pieces and sidescrapers	4	5	2	.	.
Notches and denticulates	5	3	.	1	.
Microliths	.	.	1	.	.
Backed and retouched bladelets	.	1	5	.	.
Splintered pieces	.	.	1	.	1
Points	.	2	.	.	.
Others	.	1	.	.	.
TOTAL ($n = 44$)	**12**	**18**	**11**	**1**	**2**

Table 18.14. Chalcolithic stone tool blanks and knapping debris (1996–2001)

	Level										
	4	5	5.1	5.2	5.4	6	6.1	7	7.1	7.4	7.5
0- Microburin	.	.	.	.	.	.	.	.	.	.	.
1- Non-cortical trimming flake	179	207	33	2	10	97	4	87	14	1	2
2- Cortical trimming flake	17	9	.	.	.	15	.	13	.	.	.
3- Non-cortical shatter	30	10	.	.	.	24	.	68	.	2	.
4- Cortical shatter	1	1	.	.	.	3	.	1	.	.	.
5- Plain flake	51	72	10	1	4	28	1	49	1	2	.
6- Primary decortication flake	1	3	.	1	1	2	4	4	.	.	.
7- Secondary decortication flake	10	8	1	.	2	6	.	10	1	1	1
8- Plain, whole, or proximal blade	5	3	.	.	.	6	1	6	1	2	.
9- Broken plain blade	7	.	.	.	.	5	.	8	1	.	.
10- Primary, whole, or proximal decortication blade	.	.	.	.	.	.	.	1	.	.	.
11- Secondary, whole or proximal decortication blade	.	2	.	.	.	3	.	1	.	.	.
12- Medial/distal cortical blade	1	.	.	.	.	.	.	.	.	.	.
13- Plain, whole, or proximal bladelet	18	31	3	.	1	5	3	15	.	.	1
14- Broken, plain bladelet	8	1	.	.	.	7	1	21	.	.	1
15- Whole/proximal cortical bladelet	.	1	.	1	.	.	.	1	.	.	.
16- Medial/distal cortical bladelet	.	.	.	.	.	.	.	2	.	.	.
17- Burin spall	.	1	.	.	.	.	.	1	.	.	.
18- Unidirectional crested blade	1	5	.	.	.	.	.	2	.	.	.

Table 18.14. *(continued)*

	Level										
	4	5	5.1	5.2	5.4	6	6.1	7	7.1	7.4	7.5
19- Bidirectional crested blade	.	.	.	.	.	.	.	.	.	.	.
20- Flake core	.	.	.	.	.	.	.	.	.	.	.
21- Prismatic blade core	.	.	.	.	.	.	.	.	.	.	.
22- Pyramidal blade core	.	.	.	.	.	.	.	.	.	.	.
23- Prismatic bladelet core	.	.	.	.	.	.	.	.	.	.	.
24- Pyramidal bladelet core	.	.	.	.	.	.	.	.	.	.	.
25- Mixed core	.	.	.	.	.	.	.	.	.	.	.
26- Non-cortical chunk	3	5	1	.	.	4	.	5	5	.	.
27- Cortical chunk	5	.	.	.	.	.	.	1	2	.	.
28- Platform renewal flake	.	.	.	.	.	.	.	.	.	.	.
29- Pièce esquillée	.	1	.	.	.	.	.	.	.	.	.
TOTAL LITHIC DEBRIS (1–29)	**337**	**361**	**48**	**5**	**18**	**205**	**14**	**296**	**25**	**8**	**5**

Table 18.15. Chalcolithic lithic raw materials assemblages counts by level for all lithic tools and debris

	Lithic blank and debris counts										
LEVEL	4	5	5.1	5.2	5.4	6	6.1	7	7.1	7.4	7.5
A flints[a]	79	40	9	3	4	30	5	42	13	.	2
B flints[a]	4	12	.	.	1	1	.	12	.	.	.
Other flints	42	73	4	.	1	23	.	48	3	2	1
Lutites[b]	4	2	1	.	1	1	.	6	.	1	.
Other non-flints[c]	6	4	4	.	1	3	.	7	4	.	.
Unidentified stones[d]	202	230	30	2	10	147	9	181	2	5	2
TOTAL	**337**	**361**	**48**	**5**	**18**	**205**	**14**	**296**	**25**	**8**	**5**

	Lithic tool counts										
LEVEL	4	5	5.1	5.2	5.4	6	6.1	7	7.1	7.4	7.5
A flints[a]	1	2	1	.	.	1	.	2	.	.	.
B flints[a]	.	1	.	.	.	.	.	.	.	.	.
Other flints	4	2	.	.	1	1	.	4	.	.	1
Lutites[b]	.	1	.	.	.	.	.	.	.	.	.
Other non-flints[c]	.	.	.	.	.	.	.	.	.	.	.
Unidentified stones[d]	.	.	.	.	.	.	.	.	.	.	.
TOTAL	**5**	**6**	**1**	**.**	**1**	**2**	**.**	**6**	**.**	**.**	**1**

[a]Groups of flints that (based on color, texture, inclusions) are apparently varieties of materials that can all be found at present in the Cretaceous levels outcropping along the Holocene shore of eastern Cantabria and western Vizcaya

[b]Mudstone

[c]Limestones, quartzes, quartzites, calcites, and sandstones

[d]Lithic chips or shatter too small (<1 cm) to accurately identify by raw material type

Table 18.16. Weight (g) of all artifacts (debris and tools) by major lithic groups, Chalcolithic assemblages

Level	4	5	5.1	5.2	5.4	6	6.1	7	7.1	7.4	7.5
A flints[a]	38.0	49.0	5.0	5.0	6.0	46.6	4.0	55.0	5.0	.	1.0
B flints[a]	.	8.0	.	.	.	3.0	.	3.9	.	.	.
Other flints	40.0	49.4	1.0	.	0.9	13.8	.	51.0	9.0	1.0	.
Lutites[b]	10.0	10.0	1.0	.	.	2.0	.	52.7	.	4.0	.
Other non-flints[c]	4.0	9.0	27.0	.	1.0	4.0	.	55.7	.	.	.
Unidentified stones[d]	11.0	16.0	8.0	.	.	17.8	.	18.5	.	.	.
TOTAL	103.0	141.4	42.0	5.0	7.9	87.2	4.0	236.8	14.0	5.0	1.0

[a]Groups of flints that (based on color, texture, inclusions) are apparently varieties of materials that can all be found at present in the Cretaceous levels outcropping along the Holocene shore of eastern Cantabria and western Vizcaya

[b]Mudstone

[c]Limestones, quartzes, quartzites, calcites, and sandstones

[d]Lithic chips or shatter too small (<1 cm) to accurately identify by raw material type

Table 18.17. Chalcolithic tools (1996–2001)

	Level						
TOOLS	4	5	5.1	5.4	6	7	7.5
8-Endscraper on flake	1	.	.	.	.	.	.
10-Thumbnail endscraper	1	.	.	.	.	.	.
15-Nucleiform endscraper	.		.	.	.	1	.
23-Perforator	.	.	.	.	.	1	.
24-Bec	.	1	.	.	.	.	.
28-Dejeté dihedral burin	.	.	.	.	.	1	.
30-Burin on break	.	1		.	.	.	.
48-Gravette point	.	.	.	.	1	.	.
51-Microgravette	1	.	.	.	.	.	.
55-Stemmed point	.	.	1	.			.
62-Concave truncation	.	.	.	.	1	.	.
65-Continuously retouched piece, 1 edge	.	1	.	.	.	.	.
66-Continuously retouched piece, 2 edges	.	.	.	.	.	1	.
70-Foliate point/arrowhead	.	1	.	.	.	.	.
74-Notch	.	1	.	1	.	.	.
76-Splintered piece	.	1	.	.	.	.	.
77-Sidescraper	1	.	.	.	.	.	.
85-Backed bladelet	.	.	.	.	.	2	1
87-Denticulated backed bladelet	.	1	.	.	.	.	.
TOTAL	4	7	1	1	2	6	1

Table 18.18. Blank types on which tools were made, Chalcolithic levels

	Flakes	Blades	Bladelets	Cores/chunks	Splintered pieces
Endscrapers	2	.	.	1	.
Perforators	2	.	.	.	.
Burins	1	1	.	.	.
Backed and truncated pieces	1	.	.	.	.
Continuously retouched pieces and sidescrapers	3	.	.	.	.
Notches and denticulates	2	.	1	.	.
Microliths	.	.	1	.	.
Backed and retouched bladelets	.	.	3	.	.
Splintered pieces	.	.	.	.	1
Points	.	2	.	.	.
Others	.	.	.	.	.
TOTAL (*n* = 21)	11	3	5	1	1

Note: Level 5.1, tool type 55—a stemmed point fragment—was not assigned a debris type.

Table 18.19. Bronze Age stone tool blanks and knapping debris (1996–2001)

	Level								
	1	1–2	2	2.3	3	3.2	3.3	3.4	3.5
0-Microburin	.	.	.	.	.	.	.	.	.
1-Non-cortical trimming flake	8	790	464	30	534	34	4	25	429
2-Cortical trimming flake	.	13	11	6	15	.	.	10	35
3-Non-cortical shatter	2	13	17	10	81	.	.	28	163
4-Cortical shatter	.	.	2	1	4	.	.	.	14
5-Plain flake	14	77	60	26	208	3	2	22	216
6-Primary decortication flake	.	16	4	4	9	.	.	1	11
7-Secondary decortication flake	2	12	21	3	38	.	.	3	19
8-Plain, whole, or proximal blade	1	12	2	5	18	2	.	6	17
9-Broken plain blade	1	.	1	.	18	.	.	.	18
10-Primary, whole, or proximal decortication blade	.	.	.	1	2	.	.	1	.
11-Secondary, whole or proximal decortication blade	1	.	2	.	2	.	.	.	5
12-Medial/distal cortical blade	.	.	4	1	3	.	.	1	3
13-Plain, whole, or proximal bladelet	.	132	96	3	59	3	2	9	38
14-Broken, plain bladelet	3	9	14	4	31	.	.	5	25
15-Whole/proximal cortical bladelet	.	.	.	.	3	.	.	.	4
16-Medial/distal cortical bladelet	.	.	.	.	3	1	.	.	1
17-Burin spall	.	2	5	.	4	.	.	.	.

Table 18.19. *(continued)*

	Level								
	1	1–2	2	2.3	3	3.2	3.3	3.4	3.5
18- Unidirectional crested blade	.	1	1	.	5	.	.	.	.
19- Bidirectional crested blade	1	.	.	.	.	.	.	.	.
20- Flake core	.	2	1	1	.	.	.	.	.
21- Prismatic blade core	.	.	.	.	1	.	.	.	.
22- Pyramidal blade core	.	.	.	.	.	.	.	.	.
23- Prismatic bladelet core	.	.	.	.	.	.	.	.	2
24- Pyramidal bladelet core	.	.	.	.	.	.	.	.	1
25- Mixed core	.	1	.	.	1	.	.	.	.
26- Non-cortical chunk	2	4	3	4	7	.	.	3	19
27- Cortical chunk	1	1	6	7	5	.	.	.	6
28- Platform renewal flake	.	.	2	.	1	.	1	.	.
29- Pièce esquillée	1	1	1	.	.	.	.	.	1
TOTAL LITHIC DEBRIS (1–29)	**37**	**1,086**	**716**	**106**	**1,052**	**43**	**9**	**114**	**1,027**

Table 18.20. Bronze Age lithic raw materials assemblages counts by level for all lithic tools and debris

	Lithic blank and debris counts								
LEVEL	1	1–2	2	2.3	3	3.2	3.3	3.4	3.5
A flints[a]	15	60	65	11	136	2	5	7	82
B flints[a]	4	42	26	.	32	5	.	4	11
Other flints	10	47	42	43	139	2	.	18	80
Lutites[b]	.	1	5	.	18	.	.	1	3
Other non-flints[c]	.	9	8	2	15	.	.	.	86
Unidentified stones[d]	8	927	570	50	712	34	4	84	765
TOTAL	**37**	**1,086**	**716**	**106**	**1,052**	**43**	**9**	**114**	**1,027**

	Lithic tool counts								
LEVEL	1	1-2	2	2.3	3	3.2	3.3	3.4	3.5
A flints[a]	1	6	8	.	10	1	.	.	5
B flints[a]	.	2	.	.	2	.	.	4	2
Other flints	2	3	2	.	13	.	.	4	5
Lutites[b]	.	.	.	.	.	.	.	.	1
Other non-flints[c]	.	.	.	.	1	.	.	.	.
Unidentified stones[d]	.	2	.	.	1	.	.	.	.
TOTAL	**3**	**13**	**10**	**.**	**27**	**1**	**.**	**8**	**13**

[a] Groups of flints that (based on color, texture, inclusions) are apparently varieties of materials that can all be found at present in the Cretaceous levels outcropping along the Holocene shore of eastern Cantabria and western Vizcaya

[b] Mudstone

[c] Limestones, quartzes, quartzites, calcites, and sandstones

[d] Lithic chips or shatter too small (<1 cm) to accurately identify by raw material type

Table 18.21. Weight (g) of all artifacts (debris and tools) by major lithic groups, Bronze Age assemblages

Level	1	1–2	2	2.3	3	3.2	3.3	3.4	3.5
A flints[a]	15.0	78.8	64.0	6.0	114.3	1.0	5.0	8.0	62.0
B flints[a]	3.0	15.6	21.6	.	14.2	1.0	.	0.4	38.0
Other flints	9.0	34.1	40.0	50.0	183.0	1.0	.	12.2	76.0
Lutites[b]	.	3.5	10.4	.	29.0	.	.	2.0	1.0
Other non-flints[c]	.	21.0	26.1	3.0	28.0	.	.	.	28.0
Unidentified stones[d]	2.0	46.7	36.9	.	113.2	2.0	.	8.0	90.0
TOTAL	29.0	199.7	199.0	59.0	481.7	5.0	5.0	30.6	295.0

[a]Groups of flints that (based on color, texture, inclusions) are apparently varieties of materials that can all be found at present in the Cretaceous levels outcropping along the Holocene shore of eastern Cantabria and western Vizcaya

[b]Mudstone

[c]Limestones, quartzes, quartzites, calcites, and sandstones

[d]Lithic chips or shatter too small (<1 cm) to accurately identify by raw material type

Table 18.22. Bronze Age tools (1996–2001)

	Level						
	1	1–2	2	3	3.2	3.4	3.5
1-Single endscraper	.	.	.	2	.	.	1
5-Endscraper on flake/blade	.	.	1	3	.	.	.
8-Endscraper on flake	.	.	.	2	.	.	.
10-Thumbnail endscraper	.	1	.	2	.	.	.
11-Keeled endscraper	.	.	.	1	.	.	.
15-Core-endscraper	.	.	1	1	.	.	.
21-Perforator-endscraper	.	.	.	.	.	.	1
24-Bec	.	1	.	.	.	.	1
26-Micropercoir	1	.	.	.	.	.	.
27-Straight dihedral burin	.	.	.	2	.	.	.
29-Angled dihedral burin	.	.	.	1	.	.	.
30-Burin on break	.	.	.	1	.	.	.
31-Multiple dihedral burin	.	.	.	1	.	.	.
34-Burin on straight truncation	.	.	.	.	.	.	1
51-Microgravette	.	.	.	.	.	.	1
52-Font-Yves point	.	.	.	1	.	.	.
58-Completely backed blade	.	.	.	.	.	1	.
59-Partially backed blade	.	.	1	.	.	.	1
65-Continuously retouched piece, 1 edge	.	2	.	.	.	1	.
66-Contnuously retouched piece, 2 edges		.	.	1	.	.	.
67-Aurignacian-like blade	.	.	.	.	.	.	1
70-Foliate point/arrowhead	.	.	.	1	.	.	.
74-Notch	.	.	.	.	.	.	2

Table 18.22. *(continued)*

	Level						
	1	1–2	2	3	3.2	3.4	3.5
75-Denticulate	.	.	.	.	1	.	.
76-Pièce esquillée	1	1	1	.	.	.	1
77-Sidescraper	.	.	.	1	.	.	.
79-Triangle	.	1	.	1	.	.	.
84-Truncated bladelet	1	.	.	1	.	.	1
85-Backed bladelet	.	6	3	5	.	6	2
86-Truncated backed bladelet	.	1	1	1	.	.	2
90-Retouched bladelet	.	.	1	.	.	.	.
TOTAL	**3**	**13**	**10**	**27**	**1**	**8**	**15**

Table 18.23. Blank types on which tools were made, Bronze Age levels

	Flakes	Blades	Bladelets	Cores/chunks	Splintered pieces
Endscrapers	8	3	.	4	.
Perforators	1	.	.	.	.
Burins	3	.	.	3	.
Backed and truncated pieces	1	2	.	.	.
Continuously retouched pieces and sidescrapers	3	2	.	1	.
Notches and denticulates	.	3	.	.	.
Microliths	.	.	2	.	.
Backed and retouched bladelets	1	2	28	.	.
Splintered pieces	.	.	.		.
Points	.	1	1	.	.
Others	1	.	.	.	
TOTAL (*n* = 73)	**18**	**13**	**31**	**8**	**3**

Note: Level 1, tool type 26 (a micropercoir) and Level 3, tool type 70 (a foliate point/arrowhead) were not assigned a debris type.

CHAPTER NINETEEN

The Ceramics of El Mirón Cave
Production, Morphology, and Discard

Cristina Vega Maeso
TRANSLATED BY LAWRENCE GUY STRAUS AND ELIZABETH A. STONE

The research that resulted in the work presented here was focused on Levels 1–10 in the Cabin and the equivalent levels from the Mid-Vestibule Trench, 300–303.3. These levels have been ^{14}C-dated, allowing their assignment to chronological phases. Thus, the chronology for the material discussed here can be established as follows:

- **Neolithic:** 5690 BP (Level 10) to 4680 BP (Level 8.1)
- **Chalcolithic:** 3740 BP (Level 7) to 3820 BP (Level 5)
- **Chalcolithic/Early Bronze Age:** 3700 BP (Level 3)
- **Early Bronze Age:** Levels above 3

The nature of the material determined the course of the research, since the collection is so limited and fragmented that it was impossible to carry out a traditional morphotypological study because complete vessels or sherds allowing the reconstruction were quite rare. Instead, a broader study of the assemblage was adopted, beginning with locating clay sources and continuing through the recovery of the sherds in excavation and their subsequent treatment.

Thus the investigation was organized along three axes:

1. **Abandonment:** First, a methodology was adopted that dealt with the recovery of ceramic material in the cave through the last treatments of the pieces. Later, the distribution of ceramic remains in the various excavation units of the cave, along with post-depositional transformations, will be discussed.
2. **Morphology:** Before beginning the morphological description of the Mirón ceramic assemblage, other known ceramic assemblages from the region were examined. As noted previously, the high degree of fragmentation made it impossible to reconstruct, physically or graphically, the vessel form of most of the sherds. However, the numerous ^{14}C dates allow the assemblages to be placed in stratigraphic and regional context.

3. **Manufacture:** Beginning with macroscopic study of all of the sherds, followed by a mineralogical analysis of samples by x-ray diffraction (XRF), a series of hypotheses were tested by experimentation and a component study of clays collected in the local area in order to determine the origin of the raw materials used in the ceramics along with the processing of the clays and the ceramic manufacture sequence.

The goals of this research are:

1. Determine the validity of the methodology employed
2. Characterize the El Mirón ceramic assemblage
   2.2. Identify the degree of taphonomic alteration
   2.3. Establish a typological seriation
   2.3. Identify the production sequence
3. Determine the function of vessels
4. Establish the socioeconomic implications of ceramic production
5. Define, insofar as it is possible, the role played by ceramics in the ideological realm

In any case, the ultimate goal of this work is to produce an efficient tool for the understanding of prehistoric ceramic production in the Asón Valley. It is hoped that the work will be integrated into other studies done by the interdisciplinary team working on different aspects of the El Mirón Cave project, with the hope of producing the most accurate and complete reconstruction possible of recent prehistory in the Asón valley.

Abandonment

The ceramic-bearing levels of El Mirón Cave range over a substantial period of time that covers most of what is known as "recent prehistory" (i.e., every major phase except the Iron Age), and—in addition to the handmade vessels of the Neolithic, Chalcolithic, and Bronze Age—there are even indications of medieval and modern times among the sherds found in mixed-surface sediments (e.g., common wheel-turned types, vitrified pieces, china). The recent prehistory at El Mirón (table 19.1) has been documented in a long stratigraphic sequence characterized by considerable complexity.

The horizontal layers of the post-Paleolithic period in the front and middle zones of the El Mirón Cave vestibule are cut through by many man-made holes, which in turn are often cut into by other, more recent holes. During the course of the archaeological excavations these pits were labeled as features, and generally their infillings were considered to be separate stratigraphic units. The pit fills often contained abundant ceramic sherds. The stratigraphic complexity, as well as the fact that some of the pits were archaeologically dug during different excavation campaigns, makes it difficult to define the precise relationship among the features.

In addition, after abandonment the ceramics underwent a series of surface alterations whose analysis can provide information on the nature of the contexts under which the vessels were discarded. Indeed, one of the objectives of this study is to identify the principal natural and anthropic processes that affected the spatial distribution and state of preservation of these artifacts.

For the study of the distributions of sherds among the various Holocene levels at the top of the El Mirón stratigraphic sequence, the excavation techniques used, the treatments applied in dealing with the ceramics in the field laboratory, and the post-depositional processes will all be the objects of study in this chapter to help understand the nature of the ceramic assemblages now curated in the Museo de Arqueología y Prehistoria de Cantabria. The study of the distribution of sherds among the uppermost levels of the stratigraphic sequence, excavation methods, laboratory processing, and post-depositional processes are the subjects of the following section.

Methodology

EXCAVATION METHODS

The ceramic-bearing levels in El Mirón were initially documented during the first excavation campaign, which was conducted mainly to determine whether the cave contained intact archaeological deposits. Given their stratigraphic position at the top of the sequence, it was during this campaign (1996) that many of the most important Holocene levels were found. Although the areas in which the post-Paleolithic occupations would eventually be excavated were expanded both in the Cabin area and (for the Neolithic) in the western half of the Mid-Vestibule Trench in subsequent years, many of the richest ceramic assemblages *and* most of the Chalcolithic pit features were found in 1996 in the Cabin area. Given the exploratory nature of that first campaign, the systems for processing and documentation of finds had not yet been perfected, as they would be in subsequent seasons, although all sediments were water-screened on-site and photos taken and drawings made of the features. Beginning in 1997 sediments were systematically subjected to flotation and photo grammetric methods were used to more thoroughly document the areas excavated. These methods were improved over the years and, beginning in 1999, included use of an EDM for piece-plotting of finds (González Morales and Straus 2000).

Like the other finds, all the sherds found during excavation were either individually piece-plotted in three dimensions (in the cases of lithics measuring >1 cm or ceramics having special features such as rims, lugs, angles, or decorations) or by quarter square-meter uni per excavation spit (in the cases of very small sherds, including those found in the screen). The individually piece-plotted sherds were assigned specimen numbers by meter square. Those sherds that were in fragile condition were consolidated on-site.

POST-EXCAVATION TREATMENT

Field Laboratory

Once removed from the site, the ceramics were subjected to several processes of study. The first took place in the field lab in Ramales, while others that were more elaborate and time-consuming were conducted later at the Universidad de Cantabria in Santander.

The manner in which the different sherds were recovered also in part determined their subsequent treatment, which was different for individually piece-plotted ceramics than for those sherds coming from general subsquare/spit bags or those that were found during dry screening or water flotation of the sediments. The individually recorded sherds were not washed, so as not to damage them. They were allowed to dry out until they lost all the humidity accumulated during the millennia that they had been buried, after which they were gently dry brushed. The same process was generally used for sherds recovered during the dry screening that preceded flotation. Some very small sherds that had passed through the flotation process were recovered during the posterior dry screening of residues that were subsequently sorted and from which finds were picked out from the two coarser of three screen fractions.

Every sherd was labeled with a code consisting of the letter *M*, for El Mirón, followed by the excavation year, square, subsquare, level, and inventory number (the latter including a decimal after the bag number if the sherd in question came from a general bag [e.g., M-1997-J4A-309.5]).

Final Processing

After their initial processing, the individually recorded sherds were inventoried and the data entered into a relational database for El Mirón artifacts designed by Enrique Maeso. Each sherd is represented in the database with information on its provenance (square, subsquare, level, spit, and three-dimensional coordinates) and on its type.

Ziortza San Pedro, in her conservation work on the ceramics, basically cleaned the surfaces of the sherds, only using more elaborate consolidation methods on those pieces that were in the most fragile condition. "Families" of sherds were recognized during the course of this stage of processing. Some of the similar sherds that could be refitted into the same vessels were found to have come from different levels. By refitting and gluing such groups of sherds, 18 different vessels could be partially reconstructed.

CERAMIC ANALYSIS: CLASSIFICATION

In order to study the ceramic assemblages, a general table listing all the finds was downloaded into an Excel file, which facilitated statistical analyses. Significant, information-rich pieces (partial vessels, rim sherds, decorations) were drawn and photographed. In addition, virtual reconstructions of several vessels were produced by means of Cad and 3D Studio software programs.

For the analysis of the ceramic collection from El Mirón we opted to follow two models:

Individual analysis of sherds
Grouping of sherds into vessels as determined by the refittings done by Ziortza San Pedro

Files were created for ultimate inclusion in the overall El Mirón database: Individual sherd files, including a series of variable classes that were recorded for each sherd. The sherds that were refitted to reconstruct partial vessels were also included in these individual sherd files. The data recorded include the following:

- Excavation zone within El Mirón Cave (Cabin, Mid-Vestibule Trench, Corral)
- Square and subsquare
- Level and spit

Technical characteristics of each sherd as observed macroscopically are also included in the database, as follows (Orton, Tyers, and Vince 1997):

- Firing: oxydizing, reducing, mixed, irregular
- Shaping technique: hand (M), wheel (T)
- Temper grain: fine—up to 0.5 mm (F), medium—0.5–1.0 mm (M), coarse–>1.0 mm (G)
- Temper type: organic (C), flint (F), grog (G), shell (H), iron ore (I), mica (M), sand (quartz, quartzite) (S), volcanic rock (V), other/unknown (X), no apparent temper (N) (Orton, Tyers, and Vince 1997:270)
- Paste color: following Cailleux (1981)
- Surface color: following Cailleux (1981)
- Surface finishing: coarse (1.2.1), fine/polished (1.2.2), tool burnished (1.2.3), varnished (1.2.4), slipped (1.2.5) (Llanos Ortiz and Vega Saramburu 1974)

The morphological characteristics of each sherd were analyzed according to the following attributes, using the codes indicated below:

Count: number of sherds with the same specimen number that could belong to the same vessel and with the same typological and technological characteristics

Vessel portion:

1. Rim	2. Lip	3. Neck	4. Shoulder	5. Body
6. Suspension	7. Keel/angle	8. Base	9. Pedestal	10. Clinker

Diameter: Edge and base diameters measured in millimeters

Thickness: Maximum thickness measured in millimeters for each sherd

Weight: Measured in milligrams

Fragment type (Caro 2002; Eiroa, et al. 1999):

1. Edge

Open

1.1.1 Flat lip	1.1.2 Rounded lip	1.1.3 Semi-oval-shaped lip
1.1.4 Pointed lip	1.1.5 Beveled lip	1.1.6 Swollen lip
1.1.7 Thinned lip	1.1.8 Other (as individually defined in "Observations")	

Straight

1.2.1 Flat lip	1.2.2 Rounded lip	1.2.3 Semi-oval-shaped lip
1.2.4 Pointed lip	1.2.5 Beveled lip	1.2.6 Swollen lip
1.2.7 Thinned lip	1.2.8 Other (as individually defined in "Observations")	

Closed

1.3.1 Flat lip	1.3.2 Rounded lip	1.3.3 Semi-oval-shaped lip
1.3.4 Pointed lip	1.3.5 Beveled lip	1.3.6 Swollen lip
1.3.7 Thinned lip	1.3.8 Other (as individually defined in "Observations")	

Protruding

1.4.1 Flat lip	1.4.2 Rounded lip	1.4.3 Semi-oval-shaped lip
1.4.4 Pointed lip	1.4.5 Beveled lip	1.4.6 Swollen lip
1.4.7 Thinned lip	1.4.8 Other (as individually defined in "Observations")	

Fragment type (*continued*):

2. Neck
   - 2.1 Concave
     - 2.1.1 Short — 2.1.2 Medium — 2.1.3 Long
   - 2.2 Straight
     - 2.2.1 Short — 2.2.2 Medium — 2.2.3 Long

3. Body
   - 3.1 Straight — 3.2 Curved
   - 3.3 Keeled — 3.3.1 High — 3.3.2 Mesial — 3.3.3 Low
   - 3.4 S-shaped profile

4. Base
   - 4.1 Flat — 4.2 Convex — 4.3 Conical — 4.4 Marked pedestal
   - 4.5 Angular pedestal — 4.6 Wineglass pedestal — 4.7 Lipped bowl

5. Suspension element
   - 5.1 Short handle
   - 5.1.1 Tubular — 5.1.1c Circular — 5.1.1.0 Oval
   - 5.1.2 Band form — 5.1.2e Elliptical — 5.1.2r Rectangular — 5.1.21 Lobate
     - 5.1.2b Bilobate — 5.1.2p Polylobate
   - 5.2 Lug
   - 5.3 Tongue-shaped
   - 5.4 Perforation
   - 5.5 Long Handle
   - 5.6 Pill-shaped — 5.6.1 Simple — 5.6.2 Repoussé

Vessel Form:

1. Open — 2. Straight-sided (beaker) — 3. Closed — 4. Globular

Decorations (following the typology of Llanos Ortiz and Vega Saramburu 1974):

1. Impressions:
   - Basketry — Stamped — Cord-marked — Springs — Fingerprints
   - Fingernail marks — Fingerprints + nails — Stylus — Cardium shell
2. Incisions:
   - Polishing marks — Stab-marks — Deep lines — Engravings — Grooved
   - Dashed lines — Combed — Brushed
3. Excisions:
   - 3.1 Excised
4. Plastic decorations:
   - 4.1.1 Square plain cords — 4.1.2 Rectangular plain cords
   - 4.1.3 Trapezoidal plain cords — 4.1.4 Triangular plain cords
   - 4.1.5 Rounded plain cords
   - 4.2.1 Fingerprint-marked cords — 4.2.2 Fingernail-marked cords
   - 4.2.3 Cords marked with fingerprints + nails

Decorations *(continued)*

4.2.4 Artifact-marked cords

4.3.1 Flat nipples	4.3.2 Conical nipples	4.3.3 Cylindical nipples
4.3.4 Multiple nipples	4.4.1 Bumps	

5. Painted decorations

Black	White	Red	Yellow	Blue	Green	Purple	Orange	Brown
Polychrome	Metallic	Graphite						

6. Others

6.1 Paste fill	6.2 Incrustations	6.3 Stamped

Various processes can alter certain parts of a vessel. Some variations that can be caused during the manufacture of a vessel, while others are caused after its abandonment. The former can be caused by the presence of ash, vegetal impressions, and internal burnishing. Post-depositional processes can include differential erosion and the precipitation of calcium carbonates on sherds.

For each sherd, additional observations can be noted in its record.

Each (partially) reconstructed vessel was also recorded with its own file, including the same basic provenance, technical, morphological, and alteration information as is recorded for the individual sherds. The vessels were attributed to a cultural phase, and each one is documented by photos and drawings of the actual refitted sherds, plus a virtual reconstruction of how the whole vessel might have looked before breakage. All the data were downloaded into an Excel file to be analyzed statistically. Provenance data were analyzed with Surfer software (version 8.2) to provide distribution plans drawn with Autocad. These plans show the locations of all the sherds that make up the refitted vessels relative to the site grid and to the archaeological features (pits) that were documented during the excavations. These procedures allow us to study the three-dimensional distribution of vessel elements—including evidence of cross-level "migration" of sherds due to post-depositional disturbances, notably prehistoric pit digging.

A sample of 17 sherds was submitted for X-ray diffraction analyses at the Archaeometry Laboratory of the University of Santiago de Compostela (Galicia) in order to determine the mineralogical components of the clays and the approximate firing temperatures.

GRAPHIC REPRESENTATION

Sherds that could provide some information on the shape of the original vessels or that contained some decorative or special technical element were selected for graphic representation. This consisted of both photography and drawing, in order ultimately to be able to create virtual three-dimensional reproductions showing what certain whole vessels may have looked like before breakage.

The technique of drawing allows sherds to be represented in such a manner that a great deal of information can be efficiently synthesized concerning the morphology, technological characteristics, and typology of each piece through the use of a series of internationally recognized conventions that facilitate both the understanding of each drawing and comparison among different sherds. In the specific case of El Mirón, drawing was of great importance in bringing us closer to the detailed morphology of the sherds, allowing us—despite the very small size of most—to derive at least some information about the shapes of the original vessels.

Those sherds selected for drawing included ones from rims, bases, and vessel bodies—the latter when some decorative or suspensive element was present. The first step involved determining the correct orientation of each sherd and drawing its profile on paper. In the case of a rim sherd, orientation is estimated by assuring that the majority of points along the lip coincide with a theoretically flat base, then taking a measurement of the vertical projection of the piece in order finally to be able to place it in its original position with a correct angle for the vessel mouth. Next the diameter of the mouth or of the base can be estimated, depending on which kind of sherd is being drawn. This is done by means of a simple mathematical formula or by matching the individual sherd against templates with a series of concentric circle segments. The next step in drawing involves the estimation of vessel profile, that is, a longitudinal section for each original piece, which will show its form and thickness. This is done with a calipers and shape template (Melero, León, and Torres 2003) Once all these measurements have been obtained, a final drawing is done following the standard norms for ceramic representation.

The drawings on paper were scanned into a CAD file by means of a digitizing tablet. Plainware sherds were not dotted, since the only information they supply is morphological. However, in the cases of decorated sherds or those that have special characteristics, photographs were taken and manipulated in Photoshop in order to bring out shadows and hence micro-relief. These images in turn were processed in AutoCAD in order to be able to shade them with dots to better show the decorations or other special features.

In contrast to drawing, photographic representation of ceramics is more effective in showing textures, decorations, certain technical details, and the effects of post-depositional processes on sherds. The sherd forms are, however, not as well shown by photography as by drawing, so both methods of representation are necessary in many cases.

Not all the El Mirón sherds were photographed—only those that had some attributes that allowed them to stand out, such as decorations, technological details, or some kinds of alterations. Of course, all the partially refitted vessels were photographed from different angles. First the selected sherds were individually photographed and then they were grouped by chrono-cultural phases to allow for easier comparison. The images were made with a digital camera with side lighting against a neutral backdrop set below matte glass to avoid both shadows and reflections. Later the background was manipulated in Photoshop to provide absolute uniformity among the images.

By means of the drawings and photographs it was possible to produce virtual reconstructions of some of the vessels. With the digitized profiles of sherds in CAD, using the "rotation" tool, we could produce full 360-degree profile reconstructions. The resultant image generally required some retouching to avoid giving the impression that these were perfectly symmetrical, wheel-turned vessels (which, of course, none of them were). In addition, in this final stage of image manipulation certain traits could be added to the plain vessel profile, such as nipples, incisions, and so on.

The images in CAD format were then imported into the 3D Studio program, where they could be made to look more "real" in terms of texture, light and shadow, color, and so on. The final products are JPG images of reconstructed vessels that are as realistic as possible.

The reconstruction of vessels in scaled CAD format also allowed us to work with complete objects that had been only very partially recovered from the site, without having to use methods that would have required manipulation of the original sherds. We can use the virtual reconstructions to calculate various measurements, such as diameter, volume, and so on. We can also obtain a "real" view of the vessels, allowing us to have a broader understanding of a number of questions about the ceramic technology that is difficult to obtain from just the sherds or very partial refitted vessels. In this way it is ultimately possible to group objects into assemblages of vessels that can give us an idea of the different kinds of forms that were used by the inhabitants of El Mirón Cave during different cultural phases, thereby allowing us to make comparisons through time at the site as well as between El Mirón and other sites during specific periods or across different periods.

The Ceramic Record from El Mirón Cave

SPATIAL DISTRIBUTION OF THE SHERDS

One of the goals of this research is to understand the natural and anthropogenic processes that caused the spatial distribution of the ceramic sherds. To this end we had to determine the spatial distribution of the sherds and their physical relationship to the features that were uncovered in the post-Paleolithic levels of the El Mirón deposit.

In levels dating to the early Bronze Age there are relatively few ceramics—at least in the excavated area of the cave, that is, the 11 m^2 of the Cabin excavation zone. Level 1 (which may be a mixture of Bronze Age and more recent materials) yielded ceramics only in Subsquare C of Square I3.

Level 2, which is composed of several lenses, has a somewhat more extensive distribution of sherds—

mostly concentrated in Squares I1A+B, I2A+C, and throughout Squares I4, J4, and K5. The sherds from Levels 3.4 and 3.5 are respectively concentrated within an in situ combustion feature and in Feature O—the latter an intrusive trench with mixed fill, probably including some post–Bronze Age material—in the northwest part of the Cabin excavation area. There are also possible pieces of slag, which suggest in situ metallurgy at least in Level 3.4.

In general, the sherds assigned to the transition between the Chalcolithic and early Bronze Age are concentrated in the southwestern quadrant of the Cabin excavation area. This fact may be due to the more intensive destruction of the upper part of the late prehistoric stratigraphic sequence in the northeastern area of the Cabin, which was a little higher and more exposed to recent activity (e.g., livestock stabling) in the cave. However, it could also be due to some compartmentalization of human activities in the more remote past. This differential distribution should be checked against the structural, faunal, and lithic records from these levels, since, after all, the ceramics are only a part of the whole cultural record.

The sherds found in Lenses 3.1 (travertine), 3.2 (mondmilch), and 3.3 were highly localized. The very few in 3.1 are in Squares I1 and I2. There is a great concentration of sherds in Level 3.2 in Squares H1 and H2, while Level 3.3 has a major concentration in H2. Curiously, the area nearest to the hearth features and postholes has virtually no ceramics. The combustion structures are mostly located in the eastern part of the Cabin excavation area, in the zone least affected by the humidity, which is reflected by the calcium carbonate precipitation of Levels 3.1 and 3.2.

Level 3 is a much more extensive unit, found in most of the excavation area, although best represented in the H row of squares. There is a notable area nearly devoid of sherds near pit Features A and B, which are related to Level 3 (figure 19.1).

The Chalcolithic levels are very complex and are cut by many pits. The ceramics that were recovered from one of the main levels (4) were concentrated in the northern parts of Squares I4 and J4. Level 4 is cut through by pit hearths A and B, which were dug from Level 3. At the surface of Level 4 was pit Feature F, whose function is unknown (figure 19.2).

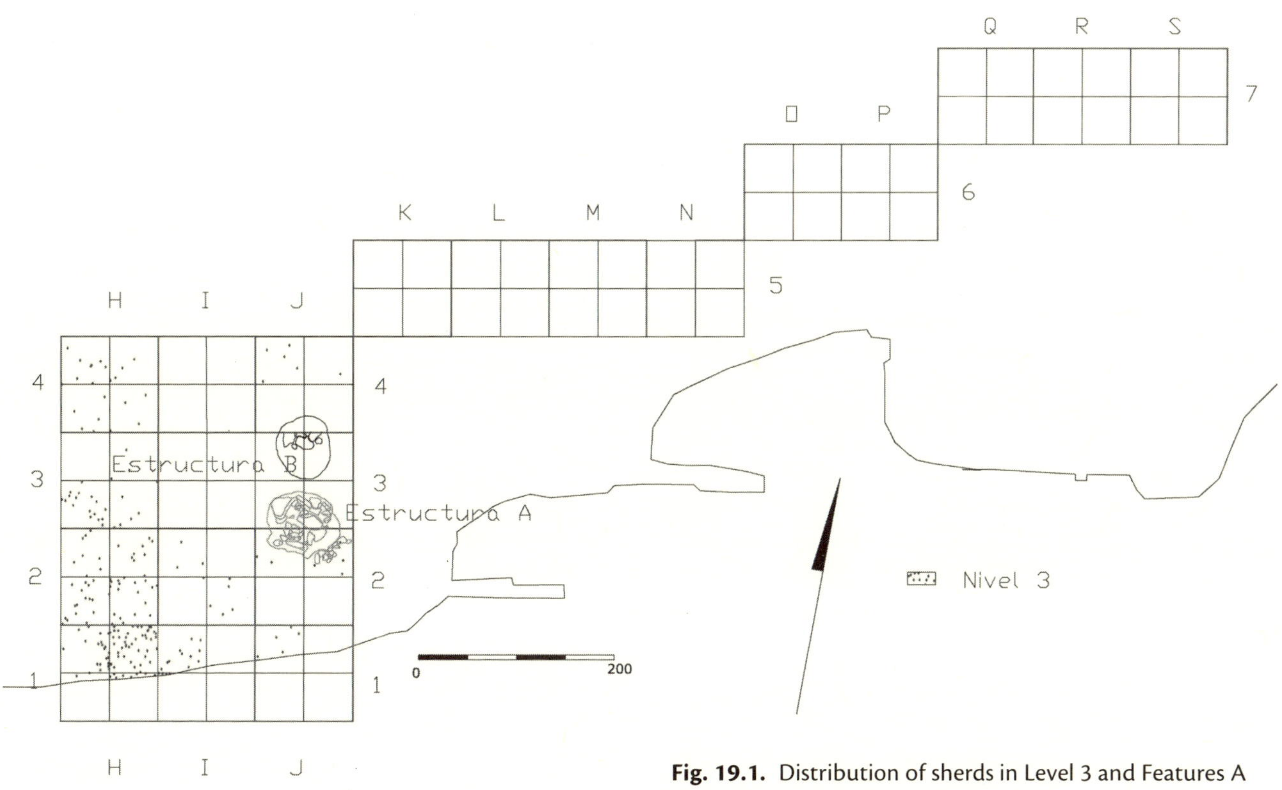

Fig. 19.1. Distribution of sherds in Level 3 and Features A and B. Drawing by Cristina Vega Maeso.

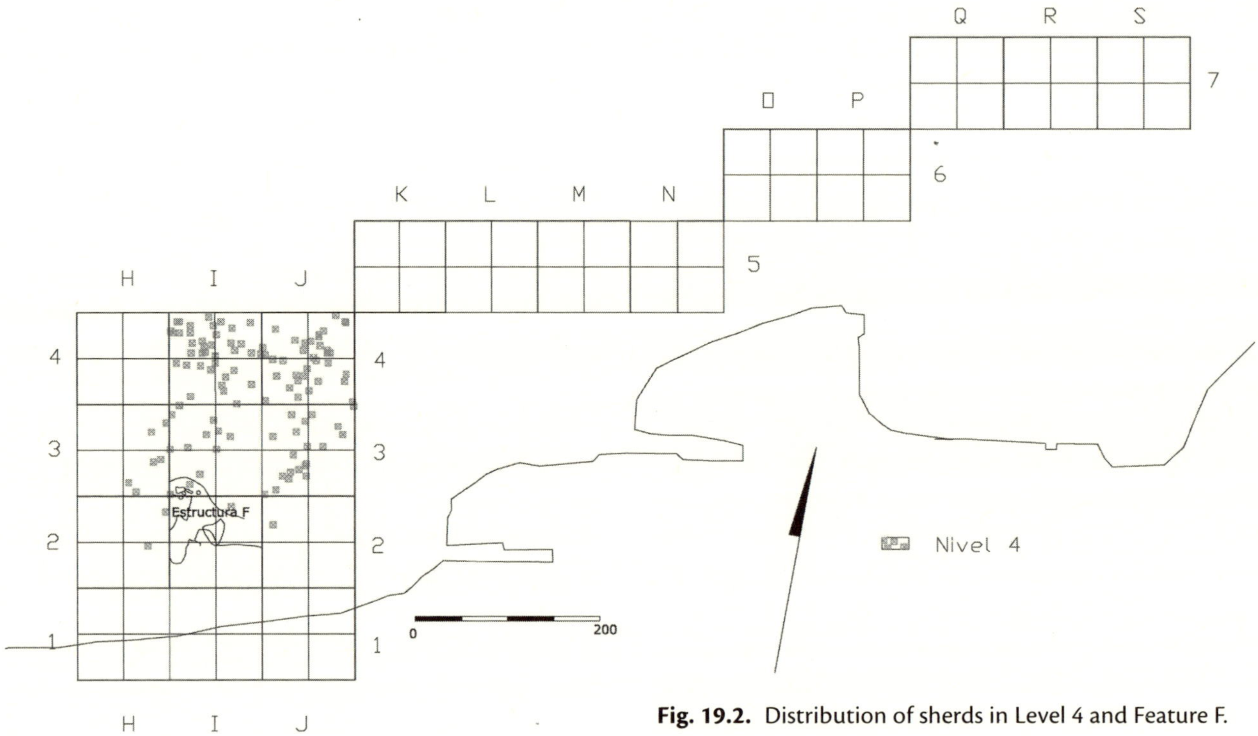

Fig. 19.2. Distribution of sherds in Level 4 and Feature F. Drawing by Cristina Vega Maeso.

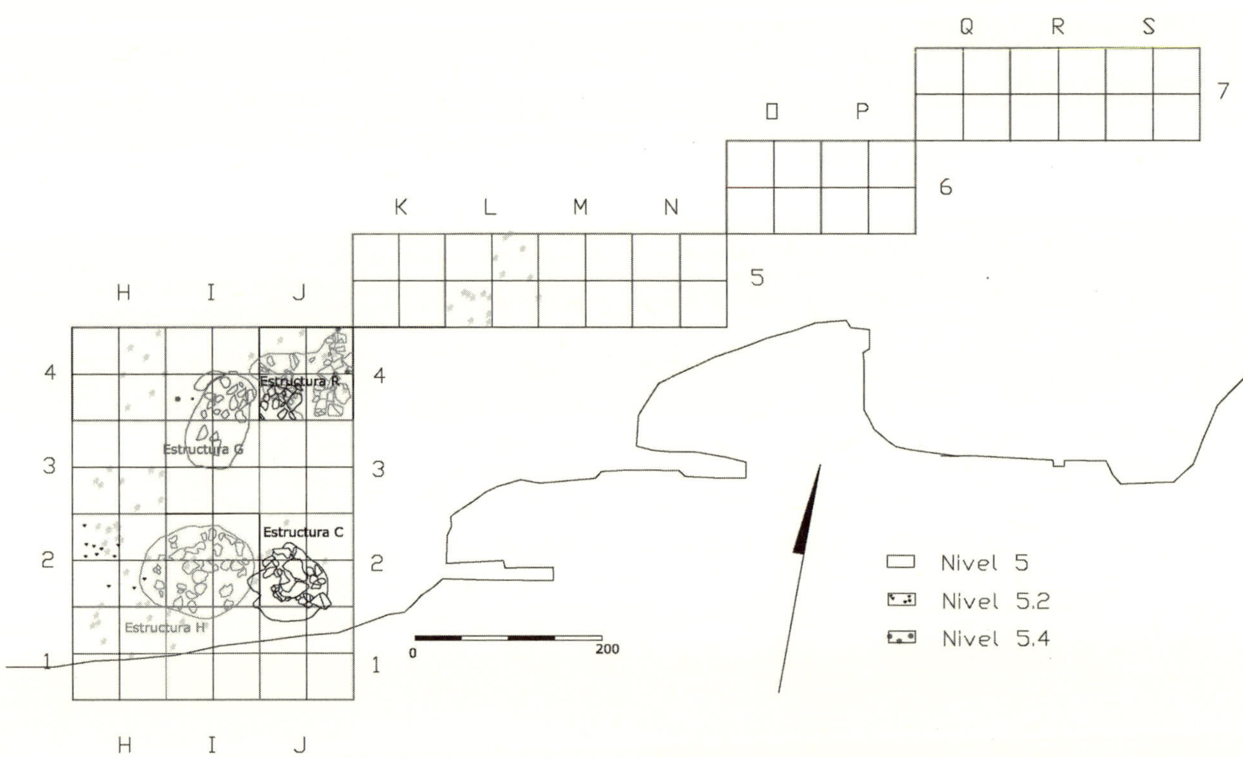

Fig. 19.3. Distribution of sherds in Levels 5, 5.2, and 5.4 and Feature H, C, G, R, and D. Drawing by Cristina Vega Maeso.

Five pits were dug from Level 5 in the Cabin excavation area, some of which contained ceramics. Most of the sherds from Level 5 came from the southwest quadrant in H1, H2, and part of H3. In contrast, the sherds from Level 6 were almost all found in the north and west of the Cabin area, with virtually none in the rest of the excavation. They are related to pit Features G, H, C, R, and D, which were dug down into Level 6 from Level 5 (figure 19.3). The Level 7 sherds are distributed homogeneously in the I and H rows, but there are none in the J row, a fact that may be related to the intrusive structures located there.

It seems clear that the areas devoid of ceramic materials in the horizontal strata were caused by posterior human activities, namely pit digging. Thus we have to recognize that the deposits that filled the pits contained sherds whose origins were in older levels that had been disturbed—hence, in secondary position. This is shown both by the voids in the ceramic distribution plans and by refits of sherds, which in some cases include items found in different levels—clearly as a result of post-depositional disturbance.

The Neolithic levels in the Cabin area range from Level 8 at the top of the sequence through Level 10 at the bottom. The distribution of sherds in Level 8 is curiously similar to that of overlying Level 7, although the former is not cut through by pits. Level 9 displays a sherd distribution that is more or less homogeneous throughout the whole Cabin area. The ceramics from Lens 9.1 correspond to the fill of pit Feature L and thus are highly localized (figure 19.4).

The sherds in Level 9.6 are found in the northwestern part of the Cabin excavation area, especially in Squares H3 and H4, while those from Levels 9.7 and 9.8 were found in I3. It is notable that these three lenses are found one on top of the other in the same highly localized area in the northwest half of the excavation area. This area was relatively little occupied during Level 10 times compared to its more southerly concentration in terms of sherds (although Level 10 has no sherds in the extreme southeastern area) (figure 19.5).

As for the Neolithic levels of the Mid-Vestibule Trench that connects the Cabin and Corral excavation areas (the latter of which has no surviving ceramic-bearing levels), these (i.e., Levels 303, 303.1, 303.2, 303.3) display no clear differences in spatial distribution. The sherds seem to be more or less evenly distributed throughout the 1 × 3 m area of the western part of the

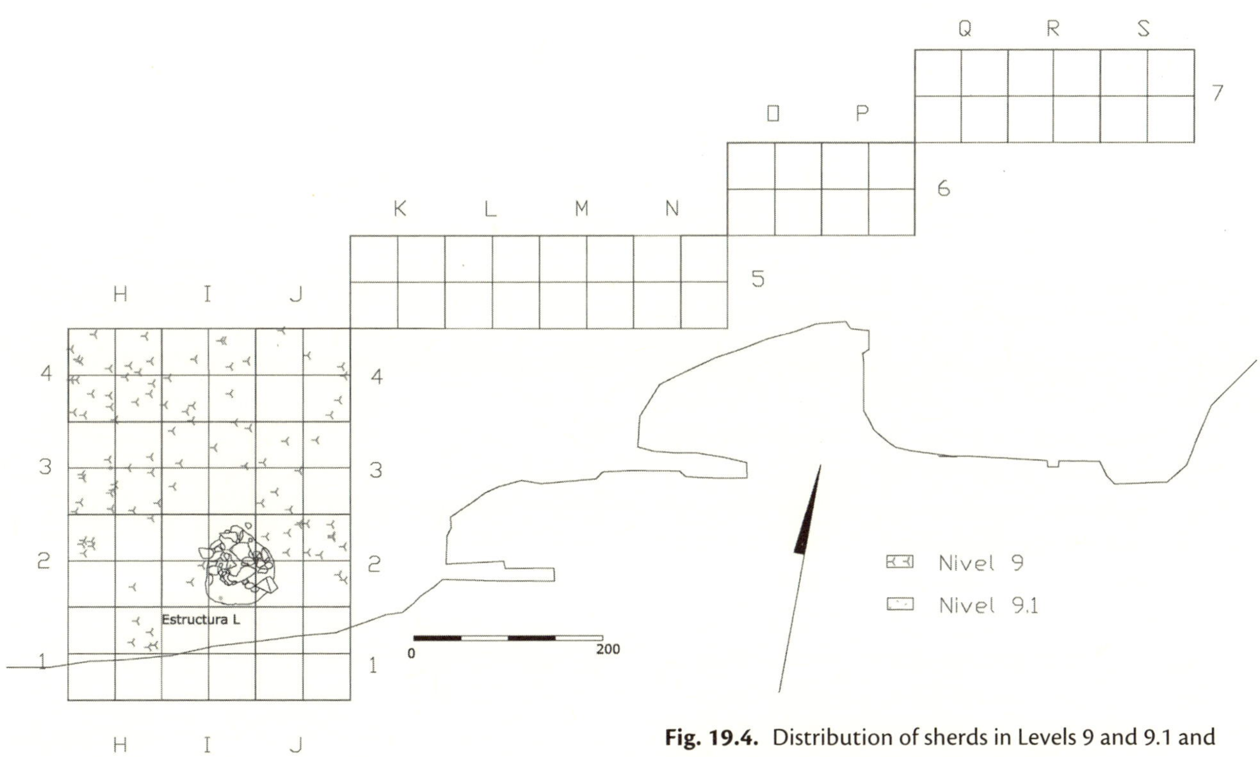

Fig. 19.4. Distribution of sherds in Levels 9 and 9.1 and Feature L. Drawing by Cristina Vega Maeso.

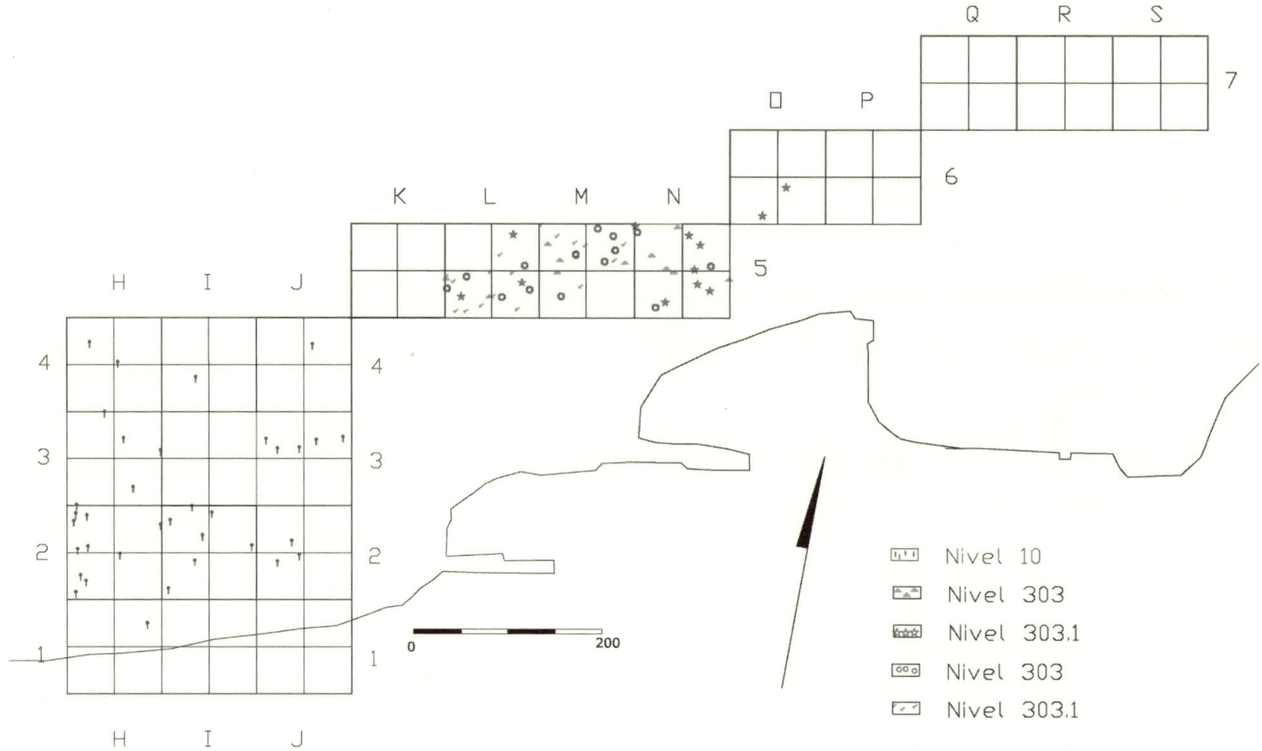

Fig. 19.5. Distribution of sherds in Levels 10 and 303–303.3. Drawing by Cristina Vega Maeso.

trench in all these levels. We also present the distribution plan for the post-Neolithic (but at least partially mixed) Levels 301, 301.1, 301.2, and 302 in the trench area. Those from 302 are almost all from squares O–P/6 in the center of the trench, while the others are more widely dispersed.

In conclusion, we can note that generally in the earlier occupations the sherds are concentrated in the northern and western parts of the excavation area (i.e., closest to the cave mouth but away from the southern cave wall). Most of the pits, in contrast, are in the eastern part of the excavation area. This tendency is somewhat reversed in the final occupations (Levels 2 and 3), with greater concentrations near the southern cave wall, especially in the southwest part of the Cabin area.

POST-DEPOSITIONAL PROCESSES

Once abandoned, the ceramic vessels underwent a series of processes that modified both their internal and external structure, beginning of course with their breakage into numerous, generally very small sherds. These processes can be caused by a number of factors, synthesized by M. Schiffer (cited in Renfrew and Bahn 1991) as cultural post-depositional processes, which include daily human activities (both deliberate and accidental), and natural post-depositional processes, namely all those natural processes that can alter the archaeological record after abandonment.

The study of these processes can provide important behavioral and paleoenvironmental information about a site. In the study of the El Mirón ceramics we took into consideration only those phenomena that could be observed visually; no other techniques were used to detect alterations in the material structure of the sherds, so only those that can be seen on the surface were analyzed. Four post-depositional processes were documented: alteration in the original physical position of the abandoned sherds, a heavy degree of breakage, erosion of many of the sherds, and the presence of carbonate concretion on some of the pieces.

ANTHROPOGENIC PROCESSES

Sherds Found in Secondary Position

Daily human activities in the cave vestibule modified the original position of many of the sherds. This is apparent

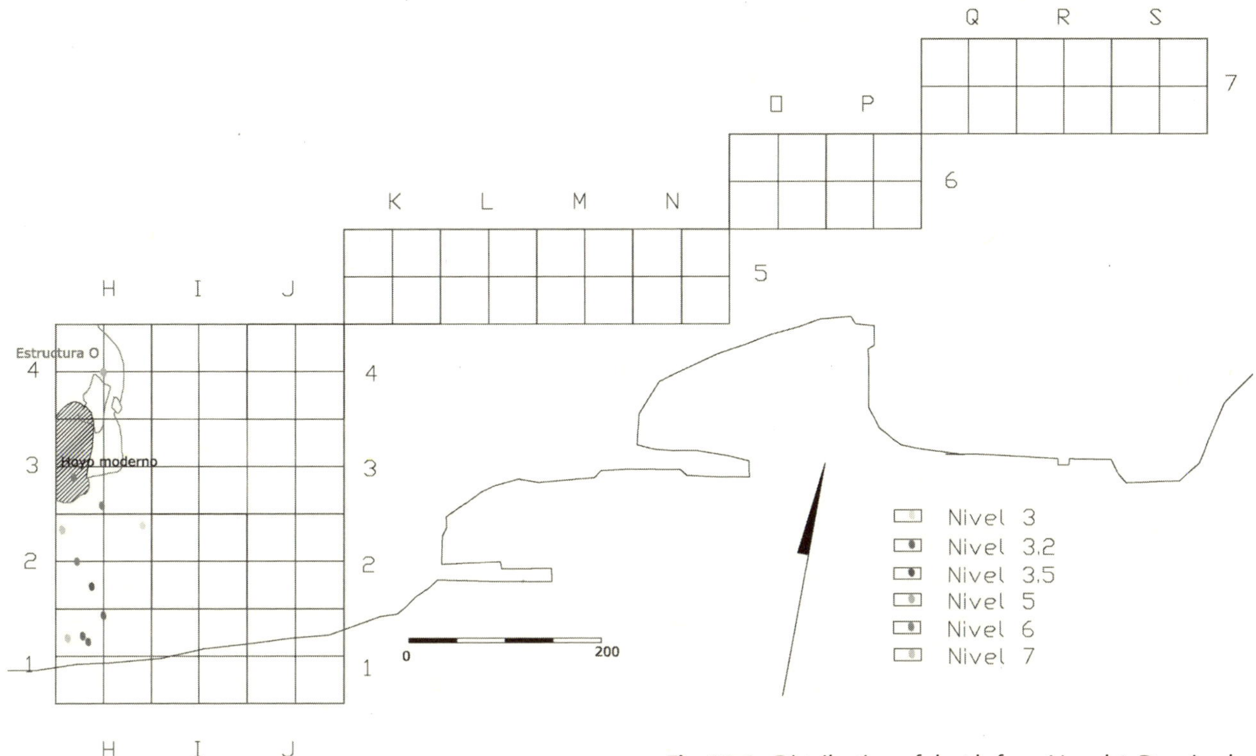

Fig. 19.6. Distribution of sherds from Vessel 1. Drawing by Cristina Vega Maeso.

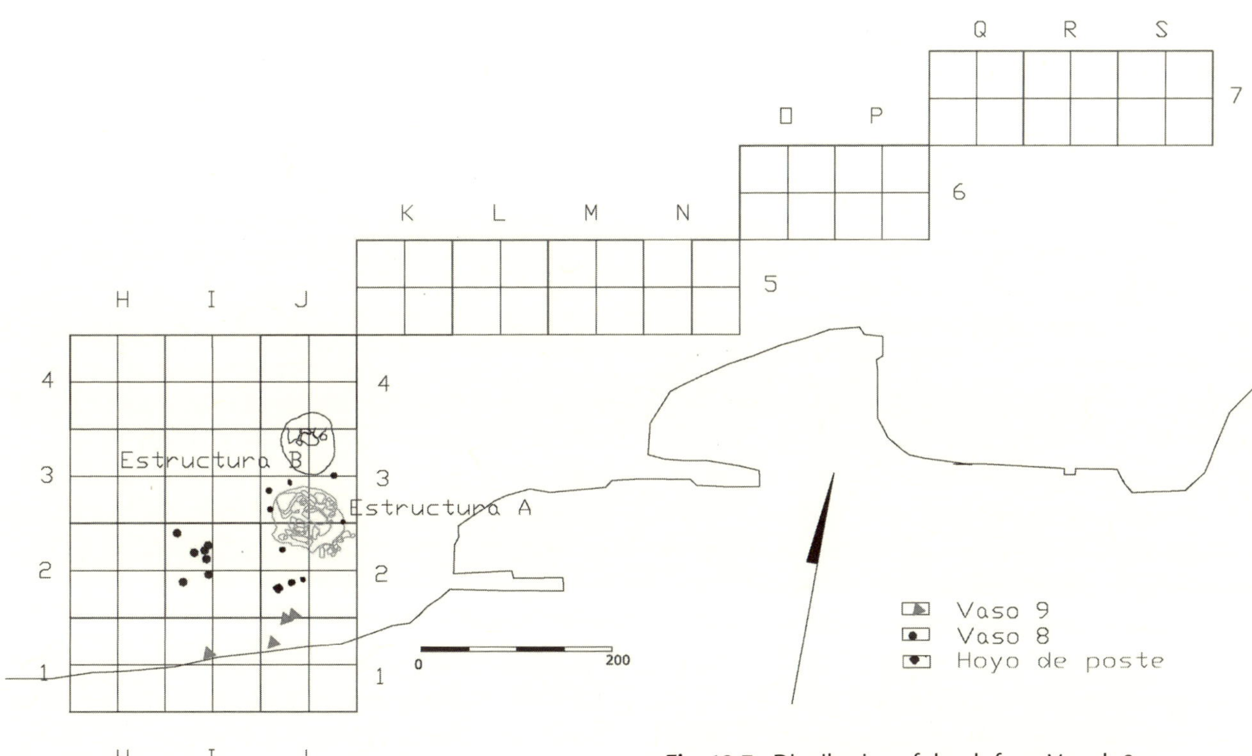

Fig. 19.7. Distribution of sherds from Vessels 8 and 9 and postholes. Drawing by Cristina Vega Maeso.

in the partially refitted vessels, the pieces of which were sometimes found in different levels. One of the most dramatic cases is Vessel 1, whose refitted sherds came from Levels 3, 3.2, 3.5, 5, 6, and even 7. Given the horizontal location of the finds, it is clear that this mixture is the result of the digging of Feature O (figure 19.6). Some of the sherds were found in the fill of this large pit (i.e., the ones attributed to Level 3.5), while others come from the levels that it cut through (Level 3 and travertine Lens 3.2).

The dispersion of sherds from Vessels 8 and 9 both horizontally and vertically is due to pit Features A, B, and E as well as the associated postholes. Sherds originally deposited in Level 5 were brought up to Level 3 (figure 19.7). Vessel 13 was dispersed in the same way; some of the sherds were found within the fill of pit Feature H, while others were found in their original place of deposition, Level 7. The distribution of sherds from Vessel 14 is perhaps the most complex, reflecting the imbrication of several pits dug in the Chalcolithic levels, which were considerably disturbed by human and animal activity.

Fragmentation

As noted above, the ceramics from El Mirón are highly fragmented. There are several causes, first among them being technological phenomena that produced rather fragile vessels that were easily broken. These include the scarcity of temper and the fact that breakage often occurred during or soon after firing. However, most breakage happened after some amount of use, causing the vessels to be abandoned, although some may have broken after people ceased to use them. Further breakage would result from the continuous trampling of humans and livestock in the cave vestibule, which was intensively utilized, especially during the Chalcolithic and the Bronze Age. In order to get a clear picture of the relative degree of fragmentation of the various samples, we determined the average weight of the sherds from each of the levels. We applied a simple mathematical formula whereby the average sherd weight for a level is the total weight of all ceramics divided by the number of sherds recovered from each layer.

The sherds from the Neolithic levels weigh on average between 12 and 26 g, with an outlier of 45 g from Level 9.1 (table 19.2), which is the fill of a hearth pit, where the sherds would have been somewhat protected from trampling.

The sherds are somewhat larger in the Chalcolithic levels, on average 24–47 g per level (table 19.3). But these are still really quite small pieces. The one exception is Level 6.1, where there was just one sherd, but it is large and came from the fill of a pit. The other associated features yielded highly diverse average sherd weights; pit fills 5.1 and 5.4 have average sherd weights within the overall range for the main Chalcolithic levels, but Lens 5.2 has an average sherd weight of 62 g. The data from the principal Chalcolithic Level—Level 4—are most notable; this is the level with the most ceramics, and they are the most highly fragmented. This is a stratigraphic unit composed of successive burning episodes manifested by ash lenses, probably the result of the use of fire to clean stabling debris (straw and dung)—phenomena that are known as "fumiers." The association of these kinds of deposits with the high degree of ceramic fragmentation point to the use of the vestibule as a place of both animal stabling and human habitation in the late Chalcolithic.

In general, the degree of trampling-related ceramic breakage seems to have gone down somewhat in the transition to the early Bronze Age, with larger average sherd weights: 35–86 g (table 19.4). It is not surprising that Lens 3.2—a travertine—has relatively larger sherds, and these are coated with precipitated calcium carbonate. The humid area (with ponded water?) of Lens 3.2 also meant that there was less trampling in this part of the cave. The superficial levels are highly varied in terms of average sherd weights; most notable is the relatively high average for Bronze Age Level 2: 77 g (table 19.5).

NATURAL PROCESSES

Calcium Carbonate Concretions

The surfaces of some of the sherds are covered with calcium carbonate precipitated from the lime-rich water of the cave. This post-depositional process is typical of karstic contexts (Calvo Trias et al. 2004). By observing the presence, absence, or degree of such concretions one can infer which levels were more or less affected by humidity or which areas of the cave witnessed more or less precipitation in the past. We thus took into account the numbers and percentages of concreted sherds per level (table 19.6). In analyzing the distribution of percentages of concreted sherds, we excluded Level 6.1, whose one sherd (which is concreted) would have unrealistically skewed the results.

The highest percentage of concreted sherds is from Level 1, with 37 percent, but much older Levels 8 and 9.7 also have very high values: 30.5 percent and 32 percent, respectively. Other Neolithic levels in the Cabin area (9 and 10) and in the Mid-Vestibule area (303, 303.1, 303.2) also have high percentages of calcium carbonate–coated sherd surfaces (16–28 percent), as do the Bronze Age Levels 2–3.3 (10–26 percent). Indeed, 83 percent of the concreted sherds come from the Bronze Age and Neolithic levels, suggesting the existence of humid conditions in the cave during these two periods. The concreted sherds of Level 3 and its underlying lenses are very localized in space, suggesting the presence of a drip from the cave ceiling. However, in the Neolithic levels the concreted sherds are much more widely distributed, suggesting generalized humidity.

Erosion

Erosion is another one of the most common phenomena of natural alteration affecting archaeological materials (Calvo Trias et al. 2004). In the present case we have documented only those sherds whose surfaces are clearly eroded, most probably by running water within the cave (table 19.7).

Such objects are not common. Most are from Level 8 and, to a lesser extent, Level 9.7: respectively, 27 percent and 10 percent of the sherds are eroded in these levels. Their abundance in these late Neolithic levels corresponds to the high frequencies of concreted sherds in the same units, corroborating the existence of very humid conditions during this period.

Morphology

THE IMMEDIATE GEOGRAPHIC CONTEXT: THE RÍO ASÓN VALLEY

Along with the typological characterization of the ceramics from El Mirón, we did a general comparison with known prehistoric ceramics from other sites in the surrounding area, namely the Asón valley. This will allow us to integrate El Mirón in its geographic context by establishing both parallels and differences with the areal ceramic record.

To do this, we set out to gather documentation on sites in the townships of Liendo, Escalante, Guriezo, Ruesga, Laredo, Santoña, Rasines, Ramales, Limpias, Ampuero, Arredondo, and Soba, whose territories lie within and/or adjacent to the Asón drainage basin. We began by consulting the *Archaeological Map of Cantabria*, which was assembled for the Department of Culture and Sports of the Cantabrian Regional Government by a team from the Department of Historical Sciences at the University of Cantabria beginning in 1998. This is, at present, the most complete inventory of archaeological sites in the region. Then we used information on the specific sites in a wide variety of publications, from general works to area survey reports, site monographs, articles in congress proceedings and journals, theses, and so on. From descriptions and illustrations in these documents we were able to obtain general information on the prehistoric ceramics of the Asón valley.

We created a database to organize the information we obtained from the literature search. Its variable fields include information on site location, type of archaeological work done at each site, site chronology, use, ceramic characteristics, other associated materials, drawings of sherds (when available), and bibliography. This database includes 103 sites known mostly from prospections and test pits dug by workmen of the provincial government in the 1950s, by the Santander Speleology group in the 1960s and 1970s (Begines and Garcia Caraves 1966), and by the CAEAP archaeological group in the 1980s (Bohigas Roldán, et al. 1984; Bohigas Roldán, Molinero, and Muñoz 1992; CAEAP 1984; Muñoz and Arozamena 1995; Muñoz, Malpelo, and Gomez 1996; Muñoz Fernández 1988, 1991, 1992, 1993, 1995), and by the British expedition to Matienzo in the 1990s (Ruiz Cobo and Smith 1997; Ruiz Cobo 1999, 2000a, b; Smith 1981; Smith 1995; Smith and Ruiz Cobo 1999) (figure 19.8).

The many poorly documented test pits in caves, without clear stratigraphic relationships between specific levels and sherds, permit only very general chronological attributions—often no more precise than that of "recent prehistory." The past use of these caves is undetermined in these kinds of archaeological interventions, being usually described as "unknown." Most of the other caves are considered to be either residential and/or funerary in function.

Fig. 19.8. Map of post-Mesolithic sites in the Río Asón basin. Map by Cristina Vega Maeso, redrafted by Ronald Stauber.

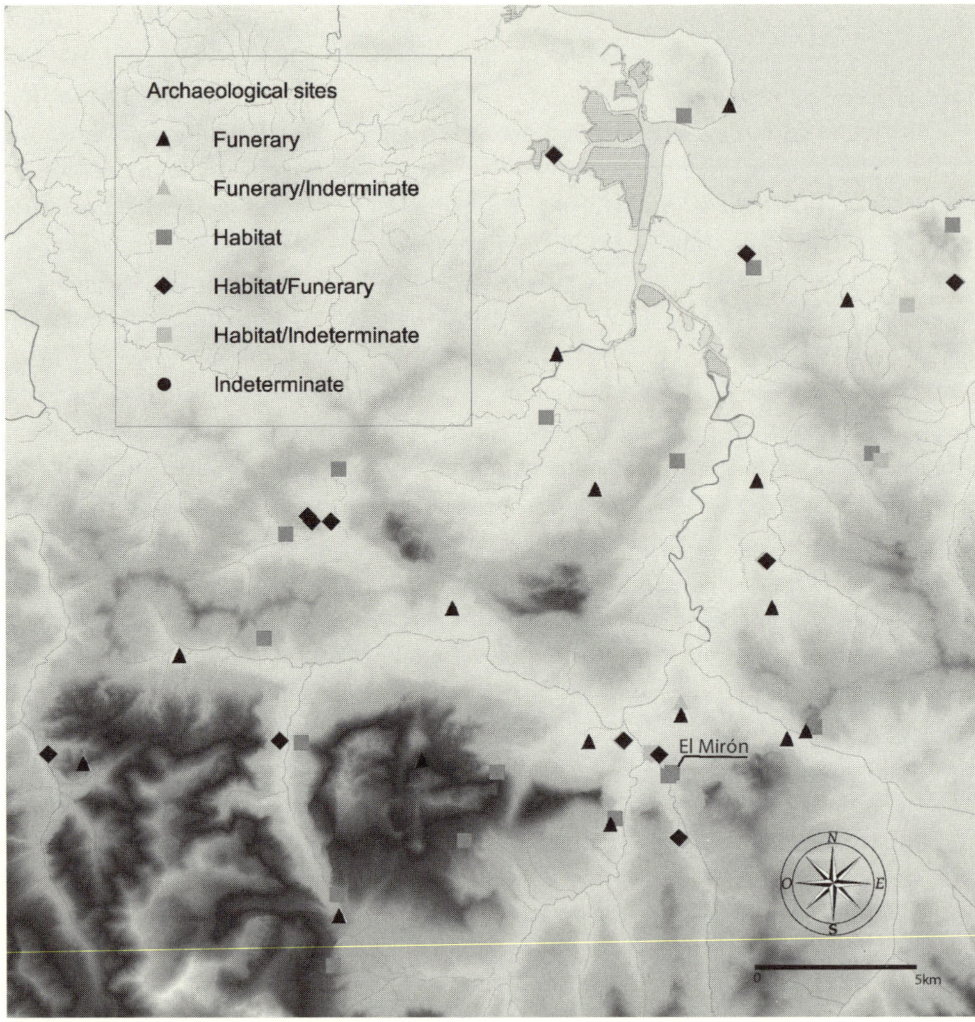

Thus, while this geographic area has been profusely investigated archaeologically by means of prospections and test pitting, the published information is often lacking in context, making it very difficult (or impossible) to ascertain site age(s) or function(s). This panorama is beginning to change because of the long-term research at El Mirón and the recent (associated) excavations by M. Fano in nearby El Horno Cave. Both have long stratigraphic sequences, and their results will aid immeasurably in the characterization of the Holocene ceramic industries of the Asón.

Among the other sites in the valley most of the ceramics are undecorated, making it very hard (in the absence of clear stratigraphic contexts and radiocarbon dates) to attribute them to specific cultural periods. Decorated sherds are usually the key to being able to make such assignments, so the situation from the limited, poorly defined finds at many of these caves makes cultural-historical classification nearly impossible in most cases. Nonetheless, among the 20 percent of the cases reviewed from the literature on the Asón valley that are not just listed as "recent prehistory," 12 percent are attributed to the Bronze Age (plus 6 percent Chalcolithic/Bronze Age) and only 2 percent Neolithic.

Decorated pottery simply seems to have been rare in the Asón area during late prehistoric periods (i.e., pre–Iron Age, pre-Roman). When there are decorations, they are mainly plastic ones (*appliqués*), with a few cases of incision.

It has traditionally been argued that there is duality in the ceramic record insofar as concerns the uses of the caves in which the vessels are found. Normally, residential use of a cave has been associated with plainware ceramics, while funerary use is said to be associated with decorated vessels. Indeed, El Mirón Cave was used residentially during recent prehistory, and its ceramic assemblages are eminently plain (undecorated).

In the immediate vicinity, in contrast, is the AER Cave in the Soba area, upstream of El Mirón in the Río Gándara drainage. This is considered to be a funerary cave dating to the Chalcolithic and the initial Bronze Age, based on the human and faunal remains and artifacts recovered from its surface in the 1960s. Unfortunately the site currently lacks any stratigraphic evidence. The ceramic assemblage includes a group of vessels dominated, according to Toledo Cañamero (1998), by ovoid-shape vessels with necks, some with a high keeled angle, others with straight sides, both decorated and plain. The decorations of the AER vessels are mainly composed of incised lines, "tear-shape" punctuations, and motifs formed by lines of dots. The absence of a stratigraphic context for the AER finds has caused there to be discrepancies in the cultural attribution thereof. The polemic over dating AER is reflected in the works of Toledo Cañamero, who agrees with Begines and Garcia Cáraves (1966) in attributing the ceramics to the Campaniform Chalcolithic, given the characteristics of the vessel profiles and the decorations. Rincón (1985), on the other hand, assigns them to the late Bronze Age based on their similarity to Urnfield ceramics. Toledo Cañamero and other authors situate the AER materials at the time of transition between the Chalcolithic and the initial Bronze Age, finding parallels to the ceramics of this age in the caves of Santimamiñe (Vizcaya), Trespando, and La Llana (Asturias).

The AER finds coincide with the record from El Mirón, where there are similar vessel types, namely ovoid-shape jars and closed-mouth pots. The El Mirón vessels are from levels dated to the late Chalcolithic–early Bronze Age, confirming the Begines–Garcia Cáraves–Toledo Cañamero hypothesis for AER, the difference between the two sites being the presence/absence of decoration. So the AER–El Mirón comparison serves to validate the expected chronology as well as the fact that funerary sites have more decorated vessels and residential ones mostly undecorated. A ritual-domestic dichotomy in cave use seems to be reflected in the upper Asón valley by the contrast between AER and El Mirón. It raises socioeconomic questions that lie beyond the scope of this chapter.

THE CERAMIC COLLECTION FROM EL MIRÓN: MORPHOTYPOLOGICAL CHARACTERIZATION

A series of questions have to be dealt with before presenting the morphological study of the El Mirón ceramics. One is the high degree of fragmentation caused by the constant trampling of people and livestock to which the ceramics have been subjected. This is indicative of periods of more or less intensive occupation of the cave, with many activities taking place on the surfaces on which the ceramics had been abandoned, thereby exposing those ceramics to continuous degradation. There was also a significant amount of movement among the sherds, especially vertically, caused by human actions (the digging of pits, hearths, postholes), such that many are found in secondary position. This shows up in the sherd refits, with some vessels having sherds that were found in several different levels.

These two phenomena—fragmentation and vertical movement—mean that the kinds of vessel reconstructions that can be of great utility when undertaking a typological study are very scarce and always very partial at El Mirón. Thus we are ignorant of the original size and shape of the majority of the vessels once present at the site. In general, however, this is a ceramic collection that is overwhelmingly composed of plainwares, with unrefined pastes and only simple technical production.

The typological analysis, therefore, is determined by the characteristics of a sample that is mainly composed of small fragments from which we can obtain hardly any information about the morphology of the original vessels. Nevertheless, El Mirón has a major stratigraphic sequence that is amply dated by radiocarbon, permitting the secure attribution of most of the sherds to specific chronological periods. There are no levels uncovered by the excavations that pertain to the late Bronze Age, although there are a few sherds from mixed surface fill that could belong to this period, such as a jar with a truncated conical, high-keeled form typical of the Middle Bronze Age and a decorated sherd with *boquique* chevron decoration, as found in the Late Bronze Age, Cogotas I Horizon.

For these reasons, and starting with a description of the sherds together with their graphic representation (figure 19.9), I will analyze the different vessel forms, the kinds of sherds, and the decorations attributable to

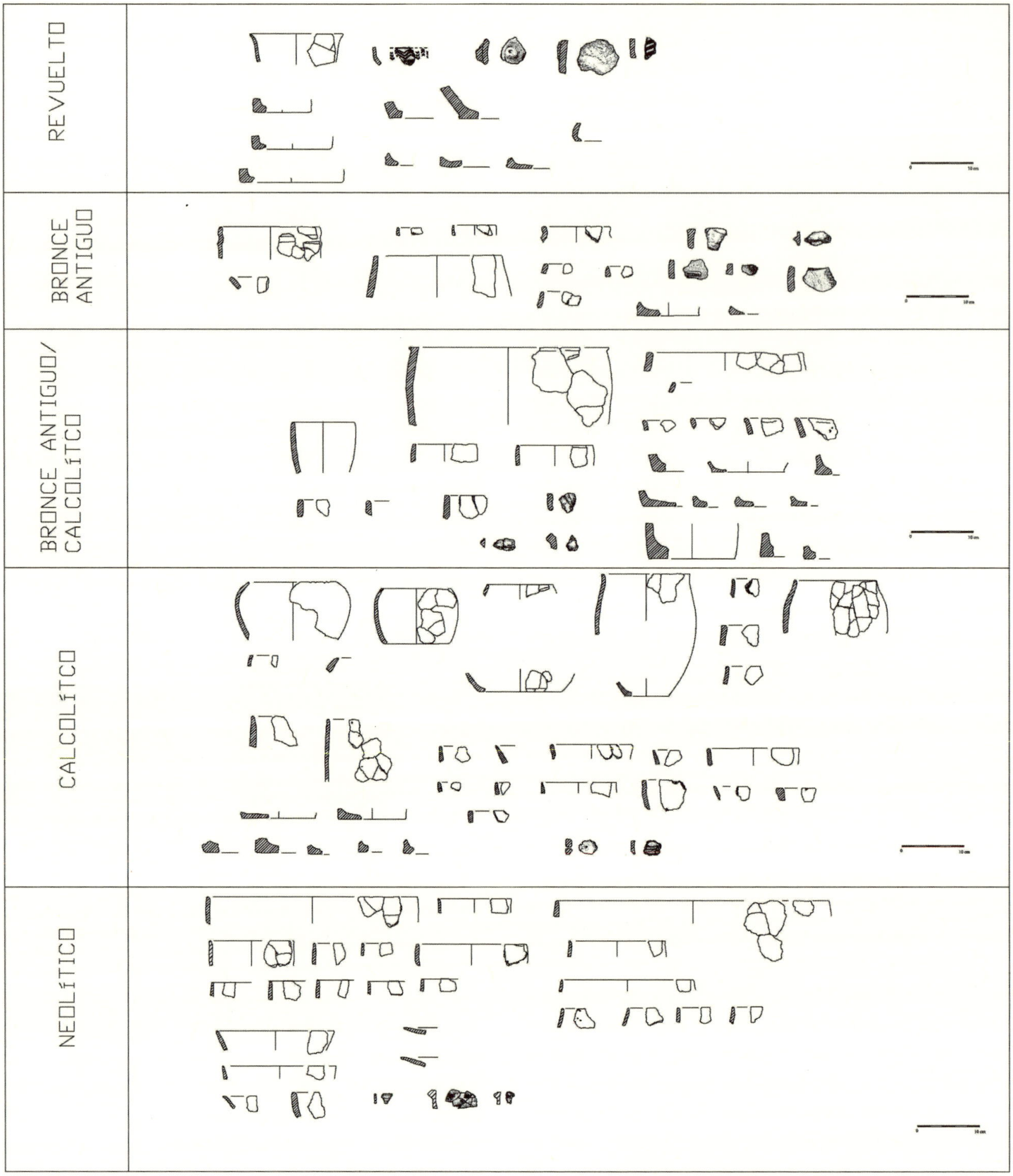

Fig. 19.9. General chart of ceramic vessel forms ("Revuelto" = disturbed surface deposits). Chart by Cristina Vega Maeso.

the different cultural phases represented in El Mirón in order to be able to establish a typological seriation, despite the high degree of ceramic fragmentation.

Neolithic

Levels 8–10 in the Cabin area and 303–303.3 in the Mid-Vestibule area are attributable to the Neolithic (table 19.8).

Most of the ceramic fragments from these units are body sherds, notably curved ones followed numerically by straight ones. There are no sherds whose profiles would suggest the presence of complex vessel forms.

There are also a smaller number of rim sherds, most with straight edges. These are followed by very similar numbers of open- and closed-lip rim sherds. There is just one outwardly protruding lip.

No definite base sherds were found in the Neolithic collections, which might mean that the vessel bottoms were rounded, such that there was no clear difference between bases and bodies of globular pots. This is a characteristic that is common in this period.

The sherds are generally rather thin, mostly ranging between 4 and 8 mm (table 19.9). Surfaces are mainly burnished (42 percent) but there are also polished (27 percent) and coarse sherd surfaces (23 percent). No decorated sherds were found in the Neolithic, with the exception of a few grouped small, conical nipples, all from a single vessel (figure 19.10).

The vessel forms are mostly open ones, some of which could have been dishes or wide bowls. There are many sherds from straight-sided jars, a few of which could have been closed in shape (table 19.10, figure 19.11).

In sum, the Neolithic vessels are rather small and have thin walls and simple profile shapes, usually with open mouths and probably rounded bases. These are plainwares, with no decoration except occasional nipples, which I have inventoried as suspension elements. There is one post-firing perforation, also probably used for suspension of the vessel. The internal surfaces of a couple of sherds bear a few incised lines that, rather than being decorations, were probably caused during the process of manufacture of the vessel. All these characteristics seem to correspond to small containers used in daily life, such as for cooking. This may be why so many sherds are burnished, since this technique reduces ceramic porosity and thus evaporation when used to contain liquids. The ensemble of plain bowls and jars was made very simply, with minimal time or effort investment. They were made to be functionally optimized as utilitarian receptacles. There is no evidence to suggest that these Neolithic ceramics were used in any ritual way at El Mirón, but rather they seem to have been made for daily use by people living in the cave.

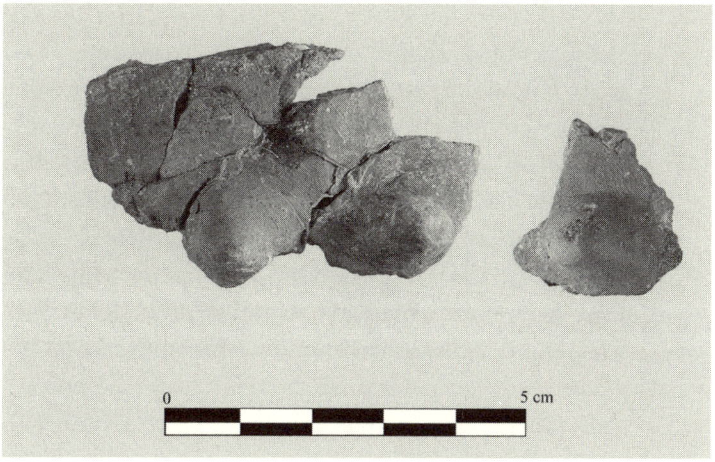

Fig. 19.10. Photo of ceramic "nipple" decorations from Neolithic levels. Photo by Cristina Vega Maeso.

Fig. 19.11. Virtual reconstruction of some Neolithic vessel forms. Image by Cristina Vega Maeso.

Chalcolithic

The radiocarbon dates lead to a Chalcolithic attribution for ceramic-yielding Cabin Levels 4.5, 5.1, 5.2, 5.4, 6, 7, and 7.4 (table 19.11).

Most of the pieces are body sherds from vessels with a predominantly curved shape (76 percent). These are followed by an also high number of straight-sided forms (22 percent) and a few keeled forms (2 percent).

Most rims are open (40 percent), followed by smaller, nearly equal proportions of straight- and closed-lip rim shapes (40 percent and 26 percent, respectively). There are a few more outwardly protruding rims than in the Neolithic, but these are still present only in trace amounts. The only bases that are represented are pedestaled (40 percent) and especially flat ones (60 percent). The sherd thickness ranges between 8 and 11 mm, with a mode at 6 mm (table 19.12).

Fig. 19.12. Decorative elements from Chalcolithic levels. Image by Cristina Vega Maeso.

Fig. 19.13. Virtual reconstruction of some Chalcolithic vessel forms. Image by Cristina Vega Maeso.

Coarse surfaces dominate the sherds (37 percent), followed by burnished and polished ones in the same quantities (30–31 percent). Quite a few sherds are so eroded that we could not determine how they were finished (2 percent).

In general, these are plainwares, although there are a few sherds with simple incised decoration, a couple with simple cord decorations, and a perforation that was probably made for functional or production reasons rather than for decoration (figure 19.12).

In sum, the Chalcolithic levels yielded mainly vessels with curved profiles, closed mouths, and flat or pedestaled bases. There are more types of forms and vessel sizes than in the Neolithic. There are closed bowls, small jars with slightly closed mouths, and for the first time here, some keeled-profile jars. Most of the vessels are plain, with no decoration, and small-to-medium size. The vessel surfaces are normally burnished, and these pastes are more refined and higher quality than in the Neolithic. The few decorated sherds do not provide much relevant information, the decoration types all being very simple (incised lines, grooves, and a few plain cords).

The fundamental characteristic of this ensemble is the greater diversity of vessel forms and sizes (figure 19.13), suggesting a wider range of functions, with the small vessels being used in cooking, liquid storage, and so on. At this time there are hints of new uses for ceramic vessels, with the larger ones possibly being storage containers. As in the Neolithic, however, most are plain, simple, and "inexpensive" in terms of the time and effort invested in their production. Utilitarian wares simply continued to be made, suggesting that the ceramics had scarce social value.

Late Chalcolithic–Early Bronze Age

The levels attributable to the transition between the Chalcolithic and Bronze Age are 3, 3.1, 3.2, and 3.3 (table 19.13).

Body sherds predominate and are mostly curved (64 percent), with very few straight-sided (35 percent) and only two keeled ones. The rim sherds are mostly straight lipped (66 percent), with smaller, equal proportions of closed- and open-lip rims (17 percent). There are very few base sherds, mostly flat (71 percent), with a very few that are pedestaled (29 percent). One lug sherd was found. The sherd thickness ranges between 6 and

Fig. 19.14. Virtual reconstruction of vessel forms from the Chalcolithic–Bronze Age transition. Image by Cristina Vega Maeso.

12 mm, with a mode at 8 mm (table 19.14). The surface finishing is mostly coarse (43 percent), followed by some polishing (34 percent) and then burnishing (23 percent).

The sherds correspond mainly to large jars with simple profiles and thick walls, flat bases, and coarse finishing. Most of the vessels are storage jars. There are, however, a few small vessels, but these are also very simple. In terms of finishing, surfaces are mostly either coarse or polished, with only 15 percent being burnished. Decorations are few and simple, with some incised lines and a few applications of simple cords and nipples.

We observe that the Chalcolithic–Bronze Age transition ceramics are rather coarser in quality and larger in size overall than in the Neolithic, with many vessels now being used for product storage (figure 19.14).

Early Bronze Age

Those levels attributable to the Early Bronze Age are 1, 2, 3.4, and 3.5 (table 19.15). Unfortunately the sample is quite small; it is dominated by body sherds that are mostly curved in profile (72 percent). There are a very few straight-sided sherds (27 percent) and just one that is keeled.

Rim sherds are mainly straight (41 percent), followed by a quarter that are open (25 percent), and then smaller proportions that are outwardly protruding and a few that are closed (17 percent). The increase in the percentage of protruding rims is notable. This is probably related to the presence of new types of vessels in the Bronze Age. There are only three base sherds (all flat). There is one ribbon suspension lug and a tab.

Sherd walls range in thickness between 3 and 17 mm, with most being between 8 and 11 mm, although there are a significant number that measure 6 mm (table 19.16). Most sherds have a coarse finish (50 percent), although there are fair numbers of polished (31 percent) and burnished (19 percent) sherds as well. Plainwares dominate (table 19.25), although there are a few impressed decorations, made either with tools, fingers, or fingernails, and a few grooves and plastic cord and nipple appliqué decorations (figure 19.15).

There is continuity in the ceramics with the early "transition" levels, with large, simple profile vessels that have thick walls and coarse finishes, mostly undecorated. The dominant types are storage jars. The decorated sherds have impressions made by fingertips or fingernails, with a few grooves and plastic decorations such as nipples and flat cords, all of which are characteristic of the Early Bronze Age in the north of the Iberian Peninsula (Arias Cabal 1996; Arias Cabal and Armendáriz 1998; Moral del Hoyo 2004) (table 19.17). This period is typified by large vessels decorated with cords and nipples that form motifs. In the El Mirón collection there is one vessel (no. 10) that stands out; it has dark paste of good quality, with a wide groove around its circumference near the rim (figure 19.16).

The overall impression is like that of the vessels from the transition levels; they are large and rather crude. The function of most of them was probably product storage.

Fig. 19.15. Photo of decorative elements from Early Bronze Age levels. Image by Cristina Vega Maeso.

Fig. 19.16. Virtual reconstruction of one of the jars from the Early Bronze Age. Image by Cristina Vega Maeso.

Material from Mixed Surface Fills

In addition to ceramics from intact archaeological levels, there are prehistoric finds from clearly mixed fill and from the present cave surface. Although they have no stratigraphic provenance, some of these sherds deserve mention because of their particular characteristics.

Some include a group of plainware sherds that come from a burnished vessel with a truncated conical, high-keeled profile (figure 19.17). There are other sherds with mostly plastic appliqué decorations, including a large nipple, a cord, and pinchmarks. Also found on the surface is a small, keeled body sherd with boquique (chevron) decoration made up of three parallel broken lines under a straight line, along which the sherd was broken (figure 19.18). The technique used as well as the decorative motifs and their placement at a keel indicate that this vessel pertained to the full-stage Cogotas I, dating to around 1500/1450–1150/1100 BC (i.e., Mid- to Late Bronze Age) (Abarquero Moras 1996, 2005). Although not found in situ, this sherd suggests that El Mirón was used by people not only in the Early Bronze Age, but also in the Mid- to Late Bronze Age, something that is also suggested by the radiocarbon age of a piece of charcoal (a torch remnant?) from the trench in the inner cave (3230±40 BP = 1400–1500 cal. BC).

If we think about the composition of the collection of sherds picked up on the surface of the El Mirón vestibule we realize that, if that was all we had from this site (as is the case with so many sites in Cantabria), we would have a very distorted vision of what the site actually contains. If we were to base our diagnoses of the site on that unsystematic surface collection, we would conclude that there were proportionally many more decorated sherds than actually exist in situ (figure 19.19). In addition, we would have simply labeled the site as being Middle or Upper Bronze Age in cultural attribution. Without the excavation, which proved the existence of major Neolithic, Chalcolithic, and Early Bronze Age levels, our diagnosis of the site's human occupation would have been limited to the very end

Fig. 19.17. Virtual reconstruction of truncated cone-shape vessel from disturbed surface context. Image by Cristina Vega Maeso.

Fig. 19.18. Photo of sherd with a chevron design found on the cave vestibule surface. Photo by Cristina Vega Maeso.

Fig. 19.19. Photo of decorated sherds from disturbed and surface contexts. Image by Cristina Vega Maeso.

of recent prehistory. Since in many other sites there are also many plainware sherds that give no precise chronological information, it is possible that the Neolithic is heavily underrepresented in the regional archaeological record, for lack of excavation and even because of the absence of culturally diagnostic surface finds.

Manufacture

Material products—in this case, ceramics—are the fruit and reflection of the societies that create them. A particular kind of object can often be made in several different ways. Nevertheless, in each society there are a series of socially acceptable choices that are made at the time of manufacture (Prieto Martínez 2001). One of our basic objectives is to define the ceramic production process in the upper Asón valley through study of the sherds uncovered in El Mirón Cave, together with the experimental replication of the technical ceramic production process, in order to help characterize the societies in which the vessels were made. Production processes and societal rules are intimately related in such a way that variability in the former denotes variability in the latter. Of course, since societies are made up of individuals there can be small variations in ceramic manufacture that can be indicative of the action of different agents who actually controlled the production. In addition, the characteristics and location of the place in which the productive process takes place can be determinant vis-à-vis the acquisition of necessary raw materials, and thus they condition the detailed conduct of manufacture.

This combination of factors can be defined as setting to work an ensemble of human knowledge skills, both conscious and unconscious, gestural and intellectual, collective and individual. These depend on relationships that exist among people as well as between humans and the laws of nature. This is what is known in French as *savoir-faire*, namely the sum total of the conscious and unconscious knowledge of an individual, translated into actions that he or she carries out during the course of his or her work (Merino 1994).

Interpretations of technological processes in ceramic production mainly come from ethnography and thus are based in the observation of tangible, living realities. However, the prehistorian has to interpret archaeological data, meaning objects that are found in the state of abandonment—no longer functioning in a living setting. The remains have also usually lost their exact original location as well as their relationship with the society that had created and used them. Thus the researcher is faced with objects that have been discarded, and yet he or she wants to try to reconstruct an entire *chaîne operatoire*, something that is not always possible in fact.

Technological processes are articulated in a series of successive, interrelated actions that together constitute a chaîne operatoire. Calvo Trias et al. (2004) structure the ceramic production chain into several levels of classification:

Level 1: Phase. The term *phase* is applied to each of the different states of the worked material and their physical properties. Specifically, it denotes how the clay loses water as the result of the actions of the potter. These processes can be classified sequentially, and the final result is the conversion of clay into ceramic.

Level 2: General technological process. This level defines in general terms the kind of process of clay manipulation to be followed by the potter as a function of minimal, universal technical requirements that are necessary for converting clay into ceramic. The operational chain is composed of several phases, which in turn can be subdivided into general technological processes. These can be broken down into detailed technological processes.

Level 3: Detailed technological process. It is at this level that one specifies the elements that make up the general process of manufacture. The following variables are analyzed: physical properties of materials, concrete technological developments, knowledge skills, and social traditions related to manufacture. At this level technological variation is significant and depends on the technological capacity of each group, their work habits, relevant aspects of social character, and so on.

Level 4: Technique. This level is determined by such aspects as the work tradition of the manufacturing group, their technological knowledge skills, and so on. This level of classification is related to the following variables: physical properties of materials, technological development, knowledge skills, social traditions, and work habits.

Level 5: Action. This has to do with the attitude of the potter vis-à-vis a particular technique. At this level one deals with the relationships among a society, its stage of technological development, its social traditions, its work habits, its artisans, and raw materials. It is related to following variables: physical properties of materials, technological development, knowledge skills, social traditions, work habits, apprenticeship, and potting experience.

The study of the ceramic remains from El Mirón permitted us to infer a series of things about the production process that we felt necessary to compare by means of realistic experiments and ethnographic examples in order to be able to reconstruct the operational chain. This is no more than a series of operations with a beginning and an end, delimited by two material states, one raw (clay) and the other a product (ceramic).

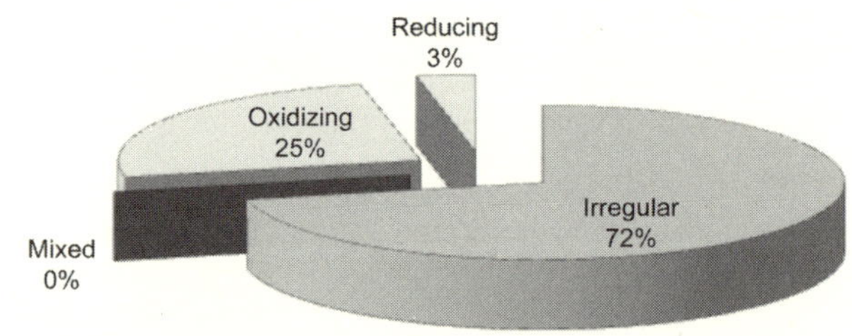

Fig. 19.20. Firing atmosphere, Neolithic ceramics. Chart by Cristina Vega Maeso.

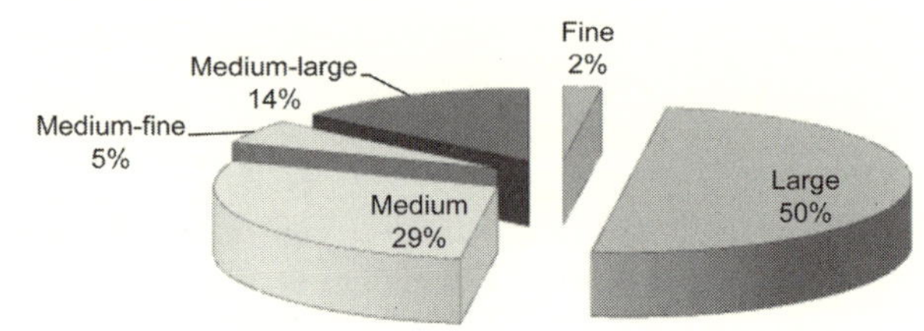

Fig. 19.21. Temper size, Neolithic ceramics. Chart by Cristina Vega Maeso.

THE CERAMIC ASSEMBLAGE FROM EL MIRÓN: TECHNICAL CHARACTERIZATION

Neolithic

The Neolithic ceramic vessels were all handmade and generally display evidence of uneven firing, mostly under reducing conditions but with some mixed or oxidizing atmosphere (table 19.18, figure 19.20). The firing was not done under conditions of controlled atmosphere, for which reason the external coloration of the sherds is not homogeneous, ranging from the grays to the pinks. The internal surfaces of the sherds have somewhat more homogeneous coloration, mostly grays (especially dark tones), which shows that during firing the passage of air was cut off to the vessel interiors. The sherds are generally fairly thin, with mostly limestone and quartzite tempers of medium to large grain size (table 19.19, figure 19.21).

Among the Neolithic sherds are three (two of which refit) that have tool marks on their interior surfaces, probably from the smoothing process during manufacture (figure 19.22).

Thus we are looking at vessels with thin walls that were formed by hand by a human group that used only a simple technological process, aimed solely at transforming clay into ceramic. The vessels were certainly fired in little pits or even in open hearths laid on the ground surface, with very little technological investment. All that was sought was to reach a temperature high enough to completely eliminate water from the clay. The blackish color of the interior sherd surfaces indicates that the vessels were fired mouth-down atop wood coals in such

Fig. 19.22. Photo of sherd with interior markings. Image by Cristina Vega Maeso.

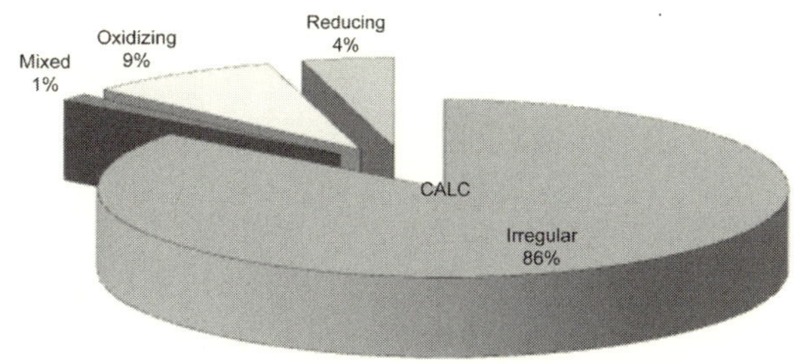

Fig. 19.23. Firing atmosphere, Chalcolithic ceramics. Chart by Cristina Vega Maeso.

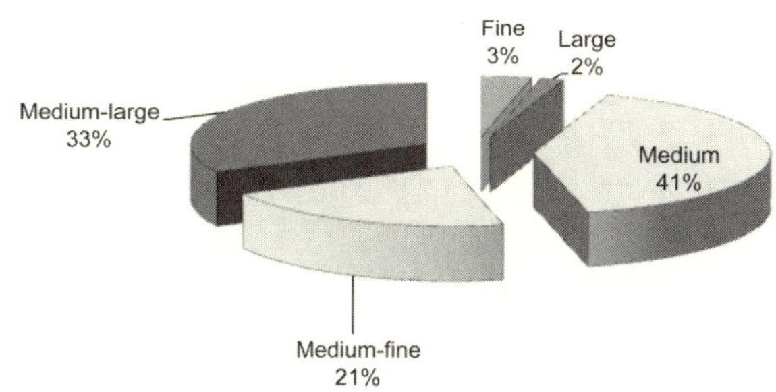

Fig. 19.24. Temper size, Chalcolithic ceramics. Chart by Cristina Vega Maeso.

a way that air did not enter them, creating a de facto reducing atmosphere in the inside of each vessel. The external parts of the vessels were directly exposed to the fire, however, such that the variability of the resultant coloration is much greater depending on how the fire acted on different sections of the vessel surfaces. To make these vessels the potters used undecanted clays in which the temper may have simply been materials that were naturally found in the clay, without even bothering to add any additional materials. The finishing, mostly by burnishing, gave the vessels less porosity, suggesting that they were mainly used for containing liquids.

Chalcolithic

All the vessels dating to the Chalcolithic were also handmade, and most (69 percent) show evidence of irregular (uncontrolled) firing (table 19.20, figure 19.23). The sherd surfaces are not homogeneous, varying widely in color: pink, grayish-pink, gray, or light red. These facts indicate that the vessels were exposed to flames and that no clearly defined firing atmosphere had been attained. The internal paste colors are mainly grays, although there are quite a few reddish-browns and pinks.

The pastes are rather unrefined and contain medium-size (41 percent) natural inclusions, mainly quartzites and limestones, although the presence of grog temper is numerically significant in the Chalcolithic ceramics (table 19.21, figure 19.24). The grog attests either to the occasionally deliberate addition of temper materials in the mixing of the clay to be shaped and fired or to the possibility that the mixing of clay and water was not done very thoroughly during the production process, leaving behind small lumps of dirt that could be interpreted as temper.

The appearance of the ceramics changes relative to the underlying Neolithic levels: average thickness increases, with the majority of the sherds being between 5 and 12 mm thick. Surface finishes are more varied and are distributed more or less equally among the burnished, polished, and coarse types. As in the Neolithic, however, the technological process used was relatively simple, with the clear goal of simply transforming clay into ceramic, without greater pretensions.

Chalcolithic to Initial Bronze Age

Most of the sherds (74 percent) were the products of irregular firing, as in the earlier periods at El Mirón, although there is an increase in the number of sherds with evidence of either oxidizing or mixed atmospheres

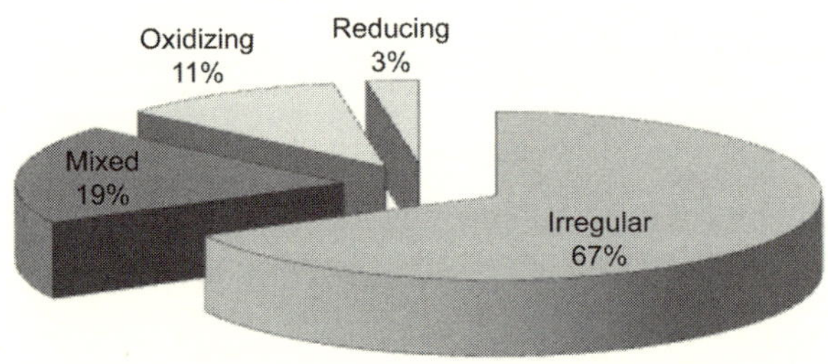

Fig. 19.25. Firing atmosphere, Chalcolithic–Bronze Age transition ceramics. Chart by Cristina Vega Maeso.

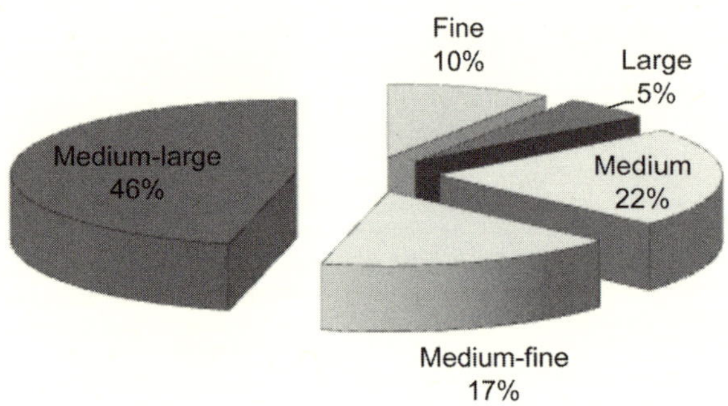

Fig. 19.26. Temper size, Chalcolithic–Bronze Age transition ceramics. Chart by Cristina Vega Maeso.

(table 19.22, figure 19.25). There is a clear predominance of the reddish, pinkish, and yellowish colorations among the sherds in this transition period. The temper materials are of medium size and are mostly natural in origin, mainly a mix of limestone and quartzite grains (table 19.23, figure 19.26).

In general the archaeological record informs us only about the paste compositions and firing temperatures, while the aspect of the ceramic technological process that concerned vessel drying generally left behind no material evidence in the sherds. However, in Level 3 there are a few base fragments that bear leaf imprints, suggesting that wet vessels were set—base-down—on leaves before firing.

The dark colors of the interior vessel walls are mainly browns, while the exterior surfaces are often pinks or yellows, with no homogeneity in external colors, marking the lack of homogeneity in firing atmospheres. On the other hand, the internal colors reflect the fact that the vessels were separated from the coals during firing, perhaps set atop rocks, thereby permitting the circulation of air within the upturned vessels during the firing process. The pastes are little refined and contain both large- and medium-size temper elements. As in the earlier periods these are mainly a mix of limestone and quartzite sands that probably occurred naturally within the clay. Vessel walls are even thicker than before: on average 9–12 mm. Among the surface finishes the coarse and polished types dominate, with only 15 percent being burnished. This might suggest some change in the functions of the vessels overall.

Early Bronze Age

Irregular firing continued to be generalized among the sherds of the Early Bronze Age occupations (table 19.24, figure 19.27). The vessel colors continue to be heterogeneous, revealing that people had still not achieved a specific firing atmosphere. The external surfaces are dominated by yellowish-orange coloration, followed by significant numbers of browns and grays. As for the internal surfaces, most are grays, although there is a nonnegligible number of yellows. The temper inclusions are

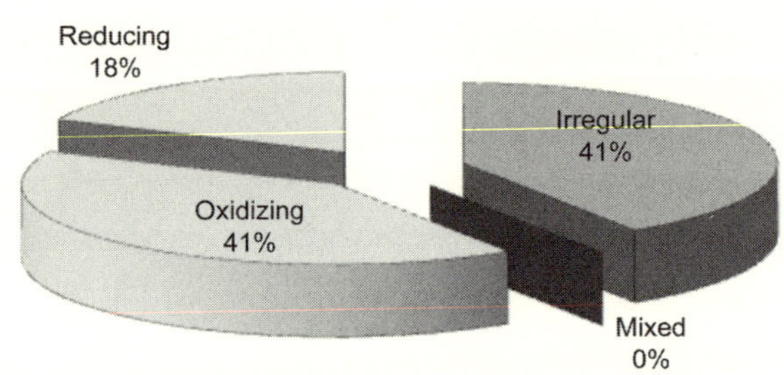

Fig. 19.27. Firing atmosphere, Bronze Age ceramics. Chart by Cristina Vega Maeso.

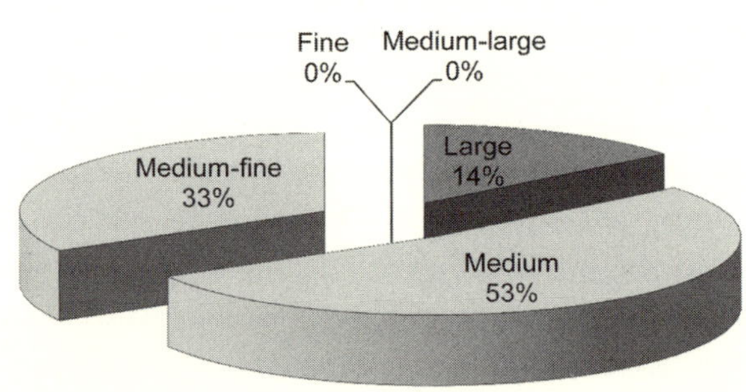

Fig. 19.28. Temper size, Bronze Age ceramics. Chart by Cristina Vega Maeso.

Fig. 19.29. Photo of bay leaf impression on a vessel base sherd. Image by Cristina Vega Maeso.

medium-size and are mostly quartzite grains or a mixture of limestone and quartzite—all probably natural to the clay (table 19.25, figure 19.28). Sherd walls are thick, as in the underlying levels. Exterior surfaces are either coarse or polished in terms of finishing.

Once again the drying process of at least one of the vessels is detectable: one sherd from Level 2 (similar to those of Level 3) has a leaf imprint; indeed, this impression is most likely attributable to a laurel (bay) leaf upon which the vessel was set to dry before firing (figure 19.29).

There is continuity with the earlier period, with the majority (75 percent) of the sherds displaying evidence of irregular firing, orange and brown exteriors and grayish interiors, generally thick walls, and abundant medium- and large-grained tempers, which, as before, are usually quartzitic sand mixed with limestone. These inclusions seem to be of natural origin, possibly coming from the clay itself as it was dug from the ground. Also as before the surface finishing of the vessel is either coarse or fine.

The Ceramic Production Process

After macroscopically analyzing the archaeological sherds, we proceeded to compare the hypotheses that had been developed through the study of mineralogical elements that constituted the ceramic pastes with the results of experiments we conducted.

The ceramic repertoire of El Mirón presents some technological characteristics that, in general terms, could be said to be rather homogeneous; there are no big differences through time. In general the site has plainwares with unrefined pastes whose inclusions are of medium size and probably were naturally present in the clays (i.e., not deliberately added). The fundamental difference that does exist is in vessel forms; there is a certain degree of variability in shapes, sizes, and finishes. Among the Neolithic levels the vessels are mostly burnished, but through time the finish types became more diversified. In the Neolithic most of the vessels are small and open, with straight walls. In the Chalcolithic the typological repertory of forms diversified, with more shape types of small bowls and various kinds of closed-mouth jars. Large storage vessels appeared during the transition to the Bronze Age. In principle there were no changes in the basic ceramic production process; rather, there were changes in the needs of the human groups, which in turn adapted their technological knowledge to the new uses by developing new vessel forms, sizes, and finishes as functionally needed.

The mineralogical analysis of the pastes and estimation of firing times were done by means of X-ray diffraction, a method that consists of X-ray bombarding of samples that have previously been pulverized (Antón Bernet 1973; Capel, Najera, and Linares 2001; Clop García 1995; Domínguez et al. 2003; García Heras and Olaetxea 1992). Seventeen samples, from several different levels of all periods, were sent to Francisco Guitián at the ceramic laboratory at the Universidad de Santiago de Compostela for X-ray diffraction analysis.

In parallel fashion, we tried experimentally to reproduce the operational chain of ceramic production used in recent prehistory at El Mirón. This type of experimental work was governed by two basic factors:

1. Our perception of reality makes us try to reproduce phenomena that we think may have happened but that may not actually have occurred.
2. Many of the actions we take could have been done differently in the past, even if they achieved the

same results, or at least results so close to the same as to be imperceptible in our studies of the remains present in the archaeological record (i.e., the problem of equifinality).

After the mineral components of the samples from the site were known, we submitted samples of clay in the natural state from three different clay deposits that, because of their nearness to El Mirón, could have been used by the prehistoric inhabitants. In order to determine changes that happened in the X-ray spectrogram after firing, we also analyzed nine other samples of experimentally fired clay. These experimentally fired samples had received different treatments, either by eliminating the coarsest fraction of natural inclusions or by adding temper materials, depending on the desired type of resultant paste. The firing was done in a basin dug in the ground, in an effort to replicate the temperatures that would have been achieved by the kinds of simple firing done by the prehistoric potters of the site. In conducting these experiments we followed the protocol proposed by Calvo Trias et al. (2004), which defines the principal phases of the operational chain; we complemented this with data we obtained through macroscopic observation and mineralogical analysis of the archaeological sherds from El Mirón. These analyses and experiments bring to light a series of results that are significant in developing our understanding of the production of ceramics during the recent prehistory of the Asón valley.

MINERALOGICAL ANALYSIS

Ceramic Pastes

Characterization of Minerals

Ceramic pastes are composed of a mixture of plastic materials—clay minerals—and non-plastic materials commonly known as temper, which can either be present naturally in a clay deposit or be added deliberately by the potter to change the physical characteristics of the clay (Capel, Najera, and Linares 1982; Druc 2001; Eygun 2001; Linares, Huertas, and Capel 1983).

The clay minerals are composed of hydrated silicates, each with a specific crystalline structure. In general one can distinguish four principal types of clays based on the reticular distance in the unions among the molecules that make up the clays:

1. Kaolinite group
2. Illite group
3. Chlorite group
4. Montmorillonite group

The most common coarser-fraction minerals naturally associated with the clay minerals are:

1. **Quartz:** Silica in the form of quartz is the most common form of temper in ceramic pastes. Quartz can occur naturally in the clay matrix or be added by the potter. The silica particles have no plasticity, so they are used as "anti-plastic" materials. Large grains of quartz help lessen the contraction of the ceramic paste during the process of drying. Under high firing temperatures, however, the fine fraction acts as a flux, thereby facilitating the appearance of newly formed minerals.
2. **Mica:** The presence of micaceous minerals such as versiculite, talcum, pirophylite, illite, and moskovite in ceramic pastes is natural, depending on the geological context and the conditions under which the clays were formed.
3. **Feldspar:** The feldspars, like the micaceous tempers, appear naturally within pastes. The most common feldspars in clay are the potassic ones, such as orthoclase and microcline, and the calcio-sodium ones, such as albite and anorthite.
4. **Calcite or calcium carbonate:** The presence of calcite or calcium carbonate in pastes could be either natural or the result of deliberate addition (Cau Ontiveros, Day, and Montana 2002). As with quartz, these tempers are used in unfired pastes as "anti-plastics," although unlike quartz they do not prevent shrinkage during the drying phase. These types of tempers are much used in the production of cooking wares.
5. **Vegetal materials:** The presence of vegetal tempers is normally the result of deliberate inclusion in the paste by potters. They disintegrate during firing, so they can be "observed" only in the form of the traces ("casts") that are left behind in the ceramics. As among limestone tempers, vegetal ones are frequently used in making cooking wares.
6. **Grog:** The presence of small fragments of older, ground-up ceramics in pastes attest to sophisticated

technical knowledge, since grog's expansion index is the same as that of the new ceramic itself. Thus it is perfect for the manufacture of cooking ware. This type of temper is very hard to identify, since, on occasions, it can be confused with lumps of the paste itself.

Temperature Effects on the Clay Minerals

The most important transformations of the whole ceramic manufacturing process occur during firing (Clop García 2001; Martín Socas et al. 1985). High temperature causes the destruction of some of the minerals originally present in the paste, while it modifies the chemical composition of others. These minerals react among themselves, giving rise to newly formed minerals. The first phase of firing, up to around 100°C, is not firing per se but the continuation of the process of ceramic drying, during which time dilations can be produced in the clay. Up to about 600°C the effects produced include the disappearance of organic temper and, usually, the elimination of water from the paste, although in unusual cases water can subside until around 900°C.

In the case of illite, dehydration generally occurs around 600–700°C, although among alumino-ferric illites this process is hastened and occurs around 450–550°C; to the contrary, it is slower among the ferro-magnesium illites. The structure of illite is maintained intact in anhydrous form until about 850–1,000°C. Above this temperature spinel appears until about 1,200°C, when it is in turn replaced by mullite.

Kaolinite loses its water around 450–600°C. Above that it is converted into metakaolinite, which in turn is transformed into spinel at around 950°C. This in turn becomes mullite above 1,000°C, and above 1,300°C the final conversion into christobalite occurs.

Chlorite dehydrates around 400–850°C, above which appears olivine. At around 1,100°C spinel and enstatite can appear. The chemical composition and degree of crystallinity of the chlorites are variable, for which reason the temperatures estimated for the formation of new chlorite phases are quite varied.

The smetics lose structural water around 500–700°C. Smetics, ferro-aluminum montmorillonites, react at low temperatures, while the magnesites require much higher temperatures. Crystalline phases are formed above about 900°C, depending a great deal on their particular mineralogical composition and structural variability.

The action of temperature also modifies the structure of the coast fraction (temper) within pastes. Thus quartz crystals change their structure at 573°C, causing changes in volume and thereby affecting the ceramic product. The fine silicious fraction favors the appearance at high temperatures of newly formed minerals such as diopside, wolastonite, ghelenite, and anorthite.

On their part, the feldspars do not react at low temperatures, but they act as fluxes when submitted to high temperatures. Their disappearance begins at about 1,000–1,100°C; above these temperatures anorthite is formed. The presence of feldspar is indicative of not having exceeded a certain temperature.

Like the feldspars, the presence of micaceous tempers is indicative of not having reached a given temperature in firing. The transformations that mica can undergo are very dependent on particle size. The process of dehydration of micas occurs around 820–1,200°C.

Calcium carbonate, under the effect of heat, is decomposed into calcium oxide around 850–950°C, giving off carbonic gas in the process. This is why it is used as temper when one wants to produce a porous ceramic. If it appears in the clay in the form of large grains, the lime formed during firing tends to carbonate and hydrate by the action of the atmosphere once the vessel is taken out of the firing kiln. Cracks may be produced in the interior of the vessel and the vessel surfaces may erode as a result of the increase in volume due to carbonation and hydration of the lime temper. If the carbonate particles are small, they can interact with philosilicates and quartz to cause the appearance of high-temperature minerals such as gehlenite, diopside, wolastonite, anorthite, and so on.

X-ray Diffraction of Ceramic Samples from El Mirón

To determine the composition of ceramic pastes, one can use two types of analyses: chemical and mineralogical. Although both methods are valid, it is sometime useful to apply both in order to compare their results. This was not possible in the case of El Mirón, so we opted for using X-ray diffraction because, in addition to providing data on clay composition, it gives estimates of firing temperatures.

The method consists of bombarding a pulverized sample with X-rays. The X-rays are diffracted in a characteristic fashion by each crystalline structure present in the sample, thereby permitting the identification of the mineral components of a paste. Atoms arranged in an orderly configuration and at distances that are typically near 10^{-10} m produce diffraction peaks. These peaks appear only when they coincide with radiation of a wavelength similar to these interatomic distances ($\lambda \approx 10^{-10}$ m). In the electromagnetic spectrum it is the radiation of the X-rays that establishes this wavelength. Thus one directs a bundle of X-rays at a determined angle (θ) at a sample and then detects the emergent radiation at the same angle. If the angle of incidence (θ) is adequate, there is a spectacular rise in the detector's radiation count, caused by constructive interference that is called a diffraction peak. The location of these peaks (after performing an angular scan) and the relative intensity among them are the hallmarks of the identified materials. Their presence can be estimated only in semi-quantitative terms, which represents one of the chief drawbacks of the method. But through it one can determine the crystalline elements that are present in the ceramics in the form of temper. It is also possible to determine firing temperature, based on the presence of nearly formed minerals that appear through the transformation of other minerals during the course of firing (García Heras and Olaetxea 1992).

The samples run at different laboratories cannot be compared, since diffraction uses only approximate, semi-quantitative models and not all labs use the same technique. Some do a count of the crystalline matter that they are trying to identify; others use a model that is known in advance. The El Mirón ceramic samples were analyzed at the Universidad de Santiago by F. Guitián, and the experimental ceramic samples were analyzed by Luis Fernández Barquín in the diffractometer in the Department of Physics at the Universidad de Cantabria. Thus we will have to avoid quantitative comparisons, since two different labs did the work and the results of a comparison would not be valid.

The 17 archaeological sherds selected for X-ray diffraction from the Cabin area are listed in table 19.26. All are small, plainware sherds with no noteworthy technical or typological characteristics, thereby justifying their destruction. They are from most of the post-Paleolithic levels, representing a broad span of time. Their apparent firing atmospheres, macroscopically identified tempers, and surface finishes also run the gamut of the kinds represented at the site. Some of the samples (nos. 4, 6, 9, 15, and 16) were coated with calcium carbonate that had precipitated in the sedimentary matrix.

The results of the spectrograms of the samples are summarized in table 19.27. Given these results, we can conclude that:

1. The main component of the clays is kaolinite, from which various minerals can be produced depending on firing temperature, duration, and conditions. Kaolinite becomes metakaolinite at 550°C, which in turn becomes spinel at 950°C and mullite at temperatures above 1,200°C.
2. In the analyzed samples kaolinite has become metakaolinite, indicating that prehistoric firing temperatures were above 550°C and below 950°C, since there are no traces of spinel or mullite.
3. The spectra for Samples 11 and 12 (Chalcolithic) reflect the presence of chlorite. Because this mineral normally disappears between 450°C and 850°C, it is hard to explain its presence in these samples. It is possible, however, that we are dealing with a form of chlorite that does not disappear until temperatures above 850°C are reached. But this could also be a newly formed chlorite derived from the conditions of humidity under which the ceramics were buried for so long.
4. Macroscopic observations showed that the most frequently found temper is quartzite. The X-ray diffraction analyses confirm this; quartzite is predominant.
5. The presence of calcium carbonate in six of the samples indicates that firing temperature was below 850°C, at least in these cases, given that above this temperature calcium carbonate decomposes, giving rise to calcium oxide.
6. Three of the samples whose spectra reveal the presence of calcium carbonate also had significant calcareous concretion on their outer surfaces. This has to be taken into consideration when interpreting the results, since the samples could have been contaminated by the concretions, and the presence of calcium carbonate in the spectra could have been due to post-depositional and not technological processes.

7. The existence of mica and feldspar is indicative that certain limits were not surpassed in firing. Feldspar melts above 980°C and mica above 1,100°C, so we can conclude that these high temperatures were not reached during firing of the sherds in the El Mirón archaeological sample.

Therefore, we are dealing with clays whose main component is kaolinite, which has been transformed by heat into amorphous metakaolinite. The predominant temper material is quartzite, along with remnants of limestone, mica, and feldspar. If we look at the temperatures at which these various minerals get transformed, we can conclude that the analyzed ceramics were fired between about 600 and 850°C. The comparison of these results with the data obtained by experimentation should permit us to ascertain the source of the raw materials and the nature of the techniques used to produce ceramics in El Mirón. In turn, this information can be the key to developing cultural interpretations beyond the technical level.

EXPERIMENTATION

The study of archaeological finds always leaves certain gaps in information that we can try to fill by experimental reproduction of the ceramic operational chain. The objective of this work is to learn about technological processes that, starting with the right combination of earth, water, air, and fire, end up transforming clay into ceramic. In order to empirically reproduce the modi operandi of the prehistoric potters, one first has to obtain information on the clay sources, water, and fuels used as well as the technical actions that were carried out at each step of the ceramic production process (Calvo 1992; Cassano et al. 1995; Clop García 2001; Garrido, Moreno, and Padial 1995; Tite and Kilikoglou 2002; Volkova 2002; Von Der Crone 2002).

As noted earlier, in conducting this series of experiments we followed the model operational chain for ceramic production laid out by Calvo Trías et al. (2004), which establishes different phases based on changes that are produced in the state of the material as the process progresses. This interpretation was complemented with data provided by relevant ethnographic literature (González Urquijo et al. 2001) as well as by direct observation of the archaeological materials from the site.

Acquisition of Ceramic Raw Materials

The experimental reconstruction of the ceramic operational chain requires a preliminary study of the immediate surroundings of the cave to determine possible sources of raw materials.

The ceramic raw materials are composed of clay pastes, temper inclusions, and water, and of course, it is necessary to have adequate fuel (wood) to fire the vessels. The location of the clay and fuel sources, the time required to obtain them, their possible storage, and so on, all have to be taken into account when studying the technological process of ceramic manufacture.

According to most ethnoarchaeological evidence, domestic ceramic production is usually done with very local clays—those found close to the places of firing. On this basis and given the results of other experimental studies, we believe that the inhabitants of El Mirón exploited clay sources close to the cave.

To determine the exploitation territory of El Mirón and thus to establish the location of the geological deposit(s) likely to have been used for the production of ceramics, we employed site catchment analysis (Davidson and Bailey 1984; Fernández Martínez 1984; Martínez Fernández and Gavilán Ceballos 1996). This kind of analysis was developed in the late 1960s to 1970s by E. Higgs and C. Vita-Finzi for the geographic/archaeological study of areas used for obtaining subsistence and raw material resources around individual sites. The standard method is to establish how far a person could walk from the site in one hour and in two hours. To calculate this without actually doing the walking, we used Naismith's formula, which establishes an average time of two hours to walk 10 km on level ground, to which one has to add an extra half hour for each rise of 300 m in elevation en route (Clop García 1995).

Inspection of the topographic and geological maps (and personal familiarity with the area around the cave) shows that the radius of action of the El Mirón inhabitants was dominated by steep slopes in an area of limestone bedrock that includes three rivers: the Asón and its tributaries the Gándara and Calera. In this predominantly limestone setting, we located two areas where the surface sediments are composed of Cretaceous clays. The first is south of the town of Ramales and very close to the cave. The second, which is larger, is east of Ramales at Guardamina. Both these deposits

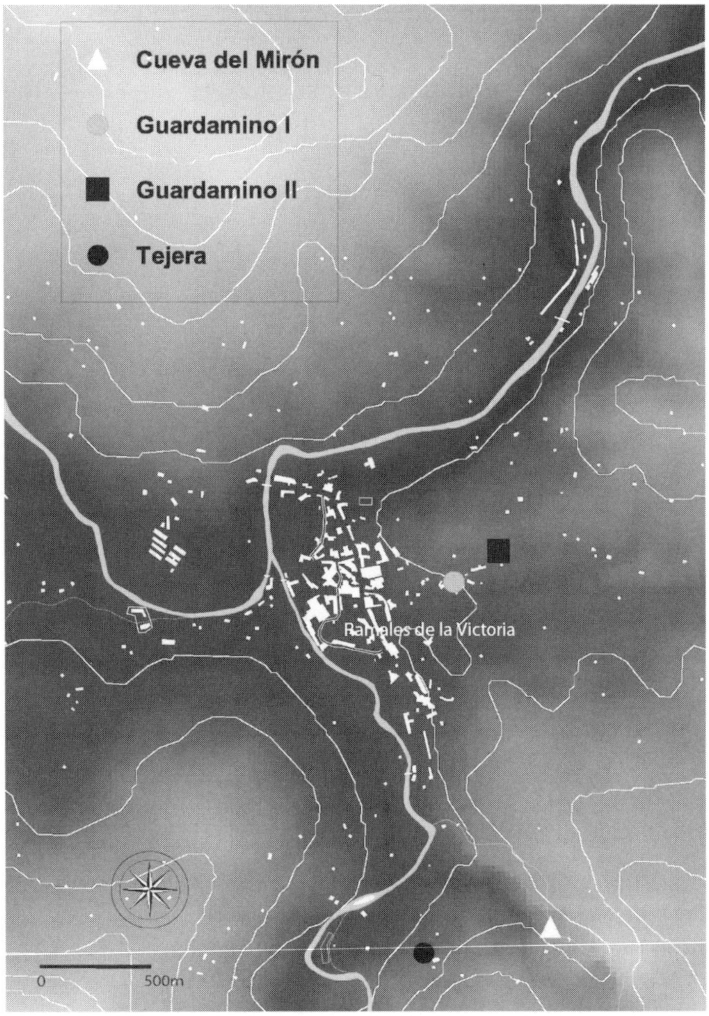

Fig. 19.30. Location of clay sources near El Mirón. Image by Cristina Vega Maeso.

had small-scale, local-production roof-tile works that functioned until the middle of the twentieth century, proving that the clays are plastic enough to be molded.

We collected samples of clay from both deposits. In the case of Guardamina, where the clay bank is much larger and we could not find the exact location of the former tile factory, we took samples in two parts of the deposit, because laterally the clays seemed to be different. In addition, we took samples of clay from the interior of the cave itself (figure 19.30).

The Ramales Tileworks

This is a place actually known as La Tejera, located directly below the mouth of El Mirón. It can be reached by going south from Ramales on the main road (N-629) in the direction of Lanestosa. About 1,300 m from the edge of the town one reaches a dirt road that leads into a private property where the old factory was located. The Lambert coordinates are X: 462753.43; Y: 478796.43. It is 700 m in a straight line from and 100 m in elevation below the cave in the valley of the Calera. It would take little more than half an hour to walk to the deposit from the cave. The clay is of excellent quality, very plastic, pure, and yellowish in color.

Guardamina I

The spot where this sample was taken is about 700 m from the edge of Ramales, in a road cut between the Guardamina neighborhood and a place known as El Castillo. The coordinates are X: 463026.17; Y: 4790004.85. It is about 1,700 m in a straight line from and 200 m below El Mirón. The estimated time to walk to this spot from the cave would be a bit less than an hour. Most of the clay is only moderately plastic; it is brown in color and contains medium-size inclusions.

Guardamina II

We also took clay samples at another place in Guardamina that is about 500 m from the edge of Ramales in a road cut along the same road that leads to El Castillo. The coordinates are X: 462942.53; Y: 4780004.85. This source is about 1,500 m from and 200 m below the cave, for a similar walking time as Guardamina I. The clay is of a fairly acceptable degree of plasticity for pottery making. It is reddish in color and contains medium-size inclusions.

El Mirón Cave Interior

Some of the sediments in the interior of the cave are clayey, so we also sampled there. However, the lack of plasticity of the sample (in reality clayey silt) caused it to fall apart during the course of experimentation.

In addition, we sought out elements in the cave area that could have been used for temper, such as sands and limestone. Naturally, there was no problem finding such materials in or near the cave: the limestone bedrock of the mountain in which El Mirón is formed and the quartzite sands in the bed of the nearly Río Calera would have provided perfectly useful temper. That or the other nearby streams (or even the seasonally more-or-less active drips from the cave ceiling) would have provided the water necessary to make the clay sufficiently plastic for ceramic production. From the wood charcoal studies of the El Mirón Holocene levels by L. Zapata we know that oak wood was most often used for fuel. Although the modern local vegetation is heavily altered as a result of centuries of heavy human exploitation, it is still possible to obtain oak wood in the area.

Preparation of the Raw Materials

Once we had gathered the necessary raw materials, we had to mix the clays to obtain a plastic material to be molded into vessels (figure 19.31). The study of the archaeological sherds does not permit us to know for sure whether the inclusions within the ceramic pastes were added intentionally by the prehistoric potters or occurred naturally within the clays. To try to get at this, we treated the experimental clays in different ways: some were fired in their natural state, from others we removed any natural inclusions before firing, and to others we added temper materials. We did this to determine whether the inclusions found in the pastes of the archaeological sherds may have been there naturally or were the result of intentional addition.

The first step involved removing impurities from both the clays and the river silts. This was done by flotation of the sediments, based on the principle that the impurities would have different densities than the clays that we wanted to keep. The clay was submerged in water and stirred; then, once the water was still, the lightest impurities floated and the heaviest ones sunk and were deposited at the bottom of the flotation tub. Once the lightest impurities were skimmed off, part of the remaining sample was screened using a mesh of less than 1 mm. In another aliquot of the sample the heavier, naturally occurring temper grains were left in the clay paste. Next the mixture of clay and water was deposited in a porous receptacle so that the excess water could be absorbed, resulting in the formation of a paste plastic enough to be molded.

This same procedure was used to extract quartzite grains from the silts of the streambed. The silts were submerged in water and the quartzite sands, being heavier, were deposited in the bottom of the tub, while the rest floated. The limestone temper elements, on the other hand, were obtained by simply crushing a limestone block until grains measuring 1–4 mm were produced. Finally, temper materials—limestone, quartzite, or both—were mixed into some of the clay samples.

At the end of this process we had several types of ceramic pastes with which we modeled various kinds of vessels whose composition could be compared systematically with the archaeological ceramic record.

The Plastic State

Once the paste was mixed to a plastic state, we hand-formed several types of simple vessels. The macroscopic study of the sherds from El Mirón provided evidence as to how the prehistoric vessels were modeled, so we used several different techniques during our replication experiments.

1. Rolled "strings" of clay were used to gradually build up the circumference of vessel walls.
2. Lumps of clay were emptied out to create cavities surrounded by vessel walls that were thinned by hand.
3. Slabs of clay were made by flattening balls of clay on a surface, and then they were put together to form vessels by hand.
4. Masses of clay were dragged between the hands to raise the walls of a vessel, while at the same time the walls were scraped and polished upward.

After the vessels were formed we added lugs and simple decorations to some of them.

First Drying Phase

The process of turning clay into ceramic is based on the removal of water from the clay. However, the first, very important drying phase generally leaves no visible traces on the resultant vessels. In the El Mirón case, however, there were prints of leaves on a few base sherds, showing that the still-wet vessels were set down on laurel leaves to dry before firing.

"Leather" Hardness

By "leather" hardness, we mean a state in which one can still manipulate the clay of a vessel; it still has enough humidity to be substantially deformed. In this phase some secondary modeling was done with the experimental vessels, such as the application of plastic decorations and the final treatment of

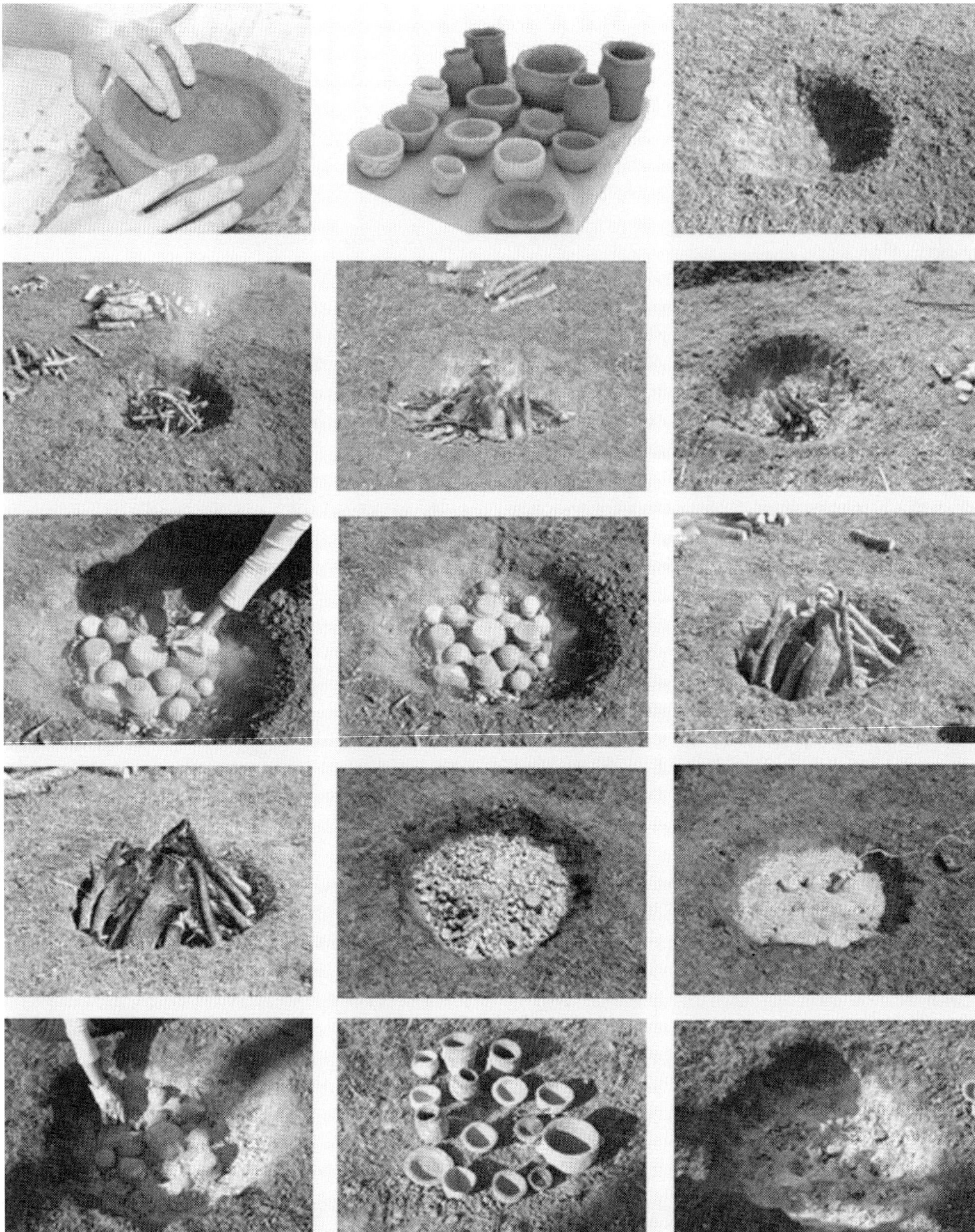

Fig. 19.31. Photomontage of the various stages of the ceramic replication experiment. Image by Cristina Vega Maeso.

the surfaces in order to improve the physical qualities of the pastes and the ultimate appearance of the vessels. In the archaeological sherds from El Mirón we observed several types of surface finishes, all of which we reproduced in our experimental replicas:

1. **Polishing:** treatment of surfaces to remove irregularities
2. **Slipping:** application of liquid clay slip to surfaces
3. **Burnishing:** use of a tool on surfaces to achieve a uniform, shiny appearance
4. **No treatment:** surfaces left coarse

Second Drying Phase

During this phase the goal is to remove as much remaining water as possible from the paste. This is very important, since the success of the firing process is in large part dependent on correct drying. The length of the second drying phase varies depending on the ambient humidity at the time and on the thickness of the vessel walls. In our case, we dried the experimental pieces for a week.

Firing

Even though vessels have dried out completely in the open air, they still cannot be considered ceramics. To attain this state they must be fired for a sufficient amount of time at a temperature that causes the destruction of the crystalline structure of the clay minerals. The elimination of all water from the composition of the clay is completed at a temperature between 500°C and 700°C, depending on the exact kind of raw materials being fired. This technological phase is the most exacting of all and requires a good knowledge of fire and of the complex transformations undergone by clays under its effects.

Firing does leave traces in the ceramic record, namely paste coloration, which can help determine the kind of atmosphere in which specific vessels were fired as well as the different components of the clays. These vary depending on the firing temperature attained. The experimental study showed that firing temperature ranged from a minimum of about 600°C and a maximum of about 800°C under irregular atmospheric conditions.

The experimental firing involved first digging a basin in the open air that was about 1 m wide by about 30 cm deep. A wood fire was built in the pit, which allows for better control of a fire, with high and stable temperatures, than would a surface bonfire, since the surrounding dirt acts as insulation. In addition, cooling is slower in a pit, reducing the risk of vessel breakage. Oak wood was burned until a layer of burning embers was formed. A few quartzite cobbles were placed on top of the embers to maintain the temperature and to isolate the vessels from direct contact with the fire. The vessels were set atop the cobbles, mouth down. Then more fuel was carefully added to cover the vessels in a tepee-like configuration. The embers progressively set the additional fuel on fire. This is the most complex phase of all,

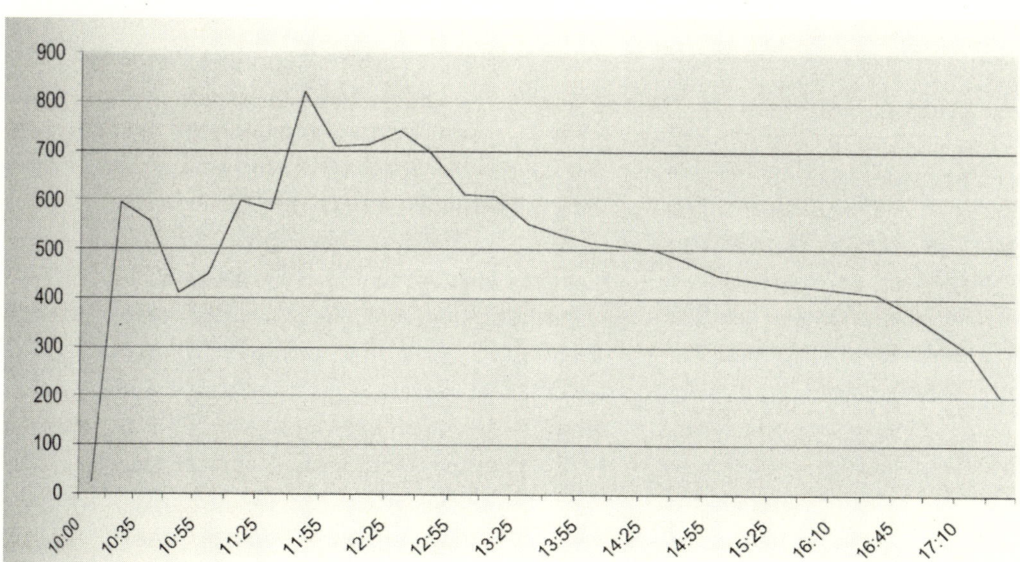

Fig. 19.32.
Firing temperatures during the experiment. Graph by Cristina Vega Maeso.

since it requires specific technological knowledge upon which the final result entirely depends. The whole firing process took five and a half hours. We took 27 temperature measurements at intervals of 10–15 minutes, using a Tes 1310, Type K thermometer (figure 19.32).

The first measurements correspond to the formation of the embers at the beginning of the process. While the wood is burning a very high temperature is reached, but the temperature falls sharply as the wood becomes embers. During the first two hours of firing we measured temperatures above 600°C in the interior of the pit, and sometimes they reached above 700°C, even up to a bit above 800°C. The temperature was maintained above 400°C for more than four hours. The firing process was considered to have ended an hour later, when the temperature had fallen to 200°C.

Cooling

This is one of the most risky phases in the pottery-making process, since the vessels can easily break. Cooling is done within the hearth to avoid brusque changes in temperature, which could break vessels through thermic stress. We removed our experimental vessels from the fire pit when the temperature reached 200°C. Two pieces broke at this point. Once fired, the surfaces of the vessels were cleaned of ash and charcoal from the fire.

X-ray Diffraction Analysis of the Experimental Samples

Once the experiment in reproducing the prehistoric production process was done, we selected a series of samples for X-ray diffraction analysis, thus to compare them with the ceramics from the site. Since some of our objectives were to determine the clay sources and the changes produced by firing in the pastes, we analyzed clay samples from the three clay deposits located near the cave that perhaps were used by its prehistoric inhabitants. To determine the changes in the X-ray spectrogram produced after firing, we also analyzed the mineralogical components of nine samples from the experiment. These had received differing manufacturing treatment, either by removing coarse-grain inclusions or by adding new minerals, depending on the kind of paste that we wanted to obtain. In all respects we had tried to emulate the conditions of firing (temperature and atmosphere) that may have been used by the prehistoric potters (table 19.28).

The X-ray diffraction analyses were conducted at the Universidad de Cantabria using a commercial instrument: a Bruker D8 Advance with a copper tube of wavelength $\lambda = 1.54 * 10^{-10}$ m. The samples were first pulverized in an agate mortar and placed in a conventional sample holder in a high-temperature chamber. The spectra were measured at 2° angles between 7° and 700° at room temperature. The object of the analysis was to compare the spectra of the experimental samples with the spectra from the sherds found in the site, so we did not do quantitative estimates for the materials but rather identified the presence or absence of minerals in the spectra. Quantitative comparisons of results obtained at different labs would not be reliable.

The samples from the three local clay deposits are generally quite similar (table 19.29). All the peaks in their spectrograms coincide, except one at 480 in sample no. 11. Perhaps the most notable observation is that the peaks at 12.50 and 180 are especially sharp. Although we do not know which mineral corresponds to the peak at 480, the peaks at 12.50 in sample no. 12 and the one at 180 in sample no. 10 both correspond to kaolinite. Otherwise, the samples are very similar and are characterized by an abundance of quartzitic and micaceous inclusions in different proportions, as well as feldspars that in the Ramales sample (no. 10) show more marked peaks than in the other samples from the Guardamina area.

On the other hand, if we look at the experimental samples, all have peaks at the same angles except samples 3, 5, and 8, which display a new phase at 29.40, corresponding to the position of limestone. These particular samples were made by adding limestone temper to the natural pastes, which is reflected in the spectra. Samples to which we added siliceous temper (2 and 7) showed no increases in the intensity of the corresponding peak in the X-ray diffraction analysis.

The clearest difference between the clay samples and the experimentally fired ones is in the presence of peaks at 12.50 and 180 among the former and the absence of peaks among the latter because of the firing effect on kaolinite, which transforms into metakaolinite with temperatures above 600°C. It is interesting to observe how the addition of quartzitic tempers did not substantially modify the spectrogram, something that the addition of limestone does do, with the appearance of a new mineral phase in samples 3, 5, and 8. As a consequence of these analytical results, we can conclude the following:

1. The main component of the clays found in the immediate vicinity of El Mirón is kaolinite, from which various other minerals are produced depending on temperature, duration, and conditions of firing. Kaolinite disappears from the samples at temperatures higher than 600°C, when it converts to amorphous metakaolinite.
2. The spectra from the samples show that the most common inclusions are quartzites, accompanied by micaceous and feldspar tempers.
3. The presence of calcium carbonate in three of the spectra is the result of the deliberate addition of crushed limestone to the pastes of these experimental samples.

COMPARATIVE RESULTS

The comparison of samples from the site and from the experiment involved only the presence and absence of minerals, regardless of their respective quantities, since quantitative comparisons would be too imprecise. Nevertheless, we can still note that:

1. The clays of both sets of samples are kaolinitic.
2. The spectra of samples 11 and 12 from the site include chlorite. This is peculiar given that chlorite disappears at temperatures between 450 and 850°C. Since these were the temperatures at which the ceramics seem to have been fired, chlorite should be absent from them. This fact, together with the absence of chlorite in the experimental samples, leads us to think that the observed clorite in the archaeological sherds might have been formed post-depositionally under the very humid conditions in which the sherds had been buried for so long.
3. In both sets of samples, there are quartz, mica, and feldspar temper grains.
4. Some of the X-ray diffraction spectra from archaeological sherds show the presence of limestone. We cannot say whether this mineral was deliberately added as a temper or not, since we did not do a microscopic analysis of these samples, which could have helped determine whether they are natural or stem from secondary calcite incrustations on the sherd surfaces.

All in all, we are able to affirm that the El Mirón ceramics were made with clays whose mineral composition was the same or very similar to those local clays we used in our experiments. Since the composition of ceramic pastes depends on the geological substrate in which the clays originate, we conclude that the El Mirón vessels were made of clays from the vicinity of the cave or at least from clays formed in a similar geological substrate.

Conclusions and Interpretation

The ceramic collection from El Mirón that analyzed was made up of two subcollections: individual sherds and reconstructed vessels. It had obviously been worked on previously (principally by Ziortza San Pedro). This artificial division of the collection caused a loss of some information on some individual sherds, although sometimes there was also duplication of information, since some were recorded on two files (as individual sherds and as elements of reconstructed vessels). If the reconstruction had been done at the same time or after cataloguing, we probably would have obtained more information from many sherds that are integrated into the restored vessels, and not merely typological information.

Characterization of this ceramic assemblage is based on macroscopic observation, a method whose utility in determining the morphology of the vessels and much of the technological process is well established. However, it would have been interesting to compare this visual characterization with observation using a binocular microscope to determine certain aspects of the technical process of ceramic production, such as paste composition, alterations, temper, and so on. The same is true for X-ray diffraction, which, even if it provides important information on the mineralogical composition of pastes and firing temperature, is not as complete as other types of analyses, such as chemical ones, which can reveal more information on clay composition in order to more precisely determine their source.

Most of the El Mirón archaeological sherds are from the Cabin area, although there are in situ finds from the Neolithic levels in the Mid-Vestibule Trench.

The distribution of these sherds shows that the Holocene deposit at the rear of the vestibule has been disturbed and exemplifies a general tendency for ceramics to have accumulated in the area near the cave mouth toward the west. It is hard to determine the cause of this tendency, but it could have been the result of the process of infilling of the cave, which resulted in an upward slope of sediment toward the cave exterior.

The distribution across several levels of sherds that made up several vessels reveals the vertical movement of materials in the past, especially in the Chalcolithic–Bronze Age transitional levels. Sherds originally abandoned in one spot when a vessel broke later became parts of the fills of pits or of the contents of other levels as a result of human (and animal) activity. These items came to be associated with chronological periods different from the one in which they had actually originated, a fact that would have caused serious errors in chrono-cultural attribution had it not been detected.

The activities of human groups that inhabited El Mirón not only altered the original location of many sherds but also caused extreme fragmentation of the pieces, notably by heavy, repeated trampling. The most significant case is that of Level 4, whose sherds are very numerous but very small. According to the stratigraphic data, these sherds come from the deposit that consists of a series of layers of ash that probably represent the repeated burning of straw and dung from the animals that were stabled in the cave. All this indicates major use of the cave as a stable (especially for cattle) in the late Chalcolithic.

Natural phenomena also altered the external appearance of the El Mirón ceramics. Many sherds are coated with calcium carbonate, an alteration provoked by the action of water, although the specific causes can be different in each case. In Levels 3.1, 3.2, and 3.3 there are many $CaCO_3$-coated sherds, which is not surprising since these layers are localized travertine lenses related to the major terminal Chalcolithic–early Bronze Age Level 3, probably caused by episodes of water dripping from the cave ceiling that created discrete areas of ponding. In the Neolithic levels there are also many such coated sherds, but they are more widely distributed throughout the excavation areas, suggesting a generalized phenomenon caused by higher humidity throughout the cave vestibule in this period (late Atlantic phase). In addition to the $CaCO_3$ coating, sherds in two Neolithic levels (8 and 9.7) also display surface erosion caused by water, thereby tending to confirm the higher-humidity hypothesis.

Typologically, the El Mirón ceramic assemblage as a whole is composed of vessels with simple forms and generally plainwares, with very few decorated sherds. We did document certain changes through time, however, such as in the treatment of surfaces. While Neolithic and Chalcolithic vessels included high percentages of burnishing, those from later periods had coarse finishing or simple polishing. There is also a shift in vessel types through time. Most of the Neolithic vessels are small, with straight or open walls. Later there is an increase in vessel diversity, with small jars, closed-mouth bowls, and larger pots, also with closed mouths. Finally, in the period of transition between the Chalcolithic and initial Bronze Age, big storage jars become quantitatively significant. These changes in ceramic types could be due to several causes. On the one hand there clearly could have been a functional differentiation among vessels, so the changes could have corresponded to cultural changes or at least to changes in the use of this cave within broader socioeconomic systems. In the Neolithic most of the ceramic vessels seem to have been destined for use in immediate daily activities, possibly involving cooking, eating, and short-term storage. Later, in the Chalcolithic, the range of uses for ceramics broadened, leading to the creation of more vessel forms. The small- and medium-size vessels were still made, however, and it seems likely that they continued to be used by small, family-type groups. A major shift occurred, however, during the Chalcolithic–Bronze Age transition, when many of the sherds found correspond to voluminous storage vessels, suggesting large-scale, long-term storage in the cave. On the other hand, the change in vessel types could be due to a change in the activities conducted in the interior of the cave vestibule and in the length of human occupations. Determining whether these changes in ceramics observed in the El Mirón post-Mesolithic sequence are reflections of wider, regional socioeconomic changes or merely evidence of changes in the human uses of this particular cave area is a major challenge for future regional research.

There are clearly implications of a socioeconomic nature. If the Neolithic ceramics are related to daily use and only short-term, small-scale storage, they reflect a society that used the cave as a simple, perhaps only periodic, residential site. During the Chalcolithic ceramics became even more important, with a wider variety of functional uses, as reflected in the diversification of vessel forms, but the changes were subtle. It is probable that the site was used residentially at a family level but perhaps over longer, continuous periods. The appearance of some larger jars suggests longer-term storage uses with a somewhat greater accumulation of agricultural surpluses. A more notable change seems to have marked the Chalcolitic–Bronze Age transition, with the appearance of more and larger storage vessels. The use of the cave and perhaps the nature of the regional socioeconomic system seem to have changed significantly, with increased agricultural production and/or new kinds of resource management. From our perspective, the changes may have been in two directions—the one not excluding the other. It is possible that human occupation of the cave became less continuous than before and, evidenced by the change in ceramic types, became a different kind of place in a production system that was more complex (and larger-scale) than it had been before. Finally, in the uppermost levels the ceramic material seems to confirm the consolidation of this trend in use of the cave, including significant use as a stable and possibly for other specialized activities, with a few decorated sherds suggestive of an early Bronze Age.

In addition, two sherds from the surface and from mixed surficial fill are suggestive of human occupations or at least visits to the site during the Final Bronze Age (as also attested by a ^{14}C date on a culturally unassociated lump of charcoal from Level VII in the trench in the dark, inner cave). The sherds in question are from a truncated, conical-shaped vessel with a high-keeled wall and from another keeled vessel decorated with chevron boquique design typical of the Cogotas I Horizon (Abarquero Moras 1996, 2005) of the Late Bronze Age in the central Iberian Peninsula. Although these pieces do not come from in situ layers, they are very interesting because they suggest a continued human use of the cave beyond the beginning of the Bronze Age. The Cogotas-type sherd in particular attests to some kind of relationship—perhaps trade—between the Cantabrian region and the Meseta Central, at least in the Final Bronze Age.

From the technical standpoint, there are no notable changes in the mineral pastes through time in the El Mirón ceramic sequence. Likewise, the temper materials remain the same: mainly quartzites, distantly followed by mixtures of quartzite sands and limestone, which are equally common in all periods. The X-ray spectrograms from the archaeological and experimental samples all show the presence of quartz, mica, and feldspar natural inclusions. The macroscopic study and the spectrograms of some samples also show the presence of limestone inclusions that were not found in the natural clay samples from the immediate vicinity of the cave. This fact, combined with the large amounts of limestone present in some of the sherds, leads us to believe that prehistoric potters intentionally added limestone in order to change the properties of the paste in some way that was specifically adapted to their needs. During the Chalcolithic there is a significant increase in the presence of what could be interpreted as grog, which indicates a high degree of technical knowledge. However, the material in question consists of small lumps of clay that are present in the paste and that, once fired, are practically impossible to distinguish from true grog. It is significant that both components are normally associated with cooking wares, such that their addition was probably done deliberately for certain vessels used in preparing foods.

The comparative mineralogical study of the archaeological and experimental samples indicates that the clays used for both are kaolinites. The spectra of pastes from archaeological samples 11 and 12 show the presence of chlorite associated with metakaolinite. Because chlorite disappears in the range of temperatures that we believe was achieved during prehistoric firing (and because chlorite is absent from the experimentally fired pieces), however, we believe that the chlorite in the archaeological samples was formed after firing, during the long period of burial of the sherds in humid sediments where it could have been precipitated.

Firing was always mostly irregular, indicating a lack of control over the atmosphere. The vessels were probably fired either in fires built in small pits or directly on the ground surface, with little technological investment, since all that was required was to reach the temperature needed to eliminate all water from the clay of the formed vessels. The Neolithic and Chalcolithic vessels have blackened interiors, from which we can deduce that they were fired mouth-down, excluding the entry of air and thus creating a reducing atmosphere in the interiors of the vessels. In contrast, the vessel exteriors, which were directly exposed to the fire, display a wider range of colors, even on individual pieces. During the transition to the Bronze Age one begins to observe a greater degree of color variation in the vessel interiors, suggesting that some material may have been placed between the vessels and the embers, which permitted air to circulate within the vessels during firing.

The X-ray diffraction analysis of the pastes shows the presence of metakaolinite, which indicates firing temperature between 550 and 980°C. In the case of ceramics with limestone temper, the maximum temperature would be reduced to 850°C.

Although the treatment of the pastes in their plastic state and the firing of the vessels did not change much through time, the external finishing of the pieces did change. The Neolithic sherds were mainly burnished. In the Chalcolithic the surface finishes diversified, with roughly equal numbers of burnished, polished, and coarse types. During the transition to the Bronze Age the coarse and polished finishes become more common, while burnished ones fall to barely 15 percent of the total sample, a trend that continued during the early Bronze Age. In many cases the treatment of the vessel surfaces was done not for mere aesthetic reasons but for functional ones. Burnishing of surfaces eliminates porosity such that very often burnished vessels are interpreted as liquid containers. Lacking residue studies to more directly determine vessel contents, burnishing can indeed be considered a relative functional indicator.

The study of the ceramic material from El Mirón has led to the definition of a series of general characteristics for the overall collection that allow us to establish the following conclusions:

1. The El Mirón assemblages belong to the domestic sphere in all periods represented. There are no other kinds of activities represented (such as funerary ones, more common in so many other caves of the Vasco-Cantabrian region).
2. From the Neolithic through the early Bronze Age the domestic ceramics of the Asón Basin lack decoration and have very simple vessel forms.
3. There is no clear evidence to indicate a lineal evolution of vessel forms. However, there is variability through time that could be due to cultural factors (norms) and/or different functions of the cave. These differences are fundamentally documented in the surface treatments and in the increase in the percentage of large vessels.
4. Thus it can be affirmed that the technological process was very simple and barely changed through time. The ceramics were made with local, unrefined clays to which limestone temper was sometimes added either to adapt vessels to cooking use or (problematically) for a lumpy form of decoration. The firing techniques did not vary much through time either, using a temperature range between 600 and 850°C to change the state of the clay. Although there were no significant ceramic technological changes, human societies adapted their ceramic skills to changes in their daily needs for cooking and storage—short- and long-term.

If we add a comparative study of this collection in relation to materials from sites in the vicinity of El Mirón, a number of other questions come to light:

1. To establish a morphotypological sequence of ceramics for the recent prehistory of the Asón Basin, it would be necessary to have a solid stratigraphic base from other sites in order to contextualize finds correctly, without having to rely only on fossil director types. Without this, any resultant patterns might be incorrect.
2. In the upper levels of El Mirón, which were affected by disturbances over a long period of time, sherds pertaining to several different eras have been found, including temporally non-diagnostic plainwares as well as items with very characteristic attributes pertaining to quite recent times. This kind of mixed record is typical of many caves in the Asón valley, leading us to believe that in many cases it is necessary to revise the chrono-cultural ascriptions made on the basis of only surface collections of sherds.
3. In the rest of the caves in the area most of the sherds are plainwares, but the dating is done on the basis of the (few) decorated sherds. Using this modus operandi, a lot of sites have been attributed to the full

and final Bronze Ages and none have been attributed to earlier periods. It is very possible that hidden among the mass of undecorated sherds are plainware materials pertaining to the Neolithic and/or Chalcolithic, which, if they are like the corresponding sherds from these periods in El Mirón, are undecorated and banal in terms of vessel form.

4. It is clear that without a stratigraphic sequence within which to contextualize ceramic finds it is impossible to produce a correct attribution. In this sense the El Mirón record represents an important corpus of evidence for synthesizing a major portion of recent prehistory. Naturally, the patterns observed there need to be corroborated by other excavations in the area, such as the ongoing work of M. Fano in the cave of El Horno, located at the base of the same mountain in which El Mirón is located.
5. El Mirón can be considered the model of residential caves in the Asón valley, serving as a counterpoint to AER Cave, which is traditionally considered to have been a funerary cave. Both caves contain ceramics from the same Chalcolithic–Bronze Age transitional period, but decorated pottery is much more abundant in AER. Nonetheless, there is a clear connection with El Mirón, namely certain vessel forms—both plain and decorated—in AER that are identical to a few that have been found in El Mirón. The most evident ones are the jars of ovoid shape with closed mouths, such as El Mirón vessel no. 1.
6. It is necessary to integrate all the various collections and to develop a synthesis of the material from the recent prehistory of eastern Cantabria, beginning with certain phenomena that tend traditionally to be studied separately but that need to be combined, such as habitation sites with plainwares, funerary caves with decorated wares, and megalithic monuments.

The study of the ceramic material from El Mirón permits us to establish a series of generalizations about the record that shed light on the types of settlement and lifeways of the inhabitants of the Upper Asón during recent prehistory. Through this material one can deduce the existence of a significant transformation not only in the archaeological record but also in prehistoric societies at the end of the Chalcolithic. This observation leads to the formulation of a number of hypotheses that could be tested with different classes of archaeological evidence from the site.

Thus there remain open many lines of investigation by other specialties, as suggested by the results of this ceramic analysis. One of our most striking conclusions is that the ceramic technology remained very conservative and stable during the whole of recent prehistory at El Mirón. It was a simple technology, requiring no very complex skills, and it was oriented toward the satisfaction of domestic needs. For the fabrication of vessels, the potters almost certainly used strictly local materials, as demonstrated by the mineralogical analyses. This is important in the study of the Neolithic, as it is relevant to the question of the origins of new technologies in this period, although well beyond the scope of this work.

It seems clear that the "Neolithic" groups that inhabited El Mirón Cave carried out an agro-pastoral exploitation of the surrounding environment, using local resources to survive. As a result of our typological study, as noted above, it became apparent that there were slight changes through time in the record that, as we understand them, had to do with changes in the needs of the local human communities. This observation poses questions that will have to be addressed by future research:

1. There is no evidence in the Neolithic of any large-scale storage (at least not in ceramic vessels). Since the existence of cereal agriculture is well documented through carpological and palynological studies, the question is: where were they storing their harvests? Even if they were eating much of the wheat soon after harvest, they still would have had to store seed grain for the next planting season.
2. There are several lines of evidence suggesting that the Chalcolithic was a dynamic period in the cave vestibule, with many structures, heavily altered surfaces, and indications of a diversification of productive activities, including the introduction of legumes, addition of new types of vessels, and so on. The question is: To what extent is this impression true and, if it is, what was the reason for this diversification? What were the new productive strategies that were in play?
3. At the end of the Chalcolithic one perceives a change in the ceramic vessels that will become consolidated

in the early Bronze Age: an increase in the capacity of the vessels to include very large storage jars. The questions that arise are: What were they storing and why? Who was managing and controlling the surpluses that now seem to have been larger? Does this represent the beginning of significant differences in wealth and power among the members of society?

In the future it would be very interesting to conduct a study of the composition of the clays from different sites throughout the Asón valley as well as from many possible clay sources in the same basin. This could provide information on the nature of ceramic production and possible exchange networks. It would also be important to do analyses of residues from the interior surfaces of sherds of different ages at El Mirón and other sites. The present work represents the beginning of a new chapter in the more detailed, more precise, and richer appreciation of long-undervalued ceramic materials from the long-understudied and heavily skewed record of recent prehistory both in eastern Cantabria and, more generally, throughout the Vasco-Cantabrian region.

Acknowledgments

This research, done for a licenciatura at the Universidad de Cantabria, was conducted as part of the El Mirón Cave Prehistoric Project, directed by Professors Manuel González Morales and Lawrence Straus, whom I thank for their assistance. I also thank many of my fellow El Mirón team members for their help and enthusiasm, especially Eduardo Carmona. Dr. Luis Fernández Barquín (Universidad de Cantabria) kindly performed the X-ray diffraction analysis of the ceramic paste mineralogy and provided much invaluable advice and information. Dr. F. Guitián of the Ceramic Institute at the Universidad de Santiago de Compostela graciously conducted the analysis of the mineralogical content of a sample of 19 sherds from Neolithic through Bronze Age levels at El Mirón at the request of Professor González Morales.

References Cited

Abarquero Moras, F. J.

1996 Rasgos de identificación de la cerámica tipo Cogotas I fuera de la Meseta. In *II Congreso de Arqueología Peninsular,* edited by R. Bolbín and P. Bueno, pp. 113–128. Fundación Rei Afonso Henriques, Zamora, Spain.

2005 *Cogotas I. La difusión de un tipo cerámico durante la Edad del Bronce.* Monografías Arqueología en Castilla y León. Junta de Castilla y León, Valladolid, Spain.

Antón Bernet, G.

1973 *Análisis por difracción de reyos X de cerámicas ibéricas valencianas.* Servicio de Investigación Prehistórica, Valencia, Spain.

Arias Cabal, P.

1996 La cronología absoluta del Neolítico y el Calcolítico de la región cantábrica. Estado de la cuestión. *Cuadernos de Sección Prehistoria-Arqueología* 6:1–21.

Arias Cabal, P., and A. Armendáriz

1998 Aproximación a la Edad del Bronce en la región cantábrica. In *A idade do Bronce en Galicia: Novas perspectivas,* edited by R. Fábregas, pp. 47–80. Ediciones do Castro, Sada, Spain.

Begines, A., and J. M. Garcia Cáraves

1966 Hallazgos del Bronce I en dos cuevas de Santander. In *Actas del IX Congreso Nacional de Arqueología (Valladolid),* pp. 122–126. Universidad de Zaragoza, Zaragoza, Spain.

Bohigas Roldán, R., A. Iriondo, J. Tocino, and F. Ruiz

1984 El Karst de Tarrueza (Laredo). *Boletín Cantabro de Espeleología* 5:48–61.

Bohigas Roldán, R., J. Molinero, and E. Muñoz

1992 Los materiales de Cueva Cofiar o de Los Trillos. In *Actas del V Congreso de Espeleología,* pp. 229–241. Santander, Spain.

C.A.E.A.P

1984 Las culturas prehistóricas con cerámicas. *Boletín Cantabro de Espeleología* 4:103–128.

Cailleux, A.

1981 *Notice sur le code des couleurs des sols.* Boubée, Paris.

Calvo, M.

1992 Experimentando con arcilla y el fuego como en la antigüedad. *Ceramología* (Alicante, Spain), pp. 39–50.

Calvo Trias, M., J. Fornés Bisquerra, J. García Roselló, V. M. Guerrero Ayuso, E. Juncosa Vecchierini, C. Quintana Abraham, and B. Salvà Simonet
2004 *La cerámica prehistórica a mano: una propuesta para su estudio.* El Tall, Mallorca, Spain.

Capel, J., F. Najera, and J. Linares
1982 Algunos aspectos del proceso de manufacturación de cerámicas neolíticas. Estudio del contenido en desgrasantes mediante lupa binocular. *Cuadernos de Prehistoria de la Universidad de Granada* 7:137–159.
2001 Aproximación al estudio de procedencia y technologia de fabricación de los cerámicas campaniformes del yacimiento de la Edad del Cobre de los Millares. *III Congreso Nacional de Arqueometría,* edited by M. Gómez and M. Pardo, pp. 207–214. Universidad de Sevilla, Seville, Spain.

Caro, A.
2002 *Ensayo sobre cerámica en arqueología.* Seville, Spain.

Cassano, S., I. Muntoni, G. Eygun, and Y. Garidel
1995 Pottery-making in Southern Italy: an experimental study. In *Estudis sobre cerámica antigua,* edited by M. Vendrell-Saz, T. Pradell, J. Mulera, and M. Garcia, pp. 11–16. Generalitat de Catalunya, Barcelona, Spain.

Cau Ontiveros, M. A., P. M. Day, and G. Montana
2002 Secondary calcite in archaeological ceramics: evaluation of alteration and contamination processes by thin section study. In *Modern Trends in Scientific Studies on Ancient Ceramics. Papers Presented at the 5th European Meeting on Ancient Ceramics,* edited by V. Kilikoglou, A. Hein, and Y. Maniatis, pp. 9–18. British Archaeological Reports S-1011. Oxford, UK.

Clop García, X.
1995 Estudio mineralógico de los materiales cerámicos del instituto de Bachillerato de Manlleu. Una aproximación a la cerámica del Calcolítico-Bronce Antiguo de Catalunya. Estudis sobre cerámica antiga. In *Studies on Ancient Ceramics. Proceedings of the European Meeting on Ancient Ceramics,* edited by M. Vendrell-Saz, T. Pradell, J. Molera, and M. Garcia, pp. 83–88. Generalitat de Catalunya. Departamento de Cultura, Barcelona.
2001 El foc i la ceràmica. *Cypsela* 13:59–72.

Davidson, I., and G. Bailey
1984 Los yacimientos, sus territorios de explotación y la topografía. *Boletín del Museo Arqueológico Nacional* 2:25–46.

Domínguez, S., J. Ramos, V. Castañeda, and M. Sánchez
2003 Análisis arqueométrico de las cerámicas prehistóricas del embarcadero del Río Palmones. In *Avances en arqueometría 2003,* edited by M. Feliu Ortega, M. Edreira, and M. Fernández, pp. 138–144. Universidad de Cádiz, Cadiz, Spain.

Druc, I.
2001 Introduction. In *Archaeology and Clays,* edited by I. Druc, pp. iii–vi. British Archaeological Reports S-942. Oxford, UK.

Eiroa, J. J., J. A. Bachiller Gil, L. Castro Pérez, and J. Lomba Maraundi
1999 *Nociones de tecnologia y tipología en Prehitoria.* Ariel, Barcelona.

Eygun, G.
2001 Clays and early Neolithic potters. In *Archaeology and Clays,* edited by I. Druc, pp. 55–70. British Archaeological Reports S-942. Oxford, UK.

Fernández Martínez, V.
1984 El análisis de territorios arqueológicos: una introducción crítica. *Arqueología Espacial* 1:55–71.

García Heras, M., and C. Olaetxea
1992 Métodos y analisis para la caracterización de cerámicas arqueológicas. Estado actual de la cuestión. *Archivo Español de Arqueología* 65:263–289.

Garrido, O., F. Moreno, and B. Padial
1995 Arqueología experimental: una forma de acercarse al pasado. *Revista de Arqueología* 166:6–13.

González Morales, M. R., and L. G. Straus
2000 La Cueva del Mirón (Ramales de la Victoria. Cantabria): excavaciones 1996–1999. *Trabajos de Prehistoria* 57(1):121–133.

González Urquijo, J., J. J. Ibáñez Estévez, L. Zapata Peña, and L. Peña Chocarro
2001 Estudio etnoarqueológico sobre la cerámica gzaua (Marruecos). Técnica y contexto social de un artesanado arcaico. *Trabajos de Prehistoria* 58(1):5–27.

Linares, J., F. Huertas, and J. Capel
1983 La arcilla como material cerámico. Característica y comportamiento. *Cuadernos de Prehistoria de la Universidad de Granada* 8:479–490.

Llanos Ortiz, A., and J. I. Vega Saramburu
1974 Ensayo de un método para el estudio y clasificación tipológica de la cerámica. *Estudios de Arqueología Alavesa* 6:265–313.

Martín Socas, D., M. D. Camalich, M. J. Tejedor, A. Rodríguez, and P. González
1985 Composición mineralógica y evaluación de las temperaturas de cocción de la cerámica de campos. *Cuadernos de Prehistoria de la Universidad de Granada* 10:131–185.

Martínez Fernández, M., and B. Gavilán Ceballos
1996 Análisis de explotación del territorio a través de la captación de arcilla por parte de una sociedad neolítica. *II Congreso de Arqueología Peninsular,* vol. 2, edited by R. Balbín and P. Bueno, pp. 77–82. Fundación Rei Afonso Henriques, Zamora, Spain.

Melero, J., A. León, and J. C. Torres
2003 Un sistema interactivo de reconstruccion y dibujo de cerámica arqueologica. In *Congreso Español de Informatica Grafica*, edited by F. Serón, A. Vinacua, and J. Regincós, pp. 321–324. Universidade da Coruña, A Coruña, Spain.

Merino, J. M.
1994 Tipología lítica. Special issue, *Munibe*, suplemento 9.

Moral del Hoyo, S.
2004 *La Cueva del Mirador*. Ediciones Sierra de Atapuerca, Burgos, Spain.

Muñoz, E., and J. Gomez Arozamena
1995 Catálogo topográfico de las cavidades con interés arqueológico: Asón–Castro Urdiales. *Boletín Cantabro de Espeleología* 11:137–145.

Muñoz, E., B. Malpelo, and J. Gomez
1996 Topografía de los cavidades con interés arqueológico. Part 9. *Boletín Cántabrio de Espeleología* 12:105–121.

Muñoz Fernández, E.
1988 *Carta Arqueológica de Cantabria*. Tantín, Santander, Spain.
1991 Los yacimientos arqueológicos del Valle de Carranza. *Arquenas* 1:89–140.
1992 Los cavidades conyacimiento arqueológico en Cantabria. In *Actas del V Congreso Español de Espeleología*, pp. 247–255. Santander, Spain.
1993 Las cavidades sepulcrales en Cantabria. In *Actas del VI Congreso Español de Espeleología*, pp. 287–322. A Coruña, Spain.
1995 Catálogo de cavidades con interés arqueológico: Arredondo-Coba. *Boletín Cántabrio de Espeleología* 11:129–135.

Orton, C., P. Tyers, and A. Vince
1997 *La cerámica en arqueología*. Crítica, Barcelona.

Prieto Martínez, M. P.
2001 *La cultura material cerámica en la prehistoria reciente de Galicia: Yacimientos al aire libre*. Trabajos de Arqueología del Paisaje, 20. Santiago de Compostela, Spain

Renfrew, C., and P. Bahn
1991 *Arqueología. Teoría, métodos y técnicas*. Akal, Madrid.

Rincón, R.
1985 Las culturas del metal. In *Historia de Cantabria: Prehistoria, edades antigua y media*, edited by M. A. García Guinea, pp. 113–209. Estudio, Santander, Spain.

Ruiz Cobo, J.
1999 La Cueva de Cofresnedo (Matienzo, Cantabria). In *Regio Cantabrorum*, edited by J. Iglesias and J. Muñiz, pp. 31–41. Caja Cantabria, Santander, Spain.
2000a La Cueva de Las Grajas (Matienzo, Ruesga). In *Actuaciones arqueológicas en Cantabria, 1984–1999*, edited by R. Ontañón, pp. 235–236. Gobierno de Cantabria, Santander, Spain.
2000b La perhistoria reciente de Matienzo. In *Actuaciones arqueológicas en Cantabria, 1984–1999*, edited by R. Ontañón, pp. 381–382. Gobierno de Cantabria, Santander, Spain.

Ruiz Cobo, J., and P. Smith
1997 El depósito cerámico de la Cueva de los Grajas (Matienzo, Ruesga). *Munibe* 49:65–76.

Smith, P.
1981 Las cuevas de Emboscados y Patatal. In *Memorias ACDPS*, pp. 45–46. ACDPS, Santander, Spain.
1995 Hallazgos de cerámica prehistoria en Matienzo (Cantabria). *Trabajos de Arqueologia en Cantabria* 3:19–23.

Smith, P., and J. Ruiz Cobo
1999 Avance al inventario arqueológico de la depresión cerrada de Matienzo. *Sautuola* 6:243–255.

Tite, M., and J. Kilikoglou
2002 Do we understand cooking pots and is there an ideal cooking pot? In *Modern Trends in Scientific Studies on Ancient Ceramics: Papers Presented at the 5th European Meeting on Ancient Ceramics*, edited by V. Kilikoglou, A. Hein, and Y. Maniatis, pp. 1–8. British Archaeological Reports S-1011. Oxford, UK.

Toledo Cañamero, M. C.
1998 Tipología y caracterización de las cerámicas de las Edad del Bronce en Cantabria. M.A. thesis. Universidad de Cantabria.

Volkova, H. V.
2002 Role of experiment in reconstruction of the Fatyanovo pottery technology. In *Modern Trends in Scientific Studies on Ancient Ceramics: Papers Presented at the 5th European Meeting on Ancient Ceramics*, edited by V. Kilikoglou, A. Hein, and Y. Maniatis, pp. 95–102. British Archaeological Reports S-1011. Oxford, UK.

Von Der Crone, M.
2002 Experimental firing of clays using salt water. *Archaeometry 98*, edited by E. Jerem and K. Biro, pp. 249–256. British Archaeological Reports S-1043(II). Oxford, UK.

Table 19.1. Ceramic-bearing levels of El Mirón Cave

Period	Level[a]	Uncal. ^{14}C age
Mixed surface fill and rev.	surface, 300, 301, 301.1, 301.2, 302(?)	Medieval
Early Bronze Age	1[b], 2, 3.4, 3.5	
Chalcolithic–Early Bronze Age (undifferentiated)	3, 3.1, 3.2, 3.3	3700±40 BP
Chalcolithic	4, 5, 5.1, 5.2, 5.4, 6, 7, 7.4	4120–3740 BP
Neolithic	8, 8.1, 9, 9.1, 9.6, 9.7, 9.8, 10, 303, 303.1, 303.2, 303.3	5790–4680 BP

[a]Units with ceramics

[b](Level 1 is probably somewhat mixed but without modern materials —LGS).

Table 19.2. Weight of sherds from Neolithic levels

Level	Count	Total weight (g)	Average sherd weight (g)
8	95	1,324	13.98
9	174	4,644	26.69
10	96	1,441	15.01
303	11	261	23.72
303.2	25	530	21.20
303.3	26	513	19.73
9.1	5	225	45.00
9.6	354	7897	22.30
9.7	79	975	12.34
9.8	50	772	15.44

Table 19.3. Weight of sherds from Chalcolithic levels

Level	Count	Total weight (g)	Average sherd weight (g)
4	438	11,079	25.29
5	122	4,003	32.80
6	118	3,833	32.48
7	16	389	24.31
5.1	24	661	27.54
5.2	11	710	64.54
5.4	4	99	24.75
6.1	1	10	10.00
7.4	2	94	47.00

Table 19.4. Weight of sherds from the transition to the Bronze Age levels

Level	Count	Total weight (g)	Average sherd weight (g)
3	592	27,748	46.87
3.1	9	777	86.30
3.2	232	16,202	69.83
3.3	81	2,843	35.09

Table 19.5. Weight of sherds from Bronze Age levels

Level	Count	Total weight (g)	Average sherd weight (g)
1–2	9	290	32.20
1	22	353	16.04
2	103	7,936	77.04
2–3	1	255	255.00
3.4	13	310	23.84
3.5	67	2,886	43.07

Table 19.6. Number and percentage of specimens with secondary calcite on their surface, by level

Level	No.	%
1	8	37.00
2	13	13.00
3	58	9.70
4	3	0.60
5	14	11.60
6	9	7.60
7	2	12.50
8	29	30.50
9	47	27.00
10	24	25.00
301	8	8.90
303	3	27.70
3.2	60	26.00
3.3	14	17.30
303.1	3	19.00
303.2	4	16.00
5.1	4	17.00
5.2	1	9.00
6.1	1	100.00
9.6	47	13.30
9.7	25	32.00
9.8	3	6.00

Table 19.7. Number and percentage of eroded sherds, by level

Level	No.	%
3	3	0.50
5	2	1.50
6	8	6.70
8	26	27.40
9	1	1.00
3.4	3	23.00
3.5	1	1.50
9.7	8	10.00

Table 19.8. Types of ceramic fragments in Neolithic levels

Level	Rim	Neck	Body	Suspension
8	3		87	5
9	13		161	
10	22		73	1
303	1		10	
303.1	1		13	2
303.2	1	1	23	
303.3	3		23	
9.1			5	
9.6	45		309	
9.7	4		75	
9.8	4		46	
TOTAL	**97**	**1**	**825**	**8**

Table 19.9. Thickness of Neolithic sherds (mm)

Level	1	3	4	5	6	7	8	9	11	13	N/A
8		1	4	34	9	13	14	17			3
9		3	28	44	26	18	37	10			8
10			11	11	42	29		1		1	1
303	1		1		4	2	1				2
303.1		1	2		7	1			1	2	2
303.2			12	7	4	2					
303.3			1	5	15	3	1				1
9.1			1		2		2				
9.6		3	50	40	124	65	53	12			7
9.7		1	16	36	19	4					3
9.8			31	5	9	4	1				
TOTAL	**1**	**9**	**157**	**182**	**261**	**141**	**109**	**40**	**1**	**3**	**27**

Table 19.10. Vessel forms in Neolithic levels

Level	Open	Straight	Closed	N/A
8	1	1		93
9	3		1	170
10	23	10	1	62
303	1			10
303.1			1	15
303.2		1		24
303.3		1	1	24
9.1				5
9.6	8	9	5	332
9.7	1	2		76
9.8	26	6		18
TOTAL	63	30	9	829

Table 19.11. Types of ceramic sherds in Chalcolithic levels

Level	Rim	Neck	Body	Base	Foot	N/A
4	22		391	23		1
5	10	1	109	2		
6	11		100	4	1	2
7	1		14		1	
5.1	2		22			
5.2			11			
5.4	1		3			
6.1			1			
7.4			2			
TOTAL	47	1	653	28	2	3

Table 19.12. Thickness of Chalcolithic sherds (mm)

Level	4	5	6	7	8	9	10	11	12	13	14	16	N/A
4		4	8	20	125	78	100	39	38	2	2	7	15
5	1	10	19	5	8	18	6	43	2	1	2		7
6		9	34	2	16	3	11	9	22			1	11
7	1	8	2		2	1	1	1					
5.1			13			3	1	7					
5.2		1			1				5	4			
5.4			1			1		2					
6.1			1										
7.4					2								
TOTAL	2	32	78	27	154	104	119	101	67	7	4	8	33

Table 19.13. Types of ceramic sherds in transition to the Bronze Age levels

Level	Rim	Body	Suspension	Base
3	20	552	1	18
3.1	1	8		
3.2	7	218		7
3.3	5	73		3
TOTAL	**33**	**851**	**1**	**28**

Table 19.14. Thickness of sherds from the transition to the Bronze Age (mm)

Level	4	5	6	7	8	9	10	11	12	13	14	15	16	18	19	N/A
3	12	4	38	31	117	145	61	71	41	25	12	25		1	1	8
3.1										1						8
3.2	3	1	3	13	17	19	21	63	29	23	25	4	1	1	5	4
3.3	1	1	6	7	18	1		14	24	6	1	2				
TOTAL	**16**	**6**	**47**	**51**	**152**	**165**	**82**	**148**	**94**	**55**	**38**	**31**	**1**	**2**	**6**	**20**

Table 19.15. Types of ceramic sherds in Bronze Age levels

Level	Rim	Galbo	Suspension	Base	N/A
1–2		6			2
1	4	18			
2	6	92	1	4	
3.4	3	10			
3.5	3	63		1	
TOTAL	**16**	**189**	**1**	**5**	**1**

Table 19.16. Thickness of Bronze Age sherds (mm)

Level	3	4	5	6	7	8	9	10	11	12	13	15	16	17	N/A
1–2				1				5							3
1				7	4	5	3	1	2						
2	1	2		3	1	10	5	24	16	11	17	1		8	4
3.4		1		1	4	2	1			1					3
3.5	1	1	6	8	1	12	15	3	10	1	1	1	4		3
TOTAL	**2**	**4**	**6**	**20**	**10**	**29**	**24**	**33**	**28**	**13**	**18**	**2**	**4**	**8**	**13**

Table 19.17. Types of decoration on Bronze Age sherds

Level	Fingerprints	Fingerprints + nails	Impression	Channel incisions	Triangular cord	Conical nipple	Smooth
1	1						11
2				1	1	1	41
3.4			1				8
3.5	1	1		4			37
TOTAL	2	1	1	5	1	1	97

Table 19.18. Firing atmosphere of Neolithic ceramics

Level	Irregular	Mixed	Oxidizing	Reducing
8	68		24	3
9	142	3	19	9
10	71	10	7	8
303	3	1	2	4
303.1	5		6	5
303.2	8	3	5	9
303.3	12	1	11	2
9.1	1	2		2
9.6	243	13	20	77
9.7	53	3	5	18
9.8	41	3	2	4
TOTAL	647	39	101	141

Table 19.19. Types of temper in Neolithic ceramics

Level	Iron	Limestone	Limestone/iron	Limestone/mica	Limestone/quartzite	Limestone/quartzite/grog	Limestone/quartzite/iron	Mica/quartzite	Quartzite	Quartzite/grog	Quartzite/iron
8		8			57	1			27	1	
9		14	4	1	77	3	3		54	10	7
303		1	1		5	1	1		2		
303.1		1		1	6	1			5		2
303.2		1			12	5			7		
303.3	1	2	1		10		4		8		
9.1					2	1			2		
9.6		30	1		175		13	1	103	5	26
9.7		2	1		47				28	1	
9.8		2	10		17	4	5		12		
TOTAL	1	61	18	2	408	16	26	1	248	17	35

Table 19.20. Firing atmosphere of Chalcolithic ceramics

Level	Irregular	Mixed	Oxidizing	Reducing	N/A
4	374	6	39	19	
5	82	5	13	22	
6	93		11	12	2
7	13		2	1	0
5.1	6		5	13	
5.2	11				
5.4	2		1	1	
6.1			1		
7.4				2	
TOTAL	**581**	**11**	**72**	**70**	**2**

Table 19.21. Types of temper in Chalcolithic ceramics

Level	Limestone	Limestone/grog	Limestone/mica/quartzite	Limestone/quartzite	Limestone/quartzite/grog	Limestone/quartzite/iron	Mica	Mica/quartzite	Mica/quartzite/iron	Quartzite	Quartzite/grog	Quartzite/iron
4	10	1		204	5	1		1		139	67	10
5	2	5		54	8	15	1		1	33	2	1
6		1	4	62	25					17	7	
7				3	1					10	2	
5.1				6						18		
5.2	4			7								
5.4				3						1		
6.1				1								
7.4				2								
TOTAL	**16**	**7**	**4**	**342**	**39**	**16**	**1**	**1**	**1**	**218**	**78**	**11**

Table 19.22. Firing atmosphere of ceramics from the transition to the Bronze Age

Level	Irregular	Mixed	Oxidizing	Reducing
3	396	113	65	15
3.1	9			
3.2	194	16	22	
3.3	76		4	1
TOTAL	**675**	**129**	**91**	**16**

Table 19.23. Types of temper in ceramics from the transition to the Bronze Age

Level	Concha/limestone	Iron/quartzite/grog	Limestone	Limestone/grog	Limestone/mica	Limestone/mica/ quartzite	Limestone/mica/ quartzite/iron	Limestone/quartzite	Limestone/quartzite/ grog	Limestone/quartzite/ iron	Limestone/quartzite/ Iron/grog	Mica	Mica/quartzite	Quartzite	Quartzite/grog
3			13	2	84	7	1	335	4	1		7	12	121	
3.1								9							
3.2		1	23		5			108	24	1	4	1	1	47	14
3.3	2			2				64						13	
TOTAL	2	1	36	4	89	7	1	516	28	2	4	8	13	181	14

Table 19.24. Firing atmosphere of Bronze Age ceramics

Level	Irregular	Mixed	Oxidizing	Reducing
1–2	2		2	5
1	9		9	4
2	81	11	9	2
3.4	13			
3.5	57	1	6	3
TOTAL	162	12	26	14

Table 19.25. Types of temper in Bronze Age ceramics

Level	Limestone	Limestone/grog	Limestone/iron/ grog	Limestone/ quartzite	Limestone/ quartzite/grog	Limestone/ quartzite/iron	Mica	Mica/quartzite	Mica/quartzite/ iron	Quartzite	Quartzite/grog	Quartzite/iron
1–2	2	5	1					1				
1										21		
2	9			57	1	3		1	1	31		
3.4				3			4			3	2	1
3.5	1			24	2	4				35	1	
TOTAL	12	5	1	84	3	7	4	2	1	90	3	1

Table 19.26. Characteristics of the sample selected for X-ray diffraction analysis

Sample	Square	Number	Sub-square	Level	Spit	Firing atmosphere	Temper grain[a]
1	H2	65	D	3.0	3	Reducing	M
2	H2	213	A	3.3	6	Reducing	F
3	H2	110	A	3	2	Oxidizing	F
4	H2	204	C	3.3.	6	Oxidizing	F
5	H3	40	D	4	7	Oxidizing	F
6	I1	7	A	5	4	Reducing	F
7	H3	94	C	6	9	Reducing	F
8	I1	7	A	5	4	Irregular	F
9	H3	107	A	6	10	Oxidizing	M/F
10	I1	7	A	5	4	Irregular	F
11	I4	245	B	7	16	Mixed	M/F
12	H4	243	B	5	9	Reducing	F
13	J3	50	B	10	14	Oxidizing	M/F
14	I2	46	C	8	11	Reducing	M/L
15	H1	158	B	5.2	4	Irregular	
16	J3	46	B	9	13	Reducing	M/F
17	I2	55	C	10	14	Reducing	M/F

(*Characteristics continued*)

Sample	Temper type[b]	Paste color	Surface color	Finish	Thickness (mm)	Post-depositional alterations
1	S	Blackened	Reddened	Engobado	3	
2	S	Blackened	Blackened	Engobado	6	
3	S	Reddened	Brown-gray	Smoothed	7	
4	S	Reddened	Whitened	Smoothed	8	Calcium carbonate
5	S	Reddened	Reddened	Smoothed	4	
6	S/M	Blackened	Brown-gray	Burnished	6	Calcium carbonate
7	S/M	Blackened	Blackened	Burnished	2	
8	S	Brown-gray	Reddened	Smoothed	10	
9	L/S	Orange	Orange	Smoothed	13	Calcium carbonate
10	S	Brown-gray	Reddened	Smoothed	6	
11	L/S	Brown-gray	Brown-gray	Burnished	6	
12	M/S	Blackened	Brown-gray	Burnished	5	
13	L/S	Brown-gray	Brown-gray	Burnished	4	
14	L/M/S	Blackened	Blackened	Smoothed	4	
15		Reddened	Grayish	Smoothed	10	Calcium carbonate
16	L/S	Blackened	Reddened	Engobado	6	Calcium carbonate
17	L/S	Blackened	Reddened	Engobado	3	

[a]F = fine; M = medium

[b]S = sand; M = mica; L = limestone

Table 19.27. Summary of the X-ray diffraction analysis of the archaeological sample

Sample No.	Quartz	Micas	Feldspar	Calcite	Kaolinite	Other
1	++	+	TRZ	—	*	—
2	++	+	TRZ	—	*	—
3	+++	TRZ	TRZ	—	*	—
4	++	+	TRZ	—	*	—
5	+++	+	TRZ	—	*	—
6	+++	TRZ	TRZ	—	*	—
7	+++	+	+	+	*	—
8	+++	—	TRZ	—	*	—
9	+++	+	+	+	*	—
10	++	++	TRZ	—	*	—
11	++	+	+	—	*	Chlorite
12	++	+	TRZ	—	*	Chlorite
13	++	+	+	++	*	—
14	+++	+	TRZ	—	*	—
15	+++	TRZ	TRZ	TRZ	*	—
16	+++	+	TRZ	+	*	—
17	++	—	TRZ	+++	*	—

+++ = Majority (30–50 percent)
++ = Very abundant (15–30 percent)
+ = Abundant (5–15 percent)
TRZ = Present (<5 percent)
— = Not present
* = Metakaolinite

Table 19.28. Characteristics of the experimental sample selected for X-ray diffraction analysis

Sample No.	Source	Temper	Temperature (°C)	Atmosphere
1	Ramales	No temper	600–800	Irregular
2	Ramales	Natural+silica	600–800	Irregular
3	Ramales	Natural+limestone	600–800	Irregular
4	Guardamino 1	Natural	600–800	Irregular
5	Guardamino 1	Natural+limestone	600–800	Irregular
6	Guardamino 1	No temper	600–800	Irregular
7	Guardamino 2	Natural+silica	600–800	Irregular
8	Guardamino 2	Natural+limestone	600–800	Irregular
9	Guardamino 2	Natural	600–800	Irregular
10	Ramales	Natural		
11	Guardamino 1	Natural		
12	Guardamino 2	Natural		

Table 19.29. Summary of the X-ray diffraction analysis of the experimental sample

Sample No.	Quartz	Micas	Feldspar	Calcite	Kaolinite
1	*	*	*		
2	*	*	*		
3	*	*	*	*	
4	*	*	*		
5	*	*	*	*	
6	*	*	*		
7	*	*	*		
8	*	*	*	*	
9	*	*	*		
10	*	*	*		*
11	*	*	*		*
12	*	*	*		*

*Presence

CHAPTER TWENTY

Conclusions

Preliminary Overview and Focus on the Post-Paleolithic Occupations of El Mirón Cave

Lawrence Guy Straus and Manuel R. González Morales

El Mirón Cave in Geographic and Archaeological Context

The Cantabrian region of northern Spain (particularly to the west of the Basque Country) is, from a geographic standpoint, structurally defined by the juxtaposition of narrow coastal lowlands and mountain ranges culminating in the cordillera that forms the northern rampart of the Castilian Meseta. This fundamental fact is the underlying constraining basis of the lifeway choices made by humans over millennia of the Upper Pleistocene and Holocene in this confined but relatively favored space. Coast and mountains, with bedrock lithology dominated by highly karstified limestone in many sectors and short, deep river valleys, provide the underlying parameters for human settlement, going well beyond the patterns of residence and subsistence to include key aspects of social relations and mobility. If geography is destiny, then in the case of Cantabrian Spain this has been true from Mousterian times to the present, and it was certainly very true among both Magdalenian hunter-fisher-gatherers and Neolithic, Chalcolithic, and Bronze Age farmer-pastoralists living in the montane setting of El Mirón Cave above the upper valley of the Río Asón. There is indeed a *longue durée* to this story.

Nonetheless, despite the underlying structure imposed by an uplifted shore and high mountain relief caused by ancient forces of orogeny, the human story in Cantabrian Spain also has been influenced significantly by dramatic changes in glaciation, sea level, vegetation, and fauna during the course of the tens of thousands of years that humans have been using El Mirón and so many other caves throughout the region. El Mirón, both as a karstic cave and as a place in an evolving natural and human landscape, has changed physically and, more importantly, in terms of its role in differing settlement systems, just as the montane zone has seen its importance wax and wane in the overall history of its use by humans. Such changes did not come to an end with the Bronze Age or even the medieval use of the cave; now the site's role is to provide information for archaeologists and to help bolster economic activity in the town of Ramales through tourism, as it is supposed to replace erstwhile dairy cattle raising and heavy industry as the town's main income sources. Not long ago the cave is said to have sheltered a French hermit and gypsies, possibly Republican guerilla fighters, and definitely, until the mid-1990s, livestock.

Never an isolate, the cave must be understood as either an integral or marginal part of changing systems of human exploitation of both the upper valley of the Asón and the whole of easternmost Cantabria. And the cave's location must been seen as a crossroads, not as a closed-in cul-de-sac, as indeed it is a major point of connection between the worlds of the Meseta and the North Atlantic shore and between those of Cantabria and the Basque Mountains, the Pyrenees, and the great beyond in France. From a potentially symbolic viewpoint (literally) the vast, gaping mouth of El Mirón is aligned with and looks out from the face of the Monte Pando cliff onto the 1,000 m high, pyramid-shaped Pico San Vicente at the eastern end of the dramatic sawtooth ridge of the Sierra del Hornijo. Among Paleolithic sites in Cantabrian Spain it is hard to find a more dramatic pair of landmarks than El Mirón and Pico San Vicente; they are rivaled only by the huge El Castillo Cave, which is on the steep, eastern face of a similarly dramatic, pyramid-shaped mountain (Monte Castillo) at the end of a coastal range in the central sector of the province.

One of many large caves in an area of eastern Cantabria, western Vizcaya, and northern Burgos that contains many of the longest and deepest karstic systems in Spain, El Mirón shares some fundamental geological attributes with other cavities that eventually would become favored loci of long-term human use, from short-term, limited-function visits to long-term, multifunction residential occupations, to sites of rupestral art activity. Formed at a time when the water table was locally far higher than today and long before great tectonic forces had left it hanging high above the floor of the present-day upper Asón valley, El Mirón was filled with alluvial sediments and subsequently eroded—perhaps multiple times, as was the case of several of the famous caves of Monte Castillo. The same ancient (early or pre-Pleistocene?) alluvial silts, sands, pebbles, and cobbles from the inner cave constantly served as the main materials with which the gaping vestibule would be (and slowly still is being) refilled. At some (as yet unknown) time in the Middle Paleolithic and until around 41,000 ^{14}C years ago, it too was occupied by Neandertals but on an apparently far less frequent basis than El Castillo Cave, for example. Similarly discrete visits continued during the early Upper Paleolithic and, though rather more spectacularly due to the presence of many foliate and shouldered stone points, in the Solutrean. The rhythm and intensity of human use seem to have changed abruptly after the Last Glacial Maximum with the onset of a new techno-cultural complex, the Magdalenian, beginning around 17,000 ^{14}C years ago.

Massive palimpsest horizons of Initial, Lower, and Middle Magdalenian were followed by levels attesting to less frequent and/or intensive occupations during the Upper Magdalenian and Azilian (far poorer in fauna, artifacts, and features than the early Magdalenian layers), apparently marking a time when low-elevation sites were favored for human residence in the Asón valley.

The range of time that is of particular concern in this monograph, the early to mid-Holocene, began in El Mirón with scant evidence of human visits during the Mesolithic and at least two major depositional hiati in the period beginning around 10,000 ^{14}C years ago. As an "interior" site (despite the relative proximity of the shore—20 km), El Mirón joins a short list of Cantabrian (*sensu lato*) sites for which there is evidence of only very modest use of the hinterland at a time when human settlement seems to have been strongly concentrated along the coast, especially at inlets and estuaries such as the mouth of the Asón (González Morales et al. 2004; Straus 2008; Straus and González Morales 2003). It would not be until less than 5,800 ^{14}C years ago (ca. 4,500 cal. BC) that humans would once again intensively use El Mirón Cave—but now as farmers and herders in the Neolithic (Peña-Chocarro, Zapata, García et al. 2005; Peña-Chocarro, Zapata, Iriarte et al. 2005). The site became a domicile not only for people but also for their livestock, and this would continue throughout the Chalcolithic and early Bronze Age. The magnitude of the human impact within the cave exploded, with massive pit digging and huge inputs of ceramics; grinding stones; bones of cattle, ovicaprines, and pigs; and fire-cracked rocks, charcoal, and ash from frequent, extensive fires. Despite continued hunting (and presumably enhanced gathering of wild plants in the lush woodlands of the Atlantic climatic phase), these agricultural occupations (here and throughout the region) caused profound environmental changes; for the first time *two* forces were at work in reshaping the vegetation of Cantabria: natural and human. Sorting out the factors at work not only in deforestation but also, more subtly, in the succession of plant taxa, makes the work of the palynologists and other paleobotanists both difficult and all the more critical, as evidenced in the relevant chapters of this volume. Likewise, the archaeology of the post-Mesolithic deposits was enormously complicated by the intensity of human occupations and activities (notably pit digging and re-digging and pit intrusions) during this relatively short period of time that is represented by a fairly limited thickness composed of many levels and discontinuous lenses. It is clear that many of the larger pits were reused and even modified over time; the complexity of the formation problems of these levels is further manifested by the many ceramic sherd refits that crosscut refits as well as by the presence of apparently erosionally "reworked" faunal remains and artifacts of Magdalenian aspect in Chalcolithic and Bronze Age layers and pits.

While the Bronze Age and Chalcolithic strata subside only in the outer vestibule area, the excavation of the Mid-Vestibule Trench revealed the continuity of both early and later Neolithic strata at least several meters to the east of the Cabin area excavation. This continuity is further supported by the stratigraphic coincidence of radiocarbon dates from the Mid-Vestibule (Levels 303.3–302 pit 98a base) and Cabin (Levels 10–8.1). The Neolithic and overlying prehistoric deposits seem to have been removed from the vestibule rear, perhaps to level it for recent livestock stabling (and as a source of fertilizer?). Below the Neolithic series of levels, however, the horizon corresponding to Mesolithic times (Preboreal, Boreal, early Atlantic) is continuous (albeit with lateral gradation in granulometry based on distance from the erosional ramp leading back to the inner cave, but always rich in calcium carbonates), sealing the long series of Azilo-Magdalenian levels that are also clearly continuous in at least a general sense throughout the El Mirón vestibule.

The Mesolithic of the Asón valley is relatively well known for sites near the then–recently formed estuary (La Fragua, El Perro) and along tributaries on the nearby low coastal plain (La Chora) (González Morales 1995, 2000; González Morales and Díaz Casado 2000; González Morales, Yudego, and Ituarte 2000). Recently, excavations in caves of the large Matienzo *polje* within the first range of coastal hills have revealed shell middens and other deposits of Mesolithic artifact typology and radiocarbon age at intermediate distances from the shore (Cosfresnedo, 6865 BP, and Cubío Redondo, 6630–5780 BP) (Ruiz Cobo and Smith 2001, 2003). These join the small, aceramic assemblage from the narrow cave of Tarrerón, about 2 km upstream of El Mirón along the Río Calera gorge, dated to 5780 BP, right at the time of transition between the Mesolithic and Neolithic in Vasco-Cantabria (Apellániz 1971).

The Neolithic record in eastern Cantabria (and indeed the whole province) is extremely scanty, consisting mainly of megalithic monuments located along ridges lining the Asón valley. The only one of these to be recently excavated and radiocarbon dated is the dolmen of Hayas: 5490 BP on charcoal from the floor of the central burial chamber (Serna 1997). Farther afield, evidence from the coastal Basque provinces to the east and from Asturias to the west suggests that the appearance of Near Eastern cultigens, domesticated livestock, and ceramic technology occurred in the period between about 6000 and 5500 BP (see Arias 1999). There seems to have been a period of coexistence between foraging and food production modes of subsistence and continued utilization of caves for human residence well after the "transition" to the Neolithic, presumably among local Mesolithic hunter-fisher-gatherers who had acquired the new knowledge and products via contacts with pioneer groups of Mesolithics-turned-Neolithics in nearby regions, such as the upper Ebro River Basin in Alava, La Rioja, and Treviño on the south side of the Cantabrian Cordillera, only about 80–100 km from El Mirón (see Alday 2005).

Holocene Environments around El Mirón Cave

In contrast to the rather dramatic environmental fluctuations of the Upper Pleistocene record of El Mirón and other sites in Vasco-Cantabria (despite the relatively low latitude and oceanic setting of the region), the Holocene record is characterized by relative stability marked only by fairly subtle changes, and many of these seem to have had human cause. As is characteristic of so much of the world's recent situation, humans, while not supplanting natural agents of climatic forcing, increasingly made their presence felt through their growing demographic density and especially because of their agro-pastoral practices and intensification. The human role in changing the face of the landscape is manifested not only in vegetation (through clearance and ecological succession) but consequently also in fauna and geomorphology. Just as native wild plants were replaced by cultigens of Near Eastern origin and mature, climax woodlands replaced by certain open-space colonizing species of trees and "weeds," so too were some terrestrial and avian animal species hurt and others favored by human actions (and competition from nonnative species). The ultimate consequences of forest clearance (which could involve burning and the use of tree foliage for livestock fodder) included accelerated erosion and even slope denudation. These major "anthropic" phenomena of course took place against the backdrop of the natural climate shifts from cold Dryas III conditions at the end of the Last Glacial to the more temperate ones of the Preboreal and Boreal, interrupted by the recently recognized cold snap around 8,000 radiocarbon years ago and followed by the Postglacial Optimum (Atlantic pollen zone) with high humidity and very warm temperatures, in turn punctuated by a "little ice age" during the late Chalcolithic–early Bronze Age around 4,000 radiocarbon years ago. One of the greatest problems facing the study of Holocene environments at El Mirón, as well as in the Cantabrian and other regions of western Europe, is the untangling of "natural" versus "human" agency in the explanation of the vegetational, faunal, and geomorphological changes observed in the various paleobiological and geological records. This question makes the interaction of archaeological and paleoenvironmental evidence and interpretation all the more critical to development of a holistic understanding of "the big picture" of human-land relations in a prehistory that, while relatively recent, is still quite remote despite being at the root of our modern "civilization."

The sedimentological analysis of El Mirón (Farrand, chap. 5, this volume) records hiati in the Holocene depositional record in the vestibule front area (Cabin), as well as the elimination of post-Neolithic levels in the mid-vestibule area and of all post-Mesolithic strata in the vestibule rear area (Corral). These latter two absences of surviving deposits, coupled with the absence of any Paleolithic or post-Paleolithic deposits on the ramp leading up into the inner cave (except for small Early Magdalenian remnant in-fillings of niches in the cave wall and remnant Azilian-age breccia below a flat flowstone corniche adhering to the cave wall at the top of the ramp), all point to massive erosion and/or removal (and downslope redeposition) of sediments during the Holocene.

The depositional hiati in the Cabin area occurred around 11,600–10,300 BP (uncal.) and around 8000–6000 BP. Despite these gaps, the rate of sedimentation in the late Holocene levels in the Cabin area was about 40 percent faster than in the Late Glacial ones, perhaps attesting to both greater humidity (and thus more running water in the cave) and higher intensity of human activity within the cave (coupled with the movements of livestock), all of which would tend to erode earlier sediments from the ramp and vestibule rear areas toward the vestibule front (Cabin area). The high humidity of the early Holocene is attested by carbonate-rich sediments deposited during Mesolithic times, when there were no more than occasional, ephemeral human visits to the cave. These levels (from rear to front: 102–101, 304, and 10.1) are more or less granular or less or more mondmilch-like with distance from the vestibule rear and ramp, respectively. Very humid conditions are also seen in the micromorphological analyses of the early Holocene sediments by Courty (pers. comm., July 2008).

In more detail, Courty and Vallverdu (2001:492–493) divide the approximately 1,000-year-long formation of Mesolithic-age horizon Level 10.1 into two stages. Following the marked cooling (and relative drying) of Dryas III, there was first a period of warm, wet climate with a high level of carbonate production caused by acidified water percolating into the cave from thick soils developed on the slopes above it. This was followed by a brief, colder episode, albeit with significant rainfall (and snowfall). The second stage was "marked by pedogenic alteration with negligible sediment supply," such that it does not appear stratigraphically distinct from the first stage within Level 10.1 (Courty and Vallverdu 2001:493).

Magnetic susceptibility analyses of the El Mirón Cabin stratigraphic sequence by Ellwood (Ellwood et al. 2001) suggest significantly warm temperatures immediately following the Dryas III cold event. However, they indicate another cold period after the first hiatus (i.e., Level 10, early Neolithic), followed by a major warming punctuated by several short, cool episodes during the early metal ages (notably one at the time of the transition between the Chalcolithic and Bronze Age).

The palynological record of El Mirón Cave is fraught with gaps (sterile samples). Nonetheless, Iriarte (chap. 8, this volume) has managed to identify four pollen zones among the Holocene deposits. During late Level 10.1 times (Mesolithic) the environment was heavily wooded, dominated by hazel, followed by English oak and Scots pine, with traces of birch, alder, and willow. (It should be noted that the arboreal pollen percentage in the Holocene levels is never below 66 percent; there were always significant woodlands in the general vicinity of the cave.) After a major hiatus, in late Neolithic Level 9 times, there was at first an even higher arboreal pollen percentage, with even more hazel but smaller percentages of English oak, Scots pine, and birch, while the percentage of ferns increased. But then, in the terminal Neolithic, hazel and oak declined as pine, grasses, shrubs, and ferns increased, suggesting the opening up of more clearings (probably due to human agropastoral activities). Scots pine is a known (re-)colonizer of open land. There is another gap in the pollen record at the base of Level 7 (early Chalcolithic), followed by evidence of the appearance of linden and holm oak in the area of the cave and fluctuations in hazel and alder, within spectra characterized by great arboreal diversity. Grasses (Poaceae) decreased somewhat, but other herbaceous plants fluctuated and ferns increased. In a fourth zone, corresponding to the final Chalcolithic and early Bronze Age, there was a jagged decline in tree pollen, with Scots pine apparently replacing many of the deciduous trees as it colonized lands that were increasingly being cleared for agriculture and/or pastoralism. Herbaceous taxa are abundant, while fern spore percentages fluctuate. The overall decline in trees (except for Scots pine) is also seen in the short spectra from the mid-vestibule. Much of this trend can be attributed to human deforestation and farming activity.

The carpological analyses by Peña-Chocarro (chap. 10, this volume) show that acorns and hazelnuts were gathered by humans and brought to the cave during the Neolithic and Chalcolithic. Other seeds include those of woodland plants and weeds, the latter in particular growing in disturbed ground such as agricultural fields and pastures. Domesticates (cereals—present since the early Neolithic) increased notably in abundance and diversity in Chalcolithic times as agriculture was intensified and more woodlands were cleared.

Wood charcoal analyses ("anthracology") by Zapata (chap. 11, this volume) show that mostly oak (together with some hazel and Rosaceae [wild fruit trees and hawthorn]) was selected for fuel by people in the Neolithic of El Mirón, meaning that these trees were growing abundantly near the cave. In some Chalcolithic levels oak charcoal decreased, but yew increased, attesting to even milder climate with abundant rain as well as to the existence of rocky areas with thin soil. Yew is toxic to most livestock, although its foliage could have been fed to goats. In the Bronze Age oak increased somewhat as yew decreased. Oak foliage could have been fed to livestock (including cattle), while the wood provided fuel. The mixed-oak forests that developed were probably caused by human disturbance of the natural (i.e., more homogeneous) oak forests of the earlier Holocene.

The ungulate faunas studied by Altuna and Mariezkurrena (chap. 16, this volume) are not very instructive concerning Holocene environments, although they are certainly all modern and either temperate or highly tolerant in their climatic preferences. Red deer is the dominant game species in the Mesolithic, Neolithic, Chalcolithic, and Bronze Age levels, although hunting clearly declined beginning with the appearance of food production around 5800 BP (uncal.). This is a species that can live in open grass- or heathlands as well as in woods (especially if there are clearings). On the other hand, roe deer and boar, present in the Mesolithic and Neolithic levels, are truer woodland denizens. Ibex and chamois (animals that prefer steep, rocky habitats, although the latter is also very much at home in woods) are present in the Neolithic and Chalcolithic (and ibex alone in the Bronze Age). There is a small representation of wild horse in the Chalcolithic and traces of aurochs throughout. Both species could have lived in woods and/or open grasslands. The possibility exists that native aurochs may have been independently domesticated in northern Spain from local wild stock, while at the same time there were also "imported" cattle.

A much more detailed picture of changing local environments during the Holocene is painted by the analyses of micromammals by Cuenca-Bescós and García Pimienta (chap. 12, this volume). Overall, the abundant samples of rodents and insectivores chronicle a decrease in woodland cover from the Mesolithic through the Bronze Age in the vicinity of El Mirón. The cave must have served as roost for raptors that daily scoured nearby areas for these small prey, especially when humans were not present, either seasonally or during longer periods of site abandonment. Humid meadows are always attested to in the record, but these wet, open areas increased markedly in the Bronze Age. There are abundant habitats along watercourses (rivers and streams) evidenced in most levels, though less so in Levels 9 (late Neolithic), 7, and 4 (early and final Chalcolithic, respectively); Level 9 was formed during a particularly dry episode, while the decrease in faunal diversity observed in Levels 7 and 4 could be due to intensified human activity in the cave at these times. There was a major increase in barren, rocky slope habitats in Level 3 (Bronze Age) as well as in Level 9 times.

Cuenca-Bescós and García Pimienta divide the Holocene sequence into six phases:

1. **Early Holocene** (Mesolithic and early Neolithic): warm and humid, with woodlands that were dotted with meadows and open, rocky slope areas.
2. **Levels 9–6** (late Neolithic and early Chalcolithic): decrease in woodlands with a corresponding increase in meadows, including wet ones, probably caused in part by human clearance.
3. **Level 5 + associated lenses** (Chalcolithic): slight increase in species abundance and diversity, with the presence of the Nordic vole indicating somewhat colder and wet conditions.
4. **Level 4** (late Chalcolithic): significant decrease in species diversity due to intensive human activity and livestock stabling in the cave.
5. **Level 3** (early Bronze Age): major decline in woodlands, replaced by fields and barren, rocky slopes, probably mainly caused by agricultural intensification, perhaps coupled with a climatic downturn, as attested elsewhere in the European Bronze Age.
6. **Levels 2 and 1** (medieval/modern): total humanization of the environment with great loss of woodlands and few remaining species of micromammals.

Of all the paleoenvironmental methods utilized at El Mirón, the micromammalian study is probably the most complete, with the most abundant data and some of the most sensitive indicators of habitats within a fairly short radius of the cave. While the sedimentological record is long and (albeit with depositional hiati) fairly complete, it tells only a muted story about external conditions, since a high proportion of the sediments in the vestibule came from the ancient alluvia of the inner cave. The palynological record is unfortunately highly fragmented and discontinuous, while the macromammalian record is not very sensitive.

The herpetology (Sanchiz, Lobo, and Bailon, chap. 14, this volume) suggests that the early and middle Holocene periods saw conditions that were somewhat colder and drier than at present. This conclusion, based on relatively small samples of reptiles and amphibians, does seem to contradict many of the other paleoenvironmental analyses, but it is interesting to note that modern conditions are considered to be the most Mediterranean-like of the sequence in El Mirón cave.

The avian faunas (Elorza, chap. 13, this volume) show the presence of woods in the cave vicinity during all sampled times, but always with greater or lesser areas of clearings and meadows.

In conclusion, after the sharp climatic downturn of Dryas III that marked the end of the Last Glacial (and of the Pleistocene), temperate, humid conditions reigned in this oceanic region at 43° North latitude. Cantabrian Spain, and the upper Asón valley specifically, was a densely wooded area, dominated by hazel, oaks, various other deciduous trees, and varying amounts of pines but with clearings and open, rocky slopes. The specific composition of the woods and the proportion of woods to meadows varied through time (as indicated by many of the methods reviewed here), but whether and in what relative proportions these variations were caused by natural fluctuations in climate (temperature and, to some degree, precipitation) or by human activity (forest clearance for agriculture, pastoralism [including the use of foliage for fodder], and fuel, followed by erosion and slope denudation) is uncertain, although there do seem to have been cold crises during both the late Mesolithic and Bronze Age of Iberia, however attenuated in the oceanic environment of northern Atlantic Spain. To what extent environmental change affected humans or humans affected the environment by the growth in their numbers, their adoption of agriculture and livestock herding, and their episodic intensification of food production activities during the Holocene in Cantabria remains an open and very important question in the study of the post-Paleolithic of this and other regions. Indeed, even the recent history of vegetation (and fauna) in the Cantabrian region has been driven by fluctuations in the nature, intensity, and venues of farming and grazing as well as industrial-scale deforestation for charcoal used in foundries and vast silviculture plantations (of nonnative trees) for paper pulp.

Human Uses of El Mirón Cave during the Early and Mid-Holocene

MESOLITHIC

As noted above, following the Azilian, so well represented both in the upper-middle Asón valley in the caves of El Valle and El Horno (as well as in El Mirón) and in the lower Asón valley in the caves of La Chora, El Perro, and La Fragua until around 9000 ^{14}C BP, human settlement seems to have concentrated heavily in the area around the newly formed estuary of the Asón between the present-day cities of Santoña and Laredo (González Morales 1990). This situation is not unusual in the Cantabrian context as a whole.

It is generally the case in the montane Vasco-Cantabrian interior that cave sites occupied throughout the Upper (and even Middle) Paleolithic saw human residence end with the Azilian, sometimes to be reused in Neolithic or more frequently Chalcolithic/Bronze Age times, often for funerary purposes. There is rarely evidence of Mesolithic occupation. El Castillo is a good instance of this phenomenon. Indeed, several caves in the montane interior were first and only (or virtually only) occupied during the Azilian (e.g., Los Azules in Asturias, El Piélago in Cantabria, Silibranka in Vizcaya, Antón Koba and Pikandita in Guipúzcoa)—and then abandoned (see Fernández-Tresguerres 2004 [2007]). On the other hand, caves near the coast (generally within one or very few kilometers of the recently formed early Holocene shore with its new inlets and estuaries) were often occupied during the Mesolithic. Sometimes these caves had been occupied during the Paleolithic and there is no "hiatus" between the Epipaleolithic Azilian culture and the Mesolithic (e.g., Santimamiñe in Vizcaya, La Riera in Asturias). In other cases caves seem to have been used for the first time during the Mesolithic, perhaps because all the other cavities in particular localities had already been filled to the ceiling with cultural deposits, culminating in previously formed Mesolithic shell middens. The best-known coastal Mesolithic "cultural tradition" of northern Spain is the Asturian, whose characteristic shell middens, dating around 9000–6500 ^{14}C BP, fill many coastal caves in the eastern third of Asturias and in the western end of Cantabria (Clark 1976; González Morales 1982; Fano 1998). Almost all are within 3 km of the shore (Fano 2004 [2007]). There are only a very few known Mesolithic sites in the montane interior inland of the dense coastal concentration of Asturian sites: Los Canes and Arangas in eastern Asturias (only 11 km from the shore but separated from it by the steep Sierra de Cuera coastal range, whose maximum elevation reaches nearly 900 m above sea level) and La Calavera (high in the Liebana valley of the Picos de Europa in western Cantabria) (Fano 2004 [2007]). The Basque Country record basically lacks interior Mesolithic sites, except for one very interesting (but unfortunately disturbed) site, Urratxa, located right at the watershed divide between the Ebro (Mediterranean) and Nervión (Atlantic) drainages on the Vizcaya-Alava border, 1,015 m high in the cordillera (Muñoz and Berganza 1997). After a late Azilian occupation (10,240 BP) there was a hiatus and then an Epipaleolithic (Mesolithic) use of the site at about 6950 ^{14}C BP. Already in Azilian times, flints from both the flysch outcrops of the Basque coast and (especially) from the Treviño sources in southern Alava (Ebro drainage) (as well as from the nearby Sierra de Urbasa) were being transported to Urratxa, attesting to human movement and contacts across the cordillera at the end of the Last Glacial and in the early Postglacial. The passes and the routes between the two worlds—Atlantic and Mediterranean—were already known and used, at least occasionally.

Nevertheless, it is clear that (unless there were many now-lost open-air sites on the steep slopes or now–deeply buried sites in the valley bottoms of the Vasco-Cantabrian interior) Mesolithic settlement of the region was heavily concentrated along the coast, while the interior was substantially abandoned. Reasons for this could include the density of the forests (and hence the difficulty of movement therein, especially given the high relief of the interior) and the relative scarcity of edible food resources on the forest floor (more solitary game that was hard to see and hard to hunt, such as roe deer and boar). The coast, with its coves and estuaries, constituted a macro-ecotone and a long, linear forest-edge "clearing," combined with rich marshes—all making for a rich environment endowed with a plethora of food sources, including terrestrial and palustrine ungulate game, fowl, and plants; marine and anadromos fishes; molluscs; and sea birds. Compared to the difficulty of getting around and obtaining food in the dense, dark, damp mixed deciduous forests of the interior, the choice of the coast for most human occupation seems logical. The result was that the interior was at best visited and/or lived in by small, rather isolated groups (as hinted at by the isotopic evidence for terrestrial diets among the humans buried in Los Canes versus the marine diets displayed among human remains from Asturian shell middens [Arias and Fano 2005]). The nature of the relationship between terminal forager sites (and hence inhabitants) along the coast and those (few) in the montane interior of the Vasco-Cantabrian region remains unclear. Were human group territories ultra-restricted (mainly narrow bands of shore and limited areas of adjacent coastal plain), or did coast-based groups continue to move residentially or send logistical parties (e.g., hunters) up into the montane zone, as they seem to have done in the late Upper Paleolithic seasonally and/or situationally? This and related questions cannot be answered satisfactorily at the present time, but future excavations and analyses like those of El Mirón will shed additional light thereon.

The Mesolithic-age evidence from El Mirón (Cabin Level 10.1, Mid-Vestibule Level 304, Corral Levels 101–102) clearly indicates that the intensity (and perhaps frequency) of human occupations of the cave in the first few millennia of the Holocene was very low: sporadic, ephemeral visits, with no evidence (at least in the areas excavated) of structures or even of defined living structures. Faunal remains are scarce, and the cultural evidence amounts to no more than a few hundred lithic artifacts (almost all knapping debris). The total of 11 retouched tools are distributed among several types, but there is scant evidence of weaponry such as would be expected if the cave had been used as a hunting camp for parties that had gone up into the mountains to procure game. (The total minimum number of individuals [MNI] for Level 10.1—the only Mesolithic stratum with a "substantial" number of bones—is a mere 5: two each of red and roe deer and one boar.) The lithic raw materials are a mix of excellent-quality flints that probably came from coastal sources and local non-flints (mudstones, quartzites). This could be interpreted to suggest that people did go into the upper Asón from the coast but then supplemented the flint they brought along with local stones during their stays in the cave and its vicinity. A definitive answer is lacking. What is clear from El Mirón (and nearby Tarrerón) is that the upper Asón was not totally abandoned for the whole Mesolithic period but rather, as with the sites of the Matienzo valley (Cubío Redondo and Cofresnedo, both closer to the shore than El Mirón), it was used perhaps as a backup resource "hinterland" for the sites clustered around the estuary.

NEOLITHIC

The use of the term *Neolithic* to contrast with *Mesolithic* carries with it the implication of a total break in lifeways, from hunting-gathering-fishing to farming-herding, with a revolution in technology, habitat, social relations, and even beliefs. The adoption of food production in the Iberian Peninsula has been the subject of abundant, intensive debate for decades, essentially revolving around the question of the relative contributions and roles of nonlocal ideas and products and/or peoples versus local groups and their long-established adaptations. Did the Neolithic literally "replace" the Mesolithic (meaning displacement of local foragers by outsider farmers), or did local foragers themselves change their means of subsistence, their technologies, and other aspects of their lives once they were exposed to the cultigens and domesticated animals of Near Eastern origin and to the organizational ideas and tools needed to most effectively exploit them? Some of these issues of "revolution" versus "continuity," "non-local" versus "local" in the Iberian Peninsula, especially in the wake of the well-known European Neolithic "demic diffusion" hypothesis of Ammerman and. Cavalli-Sforza (1984) are raised in recent publications (e.g., Lewthwaite 1986; Zilhão 1993, 2000; Zvelebil and Rowley-Conwy 1986; Lubell et al. 1994; Schmich and McClure 2009).

What interests us here in particular is the question of how, once established (by whatever mechanism[s]) along the Mediterranean coast of southeastern France and eastern Spain by around 7000 ^{14}C BP, the Neolithic technologies (notably ceramics) and lifeways spread westward into the interior of Spain (and southern France), rapidly reaching the upper basin of the Ebro River and eventually (after a considerable pause) crossing the Cantabrian Cordillera (or Pyrenees), reaching the Atlantic environment of Cantabria by less than 6,000 ^{14}C years ago (see Zapata et al. 2005). Indeed, recent excavations in Mendandia rockshelter in Treviño (an enclave of Burgos [Old Castile] within Alava, in the trans-cordilleran Basque Region), have yielded two ^{14}C dates of 7200 BP (uncal.) for a level (III sup.) with 343 ceramic sherds but with no domesticated animal remains or cultigens (Alday 2007). Prima facie evidence of agriculture and animal husbandry are still lacking in 6,400 year-old Level II, suggesting that even in this area of the upper Ebro, where food production became established very early, hunting and gathering continued to be major components of human subsistence in the Neolithic.

El Mirón has a significant part in the construction of the modern record for the early Neolithic of the whole region because of its clear stratigraphy, the completeness of its evidence, and the abundance of coherent radiocarbon dates, as well as the existence of the various different disciplinary analyses reported on in this volume. This contrasts with all the other sites of the Vasco-Cantabrian region in the 6000–5500 BP (uncal.) time range, which have partial or equivocal Neolithic records (e.g., ceramics but no cultigens or domesticates, cereal pollen but no ceramics, problematic stratigraphy, debatable evidence of domestication of otherwise local aurochsen, or attribution to the Neolithic solely on the basis of radiocarbon dates or the presence of beveled retouch geometric microliths that, in fact, are also known in Mesolithic contexts both in the Ebro Valley and the Cantabrian region).

The research of Vega Maeso (2005; chap. 19, this volume) shows that small, simple, generally undecorated ceramic jars are found throughout the Neolithic levels (10–8 in the Cabin and 303.3–303 in the Mid-Vestibule Trench) in quantities that are always modest but that increased in the late Neolithic occupations. Judging from experiments she did, Vega Maseo concludes that the vessels were made with local clays and rock temper and fired by placing them mouth-down on the coals of simple fires. The ceramists fully understood the nature of the processes involved and were making vessels apparently for everyday domestic use. Presumably the techniques were learned from people who had already been making ceramics for some time in other regions. But decoration of pots was not in the agenda of those inhabitants of El Mirón who had somehow acquired knowledge of the basic methods.

The lithic artifact assemblages from the Neolithic levels are small and heavily dominated by debitage. The latter are dominated by local non-flint materials, but the few retouched tools are generally made on high-quality flints, probably from coastal sources. The tools include small numbers of endscrapers, denticulates, notches, backed blade(let)s, and a few geometric microliths, including a circle segment with beveled ("Helwan") retouch and other geometrics with backing retouch. These kinds of Helwan microliths could be seen as providing evidence of human connections with the early Neolithic of the upper Ebro Basin and the Mediterranean world beyond. They are found, of course, not only at El Mirón and in other similarly dated (but aceramic) sites in the upper Asón (Tarrerón, Cubío Redondo, Cofresnedo) but also at a number of "terminal Mesolithic," "initial," or "aceramic Neolithic" sites in the coastal Basque provinces (e.g., Herriko Barra), Cantabria, and Asturias (e.g., Los Canes).

The faunal assemblages from the early Neolithic levels of El Mirón (Levels 10, 303.3–303) are heavily dominated by ovicaprines, according to the analyses of Altuna and Mariezkurrena (chap. 16, this volume). There are small numbers of domesticated cow remains (increasing among the Mid-Vestibule levels to reach a significant presence in Level 303) as well as domesticated pig. The later Neolithic levels (9.8–8) continue to be dominated by ovicaprines but cattle increase, as do pigs. Only in these levels are there traces of dog. Red deer continued to be hunted during all the Neolithic occupations, together with very small amounts of red deer, boar, ibex, chamois, and aurochs, as well as small carnivores and hedgehogs. The continued evidence of hunting is typical of Neolithic occupations in the Vasco-Cantabrian region (Altuna 1980). For example, Ontañón (2005) presents preliminary evidence of the progressive proportional increase of domesticated animals, albeit in the context of significant continuing dependence on hunting, in the Neolithic levels of Los Gitanos Cave in easternmost coastal Cantabria, together with cereal pollen, ceramics, and geometric microliths with Helwan retouch.

Similarly, wild plants were probably exploited not only for fuel and construction material but also for food (notably acorns, hazelnuts, and wild cherries), as suggested by Peña-Chocarro and Zapata (chaps. 10 and 11, respectively, this volume). However, the dramatic change vis-à-vis the Mesolithic is the appearance of emmer wheat at the base of the Neolithic sequence, followed in higher levels by einkorn wheat and free-threshing wheats, but no barley. Barley is found, however, in some of the early Neolithic sites of the coastal Basque provinces. The pollen analyses of Iriarte Chiapusso (chap. 8, this volume) indicate the presence of clearings in the forest, suggesting that people had begun the process of deforestation for agricultural fields and/or pasturelands. Of course, human creation of clearings (often by burning) is also known to have happened, at least in other regions of Europe, in the Mesolithic, to provide for deer grazing and to facilitate hunting. Despite the simultaneous appearance of "the whole Neolithic package" in El Mirón somewhere around 4500 cal. BC, the ancient ways of the Cantabrians, long centered on the inhabitation of caves and deer and ibex hunting, supplemented—in Post-Glacial times—by the use of wild plant foods, did not end then. Indeed, it continued for many generations to come—in sharp contrast to the trajectory of cultural change in most other regions of the Iberian Peninsula, where Neolithic villages based heavily on extensive agriculture and pastoralism were rapidly established and extended across the landscape. Nonetheless, it is apparent that a new conception of territory and land use took over in Cantabrian Spain (and indeed in the Asón Basin), if not immediately then very shortly after the arrival of ceramics, cultigens, and domesticated animals, as shown by the renewed intensive human occupation of the montane interior and as suggested by the construction of megalithic monuments beginning around 5500 BP (González Morales 1992; Serna 1997). Whatever their exact function(s), the (albeit relatively modest) dolmens, tumuli, and menhirs of the region are often located on ridgetops or hilltops, locations that (once the woods had been cleared) would have provided dominant, controlling views of wide areas that were in the process of being claimed and "humanized" for the purposes of an economy in which the means of production most likely were increasingly being apportioned unequally among the members of the society, who were now more bound to particular, delimited areas of land. A reorientation of the relationship between humans and the land was underway, even if some aspects of culture remained superficially similar to the Mesolithic status quo.

CHALCOLITHIC

The connotation of a Chalcolithic (Copper Age) or "Eneolithic" period in Cantabrian Spain carries with it a far lesser sense of a quantum leap to socioeconomic complexity than it does in much of the rest of the peninsula (or in other regions of Europe). Indeed, the lack of villages and towns (or, indeed, even open-air sites), giant burial tumuli, or other vast monuments in northern Atlantic Spain (in contrast to such regions as Andalucía or Portugal) has meant that there is simply not a major archaeological record for the Chalcolithic (or for the early Bronze Age, from which it is poorly distinguished) in this region. Most known Chalcolithic sites are caves that were used for funerary purposes or as ceramic deposits ("caches"). Indeed, many well-known Paleolithic cave sites saw their last significant human use as burial places in the Chalcolithic, yielding grave goods that are often out of their original functional context. The other unfortunate fact is that, being at the top of most stratigraphic sequences (sometimes right on the surface), Chalcolithic materials have very often been disturbed by modern human activity, from the passage of recent cave visitors or livestock, to deliberate trenching by spelunkers seeking to penetrate inner cave chambers, to surficial but highly destructive pothunting. Very little in the way of actual archaeological excavation of intact Chalcolithic sites of any kind (let alone residential as opposed to funerary ones) has taken place in north Atlantic Spain in general or, especially, Cantabria in particular, and there had been no radiocarbon dating of any Cantabrian Chalcolithic sites until the inception of the El Mirón Cave project (Ontañón 1996). The distinctions between late Neolithic and Chalcolithic on the one hand and Chalcolithic and early Bronze Age on the other in Vasco-Cantabria are tenuous and vague, a fact that is probably exacerbated here by the rural nature of El Mirón as a residence for humans and their livestock.

The Chalcolithic levels (7–4) of the El Mirón outer vestibule (Cabin area) are characterized by the presence of numerous pits, many of which are very large (±1 m in diameter and 20–25 cm deep). Of the 19 well-defined features in the Holocene levels of the 9 m^2 Cabin area, 12–14 are of Chalcolithic age (2 could be assigned to either the terminal Chalcolithic or initial Bronze Age). These pits really define the Chalcolithic occupations, along with extensive ash lenses. The presence of abundant fire-cracked rocks suggests extensive roasting and/or stone boiling. Many of the cobbles may originally have been used as anvils or grinding stones. Possible postholes associated with some of the pits imply the existence of other structures (racks, screens, partitions, etc.). The abundance of structures is in sharp contrast with the Neolithic layers, which have plenty of fire evidence but few pits, at least in the areas excavated (Cabin and western half of the Mid-Vestibule Trench). The possibility exists that some of the large pits could have been dug for storage, although their final use might have been for cooking, as indicated by the presence of fire-cracked rocks.

In the Chalcolithic levels the abundant ceramics continue to be simple, undecorated wares, almost certainly made locally with local clays and tempers. The diversity of vessel forms and sizes is greater than in the underlying Neolithic levels. Some of the larger jars could have been used for storage, while smaller bowls and jars could have been used for cooking and eating. The ceramic assemblage is clearly domestic (i.e., non-funerary) in nature.

As in the Neolithic level, lithic artifacts in the Chalcolithic units (levels, lenses, pit fills) are not abundant. Lithic raw materials seem to have continued to be procured in the same local (non-flint) and (especially) non-local (good-quality flints) sources as during the Neolithic. The coastal flint outcrops continued to be preferred, and among the artifacts made from such excellent materials are the two diagnostic arrowheads from Levels 5 and 5.1. Although there is relatively much microdebitage, no cores were recovered. Larger flakes and especially bladelets and blades are not very abundant. The scarcity of cortical items also suggests that primary knapping was not going on here, at least in the area excavated by us. Retouched tools are mainly made on flakes and include a very small but typologically diverse collection from Levels 7–4 combined. In addition to the bifacially worked arrowheads there are a couple of backed points plus very small numbers of endscrapers, perforators, burins, and miscellaneous retouched, truncated, and backed pieces. This collection is a far cry from the arrowhead- and knife-rich assemblages that are found in some Chalcolithic funerary cave contexts in the region. Possible sickle elements are not apparent. In addition to the chipped stone artifacts there are a few grinding stones and a couple of bone awls. All in all, despite the relative abundance of domesticated faunal remains and cereal grains and the spectacular presence of many pits and fire lenses (all indicating intensive, repeated, probably long-term human occupation of the cave during the Chalcolithic), the artifactual record is surprisingly poor, especially when compared to the grave goods deposited in specialized ossuary caves during the same period in the Vasco-Cantabrian region.

Subsistence seems to have been increasingly based on cattle pastoralism during the Chalcolithic at El Mirón, compared to the Neolithic. Although sheep and goats continued to be important, there was also an increase in pigs. There are hints from the maintenance of some older ovicaprines and cows for an ongoing "secondary products revolution" that had begun in the late Neolithic. (There is only one Chalcolithic dog bone—in Level 7.) A variety of wheats (free-threshing, naked, and hulled) are present, along with trace presence of pea (a pulse), but there are also remains of edible wild plants, notably acorns and hazelnuts. As in the Neolithic levels (and surprisingly so), barley is absent. Pollen and micromammal evidence suggest stepped-up forest clearance (apparently with some regrowth in abandoned fields, indicated by an upturn in the presence of Scots pine, perhaps the result of swidden-like shifts in field use). However, there may also have been a decrease in temperature and increase in humidity in the late Chalcolithic (Level 5), based on the appearance of the Nordic vole. The absence of obvious lithic sickle elements suggests the use of wooden "tongs" or simple hand picking for cereal harvesting. The presence of some large ceramic jars is indicative of storage of—among other things—cereal grains but also liquids, including milk. However, no cheese strainers have been found among the ceramic sherds. Wild game is still present, but only in small quantities (mainly red deer).

The transition between the Chalcolithic and Bronze Age is particularly poorly defined in this region and indeed seems rather arbitrary in the (scanty) literature thereon. Thus the specific attributions of Levels 4 and 3 are ambiguous, especially given the inconsistencies of the few radiocarbon dates for the top of the El Mirón sequence. Certainly, the kinds of ceramics (beaker-like jars) and lithics (arrowheads) seem to straddle the transition, which, after all, like the Chalcolithic and Bronze Age periods themselves, are archaeological constructs. These are not clearly distinguished in a region so peripheral to the core areas of the development of socioeconomic complexity (and metallurgy) in more southerly regions of Iberia.

EARLY BRONZE AGE

Two of the Bronze Age levels (3 and 2) each yielded slightly over 1,000 lithic artifacts—mostly microdebitage. There is more evidence of the full gamut of in situ flint-knapping activities now than in the Chalcolithic levels. The flints are dominated by good-quality materials, probably from coastal sources, with relatively little poorer-quality local (mainly non-flint) stones. Cores, cortical debris, platform renewal flakes, and crested blades, all testify to primary knapping in the cave.

There is a pair of hammerstones or anvils. Stone tools are relatively abundant (but still fewer than 65 total) compared to the Chalcolithic assemblages. Projectile tips are scarce (at most three), but backed bladelets together with endscrapers dominate the retouched pieces, with scatterings of other types, including truncated and retouched bladelets. The tools are, in fact, mainly made on flakes and bladelets. But oddly, at least in the area excavated, there were no bladelet cores. Clear sickle elements are no more in evidence than in the Chalcolithic levels. As in the Chalcolithic, the Bronze Age lithic collection is also very poor in spectacular items (arrowheads, daggers) more commonly found in burial contexts. There is, however, an awl or pin made of arsenical copper in Level 3. This is really the only "luxury" item among the Bronze Age artifacts from the site. Contact with the coast is confirmed by the presence of a few marine shells.

There are very few Chalcolithic–early Bronze Age sites in the area with which to compare the El Mirón occupations. As noted earlier, there are many burial caves, but almost all of these were either looted or excavated by spelunkers with minimal provenance control or publication. One of the most famous of these is the AER Cave in the Soba (Río Calera) Valley upstream of El Mirón, explored by the local speleological association (Agrupación Espeleológica Ramaliega) in the 1960s (García Cáraves 1966). This site yielded human skeletal remains and large numbers of ceramic vessels, many decorated. Another, much more recently excavated, is Cofresnedo Cave in the karstic valley of Matienzo within the first cordilleran front range, northwest of El Mirón (Ruiz Cobo and Smith 2003). Again, the human remains are generally associated with numerous ceramic vessels (some quite large and many decorated), as well as a copper awl and arrowhead, plus ornaments. There is some evidence that parts of Cofresnedo may also have been used for occasional habitation (by humans and livestock), particularly in the Bronze Age. Other caves in the immediate area of El Mirón have yielded "caches" of Chalcolithic–early Bronze Age ceramics, such as Las Grajas in Matienzo, with a date of 3710±60 BP (Ruiz Cobo and Smith 1997). Currently, the best-published of the funerary caves of Chalcolithic age in the area is Pico Ramos in westernmost Vizcaya, with an astounding collection of metal artifacts (including a copper pin very similar to the one from El Mirón Level 3); shell, bone, and stone ornaments; polished stone axes; and chipped stone arrowheads, as well as a few ceramic vessels (see Zapata 1995 and related articles in *Munibe*, vol. 47). Even nearer to El Mirón, on the Cantabria-Vizcaya border, is the funerary cave of Las Pajucas, dated to about 3700 BP, like El Mirón Level 3 (Apellániz 1967). Near it is Tarrerón Cave, with a burial deposit attributed to the Bronze Age (Apellániz and Nolte 1979). A bit farther away, on the coastal plain near Santander, is the cave of El Ruso, which underwent a confused series of lootings and poorly documented "excavations," the overall results of which were published by Gavelas (1986). The topmost (and most disturbed) deposit, Level 1, yielded a single human burial, attributed to the Chalcolithic on the basis of the associated bell beaker pottery, some of it with complex incised decorations. Arrowheads also accompanied this burial.

Naturally, such funerary cave sites lack structures, combustion lenses or features, or the abundance of faunal remains that are characteristic of the Chalcolithic and early Bronze Age strata in El Mirón. El Mirón Cave was apparently one of the major residential sites during the Chalcolithic and early Bronze Age in the montane upper valley of the Asón, a territory that, as noted above, is marked by numerous other caves used for burial of the dead or caching of ceramics. As a capacious cave overlooking the confluence of the Calera and Gándara rivers with the Asón at the eastern end of the relatively broad Ruesga Valley sector of the Asón (well suited for agriculture), with ample shelter and a warm westward orientation, surrounded by slopes that would have been good goat pastureland, El Mirón would have been an ideal location for human habitation and livestock stabling. Use of a cave for such residential and pastoral purposes would have been practical, avoiding the necessity of building structures. One can imagine that some of the funerary caves in the area were the final resting places of people who might have resided in El Mirón, either occasionally or for long periods of time. However, the evidence from El Mirón that Chalcolithic and early Bronze Age people used caves in this residential way and not only for burial does not prove that they did not also have more normal, open-air villages or even towns somewhere, especially in the latter period. That none have (yet) been clearly identified in the Vasco-Cantabrian region is, however, suggestive of the continuing emphasis on the multifunctional use of caves even in these times of developing socioeconomic complexity. The pluvial climate and cool temperatures of the region may well have favored continued use of caves for human residence, at least until appropriate building practices were acquired and/or local population levels became so high as to permit or necessitate abandonment of caves as living sites. It was in the Iron Age that open-air settlements finally became major features of the landscapes of northern Atlantic Spain, in the form of the famous *castros* (defensible, walled hill towns) (e.g., Sastre 2008). Even so, caves continued to serve as livestock stables and human refuges in times of trouble or need up to the modern era.

Bronze Age environments were increasingly humanized, with major losses of woodlands due to clearance for agricultural fields and pastures. Paleobotanical evidence continues to show the presence of various wheats as well as wild edible plants (notably acorns and hazelnuts), but no barley. While cattle continued to be important, there seems to have been some decline in favor of sheep and goats. One could speculate as to whether this shift back to ovicaprines (if proven to be generalized, and not just an El Mirón phenomenon) might have been related to the well-known cooling climatic event in the early Bronze Age of Europe that is hinted at by the micromammalian record in El Mirón. Pigs are also represented among the site's domesticated fauna in the Bronze Age. Wild game animals are present in only modest quantities. There is still no direct evidence of cheese making.

Epilogue: El Mirón Cave after the Early Bronze Age

Following the Early Bronze Age there are only traces of human use of the cave until one reaches the historic-era structures (the cabins) whose foundation stones lie directly atop the last Bronze Age level. The uppermost prehistoric deposits (Level 2) in the Cabin excavation area (including the fill of the large pit O, called Level 3.5) are manifestly mixed, as is Level 302 in the Mid-Vestibule Trench, but lack modern artifacts. Mixed surface sediments did yield a single prehistoric sherd with a deeply incised, multiple zigzag decoration as well as a series of sherds from a polished bowl with a long, truncated conical base, both of which, according to Vega Maseo (chap. 19, this volume), are similar to pieces from the Late Bronze Cogotas culture of Old Castile, a few hundred years later than the most recent of the Early Bronze Age radiocarbon dates from an intact level in the El Mirón vestibule (Level 3: 3700 BP). That there were at least visits to the cave in the Late Bronze Age is also suggested by a radiocarbon date of 3230 BP on a chunk of charcoal (torch remnant?) from Level VII of the Inner Cave Trench, dug in the 1950s and whose sections were straightened and cut back by us in 1996.

Visits to the cave also took place at least in the eleventh to twelfth and in the fourteenth to fifteenth centuries based on (and only on) one charcoal radiocarbon date each from the inner cave and the vestibule rear, respectively. El Mirón, like many other caves in the region, was probably a place of refuge during troubled times. But no apparent medieval artifacts were found in the mixed surficial sediments, at least in the areas of the vestibule that we cleared prior to formal excavation. These goat excrement–rich sediments did, however, yield modern artifacts, from horseshoes, glass, and china to an automatic pistol clip full of bullets, as well as a few human child bones only barely buried between a rock and the cave wall near the vestibule rear. Now added to the record will be items placed or lost in the cave by archaeologists from Spain, the United States, and numerous other European and American countries in the late twentieth and early twenty-first centuries. *Sic transit gloria mundi.*

References Cited

Alday, A.

2005 The transition between the last hunter-gatherers and the first farmers in southwestern Europe: the Basque perspective. *Journal of Anthropological Research* 61:469–494.

2007 Mésolithique et Néolithique au Pays Basque d'après l'abri de Mendandia (8500–6400 BP). *L'Anthropologie* 111:39–67.

Altuna, J.

1980 Historia de la domesticación animal en el País Vasco desde sus origenes hasta la romanización. *Munibe* 32:1–163.

Ammerman, A., and L. L. Cavalli-Sforza

1984 *The Neolithic Transition and the Genetics of Population in Europe.* Princeton University Press, Princeton, NJ.

Apellániz, J. M.

1967 Cuevas sepulcrales de Vizcaya: excavación, estudio y datación por el C-14. *Munibe* 19:159–226.

1971 El Mesolítico de la Cueva de Tarrerón y su datación por el C-14. *Munibe* 27:91–104.

Apellániz, J. M., and E. Nolte

1979 Memoria de las excavaciones de las cuevas de Tarrerón (Santander), Cuestalaviga (Vizcaya) y Orejones de Mostescusu (Burgos). *Kobie* 9:73–101.

Arias, P.

1999 The origins of the Neolithic along the Atlantic coast of continental Europe: a survey. *Journal of World Prehistory* 13:403–464.

Arias, P., and M. A. Fano

2005 Le rôle des ressources marines dans le Mésolithique de la région Cantabrique (Espagne): l'apport des isotopes stables. *Mémoires de la Société Préhistorique Français* 36:173–188.

Clark, G. A.

1976 *El Asturiense Cantábrico.* Bibliotheca Praehistorica Hispana 13, Madrid.

Courty, M. A., and J. Vallverdu

2001 The microstratigraphic record of abrupt climatic changes in cave sediments of the western Mediterranean. *Geoarchaeology* 16:467–500.

Ellwood, B., F. Harrold, S. Benoist, L. Straus, M. González Morales, K. Petruso, N. Bicho, J. Zilhão, and N. Soler

2001 Paleoclimate and intersite correlations from Late Pleistocene/Holocene cave sites: results from southern Europe. *Geoarchaeology* 16:433–463.

Fano, M. A.

1998 *El Habitat Mesolítico en el Cantábrico Occiental.* British Archaeological Reports S-732. Oxford, UK.

2004 [2007] Un nuevo tiempo: el Mesolítico en la región cantábrica. In *Las sociedades del Paleolítico en la región cantábrica*, edited by M. A. Fano, pp. 337–402. Kobie. Anejo 8. Diputación Foral de Bizkaia, Bizkaiko Foru Aldundia, Bilbao, Spain.

Fernández-Tresguerres, J.

2004 [2007] El final del Paleolítico en los espacios cantábricos: el Aziliense. In *Las sociedades del Paleolítico en la región cantábrica*, edited by M. A. Fano, pp. 309–336. Kobie. Anejo 8. Diputación Foral de Bizkaia, Bizkaiko Foru Aldundia, Bilbao, Spain.

García Cáraves, J. M.

1966 Hallazgos antropológicos en la Cueva del AER. In *Actas del IX Congreso Nacional de Arqueología (Valladolid).* Universidad de Zaragoza, Zaragoza, Spain.

Gavelas, A. J.

1986 El enterramiento con cerámica campaniforme de la Cueva del Ruso I (Igollo de Camargo, Cantabria). *Boletín del Instituto de Estudios Asturianos* 118:563–588.

González Morales, M.

1982 *El asturiense y otras culturas locales.* Monografías 7. Centro de Investigación y Museo de Altamira, Santander, Spain.

1990 La prehistoria de las Marismas: excavaciones en el Abrigo de la Peña del Perro (Santoña, Cantabria). *Cuadernos de Trasmiera* 2:13–28.
1992 Mesolíticos y megalíticos: la evidencia arqueológica de los cambios en las formas productivas en el paso al megalítico en la Costa Cantábrica. In *Elefantes, ciervos y ovicaprinos*, edited by A. Moure, pp. 185–202. Universidad de Cantabria, Santander, Spain.
1995 La transición al Holoceno en la región cantábrica: el contraste con el modelo del Mediterráneo español. In *Los ultimos cazadores*, edited by V. Villaverde, pp. 63–78. Instituto de Cultura Juan Gil-Albert, Alicante, Spain.
2000 La prehistoria de las Marismas: excavaciones en la Cueva de La Fragua (Santoña). Excavaciones de 1990, 1991, 1993, 1994 y 1996. In *Actuaciones arqueológicas en Cantabria, 1984–1999*, edited by R. Ontañón, pp. 177–180. Gobierno de Cantabria, Santander, Spain.

González Morales, M., and Y. Díaz Casado
2000 La prehistoria de las Marismas: excavaciones arqueológicas en los abrigos de la Peña del Perro. In *Actuaciones arqueológicas en Cantabria, 1984–1999*, edited by R. Ontañón, pp. 93–96. Gobierno de Cantabria, Santander, Spain.

González Morales, M., L. G. Straus, A. Diez, and J. Ruiz
2004 Postglacial coast and inland: the Epipaleolithic-Mesolithic-Neolithic transitions in Vasco-Cantabrian Spain. *Munibe* 56:61–78.

González Morales, M., C. Yudego, and C. Ituarte
2000 La prehistoria de las Marismas: prospección arqueológica de la zona del bajo Asón y marismas de Santoña y toma de nuestras en las yacimientos de las cuevas de Otero, La Chora y El Valle. *Actuaciones arqueológicas en Cantabria 1841–1999*, edited by R. Ontañón, pp. 151–153. Gobierno de Cantabria, Santander, Spain.

Lewthwaite, J.
1986 The transition to food production: a Mediterranean perspective. In *Hunters in Transition*, edited by M. Zvelebil, pp. 53–66. Cambridge University Press, Cambridge, UK.

Lubell, D., M. Jackes, H. Schwarcz, and M. Knyf
1994 The Mesolithic-Neolithic transition in Portugal, part 1: isotopic evidence of diet. *Journal of Archaeological Science* 21: 201–216.

Muñoz, M., and E. Berganza
1997 *El yacimiento de la cueva de Urratxa (Orozko, Bizkaia)*. Universidad de Deusto, Bilbao, Spain.

Ontañón, R.
1996 Las industrias líticas del Neolítico final—Calcolítico en Cantabria. *Munibe* 48:15–51.
2005 La secuencia de la Cueva de Los Gitanos (Castro Urdiales, Cantabria) y el Neolítico cantábrico. In *Actas del III Congreso del Neolítico en la Península Ibérica*, edited by P. Arias, R. Ontañón, and C. García-Moncó, pp. 1035–1043. Universidad de Cantabria, Santander, Spain.

Peña-Chocarro, L., L. Zapata, J. García, M. González Morales, J. Sesma, and L. G. Straus
2005 The spread of agriculture in northern Iberia: new archaeobotanical data from El Mirón Cave (Cantabria) and the open-air site of Los Cascajos (Navarra). *Vegetation History and Archaeobotany* 14:268–278.

Peña-Chocarro, L., L. Zapata, M. J. Iriarte, M. González Morales, and L. G. Straus
2005 The oldest agriculture in northern Atlantic Spain: new evidence from El Mirón Cave (Ramales de la Victoria, Cantabria). *Journal of Archaeological Science* 32:579–587.

Ruiz Cobo, J., and P. Smith
1997 El depósito cerámico de la Cueva de los Grajas (Matienzo, Ruesga). *Munibe* 49:65–76.
2001 El yacimiento del Cubío Redondo (Matienzo, Ruesga): una estación mesolítica de montaña en Cantabria. *Munibe* 53:31–55.
2003 *La Cueva de Cofresnedo en el Valle de Matienzo: Actuaciones arqueológicas 1996–2001*. Gobierno de Cantabria, Santander, Spain.

Sastre, I.
2008 Community, identify and conflict: Iron Age warfare in the Iberian Northwest. *Current Anthropology* 49:1021–1051.

Schmich, S., and S. McClure
2009 The Iberian Pleistocene-Holocene transition. Special issue, *Journal of Anthropological Research* 65(2).

Serna, M.
1997 Ocupacion megalitica y proceso de neolitización en la Cornisa Cantabrica. In *O Neolitico Atlantico e as orixes do Megalitismo*, edited by A. Rodriguez Casal, pp. 353–369. Universidade de Santiago de Compostela, Santiago, Spain.

Straus, L. G.
2008 The Mesolithic of Atlantic Iberia. In *Mesolithic Europe*, edited by G. N. Bailey and P. Spikins, pp. 302–327. Cambridge University Press, Cambridge, UK.

Straus, L. G., and M. R. González Morales
2003 The Mesolithic of the Cantabrian interior: fact or fantasy? In *Mesolithic on the Move*, edited by L. Larsson et al., pp. 359–369. Oxbow Books, Oxford, UK.

Zapata, L.
1995 La excavación del depósito sepulcral calcolítico de la cueva Pico Ramos (Muskiz, Bizkaia). La industria ósea y los elementos de adorno. *Munibe* 47:35–90.

Zapata, L., L. Peña-Chocarro, G. Pérez, and H. P. Stika
2005 Difusión de la agricultura en la Península Ibérica. In *Actas del III Congreso del Neolítico de la Península Ibérica*, edited by P. Arias, R. Ontañón, and C. García-Moncó, pp. 103–113. Universidad de Cantabria, Santander, Spain.

Zilhão, J.
1993 The spread of agro-pastoral economies across Mediterranean Europe: a view from the far west. *Journal of Mediterranean Archaeology* 6:5–63.

2000 From the Mesolithic to the Neolithic in the Iberian Peninsula. In *Europe's First Farmers*, edited by T. D. Price, pp. 144–182. Cambridge University Press, Cambridge, UK.

Zvelebil, M., and P. Rowley-Conwy
1986 Foragers and farmers in Atlantic Europe. In *Hunters in Transition*, edited by M. Zvelebil, pp. 67–93. Cambridge University Press, Cambridge, UK.

CONTRIBUTORS

JESÚS ALTUNA and KORO MARIEZKURRENA.
Centro de Custodia de los Materiales Arqueológicos y Paleontológicos de Gipuzkoa, Paseo Zarategi 84–88, 20015 San Sebastián, Spain.

ÁNGEL CARRANCHO and JUAN JOSÉ VILLALAÍN.
Laboratorio de Paleomagnetismo, Departamento de Física, Universidad de Burgos, Avenida de Cantabria s/n, 09006 Burgos, Spain.

GLORIA CUENCA BESCÓS and JUAN CARLOS GARCÍA PIMIENTA.
Area de Paleontología, Departamento de Ciencias de la Tierra, Universidad de Zaragoza, Calle Pedro Cerbuna 12, 50009 Zaragoza, Spain.

MIKELO ELORZA.
Sociedad de Ciencias Aranzadi, Zorroagagaina 11, 20014 San Sebastián, Spain.

WILLIAM R. FARRAND (deceased).
Museum of Anthropology, University of Michigan, Ann Arbor, MI 48109–1079, USA.

MANUEL R. GONZÁLEZ MORALES.
Instituto Internacional de Investigaciones Prehistóricas, Universidad de Cantabria, Avenida de los Castros s/n, 39005 Santander, Spain.

MARÍA JOSE IRIARTE CHIAPUSSO.
Area de Prehistoria, Facultad de Historia, Universidad del País Vasco, Calle Tomás y Valiente s/n, 01006 Vitoria, Spain.

LEONOR PEÑA-CHOCARRO.
Grupo de Investigación Arqueobiológica, Instituto de Historia, Centro de Ciencias Humanas y Sociales, Calle Albasanz 26–28, 28037 Madrid, Spain.

MANUEL PÉREZ RIPOLL and MARÍA DOLORES LÓPEZ GILA.
Departament de Prehistori i Arqueología, Universitat de Valencia, Avenida Vicente Blasco Ibañez 28, 46010 Valencia, Spain.

BORJA F. SANCHIZ, JORGE M. LOBO, and SALVADOR BAILON.
Departamento de Paleontología, Museo Nacional de Ciencias Naturales, Calle J. Gutierrez Abascal 2, 28006 Madrid, Spain.

LAWRENCE GUY STRAUS and JOHN D. RISSETTO.
Department of Anthropology MSC01 1040, University of New Mexico, Albuquerque, NM 87131–0001, USA.

CRISTINA VEGA MAESO.
Calle Eduardo Martínez del Campo 3, 5B, 09003 Burgos, Spain.

LYDIA ZAPATA.
Area de Prehistoria, Facultad de Historia, Universidad del País Vasco, Calle Tomás y Valiente s/n, 01006 Vitoria, Spain.

DÉBORA ZURRO.
Laboratorio de Arqueología y Antropología, Instituto Mila I Fontanals, IMF-CSIC, Calle Egipciaques 15, 08001 Barcelona, Spain.